Fundamentals of Enzymology

The Cell and Molecular Biology of Catalytic Proteins

THIRD EDITION

Nicholas C. Price

and

Lewis Stevens

Department of Biological Sciences,
University of Stirling

OXFORD

UNIVERSITY PRESS

OXFORD

UNIVERSITY PRESS

Great Clarendon Street, Oxford OX2 6DP
Oxford University Press is a department of the University of Oxford.
It furthers the University's objective of excellence in research, scholarship,
and education by publishing worldwide in

Oxford New York

Auckland Cape Town Dar es Salaam Hong Kong Karachi
Kuala Lumpur Madrid Melbourne Mexico City Nairobi
New Delhi Shanghai Taipei Toronto

With offices in
Argentina Austria Brazil Chile Czech Republic France Greece
Guatemala Hungary Italy Japan Poland Portugal Singapore
South Korea Switzerland Thailand Turkey Ukraine Vietnam

Oxford is a registered trade mark of Oxford University Press
in the UK and in certain other countries

Published in the United States
by Oxford University Press Inc., New York

First published 1982
Second edition 1989
Third edition 1999
Reprinted 2000, 2001 (twice), 2003, 2005, 2006, 2009 (twice), 2011, 2012, 2014

British Library Cataloguing in Publication Data
Data available

Library of Congress Cataloging in Publication Data
Price, Nicholas C.
Fundamentals of enzymology: the cell and molecular biology of
catalytic proteins / Nicholas C. Price and Lewis Stevens. — 3rd ed.
Includes bibliographical references and index.
1. Enzymes. I. Stevens, Lewis. II. Title.
[DNLM: 1. Enzymes. QU 135 P946F 1999]
QP601.P83 1999
572'.7—dc21
DNLM/DLC
for Library of Congress   99-25876 CIP

ISBN 13: 978-0-19-850229-6
ISBN 10: 0-19-850229-X

14

Typeset by EXPO Holdings, Malaysia

Printed in Great Britain
on acid-free paper by
CPI Group (UK) Ltd,
Croydon, CR0 4YY

Fundamentals
of Enzymology

For
Margaret, Jonathan, Rebekah, and Naomi, and
Evelyn, Rowena, Catherine, and Andrew.

Preface
to the third edition

In the preface to the second edition written some ten years ago, we drew attention to the enormous potential afforded by the application of recombinant DNA technology to enzymology. Over the last decade much of this potential has been realized. In parallel there has been a dramatic increase in the rate at which three-dimensional structures of proteins have been solved. The explosive growth in structural information about proteins has led to the development of large databases; these in turn have allowed detailed comparisons between proteins to be made at both the sequence and three-dimensional structural levels. The growth of the new science, bioinformatics, which has evolved to develop and exploit the information in the databases, has been catalysed by developments in computing. These have meant that access to even a modest networked desktop computer allows a scientist to gain access to structural information about proteins. Given the rapid pace of change in the subject, it is worthwhile asking whether a textbook is still a useful tool for the teachers and students of enzymology. We feel that the answer is yes, mainly because a book is still the best way of giving access to the core of the subject, and we hope that the changes we have made in the preparation of this new edition will convince the readers that this is the case. In order to allow readers to keep up with the developments, we have given a number of key references to important compilations of structural and other information about enzymes on the World Wide Web; these sites are being continually updated and expanded and should be consulted for recent work.

Stirling
February 1999

N. C. P. and L. S.

Preface
to the second edition

In preparing the second edition, we have attempted to highlight the major developments in enzymology in the seven years since the first edition appeared. Undoubtedly the most significant advances have been in the application of recombinant-DNA technology to the production of enzymes (Chapters 2 and 11), to the indirect determination of amino-acid sequences (Chapter 3) and to the analysis of enzyme structure and function using site-directed mutagenesis (Chapters 3 and 5). In many ways these advances can be said to have ushered in a 'golden age' of enzymology.

Other major advances have been made in the techniques of enzyme purification and analysis (Chapters 2 and 3), in the identification of catalytic domains within multienzyme proteins (Chapter 7), in understanding the organization of enzymes within membranes (Chapter 8), and in the application of purified enzymes in clinical analyses (Chapter 10).

We have been greatly encouraged by the response of colleagues to the first edition and hope that the new edition will continue to serve as a valuable guide for students and teachers of enzymology.

Stirling   N.C.P.
August 1988   L.S.

Preface
to the first edition

Over the last few years a number of books that have dealt comprehensively with the kinetics, structures, and mechanisms of enzymes have appeared. In general, the emphasis of these books has been on understanding the behaviour of isolated enzymes and much less attention has been focused on enzymes in their natural environment in an organism. This book aims to restore the balance to some extent by discussing the properties of enzymes in systems of increasing complexity from isolated enzymes to enzymes in the cell. It should be of value to undergraduate students of biochemistry and related biological sciences. We have assumed that the reader will have an elementary knowledge of biochemistry such as might be gained from an introductory course at university.

In order to kepp the lenght of the book within manageable limits, certain topics have been discussed only in outline. However, we have aimed to explain the principles involved in these topics in sufficient detail so that a student can pursue them if he or she wishes in the many books, research articles, and reviews referred to at the end of each chapter. We feel that it is important for undergraduates to make use of the original literature because biochemistry is such a rapidly expanding research-based subject.

In this book we have adopted the SI system of units and followed, as far as possible, the recommendations of the International Union of Biochemistry regarding the nomenclature of enzymes and substrates. Some of the recommended names may not be familiar or many biochemist, e.g. the enzyme RNA nucleotidyl transferase is more widely known as RNA polymerase. In such cases the more commonly used name is given as wellas the recommended name in the Appendix at the end of the book, which also gives reactions catalysed and Enzyme Commission numbers of all the enzymes mentioned in the book.

Stirling   N.C.P.
January 1981   L.S.

Acknowledgements

It is particularly appropriate in a book on biological catalysts to thank the many people who helped to bring this work to fruition. We should like to thank our colleagues in the department for many useful discussions and especially to pay tribute to Dr Evelyn Stevens who improved the style of much of the book. Professors Michael Rossmann and David Blow and Drs Heiner Schiemer and Georg Schulz generously supplied original photographs of enzyme structures. Dr Hal Dixon gave us some valuable advice on nomenclature, and Dr Linda Fothergill helped with some points relating to amino-acid sequencing. Dr Neil Paterson gave some very helpful comments and advice on current medical applications of enzymology. We should also like to thank Mrs May Abrahamson for her excellent typing. In one sense, all these people acted as more than catalysts; their intervention improved the book, rather than leaving it unchanged. The errors are, of course, our responsibility.

Acknowledgements
for second edition

We would like to thank Khlayre Mullin, June Watson, and Marianne Burnett for their help in preparing the second edition, and Professor David Blow for supplying a photograph of the structure of tyrosyl-tRNA synthase.

Acknowledgements
for third edition

We would like to thank Drs Lindsay Sawyer, Adrian Lapthorn, and Xiaodong Cheng for supplying photographs of the structures of type I dehydroquinase, type II dehydroquinase, and DNA methyltransferase, respectively, and Dr Tim Wess for a number of helpful discussions on structural and graphics programs. Liz Owen and her colleagues at Oxford University Press have helped greatly in bringing this third edition to fruition.

Contents

The plates fall at the back of the book.

A note on units

In this book we have tried, wherever possible, to use SI units. These are based on the metre-kilogram-second system of measurement. Multiples of the basic units by powers of 10 are as follows.

Prefix		Abbreviation	Prefix		Abbreviation
$10(10^1)$	deca	da	10^{-1}	deci	d
10^2	hecto	h	10^{-2}	centi	c
10^3	kilo	k	10^{-3}	milli	m
10^6	mega	M	10^{-6}	micro	μ
10^9	giga	G	10^{-9}	nano	n
10^{12}	tera	T	10^{-12}	pico	p

We do not use compound prefixes, thus 10^{-9} metre (m) = 1 nm, not 1 mμm. SI units for the various physical quantities mentioned in this book are listed below.

Quantity	SI unit
Time	*second* (s).
Length	*metre* (m). (Supplementary units retained for convenience are dm, cm)
Mass	*kilogram* (kg). Note that multiples are based on 1 gram (g), i. e. mg, μg rather than μkg, nkg (see above compound prefix rule)
Volume (given in units of length cubed)	*cubic metre* (m^3). For convenience the litre ($1l = 1dm^3$) and millilitre (1 ml = 1 cm^3) are retained
Amount of substance	*mole* (mol). This quantity contains 1 Ava-gadro number of basic units (e. g. electrons, atoms, or molecules)
Concentration	*mol dm^{-3}* used instead of molar (M) *mol kg^{-1}* used instead of molal (m)
Temperature	*kelvin (K)* note 0°C = 273.15 K

Quantity	SI unit
Force	*newton* (N) (1 m kg s^{-2})
Pressure	*pascal* (Pa) (1 Nm^{-2}). In this text, however, we have retained the atmosphere unit (atm) since the definition of standard states is then less cumbersome, being under a pressure of 1 atm rather than 101.325 kPa. (1 atm = 760 mm Hg = 101.325 kPa)
Energy	*joule* (J) (1 m^{2}kg s^{-2}) One calorie = 4.18 J. The gas constant, R = 8.31 J K^{-1} mol^{-1} (1 electron volt = 96.5 kJ mol^{-1})
Electric charge	*coulomb* (C) (1 amphere second). Faraday constant = 96 500 C mol^{-1}
Frequency	*hertz* (Hz) (1 s^{-1})
Viscosity	the units are kg m^{-1} s^{-1} (Note that 1 centipoise = 10^{-3} kg m^{-1} s^{-1}).

The main implications of these units for biochemists are as follows.

(i) Temperatures are quoted in kelvin (or absolute). Thus 25°C is 298.15 K (in practice, this is given as 298 K).

(ii) Enthalpy, internal energy, and free-energy changes are given in J mol^{-1}.

(iii) Entropies are quoted in J K^{-1} mol^{-1}.

(iv) Molar concentrations are given in terms of mol dm^{-3} rather than M.

(v) Enzyme activities are quoted in terms of katals (the amount of enzyme catalyzing the transformation of 1 mol substrate per second). Specific activities are quoted in terms of katal kg^{-1}.

(vi) The curie (Ci) is redundant as a unit of radioactivity; 1 Ci = 3.7 × 10^{10} disintegrations s^{-1}. (The unit disintegrations s^{-1} is given the name *becquerel* (Bq)).

For fuller discussions of SI units the following may be consulted:

Quantities, units, and symbols (2nd edn). The Royal Society, London (1975).
Physicochemical quantities and units (2nd edn). M. L. McGlashan. Royal Institute of Chemistry, London (1971).

Abbreviations

The following standard abbreviations, which are in common use in biochemistry, are used in the text. Any others are defined where used.

A	absorbance
ACP	acyl-carrier protein
ADP	adenosine 5′-pyrophoshate (adenosine 5′-diphosphate)
AMP	adenosine 5′-phosphate (adenosine 5′-monophosphate)
ATP	adenosine 5′-triphosphate
cAMP	3′:5′ cyclic AMP
cDNA	complementary DNA
CoA	coenzyme A
CTP	cytidine 5′-triphosphate
dATP	2′-deoxyadenosine 5′-triphosphate
dGDP	2′-deoxyguanosine 5′-pyrophosphate (2′-deoxyguanosine 5′-diphosphate)
dGTP	2′-deoxyguanosine 5′-triphosphate
DNA	deoxyribonucleic acid
dTMP	thymidine 5′-phosphate (thymidine 5′-monophosphate or 2′-deoxyribosylthymine 5′-monophosphate)
dTTP	thymidine 5′-triphosphate (2′-deoxyribosylthymine 5′-triphosphate).
dUMP	2′-deoxyuridine 5′-phosphate (2′-deoxyuridine 5′-monophosphate)
EDTA	ethylenediaminetetra-acetate
FAD	flavin-adenine dinucleotide (oxidized form)
$FADH_2$	flavin-adenine dinucleotide (reduced form)
g	gravitational field, unit of (9.81 m s^{-2})
GDP	guanosine 5′-pyrophosphate (guanosine 5′-diphosphate)
GTP	guanosine 5′-triphosphate
h.p.l.c.	high-performance liquid chromatography
i.r.	infrared
ITP	inosine 5′-triphosphate
mRNA	messenger RNA
M_r	relative molecular mass
NAD$^+$	nicotinamide-adenine dinucleotide (oxidized form)

NADH	nicotinamide-adenine dinucleotide (reduced form)
$NADP^+$	nicotinamide-adenine dinucleotide phosphate (oxidized form)
NADPH	nicotinamide-adenine dinucleotide phosphate (reduced form)
NMN	nicotinamide mononucleotide
n.m.r.	nuclear magnetic resonance
RNA	ribonucleic acid
r.p.m.	revolutions per minute
rRNA	ribosomal ribonucleic acid
tris	2-amino-2-hydroxymethyl-propane-1, 3-diol (tris(hydroxymethyl) methylamine)
tRNA	transfer RNA
UDP	uridine 5′-pyrophosphate (uridine 5′-diphosphate)
UMP	uridine 5′-phosphate (uridine 5′-monophosphate)
UTP	uridine 5′-triphosphate
u.v.	ultraviolet

Nucleotide sequences

A sequence of nucleotides can be written as, e. g., A C G C U C where each letter signifies a nucleotide.* The convention is to write the order so that the phosphodiester link runs 3′ to 5′ from left to right. A nucleoside 5′-phosphate is written as pX, and a 3′-phosphate as Xp. The sequence above could also be written as ApCpGpCpUpC. To indicate that there is a phosphate, pyrophosphate or triphosphate group at the 5′-end, we would write

pA C G C U C	5′-phosphate
ppA C G C U C	5′-pyrophosphate
pppA C G C U C	5′-triphosphate

* The nucleotide is taken to be a ribonucleotide unit, unless the context makes it clear that it is a deoxyribonucleotide (e. g. in DNA). In a DNA sequence, the letter T refers to a thymidine phosphate (2′-deoxyribosylthymine phosphate) unit.

1

Introduction

1.1 Aims of this book

In this book we have tried to give a broad account of enzymology and have aimed to put current knowledge into perspective. Studies of enzymes have as their ultimate goal an understanding of the crucial role that these catalysts play in the metabolic processes of living organisms. Because of the complexity of such processes, it is at least necessary to gain an insight into the properties of enzymes in simpler systems, i.e. as isolated entities studied in the test-tube or spectrophotometer cuvette. The chapters in the book follow a progression from the properties of isolated enzymes to the behaviour of enzymes in increasingly complex systems, leading up to the cell. We have included some discussion of the importance of enzymes in medicine and industry to emphasize that enzymology is not a purely academic subject but has increasingly wide applications.

1.2 Historical aspects

Enzymes are catalysts (i.e. they speed up the rates of reactions without themselves undergoing any permanent change). Each reaction taking place in the cell is catalysed by its own particular enzyme, so that in a given cell there are a large number of enzymes. It is difficult to make a precise estimate of the number of different enzymes in each cell, but clues are beginning to emerge from the results of genome sequencing projects. Thus the bacterium *Escherichia coli* has the coding potential for 4288 proteins, 2656 (62%) of which are characterized; the remaining 1632 are categorized as hypothetical, unclassified, or unknown. Of the characterized proteins, 1701 (64%) are enzymes.[1,2] In higher eukaryotes, the number of different enzymes is likely to be an order of magnitude higher.

In the absence of enzymes most of the reactions of cellular metabolism would not occur even over a time period of years, and life as we know it could not exist.

The word *enzyme* is derived from the Greek meaning 'in yeast' and was first used by Kühne in 1878.*

At the time it was used to distinguish between what were referred to as 'organized ferments' (meaning whole microorganisms) and 'unorganized ferments'

* It was pointed out by Fruton[3] that the term was used as far back as the twelfth century by the Armenian philosopher, Theorianus—although obviously not in a biochemical context.

(meaning extracts or secretions from whole organisms). The term *enzyme* was thus intended to emphasize that catalytic activity was 'in yeast', i.e. a manifestation of an extract or a secretion rather than of the whole organism. Although the term specified yeast, it was to be used for all 'unorganized ferments'. Of course, the catalytic activity of enzymes in microorganisms has been utilized by mankind for many thousands of years in processes such as fermentation and cheese-making, but this was very much purely a practical use. It was only when it was shown that enzyme activity could be expressed without the need for an intact cellular structure that the study of enzymes could proceed along the paths already established in the study of chemistry. In this respect Büchner's demonstration (1897) that filtrates of yeast extracts could catalyse fermentation was highly significant.

Emil Fischer in 1894 had performed some classical studies on carbohydrate metabolizing enzymes in which he demonstrated the specificity shown by an *enzyme* for its *substrate* (the molecule acted on by the enzyme). On the basis of his experiments, Fischer proposed the 'lock and key' hypothesis to describe this interaction and for many years this proved to be a fruitful way to picture the binding of enzyme to substrate (Fig. 1.1).

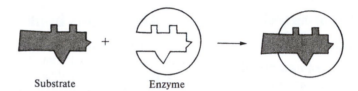

Substrate   Enzyme

Fig. 1.1 Fischer's 'lock and key' hypothesis to explain enzyme specificity.

At this time (1890s) the chemical nature of enzymes was not clear. In fact this point was only established many years later after a number of enzymes had been crystallized and shown to consist entirely of protein (i.e. made up of amino acids linked by amide bonds, see Chapter 3, Section 3.3). The first enzyme to be crystallized (in 1926 by Sumner) was urease, which catalyses the hydrolysis of urea to yield carbon dioxide and ammonia. In fact, if Sumner had had more sensitive methods of analysis he would have found that the preparation contained a small amount of nickel (approximately 0.1% by weight, two Ni^{2+} per active site) which is essential for catalysis (Fig. 1.2).[4,5] Perhaps in retrospect it was fortunate that he did not have such sensitive methods, since otherwise the nature of enzymes might not have been settled for many more years.

The development of the ultracentrifuge by Svedberg (also in the 1920s) allowed very high centrifugal fields capable of sedimenting macromolecules* to be generated. These studies showed that proteins in solution generally consist of homogeneous molecules of definite M_r (in the case of enzymes the M_r values range between about 10^4 and 10^7) rather than of colloidal suspensions. The description of enzyme structures in precise chemical terms was then a realistic possibility: this was first achieved in 1960 when the amino-acid sequence of ribonuclease[7] (an enzyme catalysing the hydrolysis of ribonucleic acid) was deduced. In 1965 the three-dimensional structure of lysozyme (an enzyme cleaving certain bacterial cell walls) was deduced by the technique of X-ray crystallography (see Chapter 3, Section 3.4.1), and for the first time a mechanism of action could be postulated in precise chemical and structural terms. We now know the amino-acid sequence and three-dimensional structure of many thousands of enzymes and generalizations and comparisons between them can be undertaken (see Chapter 3, Sections 3.3.2.11 and 3.4.4.6).

* The term macromolecule has been generally reserved for molecules of relative molecular mass (M_r) > 10 000; however, the smallest known enzyme polypeptide chain (that of 4-oxalocrotonate tautomerase, with an M_r of 6811) would still be regarded as a macromolecule.[6]

During the late 1950s and the 1960s a number of observations were made that suggested that enzymes show considerable flexibility. In 1958 Koshland[8] proposed the 'induced fit' theory to account for the catalytic power and specificity shown by enzymes (see Chapter 5, Section 5.6). It also became clear that the catalytic activity of certain enzymes could respond to changes in physiological conditions. Monod and his colleagues[9,10] put forward their 'allosteric model' to explain in a quantitative way how the activity of certain enzymes can be regulated by the binding of small molecules (*effectors*) and this provided a basis for understanding many features of the control of enzymes in the cell. An important feature of models for allosteric enzymes in general is that they postulate that the binding of effectors to the enzymes induces structural changes in the enzymes (Chapter 6, Section 6.2.2).

The first chemical synthesis of an enzyme (ribonuclease) from amino-acid precursors was reported in 1969.[11] Although this represented a considerable achievement, it should be noted that both the chemical purity and catalytic activity of the preparation were rather low. The chemical synthesis can, perhaps, be said to represent the final proof that enzymes are no different qualitatively from other non-biological catalysts.

The application of recombinant DNA techniques to the study of enzymes has produced some remarkable new insights (see Chapter 3, Section 3.3.1, and Chapter 5, Sections 5.4.5 and 5.5). It has proved possible to alter catalytic activity and specificity in a rational manner by introducing mutations at defined positions using site-directed mutagenesis. This has helped in understanding the mechanism of enzyme action and has also opened the prospect of designing enzymes with specific required properties.[12] For example, the specificity of lactate dehydrogenase has been changed to that of malate dehydrogenase by introducing three particular mutations at the active site (see Chapter 5, Section 5.5.4.5). Attempts are being made to manipulate the specificity of restriction endonucleases, so that the number of nucleotides recognized at the catalytic site might be increased (e.g. from six to eight).[13]

Recently, it has become evident that catalytic activity can be shown to a limited extent by biological molecules other than the 'classical' enzymes. In the process of enzyme catalysis the substrates are converted via a high-energy transition state to the eventual products (see Chapter 5, Fig. 5.1). A number of experiments have shown that antibodies raised to stable analogues of the transition states of a number of enzyme-catalysed reactions (see Chapter 5, Section 5.3.8) can act as effective catalysts for those reactions.[14–16] In 1986 Cech discovered that RNA can also act as a catalyst for reactions involving hydrolysis of RNA and that rate enhancements of up to 10^{11} can be achieved.[17,18] Since the initial discovery a number of different types of RNA-catalysed reactions have been discovered and the RNA catalysts are known as ribozymes. Ribozymes catalyse self-splicing of introns, tRNA processing, and splicing of RNA viral genomes. All naturally occurring ribozymes known so far catalyse either transphosphoesterification or phosphate ester hydrolysis,[18] but ribozymes have been engineered that have peptidyl transferase activity similar to that of ribosomes.[19] The RNA fragments involved in the catalysis have been identified, and all have a metal ion requirement. Ribonuclease P is a ribozyme catalysing the hydrolysis of a phosphodiester in precursor tRNA. It comprises a 377-nucleotide RNA component and a 119 amino-acid residue protein component. The catalytic site is on the RNA component.[20,21] Recently, attention has focused on a group of small ribozymes known as hammerhead ribozymes. These are composed of approximately 30 nucleotides folded in the shape of a hammerhead and able to induce the site-specific cleavage of a phosphodiester bond (Fig. 1.3).

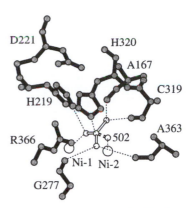

Fig. 1.2 Urease active site, showing the positions of the substrate, urea, the amino-acid residues (indicated using the single letter code, see Chapter 3, Table 3.1) forming the active site, and the two Ni atoms and oxygen atom of H_2O (502) hydrating Ni-2.[5]

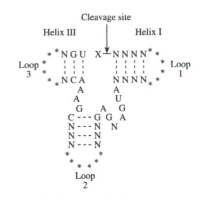

Fig. 1.3 Two-dimensional representation of the hammerhead ribozyme. The arrow indicates the site of phosphodiester cleavage. N is any nucleotide, X is A, C, or U, and the asterisks denote loops of RNA which can be 100–200 nucleotides in length.[20]

The three-dimensional structure has been determined by X-ray diffraction and comprises three base-paired helices connected by two single-strand regions, and the cleavage site has been identified.[22] Because of their small size the hammerhead ribozymes are ideal for the design of new ribozymes and have therapeutic potential. The observation of these 'ribozymes' has important implications for theories concerning the evolution of catalytic function.[23]

For further details on the history of enzymology see references 24 and 25.

1.3 Remarkable properties of enzymes as catalysts

Enzymes display a number of remarkable properties when compared with other types of catalyst. The three most important are their high catalytic power, their specificity, and the extent to which their catalytic activity can be regulated by a variety of naturally occurring compounds. These three properties will be illustrated below.

1.3.1 Catalytic power

Enzymes may increase the rate of a reaction by as much as 10^{17}-fold. There are not many examples where a direct comparison can be made between the rates of an enzyme-catalysed reaction and the reaction occurring under similar conditions of temperature, pH, etc., but in the absence of enzyme. This is because in the absence of an enzyme the rates may be too low to be measured easily. Where the comparison has been made, very high rate enhancements have been found,[26,27] e.g. hexokinase $> 10^{10}$, phosphorylase $> 3 \times 10^{11}$, alcohol dehydrogenase $> 2 \times 10^8$, triosephosphate isomerase $> 10^9$, carboxypeptidase A $> 10^{11}$, and urease $> 10^{14}$. In other instances, where enzymatic and non-enzymatic catalysts are compared, the former catalyse much higher rates and in some cases do so at significantly lower temperatures (Table 1.1). The optimum conditions for enzyme catalysis are usually moderate temperatures and pH values close to neutrality, although enzymes from a number of archaebacteria function under more extreme conditions.[28] The contrast between an enzyme-catalysed reaction and a reaction catalysed by a non-enzymatic catalyst is well illustrated by the process of nitrogen fixation (i.e. reduction of N_2 to ammonia). Nitrogenase* catalyses this reaction at temperatures around 300 K and at neutral pH. The enzyme is a complex system comprising two dissociating protein components, one of which contains iron and the other iron and molybdenum.[29] Several molecules of ATP are hydrolysed during the reduction, although the exact stoichiometry is still uncertain. By

* Only certain prokaryotes can carry out nitrogen fixation and they may be either symbionts or non-symbionts. In terrestrial ecosystems, symbiotic fixation appears to exceed non-symbiotic and the most outstanding example of symbiotic fixation is that between bacteria of the genus *Rhizobium* and the roots of leguminous plants.

Table 1.1 Examples of the catalytic power of enzymes

Substrate	Catalyst	Temperature (K)	Rate constant k (mol dm^{-3})$^{-1}$s^{-1}
Amide (hydrolysis)			
benzamide	H+	325	2.4×10^{-6}
benzamide	OH-	326	8.5×10^{-6}
benzoyl-L-tyrosinamide	α-chymotrypsin	298	14.9
Urea (hydrolysis)	H+	335	7.4×10^{-7}
	urease	294	5.0×10^6
$2H_2O_2 \rightarrow 2H_2O + O_2$	Fe^{2+}	295	56
	catalase	295	3.5×10^7

Data taken from reference 31.

contrast, in the industrial synthesis of ammonia from nitrogen and hydrogen the conditions used are as follows: temperatures between 700 and 900 K, pressures between 100 and 900 atmospheres, and the presence of an iron catalyst, often promoted by traces of oxides of other metals.[30] The basis of catalytic power is discussed in Chapter 5, Section 5.3.

1.3.2 Specificity

Most enzymes are highly specific both in the nature of the substrate(s) they utilize and also in the reaction they catalyse. The range of specificity varies between enzymes. There are some enzymes that have relatively low specificities (*bond specificity*), e.g. certain peptidases, phosphatases, and esterases, which will utilize a wide range of substrates provided they contain the required chemical bond, i.e. peptide, phosphate ester, and carboxylate ester, respectively, in these three examples. Low specificity is more commonly encountered with degradative enzymes but is only very rarely observed with biosynthetic enzymes. The role of the former may be that of digestion, where wider specificity would be more economical. An intermediate set of enzymes show *group specificity*, e.g. hexokinase. This enzyme will catalyse the phosphorylation of a variety of sugars provided they are aldohexoses. However, many enzymes show *absolute* or *near-absolute specificity*, in which they will only catalyse at an appreciable rate the reaction with a single substrate (or a single pair of substrates in a bimolecular reaction), e.g. urease will only catalyse the reaction with urea, or with very similar analogues at a very much lower rate—thus its next best substrate is semicarbazide which it degrades at $> 10^3$ slower rate.[5] The quantitative definition of specificity in kinetic terms is described in Chapter 4, Section 4.3.1.3.

The terms *group specificity* and *absolute specificity* can be readily appreciated in relation to low-M_r substrates, but when considering macromolecular substrates the position may be somewhat different in the sense that the active site of the enzyme can only interact with a part of the macromolecule. This is simply due to the relative dimensions of the active site and the macromolecule. A group of enzymes that has been studied extensively in the past decade is the restriction endonucleases. These enzymes generally recognize a sequence of four to six base pairs in DNA and then cleave the phosphodiester links in both strands but not necessarily in opposite positions. There are now known to be at least 1900 of these enzymes exhibiting a wide range of sequence specificity.[32] In a sense, each enzyme shows absolute specificity for the region of the substrate that is in contact with the active site, although it can act on any DNA molecule or fragment that contains the appropriate sequence.

Another distinct feature of many enzyme-catalysed reactions is their stereospecificity; this is well illustrated in the case of NAD^+- and $NADP^+$-requiring dehydrogenases. It has been demonstrated by use of suitably labelled substrates that dehydrogenases catalyse the transfer of hydrogen from the substrate on to a particular side of the nicotinamide ring; these are designated A-side and B-side dehydrogenases (Fig. 1.4). In addition, almost all dehydrogenases act on either NAD^+ or $NADP^+$. The basis of these specificities is clear in the case of those dehydrogenases whose three-dimensional structures are known, e.g. liver alcohol dehydrogenase, lactate dehydrogenase, and glyceraldehyde-3-phosphate dehydrogenase,[34] and is discussed further in Chapter 5, Section 5.5.4.3.

There are a number of other examples of prochirality in enzyme-catalysed reactions.[35] These reactions occur in such a manner that although the reactant does not have a chiral* centre, the reaction occurs in a stereospecific manner. A good

* A compound is said to be chiral if it cannot be superimposed on its mirror image.

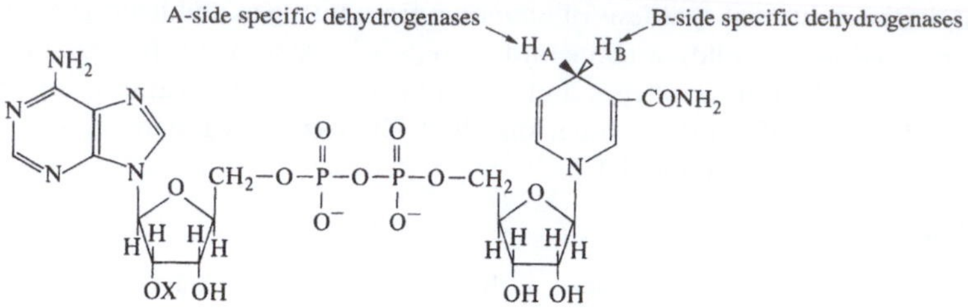

A-side specific dehydrogenases   B-side specific dehydrogenases

H_A H_B

$CONH_2$

NH_2

$CH_2-O-\overset{\overset{\displaystyle O}{\|}}{P}-O-\overset{\overset{\displaystyle O}{\|}}{P}-O-CH_2$

O^- O^-

OX OH   OH OH

Reduced nicotinamide adenine dinucleotide (NADH), X = H, and reduced nicotinamide adenine dinucleotide phosphate (NADPH), X = phosphate.

A-side dehydrogenases	**B-side dehydrogenases**
Alcohol dehydrogenase (NAD+)	Glycerol 3-phosphate dehydrogenase (NAD+)
Lactate dehydrogenase (NAD+)	3-Hydroxybutyrate dehydrogenase (NAD+)
Malate dehydrogenase (NAD+)	Glucose dehydrogenase (NAD+)
Cytochrome b_5 reductase (NAD+)	Glyceraldehyde-phosphate dehydrogenase (NAD+)
Shikimate dehydrogenase (NADP+)	Homoserine dehydrogenase (NADP+)
3-Oxyacyl [acyl-carrier-protein] reductase (NADP+)	Glucose 6-phosphate dehydrogenase (NADP+)

Fig. 1.4 Stereospecificity of NAD+- and NADP+-requiring enzymes. For a more complete list see reference 33.

example is the reaction catalysed by fumarate hydratase in which fumarate is converted to malate. It can be shown by carrying out the reaction in the presence of tritiated water (^3H_2O) that ^3H addition occurs in a stereospecific manner.

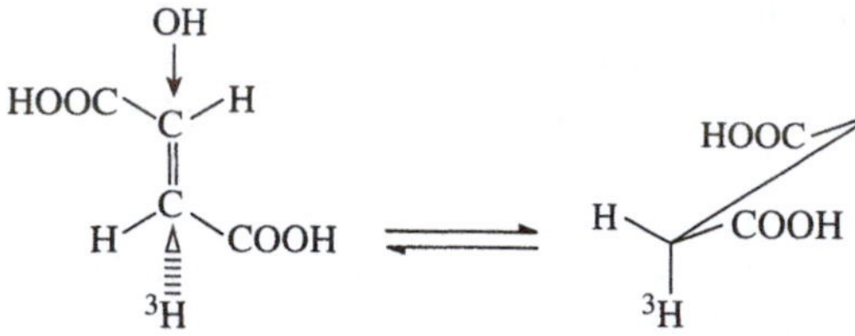

OH

HOOC–C–H

C

H   COOH

^3H

⇌

OH

HOOC   H

H   COOH

^3H

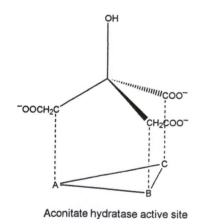

Aconitate hydratase active site

Fig. 1.5 Diagram of a three-point attachment to the enzyme surface. Citrate lacks a chiral centre, but when it makes a three-point attachment to the asymmetrical enzyme surface (A, B, C) of aconitate hydratase the two methylene groups can be distinguished.[36]

There are several other examples of this phenomenon, but historically the most important is that of aconitate hydratase, the enzyme catalysing the conversion of citrate into *cis*-aconitate. Citrate, a molecule without a chiral centre, is the first intermediate of the citric acid cycle. In 1941 it was shown that [^{13}C]-labelled citrate gave rise to 2-oxoglutarate, labelled only in the C_2 and not the C_4 position. This finding led to the suggestion that citrate was not in fact an intermediate of the citric acid cycle, which was accordingly named the tricarboxylic acid cycle. The situation was resolved in 1948 when Ogston pointed out that although citrate had no chiral centre, if it made a three-point attachment (Fig. 1.5) with aconitate hydratase the two methylene ($-CH_2-$) groups on citrate could be distinguished. This could account for the observed differential labelling of 2-oxoglutarate and so 'citrate returned to the tricarboxylic acid cycle'.[36]

A further aspect of enzyme specificity has been recognized as important in DNA replication and protein synthesis. It is known that the error rate in DNA replication *in vivo* is as low as one mistake in 10^{10} nucleotides polymerized. This is far lower than can be accounted for in terms of the preference of hydrogen bonding in Watson–Crick base pairs (Adenine with Thymine and Guanine with Cytosine) which would lead to an error rate of one base in about 10^4. The much lower rate observed is due to the existence of a proofreading, or editing, mechanism in DNA polymerase which examines the base which has just been added to the growing strand, and removes any incorrectly added base by hydrolysis (this is known as 3′-5′ exonuclease activity). The hydrolysis occurs at an active site distinct from the polymerase site.[37] A proofreading mechanism also operates in the formation of aminoacyl-tRNAs catalysed by some aminoacyl-tRNA ligases.[38] The charging reaction occurs in two steps

amino acid + ATP + enzyme → enzyme–aminoacyl-AMP + pyrophosphate

enzyme–aminoacyl-AMP + tRNA → aminoacyl-tRNA + AMP + enzyme

The enzyme has to recognize a specific amino acid and also a specific tRNA. The latter is recognized very precisely since it is a large molecule and thus makes many contacts with the enzyme. The amino acids, which are very much smaller, are not selected so precisely. It is possible to monitor both the first and second steps in the reaction and this shows that significant errors do occur in the formation of the aminoacyl-adenylate intermediates but not in the formation of aminoacyl-tRNAs. However, any incorrect aminoacyl-adenylates formed rapidly become hydrolysed; this editing occurs at a second site, distinct from that on which the synthesis occurs.[39]

1.3.3 Regulation

The third important property of enzymes is that their catalytic activity may be regulated by small ions or other molecules, or by small changes in their covalent structure. A good example of these effects is provided by the enzyme phosphorylase, which catalyses the first step in the breakdown of glycogen in skeletal muscle. This is an important enzyme that enables the carbohydrate reserves to be degraded in order to generate ATP required for muscle contraction. The onset of muscle contraction is triggered by release of Ca^{2+} from the sarcoplasmic reticulum (see Chapter 8, Section 8.4.4) and this also brings about the activation of phosphorylase (by the phosphorylation of a single serine side-chain on each subunit of the enzyme) to ensure the continued production of ATP. The mechanisms of regulation of phosphorylase both by Ca^{2+} and also by hormones such as adrenalin involve a complex sequence of events and are discussed in Chapter 6, Section 6.4.2.

The phenomenon of feedback inhibition is common in many biosynthetic pathways. For example, in the biosynthetic pathway leading to the synthesis of pyrimidine nucleotides, the end-products UTP and CTP are able to inhibit the first enzyme in the pathway; thus they are able to limit the flow of metabolites into that pathway and so regulate their own biosynthesis. This regulation is effected through changes in the catalytic activity of early enzymes in the pathway, carbamoyl-phosphate synthase and aspartate carbamoyltransferase; this topic is discussed in Chapter 6, Sections 6.2.2.1 and 6.3.1.

1.4 Cofactors

The demonstration of enzyme crystallinity and analysis of these crystals have shown that enzymes are proteins with chain lengths generally of 100–2500 amino acids. Enzymes such as chymotrypsin (a protein-hydrolysing enzyme) or triosephosphate isomerase (an enzyme catalysing the interconversion of two phosphorylated three-carbon compounds) are active without needing any other factor present. However, many enzymes require a non-protein component for activity; this is termed a *cofactor*. One group of cofactors comprises metal ions. Thus,

Table 1.2 Examples of the metal ion requirements of some enzymes

Metal	Enzyme
Na	Intestinal sucrose α-D-glucohydrolase
K	Pyruvate kinase (also requires Mg)
Mg	Kinases (e.g. hexokinase, pyruvate kinase), adenosinetriphosphatases (e.g. myosin adenosinetriphosphatase)
Fe	Catalase, peroxidase, nitrogenase
Zn	Alcohol dehydrogenase, carboxypeptidase, carbonic anhydrase
Mo	Xanthine oxidase, nitrogenase
Cu	Cytochrome c oxidase, amine oxidase
Ni	Urease, hydrogenase, carbon monoxide dehydrogenase, superoxide dismutase
Mn	Histidine ammonia lyase
V	Nitrogenase

Further information on the metal ions in enzymes can be found in references 40–44.

EDTA

$Mg\text{-}ATP^{2-}$

* $NAD(P)^+$ is often listed as a cofactor, and is generally known as a *coenzyme*. In fact it functions as a normal substrate for dehydrogenases and in our view is best considered as a substrate. It does, however, differ from most cell metabolites in that it becomes alternately reduced and oxidized. In this respect it does resemble redox cofactors. The same arguments apply to coenzyme A which can also be regarded as a substrate. Pantotheine, however, which is covalently attached to certain fatty-acid synthase systems (see Chapter 7, Section 7.11.1), is a cofactor. The problems of nomenclature of cofactors, coenzymes, and prosthetic groups are discussed further in reference 45.

† Old yellow enzyme is now termed NADPH: acceptor oxidoreductase, catalase is hydrogen peroxide: hydrogen peroxide oxidoreductase, papain and trypsin do not have systematic names because of the wide variety of substrates acted upon, and rhodanese is termed thiosulphate sulphurtransferase.

carboxypeptidase (an enzyme catalysing the hydrolysis of proteins from the C-terminal end) requires zinc for activity. Removal of zinc (e.g. by EDTA) leads to inactivation of the enzyme: activity can be restored by addition of zinc or, to a lesser extent, by other metals. In this case, it appears that zinc forms an integral part of the enzyme structure. The kinases (enzymes that catalyse the transfer of the γ-phosphoryl group of ATP to some acceptor molecule) have a requirement for magnesium ions; in fact the metal ion is actually bound to substrate rather than to the enzyme so that the true substrate is $Mg\text{-}ATP^{2-}$ rather than ATP. Some examples of metal ion requirements among enzymes are given in Table 1.2.

The second major class of cofactors comprises organic cofactors (many of which are derivatives of B vitamins). Thus most carboxylases (enzymes involved in the incorporation of CO_2) require biotin, which is covalently linked to its enzyme. Similarly, in the case of aminotransferases (enzymes involved in inter-conversions of oxo-acids and amino acids), pyridoxal phosphate is bound to the enzyme as a Schiff base. Some examples of organic cofactors are listed in Table 1.3.*

Tightly bound cofactors (which cannot be removed by dialysis) are often termed *prosthetic* groups. An enzyme containing a cofactor or prosthetic group is termed a *holoenzyme:* when the cofactor is removed, the resulting protein is termed an *apoenzyme*. Any small molecule or other species which can reversibly bind to an enzyme is termed a *ligand* (this can include substrates, inhibitors, metal ions, etc.).

1.5 Nomenclature and classification of enzymes

1.5.1 General classification

As in the development of organic chemistry, many enzymes were given 'trivial' names before any attempt was made at a system of nomenclature. In general, the 'trivial' names consisted of the suffix '-ase' added to the substrate acted on (as in the case of urease), or implied something about the reaction catalysed (as in the case of lactate dehydrogenase, which catalyses the dehydrogenation of lactate to yield pyruvate). However, some trivial names are distinctly unhelpful in this respect, e.g. old yellow enzyme, catalase, papain, trypsin, rhodanese, etc.†

Table 1.3 Examples of organic cofactors

Cofactor	Linkage to apoenzyme	Type of reactions catalysed by holoenzyme
Pyridoxal phosphate 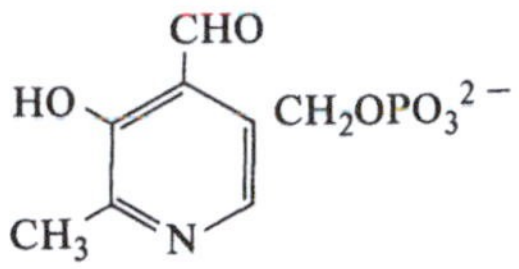	Usually Schiff base to lysine residue	Transamination, decarboxylation, racemization
Biotin 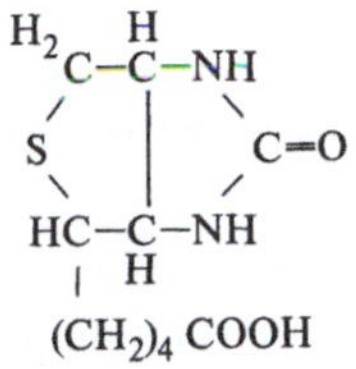	Amide bond to lysine residue	Carboxylation reactions, e.g. acetyl-CoA carboxylase, pyruvate carboxylase
Lipoic acid 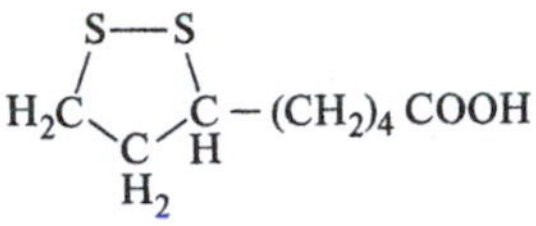	Amide bond to lysine residue	Acyl transfer, e.g. pyruvate dehydrogenase and 2-oxoglutarate dehydrogenase systems
Thiamin diphosphate 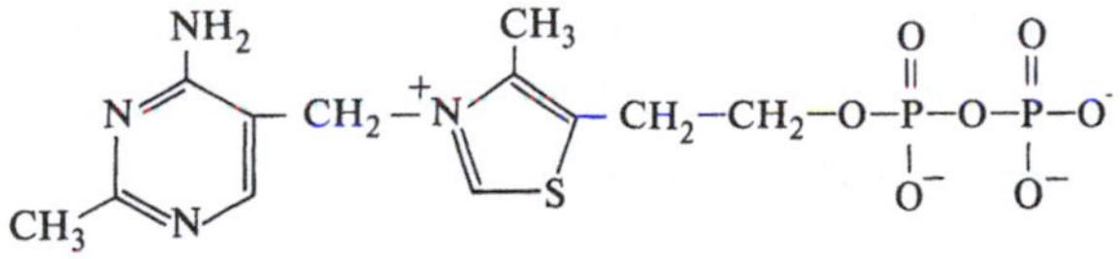	Non covalent binding dissociation constant $\approx 10^{-6}$ mol dm^{-3}	Decarboxylation of 2-oxo-acids, e.g. pyruvate dehydrogenase and 2-oxoglutarate, transketolase dehydrogenase systems
Flavin nucleotides: flavin adenine dinucleotide (FAD) and flavin mononucleotide (FMN) 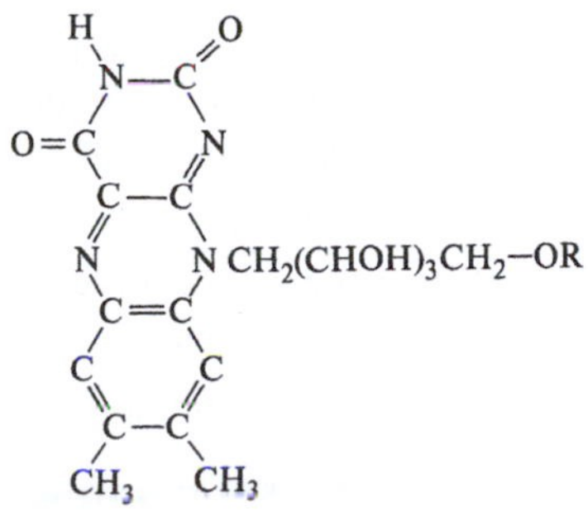	Non-covalent in, e.g. amino acide oxidase; covalent link in succinate dehydrogenase	Redox reactions, e.g. xanthine oxidase (FAD), succinate dehydrogenase (FAD), glucose oxidase (FAD), NADPH-cytochrome reductase (FMN)

Pyridoxal phosphate structure:

CHO

HO — CH$_2$OPO$_3{}^{2-}$

CH$_3$ — N

Biotin structure:

H$_2$C–C–NH

S   C=O

HC–C–NH

H

(CH$_2$)$_4$ COOH

Lipoic acid structure:

S—S

H$_2$C   C – (CH$_2$)$_4$ COOH

C H

H$_2$

Thiamin diphosphate structure:

NH$_2$ CH$_3$ O   O

N — CH$_2$—$^+$N   CH$_2$—CH$_2$—O–P–O–P–O$^-$

CH$_3$ — N   S   O$^-$ O$^-$

Flavin structure:

H   O

N—C

O=C   N

C—C

N   N CH$_2$(CHOH)$_3$CH$_2$—OR

C=C

C   C

C—C

CH$_3$ CH$_3$

(FMN, R = phosphate)
(FAD, R = ADP)

* An Enzyme Nomenclature Database can be accessed through the internet (http://www.expasy.ch). From this database an enzyme can be accessed by its official name, alternative name, EC number, its class, a reactant, or cofactor. Each entry gives details of the reaction catalysed, its position in a metabolic pathway, and, if appropriate, any human deficiency diseases in which it has been implicated.

The present-day accepted nomenclature of enzymes is that recommended by the Enzyme Commission, which was set up in 1955 by the International Union of Biochemistry (now known as the International Union of Biochemistry and Molecular Biology) in consultation with the International Union of Pure and Applied Chemistry. Enzymes are named according to certain well-defined rules (for details see references 46 and 47).*

The six major types of enzyme-catalysed reactions are:

1. oxidation–reduction reactions, catalysed by *oxidoreductases*;

2. group transfer reactions, catalysed by *transferases*;

3. hydrolytic reactions, catalysed by *hydrolases*;

4. elimination reactions in which a double bond is formed, catalysed by *lyases*;

5. isomerization reactions, catalysed by *isomerases*;

6. reactions in which two molecules are joined at the expense of an energy source (usually ATP), catalysed by *ligases*.

The full systematic name of an enzyme not only shows the type of reaction catalysed but describes the substrate(s) acted on and any other important information. One of the enzymes from each major group will now be given to illustrate the classification scheme.

Group 1, e.g. lactate dehydrogenase. The full systematic name is (*S*)-lactate: NAD^+ oxidoreductase, which denotes the fact that (*S*)-lactate acts as the electron donor and NAD^+ as the electron acceptor. It should be noted that the substrate is usually designated as L-lactate in the biochemical literature, so the enzyme is referred to as L-lactate dehydrogenase:

$$(S)\text{-Lactate} + NAD^+ = \text{pyruvate} + NADH.$$

Group 2, e.g. hexokinase. The full systematic name is ATP: D-hexose 6-phosphotransferase, indicating that ATP is the phosphate donor, a D-hexose the phosphate acceptor, and that transfer is to the hydroxyl on the 6-carbon atom of the hexose:

$$ATP + \text{D-hexose} = ADP + \text{D-hexose 6-phosphate}.$$

Group 3, e.g. adenosinetriphosphatase. The full systematic name is ATP phosphohydrolase, indicating that ATP, the substrate, is hydrolysed so as to split the bond which allows release of orthophosphate:

$$ATP + H_2O = ADP + \text{orthophosphate}.$$

Group 4, e.g. fructose-bisphosphate aldolase. The full systematic name is D-fructose 1,6-bisphosphate D-glyceraldehyde 3-phosphate-lyase, which indicates that the substrate D-fructose 1,6-bisphosphate is cleaved in such a way as to yield D-glyceraldehyde 3-phosphate as one of the products:

$$\text{D-Fructose 1,6-bisphosphate} = \text{glycerone phosphate} + \text{D-glyceraldehyde}$$
$$3\text{-phosphate}.$$

(It should be noted that glycerone phosphate is more usually known as dihydroxyacetone phosphate.)

Group 5, e.g. triosephosphate isomerase. The full systematic name is D-glyceraldehyde 3-phosphate ketol-isomerase, indicating that the aldose D-glyceraldehyde

3-phosphate is isomerized to the ketose glycerone phosphate (dihydroxyacetone phosphate):

$$\text{D-Glyceraldehyde 3-phosphate} = \text{glycerone phosphate.}$$

Group 6, e.g. isoleucine-tRNA ligase. The full systematic name is L-isoleucine: tRNA^{Ile} ligase (AMP-forming) and it indicates that L-isoleucine becomes bonded to a specific tRNA acceptor (tRNA^{Ile}) and that in the process ATP is split to AMP and pyrophosphate:

$$\text{ATP + L-isoleucine + tRNA}^{Ile} = \text{AMP + pyrophosphate + L-isoleucyl-tRNA}^{Ile}.$$

The Enzyme Nomenclature Committee of the IUBMB has recommended that Group 6 enzymes be termed ligases and that the use of the term synthetase be discouraged. The term synthase is used wherever it is desired to emphasize the synthetic nature of the reaction. This includes enzymes in both groups 4 and 6. For example, the substrates of tryptophan synthase (EC 4.2.1.20) are serine and (indol-3-yl)glycerol 3-phosphate and the main product is L-tryptophan but there is no concomitant hydrolysis of the pyrophosphate bond in ATP or similar triphosphate (for details see Chapter 7, Section 7.9). The term ligase is used in conjunction with the reactants in an enzyme-catalysed synthesis, e.g. acetate-CoA ligase (EC 6.2.1.1) in which acetyl-CoA is formed from acetate, ATP, and CoA. However, in some group 6 enzymes it is less cumbersome and preferable to emphasize the product rather than the reactants, e.g. carbamoyl-phosphate synthase (EC 6.3.5.5). In this case the reactants are glutamine, carbon dioxide, water, and ATP, and the main product is carbamoyl-phosphate (for details see Chapter 7, Section 7.10).

Many of the systematic names of enzymes are rather cumbersome and for everyday use shorter names can be used (e.g. lactate dehydrogenase, hexokinase, etc.). These shorter names are known as *recommended names*.

Enzymes are further classified by being assigned an Enzyme Commission (EC) number consisting of four parts (a, b, c, d).

The first number (a) indicates the type of reaction catalysed and can take values from 1 to 6 according to the classification of reaction types given above.

The second number (b) indicates the subclass, which usually specifies the type of substrate or the bond cleaved more precisely. In the case of oxidoreductases, for example, this shows the type of chemical grouping acting as an electron donor, whereas with hydrolases it would show the type of bond that is being broken.

The third number (c) indicates the sub-subclass, allowing an even more precise definition of the reaction catalysed in terms of the type of electron acceptor (oxidoreductases) or the type of group removed (lyases), etc.

The fourth number (d) indicates the serial number of the enzyme in its sub-subclass.

Following these rules the EC numbers of lactate dehydrogenase, hexokinase, adenosinetriphosphatase, fructose-bisphosphate aldolase, triosephosphate isomerase, and isoleucine-tRNA ligase are 1.1.1.27, 2.7.1.1, 3.6.1.3, 4.1.2.13, 5.3.1.1, and 6.1.1.5, respectively. All the enzymes mentioned in this book are given their recommended names, together with their EC numbers, and the reactions catalysed in the Appendix at the end of the book. (There are a few enzymes that have not yet been assigned EC numbers because the reactions they catalyse have not been precisely determined.)

It should be noted that the system of nomenclature and classification of enzymes is based only *on the reaction catalysed* and takes no account of the origin of the enzyme (i.e. from the species or tissue it derives). Enzymes catalysing the same reaction but isolated from different species may well have substantially different amino-acid sequences and may even act via different catalytic mechanisms, but these points will not be distinguished in the classification system. For example, the adenosinetriphosphatases from the inner mitochondrial membrane and from the sarcoplasmic reticulum are not distinguished in this classification scheme, although the former is concerned with the transport of protons across the inner mitochondrial membrane and the latter with the transport of Ca^{2+} across the sacroplasmic reticulum (for details see Chapter 8, Section 8.4.4). Both catalyse hydrolysis of ATP. Another example is the two classes of dehydroquinase discussed in Chapter 5, Section 5.5.3. Therefore, for a more exact specification of an enzyme the source should also be mentioned, and if necessary the isoenzyme type. The meaning of the term isoenzyme is discussed below.

1.5.2 Isoenzymes

Within a single species there may exist several different forms of enzyme catalysing the same reaction. These could differ from one another in terms of amino-acid sequence, some covalent modification (e.g. phosphorylation of serine hydroxyl groups), or possibly in terms of three-dimensional structure (conformational changes), etc. The term *isoenzyme* should be restricted to those forms of an enzyme that arise from *genetically determined differences in amino-acid sequence* and should not be applied to those derived by modification of the same amino-acid sequence.*

* See reference 48 for recommendations on nomenclature of isoenzymes.

Thus, for example, heart muscle malate dehydrogenase (which catalyses the oxidation of malate to oxaloacetate) occurs both in the cytoplasm and in the mitochondria; these two forms of the enzyme are termed *isoenzymes* of malate dehydrogenase. Similarly, in serum or in muscle extracts, lactate dehydrogenase occurs in a variety of forms (see Chapters 5 and 10, Sections 5.5.4.1 and 10.2); these are also referred to as isoenzymes. It is recommended that the naming of isoenzymes should be based on the extent of migration in an applied electric field using the technique of *electrophoresis* (see Chapter 2, Section 2.6.2.2) rather than on the basis of tissue distribution (e.g. brain type, muscle type, etc.) since this distribution can vary between different species (or in a single species or tissue with the stage of development). Electrophoresis is recommended as the basis of classification because it is a widely used technique with high resolving power. Isoenzymes are numbered starting with the species having highest mobility towards the anode (see for example Fig. 10.1).

1.5.3 Multienzyme systems

Multienzymes are proteins that exhibit more than one catalytic activity and are described in some detail in Chapter 7. For the purposes of nomenclature, the Enzyme Commission recommendation is that where more than a single catalytic activity is to be ascribed it should be referred to as a *system*. Thus, for example, the unit that contains all the catalytic entities for the synthesis of fatty acids is referred to as the fatty-acid synthase system. However, since each enzyme-catalysed reaction has an EC number and a recommended name, it follows that multifunctional enzymes will have more than one EC number and position in the classification scheme. Two examples that are discussed later in the book will serve to illustrate this. The debranching enzyme (see Chapter 6, Section 6.4.2) that catalyses the removal of 1,6-branches in glycogen is a single polypeptide chain having two catalytic activities, amylo-1,6-glucosidase and 4-α-glucanotransferase, and thus appears twice in the classification scheme as EC 3.2.1.33 and EC 2.4.1.25. Similarly, mammalian fatty acid synthase (see Chapter 7, Section 7.11.1) is a multi-enzyme polypeptide catalysing reactions of EC 2.3.1.38, EC 2.3.1.39, EC 2.3.1.41, EC 1.1.1.100, EC 4.2.1.61, and EC 1.3.1.

A further problem arises with the nomenclature of multienzyme systems, namely, that some consist of several polypeptide chains each having a distinct catalytic activity and associated with one another by non-covalent bonds, whilst others, like the two examples just described, exist as single polypeptide chains having multiple catalytic sites. Some have a combination of both. The Nomenclature Committee of the IUBMB (1992) recommends that the term multienzyme complex is used for multienzymes with catalytic domains* on more than one polypeptide chain, and multienzyme polypeptide for polypeptide chains containing at least two types of catalytic domains.[49]

> * A domain represents an independently folded, globular, unit of a polypeptide chain (see Chapter 3, Section 3.4.4.5).

1.6 The contents of this book

In the next five chapters we shall be discussing the behaviour of isolated enzymes, dealing in turn with isolation methods (Chapter 2), structural characterization (Chapter 3), kinetics (Chapter 4), catalytic action (Chapter 5), and control of activity (Chapter 6). These are areas in which there is now a considerable body of knowledge and understanding. The methods for isolation and characterization of enzymes are now well-established procedures, so the rate at which three-dimensional structures and mechanisms are being determined is increasing dramatically. Ultimately, of course, we want to know how enzymes behave in living cells. This involves in part a synthesis of the information obtained from the study of isolated enzymes, but it also requires detailed knowledge of the molecular morphology of the cell, which in turn requires methods for making measurements on intact cells. There are a large number of unanswered questions in this area, such as, what is the state of proteins in the cytosol? Are they just part of an intracellular soup or do they have an organization that so far has not been characterized? What are the concentrations of free ligands such as metal ions in cells? It is relatively straightforward to determine the total concentration of a ligand in cells, but very difficult to ascertain the extent to which it is bound to different macromolecules. Many of these questions require new methods of investigation. Some of these aspects are considered in Chapters 7 and 8 on multienzyme systems and enzymes in the cell, respectively.

As already mentioned, there are a large number of different enzymes in each cell. The amounts of certain enzymes are relatively constant throughout the life of a cell, whereas others vary with physiological conditions. How these amounts are controlled is the topic of Chapter 9 on enzyme turnover.

Of the 3196 different enzyme activities listed in the 1992 classification scheme, over 500 are commercially available in a purified or partially purified state. The number of restriction endonucleases isolated from a number of sources now exceeds 1900, and about 300 are commercially available.[32] This has meant that many enzymes can be used as reagents on either an analytical or a preparative scale, a process which has been greatly assisted by developments in recombinant DNA technology. The study and application of enzymes have assumed increasing importance both in medicine and in industry and a discussion of these aspects is therefore included in the final two chapters of the book.

References

1. Blattner, F. R., Plunkett, G., Bloch, C. A., Perna, N. T., Burland, V., Riley, M. *et al.*, *Science* **277**, 1453 (1997).

2. Clayton, R. A., White, O., Ketchum, K. A., and Venter, J. C., *Nature* **387**, 459 (1997).

3. Fruton, J. S., *Trends Biochem. Sci.* **3**, N281 (1978).

4. Dixon, N. E., Gazzola, C., Blakeley, R. L., and Zerner, B., *J. amer. chem. Soc.* **97**, 4131 (1975).

5. Karplus, P. A., Pearson, M. A., and Hausinger, R. P., *Acc. Chem. Res.* **30**, 330 (1997).

6. Chen, L. H., Kenyon, G. L., Curtin, F. *et al.*, *J. biol. Chem.* **267**, 17716 (1992).

7. Hirs, C. H. W., Moore, S., and Stein, W. H., *J. biol. Chem.* **235**, 633 (1960).

8. Koshland, D. E. Jr, *Proc. natn. Acad. Sci. USA* **44**, 98 (1958).

9. Monod, J., Changeux, J.-P., and Jacob, F., *J. molec. Biol.* **6**, 306 (1963).

10. Monod, J., Wyman, J., and Changeux, J.-P., *J. molec. Biol.* **12**, 88 (1965).

11. Gutte, B. and Merrifield, R. B., *J. amer. chem. Soc.* **91**, 509 (1969).

12. Wells, J. A. and Estell, D. A., *Trends Biochem. Sci.* **13**, 291 (1988).

13. Pingoud, A. and Jeltsch, A., *Eur. J. Biochem.* **246**, 1 (1997).

14. Lerner, R. A. and Tramontano, A., *Trends Biochem. Sci.* **12**, 427 (1987).

15. Lerner, R. A., Benkovic, S. J., and Schultz, P. G., *Science* **252**, 659 (1991).

16. Wade, H. and Scanlon, T. S., *A. Rev. Biophys. Biomol. Struct.* **26**, 461 (1997).

17. Cech, T. R., *A. Rev. Biochem.* **59**, 543 (1990).

18. Narlikar, G. J. and Herschlag, D., *A. Rev. Biochem.* **66**, 19 (1997).

19. Zhang, B. and Cech, T. R., *Nature* **390**, 96 (1997).

20. James, H. A. and Turner, P. C., *Essays Biochem.* **29**, 175 (1995).

21. Scott, W. G. and Klug, A., *Trends Biochem. Sci.* **21**, 220 (1996).

22. Birikh, K. R., Heaton, P. A., and Eckstein, F., *Eur. J. Biochem.* **245**, 1 (1997).

23. North, G., *Nature, Lond.* **328**, 18 (1987).

24. Dixon, M., in *Chemistry of life* (Needham, J., ed.), p. 15. Cambridge University Press (1971).

25. *FEBS Lett.* Vol. 62, Suppl., Enzymes: 100 years (1976).

26. Koshland, D. E., *J. cell. comp. Physiol.* **47**, Suppl. **11**, 217 (1956).

27. Radzicka, A. and Wolfenden, R., *Science* **267**, 90 (1995).

28. Jaenicke, R., Schurig, H., Beaucamp, N., and Ostendorp, R., *Adv. Prot. Chem.* **48**, 181 (1996).

29. Burris, R. H., *J. biol. Chem.* **266**, 9339 (1991).

30. Shreve, R. N. and Brink, J. A., *Chemical process industries* (4th edn), pp. 276–80. McGraw-Hill, New York (1977).

31. Laidler, K. J. and Bunting, P. S., *The chemical kinetics of enzyme action* (2nd edn), p. 256. Clarendon Press, Oxford (1973).

32. Bhagwat, A. S., *Methods Enzymol.* **216**, 199 (1992).

33. Branden, C.-I. and Ekland, H., in *Dehydrogenases requiring nicotinamide coenzymes* (Jeffrey, J., ed.), p. 63. Birkhauser, Basel (1980).

34. You, K., Arnold, L. J., Allison, W. S., and Kaplan, N. O., *Trends Biochem. Sci.* **3**, 265 (1978).

35. Goodwin, T. W., *Essays Biochem.* **9**, 103 (1973).

36. Barry, M. J., *Trends Biochem. Sci.* **22**, 228 (1997).

37. Joyce, C. M. and Steitz, T. A., *A. Rev. Biochem.* **63**, 777 (1994).

38. Fersht, A. R., *Trends Biochem. Sci.* **5**, 262 (1980).

39. Schimmel, P., *A. Rev. Biochem.* **56**, 125 (1987).

40. Fraústo da Silva, J. J. R. and Williams, R. J. P., *The biological chemistry of the elements: the inorganic chemistry of life*. Oxford University Press (1991).

41. Kaim, W. and Schwerderski, B., *Bioinorganic chemistry: inorganic elements in the chemistry of life: an introduction and guide*. Wiley, Chichester (1994).

42. Fridovich, I., *A. Rev. Biochem.* **64**, 97 (1995).

43. Kisker, C., Schinderlin, H., and Rees, D. C., *A. Rev. Biochem.* **66**, 233 (1997).

44. Hales, B. J., Case, E. E., Morningstar, J. E., Dzeda, M. F., and Mauterer, L. A., *Biochemistry* **25**, 725 (1986).

45. Bryce, C. F. A., *Trends Biochem. Sci.* **4**, N62 (1979).

46. Dixon, M., Webb, E. C., Thorne, C. J. R., and Tipton, K. F., *Enzymes* (3rd edn). Longman, London (1979).

47. *Enzyme nomenclature*, Recommendations (1992) of the Nomenclature Committee of the International Union of Biochemistry and Molecular Biology. Academic Press, London (1992).

48. Nomenclature of multiple forms of enzymes, *Biochem. J.* **171**, 37 (1978).

49. *Enzyme nomenclature*, Recommendations (1992) of the Nomenclature Committee of the International Union of Biochemistry and Molecular Biology, pp. 562–3. Academic Press, London (1992).

2

The purification of enzymes

2.1 Introduction

In this chapter we shall discuss the isolation of enzymes, concentrating initially on the principles underlying the more important separation methods employed in this work. We shall then illustrate how these methods are used in practice by considering seven specific examples of purification of enzymes; the examples have been chosen to give a broad coverage of the methods and problems involved in such procedures. However, it is necessary to deal first with some more general questions regarding the need for, and strategy underlying, purification procedures.

2.2 Why isolate enzymes?

It is not difficult to appreciate that if we hope to gain a detailed understanding of the behaviour of an enzyme in a complex system (be it a subcellular organelle such as the mitochondrion, a cell, or a whole organism) we must first try to understand its properties in as simple a system as possible. In many cases this simple system would consist of a solution of enzyme in a medium containing only small ions, buffer molecules, cofactors, etc. However, in some cases, e.g. enzymes that are bound to cell membranes (see Chapter 8, Section 8.4), the isolated enzyme may be inactive in the absence of phospholipid or detergent, and the simplified system would need to contain these additional components.

From studies of the isolated enzyme we can learn about its specificity for substrates, kinetic parameters for the reaction (see Chapter 4), and possible means of regulation (see Chapter 6). All these features would be useful in understanding the role of the enzyme in the more complex systems (see Chapters 7 and 8). In addition, enzymes pose some extremely intriguing questions about the structure of large molecules (see Chapter 3) and mechanisms of catalysis (see Chapter 5). Detailed studies of these aspects are only possible if we have been able to purify enzymes so as to remove contaminating enzymes and other large molecules. The ready availability of isolated enzymes has been of considerable value in a number of medical and industrial applications (see Chapters 10 and 11, respectively).

2.3 Objectives and strategy in enzyme purification

2.3.1 Objectives

The aim of a purification procedure should be to isolate a given enzyme with the *maximum possible yield*, based on the percentage recovered activity compared with the total activity in the original extract. In addition, the preparation should possess the *maximum catalytic activity*, i.e. there should be no degraded or other inactivated enzyme present, and it should be of the *maximum possible purity*, i.e. it should contain no other enzymes or large molecules. In the early days of enzyme purification, crystallinity was taken to be proof of purity but there are now known to be a number of crystalline enzymes[1] that are impure, and indeed the majority of present-day purification procedures do not involve a crystallization step (e.g. RNA polymerase; Section 2.8.3). Of course, crystallization is necessary if the three-dimensional structure of an enzyme is to be elucidated by X-ray crystallography (see Chapter 3, Section 3.4.1). Although the conditions which lead to growth of suitable crystals are not always predictable,[2] it is clear that successful crystallization requires enzyme preparations of very high purity. In addition, if the enzyme is to be used for therapeutic purposes (see for example Chapter 10, Section 10.9), it is essential that the preparation is of extremely high purity and free from pyrogens for instance.

The catalytic activity of a preparation is determined by a suitable assay procedure (Chapter 4, Section 4.2) in which the rate of disappearance of substrate or the rate of appearance of product is determined under defined conditions of substrate concentration, temperature, pH, etc. The units of activity are usually expressed either as μmol substrate consumed, or product formed, per minute ('units' or 'international 'units') or as mol substrate consumed, or product formed, per second ('katal' in the SI system). There is no way of predicting the catalytic activity of a purified enzyme under a given set of conditions and so purification is carried out until the *specific activity* of the preparation (i.e. units per mg or katal per kg*) increases to reach a constant value that is not increased by further purification steps. The criteria used in assessing the purity of a preparation involve the application of analytical methods to detect contaminants: these are dealt with in more detail in Section 2.7.

The rapid development since the early 1980s of recombinant DNA technology has changed the emphasis of much enzyme purification work. In many cases the amount of enzyme which can be isolated using conventional sources and methods is far too small to permit detailed structural and functional investigations to be undertaken since these usually require tens to hundreds of milligrams. Under such circumstances, the initial objective is therefore to isolate only a very small amount (of the order of a few micrograms) of enzyme, perhaps even as a band on a gel, from which sufficient amino-acid sequence information can be obtained to permit the design of an oligonucleotide probe. The probe can then used to identify the gene of interest in a suitable DNA library; the cloned gene can then be overexpressed in a suitable host to allow the production of much larger quantities of the enzyme (Section 2.4.1). In many cases the enzyme of interest can constitute a very high percentage (even up to 50 per cent) of the total cell protein, so the production of very pure enzyme is then relatively straightforward, perhaps involving only one or two steps.

2.3.2 Strategy

The steps involved in the purification of an enzyme can be discussed in terms of the type of 'flow sheet' shown in Fig. 2.1. Within this general scheme, the pro-

* It should be noted that 60 units per mg correspond to 1 katal per kg.

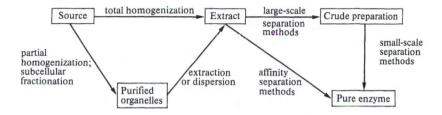

Fig. 2.1. Steps involved in the purification of enzymes.

cedure to be adopted for a given enzyme will involve choices of: (i) source of enzyme; (ii) methods of homogenization; and (iii) methods of separation. These choices are discussed in the following sections (2.4 to 2.6). The progress of a purification is recorded in a purification table (see Section 2.8.1). It should be noted that the purification of overexpressed recombinant enzymes usually involves small-scale or affinity methods.

2.4 **Choice of source**

A number of factors influence the choice of starting material in an enzyme purification. Among the most important of these are the following.

2.4.1 **Abundance of enzyme**

The classical approach to enzyme purification involves choosing a source in which the required enzyme occurs in large amounts. Thus, lactating mammary gland is an excellent source of enzymes such as acetyl-CoA carboxylase involved in fatty-acid biosynthesis,[3] and kidney is a good source of hydrolytic enzymes such as alkaline phosphatase, an enzyme that catalyses the hydrolysis of phosphate esters at alkaline pH. It had also been recognized for many years that the levels of expression of many enzymes could be manipulated by changes in the culture conditions; thus synthesis of β-D-galactosidase can be induced in *E. coli* by growing the bacteria in a medium containing lactose as the sole source of carbon.[4] While the use of wild-type organisms is still important in many cases, particularly where an abundant enzyme is required as a reagent (Chapters 10 and 11), it is becoming increasingly true that the power of recombinant DNA technology can be harnessed to increase the production of an enzyme of interest whatever its source.[5] An early example of the usefulness of the recombinant-DNA approach involved studies[6] on the enzyme 3-phosphoshikimate-1-carboxyvinyl transferase (PSCV transferase)* in *E. coli*, which is involved in the biosynthesis of aromatic amino acids. The gene for this enzyme was isolated and inserted into the multicopy plasmid pAT153. The modified plasmid was used to transform *E. coli* to give a new strain that produced approximately 100-fold more PSCV transferase than the wild-type strain; 4.8 mg pure enzyme could be isolated from 18 g cells as a result of a 50-fold purification.[6] Other examples of the recombinant-DNA approach are provided by a number of the examples discussed in Section 2.8.

   The basic requirements for the successful production of recombinant enzyme are the isolation of the gene encoding that enzyme and the development of a suitable expression system for the gene. While details of these techniques can be found in a number of books,[5,7,8] it is worthwhile mentioning briefly some of the factors involved in choice of a suitable expression system. In general, expression is more likely to be successful when the host organism is either the same as, or is

* This enzyme is also known as 3-enolpyruvoylshikimate-5-phosphate synthase

closely related to, the organism whose gene is being expressed. In every case, it is necessary to check that the level of expression in the non-transformed organism is very much smaller than that of the transformed strain. Production of a null mutant strain in which the gene encoding the wild-type enzyme has been deleted may offer considerable advantages.[9]

Prokaryotes offer potentially great advantages as host organisms in terms of their rapid growth and relatively simple nutritional requirements. The gene to be overexpressed is generally incorporated in a plasmid (extra-chromosomal DNA) under the control of a strong promoter, so that the expression of the gene can be induced by addition of an inducer or some other change in the culture medium. Disadvantages of bacteria when used to express eukaryotic proteins include the fact that they lack the correct machinery to carry out post-translational modifications such as glycosylation. In addition, many overexpressed proteins (particularly large multidomain proteins) form insoluble inclusion bodies in prokaryotes;[10] see the example of prochymosin (Section 2.8.7). Formation of inclusion bodies results from misfolding of the polypeptide chain; this has been shown to be a consequence of the fact that in prokaryotes the folding of the chain occurs predominantly after translation is complete. (By contrast, in eukaryotes, the chain appears to fold co-translationally, thus allowing the separate domains to fold sequentially to form properly folded enzyme.[11]) If formation of inclusion bodies occurs, it is necessary to attempt to recover active enzyme by solubilization of the protein and refolding under controlled conditions;[10] this process is not always successful. Nevertheless, particularly for large-scale production of enzymes, formation of inclusion bodies can be a useful way of concentrating the protein and facilitating subsequent recovery and purification.[12]

For expression of eukaryotic proteins, a variety of host organisms can be employed. Lower eukaryotes such as yeasts have proved popular, because of their good growth rates on simple media and because they are well understood at a genetic level. A comprehensive review of the factors involved in the use of yeast expression systems has been given.[13] There is relatively little evidence for the accumulation of misfolded overexpressed intracellular proteins in baker's yeast (*Saccharomyces cerevisiae*); however, problems can arise in the case of naturally secreted proteins, many of which are found to accumulate in an insoluble form. By incorporation of suitable signal sequences derived from proteins of *S. cerevisiae* or other organisms, it is however often possible to cause such proteins to be secreted efficiently (see the example of prochymosin, Section 2.8.7). (It should be noted that purification of the secreted protein from the growth medium can often present problems.) *S. cerevisiae* is not always an ideal host organism, in particular it can be difficult to grow to high cell densities in continuous culture and it has a tendency to hyperglycosylate proteins. A number of other yeast strains, such as *Kluyveromyces lactis* (which is fairly similar to *S. cerevisiae*) and the methylotroph *Pichia pastoris*, appear to offer considerable advantages in these respects and they are being increasingly used as host organisms. In the case of *P. pastoris* high levels of expression can be obtained using the promoter from the tightly regulated *AOX1* gene which encodes alcohol dehydrogenase. For some applications, baculovirus-driven expression in insect cells appears to be the system of choice. Because the baculovirus vector employs many of the protein modification, processing, and transport systems of higher eukaryotic cells, many functional recombinant proteins can be expressed to high levels.[5,8]

An alternative approach to the expression of proteins is to express them as 'fusion proteins', in which the gene for the protein is linked to that for another component such as glutathione-*S*- transferase or the maltose-binding protein.[14] The advantage of this type of system is that the purification of the expressed protein is greatly facilitated. Thus in the case of the glutathione-*S*-transferase fusion protein, the fusion protein can be purified by affinity chromatography (Section 2.6.4.1) on immobilized glutathione, followed by elution with glutathione. When the fusion protein has been purified, the protein of interest can be liberated if necessary by proteolysis with thrombin (see Chapter 6, Section 6.2.1.1). A further type of fusion protein involves the attachment of the hexa-histidine tag, permitting easy purification by immobilized metal ion (Ni^{2+}) affinity chromatography (Section 2.6.4.1).

2.4.2 Availability

A source with reasonable abundance of the chosen enzyme might not be readily available for various reasons (economic, geographical, etc.). Thus, bioluminescent species[15,16] are often found in far away 'exotic' locations, and human tissue or calf tissue (Section 2.8.7) is not readily available in many countries. Clearly, in such cases it may be necessary to make some compromise between abundance and availability. These considerations also provide an impetus for the development of suitable expression systems (Section 2.4.1) in order to increase the amount of enzyme available.

2.4.3 Comparative studies

It may be the case that a certain enzyme has been studied in one species or within one tissue in a species; in such cases it can be highly profitable to examine the corresponding enzyme in another species or another tissue. In Chapter 3, Section 3.4.9.5, we shall see how comparisons of a given enzyme from different species can be used to explore aspects of enzyme evolution. A knowledge of the properties of the different isoenzymes (see Chapter 1, Section 1.5.2) of lactate dehydrogenase is of use in discussing the role of these enzymes in heart and muscle.[17]

2.4.4 Subcellular location

If the enzyme catalysing a given reaction occurs at only one location within the cell* (e.g. succinate dehydrogenase and some other tricarboxylic acid-cycle enzymes occur only in the mitochondria), then the whole tissue can be homogenized or extracted in the purification procedure (Fig. 2.1), and subcellular fractionation is not essential. In some cases, however, a given enzyme may exist in a number of locations in the cell and some subcellular fractionation may well be necessary before the purification can proceed. Thus, in the case of mitochondrial adenosinetriphosphatase it is important to obtain a reasonably pure preparation of mitochondria from the chosen source (beef heart, a tissue rich in mitochondria) in order to avoid contamination from other adenosinetriphosphatases, for instance that involved in the transport of Na^+ and K^+ across the plasma membrane of the cell. Subcellular fractionation to prepare the constituent organelles of a cell is normally achieved by a succession of centrifugation steps with the larger organelles (nuclei, mitochondria) being sedimented first (see Chapter 8, Section 8.2.1).

Having taken these and possibly other factors into account, the choice of source of enzyme can be made. We now proceed to look at the steps in the purification

* A discussion of the subcellular locations of enzymes will be found in Chapter 8, Section 8.2.

procedure outlined in Fig. 2.1. Further details of the various methods can be found in reference 18.

2.5 Methods of homogenization

A number of methods for breaking open cells and extracting the contents are available. The choice of method usually depends on the type of tissue or organism which is used as the source of enzyme.

2.5.1 Mammalian tissue

* An isotonic solution is one which exerts the same osmotic pressure as the cell contents, containing approximately 0.3 mol dm^{-3} dissolved species (e.g. 0.3 mol dm^{-3} sucrose or 0.15 mol dm^{-3} KCl).

The lack of a rigid cell wall makes homogenization of most mammalian tissues relatively easy. The tissue is often cut up into small pieces or minced prior to homogenization by either a rotating Teflon pestle in a glass mortar (the Potter–Elvehjem homogenizer) or a high-speed blender. Extraction is performed with isotonic* solutions in those cases where it is important to avoid the rupture of subcellular organelles. This is particularly important in the case of cells from tissue such as liver which possess a large number of lysosomes; these are subcellular organelles containing a number of hydrolytic enzymes such as proteases (see Chapter 8, Section 8.2.4). In other cases a medium of low ionic strength such as 0.01 mol dm^{-3} KCl is often used for extraction. It may also be necessary in particular cases to add protease inhibitors or other substances (e.g. the reducing agent dithiothreitol) to prevent damage to the enzyme of interest during the extraction process.

The use of the pressure homogenization technique, in which a large pressure difference is created between the interior of a cell and its surroundings, has been discussed by Avis.[19]

2.5.2 Plant, fungal, and bacterial material

The rigid cell wall surrounding the cells in these types of material usually necessitates the use of harsher methods of extraction, e.g. grinding with abrasives such as alumina or sand, freezing and thawing, long periods of blending, or the addition of glass beads during blending. Alternatively, a French press, in which the cells are forced at high pressure through a narrow orifice, can be used. In this case, the mechanical shear forces lead to cell rupture.

However, it is often possible to disrupt the cell walls of plants, fungi, and bacteria by the use of appropriate hydrolytic enzymes. Protoplasts (cells completely or partially lacking a cell wall) may be prepared from Gram-positive bacteria, e.g. *Bacillus subtilis*, by incubation with lysozyme.[20] By a modification of this method that includes the addition of EDTA, protoplasts have also been prepared from the Gram-negative bacterium *E. coli*.[21] Protoplasts have been prepared from fungi, e.g. yeast, *Neurospora* and *Aspergillus*, using enzyme extracts that generally contain a mixture of chitinase and 3-glucanases.[22] An alternative method may be to use a fungal mutant that lacks the normal cell wall, e.g. the slime mutant of *Neurospora*,[23] and which can therefore be readily disrupted by solutions of lower osmotic pressure.

Particular problems can arise with plant tissues,[24] because during homogenization the release of the contents of the vacuoles (usually acidic and containing proteases) can cause damage to enzymes. Addition of buffer can protect against such damage. In addition, plant tissues often contain certain phenolic compounds that are readily oxidized to form dark pigments that can be harmful to enzymes. The pigments can be removed by adsorption on polymers such as poly(vinylpyrrolidone) (Section 2.8.2).

2.5.3 Extraction of membrane-bound enzymes

A number of enzymes are membrane-bound (integral membrane proteins; see Chapter 8, Section 8.4.1). The study of these enzymes presents special problems. In order to preserve the enzyme in its functional state, it is necessary to provide an environment which resembles that provided by the membrane. Detergents have generally been found to be able to provide this environment, but the choice of a suitable detergent for extraction from the large range of cationic, anionic, and neutral detergents available[25] is not easily made without adopting a largely empirical approach. The overriding criterion is the retention of biological activity on extraction from the membrane. In the case of enzymes which catalyse vectorial processes, such as the Na^+, K^+ adenosinetriphosphatase which transports these ions in opposite directions across a membrane at the expense of ATP hydrolysis, it may only be possible to demonstrate retention of activity in a system such as a phospholipid vesicle in which the enzyme is oriented in a correct fashion. One factor to be borne in mind is that the nature of the detergent can influence the choice of the purification methods to be used subsequently.[25] Thus, for instance, the use of the anionic detergent sodium dodecylsulphate or cholate will rule out the use of ion-exchangers such as DEAE-cellulose (Section 2.6.2.1), since they would adsorb the detergent. Non-ionic detergents such as dodecylmaltoside (see Section 2.8.6) exclude the use of ammonium sulphate fractionation (Section 2.6.3.2), since the high salt concentrations promote phase separation, which can lead to denaturation of proteins.

2.6 Methods of separation

A successful separation of one substance from another depends on there being some property by which the substances may be distinguished. The principal properties of enzymes that can be exploited in separation methods are size, charge, solubility, and the possession of specific binding sites. Methods that exploit differences in these properties are listed in Table 2.1, and are described in the following sections (2.6.1 to 2.6.4).

2.6.1 Methods that depend on size or mass

2.6.1.1 Centrifugation[26]

Large molecules such as enzymes can be sedimented by the high centrifugal fields (up to 300 000 times gravity) generated by an ultracentrifuge. Although the rate at which any particular enzyme will sediment depends on a variety of factors including the size and shape of the molecule and the viscosity of the solution, it is found that in general the higher the M_r value the greater the rate of sedimentation. This method is not widely used in purification procedures to separate one enzyme from another because only small volumes (a few cm^3) can be dealt with in ultracentrifuges that operate at high centrifugal fields. However, centrifugation is very widely used to remove precipitated or insoluble material in the course of an isolation, e.g. to remove cell debris after homogenization or to collect enzyme that has been precipitated by the addition of ammonium sulphate (see Section 2.6.3.2). When centrifugation is being used in this way, only relatively low centrifugal fields (5000 to 50 000g) are required and consequently volumes up to several dm^3 can be handled.

Table 2.1 Principal separation methods used in purification of enzymes

Property	Method	Scale
Size or mass	Centrifugation	Large or small
	Gel filtration	Generally small
	Dialysis; ultrafiltration	Generally small
Polarity		
(a) Charge	Ion-exchange chromatography	Large or small
	Chromatofocusing	Generally small
	Electrophoresis	Generally small
	Isoelectric focusing	Generally small
(b) Hydrophobic character	Hydrophobic chromatography	Generally small
Solubility	Change in pH	Generally large
	Change in ionic strength	Large or small
	Decrease in dielectric constant	Generally large
Specific binding sites or structural features	Affinity chromatography	Generally small
	Immobilized metal ion chromatography	Generally small
	Affinity elution	Large or small
	Dye–ligand chromatography	Large or small
	Immunoadsorption	Generally small
	Covalent chromatography	Generally small

The term 'large-scale' is used to indicate that amounts of protein greater than about 100 mg can be readily handled at that particular step in the purification procedure.

The use of differential centrifugation in the preparation of subcellular organelles is described in Chapter 8, Section 8.2.1.

2.6.1.2 Gel filtration[27]

In gel filtration, the separation between molecules of different sizes is made on the basis of their ability to enter the pores within the beads of a beaded gel. The most widely used types of gel are Sephadex (cross-linked dextrans), Bio-Gel P (cross-linked polymers of acrylamide), Sephacryl (cross-linked polymers of dextran and acrylamide), and Sepharose (agarose, which can be cross-linked for greater rigidity). Small molecules that can enter the pores of the beads are retarded as they pass down a column containing the gel; large molecules that are unable to enter the pores pass through the column unimpeded. By varying the size of the pores (which is controlled by the degree of cross-linking in the preparation of the beaded gel), it is possible to change the range of M_r values which can be fractionated. Sephadex G-100, for example, can fractionate globular proteins in the M_r range 4000–150 000.* If the shape of the enzyme is markedly non-spherical, it will be eluted from the column in an unexpected position, i.e. not at that expected on the basis of its M_r (see Chapter 3, Section 3.2.2). In order to minimize any non-specific interactions between the enzyme and the gel, it is important that extremes of ionic strength should be avoided.[27] Gel filtration can be carried out on a large scale, but since large columns are rather time-consuming to run and the gels required to fill them are expensive, the method finds most application in the later (small-scale) stages of enzyme purifications. A number of gel filtration media are now available which can be employed under high performance chromatography conditions; thus, for instance, Superose 12 can fractionate in the M_r range 1000–300 000 (Section 2.8.5); these offer considerable advantages over conventional gel filtration in terms of resolution and speed of operation, but are generally used for analytical or small-scale preparative purposes.

* The corresponding ranges for Sephacryl S-300 and Sepharose 6B are 10 000 to 800 000 and 10 000 to 4 000 000, respectively.

2.6.1.3 **Dialysis**[28,29] **and ultrafiltration**[30,31]

A dialysis membrane such as cellophane acts as a sieve with holes large enough to permit the passage of globular proteins with M_r values up to about 20 000 but not of larger molecules. It is possible to change the pore sizes by various mechanical and chemical treatments.[32] Although dialysis is not generally useful for the separation of enzymes from each other, it is widely used during purification procedures to remove salts, organic solvents, or inhibitors of low M_r from solutions of enzymes.

In ultrafiltration, small molecules and ions pass through the dialysis membrane under the influence of an applied pressure (generally N_2 gas at about 4 atm pressure). This leads to concentration of an enzyme solution, which can be useful in reducing the volume of a sample during the purification procedure. An alternative approach involves the use of centrifuge tubes which consist of two compartments separated by a dialysis membrane. The solution to be concentrated is placed in the upper chamber; as the tube is subjected to centrifugation, small molecules (buffer components, water, etc.) are forced through the membrane, whereas the enzyme is retained in the upper chamber.

2.6.2 Methods that depend on polarity

2.6.2.1 Ion-exchange chromatography[33,34]

Ion-exchange chromatography depends on the electrostatic attraction between species of opposite charge. Ion exchangers usually consist of modified derivatives of some support material such as cellulose, Sephadex, etc., as shown in Fig. 2.2.

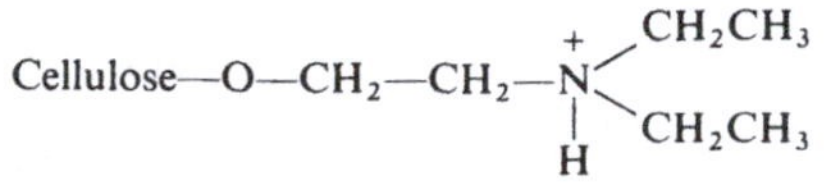

$$Cellulose-O-CH_2-CH_2-\overset{+}{\underset{H}{N}}\overset{CH_2CH_3}{\diagdown}_{CH_2CH_3}$$

DEAE-cellulose

(diethylaminoethyl-cellulose)

possesses a $pK_a \approx 10$, will bind negatively charged species and is therefore an *anion exchanger*

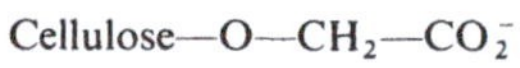

$$Cellulose-O-CH_2-CO_2^-$$

CM-cellulose

(carboxymethyl-cellulose)

possesses a pK_a of ≈ 4, will bind positively charged species, and is therefore a *cation exchanger*

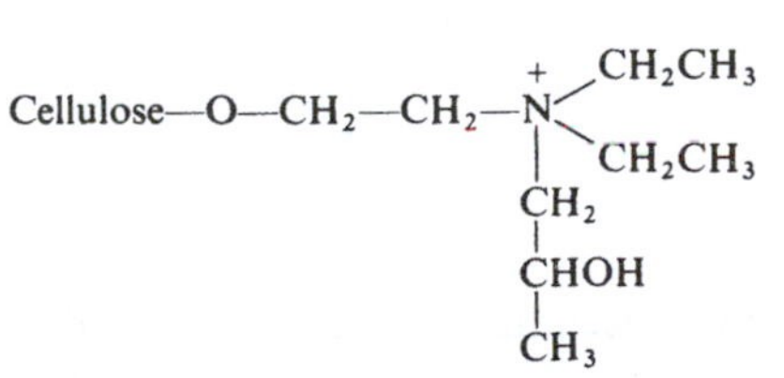

$$Cellulose-O-CH_2-CH_2-\overset{+}{N}\overset{CH_2CH_3}{\underset{\underset{\underset{CH_3}{|}}{\underset{CHOH}{|}}{CH_2}}{\diagdown}_{CH_2CH_3}}$$

QAE-cellulose
(quaternary amino ethyl-cellulose or, more correctly, diethyl [2-hydroxypropyl]-aminoethyl-cellulose), a strongly basic ion-exchanger that acts as an *anion exchanger*

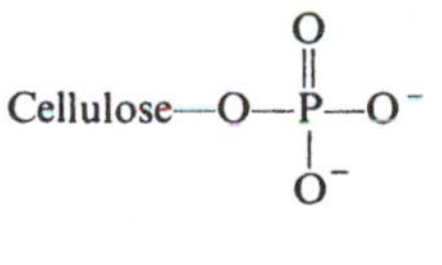

$$Cellulose-O-\overset{O}{\overset{\|}{\underset{\underset{O^-}{|}}{P}}}-O^-$$

Phosphocellulose, possesses a second $pK_a \approx 6.5$, will act as a *cation exchanger*, but in some respects acts also as an affinity adsorbent[27] (see Section 2.6.4.1)

Fig. 2.2 Ion exchangers commonly used in purifications of enzymes.

During a purification procedure, the enzyme is usually applied to an ion exchanger in a solution of low ionic strength and at a pH at which the appropriate interaction will occur (i.e. the enzyme and the ion exchanger have opposite charges). Desorption of the bound species can be brought about either by changing the pH, and thus altering the charges, or, more commonly, by increasing the ionic strength of the solution so that the increased concentration of cations or anions will compete with the enzyme for the binding sites on the ion exchanger. Use of a gradient of increasing ionic strength permits the separation of proteins in a mixture on the basis of their ability to bind to the ion exchanger. When modified forms of Sephadex (such as DEAE-Sephadex) are used, separation is made on the basis of charge and (to a limited extent) size (see Section 2.6.1.2). It should be pointed out that DEAE-Sephadex shrinks as the ionic strength of the solution increases; this can cause problems during ion-exchange chromatography. The agarose-based ion exchangers such as DEAE-Sepharose do not suffer from this problem.

Ion-exchange chromatography can be performed on a large or a small scale. On a large scale it is often convenient to work in a batchwise manner, i.e. to adsorb the enzyme by adding the ion exchanger to a solution and then to pour the material into a column for controlled desorption using an ionic-strength gradient. On a small scale, both adsorption and desorption are performed in a column.

The degree of purification effected by an ion exchange step is generally up to about 10-fold (as judged by the specific activity), but examples are known where much better results have been achieved[35]. Ion-exchange chromatography finds very wide application in present-day purification procedures. High performance chromatographic ion exchangers are now available (e.g. monoQ, an anion exchanger); these are finding considerable use in purification procedures. In Section 2.6.4.2 we shall describe the technique of affinity elution from ion exchangers.

In general, the desorption of proteins from ion exchangers by changes in pH has not been of great use in protein purification, partly because the buffering properties of the bound proteins distort the shape of the applied pH gradient. However, the technique of 'chromatofocusing' overcomes many of these problems.[36] In chromatofocusing, the pH gradient is formed by using mixtures of ampholyte-type buffers (see Section 2.6.2.3) that are of high buffering capacity. The ion exchanger is poly(ethyleneimine) agarose, which has a wide range of titrable groups of differing pK_a values. The protein mixture is adsorbed on the ion exchanger at the upper end of the range of pH and the pH gradient is generated on the column by addition of the acid form of the ampholyte buffer. As the pH falls, proteins are eluted from the column roughly in the order of their isoelectric points (see Section 2.6.2.3) and the technique is claimed to be of high resolving power. The costs of the materials involved make working on a large scale expensive, and thus chromatofocusing is probably most useful at the later stages of a purification procedure.

2.6.2.2 Electrophoresis[37,38]

Electrophoretic separation is based on the differential movement of charged molecules under the influence of an applied potential difference. The rate of movement of a species is governed by the charge it carries and also by its size and shape. In order to minimize convection currents, the buffer (electrolyte) solution is soaked into a support (paper, cellulose powder, starch, or polyacrylamide gels). The position of a protein on the support can be determined by use of a stain such as Coomassie Blue or silver.[38] Although the technique is normally performed on a small, analytical scale (a few milligrams or less), it is possible to use the method on a large preparative scale.[39] After preparative electrophoresis, the separated enzymes can be eluted out of the support or run off the bottom of a column in sequence.

Capillary electrophoresis is finding increasing application in the separation and analysis of protein mixtures.[40,41] The distinctive feature of the technique is that the electrophoretic separation takes place in solution in a capillary tube of very small cross-section (typically 100 μm) so that temperature control during separation is readily achieved. The capillary which is normally about 30 cm in length is filled with the running buffer and the sample is usually introduced by dipping one end into the sample and applying the electric field; the components of the mixture are detected as they pass a window at the far end of the capillary. A number of different types of electrophoretic separation using this approach have been developed.[40]

2.6.2.3 Isoelectric focusing[42]

Isoelectric focusing is based not on the rate at which charged species move in an electric field but rather on their equilibrium position in a pH gradient. The pH gradient is established by applying a potential difference to a gel soaked with a solution containing a mixture of ampholytes that are polyamino acids with different charge properties and hence different isoelectric* points (pI values) (Fig. 2.3). When the potential difference is applied, a negatively charged molecule will migrate towards the anode until it encounters the acid; the acid will tend to neutralize the charge so the molecule will stop. The reverse argument holds for positively charged species. By this means the ampholytes will arrange themselves in the gel so that there is a gradual increase of pH from anode to cathode.

An enzyme applied to the gel will then migrate to the position in the gel where the pH equals its isoelectric point. For example, if the enzyme were at a position where the pH is above its pI, then its net negative charge would ensure that it moved towards the anode until it reached its pI. It is possible to separate species that differ only slightly (as little as 0.01 pH units) in isoelectric points, and thus this method can be a very powerful tool in purification procedures. The method is generally used on an analytical scale, although it is possible to scale up for preparative work (up to gram quantities). Isoelectric focusing can also be performed using capillary electrophoresis[40].

* The isoelectric point of a compound is the pH at which it carries no net charge and hence does not move under an applied potential difference.

2.6.2.4 Hydrophobic interaction chromatography

In an aqueous environment, hydrocarbons and other non-polar molecules will tend to associate with each other rather than with water molecules (see Chapter 3, Section 3.5.5.4, for a further discussion of this 'hydrophobic' effect). The surfaces of proteins contain a high proportion of charged and polar amino acid side-chains (see Chapter 3, Section 3.5.4.5) but usually contain at least some non-polar

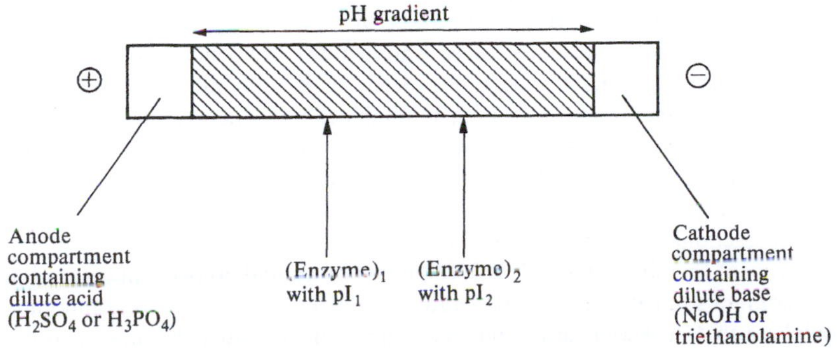

Fig. 2.3 Separation of enzymes by isoelectric focusing.

side-chains. It is the presence of these non-polar side-chains that accounts for the ability of many proteins to bind to non-polar molecules. Such interactions are exploited in the technique of hydrophobic-interaction chromatography.[43,44] In this technique, proteins are adsorbed on matrices such as octyl- or phenyl-Sepharose, usually at high ionic strength at which the hydrophobic interactions are stronger. Desorption of proteins can be brought about by decreasing the ionic strength and, if necessary, by addition of organic solvents or non-ionic detergents to weaken the hydrophobic interactions further. In the purification of 3-phosphoshikimate-1-carboxyvinyl transferase from *E. coli* (Section 2.4.1), hydrophobic-interaction chromatography on phenyl-Sepharose was employed to give a 5-fold purification with 80 per cent recovery of activity. The enzyme was adsorbed at 0.8 mol dm^{-3} $(NH_4)_2SO_4$ in a 0.1 mol dm^{-3} Tris–HCl buffer at pH 7.5 and eluted using a decreasing gradient of $(NH_4)_2SO_4$ (0.8–0.0 mol dm^{-3}) in this buffer.[6]

2.6.3 Methods based on changes in solubility

The solubility of a compound in a given solvent depends on the balance of the forces between solute and solute and those between solute and solvent. If the former predominate the compound will be insoluble, whereas if the latter are predominant the compound will be soluble. In the course of enzyme purifications it is possible to alter the balance between these forces and hence precipitate the enzyme of interest or remove contaminating enzymes. Three of the most important ways of changing the solubility of enzymes are (i) to change the pH; (ii) to change the ionic strength; (iii) to decrease the dielectric constant. These methods can be applied on a large scale and are often used in the initial stages of a purification procedure.

2.6.3.1 Change in pH[45]

An enzyme is least soluble at its isoelectric point, since at this pH there will be no repulsive electrostatic forces between the enzyme molecules. Adjustment of the pH to the appropriate value can therefore be used to precipitate an enzyme; a change in pH can also be used to precipitate unwanted enzymes and proteins (see the case of adenylate kinase; Section 2.8.1). It is important to check that the enzyme of interest is not inactivated by exposure to these changes in pH.

2.6.3.2 Change in ionic strength[46]

Large charged molecules are generally only slightly soluble in pure (deionized) water; addition of ions promotes solubility by helping to disperse the charge carried by the large molecule. This phenomenon is known as salting in. However, if the ionic strength is increased beyond a certain point, the charged molecule will be precipitated (salting out). Although the theoretical basis for this behaviour is not too well understood, it is probable that a major factor is that at very high concentrations of salt, the concentration of water is significantly decreased, leading to a decrease in solute–solvent interactions and hence in solubility. The salt of choice in enzyme purification procedures is ammonium sulphate, which has the advantages of low cost, high solubility in water (a saturated solution at 298 K (25°C) has a molarity of approximately 4 mol dm^{-3}), and lack of harmful effects on most enzymes.* Each enzyme will usually begin to precipitate at a certain concentration of ammonium sulphate and this forms the basis of an initial fractionation procedure. For instance, if it is known from a preliminary experiment that an enzyme, X, is precipitated at around 50 per cent saturation† ammonium sulphate, then the procedure might involve addition of ammonium sulphate to 45 per cent saturation, centrifugation to remove the unwanted pre-

* It should be noted that ammonium sulphate is a weak acid. When it is added to a weakly buffered solution, ammonia or another base should be added to maintain the pH at 7 or above.

† Concentrations of ammonium sulphate are usually expressed as percentages of the concentration required to saturate the solution.

cipitated protein followed by addition of more ammonium sulphate to the supernatant to bring the concentration up to 55 per cent. The precipitate containing X
is removed by centrifugation.

In general, the degree of purification that can be achieved by ammonium sulphate fractionation is less than 10-fold, but the method is often used in the initial
stages of a purification to reduce the volume of solution to be dealt with since the
precipitated enzyme can be redissolved in a small volume of buffer. In some cases,
however, such as fructose-bisphosphate aldolase[47] from rabbit muscle, it is possible
to obtain pure enzyme by ammonium sulphate fractionation of a crude extract.

Ammonium sulphate is also commonly used in crystallization of enzymes.[2]
The concentration of salt is raised until slight turbidity is observed; when the
solution is left to stand, crystals of the enzyme should form. Some of the larger
crystals (with dimensions of the order of 0.5 mm) may be suitable for X-ray crystallography (see Chapter 3, Section 3.4.1).

2.6.3.3 Decrease in dielectric constant[48]

Addition of a water-miscible organic solvent (such as ethanol or acetone) will
decrease the dielectric constant of a solution and hence increase electrostatic
forces. Although there can be rather complex effects on both solute–solute and
solute–solvent forces, in general the net effect is to cause precipitation of large,
charged molecules such as enzymes. This method can be used on a large scale as
an initial fractionation procedure. It must be remembered, however, that addition
of organic solvents can sometimes lead to the inactivation of enzymes (see
Chapter 3, Section 3.6.1) and it is normally important to work at low temperatures to minimize such inactivation.

An alternative method of precipitating proteins is to add the neutral water-
soluble polymer poly(ethylene glycol). The polymer is thought to act by removing
water from the hydration spheres of proteins.[48] Poly(ethylene glycol) preparations
of average M_r 4000 or 6000 are commonly used and can give highly successful
purifications (see Section 2.8.2). Difficulties can arise in removing residual traces
of poly(ethylene glycol) from protein preparations but this is often unnecessary.[48]

2.6.4 Methods based on the possession of specific binding sites or structural features

Enzymes normally display highly specific interactions with their substrates and
other ligands (see Chapter 1, Section 1.3.2). Advantage can be taken of this
specificity in the so-called affinity separation methods of enzyme purification
discussed below.

2.6.4.1 Affinity chromatography[49–51]

In affinity chromatography, a molecule such as a substrate or competitive
inhibitor (see Chapter 4, Section 4.3.2.2) that interacts specifically with the
enzyme of interest is linked covalently to an inert matrix (such as agarose). When
a mixture is passed down a column containing the affinity matrix, only this
enzyme is retained and other enzymes and proteins are washed away. The bound
enzyme can be desorbed by a pulse of substrate, which will compete for the
binding sites on the enzyme, or by changing the pH or ionic strength of the solution in such a way as to weaken the binding of enzyme to the column (Fig. 2.4).

This simple method is, in principle, capable of purifying an enzyme in a single
step from a crude extract and a number of such purifications have been reported
(see the example of Staphylococcal nuclease[52,53] outlined in Fig. 2.5).

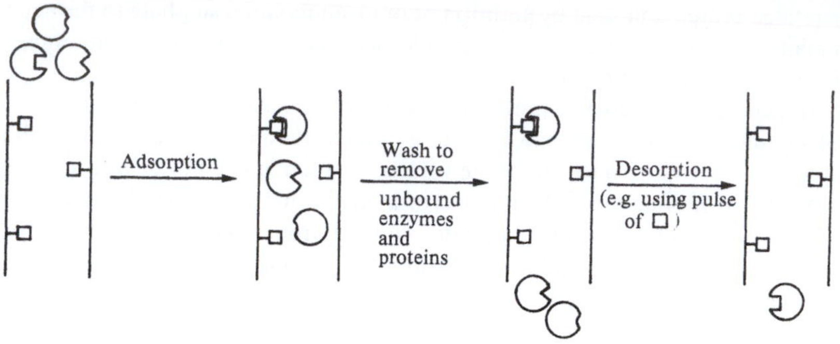

Fig. 2.4 Schematic illustration of enzyme purification using affinity chromatography.

There are, however, a number of problems associated with the use of affinity chromatography in enzyme purifications; these include

1. Attaching a suitable substrate analogue or inhibitor to the matrix can be a difficult task and the reactions involved in coupling (such as those illustrated in Fig. 2.5) are not always completely characterized.

2. Linking of the ligand to the matrix may interfere with the binding to the enzyme and lead to a loss of specificity in the interaction. In this connection, the need to interpose spacer arms between ligand and matrix to prevent interference by the matrix in the interaction has been recognized for some time.[54] However, the most widely used spacer (1,6-diaminohexane) can itself

Staphylococcal nuclease acts on substrates of general type pTp-X and is completely inhibited by pdtp. The compound

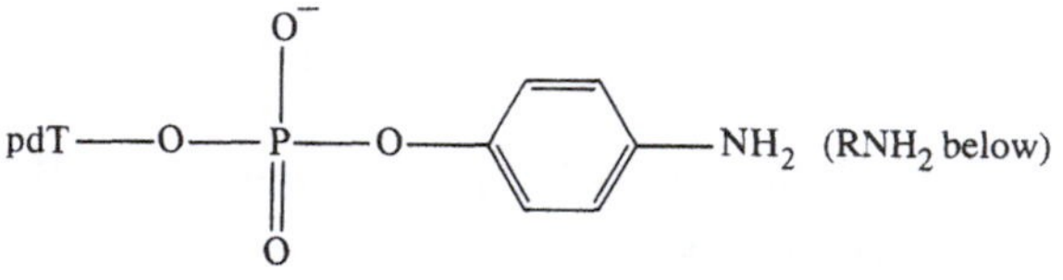

which acts as a powerful competitive inhibitor of the enzyme was attached to a cyanogen bromide-activated agarose gel via the free amino group:

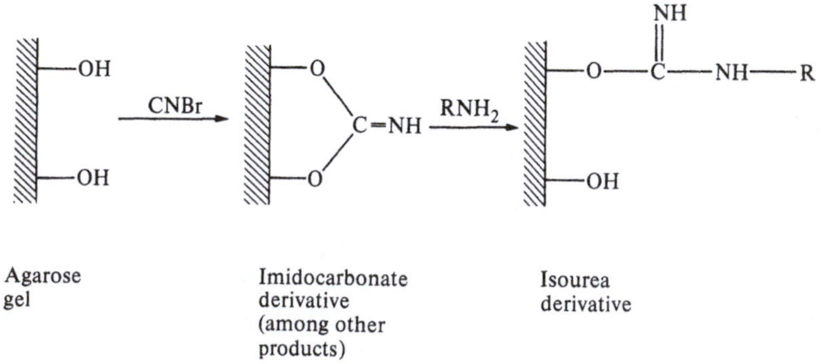

Fig. 2.5 Use of affinity chromatography in the purification of Staphylococcal nuclease.

The nuclease was retained by the affinity column and could be subsequently eluted at low pH. Purification could be achieved in a single step from a crude cell extract.[53]

participate in non-specific hydrophobic interactions (see Chapter 3, Section 3.5.5.4) with enzymes, so the use of alternative hydrophilic spacers has been suggested.[55]

3. For affinity chromatography to work successfully the strength of the interaction between matrix-bound ligand and the enzyme must be in the correct range.[56] If it is too weak, with a dissociation constant greater than about 2 mmol dm^{-3} , the enzyme will not be retarded by the column. However, if the interaction is too strong (dissociation constant smaller than about 10^{-8} mol dm^{-3}), removal of the bound enzyme may only be possible under harsh conditions that lead to inactivation of the enzyme. Information obtained from inhibition studies of an enzyme in solution can be of considerable value in the successful design of affinity chromatography systems.

4. Special problems are posed by enzymes that catalyse reactions involving more than one substrate. Thus it would be expected that all NAD$^+$-dependent dehydrogenases would 'recognize' matrix-bound NAD$^+$. The differences between the various dehydrogenases occur in the binding of the second substrate, and in order to exploit these differences we normally have to pay particular attention to the method of elution from the column. In the case of horse liver alcohol dehydrogenase, for example, specific elution of the enzyme from a column containing a matrix-bound AMP analogue could be obtained with a mixture of NAD$^+$ plus pyrazole, a competitive inhibitor of the enzyme.[57]

In spite of these problems, affinity chromatography has made a very significant contribution to the purification of enzymes and also of other biological molecules such as antibodies that possess specific binding sites. The technique is most likely to be successful when full use is made of information from solution studies concerning the interaction between the ligand and the enzyme.

Affinity chromatography has also found considerable use in the purification of fusion proteins[14] (Section 2.4.1). In these systems, the protein of interest is expressed as a larger molecule in which it is tagged to a second moiety which can act as a marker to facilitate purification. Common fusion tags include glutathione-*S*-transferase, maltose binding protein and the hexahistidine moiety discussed below. Fusion proteins involving glutathione-*S*-transferase can be readily purified by application to an affinity column in which glutathione is immobilized on a support such as Sepharose; if the maltose binding protein were the tag, the affinity ligand would be immobilized maltose. In each case, the purified fusion protein would be eluted by a solution of the appropriate ligand.

A special case of a type of affinity chromatography is provided by immobilized metal affinity chromatography.[58] In this technique, the absorbent is a metal ion complexing agent attached to a matrix, such as Ni^{2+} coordinated to nitrilotriacetic acid (NTA) which is linked to agarose. The vacant coordination sites on the metal can be taken up by certain side-chains in proteins, particularly the uncharged histidine side-chain. Although the technique could in theory be used to purify a number of proteins, it is usually difficult to achieve a high degree of specificity. However, the technique is particularly useful in the purification of His-tag fusion proteins. In this case the tag (a sequence of six histidines) is produced by addition of an oligonucleotide to the gene encoding the protein of interest; a thrombin cleavage site is also included in the tag. The (His)$_6$ sequence provides a powerful ligand for the Ni^{2+}-NTA agarose complex. The bound fusion protein can be eluted by addition of imidazole,* and the histidine tag released by treatment with

* The added imidazole will compete with the imidazole group on histidine side-chains for immobilized Ni^{2+}.

catalytic amounts of thrombin. Examples of the use of immobilized metal affinity chromatography are given in Sections 2.8.3 and 2.8.6.

2.6.4.2 Affinity elution[59,60]

Affinity elution is complementary to affinity chromatography in that in the former technique the specificity of interaction is at the stage of *desorption* from the chromatographic support material, whereas in the latter the specificity occurs at the stage of *adsorption*. In affinity elution, the enzyme of interest (along with others) is adsorbed onto an ion exchanger (e.g. DEAE-cellulose, see Section 2.6.2.1), and then is eluted specifically by the appropriate substrate. The advantages over affinity chromatography are mainly of a practical nature; namely (i) the often complex task of designing and attaching the appropriate ligand to the matrix is not required, and (ii) columns of high capacity are more readily available, since ion exchangers cost much less than affinity matrices. Affinity elution was used in the early 1960s in the purification of fructose bisphosphatase from rabbit liver,[61] but the technique has been extensively developed. For instance, Scopes[62,63] devised a method for the separation of all the enzymes involved in the glycolytic pathway (see Chapter 6, Section 6.4.1) in a single multiple-operation scheme. Thus, for instance, a mixture of glycolytic enzymes already partially fractionated by ammonium sulphate (those precipitating between 45 and 65 per cent saturation) was loaded on to a CM-cellulose column at pH 6.5; elution with phosphoenolypyruvate (0.5 mmol dm^{-3}) and D-fructose 1,6-bisphosphate (0.2 mmol dm^{-3}) yielded in turn pyruvate kinase and fructose-bisphosphate aldolase.[62,63] An analogous procedure has been reported for purification of the tricarboxylic acid-cycle enzymes from beef heart.[64]

2.6.4.3 Dye–ligand chromatography[65–67]

In the mid 1970s it became clear that the dye Cibacron Blue F3G-A (Fig. 2.6) (which had been used to label high-M_r dextran preparations to provide a

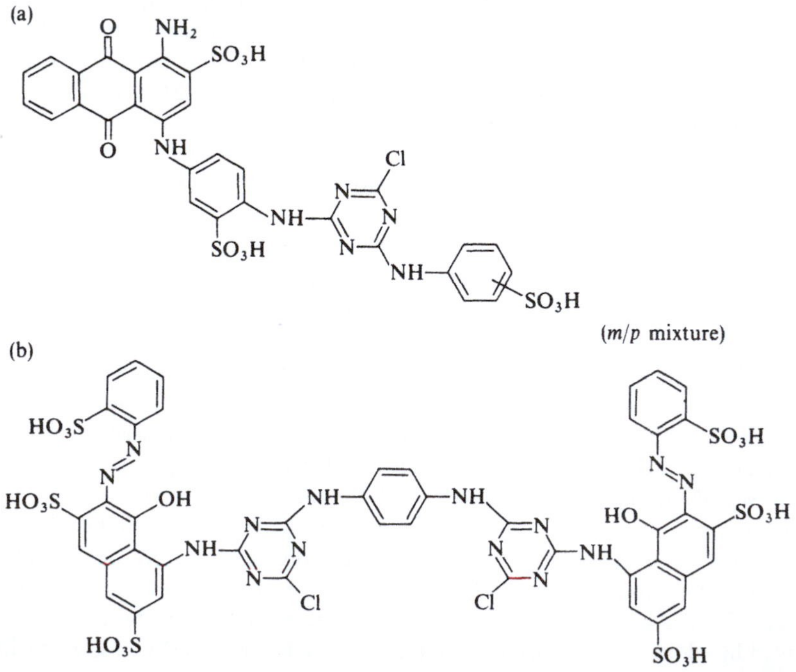

Fig. 2.6 Structures of two dyes used in dye–ligand chromatography.
(a) Cibacron Blue F3G-A; (b) Procion Red HE-3B. These dyes can be linked to matrices such as agarose by reaction of the triazinyl chloride groups.

calibration for gel-filtration columns) could bind to a number of enzymes such as dehydrogenases and kinases.[68] These enzymes possess a common structural feature known as the nucleotide-binding domain (see Chapter 3, Section 3.4.3.3). Although the molecular basis for the apparent specificity has not been completely clarified,[65-67] the interaction has proved a useful tool in protein purification. The dye can conveniently be coupled to an agarose matrix via the triazinyl chloride group and the enzymes can be eluted by addition of appropriate substrate or ligand or merely by increasing the ionic strength.[68]

Since the early work on Cibacron Blue, a large number of triazine dyes have been examined as potential tools for protein purification. The Procion dye series (produced by Imperial Chemical Industries) have been particularly well studied* and many of them show high degrees of selectivity for particular classes of proteins.

* It should be pointed out that structural information about a number of these dyes is not readily available, often for commercial reasons.

Thus Procion Red HE-3B (Fig. 2.6) linked to Sepharose binds to $NADP^+$-dependent dehydrogenases (such as glucose-6-phosphate dehydrogenase) in preference to NAD^+-dependent dehydrogenases (such as malate dehydrogenase) and can be used for purification of the former type from crude extracts of yeast.[69] By contrast, NAD^+-dependent dehydrogenases are bound preferentially by Cibacron Blue F3G-A. Hey and Dean[70] have advocated the use of a 'tandem' strategy for protein purification in which a crude extract is first passed through a dye column that does not retain the protein of interest, and then through a second dye column that does retain the protein. The first column thus acts to remove a number of contaminating proteins that might interfere with the second step.

As well as varying the structure of the dye, the conditions of adsorption and elution can be varied.[65] In particular, it is important to note that under certain conditions the immobilized dyes can act as effective cation exchangers. To minimize these effects, it is recommended that the operations should be carried out at an ionic strength ≥ 0.1 and at pH ≥ 7.[65] Examples of the use of dye–ligand chromatography in protein purification are given in Sections 2.8.4 and 2.8.5.

2.6.4.4 Immunoadsorption chromatography[71]

The high specificity shown in antibody–antigen interactions can often provide a suitable basis for purification procedures. If a small amount (ca. 0.1 mg) of a pure enzyme from one species is available, it can be used to raise antibodies in another species (usually a rabbit, goat, or mouse). A range of antibodies will be produced that vary in affinity for the enzyme because they recognize different structural features (epitopes) of the enzyme. After purification, the antibodies can be coupled to a matrix such as CNBr-activated Sepharose (see Fig. 2.5) and subsequently used to separate the enzyme antigen from a complex mixture. Desorption can be brought about by a change in pH, an increase in ionic strength, or other treatments which weaken the antibody–antigen binding. (The desorption step can be the most difficult part of the entire procedure because the enzyme activity can be destroyed during the harsh conditions required to weaken the strong interactions.) The application of monoclonal antibody techniques[72,73] has helped to overcome a number of the problems previously encountered; thus it is often possible to choose antibodies that have an appropriate affinity for the enzyme antigen and to produce an indefinite supply of such antibodies. The monoclonal-antibody technique has been used to purify aromatic L-amino acid decarboxylase from bovine brain.[74] In this case the previously purified enzyme from bovine adrenal medulla was used as the antigen for monoclonal-antibody production.

The antibody was immobilized on an activated cross-linked agarose matrix and used to provide a 25-fold purification of enzyme from extracts of bovine brain which had been already subjected to ion-exchange chromatography. Elution of enzyme from the immobilized antibody was brought about by washing with a solution containing 50 mmol dm^{-3} acetic acid and 10 per cent (v/v) ethylene glycol.

2.6.4.5 Covalent chromatography[75]

In covalent chromatography, a covalent bond is formed between molecules in the mobile phase and the stationary-phase matrix, in contrast to the non-covalent interactions described earlier in this section. The principal applications have been in the separation and purification of cysteine-containing peptides and proteins, via the disulphide exchange reactions shown in Fig. 2.7(a).

A commonly used matrix is activated thiol-Sepharose 4B in which a 2-thiopyridyl group is attached via a glutathione spacer arm in order to minimize steric interference to the disulphide exchange reactions[76] (Fig. 2.7(b)). The technique has been used to purify the cysteine-containing protease, papain, from dried papaya latex[76] and can also be adapted to separate a mixture of cysteine-containing proteins. By eluting with successively more powerful reducing agents (e.g. L-cysteine, reduced glutathione, 2-mercaptoethanol, and finally dithiothreitol), separation of protein-disulphide isomerase from glutathione-insulin transhydrogenase could be achieved.[77] Activated thiol-Sepharose can also be used to immobilize cysteine-containing proteins for affinity chromatography; for instance, immobilized cytochrome *c* has been used in the purification of cytochrome *c* oxidase.[78] A further application of covalent chromatography is mentioned in Chapter 3, Section 3.3.2.9, for the purification of a cysteine-containing peptide.

2.6.5 Other methods

In Sections 2.6.1 to 2.6.4 we have discussed the principal methods employed in enzyme purification procedures. There are, however, a number of other methods, including heat denaturation,[79,80] fractional adsorption on calcium phosphate gels

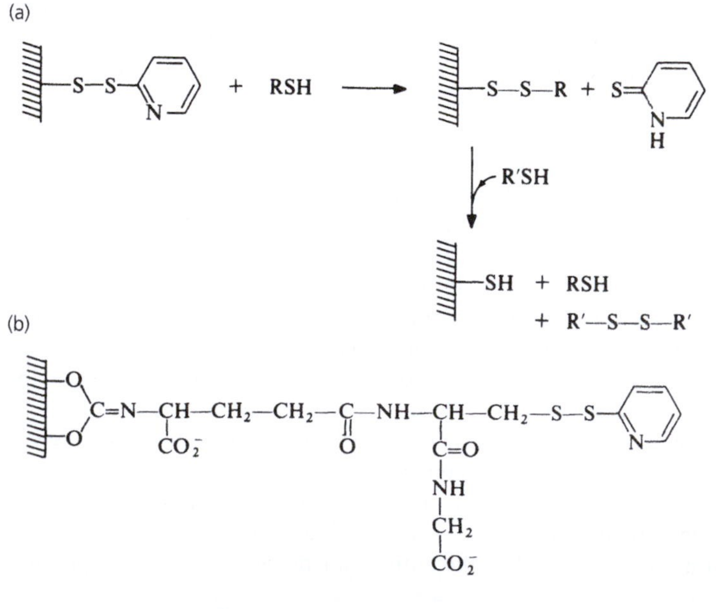

Fig. 2.7 Covalent chromatography. (a) Reaction scheme. In the first step a protein or peptide containing a cysteine group (RSH) reacts with the immobilized 2-thiopyridyl group to form a mixed disulphide. In the second step, a thiol reducing agent such as 2-mercaptoethanol or dithiothreitol (R'SH) is added to liberate RSH. (b) Partial structure of activated thiol-Sepharose.

and hydroxyapatite,[81–83] and concentration by freeze-drying (lyophilization).[84] Further details about these methods can be found in the references given.

2.6.6 **Choice of methods**

Having described some of the advantages and disadvantages of the various separation methods, it is necessary to consider some of the factors that influence the choice of methods and the order of their application in a particular purification procedure. At the outset it should be emphasized that there is rarely only one method or combination of methods that can be used to purify a given enzyme. The actual sequence of methods employed will depend on a variety of factors such as (i) the scale of the preparation and the yield of enzyme required; (ii) the time available for the preparation; and (iii) the equipment and expertise available in the laboratory.

In general, methods based on changes in solubility (Section 2.6.3) are more suitable in the earlier (large-scale) stages of a purification, whereas methods involving column chromatography (e.g. ion-exchange chromatography) or electrophoresis are more appropriate in the latter (small-scale) stages.

In some cases, especially if there is the possibility of proteolysis during the procedure, it may be desirable to aim for as quick a purification as possible at the expense of a lower yield. If so, the use of methods based on solubility would be favoured because these are more rapid than those that involve column chromatography. Recently, matrices have been developed to allow column chromatography procedures to be carried out more rapidly under high-performance liquid chromatography (h.p.l.c.) conditions. Gel filtration can be performed using matrices such as Superose, a rigid cross-linked agarose-based matrix (Section 2.8.5). Other beaded hydrophilic matrices are available for ion-exchange chromatography and chromatofocusing (Section 2.6.2.1). Although these matrices can give superior resolution compared with conventional materials, it should be noted that they are expensive and require specialized apparatus. They are thus generally used in the final stages of purification procedures.

Despite the great advances made in recent years by the advent of affinity techniques (Section 2.6.4), there are a number of enzymes that are still conveniently prepared by 'classical' techniques such as precipitation by ethanol or ammonium sulphate (Section 2.6.3). In these cases, which are generally enzymes present in large amounts, it is probably not worth the trouble to devise new procedures such as affinity chromatography unless there is evidence that the enzyme prepared by the traditional methods is heterogeneous or partially inactivated.

Whatever method or combination of methods is finally decided upon, the progress of the purification should be recorded in a purification table (see the example given in Section 2.8.1).

As stated in Section 2.3.1, developments in recombinant DNA technology have led to a change in the emphasis in enzyme purification. Thus, the initial purification of the enzyme may only aim at isolating a small amount of pure enzyme, sufficient to obtain sequence data to be able to design an oligonucleotide probe to isolate and clone the gene encoding the enzyme. Once a suitable expression system for the gene has been devised (Section 2.4.1), the purification of the enzyme may be a relatively straightforward task, since the degree of purification required is often only 10–100-fold. It must be remembered, however, that for successful crystallization, a prelude to structure determination, enzyme of very high purity is usually required; in these cases it is usual to employ high-performance chromatography or affinity techniques at the final stages of the procedure.

2.7 How to judge the success of a purification procedure

As explained in Section 2.3.1, the aim of a purification procedure is to obtain enzyme of the *maximum possible purity and maximum catalytic activity.* We shall now consider some methods used to test these properties.

2.7.1 Tests for purity

Many of the methods of separation described in Section 2.6 can be used on an analytical scale to check that an enzyme preparation is homogeneous, although we should remember that an analytical test can only really be used to demonstrate the presence of impurities rather than prove their absence. Some of the more commonly employed analytical methods are listed in Table 2.2.

In using these methods we should be aware that the impurities may only constitute a small amount (1 per cent or so) of the total protein and may be missed altogether. It is often easier to test for the presence of contaminating enzymes by measurements of catalytic activity (e.g. testing for the presence of lactate dehydrogenase in a preparation of pyruvate kinase) than to use analytical methods to

Table 2.2 Some commonly employed analytical methods to check the purity of enzyme preparations

Method	Comments
Ultracentrifugation (Chapter 3, Section 3.2.1)	Not very satisfactory for detecting impurities at the ≤5% level. Problems can arise from associating–dissociating systems (Chapter 3, Section 3.2.5)
Electrophoresis (Section 2.6.2.2)	A good method for examining enzymes composed of non-identical subunits (Chapter 3, Section 3.5.1.5)
Electrophoresis in the presence of sodium dodecylsulphate (Chapter 3, Section 3.2.3)	A good method for detecting impurities that differ in terms of subunit M_r; excellent for detecting proteolytic damage. Problems arise from enzymes composed of non-identical subunits, which give rise to multiple bands (Chapter 3, Section 3.5.1.5)
Capillary electrophoresis (Section 2.6.2.2)	A powerful analytical technique which can be used in a variety of modes, including isoelectric focusing. Equipment required is specialized and relatively expensive
Isoelectric focusing (Section 2.6.2.3)	A very sensitive method for detecting impurities. Artefacts can arise suggesting apparent heterogeneity[85–87]
N-terminal analysis (Chapter 3, Section 3.3.2.4)	Should indicate the presence of a single polypeptide chain. Some enzymes have a blocked N-terminus (Chapter 3, Section 3.3.2.7); others consist of multiple polypeptide chains held together by disulphide bonds (e.g. chymotrypsin)
Mass spectrometry (Chapter 3, Section 3.2.4)	A very powerful but specialized technique. Subunit M_r values can be obtained very precisely, confirming the authenticity of the primary structure (Chapter 3, Section 3.3.2.10). Post-translational modifications can be identified

try to indicate their absence. However, if the preparation of the enzyme of interest appears to be homogeneous in a number of these analytical tests we can be
reasonably confident that it is pure.

2.7.2 Tests for catalytic activity

In testing for the catalytic activity of a preparation it is important to check that
the assay conditions (see Chapter 4, Section 4.2) are optimal, i.e. that any
necessary activators or cofactors are present and that inhibitors are absent. It is.
also worthwhile to investigate the conditions under which the enzyme is stable
during storage; in some cases a reducing agent such as dithiothreitol or 2-
mercaptoethanol may be necessary to maintain cysteine side-chains in a reduced
state; in others, storage at low temperatures, e.g. in 50 per cent (v/v) glycerol
solution at 255 K (– 18°C), may serve to minimize processes leading to inactivation. The possibility of degradation (e.g. caused by traces of proteases) during
long-term storage should also be borne in mind (see the example in Section
2.8.4).

2.7.3 Active site titrations

Even if an enzyme preparation is shown to be homogeneous by the various analytical methods listed in Table 2.2, there is still a possibility that some of the
enzyme may be in an inactive form. In a number of cases it has proved possible
to estimate the amount of active enzyme by a technique known as active-site
titration.[88,89] In this method we observe the rapid release of a product of reaction
between the enzyme and a substrate (or pseudosubstrate), when the intermediate
breaks down slowly, if at all, to regenerate enzyme (Fig. 2.8). Inactive enzyme
does not give rise to any product, so the concentration of product formed gives
the concentration of active enzyme.

Using the reagent p-nitrophenyl-p'-guanidinobenzoate which forms a stable
acyl enzyme intermediate with trypsin, it was possible to show that commercial
preparations of the enzyme contained only 60–70 per cent of active enzyme;[90]
the inactive contaminants could be removed by affinity chromatography. Active-
site titration methods have been developed for a number of other enzymes,
including chymotrypsin,[88] papain,[91] isoleucyl-tRNA synthetase,[92] and ornithine
decarboxylase.[93]

General scheme

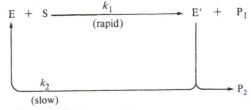

If $k_2 = 0$, the concentrationof P_1 produced gives the concentration of active enzyme directly.
If k_2 is less than k_1, but not equal to zero, it is possible to calculate the concentration of
active enzyme (see reference 94)

Fig. 2.8 Active-site titration of an
enzyme.

2.8 Examples of purification procedures

In this section, seven examples of enzyme purification procedures will be outlined to show how the various separation methods described in Section 2.6 are applied in actual cases. The examples illustrate the wide variety of sources, methods, and problems encountered.

2.8.1 Adenylate kinase from pig muscle[95]

Adenylate kinase catalyses the reaction between adenine nucleotides

$$\text{ATP} + \text{AMP} \xrightleftharpoons{\text{Mg}^{2+}} 2\text{ADP}$$

and is very widely distributed in nature. Pig muscle was chosen as the source because this enzyme was found to give crystals suitable for X-ray diffraction. The procedure is summarized in the flow sheet shown in Fig. 2.9; the progress of the purification is recorded in a purification table (Table 2.3).

Particular points to note in this purification scheme are:

1. The cells are easily disrupted and the enzyme is extracted in solutions of low ionic strength.

2. Advantage is taken in step 2 of the exceptional stability of adenylate kinase at low pH; this enables many contaminating proteins to be removed. The stability of the enzyme is probably associated with the large amount of regular secondary structure in the enzyme; see Chapter 3, Section 3.4.6.4.

3. Chromatography on phosphocellulose (step 3) is made very effective by the use of affinity elution (see Section 2.6.4.2) using AMP.

4. Gel filtration is used (step 4) to remove a contaminant of M_r 60 000 from adenylate kinase (M_r 21 000). Sephadex G-75 fractionates in the M_r range 3000–70 000.

The success of the isolation procedure was judged by the homogeneity of the product during ultracentrifugation and gel electrophoresis. Subsequent structural studies[95] confirmed the absence of any significant contaminating proteins. The enzyme has also been purified from a number of other animal sources,

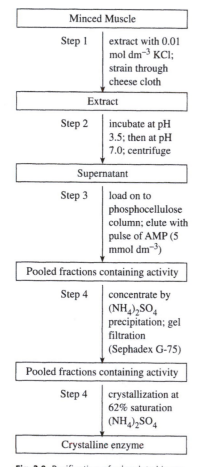

Fig. 2.9 Purification of adenylate kinase from pig muscle. All steps are performed at 273–278 K (0–5°C).

Table 2.3 Purification table for adenylate kinase (see Fig. 2.9) starting with 6 kg muscle

Step	Total volume (cm³)	Total protein (mg)	Total activity (katal)	Specific activity (katal kg⁻¹)	Yield (%)	Purification factor per step
1 (extraction)	16600	435000	0.0413	0.095	(100)	(1.0)
2 (pH)	15700	112000	0.0365	0.325	88.3	3.42
3 (phosphocellulose)	1380	1716	0.0223	13.02	54.0	40.0
4 (gel filtration)	211	462	0.0200	43.17	48.4	3.32
5 (crystallization)	—	344	0.0160	46.5	38.7	1.08

The yield is calculated from the amount of activity at each step relative to the amount in the initial extract.

The purification factor of a step is calculated on the basis of the increase in specific activity after that step (e.g. 0.325 to 13.02 in Step 3 is a 40-fold increase).

including rabbit, calf, and chicken. The gene encoding the chicken enzyme has been cloned and overexpressed in *E. coli*[96] and this has allowed the production of the wild-type enzyme and a number of mutant enzymes (see Chapter 5, Section 5.4.5) designed to explore the role of conserved amino acid side-chains. The over-expressed enzymes were purified from cell lysates by ion exchange on phospho-cellulose, and, after concentration by dialysis against poly(ethylene glycol), gel filtration on Sephadex G-100.

2.8.2 Ribulosebisphosphate carboxylase from spinach

Ribulosebisphosphate carboxylase catalyses the CO_2-fixation step in photo-synthesis:

$$\text{D-Ribulose-1,5-biphosphate} + CO_2 \overset{Mg^{2+}}{\rightleftharpoons} 2 \text{ 3-phospho-D-glycerate.}$$

The enzyme also catalyses an oxygenase reaction:

$$\text{D-Ribulose-1,5-biphosphate} + O_2 \overset{Mg^{2+}}{\rightleftharpoons} \text{3-phospho-D-glycerate} + \text{2-phosphoglycollate}$$

and is thus often referred to as ribulosebisphosphate carboxylase/oxygenase. It has been claimed that the enzyme, which occurs at high concentrations in chloro-plasts, is the most abundant protein in the world.[97] The enzyme is composed of large and small subunits (see Chapter 3, Table 3.5) that are encoded by chloro-plast and nuclear genes, respectively, and is part of a multienzyme complex (see Chapter 7, Section 7.6).[97,98]

Purification of the enzyme from spinach had been previously reported using procedures involving rather time-consuming chromatographic and sucrose density-gradient centrifugation steps; however, the procedure devised by Hall and Tolbert[99] is considerably quicker and is summarized in Fig. 2.10.

In step 1, the spinach leaves are ground in buffer (50 mmol dm^{-3} *N,N*-bis(2-hydroxyethyl)glycine, 1 mmol dm^{-3} EDTA, and 10 mmol dm^{-3} 2-mercaptoethanol adjusted to pH 8.0 with KOH) containing 2 per cent (w/v) insoluble poly(vinyl polypyrrolidone) to adsorb oxidized phenolic substances (Section 2.5.2). Addition of poly(ethylene glycol) 4000 in step 2 precipitates nucleic acids, chlorophyll, and other pigments. Precipitation of ribulosebisphosphate carboxylase is achieved in step 3 by addition of Mg^{2+} to a final concentration of 20 mmol dm^{-3}. The precip-itate is redissolved in grinding buffer and subjected to final purification on DEAE-cellulose (step 4). The poly(ethylene glycol) is not retained by the ion exchanger and is washed through before the sodium bicarbonate gradient is applied. The final product is > 95 per cent homogeneous on polyacrylamide gel electrophoresis and has a specific activity (1.4 μmol CO_2 fixed/mg protein/min) comparable with earlier preparations. On polyacrylamide gel electrophoresis in the presence of sodium dodecylsulphate, two bands are seen, corresponding to the two types of polypeptide chain (M_r values 56 000 and 14 000).

Ribulosebisphosphate carboxylase from plants has proved to be a very difficult enzyme to produce by reconstitution of the separately expressed subunits. The reason for this can be understood when the assembly process of the enzyme *in vivo* is considered. The large subunit is synthesized in the chloroplast, but is then bound to a special binding, chaperone, protein (see Chapter 3, Section 3.6.2)

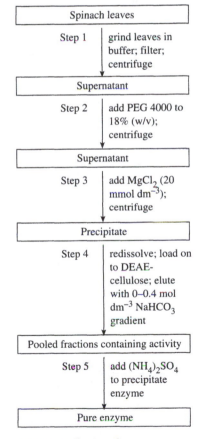

Fig. 2.10 Purification of ribulosebisphosphate carboxylase from spinach. All steps are carried out at 277 K (4°C). PEG 4000 refers to polyethylene glycol of average M_r 4000.

and is only released in a controlled fashion to assemble with the imported small subunits.[100] Reconstitution of ribulosebisphosphate carboxylase from the photo-synthetic bacterium *Rhodospirillum rubrum* (which is a simpler enzyme being a dimer of large subunits only) has been achieved when the necessary chaperone proteins are included.[101]

2.8.3 RNA polymerase from *E. coli*

RNA polymerase catalyses the DNA-dependent incorporation of the XMP moiety (X = nucleoside) of ATP, UTP, GTP, and CTP into high-M_r RNA which is insoluble in acid. (For a more detailed discussion of the properties of the enzyme, see Chapter 7, Section 7.3.) The source of the enzyme used in the classical procedure of Burgess[102] summarized in Fig. 2.11 is *E. coli* strain K-12; the cells are grown, harvested by centrifugation at 75 per cent of maximum growth, and frozen at 253 K (– 20°C).

This procedure illustrates the harsh measures (step 1) needed to disrupt bacterial cell walls. It is possible that enzymatic methods could be used for this purpose (Section 2.5.2) but these are not easy to apply on a large scale. Because of the absence of subcellular organelles (except ribosomes) in bacteria, the extract contains significant amounts of DNA and is therefore highly viscous. Addition of deoxyribonuclease (step 1) reduces the viscosity of the medium, which is important in the later chromatographic steps. It also prevents formation of the high-M_r complex DNA–RNA–RNA polymerase, which would be sedimented along with ribosomes in step 2. In step 3 the oligonucleotides produced by the deoxyribonuclease treatment in step 1 are re-extracted into $(NH_4)_2SO_4$ at 42 per cent saturation. In step 6 gel filtration on a column of Biogel A–1.5m (an agarose gel which fractionates in the M_r range 10 000–2 000 000) is used to remove a minor contaminant of higher M_r than the RNA polymerase. The overall yield of enzyme is 54 mg from 200 g of frozen cells corresponding to 56 per cent of the activity in the initial extract.

Purification factors for steps 2, 3, 4, 5, and 6 are 1.9-, 4.4-, 5.9-, 5.1-, and 1.1-fold, respectively, giving an overall purification of 260-fold. The homogeneity of the preparation is judged by ultracentrifugation (single peak) and by electro-phoresis in 8 mol dm^{-3} urea (the four bands correspond to the correct amounts of the four subunits of different size). There is no detectable deoxyribonuclease, ribonuclease, or polynucleotide nucleotidyl transferase activity in the preparation.

While the preparation procedure of Burgess[102] proved adequate to examine the gross structural features of the enzyme and the outline pathway of its assembly, further detailed analysis of RNA polymerase and its interaction with DNA requires the reconstitution of the intact enzyme from the cloned and overex-pressed subunits, which has been achieved.[103] All the core subunits (α, β, and β') and the major σ subunit can be expressed to high levels in *E. coli* cells by using the T7 RNA polymerase-T7 promoter system. After cell disruption by sonica-tion, β, β', and σ are recovered in insoluble inclusion bodies, but can be readily dissolved in high concentrations of denaturing agents such as urea or guani-dinium chloride. The α subunit is recovered in a native form either in the soluble fraction after cell lysis or by salt extraction of the insoluble material. The core enzyme can be readily reconstituted in high yield by mixing the core subunits (α, β, and β') in the presence of denaturing agents (6 mol dm^{-3} urea or guani-dinium chloride) and then removing the denaturant by dialysis against buffer. In order to avoid non-specific aggregation of the subunits it is necessary to include

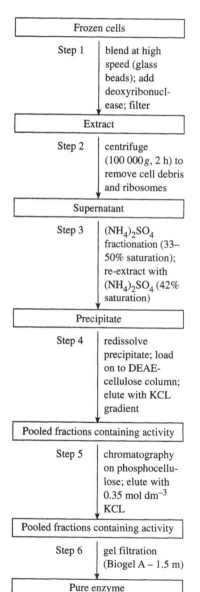

Fig. 2.11 Purification of RNA polymerase from *E. coli*. All steps are performed at 273–278 K (0–5°C).

glycerol (10–20% (v/v)) and salt (0.2–0.3 M KCl), and to maintain the protein concentration below about 0.5 mg ml^{-1}. The core enzyme complex is separated from free subunits and other contaminants by ion-exchange chromatography on DEAE-cellulose. The intact enzyme can then be prepared by adding a small (2–4)-fold molar excess of the σ subunit. A more rapid procedure for reconstitution of the intact enzyme from the recombinant subunits which does not involve large numbers of column chromatography steps and which can be scaled up to yield several hundred milligrams of enzyme has been reported.[104] The novel feature of this procedure is that the recombinant α subunit is produced in a hexa-histidine-tagged form. In the reconstitution procedure this tagged α subunit is mixed with excesses of crude preparations of the other subunits, and the intact enzyme is recovered by batchwise adsorption on an immobilized nickel matrix (Ni^{2+}-NTA-agarose; Section 2.6.4.1), followed by elution with 5 mmol dm^{-3} imidazole, to yield the enzyme.

2.8.4 The *arom* multienzyme protein from *Neurospora*

This multienzyme protein* from *Neurospora crassa* contains five distinct enzyme activities, which catalyse adjacent steps in the biosynthesis of chorismic acid, the common precursor of the three aromatic amino acids tyrosine, phenylalanine, and tryptophan. Although purification of this protein had previously been reported, there was considerable doubt about the M_r of the protein and its subunit composition (see Chapter 3, Section 3.5). It was realized that these difficulties were probably due to proteolysis during the purification procedure and accordingly an isolation procedure has been developed[105] in which stringent precautions are taken to minimize the effects of proteases. These precautions appear to be especially necessary when purifying enzymes from fungi,[106] since they contain large amounts of proteases. The procedure (outlined in Fig. 2.12) starts from cells that have been harvested, dried, stored at 253 K (– 20°C), and powdered in a blender.

Protease inhibitors are added at each step of this procedure. In the extraction step (step 1), phenylmethanesulphonyl fluoride (PMSF) is included; this irreversibly inactivates 'serine' proteases (see Chapter 5, Section 5.5.1.2). Most of the intracellular proteases are removed by adsorption on DEAE-cellulose under the conditions used in step 2. (It should be noted that the *arom* complex does not bind to the ion exchanger under these conditions, since 75 mmol dm^{-3} KCl is included in the buffer.) In step 3 (ammonium sulphate fractionation) the inhibitor benzamidine is added; this is an effective reversible inhibitor of proteases with trypsin-like specificity. It is added at this step because, unlike PMSF, it is not salted out of aqueous solutions by high concentrations of ammonium sulphate. After ion-exchange chromatography on DEAE-cellulose (step 4), the final step involves chromatography on blue dextran-Sepharose, with the enzyme eluted at high ionic strength (1.5 mol dm^{-3} KCl). The final yield of enzyme is about 4 mg from 100 g powdered cells, representing a yield of some 25 per cent. Purification factors for steps 2, 3, 4, and 5 are 1.9-, 5.3-, 3.7-, and 19.3-fold, respectively, giving an overall purification of 730-fold. The purified material appeared to be homogeneous on polyacrylamide gel electrophoresis in the presence of sodium dodecylsulphate and on native polyacrylamide gel electrophoresis. There was no trace of species of low M_r (which might be formed by proteolysis. The material could be stored satisfactorily for long periods at 253 K (–20°C) in the presence of 50 per cent glycerol, benzamidine (1 mmol dm^{-3}), and dithiothreitol (0.4 mmol dm^{-3}).

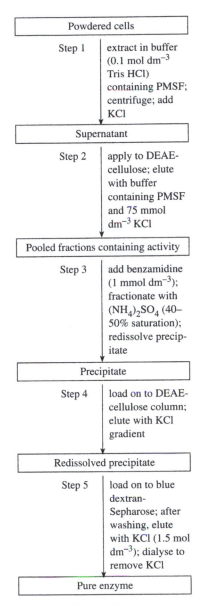

Fig. 2.12 Purification of the *arom* complex from *N. crassa*.

* The definition of multienzyme proteins is discussed in Chapter 7, Section 7.2.

The *arom* multienzyme protein consists of two identical subunits each of M_r 165 000 and each possessing five distinct enzyme activities.[105] This protein is thus a multienzyme polypeptide (see the discussion in Chapter 7, Section 7.11.2), in which the polypeptide chain is folded into a number of globular domains each with an active site; proteolytic cleavage could occur at the linking portions of the polypeptide chain. From an alignment of sequences of the *arom* multienzyme proteins from various fungi with those of the individual mono-functional enzymes from *E. coli*, it has been possible to identify these domains in detail.[105]

2.8.5 Glutathione reductase from *E. coli*

Glutathione reductase catalyses the NADPH-dependent reduction of oxidised glutathione* (GSSG):

$$\text{GSSG} + \text{NADPH} + \text{H}^+ \rightleftharpoons 2\text{GSH} + \text{NADP}^+.$$

Glutathione reductase belongs to the flavoprotein oxidoreductases, a family of enzymes (including dihydrolipoamide dehydrogenase (see Chapter 7, Section 7.7.1)) that possess a disulphide bond that is alternately oxidised and reduced as part of the catalytic mechanism. The enzyme had been previously prepared from *E. coli* using a complex two-stage affinity-chromatography procedure.[107] However, for more detailed investigations of the enzyme it was necessary to develop more a rapid procedure that could be scaled up. The preparation described by Scrutton *et al.*[108] made two important modifications. First, by recombinant DNA techniques a strain of *E. coli* was constructed that produced 100–200 times more enzyme than the wild-type strain. (The fragment of the plasmid pGR containing the gene for the enzyme was inserted into the plasmid pKK 233-3 to form a new plasmid pKGR which was used to transform *E. coli* strain JM10l.) Second, dye–ligand chromatography (Section 2.6.4.3) was used to give a substantial degree of purification. The procedure is summarized in Fig. 2.13.

The procedure uses a French press (Section 2.5.2) to disrupt the cell suspension followed by centrifugation to remove cell debris. During the addition of ammonium sulphate (step 2) the pH is maintained at 7–7.5 by addition of 2 mol dm^{-3} K_2HPO_4 as necessary. The precipitate is redissolved in a low-ionic-strength buffer (5 mmol dm^{-3} potassium phosphate, pH 7.0, containing 1 mmol dm^{-3} EDTA and 1 mmol dm^{-3} 2-mercaptoethanol) and dialysed against three changes of this buffer, before being applied to a column of Procion Red HE-7B linked to cross-linked agarose. As mentioned previously (Section 2.6.4.3), Procion dyes can show considerable specificity for different classes of enzymes and this dye had been found in preliminary work to bind glutathione reductase strongly. After washing with buffer containing 0.1 mol dm^{-3} KCl to remove weakly bound proteins, glutathione reductase is eluted by buffer containing 0.2 mol dm^{-3} KCl (step 3). In a final step the enzyme is concentrated by ultrafiltration (Section 2.6.1.3) and subjected to gel filtration on Superose 12 (which fractionates in the M_r range 1000–300 000) under h.p.l.c. conditions (Section 2.6.6). The overall yield is 11.6 mg enzyme from 100 cm^3 of cell suspension, in turn derived from 10 dm^3 of initial cell culture. The purification factors for steps 2, 3, and 4 are 1.6-, 33.6-, and 1.08-fold respectively, giving an overall purification of 58.3-fold. The enzyme is homogeneous on polyacrylamide gel electrophoresis in the presence of sodium dodecylsulphate and is identical with the enzyme prepared from the wild-type

* Glutathione (GSH) is a tripeptide γGluCysGly that acts as an intracellular reducing agent. In its oxidized form (GSSG), two molecules of glutathione are linked by a disulphide bond.

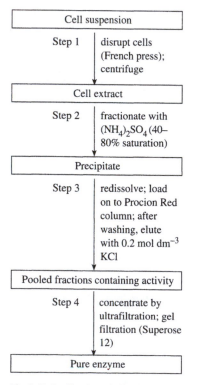

Fig. 2.13 Purification of glutathione reductase from *E. coli*. All steps are carried out at 277 K (4°C).

strain of *E. coli*[107] by a number of criteria (specific activity, M_r (97 000) and N-terminal sequence analysis).

2.8.6 Adenylate cyclase

Adenylate cyclase (EC 4.6.1.1) is a membrane-bound enzyme which catalyses the formation of 3′,5′-cyclic AMP from ATP and plays a key role in signal transduction (Chapter 6, Section 6.4.2.1, and Chapter 8, Section 8.4.5).

$$ATP \rightleftharpoons 3′:5′ \text{ cyclic AMP} + \text{pyrophosphate}$$

The activities of adenylate cyclases are regulated by a large number of factors including Ca^{2+} ions via the calmodulin system and the α and $\beta\gamma$ subunits of the heterotrimeric guanine-nucleotide binding regulatory proteins (G-proteins). A number of different forms of the enzyme have been purified, but a more detailed understanding of their regulatory properties has been hampered by the difficulty in isolating the pure isoforms of the enzyme. A recombinant baculovirus-driven (see Section 2.4.1) cell system has been used to express all six known isoforms of the enzyme.[109] The type I enzyme (activated by the α subunit of the G-protein and by Ca^{2+}/calmodulin) modified to include a hexahistidine tag at the N-terminus was isolated after expression in Sf9 (*Spodoptera frugiperda*, fall armyworm ovary) as outlined in Fig. 2.14.

In step 1, 4 litres of Sf9 cell culture are infected with the recombinant baculovirus encoding the histidine-tagged type I enzyme. Cells were harvested about 50 h after infection, suspended in a Na-HEPES buffer containing NaCl, EDTA, EGTA, dithiothreitol, and protease inhibitors, and lysed by nitrogen pressure at 40 atm. After removing the nuclei by centrifugation, the membranes were harvested by centrifugation at 70 000*g* for 30 min and resuspended in buffer, before being stored at −80°C. (The time for harvesting the cells after infection had been previously gauged by immunoblotting to give maximum expression of adenylate cyclase.)

In step 2, the type I enzyme is extracted by incubation of the membranes with the non-ionic detergent dodecylmaltoside in the presence of glycerol and NaCl. This combination gives about 90 per cent extraction, which is rather better than afforded by a number of other detergents tried including Lubrol, cholate, digitonin, etc. The detergent is added dropwise with stirring to the membrane suspension to give a final concentration of 0.8 per cent (w/v) and after homogenization (Teflon homogenizer) and centrifugation (150 000 *g*; 30 min) to remove unextracted membranes the supernatant is retained to the next stage.

In step 3, the supernatant from step 2 is subjected to affinity chromatography on forskolin-Sepharose (forskolin* is a diterpene isolated from the roots of *Coleus forskohlii*, which is a powerful activator of the enzyme). After application to the column, it is washed successively with buffer containing a lower concentration of detergent and high salt (2 mol dm^{-3} NaCl) and buffer containing detergent and the dipolar solvent dimethyl sulphoxide, in order to remove weakly bound contaminants. Finally, adenylate cyclase is eluted from the column by buffer containing detergent (0.2% (w/v)) and forskolin (200 μmol dm^{-3}).

In step 4, the material eluted by forskolin is further purified by chromatography on immobilized Ni^{2+} (Ni^{2+}-NTA-agarose; see Section 2.6.4.1). The enzyme is eluted by buffer containing 0.1% (w/v) detergent and 100 mmol dm^{-3} imidazole.

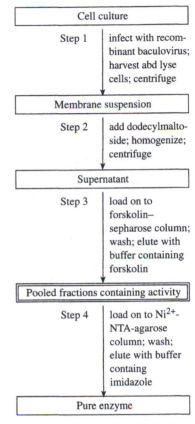

Fig. 2.14 Purification of type I adenylate cyclase from baculovirus-infected cells of *Spodoptera frugiperda*. All steps are carried out at 277 K (4°C).

* The structure of forskolin is given in Chapter 8, Section 8.4.5.

The procedure outlined gives a 10% recovery of the type I enzyme with an overall degree of purification of 2100-fold from the membrane suspension. The forskolin-Sepharose and immobilized Ni^{2+} steps lead to 180-fold and 4.9-fold purification respectively. On sodium dodecylsulphate–polyacrylamide gel electrophoresis a single band was observed at the expected molecular mass (M_r 110 000).

2.8.7 Chymosin

Chymosin (EC 3.4.23.4) is the active ingredient of rennet, the extract of calf stomach, which is used to curdle milk as the first step in cheese manufacture (see Chapter 11, Section 11.4.3.3). The enzyme specifically cleaves the bond between Phe 105 and Met 106 in κ-casein, one of the major phosphoprotein constituents of milk. Cleavage of this bond leads to a Ca^{2+}-induced aggregation of the casein micelles to form a gel. Because the demand for cheese has grown and the supply of calf stomachs declined, there has been increasing interest over the last 30 years or so for alternatives to rennet. Pepsins and other aspartic proteinases (see Chapter 5, Section 5.5.1.7) from various sources have been used but have generally not proved wholly satisfactory because of their different actions on casein, which lead to alterations in the flavour and/or texture of the product. It was recognized in the early 1980s that recombinant DNA technology could clearly be used to increase the supply of chymosin required by the industry.

Chymosin is first synthesized as preprochymosin, which has a 16 amino acid leader peptide sequence. When this is removed the zymogen prochymosin is formed; under the acid conditions of the stomach this is in turn converted to chymosin by removal of the 42 amino acid propeptides from the N-terminus. In order to produce recombinant enzyme, it was decided that the best approach for expression in a heterologous host was to express the gene for prochymosin[110] in order to avoid possible problems caused by incorrect processing of the leader sequence. Active chymosin could then be formed *in vitro* by incubation of the prochymosin at pH 6. The gene for prochymosin was expressed in *E. coli*, using the constructed plasmid pCT70 which contained the *trp* promoter, allowing expression of the gene to be induced by lack of tryptophan in the growth medium.[110] From the results of sodium dodecylsulphate–polyacrylamide gel electrophoresis of extracts of the transformed *E. coli* cells it was estimated that between 1 and 5 per cent of the total cell protein was expressed as prochymosin, but this was present in an insoluble aggregated form. The insoluble material was solubilized in 9 mol dm^{-3} urea, and after dialysis to remove the denaturing agent, the soluble enzyme was purified by ion-exchange chromatography on DEAE-cellulose (Section 2.6.2.1) followed by acidification. The resulting chymosin was shown to possess milk-clotting activity comparable with authentic chymosin. Marston and colleagues investigated the optimum conditions for solubilization of the inclusion bodies and their subsequent refolding.[111] The best results were obtained with solubilization of the Triton/EDTA washed inclusion bodies in 10 volumes 8 mol dm^{-3} urea, incubation at room temperature for 1 h, followed by a 10-fold dilution of this extract into phosphate buffer at pH 10.7. After 30 min incubation at the high pH, the pH was lowered to 8.0. The final step involved ion-exchange chromatography on DEAE-cellulose. Purified prochymosin was then activated by incubation at pH 6. Using this procedure the yield of prochymosin in the urea/alkali extract was 54 per cent of that

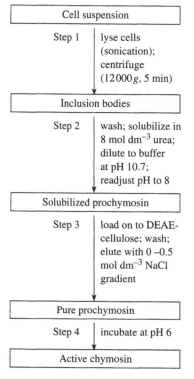

Fig. 2.15 Purification of recombinant chymosin after expression of prochymosin in *E. coli*.

in the total cell lysate, and the overall yield after ion exchange chromatography 22 per cent. It was later shown that the refolding of the solubilized inclusion body material was assisted by the addition of protein disulphide isomerase (Chapter 3, Section 3.6.2) which assists in the formation of the correct set of disulphide bonds in the zymogen.[112]

Because of the problems caused by formation of inclusion bodies when the gene for prochymosin is expressed in *E. coli*, alternative expression systems have been explored, including the yeast *Kluyveromyces lactis*.[113] A number of leader sequences were investigated to assess their ability to cause the prochymosin to be secreted into the growth medium. The best results were found with the prepro region of the α-factor gene from *Saccharomyces cerevisiae*; using this in the plasmid pKS105, more than 95 per cent of the prochymosin was secreted into the growth medium. Electrophoretic analysis of the growth medium indicated that prochymosin was the major protein present; this could be converted to active chymosin by incubation at pH 6. This type of approach has a number of advantages for the cheesemaking industry. First, the process can be scaled up relatively easily; fermentations on the scale of 40 000 litres have been run successfully.[113] By contrast, the batchwise procedure involving solubilization and renaturation of inclusion bodies in *E. coli* is difficult to scale up. Second, the purification of prochymosin from the culture medium is relatively straightforward. Third, because *Kluyveromyces* is a food yeast, its use as a host for overexpression of cloned genes is permitted by the regulatory authorities in many (but not all) countries. Recombinant chymosin has been widely used in the cheesemaking industry during recent years. There is also considerable current interest in the use of recombinant DNA technology to produce mutant forms of chymosin which may have different specificities and thus give rise to cheese products with novel flavours or textures.

2.9 Conclusions from the examples of enzyme purification

The seven examples of enzyme purification discussed in Section 2.8 illustrate the range of problems faced in this type of work and some of the solutions available. It can be seen that many of the earlier examples contain steps involving ammonium sulphate fractionation, ion-exchange chromatography, and gel filtration, whereas the more recent examples place greater emphasis on recombinant DNA technology together with affinity methods involving the recognition of particular structural features in the enzymes of interest. These developments have made it feasible to purify enzymes normally present in very small amounts, which would be difficult, if not impossible, to purify by the classical methods. The choice of methods to be used in any particular case depends on a number of factors (Section 2.6.6). If *yield* is the primary objective, then the number of steps in a purification should be kept to a minimum, since losses inevitably occur at each stage. (Of course the number of steps has to be sufficient to give a homogeneous product.) If *speed* is the primary concern, it should be remembered that methods based on changes in solubility are in general more rapid than those involving column chromatography, e.g. ion-exchange chromatography or gel filtration, although recent developments in h.p.l.c. techniques have made such chromatographic procedures much more rapid (Section 2.6.6).

References

1. Dixon, M., Webb, E. C., Thorne, C. J. R., and Tipton, K. F., *Enzymes* (3rd edn). Longman, London (1979). See p. 40.

2. Sawyer, L., in *Proteins Labfax* (Price, N. C., ed.), Ch. 15. Bios Scientific Publishers, Oxford (1996).

3. Hardie, D. G. and Cohen, P., *FEBS Lett.* **91**, 1 (1978).

4. Roberts, T. M. and Lauer, G. D., *Methods Enzymol.* **61**, 473 (1979).

5. Old, R. W. and Primrose, S. B., *Principles of gene manipulation* (5th edn). Blackwell, Oxford (1994).

6. Duncan, K., Lewendon, A., and Coggins, J. R., *FEBS Lett.* **165**, 121 (1984).

7. Brown, T. A., *Gene cloning: an introduction* (3rd edn). Chapman and Hall, London (1995).

8. Walsh, G. and Headon, D. (eds), *Protein biotechnology.* Wiley, Chichester (1994).

9. White, M. F. and Fothergill-Gilmore, L. A., *Eur. J. Biochem.* **207**, 709 (1992).

10. Marston, F. A. O., *Biochem. J.* **240**, 1 (1986).

11. Netzer, W. J. and Hartl, F. U., *Nature* **388**, 343 (1997)

12. Thatcher, D. R., in *Proteins Labfax* (Price, N. C., ed.), Ch. 12. Bios Scientific Publishers, Oxford (1996).

13. Romanos, M. A., Scorer, C. A., and Clare, J. J., *Yeast* **8**, 423 (1992).

14. Hornby, D. P., in *Proteins Labfax* (Price, N. C., ed.), Ch. 11. Bios Scientific Publishers, Oxford (1996).

15. Herring, P. J. (ed.), *Bioluminescence in action.* Academic Press, London (1978).

16. *Methods Enzymol.* **133** (1986).

17. Newsholme, E. A. and Leech, A. R., *Biochemistry for the medical sciences.* Wiley, Chichester (1983). See p. 206.

18. Scopes, R. K., *Protein purification: principles and practice* (3rd edn). Springer, New York (1994).

19. Avis, P. J. O., in *Subcellular components* (2nd edn) (Birnie, G. D., ed.), pp. 1–13. Butterworths, London (1972).

20. Freese, F. and Oosterwyk, J., *Biochemistry* **2**, 1212 (1963).

21. Repaske, R., *Biochim. biophys. Acta* **30**, 225 (1958).

22. Villaneuva, J. R. and Acha, I. G., in *Methods in microbiology* (Booth, C., ed.), Vol. 4, p. 665. Academic Press, New York (1971).

23. Emerson, S., *Genetica* **34**, 162 (1963).

24. Scopes, R. K., *Protein purification: principles and practice* (3rd edn). Springer, New York (1994). See p. 29.

25. Cogdell, R. J. and Lindsay, J. G., in *Proteins Labfax* (Price, N. C., ed.), Ch. 10. Bios Scientific Publishers, Oxford (1996).

26. Birnie, G. D. and Rickwood, D. (eds), *Centrifugal separations in molecular and cell biology.* Butterworths, London (1978).

27. Scopes, R. K., *Protein purification: principles and practice* (3rd edn). Springer, New York (1994). See p. 238ff.

28. McPhie, P., *Methods Enzymol.* **22**, 23 (1971).

29. Scopes, R. K., *Protein purification: principles and practice* (3rd edn). Springer, New York (1994). See p. 17.

30. Blatt, W. F., *Methods Enzymol.* **22**, 39 (1971).

31. Scopes, R. K., *Protein purification: principles and practice* (3rd edn). Springer, New York (1994). See p. 267.

32. Craig, L. C., *Methods Enzymol.* **11**, 870 (1967).

33. Scopes, R. K., in *Proteins Labfax* (Price, N. C., ed.). Bios Scientific Publishers, Oxford (1996). See p. 35.

34. Scopes, R. K., *Protein purification: principles and practice* (3rd edn). Springer, New York (1994). See Ch. 6.

35. Dixon, M., Webb, E. C., Thorne, C. J. R., and Tipton, K. F., *Enzymes* (3rd edn). Longman, London (1979). See p. 6.

36. Sluyterman, L. A. A. and Wijdenes, J. J., *Chromatogr.* **206**, 441 (1981).

37. Scopes, R. K., *Protein purification: principles and practice* (3rd edn). Springer, New York (1994). See p. 250ff.

38. Patel, D. and Rickwood, D., in *Proteins Labfax* (Price, N. C., ed.), Ch. 9. Bios Scientific Publishers, Oxford (1996).

39. Shuster, L., *Methods Enzymol.* **22**, 412 (1971).

40. Kemp, G., *Biotechnol. Appl. Biochem.* **27**, 9 (1998).

41. Hjertén, S., *Methods Enzymol.* **270**, 296 (1996).

42. Scopes, R. K., *Protein purification: principles and practice* (3rd edn). Springer, New York (1994). See p. 299ff.

43. Scopes, R. K., *Protein purification: principles and practice* (3rd edn). Springer, New York (1994). See p. 175.

44. Scopes, R. K., in *Proteins Labfax* (Price, N. C., ed.). Bios Scientific Publishers, Oxford (1996). See p. 40.

45. Scopes, R. K., *Protein purification: principles and practice* (3rd edn). Springer, New York (1994). See p. 74.

46. Scopes, R. K., *Protein purification: principles and practice* (3rd edn). Springer, New York (1994). See p. 76ff.

47. Lai, C. Y. and Horecker, B. L., *Essays Biochem.* **8**, 149 (1972).

48. Scopes, R. K., *Protein purification: principles and practice* (3rd edn). Springer, New York (1994). See p. 85ff.

49. Lowe, C. R., in *Laboratory techniques in biochemistry and molecular biology* (Work, T. S. and Work, E., eds), Vol. 7, p. 267. North-Holland, Amsterdam (1979).

50. *Affinity chromatography: a practical approach* (Dean, P. D. G., Johnson, W. S., and Middle, F. A., eds). IRL Press, Oxford (1985).

51. Scopes, R. K., in *Proteins Labfax* (Price, N. C., ed.). Bios Scientific Publishers, Oxford (1996). See p. 43.

52. Cuatrecasas, P. and Anfinsen, C. B., *Methods Enzymol.* **22**, 355 (1971).

53. Cuatrecasas, P., Wilchek, M., and Anfinsen, C. B., *Proc. natn. Acad. Sci. USA* **61**, 636 (1968).

54. Cuatrecasas, P. and Anfinsen, C. B., *A. Rev. Biochem.* **40**, 259 (1971).

55. Lowe, C. R., in *Laboratory techniques in biochemistry and molecular biology* (Work, T. S. and Work, E., eds), Vol. 7, p. 323. North-Holland, Amsterdam (1979).

56. Scopes, R. K., *Protein purification: principles and practice* (3rd edn). Springer, New York (1994). See p. 196.

57. Andersson, L., Jörnvall, H., Åkeson, Å., and Mosbach, K., *Biochim. Biophys. Acta* **364**, 1 (1974).

58. Scopes, R. K., in *Proteins Labfax* (Price, N. C., ed.). Bios Scientific Publishers, Oxford (1996). See p. 52.

59. Scopes, R. K., *Protein purification: principles and practice* (3rd edn). Springer, New York (1994). See p. 226ff.

60 Scopes, R. K., in *Proteins Labfax* (Price, N. C., ed.). Bios Scientific Publishers, Oxford (1996). See p. 46.

61. Pogell, B. N., *Biochem. Biophys. Res. Commun.* **7**, 225 (1962).

62. Scopes, R. K., *Biochem. J.* **161**, 253 (1977).

63. Scopes, R. K., *Biochem. J.* **161**, 265 (1977).

64. Davies, J. R. and Scopes, R. K., *Analyt. Biochem.* **114**, 19 (1981).

65. Scopes, R. K., *Protein purification: principles and practice* (3rd edn). Springer, New York (1994). See p. 210ff.

66. Scopes, R. K., in *Proteins Labfax* (Price, N. C., ed.). Bios Scientific Publishers, Oxford (1996). See p. 48.

67. Lowe, C. R. and Thomas, J. A., in *Enzymology Labfax* (Engel, P. C., ed.). Bios Scientific Publishers, Oxford (1996). See p. 9.

68. Thompson, S. T., Cass, K. H., and Stellwagen, E., *Proc. natn. Acad. Sci. USA.* **72**, 669 (1975).

69. Watson, D. H., Harvey, M. J., and Dean, P. D. G., *Biochem. J.* **173**, 591 (1978).

70. Hey, Y. and Dean, P. D. G., *Biochem. J.* **209**, 363 (1983).

71. Scopes, R. K., *Protein purification: principles and practice* (3rd edn). Springer, New York (1994). See p. 204ff.

72. Köhler, G. and Milstein, C., *Nature* **256**, 495 (1975).

73. Eisenbarth, G., *Analyt. Biochem.* **111**, 1 (1981).

74. Nishigaki, I., Ichinose, H., Tamai, K., and Nagatsu, T., *Biochem. J.* **252**, 331 (1988).

75. Brocklehurst, K., in *Proteins Labfax* (Price, N. C., ed.), Ch. 7. Bios Scientific Publishers, Oxford (1996).

76. Brocklehurst, K., Carlsson, J., Kiersten, M. P. J., and Crook, F. M., *Biochem. J.* **133**, 573 (1973).

77. Hillson, D. A., and Freedman, R. B., *Biochem. J.* **191**, 373 (1980).

78. Bill, K., Casey, R. P., Broger, C., and Azzi, A., *FEBS Lett.* **120**, 248 (1980).

79. Dixon, M., Webb, E. C., Thorne, C. J. R., and Tipton, K. F., *Enzymes* (3rd edn). Longman, London (1979). See p. 30.

80. Scopes, R. K., *Protein purification: principles and practice* (3rd edn). Springer, New York (1994). See p. 96.

81. Dixon, M., Webb, F. C., Thorne, C. J. R., and Tipton, K. F., *Enzymes* (3rd edn). Longman, London (1979). See pp. 33–5.

82. Dixon, M., Webb, F. C., Thorne, C. J. R., and Tipton, K. F., *Enzymes* (3rd edn). Longman, London (1979). See p. 35.

83. Scopes, R. K., *Protein purification: principles and practice* (3rd edn). Springer, New York (1994). See p. 173.

84. Everse, J. and Stolzenbach, F. E., *Methods Enzymol.* **22**, 33 (1971).

85. Wrigley, C. W., *Methods Enzymol.* **22**, 559 (1971).

86. Hare, D. L., Stimpson, D. I., and Cann, J. R., *Archs. Biochem. Biophys.* **187**, 274 (1978).

87. Scopes, R. K., *Protein purification: principles and practice* (3rd edn). Springer, New York (1994). See p. 262.

88. Kédzy, F. J. and Kaiser, E. T., *Methods Enzymol.* **19**, 3 (1970).

89. Brocklehurst, K., in *Enzymology Labfax* (Engel, P. C., ed.). Bios Scientific Publishers, Oxford (1996). See p. 59.

90. Price, N. C., *Analyt. Biochem.* **73**, 447 (1976).

91. Baines, B. S. and Brocklehurst, K., *Biochem. J.* **173**, 345 (1978).

92. Fersht, A. R. and Kaethner, M. M., *Biochemistry* **15**, 818 (1976).

93. Pösö, H. and Pegg, A. E., *Biochim. biophys. Acta* **747**, 209 (1983).

94. Fersht, A. R., *Structure and Mechanism in Protein Science*. Freeman, New York (1999). See p. 155.

95. Heil, A., Müller, G., Noda, L. *et al.*, *Eur. J. Biochem.* **43**, 131 (1974).

96. Tian, G., Sanders, II, C. R., Kishi, F., Nakazawa, A., and Tsai, M. D., *Biochemistry* **27**, 5544 (1988).

97. Ellis, R. J., *Trends Biochem. Sci.* **4**, 241 (1979).

98. Roy, H. and Cannon, S., *Trends Biochem. Sci.* **13**, 163 (1988).

99. Hall, N. P. and Tolbert, N. E., *FEBS Lett.* **96**, 167 (1978).

100. Ellis, R. J. and van der Vies, S. M., *A. Rev. Biochem.* **60**, 321 (1991).

101. Goloubinoff, P., Christeller, J. T., Gatenby, A. A., and Lorimer, G. H., *Nature* **342**, 884 (1989).

102. Burgess, R. R., *J. biol. Chem.* **244**, 6160 (1969).

103. Fujita, N. and Ishihama, A., *Methods Enzymol.* **272**, 121 (1996).

104. Tang, H., Kim, Y., Severinov, K., Goldfarb, A., and Ebright, R. H., *Methods Enzymol.* **272**, 130 (1997).

105. Coggins, J. R., Boocock, M. R., Chaudhuri, S. *et al.*, *Methods Enzymol.* **142**, 325 (1987).

106. Pringle, J. R., *Methods cell. Biol.* **12**, 149 (1975).

107. Mata, A. M., Pinto, M. C., and Lopez-Barea, J., *Z. Naturforsch. C: Biosci.* **39**, 908 (1984).

108. Scrutton, A. S., Berry, A., and Perham, R. N., *Biochem. J.* **245**, 875 (1987).

109. Taussig, R., Tang, W.-J., and Gilman, A. G., *Methods Enzymol.* **238**, 95 (1994).

110. Emtage, J. S., Angal, S., Doel, M. T. *et al.*, *Proc. natn. Acad. Sci. USA* **80**, 3671 (1983).

111. Marston, F. A. O., Lowe, P. A., Doel, M. T., Schoemaker, J. M., White, S., and Angal, S., *Bio/Technology* **2**, 800 (1984).

112. Tang, B., Zhang, S., and Yang, K., *Biochem. J.* **301**, 17 (1994).

113. van den Berg, J. A., van der Laken, K. J., van Ooyen, A. J. J. *et al.*, *Bio/Technology* **8**, 135 (1990).

3

The structure of enzymes

* The term *relative molecular mass* (M_r) is now used in place of molecular weight. M_r is a dimensionless number and is the ratio of the molecular mass of a molecule to 1/12 the mass of one atom of ^{12}C. The latter value is known as a *dalton* (1.663×10^{-24} g). Molecular masses are often quoted in daltons or kilodaltons (abbreviated as Da or kDa, respectively).

† The term subunit refers to the smallest covalent unit. A subunit may consist of one polypeptide chain or two or more chains linked by disulphide bonds. Thus the subunit of insulin consists of two chains; six of these subunits can aggregate to form a hexameric species.

3.1 Introduction

In this chapter we shall consider various aspects of enzyme structure. The ultimate aim of an investigation is to establish the complete three-dimensional structure of an enzyme at atomic (or near atomic) resolution. This information provides the necessary basis not only for understanding the detailed properties of the enzyme, especially its catalytic activity (see Chapter 5), but also for a number of applications, including the rational design of drugs (Chapter 10) and the exploitation of the enzyme in commercial processes (Chapter 11).

The determination of this detailed structure has become considerably easier in recent years as a result of several technical developments but still remains a major task for most enzymes, requiring access to specialized resources and (usually) a few years of work. Assuming that a supply of purified enzyme is available, the principal stages of the work can be listed as:

(1) determination of relative molecular mass, M_r (Section 3.2);*

(2) determination of primary structure and amino acid composition (Section 3.3);

(3) determination of secondary and tertiary structure (Section 3.4);

(4) determination of quaternary structure (Section 3.5);

The term primary structure refers to the sequence of amino acids in a polypeptide chain; it contains only one-dimensional information and tells us little about the three-dimensional structure. The terms secondary and tertiary structure refer to different aspects of the three-dimensional structure; secondary structure refers to regular elements of structure such as α-helix, β-sheet, etc. (Section 3.4) in which interactions between regions of the enzyme in close proximity in the sequence are involved. Tertiary structure refers to the folding of a chain, by which portions of the molecule well separated in the sequence are brought into close contact with each other. The term quaternary structure refers to the arrangement of the individual subunits† in an enzyme consisting of more than one subunit.

We shall discuss the various aspects of structure in the order given above, but it should be emphasized that structural investigations do not necessarily proceed in this order. It is often found that information from one type of investigation can be useful in another line of work; thus the definitive M_r value is usually established once the primary structure of the enzyme is known (Section 3.3.2.10) and the primary structure is needed to help in the elucidation of the three-dimensional structure by X-ray crystallography (Section 3.4.1). In each of the sections, the main experimental approaches involved will be indicated, together with a description of the principal results obtained and the application of these results.

There has been explosive growth in the amount of structural information available over the past decade. For instance, the number of amino acids in protein sequences in the SWISS-PROT database[1] has grown over 4-fold in the period from 1990 to 1996 and the number of structures determined by X-ray crystallography deposited in the Protein Data Bank (see the Appendix to this chapter) by nearly 10-fold over this period.[2] The storage of this information and its use has given rise to a new area of *bioinformatics*. Developments in computing now allow for access and use of the databases to be made from networked computers and workstations in offices and teaching locations. Because it is important for students to become familiar with some of the more important databases and programs involved, the Appendix to this chapter contains some useful web sites and other means of getting started. Two recent volumes of *Methods in Enzymology*[3,4] have been devoted to this area.

3.2 The determination of M_r

Enzymes are macromolecules with M_r values ranging from about 10 000 to several million. A number of methods which are appropriate for this range have been developed; most present-day determinations of M_r values of enzymes are performed by use of one or more of the following techniques:

(1) ultracentrifugation;

(2) gel filtration;

(3) sodium dodecylsulphate polyacrylamide gel electrophoresis;

(4) mass spectrometry.

Of these methods, (2) and (3) are 'semi-empirical' and comparisons are made with standard molecules of known M_r. In (1), however, the M_r can be calculated using equations derived from first principles. The relatively new mass spectrometric techniques (4) can give extremely accurate M_r values.

3.2.1 Ultracentrifugation

An ultracentrifuge is capable of generating intense centrifugal fields; in a typical machine a rotor speed of 65 000 r.p.m. corresponds to a field of the order of 300 000 times that of gravity. Under these conditions macromolecules will have a tendency to sediment, provided that the density of the macromolecule is greater than that of the solution (this generally holds for enzymes in aqueous solution). The ultracentrifuge can be used in two main ways to determine M_r values: sedimentation velocity and sedimentation equilibrium.[5]

3.2.1.1 Sedimentation velocity

In this type of experiment the ultracentrifuge is operated at high speeds to generate centrifugal forces that are sufficiently intense to sediment the macromolecules. The sedimentation of an enzyme can be monitored by suitable optical means (including, for instance, changes in refractive index or absorption of 280 nm radiation) and from these measurements the sedimentation coefficient, s, can be calculated. This coefficient gives a measure of the sedimentation velocity in a unit gravitational field. The sedimentation coefficient cannot by itself be used to calculate the M_r of the enzyme, since the rate of sedimentation will depend on other factors such as the shape of the macromolecule. However, if we have other information, such as the value of the diffusion coefficient (D) of the macromolecule, its partial specific volume ($\bar{v}$) and the density of the solution (ρ), the M_r can be calculated[6] from the formula:

$$M_r = \frac{RTs}{D(1 - \bar{v}\rho)},$$

where R is the gas constant, and T is the temperature.

The partial specific volume ($\bar{v}$) of a solute is defined as the volume change upon addition of 1 g of that solute to a large volume of solution, keeping all other parameters (temperature, pressure, etc.) constant. The value of ($\bar{v}$) can be determined from very accurate measurements of the densities of solutions[7] or calculated from the amino-acid composition of the enzyme[8] by using the known values of the molar volumes of each increment of structure of the various amino acids. The diffusion coefficient, D, can be measured in a separate experiment in which the ultracentrifuge is operated at low speeds so that no detectable sedimentation occurs.

3.2.1.2 Sedimentation equilibrium

If the rotor speed is not great enough to cause complete sedimentation of the macromolecule, then after a while an equilibrium state is reached in which the tendency of the macromolecules to be sedimented is balanced by their tendency to diffuse from the region of high concentration at the bottom of the ultracentrifuge cell to the region of low concentration towards the meniscus of the solution. From measurements of the distribution of concentration, c, of the macromolecule as a function of distance, r, along the cell (i.e. distance from the axis of rotation), the M_r can be calculated[6] provided that we also know $\bar{v}$ and ρ, as in the sedimentation velocity experiment (Section 3.2.1.1):

$$M_r = \frac{2RT}{(1 - \bar{v}\rho)\omega^2} \frac{\mathrm{d}\ln c}{\mathrm{d}r^2}.$$

where R is the gas constant, T is the temperature, and ω is the angular velocity in rad s^{-1}. The value of ($\mathrm{d}\ln c/\mathrm{d}r^2$) is obtained from the slope of the plot of $\ln c$ against r^2.

The advantage of the sedimentation equilibrium method is that a system is being studied at equilibrium and thus there is no dependence on the shape of the solute or the viscosity of the solution. Using a short column of solution (0.15–0.30 cm) the time required to reach equilibrium is quite short (4–8 hours). Under favourable circumstances M_r can be determined to within a few per cent of the true value (i.e. the value calculated subsequently from the primary structure). For further details of these methods, a number of articles can be consulted.[5,9] Provided that the appropriate values of ($\bar{v}$) and ρ are used it is also possible to calculate the M_r of the enzyme under denaturing conditions (Section 3.6.1).

Ultracentrifugation is a versatile technique as it can be used to study enzymes under a wide variety of conditions. Thus the effects of ligands, cofactors, etc. on the M_r can be determined. A special application of the technique is active enzyme centrifugation (Section 3.2.5) which allows the M_r of the active enzyme unit to be determined.[10]

3.2.2 Gel filtration

As described in Chapter 2 (Section 2.6.1.2) certain cross-linked polymers such as Sephadex, Sephacryl, and Bio-Gel are able to separate molecules according to size. Large molecules are unable to penetrate the pores of the cross-linked polymer and are eluted from a Sephadex column ahead of small molecules. For molecules of a similar shape (e.g. a series of globular proteins) there has been found to be a linear relationship between the logarithm of M_r and the distribution coefficient, K_d, over a certain range of M_r values.[11,12]

The value of K_d for a given solute is defined by the relationship

$$K_d = \frac{V_e - V_o}{V_i - V_o},$$

where V_e is the elution volume of the molecule of interest, V_o is the elution volume of a molecule completely excluded by the column (e.g. dextran blue with an M_r of 2×10^6) and V_i is the elution volume of a small molecule (e.g. glucose) which is totally included by the column. Thus $K_d = 0$ for solutes which are totally excluded and $K_d = 1$ for solutes which are totally included.

For Sephadex G-200 the 'linear' range of M_r values is from about 40 000 to about 200 000. By reference to the standard curve, the M_r of the enzyme of interest can be determined (Fig. 3.1). Other types of cross-linked polymers can be used over different ranges of M_r values.

While the gel filtration method is extremely simple to use it must be emphasized that the M_r value obtained should only be used as a guide until more definitive evidence is available. It has been known for some time[13] that glycoproteins, particularly those rich in carbohydrate, show anomalous behaviour on gel filtration, and this will also be the case for proteins that are not globular in shape. Thus, the heat-stable phosphatase inhibitor protein isolated from skeletal muscle (Chapter 6, Section 6.4.2.4) appears to have an M_r of 60 000 on gel filtration through Sephadex G-100, whereas the true M_r is close to 20 000.[14] Fortunately for our purposes it does appear that the vast majority of enzymes are globular proteins and are well behaved on gel filtration, so that M_r values accurate to within 5–10 per cent are readily obtained.[11] It is also possible to measure M_r values under denaturing conditions (see Section 3.6.1) provided appropriate standard curves are constructed.[11,15] Gel filtration can also be performed under high-performance liquid chromatography (h.p.l.c.) conditions using special column support materials, such as Superose 12 (which has a fractionation range in the M_r

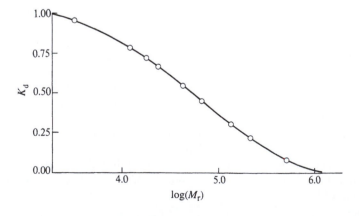

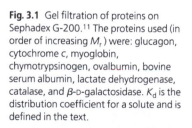

Fig. 3.1 Gel filtration of proteins on Sephadex G-200.[11] The proteins used (in order of increasing M_r) were: glucagon, cytochrome c, myoglobin, chymotrypsinogen, ovalbumin, bovine serum albumin, lactate dehydrogenase, catalase, and β-D-galactosidase. K_d is the distribution coefficient for a solute and is defined in the text.

1000–300 000 range). Apart from the higher resolution available using such methods, the M_r can be determined in a considerably shorter time (typically 30 min) than by conventional gel filtration.[16]

3.2.3 Sodium dodecylsulphate polyacrylamide gel electrophoresis

The mobility of a charged molecule in an electric field is normally a function of various factors such as the size and shape of the molecule and the charge it carries and it would therefore be expected that electrophoresis would not normally give any reliable estimates of M_r.* However, in the case of proteins we can overcome this difficulty by addition of the detergent sodium dodecylsulphate, SDS, which has the structure $CH_3(CH_2)_{11}OSO_3^- Na^+$. (SDS is also known as sodium lauryl sulphate.) A reducing agent such as 2-mercaptoethanol is also added to break disulphide bonds (see Section 3.3.1.2). Addition of the detergent has two principal effects.[18]

1. Nearly all proteins bind SDS in a more or less constant ratio, 1.4 g SDS per gram of protein. Since the negative charge carried by the SDS overwhelms any charge carried by the protein, the protein–SDS complex has a constant charge/mass ratio.

2. The native three-dimensional structure of the protein is lost and the protein–SDS complex adopts a uniform shape (thought to be a rod shape) with dimensions proportional to the M_r of the protein.

Since the charge and hydrodynamic properties of the protein–SDS complex are both simple functions of the M_r, it is not particularly surprising that the mobility on electrophoresis is a function of M_r alone. The fact that larger molecules have lower mobilities means that the hydrodynamic effects (i.e. sieving) predominate over the charge effects. In practice it is found that a graph of the logarithm of the M_r against mobility is linear (Fig. 3.2) and the M_r of an unknown protein can be determined by reference to the standard line. Different ranges of M_r can be examined by the use of gels of different polyacrylamide concentration or by the use of gradient gels;[19] thus, 10 per cent gels give good separation in the range 10 000 to 70 000 and 5 per cent gels are satisfactory in the range 25 000 to 200 000. The accuracy of the M_r obtained is estimated to be 5 per cent or better.[18,20]

* Hedrick and Smith[17] have shown that the M_r of a protein can be estimated by measuring its mobility as a function of the concentration of acrylamide used to form the polyacrylamide gels. However, this method is only reliable if the standard proteins for calibration have the same shape, degree of hydration, and partial specific volume as the unknown protein.

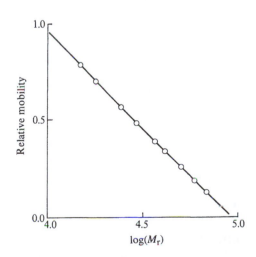

Fig. 3.2 Sodium dodecylsulphate polyacrylamide gel electrophoresis of proteins.[18] Ten per cent polyacrylamide gels were used and mobilities are expressed relative to bromophenol blue. The proteins used (in order of increasing subunit M_r) were: lysozyme, myoglobin, trypsin, carbonate dehydratase, glyceraldehyde-phosphate dehydrogenase, fructose-bisphosphate aldolase, fumarate hydratase, catalase, and bovine serum albumin.

The comments made in Section 3.2.2 about gel filtration also apply to SDS polyacrylamide gel electrophoresis; namely that the M_r obtained should be checked by other methods. There are several cases known in which anomalous behaviour is observed on electrophoresis. Thus, histones, which are highly positively charged, contribute significantly to the charge of the protein–SDS complex and give rise to rather low electrophoretic mobilities.[18] In addition, glycoproteins often show impaired binding of SDS and display low mobilities.[18] Additional complications can arise when proteins possess highly flexible regions (see the example of the lipoyl domains in the E2 component of the pyruvate dehydrogenase complex, Chapter 7, Sections 7.7.1 and 7.7.2).

Finally, we should note that addition of SDS dissociates a protein into its constituent subunits. If the subunit contains disulphide bonds, the addition of 2-mercaptoethanol ensures that these are broken, so that the molecular weight(s) we obtain by this method will be that/those of the individual polypeptide chain(s).

3.2.4 Mass spectrometry

Until the 1980s the applications of mass spectrometry to biological systems were limited to the study of a number of small molecules which could be volatilized relatively easily following chemical modification, e.g. by permethylation of the amide nitrogen atoms and modification of both N- and C-termini[21] (Fig.3.3).

In the early 1980s, the development of fast-atom bombardment (f.a.b.) mass spectrometry overcame a number of the difficulties associated with conventional mass spectrometry.[22] In this technique there is no need for prior derivatization of the sample. The peptide is dissolved in glycerol and bombarded with high-energy atoms (Xe or Ar). The transfer of momentum from the atoms to the peptide leads to the production of gaseous ions. Development of the technique has allowed peptides of up to about 50 amino acids in length to be analysed.[23] However, the key breakthroughs made in the late 1980s were the development of 'soft' ionization techniques capable of producing ions of low energy, since these allowed ions to be generated from biological macromolecules. This has led to a very large growth in the application of mass spectrometry to the study of proteins of M_r up to several hundred thousand.[24–27] In MALDITOF (matrix-assisted laser desorption ionization time of flight) mass spectrometry, a small volume (1–10 microlitres) of solution of the protein plus an organic 'matrix' such as cinnamic acid is dried on to a support. An intense pulsed UV laser beam is then directed on to this area of the support; the energy of the laser is absorbed by the matrix which then rapidly vaporizes and carries a stream of protein ions into the gas phase. The ions are accelerated to a constant kinetic energy prior to analysis, which is performed on the basis of the time of flight along a hollow tube (1 m or so in length) from the laser to a detector. The time of flight is related to the mass to charge ratio of the ion. After calibration in the appropriate mass range, the M_r value of the protein can be determined.

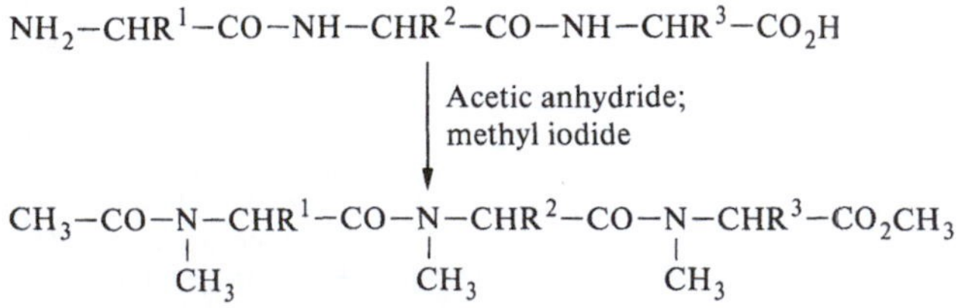

$$NH_2-CHR^1-CO-NH-CHR^2-CO-NH-CHR^3-CO_2H$$

Acetic anhydride; methyl iodide

$$CH_3-CO-\underset{\underset{CH_3}{|}}{N}-CHR^1-CO-\underset{\underset{CH_3}{|}}{N}-CHR^2-CO-\underset{\underset{CH_3}{|}}{N}-CHR^3-CO_2CH_3$$

Fig. 3.3 Modification of a peptide by acetylation and permethylation in order to prepare a volatile derivative for mass spectrometry.

In ESI (electrospray ionization) mass spectrometry, a stream of ions is formed continuously. A solution of the protein (usually in 1% (v/v) formic acid) together with a volatile organic solvent such as methanol or acetonitrile is pumped continuously into the source through a capillary needle which is maintained at 4 kV potential relative to an adjacent sampling plate. This leads to a fine spray of small droplets with high surface charge. After evaporation of the carrier solvent, the droplets break up to give eventually ions of individual molecules. A quadrupole mass analyser operated at low pressure (of the order of 10^{-8} atm) is used to measure the mass to charge ratio of the ions. It is possible to determine the M_r of the protein by applying appropriate computer-based analysis to the observed envelope of peaks obtained in the analyser.

The two mass spectrometric techniques are complementary to one another. MALDITOF–MS is particularly suitable for high molecular mass species (> 250 000) and in the analysis of mixtures, but is of lower accuracy than ESI–MS. The latter technique is capable of giving masses up to 100 000 with 0.002 per cent accuracy (± 2 units in 100 000). This level of accuracy means that a number of important tasks can be undertaken: (i) the observed mass can be compared with that calculated on the basis of the amino-acid sequence (Section 3.3.2.9) — this provides a very useful check on the accuracy of the latter; (ii) the observed mass will also indicate the occurrence of any post-translational modification — for instance the acetylation of the N-terminus (Section 3.3.2.7) will add 42 to the M_r;[24] and (iii) the mass will confirm the nature of any mass change introduced by site-directed mutagenesis (see Chapter 5, Section 5.4.5) — for instance the replacement of histidine by glutamine would lead to a decrease of 9 in the M_r.[28]

ESI–MS would normally give the subunit M_r, but under favourable conditions is capable of giving the intact molecular mass of an oligomeric protein.[29]

3.2.5 Association–dissociation systems

In the previous sections (3.2.1, 3.2.2, and 3.2.3) we have assumed that the enzyme in question has a unique M_r value, i.e. that it is *monodisperse*. While this is a reasonable assumption for most purified enzymes, there are systems known that are capable of undergoing association and dissociation. A good example is provided by glutamate dehydrogenase from beef liver, where the fundamental unit of M_r 336 000 (itself made up of six polypeptide chains) can form an aggregate:

$$nA \rightleftharpoons A_n$$

where A is the hexamer of M_r 336 000 and A_n is the aggregate of M_r up to at least 2×10^6.

The position of equilibrium in this system depends on the concentration of enzyme with the aggregate form being favoured at high concentrations.[30] At any given concentration the M_r will contain contributions from the various species present and the measured value of this average M_r will depend on the technique employed. For instance, from osmotic pressure measurements we would determine a 'number average' M_r ($\overline{M}_n$) defined by

$$\overline{M}_n = \frac{\sum n_i m_i}{\sum n_i},$$

where n_i is the relative number of the ith species, and m_i is its relative molecular mass.

However, from sedimentation equilibrium or light-scattering measurements a 'weight average' M_r ($\overline{M}_w$) is determined:

$$\overline{M}_w = \frac{\sum n_i m_i^2}{\sum n_i m_i}$$

Other examples of association–dissociation systems include pyruvate carboxylase in which the tetrameric enzyme dissociates on dilution,[31] and acetyl-CoA carboxylase, in which extensive polymerization can occur to give large filamentous aggregates.[32]

It is possible to adapt some of the methods described to determine the M_r of the active unit of the enzyme when this may be in doubt, e.g. in an association–dissociation system. Cohen and Mire[33,34] described the technique *of active enzyme centrifugation* in which enzyme at very low (catalytic) concentrations is sedimented through a solution of substrate. The rate of sedimentation down the ultracentrifuge cell is determined by monitoring the progress of conversion of substrate to product using absorption spectrophotometry or other convenient means of following the reaction. In addition, gel filtration can be performed at low concentrations of enzyme if the measurement of catalytic activity is used to monitor the position of elution of enzyme from the column. Using techniques such as these it has been possible to show, for instance, that the active unit of beef-liver glutamate dehydrogenase is the hexamer of M_r 336 000,[34,35] the active form of yeast phosphoglycerate mutase is the tetramer,[36] and the solubilized monomeric forms of the Na^+/K^+ ATPase and Ca^{2+} pumps are active.[10]

3.2.6 Uses of M_r information

The M_r of an enzyme is a fundamental piece of information because it allows us to convert the concentration of a solution from units of mass per volume (e.g. mg cm^{-3}) to units of molarity. This information can then be used in a whole variety of ways such as in considerations of composition (how many amino acids of a given type are present per molecule of enzyme?), catalytic activity (how many molecules of substrate are transformed by one molecule of enzyme per second?), and ligand binding (how many molecules of ligand are bound per molecule of enzyme?). Many of these applications will be described in the course of this and the succeeding three chapters.

Measurements of M_r made in the absence and presence of denaturing agents (see Section 3.6.1) will show whether or not the enzyme is composed of subunits and may indicate the number of such subunits (see Section 3.6.1.1). Lactate dehydrogenase from yeast has an M_r of 140 000 on gel filtration, but on SDS–polyacrylamide gel electrophoresis the M_r is close to 35 000, indicating that the enzyme is a tetramer, composed of four subunits.

3.3 The determination of amino-acid composition and primary structure

These two aspects of structure determination have been included in the same section of this chapter to reflect the fact that with the greatly increased rate of sequencing as a result of the development of DNA-based sequencing methods (Section 3.3.2.1) most of the information on the amino-acid composition data is now deduced from the amino-acid sequence itself. First, however, we shall give a brief account of the direct determination of composition from analysis of the enzyme itself.

3.3.1 Amino acid composition

3.3.1.1 The amino acids

As mentioned in Chapter 1 (Section 1.2) virtually all enzymes are proteins and are thus made up of α-amino acids linked together. In some cases additional components such as metal ions, cofactors, or carbohydrate may be present. The composition of the protein part is defined in terms of the numbers of each of the 20 amino acids that occurs in each molecule of enzyme. These amino acids are shown in Table 3.1 in the form that is predominant at neutral pH, i.e. in the *zwitterion* form $^+NH_3\text{--}CHR\text{--}CO_2^-$ rather than the uncharged form $NH_2\text{--}CHR\text{--}CO_2H$.* It must be pointed out that different ionized forms of amino acids from those shown in Table 3.1 are important in certain properties, e.g. in the reaction of cysteine with an electrophile such as iodoacetamide it is the $-CH_2\text{--}S^-$ form of the cysteine side-chain that is the reactive species even though fewer than 10 per cent of the molecules are in this form at pH 7. In general, therefore, we shall, for convenience, depict amino acids as being *in their uncharged forms*, and leave it to the reader to ascertain the predominant or most appropriate ionized form under a given set of circumstances.

* Since the pK_a values of the carboxyl and amino groups attached to the α-carbon atom in the general formula are approximately 2 and 10 respectively, the 'zwitterion' form will predominate at neutral pH.

Table 3.1 The amino acids that are normally found in enzymes

α-Amino acids are of the general formula

$$^+NH_3\text{--}CH\text{--}CO_2^-$$
$$|$$
$$R$$

(except for proline). Different amino acids differ in the nature of the R group, or side-chain. One type of classification of the amino acids can be made on the basis of the chemical nature of the R group. The pK_a values of the carboxyl and amino groups attached to the α-carbon atom are approximately 2 and 10 respectively.

Amino acid	Abbreviations	Side-chain	pK_a of side-chain
Non-polar R group			
Alanine	Ala (A)	$-CH_3$	—
Valine	Val (V)	$-CH(CH_3)CH_3$	—
Leucine	Leu (L)	$-CH_2-CH(CH_3)CH_3$	
Isoleucine	Ile (I)	$-CH(CH_3)-CH_2-CH_3$	
Phenylalanine	Phe (F)	$-CH_2-C_6H_5$	
Tryptophan	Trp (W)	$-CH_2$ (indole)	
Methionine	Met (M)	$-CH_2-CH_2-S-CH_3$	
Proline (see foot of table)			

Table 3.1 (Contd.)

Amino acid	Abbreviations	Side-chain	pK_a of side-chain
Polar, uncharged R group			
Glycine	Gly (G)	—H	—
Serine	Ser (S)	—CH_2—OH	≈ 14.0
Threonine	Thr (T)	—CH—CH_3 with OH	≈ 14.0
Cysteine	Cys (C)	—CH_2—SH	≈ 8.3
Tyrosine	Tyr (Y)	—CH_2—〈ring〉— OH	≈ 10.1
Asparagine	Asn (N)	—CH_2—C(=O)—NH_2	
Glutamine	Gln (Q)	—CH_2—CH_2—C(=O)—NH_2	
Acidic R group			
Aspartic acid	Asp (D)	—CH_2—CO_2^-	≈ 3.9
Glutamic acid	(Glu (E)	—CH_2—CH_2—CO_2^-	≈ 4.3
Basic R group			
Lysine	Lys (K)	—$(CH_2)_4$—$^+NH_3$	≈ 10.5
Arginine	Arg (R)	—$(CH_2)_3$—NH—C(=$^+NH_2$)—NH_2	≈ 12.5
Histidine	His (H)	—CH_2—C=CH (imidazole: HN, N, CH)	≈ 6.0

Proline (abbreviated Pro or P) is an imino acid with a secondary nitrogen atom. The predominant form at pH 7 is

$$CH_2—CH_2—CH_2—^+NH_2—CH(CO_2^-)$$

In this book we shall use the three-letter abbreviations for the amino acids. The one-letter code is frequently used in displays of amino-acid sequences to save space.

Apart from glycine, all the amino acids have a chiral centre; indeed, isoleucine and threonine have two such centres. They are thus capable of existing in more than one enantiomeric form. It is found that in enzymes all the amino acids occur in the L-form in which the three-dimensional configuration about the α-carbon atom is as shown in Fig. 3.4(a).

D-Amino acids, which have the opposite configuration (see Fig. 3.4(b)), do occur occasionally in some small peptides; thus D-phenylalanine occurs in the antibiotic gramicidin S. The mode of insertion of D-amino acids into such systems occurs in a sequence of reactions[37] and is thus distinct from the normal mode of protein synthesis which is mediated by ribosomes, following the information contained in the messenger RNA sequence.

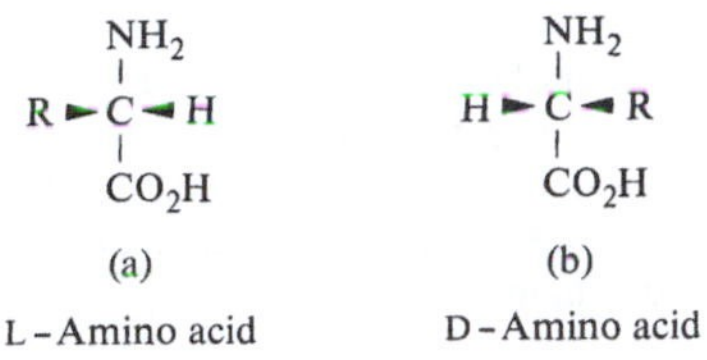

3.3.1.2 Linkages between amino acids

The bond that joins amino acid units together to form a chain is known as an *amide* or a *peptide bond*. We can view the bond as arising by the elimination of water as in the reaction:

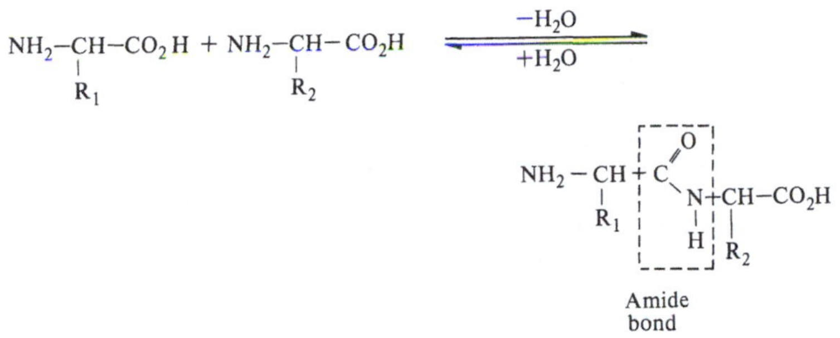

$$NH_2-CH-CO_2H + NH_2-CH-CO_2H \quad \underset{+H_2O}{\overset{-H_2O}{\rightleftharpoons}}$$
$$\quad\quad R_1 \quad\quad\quad\quad\quad R_2$$

NH$_2$ NH$_2$

R►C◄H   H►C◄R

CO$_2$H   CO$_2$H

(a)   (b)

L – Amino acid   D – Amino acid

Fig. 3.4 Structures of (a) an L-amino acid, and (b) a D-amino acid. In these diagrams, –R and –H project towards the reader; –NH$_2$ and –CO$_2$H project away from the reader.

Amide bond

In order to form this bond it is first necessary to activate the carboxyl group. During biosynthesis of proteins this is achieved by forming aminoacyl-AMP derivatives by reaction of the amino acid with ATP; the pyrophosphate group is split out.

When a number of amino acids are linked together in this way we build up a peptide chain. The term polypeptide or protein is usually applied to chains consisting of 50 or more amino-acid units. By convention, the N-terminal amino acid is written at the left-hand side and the C-terminal one at the right. The amino-acid units are numbered from the N-terminal end. Figure 3.5 shows the tetrapeptide, alanylleucyltyrosylmethionine, which would be abbreviated as Ala-Leu-Tyr-Met or ALYM in the one-letter code.[38]

Evidence for the presence of amide bonds in proteins comes from a variety of sources, e.g. UV and IR spectroscopy, the action of enzymes which are known to cleave amide bonds specifically, and direct structure determinations using X-ray crystallography.

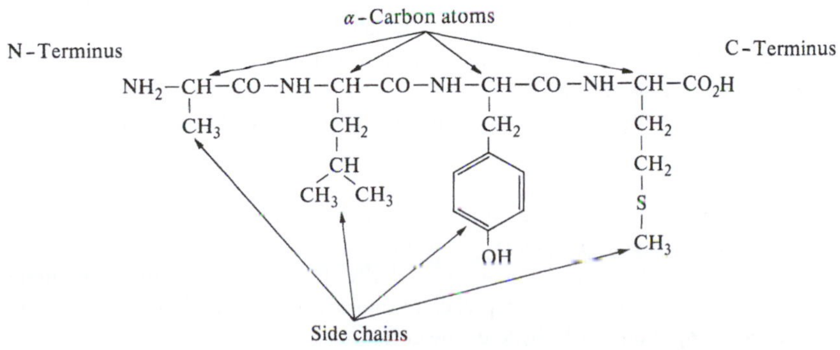

Fig. 3.5 The structure of alanylleucyltyrosylmethionine. In naming the peptide, all amino acids except the C-terminal one are given the suffix -yl.[38]

* In some proteins other types of
covalent bonds are found. For
instance, certain collagens have
cross-links formed by condensation
reactions involving the deaminated
side-chains of lysine.[39]

HS – CH$_2$ – CH$_2$ – OH

2-Mercaptoethanol

CH$_2$– SH
CHOH
CHOH
CH$_2$– SH

Dithiothreitol

Apart from the amide bond, the only other type of covalent bond involved
in linking amino-acid units in enzymes* is the disulphide (–S–S–) bond, which
can be formed between two cysteine side-chains under oxidizing conditions.
The disulphide bond can be broken by the addition of reducing agents such as
2-mercaptoethanol or dithiothreitol.

$$2\ NH_2-CH-CO_2H\ +\ \tfrac{1}{2}\ O_2 \rightleftharpoons NH_2-CH-CO_2H + H_2O$$

CH$_2$ CH$_2$
SH   S
S
CH$_2$
NH$_2$ – CH – CO$_2$H

2 Cysteine   Cystine

In some enzymes, especially the small extracellular hydrolases such as
lysozyme, ribonuclease, and chymotrypsin, there are no free cysteine side-chains,
i.e. all are paired to form disulphide bonds. By contrast, in most intracellular
enzymes such as lactate dehydrogenase the cysteine side-chains are free and there
are no disulphide bonds.

3.3.1.3 Hydrolysis of amide bonds

Hydrolysis of the amide bonds in proteins can be brought about by a variety of
reagents, but the most commonly used conditions are 6 mol dm^{-3} HCl at 383 K
(110°C) for 24 hours *in vacuo*. However, under these conditions some amino
acids, e.g. tryptophan, are partially or wholly destroyed (see Section 3.3.1.5). For
determination of the tryptophan content of a protein, hydrolysis is carried out
using *p*-toluene-sulphonic acid[40] or sodium hydroxide[41] in place of HCl. A single
hydrolysis method claimed to be suitable for all amino acids (including cysteine
and tryptophan) has been described.[42] It is based on alkylation of cysteine, followed by hydrolysis by methanesulphonic acid in the presence of tryptamine,
which serves as an amino-acid protectant.

3.3.1.4 Analysis of amino-acid mixtures

After hydrolysis of a protein according to one of the procedures described in
Section 3.3.1.3, the sample is dried and then analysed for the component amino
acids. A 'classical' amino-acid analyser of the type developed by Stein and Moore
in the 1960s consists essentially of an ion-exchange column, normally a
sulphonated polystyrene resin.[43] Samples to be analysed are dissolved in buffer at
pH 2.2 and eluted from the column by defined volumes of citrate buffers of
increasing pH, usually 3.25, 4.25, and 5.28, with the pH 5.28 buffer being of
greater ionic strength than the first two. The order of elution of the amino acids
(Asp, Thr, Ser, Glu, Pro, Gly, Ala, Val, Met, Ile, Leu, Tyr, Phe, Lys, His, Arg) can
be seen to be broadly in line with expectations, i.e. the acidic amino acids are not
retained by the negatively charged column, the neutral amino acids are retained
to some extent, and the basic amino acids are strongly retained. Other types of
interactions between the amino acids and the column are involved, since the aromatic amino acids Tyr and Phe are eluted after the aliphatic amino acids such as
Ala; this is presumably a consequence of interactions between the aromatic rings
of these compounds with those of the column.[44]

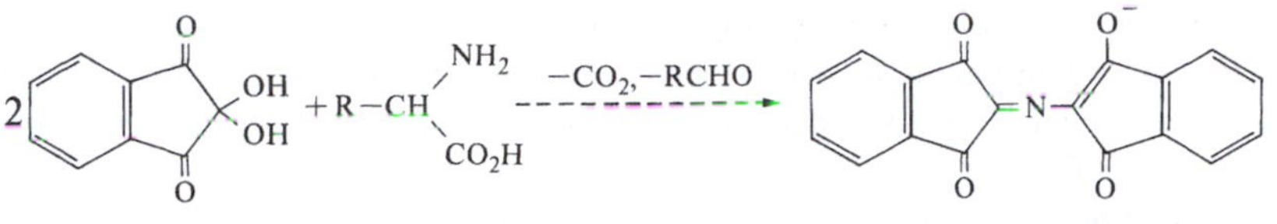

Ruhemann's purple

Fig. 3.6 The reaction of amino acids with ninhydrin. Proline does not give this reaction (see Fig. 3.7).

As the amino acids are eluted from the column they are detected by reaction with ninhydrin (Fig. 3.6) which gives rise to a purple derivative (Ruhemann's purple),[45] which can be monitored at 570 nm; with the imino acid proline a yellow product (Fig. 3.7) is formed which can be monitored at 440 nm. From the colour yield the amount of the amino acid can be evaluated and hence the composition of the protein determined. A more sensitive detection method involves the reaction of the amino group of the amino acids with the reagents fluorescamine[46] or phthalaldehyde (Fig. 3.8)[47,48] which produce highly fluorescent compounds. With these reagents amino-acid analyses can be carried out successfully on as little as 10 ng of hydrolysed protein. Phthalaldehyde has some advantages over fluorescamine in terms of stability and sensitivity.[49] For detection by reaction with phthalaldehyde, proline must first be reacted with alkaline sodium hypochlorite solution.[50]

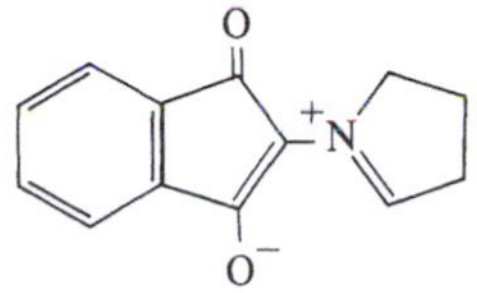

H.p.l.c. has been applied successfully to the analysis of amino-acid mixtures.[50] The mixture is reacted with reagents such as dimethylaminoazobenzenesulphonyl chloride or phthalaldehyde and the derivatives separated by reverse-phase h.p.l.c. using an acetonitrile gradient. The detection of the derivatives is by absorbance or fluorescence, respectively. Analyses can be completed in under 40 min.

Product formed in reaction between proline and ninhydrin

Fig. 3.7 The proline–ninhydrin adduct structure.

3.3.1.5 Problems encountered in amino-acid analysis

As mentioned in Section 3.3.1.3 a number of amino acids are subject to destruction during acid hydrolysis of proteins and certain precautions must be taken to obtain reliable results. A full discussion of these precautions is described in a review[51] that also outlines procedures for the analysis of the carbohydrate components of glycoproteins, but we shall mention some of the more commonly encountered problems below.

The hydroxyamino acids serine, threonine, and tyrosine are partially destroyed during acid hydrolysis (up to about 10 per cent in 24 hours). In these cases, it is usual to perform hydrolyses for different lengths of time, e.g. 24, 48, and 72 hours, and to extrapolate the results of the analyses back to zero time to obtain the true content of these amino acids.

Certain amide bonds in proteins, especially those between the branched side-chains of valine and isoleucine, are rather resistant to hydrolysis, e.g. the Ile–Ile bond is only cleaved to the extent of about 50 per cent in 24 hours. In order to obtain reliable values of the contents of these amino acids, extended periods of hydrolysis (up to 120 hours) may be necessary.

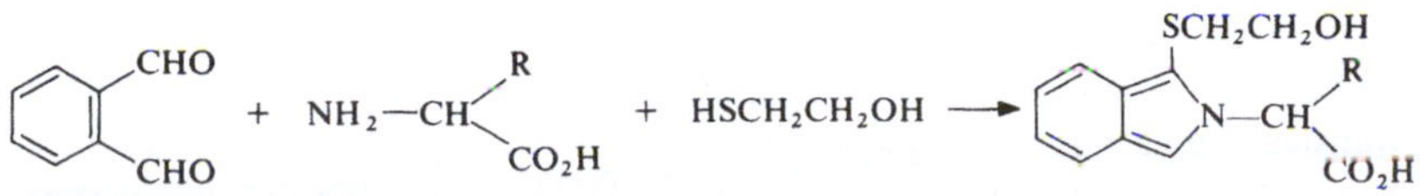

Fig. 3.8 The reaction of amino acids with phthalaldehyde in the presence of 2-mercaptoethanol to give the highly fluorescent 1-alkylthio-2-substituted isoindole product.

The amide amino acids asparagine and glutamine are hydrolysed to yield aspartic acid and glutamic acid, respectively. Thus the content of glutamic acid determined in the amino-acid analyser will correspond to the sum of the contents of Glu plus Gln in the original protein. In a sequence determination, the decision as to whether a particular amino acid in the sequence of the protein is in fact Glu or Gln is normally made on the basis of the chromatographic properties of the phenylthiohydantoin derivatives following Edman degradation (Section 3.3.2.4). If there is a doubt about the identity, the amino acid is designated as Glx (Z), or Asx (B) in the case of a choice between aspartic acid and asparagine.

The sulphur-containing amino acids cysteine, cystine, and methionine are especially susceptible to oxidation even when care has been taken to remove oxygen prior to hydrolysis. These amino acids can be determined reasonably reliably as oxidized derivatives if the protein is oxidized with performic acid prior to hydrolysis.[52]

3.3.1.6 Uses of amino-acid composition data

First, the composition of a protein can be used as a 'sum total' of amino acids to which the composition of derived peptide fragments (Section 3.3.2.3) must be related, since the total composition of the various fragments must equal the composition of the whole protein. This type of comparison could be used to show, for instance, that a particular fragment had not been accounted for during the purification of the fragments.

Second, inspection of the amino-acid composition data can give a general idea of some of the properties of an enzyme. Thus, the protease pepsin has an overwhelming preponderance of acidic amino acids (Glu + Asp = 43) over basic amino acids (Lys + His + Arg = 4) in the enzyme isolated from the pig;[53] this can be related to the fact that the enzyme is active at the acid pH values (1–5) found in the stomach. Hatch evaluated the ratio of polar to non-polar amino acids in a number of proteins and found that this ratio varied between about 1 and about 3. Proteins with a low ratio, i.e. a relatively high content of non-polar amino acids, appeared to have a high content of helical structure and vice versa.[54] An extreme example of an enzyme with a very high content of non-polar amino acids (ratio = 0.6) is the C_{55}-isoprenoid alcohol phosphokinase from *Staphylococcus aureus*.[55] This extremely non-polar enzyme will preferentially dissolve in the butanol layer of a butanol–water system.

Third, the amino-acid composition of a protein can be used to calculate[8] its partial specific volume, $(\bar{v})$, which is used in determinations of molecular weight using the ultracentrifuge (see Section 3.2.1).

Finally, quantitative amino acid analysis data can be used to evaluate the concentration of a solution of a purified protein; this value can then be used as a definitive standard with which to calibrate a routinely used method.[56]

3.3.2 The determination of primary structure

3.3.2.1 Strategies underlying the determination of primary structure

Two distinct strategies are employed to determine the primary structure of a protein or an enzyme. The 'direct' method operates at the protein level and dates back to 1953 when Sanger published the amino-acid sequence of insulin.[57] The 'indirect' method operates at the gene (DNA) level and uses the genetic code to translate the nucleic acid sequence into that of the corresponding product. It can be traced back to the development of rapid DNA-sequencing techniques in the mid 1970s.[58,59] The strategies underpinning the two approaches are shown in Figs. 3.9 and 3.10.

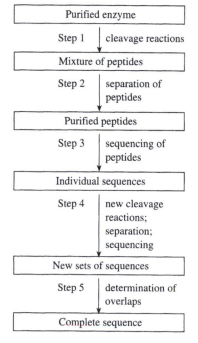

Fig. 3.9 Outline scheme for the 'direct' determination of the amino-acid sequence of a protein.

In this section we shall outline the various steps in the 'direct' sequencing method according to the general scheme shown in Fig. 3.9. Further details can be found in the review by Allen.[60] The determination of a sequence in this way usually requires 10–100 nmol of enzyme, i.e. 0.2–2 mg of an enzyme of M_r 20 000. We shall then describe an example of sequence determination (phosphoglycerate kinase from baker's yeast) in which the two approaches were used in a complementary fashion. The section will then examine the relative contributions of the two methods.

The details of the 'indirect' method are given in a number of books and reviews.[61–63] With reference to Fig. 3.10, step 1 involves determining a small amount of sequence information of the protein (15 to 20 amino acids is sufficient) using the Edman technique (see Section 3.3.2.4). If the N-terminus is blocked, the protein is fragmented using a protease or a specific chemical reagent (Section 3.3.2.2), and the sequence information is obtained from one of the purified peptides. In step 2, the genetic code is used (taking degeneracy into account) to design a suitable oligonucleotide probe. The probe is synthesized using an automated apparatus and is then labelled either with ^{32}P to permit detection by autoradiography or with one of a variety of reagents such as digoxigenin or biotin for which chemiluminescent detection methods have been developed. In step 3 a library (either genomic or based on complementary DNA (cDNA)) is screened using Southern blotting in which the probe hybridizes to (i.e. forms base pairs with) the fragment of interest and can be detected by autoradiography or chemiluminescence. The DNA sequencing method (step 4) most commonly used relies on the interruption of enzymatic replication of DNA by DNA polymerase using 2',3'-dideoxynucleotide triphosphates.[59] The products obtained are separated by electrophoresis on high-resolution gels (6–20% acrylamide in 7 mol dm^{-3} urea), which separate nucleic-acid fragments on the basis of their length. By inspection of the patterns obtained from a number of fragments obtained by causing interruption at the different bases, it is possible to deduce the sequence of the DNA.[59] This process has been automated using fluorescent labelling of the dideoxynucleotides and chromatographic separation of the products. Sequences of several hundred bases can be read off in this way, and it is thus not difficult to appreciate that the sequencing of DNA can be very rapid indeed.[64]

A number of the early applications of DNA sequencing involved addressing problems raised during the direct determination of sequences of enzymes. This is well illustrated in the case of penicillinase from *E. coli*.[65,66] The 'direct' method took about four years, whereas the nucleotide sequence of the DNA was deduced in about six months and was used to resolve some of the ambiguities encountered in the 'direct' method. More recently, the increasing availability of purified genes and the speed of DNA sequencing have combined to produce an explosive growth in the amount of sequence information available, leading in turn to the development of sophisticated databases. The Appendix to this chapter should be consulted for details of some of the databases. To give some idea of the scale of growth it should be noted that the number of amino acids in sequences deposited in the SWISSPROT protein sequence database[1] grew fourfold between 1990 and 1996 and at the October 1996 release numbered over 21 million amino acids in over 59 000 separate entries. As a result of a number of major international genome sequencing projects, the complete sequences of a number of prokaryotes such as *Escherichia coli*,[67] *Helicobacter pylori*,[68] and *Bacillus subtilis*,[69] and of the eukaryote *Saccharomyces cerevisiae*[70] have been published. For each of these organisms, about 60 per cent of the open reading frames (i.e. where start and stop

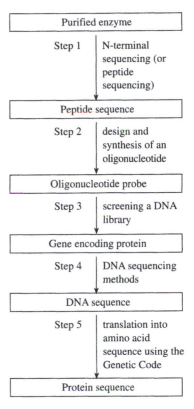

Fig. 3.10 Outline scheme for the 'indirect' determination of the amino-acid sequence of a protein.

codons have been identified) have been identified as corresponding to proteins of identified functions, but in each case a significant minority have no known analogues in the entire sequence database.[71]

3.3.2.2 Cleavage of the polypeptide chain (step 1 in Fig. 3.9)

The polypeptide chains of enzymes have been found to contain between about 100 and about 2500 amino-acid units. Even the smallest chains are too large to be sequenced directly from one end to the other (Section 3.3.2.4). It is therefore necessary to split the chain into a number of fragments that can each be sequenced before the sequence of the whole chain can be deduced by piecing together the sequences of the fragments. This piecing together will be much easier if we can cleave the polypeptide chain at a restricted number of points and so produce a limited number of well-defined fragments rather than cleave it randomly and so produce a collection of fragments of all sorts of lengths. Specific cleavages can be brought about by the use of certain proteases, since many of these possess high degrees of specificity; some of the more commonly used proteases are listed in Table 3.2. It is particularly important to ensure that pure preparations of proteases are used, since otherwise non-specific cleavages will occur; in Chapter 5, Section 5.5.1.4, we describe the use of an affinity label to ensure that trypsin preparations are free from contamination by chymotrypsin. In order to obtain complete cleavage of the chain, e.g. on the C-terminal side of every lysine and arginine in the case of trypsin, the enzyme should be treated with a denaturing agent such as guanidinium chloride or urea (Section 3.6.1) so that the compact three-dimensional structure of the enzyme is destroyed and the susceptible bonds are exposed to the protease.[60]

The specificity of the proteolytic cleavages can sometimes be altered to advantage. For instance, lysine side-chains can be reacted with an anhydride such as

Table 3.2 Some proteases commonly employed in sequencing work for cleaving polypeptide chains

Protease	Nature of amino acid side-chains at cleavage site in polypeptide chain		
	amino ——————— X Y ↓ Z ——————— carboxyl		
	X	Y	Z
Trypsin[72]	—	Basic side-chain (Lys, Arg)	—
Clostripain[73] (from *Clostridium histolyticum*)	—	Arg	—
Chymotrypsin[74]		Aromatic (Trp, Tyr, Phe) or hydrophobic (e.g. Leu, Met) side-chain	
Elastase[75]	—	Small hydrophobic side-chain (e.g. Ala)	—
Papain[76,77]	Phe (preference)	Preference for Lys over Ala	Leu or Trp (preference)
Thermolysin[78] (from *Bacillus thermoproteolyticus*)	—	—	Bulky hydrophobic side-chain (e.g. Val, Leu, Ile, Phe)
Endoprotease Glu-C[79] (from *Staphylococcus aureus* V8)	—	Glu	—
Pepsin[80]	—	Phe (preference)	Phe, Trp, Tyr (preference)

maleic anhydride[81] or citraconic anhydride;[82] this prevents attack by trypsin at these points and limits cleavage to the C-terminal side of arginine side-chains. Another possible variation involves modification of cysteine side-chains by reaction with 2-bromoethylamine to produce a side-chain analogous to that of lysine and which is in fact 'recognized' by trypsin.[83] This modification reaction can generate additional cleavage sites and has sometimes proved useful in generating fragments from large polypeptides where successive lysine or arginine groups are well separated in the chain.

Polypeptide chains can also be cleaved specifically by the use of certain chemical reagents. The most commonly used reagent is cyanogen bromide, CNBr, which under acidic conditions will cleave a polypeptide chain on the C-terminal side of methionine side-chains[84] (Fig. 3.11). This method of cleavage is important since methionine is a relatively uncommon amino acid in proteins; hence CNBr will generally cleave a protein into a smaller number of fragments than will be produced with attack by trypsin. For instance, in the case of adenylate kinase from pig muscle, cleavage with CNBr produces six fragments, whereas attack by trypsin would yield 32 fragments.[85] It is obviously easier to separate and purify the components of a mixture, the smaller the number of such components.

Chemical methods for bringing about selective cleavage at other types of amino-acid side-chains including cysteine (e.g. by incubation at pH 9 after cyanylation[86]) and tryptophan (e.g. by incubation with *o*-iodosobenzoic acid under acid conditions,[87] see Section 3.3.2.9) have been described. For reviews of work in this area, the articles by Spande *et al.*[88] and Allen[60] should be consulted.

3.3.2.3 Separation of the fragments (step 2 in Fig. 3.9)

The fragments produced by the various types of cleavage reactions described in Section 3.3.2.2 would be expected to differ in terms of their size, charge properties, chemical characteristics, etc. We could therefore utilize some of the methods, e.g. gel filtration, ion-exchange chromatography, or electrophoresis described in Chapter 2 (Section 2.6), to separate and purify the various fragments present in the mixture. However, since the early 1980s the method of choice has been based on reverse-phase h.p.l.c., in which the stationary phase consists of silica microspheres coated with long-chain (C_8 or C_{18}) alkyl groups and the mobile phase is a gradually increasing proportion of acetonitrile or propan-2-ol in aqueous solution. Detection of peptides in the eluent is usually by measurements of absorbance

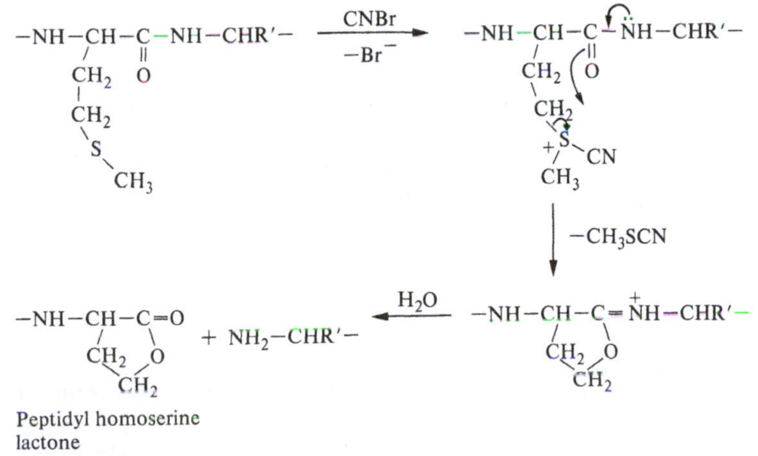

Fig. 3.11 Cleavage of a polypeptide chain on the C-terminal side of methionine by reaction with cyanogen bromide.

at 214 nm (where the peptide bond absorbs strongly). The h.p.l.c. method is capable of rapid, high resolution of peptide mixtures[89] (see Section 3.3.2.9).

3.3.2.4 Sequencing of the purified peptides (step 3 in Fig. 3.9)

The N-terminal amino acid of a peptide can be determined by reaction with dansyl chloride and subsequent identification of the dansyl derivative by electrophoresis or chromatography[90] (Fig. 3.12).

$$NH_2-CHR^1-CO-NH-CHR^2-CO-NH-CHR^3-CO_2H$$

$$\downarrow DNS-Cl$$

$$DNS-NH-CHR^1-CO-NH-CHR^2-CO-NH-CHR^3-CO_2H$$

$$\downarrow \text{Acid hydrolysis}$$

$$DNS-NH-CHR^1-CO_2H \quad + \quad NH_2-CHR^2-CO_2H \quad + \quad NH_2-CHR^3-CO_2H$$
(fluorescent)

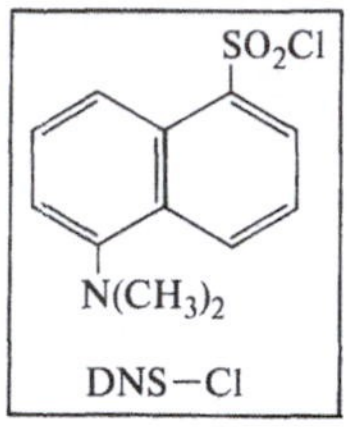

Fig. 3.12 The labelling of a tripeptide by reaction with dansyl chloride (DNS–Cl). The DNS–amino acid bond is stable to acid.

The C-terminal amino acid can be determined by treatment with hydrazine which converts all the amino acids in the peptide to their hydrazides except the C-terminal acid (Fig. 3.13); this can be separated and identified in the amino-acid analyser.[91] An alternative method for the determination of the C-terminal amino acid involves the use of carboxypeptidase, which removes amino acids one at a time from the C-terminal end of a polypeptide chain (see Chapter 5, Section 5.5.1.7). The amino acid released can be identified using an amino-acid analyser. The application of this method has been discussed in reviews.[92,93]

In order to obtain more detailed sequence information, it is necessary to be able to degrade the peptide chain in a sequential manner under carefully controlled conditions. Such a method was developed by Edman[94] and is based on the reaction of the N-terminal amino acid with phenyl isothiocyanate (Fig. 3.14).*

The initial adduct, a phenylthiocarbamyl derivative, A, rearranges under anhydrous acidic conditions, cleaving the adjacent peptide bond and giving a heterocyclic anilinothiazolinone derivative, B. Derivative B is separated from the residual peptide by extraction with an organic solvent and converted to the more stable phenylthiohydantoin (PTH) derivative, C, before identification. The Edman reaction permits the removal of amino acids one at a time from the N-terminus of a peptide chain. Although the conditions can be arranged so that the

* Methods for sequential degradation of a polypeptide chain from the C-terminus have not generally been successful. Formation of a thiohydantoin at the C-terminus followed by cleavage with thiocyanate under acid conditions can be used for the sequential removal of up to five amino acids from the C-terminus.[95]

$$NH_2-CHR^1-CO-NH-CHR^2-CO-NH-CHR^3-CO_2H$$

$$\downarrow N_2H_4$$

$$NH_2-CHR^1-CO-NH-NH_2 \; + \; NH_2-CHR^2-CO-NH-NH_2 \; + \; NH_2-CHR^3-CO_2H$$
(separated and identified)

Fig. 3.13 Hydrazinolysis of a tripeptide. The C-terminal amino acid is separated and identified.

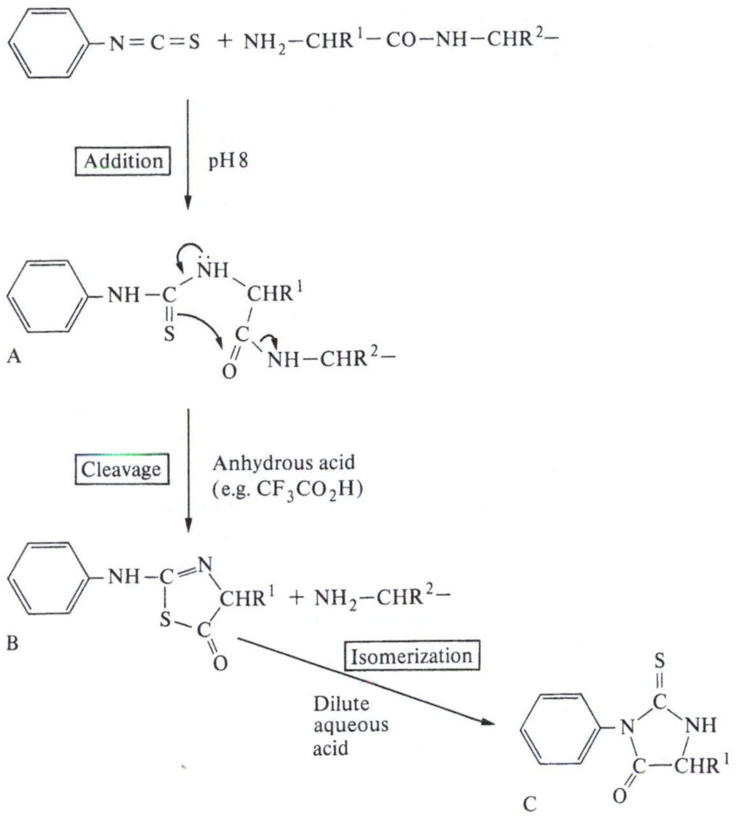

Fig. 3.14 Sequential degradation of a peptide chain by reaction with phenyl isothiocyanate.

reactions involved take place with close to 100 per cent efficiency,[94] there is a gradual accumulation of chains of differing length resulting from incomplete removal of the N-terminal amino acid at each cycle. In practice, the useful upper limit is around 30–40 cycles of degradation. Identification of the amino acid released in each cycle of Edman degradation can be achieved directly by identification of the PTH derivative (C in Fig. 3.14), using chromatographic procedures.[96] The sequential degradation and identification process can be automated in a 'sequenator' by immobilizing the peptide as a thin film in a spinning cylindrical glass cup and passing the necessary reagents and solvents over the peptide in a programmed cycle. The PTH–amino acid formed after each cycle of the Edman degradation can be identified by thin-layer chromatography[94] or h.p.l.c.[97]

An alternative approach has been the 'gas phase' sequenator in which the reagents for the Edman degradation are delivered in a stream of argon or nitrogen and the protein or peptide is immobilized by adsorption onto a glass-fibre disc. The repetitive yields have been claimed to be better than with the 'liquid phase' instrument.[98]

Although there are a number of problems associated with the use of the sequenator,[97] including the corrosive nature of some of the reagents, less than 100 per cent yields at each cycle, internal peptide bond cleavage, and hydrolysis of the side-chains of glutamine and asparagine, the instrument has made an outstanding contribution to the direct determination of protein sequences. At present, it would be unrealistic to suppose that the sequence of a polypeptide chain of, say,

150 amino acids could be determined 'in one go'; the sequenator is best employed to determine the sequences of fragments (up to about 50 amino acids in length) produced by the cleavage reactions described in Section 3.3.2.2 (see the example of phosphoglycerate kinase in Section 3.3.2.9). It should be noted that the technique of fast-atom bombardment mass spectrometry (see Section 3.2.4) has been successfully applied to the determination of sequences of peptides up to about 50 amino acids in length.[22,23] The mass spectrometric technique has a major advantage over the conventional Edman method in that it is not necessary to work with purified peptides.

3.3.2.5 Preparation and analysis of new fragments (step 4 in Fig. 3.9)

Once the peptide fragments have been purified and sequenced as described in Sections 3.3.2.2 to 3.3.2.4, the next task is to position them correctly so as to build up the overall sequence of the protein. This is usually done with the aid of a new set of fragments produced from the original protein by use of a different cleavage procedure (see Section 3.3.2.6). Thus, if trypsin were used to produce the first set of fragments, we might use chymotrypsin to produce the new set. These new fragments are purified and sequenced using the procedures already described (Sections 3.3.2.3 and 3.3.2.4).

3.3.2.6 Alignment of peptide sequences and determination of the overall sequence (step 5 in Fig. 3.9)

When the peptide sequences of the fragments produced by the various cleavage procedures are known, it is possible to position the peptides in the overall sequence by determining the regions of overlap of sequence. For instance, in the case of adenylate kinase from pig muscle,[85] two of the cyanogen bromide fragments CBc and CBd could be positioned with respect to each other by the

Leu-Ser-Glu-Ile-Met   Glu-Lys-Gly-Glu . . . Met
CBc   CBd

isolation of a peptide produced by the action of trypsin on the enzyme which had the sequence:

Met-Leu-Ser-Glu-Ile-Met-Glu-Lys.

Thus the overall sequence of this portion of the polypeptide chain is

Met-Leu-Ser-Glu-Ile-Met-Glu-Lys-Gly-Glu . . . Met

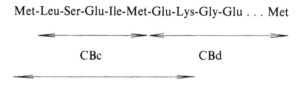

CBc   CBd

Tryptic peptide

This conclusion was confirmed by the analysis of fragments produced by the action of thermolysin on the enzyme. Assuming that amino-acid sequences are entirely random (which is almost certainly not the case!), the probability of occurrence of a given sequence of three amino acids, e.g. Asp-Gly-Leu, will be 1/8000 ($1/20^3$). It is clear that a given sequence of three amino acids is unlikely to be repeated in a polypeptide chain of length 100 to 2500 amino acids and hence an overlap of this size should serve to position peptide fragments fairly unambiguously in the overall sequence.

The positioning of fragments in the overall sequence of a protein is greatly helped if the sequence of a related protein is available. For instance, determination of the sequence of elastase was considerably speeded up because this enzyme shows a high degree of similarity to chymotrypsin, the sequence of which was already available.[99] The positioning of peptides can also be aided by information from DNA sequencing as in the example of phosphoglycerate kinase (Section 3.3.2.9).

3.3.2.7 Problems encountered in the determination of amino-acid sequences

Using the procedures outlined in the previous sections (3.3.2.2 to 3.3.2.6), the sequence of a protein can usually be deduced, provided that sufficient care is taken in the purification and characterization of the various fragments. It is important to check that the sequence is in accordance with any DNA-derived sequence data available (Section 3.3.2.8). Any discrepancy between the various sets of data could indicate that a fragment has not been accounted for during the separation procedures. Among the problems which arise in the direct determination of sequence, three of the most common are mentioned below.

Blocked N-terminal amino acid Some proteins do not appear to possess an N-terminal amino acid when subjected to the dansyl chloride or the Edman reactions (Section 3.3.2.4). This can be because the amino group has been modified post-translationally, e.g. by acetylation, formylation, or formation of a pyroglutamyl moiety.[27,100,101] Examples of an acetylated N-terminal amino acid and an N-terminal pyroglutamyl group are provided by carbonate dehydratase from human erythrocytes[100] and peroxidase from horseradish[102] respectively. Myristic acid ($C_{13}H_{27}CO_2H$) is the N-terminal blocking group in the catalytic subunit of cAMP-dependent protein kinase[103] (see Chapter 6, Section 6.4.2.1).

Allen[60] has given a general discussion of the problems of structure determination in those cases where post-translational modification has occurred. Mass spectrometry (Section 3.2.4) is now proving to be extremely useful for determining the nature of the post-translational modification of proteins.[24,27,101]

The pyroglutamyl group in which an N-terminal glutamic acid has formed an internal amide bond.

Amide assignments In Section 3.3.1.5 it was mentioned that the 'amide' amino acids asparagine and glutamine are converted to aspartic acid and glutamic acid, respectively, during acid hydrolysis of a protein or peptide fragment. A correct assignment can usually be made on the basis of the electrophoretic mobility at pH 6.5 of a fragment containing the amino acid in question[104] since the side-chain of aspartic acid (glutamic acid) will carry a negative charge at this pH, whereas the side-chain of asparagine (glutamine) will carry no charge. In order to prepare the fragment of the enzyme it is necessary to perform the hydrolysis under conditions in which the amide bond in the side-chain of asparagine or glutamine will not be hydrolysed, e.g. by the use of various proteases (see Table 3.2). Correct assignments can usually be made on the basis of Edman degradation, since the phenylthiohydantoin derivatives (Section 3.3.2.4) can readily be distinguished. Reference to the nucleotide sequence (if available) would confirm these assignments.

Location of disulphide bonds If a protein is found to contain disulphide bonds, formed between pairs of cysteine side-chains, it is generally necessary to break these bonds either by reduction (Section 3.3.1.2) followed by reaction with iodoacetate or by oxidation with performic acid (Section 3.3.2.5) in order to form derivatives that are stable during the procedures involved in sequence determination. Analysis by mass spectrometry of proteolytic fragments prior to and after reduction of the disulphide bonds can also be used to assign the positions of the disulphide bonds in a protein.[27,101]

3.3.2.8 The complementary roles of direct and indirect sequencing of proteins

The very impressive developments in DNA sequencing might appear to have rendered the direct sequencing of proteins almost entirely redundant. However, it is important to note that direct sequencing still has an important role to play in studies of protein structure. In some cases, for instance, the enzyme of interest may well be available in a pure form in sufficient quantities to complete the direct sequencing relatively quickly. This would be the case if the enzyme were not too large (say about 200 amino acids in length) and could be split into readily separable fragments. Of course, if DNA sequencing, overexpression of the gene, or the production of site-directed mutants (see Chapter 5, Section 5.4.5) were to be undertaken it would be necessary to invest the time and effort involved in identifying and cloning the gene or cDNA encoding the enzyme.

Some limited direct sequence information (typically up to about 20 amino acids) on the purified protein is necessary to design an oligonucleotide probe (Section 3.3.2.1) in order to identify and clone the relevant gene. With additional sequence information from distinct parts of the polypeptide chain, sets of primers for the polymerase chain reaction can be designed; this would permit larger more specific probes for the gene to be synthesized.

That the DNA-derived sequence does in fact correspond to the protein under study should be checked by comparing it with the direct sequence information available, and by checking that the predicted mass agrees with that determined by mass spectrometry. On the other hand, the DNA-derived sequence can be used to resolve minor discrepancies or uncertainties in the directly determined sequence, e.g. in the assignment of Glu or Gln (Section 3.3.2.1).

Finally, we should note that the indirect sequencing of proteins does not give information about the occurrence of post-translational modifications, although it may indicate that consensus motifs or sequences for such modifications are present (Section 3.3.2.10). In order to characterize such modifications it is necessary to carry out direct analysis of the mature protein product. Modifications such as glycosylation, phosphorylation, formation or N-terminal blocking by acetylation, myristoylation, etc. (see Section 3.3.2.7) can be identified by direct analysis[60] coupled with mass spectrometry.[24,27,101]

3.3.2.9 An example of sequence determination

The various methods described in Sections 3.3.2.2 to 3.3.2.7 have been used to determine the sequences of a large number of proteins. The example we have chosen to discuss is that of phosphoglycerate kinase from baker's yeast, since this work nicely illustrates the way that the 'direct' and 'indirect' sequencing techniques can complement each other.[105]

The sequence of yeast phosphoglycerate kinase (a monomer of M_r 44 500) was required to permit detailed interpretation of the electron-density map determined previously by X-ray crystallography.[106] Phosphoglycerate kinase was isolated by a procedure involving cell breakage using ammonia solution, ammonium sulphate fractionation, and affinity elution from CM-cellulose with 3-phosphoglycerate.[107,108] Traces of a contaminating protein were removed by gel filtration on Sephadex G-75.

The fragmentation steps involved in the 'direct' sequencing are illustrated in Fig. 3.15.

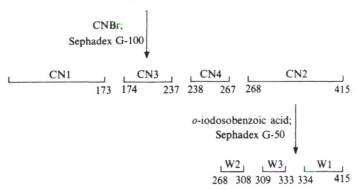

Fig. 3.15 The digestion of phosphoglycerate kinase to prepare fragments for sequencing. The numbering of fragments is described in the text.

Amino-acid analysis revealed that the molecule possesses three methionine side-chains. Cleavage with CNBr (Section 3.3.2.2) yielded four fragments that could readily be separated by gel filtration on Sephadex G-100. The numbering of the fragments (CN1–4) is on the basis of their order of elution on gel filtration, CN1 being eluted first. The fractions pooled as 'CN1' were found to be heterogeneous on polyacrylamide gel electrophoresis in the presence of sodium dodecylsulphate; further purification was achieved by covalent chromatography on activated thiol-Sepharose (see Chapter 2, Section 2.6.4.5), taking advantage of the fact that CN1 contained the single Cys in the molecule.

Fragments CN1, CN3, and CN4 were found by amino-acid analysis to contain homoserine, whereas CN2 did not. This placed CN2 as the C-terminal fragment. Fragments CN2, CN3, and CN4 showed free N-terminal amino groups by reaction with dansyl chloride (Section 3.3.2.4). CN1 did not show any free amino group and was therefore positioned at the N-terminal end of the molecule (see Fig. 3.15). (It had been previously established[109] by analysis of small peptides produced by proteolytic digestion that the N-terminus of yeast phosphoglycerate kinase was *N*-acetylserine.) The order of CN3 and CN4 was determined by nucleotide sequencing as described below.

The large CN2 fragment was further digested by treatment with *o*-iodosobenzoic acid; under the appropriate conditions, cleavage occurs specifically at the two Trp side-chains.[87] (Since Trp is abbreviated W in the one-letter code (Table 3.1) the fragments are denoted W1–3.) W1 was eluted first on gel filtration on Sephadex G-50.

The sequences of the various fragments in Fig. 3.15, i.e. CN2, CN3, CN4, W1, W2, and W3, were established by automated Edman degradation (Section 3.3.2.4) and by manual Edman sequencing of peptides derived by digestion of these fragments with various proteolytic enzymes including trypsin, chymotrypsin, thermolysin, pepsin, and V8 protease (endoprotease Glu-C) from *Staphylococcus aureus*. The resulting mixtures of peptides were separated by various techniques, such as high-voltage electrophoresis, ion-exchange chromatography and high-performance liquid chromatography (h.p.l.c.) (Section 3.3.2.3). A typical h.p.l.c. profile showing the separation of tryptic peptides from fragment CN3 is shown in Fig. 3.16.

The order of the fragments W1, W2, and W3 in CN2 was determined by automated Edman sequencing from the N terminus of CN2 and by the isolation of overlapping peptides, e.g. peptide S29 from the V8 protease digest of CN2 (Fig. 3.17).

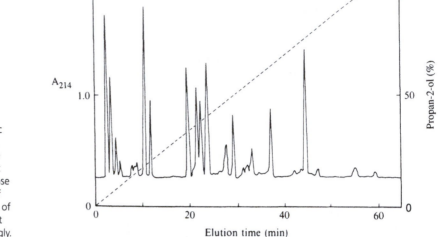

Fig. 3.16 H.p.l.c. separation of tryptic peptides from fragment CN3. Approximately 0.2 mg of the peptide mixture in 0.1 per cent trifluoroacetic acid was applied to a C$_{18}$ reverse-phase column and eluted with a gradient of propan-2-ol as indicated. The elution of peptides was monitored at 214 nm at which the amide bond absorbs strongly.

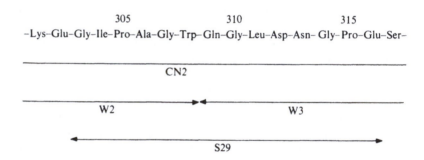

```
                 305                    310                    315
 –Lys–Glu–Gly–Ile–Pro–Ala–Gly–Trp–Gln–Gly–Leu–Asp–Asn– Gly– Pro–Glu–Ser–
```

Fig. 3.17 An overlap peptide (S29) to establish the order of fragments W2 and W3 in the CN2 fragment.

Amide assignments (Gln vs. Glu; Asn vs. Asp) were made from direct identification of the phenylthiohydantoin derivatives during Edman degradation (Section 3.3.2.4).

At this point, the sequences of the fragments CN3, CN2, and CN4 and the fact that CN4 was C-terminal had been established. The remaining parts of the protein sequence were established by complementary work being undertaken on the structure of the gene coding for phosphoglycerate kinase in yeast. It had been previously established[110] that this gene (the *PGK* gene) was located on a 2900-base (2.9-kb) fragment resulting from digestion of the vector pMA3 with the restriction endonuclease Hind III.* By treatment of the pMA3 vector† with the restriction endonuclease Sal I followed by an exonuclease (BAL 31), a series of overlapping gene fragments was prepared. After appropriate selection, these could be sequenced (see Fig. 3.18), providing a complete sequence of the first 624 bases of the *PGK* gene (starting from ATG as the initiation codon).

By use of the genetic code, the DNA sequence could then be expressed as a protein sequence (amino acids 1 to 207).

In addition, nucleotide sequence data were obtained for a part of the *PGK* gene around the EcoRI restriction endonuclease site (Fig. 3.18); this corresponds to amino acids 281 to 415 (C-terminus). The nucleotide sequence data not only confirmed the amino-acid sequence determined directly for fragment CN2

* A complete understanding of the concepts involved in the DNA cloning and sequencing methods is not important for the purposes of this discussion. An excellent account of this area is given in references 62 and 63.

† A vector is a piece of DNA that carries genetic information and can replicate autonomously in an appropriate host. pMA3 is an artificially constructed 'shuttle vector' that can carry genetic information between different organisms, in this case *E. coli* and yeast.

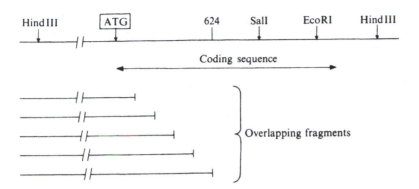

Fig. 3.18 General structure (not to scale) of the *PGK* gene in the pMA3 vector. The initiation codon (ATG) is shown together with the sites of cleavage by various restriction endonucleases (see references 62 and 63 for details of these). The overlapping fragments were sequenced in the left ← right direction, so that an unambiguous DNA sequence of the first 624 bases could be established.

(Fig. 3.15), but also helped to resolve some small ambiguities in the latter sequence where good overlap peptides could not readily be obtained.

The final sequence of 415 amino acids[105] was thus obtained by the complementary approaches of protein and DNA sequencing and was consistent with all the available data.

3.3.2.10 Uses of sequence information

The information contained in the primary structure of an enzyme can be used in a large number of ways. At a very basic level it will provide the amino-acid composition and thus give a guide to the overall properties of the enzyme (Section 3.3.1.5). Four of the more important applications of sequence information are discussed below. The use of sequences to establish relationships between enzymes is discussed separately in Section 3.3.2.11.

To calculate the M_r ***of an enzyme***   There are a number of experimental methods (Section 3.2) for determining the M_r of an enzyme. The definitive M_r can be obtained when the covalent structure, i.e. sequence of the enzyme, is established, and this calculated M_r has often been used as a check on the validity of some of the earlier experimental methods, such as gel filtration and ultracentrifugation. In the case of β-D-galactosidase from *E. coli*, the M_r of the subunit had been measured as 135 000, but when the sequence became available[111] it was clear that this value was some 15 per cent too high and that the correct value was 116 300. The discrepancy seems to have arisen from the use of an incorrect value of the partial specific volume of the enzyme (Section 3.2.1.1) and from errors in the interpretation of ultracentrifuge data. A second example is afforded by the work on penicillinase from *E. coli* in which the calculated M_r was some 20 per cent higher than that measured by physical chemical methods.[65] To a large extent the situation has now been reversed since the development of the extremely accurate mass spectrometric techniques for the determination of M_r; these now provide a very exacting check on the deduced sequence.[24,25] The importance of an accurate knowledge of the M_r of an enzyme has been discussed in Section 3.2.5.

To locate a particular amino acid in an enzyme   The primary structure of an enzyme provides a framework within which the importance of particular amino-acid side-chains can be discussed. In Chapter 5 (Section 5.4.4) we shall describe how chemical modification of an amino-acid side-chain can be used to indicate the involvement of that side-chain in the catalytic mechanism of an enzyme. For instance, the reagent diisopropyl fluorophosphate modifies one serine side-chain in chymotrypsin leading to inactivation; analysis of the peptide fragment containing this modified amino acid shows that this serine is number 195 in the sequence (see Chapter 5, Section 5.5.1.2). It is essential to have this type of information in conjunction with three-dimensional structural data (Section 3.4) to formulate detailed proposals concerning the mechanism of action of an enzyme.

To interpret data from X-ray crystallography X-ray crystallographic studies of an enzyme (Section 3.4) yield a three-dimensional map of the electron density in the molecule. In order to interpret this pattern of electron density in terms of the arrangement of atoms it is normally necessary to know the sequence of the enzyme so that the path of the chain of α-carbon atoms and amide bonds can be traced and the positions of the side-chains located.

The X-ray structure of phosphoglycerate mutase from yeast was determined in 1974[112] to a resolution of 0.35 nm. At this level of resolution some of the main structural features such as α-helices and β-sheets (Section 3.4.4) were clearly visible. However, it was not until sequence information on the enzyme became available that the higher-resolution electron-density maps could be interpreted in sufficient detail to allow a plausible mechanism of action to be proposed.[113]

To predict the three-dimensional structure of an enzyme It has been demonstrated that the primary structure contains the information necessary to specify the three-dimensional structure of an enzyme, although a number of other protein factors (chaperones) may be involved in assisting the folding process (Section 3.6.2). It would therefore be appropriate to ask whether it is possible to use sequence data to predict the three-dimensional structure. The answer to this question is a reasonably encouraging 'yes', especially when the task is limited to predicting which parts of the amino-acid sequence might occur in regions of distinct secondary structure (helices, sheets, etc.). This type of approach is described in Section 3.6.3.

To identify functional regions of enzymes In some cases, inspection of the sequence allows us to identify parts of an enzyme that may have a distinct functional role. In the case of phospholipase A_2, for instance, it is likely that the non-polar N-terminal portion of the polypeptide chain Ala-Leu-Trp-Gln-Phe is involved in the interaction of the enzyme with the non-polar portions of phospholipid substrates. This conclusion is supported by the results of chemical modification of amino-acid side-chains in the enzyme.[114] It is well recognized that certain types of N-terminal sequences are involved in directing newly synthesized proteins to particular destinations, e.g. into the lumen of the endoplasmic reticulum prior to export[115] or to the various compartments of the mitochrondrion.[116] In a membrane-associated protein, an analysis of the polarity of the side-chains of 20-amino-acid stretches in the sequence can be used to identify possible membrane-spanning helices. A 'hydropathy plot' for the protein is obtained by plotting the free energy for transfer from the membrane interior to water of such a helix against the number of the first amino acid in the helix; a peak above $+ 80$ kJ mol^{-1} is considered to indicate where a transmembrane helix (see Chapter 8, Section 8.4) is likely to occur.[117] From the large number of sequences now available in the databases it is possible to identify 'sequence motifs' characteristic of particular functions, e.g. the sequence -Gly-Xaa-Xaa-Xaa-Gly-Lys- is found in many nucleotide-binding proteins such as kinases. A detailed list of consensus sequences has been compiled.[101] Using programs such as PROSITE[118] (see Appendix 3.1) the databases can be searched for specified consensus sequences. Finally, we should note that certain types of amino-acid sequence are associated with high degrees of flexibility, e.g. the Ala/Pro-rich sequences in the lipoyl domains of the E2 component of the pyruvate dehydrogenase complex (see Chapter 7, Section 7.7.2).

3.3.2.11 **To explore relationships between enzymes**

The development and growth of databases of sequence and other structural information on proteins and of computer-based methods of searching them have given enormous impetus to the search for patterns underlying protein sequences. Two volumes of *Methods in Enzymology*[3,4] are devoted to this topic and can be consulted for further details.

Assessing the degree of similarity between enzymes A common way in which clues to the function of an uncharacterized protein (e.g. an open reading frame from a genome sequencing project) have been gained is to compare the observed ('test') sequence against all the protein sequences in the database. The best matches with proteins of known function should indicate possible functions of the protein of interest. This type of exercise is not as appropriate for enzymes which have been purified and sequenced, because the isolation procedure itself will have utilized an assay method based on the function (catalytic activity) of the enzyme. However, a good example of a case where such database searching led to an entirely unexpected finding was provided by the case of peptidyl-prolyl isomerase, an enzyme which catalyses the *cis/trans* isomerization of peptide bonds to proline, and plays a part in protein folding (Section 3.6.2).

In 1989, two groups had independently purified the enzyme from porcine kidney.[119,120] When the directly determined sequence of the enzyme was compared with those in the databases, in was clear that the enzyme was in fact identical to the protein cyclophilin. Cyclophilin had previously been discovered and characterized as a protein which bound to the immunosuppressive agent cyclosporin A, a cyclic undecapeptide of fungal origin. The identity of the two proteins suggested a possible relationship between the processes of protein folding and immunosuppression; this idea gained further support when another type of protein (the FK506-binding protein) known to bind an immunosuppressant was also found to possess peptidyl-prolyl isomerase activity.[121] Although further work has indicated that the link between the two processes is not a simple one, this example serves to illustrate the potential usefulness of database searches in linking structure to function.

The fundamental task of any search program is to calculate the probability that similarities between sequences have occurred by chance. By choosing a threshold level of probability, it is possible to assign the statistical significance of the match. It is important to note that the threshold level of probability will depend on the lengths of sequences being compared. Thus matches of > 50 per cent identity in a 20 to 40 amino-acid sequence occur frequently by chance and do not necessarily indicate meaningful similarities (homologies) between proteins. Homologous sequences are usually similar over an entire sequence or domain, typically showing 20–25 per cent (or greater) identity over a range of 200 or more amino acids. When the degree of identity between two sequences of this type of length falls within the 15–20 per cent range it is not clear whether the similarity is significant; this is termed the 'twilight zone' of comparison. A number of programs have been developed to compare a test sequence against the databases; details about these can be found from the WWW (World Wide Web) site of the European Bioinformatics Institute (see the Appendix to this chapter) or in a number of articles in volumes 183 and 266 of Methods in Enzymology. Three of the most widely used programs which are available through the EBI (European Bioinformatics Institute) are BLAST,[122] FASTA,[123] and BLITZ.[124] One of the most difficult parts of the search procedure is to know how to treat gaps in the sequences which are introduced to give optimum alignments; it is a useful exercise to study the effects of varying the penalties to be paid of introducing such gaps.[125]

In addition to the overall comparisons, the database searches allow strictly conserved amino acids to be identified. Such conserved amino acids may well be functionally important, e.g. in the catalytic mechanism of an enzyme (see Chapter 5, Section 5.5), and would serve as potential targets for site-directed mutagenesis studies (Chapter 5, Section 5.4.5).

Using similarities between enzymes to establish evolutionary relationships between organisms A change in the primary structure of an enzyme will arise from a change in the base sequence of the DNA coding for that enzyme; two of the most important types of genetic modification that can occur are point mutations and gene duplication. A point mutation refers to the replacement of a single nucleotide in a triplet codon in DNA. Comparison of the sequences of enzymes that catalyse the same reaction in different organisms shows that most, if not all, of the changes in sequence can be explained by point mutations.[126,127] Examination of the corresponding DNA sequences confirms that the steady evolution of proteins results from the steady evolution of DNA, when the effects of different codon usage in different organisms and of removal of intervening sequences in eukaryotic DNA must be borne in mind when making these comparisons.*

* Comparisons of three-dimensional structures of enzymes can also be used to examine these evolutionary processes (Section 3.4.3.3). It should be noted that a number of different amino-acid sequences can give rise to very similar three-dimensional arrangements, so in functional terms there may be much less restraint on amino-acid sequences.

Presumably, the pattern of amino-acid substitutions in enzymes reflects a combination of two processes: *mutation* which can occur randomly along the sequence, and *survival* of the mutants which is not a random process. Replacement of one amino acid by another with different characteristics (e.g. the basic side-chain of lysine replaced by the acidic side-chain of glutamic acid) would produce a non-functional enzyme if the lysine side-chain were involved in the catalytic mechanism or in the maintenance of the structure of the enzyme. The operation of natural selection would serve to ensure that organisms with such non-functional enzymes did not survive.

By aligning sequences of the enzymes, it is possible to construct a phylogenetic tree linking the organisms concerned. The tree indicates how different sequences have diverged, where the lengths of the branches reflect evolutionary distances between the organisms and can be valuable in helping to resolve questions of classification. An initial pairwise comparison between the sequences is used to identify the closest pair. The other more distantly related sequences are then included in turn. The various types of program used (e.g. TREE and PHYLIP) allow both identical and functionally similar amino acids to be matched and introduce variable gap penalties in constructing the tree. Using these procedures a tree based on a number of lactate dehydrogenase sequences in organisms ranging from archaebacteria to human could be constructed.[128] Detailed discussions of the construction and interpretation of phylogenetic trees are given in reference 129.

When enzymes catalysing similar reactions within an organism are compared it is often possible to find evidence for the occurrence of gene duplication, in which two copies of a particular gene have been produced, followed by independent evolution of these two copies by processes such as point mutation. This type of duplication process would mean that an organism would be capable of producing enzymes of differing specificity allowing a greater variety of foodstuffs to be utilized and thereby enhancing the survival prospects of the organism. The high degree of sequence identity of various proteases of differing specificity, e.g.

Table 3.3 Amino-acid sequences around the essential serine, histidine, and aspartic acid groups in various proteases[134]

Enzyme	Amino-acid sequences		
	Essential serine	Essential histidine	Essential aspartic acid
Trypsin (bovine)	Gly-Asp-SER-Gly-Gly	Ala-Ala-HIS-Cys-Tyr	Asn-Asn-ASP-Ile-Met
Chymotrypsin (bovine)	Gly-Asp-SER-Gly-Gly	Ala-Ala-HIS-Cys-Gly	Asp-Asn-ASP-Ile-Thr
Elastase (porcine)	Gly-Asp-SER-Gly-Gly	Ala-Ala-HIS-Cys-Val	Gly-Tyr-ASP-Ile-Ala
Thrombin (bovine)	Gly.Asp-SER-Gly-Gly	Ala-Ala-HIS-Cys-Leu	Asp-Arg-ASP-Ile-Ala
Subtilisin BPN′	Gly-Thr-SER-Met-Ala	Asn-Ser-HIS-Gly-Thr	Val-Ile-ASP-Ser-Gly

trypsin, chymotrypsin, elastase, and thrombin, especially around those amino acids known to be involved in the catalytic mechanism (Table 3.3), suggests very strongly that these enzymes evolved from some common precursor protease by a process of gene duplication, followed by subsequent independent evolution. These types of similarities have allowed the identification of 'superfamilies' of enzymes which are presumed to have diverged from common ancestral sources.[130] Other examples of this type of 'divergent' evolution are provided by the group of thiol proteases papain, ficin, bromelain, and actinidin,[131] the normal and acid-resistant lysozymes in ruminants,[132] and several of the enzymes involved in the blood coagulation pathway.[133]

On the other hand, comparison between certain other enzymes suggests that a process of 'convergent' evolution may have occurred in which a similar type of functional unit has been evolved starting from quite different precursor polypeptide chains. For example, there is very little similarity in primary structure (Table 3.3) or in three-dimensional structure between subtilisin BPN′, a protease isolated from *Bacillus amyloliquefaciens*, and bovine chymotrypsin, suggesting that these two enzymes did not arise by a process of divergent evolution from a common precursor. However, both enzymes possess the 'charge relay system' of three precisely positioned side-chains, Ser . . . His . . . Asp (see Chapter 5, Section 5.5.1.2), of amino acids well separated from each other in the sequence. This system is essential for the catalytic mechanism and its occurrence in the two quite different enzymes suggests that a process of 'convergent' evolution has occurred.[127] Other examples of 'convergent' evolution include the two classes of aldolases and the two classes of dehydroquinases (see Chapter 5, Section 5.5.3.1).[135] (Convergent evolution is sometimes referred to as 'parallel' evolution, since it is not clear that the enzymes are 'converging' towards one another with the passage of time!)

3.4 The determination of secondary and tertiary structure

Knowledge of the primary structure of an enzyme does not allow us to explain properties such as catalytic power and specificity. We must also consider how the polypeptide chain is folded up so that different parts of the chain are brought into close proximity with each other to create binding sites for substrates and unusual environments which facilitate catalysis (see Chapter 5, Section 5.3). X-ray crystallography has been by far the most widely applied technique for the determination of the three-dimensional structure of enzymes. However, high-resolution nuclear magnetic resonance (NMR) has been extensively developed over the last

15 years and is now capable of giving a great deal of information about the structure and dynamics of proteins, especially those of M_r up to about 25 000. We shall first outline the technique of X-ray diffraction before considering the relationship between the structure in the crystal and that in solution. A number of reviews of X-ray crystallography can be consulted for further details.[2,136–140]

3.4.1 X-ray crystallography

X-ray crystallography relies on the scattering of electromagnetic radiation of suitable wavelength by electrons belonging to the atoms in a molecule. In the case of a regularly arranged array of atoms, such as is present in a crystal lattice, we can have constructive or destructive interference between the scattered waves; only constructive interference will give rise to a detectable signal. X-rays (which are emitted when an electron falls to a lower orbital of an atom from a higher occupied orbital) provide suitable radiation to bring about these diffraction effects since their wavelength is comparable with the interatomic distances in a molecule.* Most protein structures have been determined using the K_α radiation emitted by Cu; this is of wavelength 0.154 nm. Since the early 1980s, increasing attention has been paid to the possibilities of using synchrotron radiation sources (see Chapter 5, Section 5.4.3). The intensities of these (highly specialized) sources are much higher than those of conventional X-ray sources, thereby shortening the time required for data collection.[2,137]

The positions and intensities of diffracted rays (in which constructive interference has occurred) are measured using a diffractometer that can measure intensities directly. From the pattern of spots (Fig. 3.19) it is possible to deduce certain features such as the overall symmetry of the crystal and the dimensions of the repeating unit, but additional information is required in order to determine the three-dimensional structure of the enzyme, as described below.

If we know the three-dimensional structure of a molecule, it is a relatively straightforward task to calculate the positions and intensities of the diffracted rays. Proceeding in the reverse direction from the diffraction pattern to the three-dimensional structure is difficult because we need to have information on the phases of the scattered X-rays, i.e. how the waves are arriving at the detector related in terms of numbers of wavelengths from a common origin. This information on phases is lost when only the positions and intensities of the diffracted rays are recorded. However, a number of methods have been developed to solve the phase problem. The most widely used of these methods is that of *multiple isomorphous replacement*, in which derivatives of the enzyme are made in which a limited number of heavy atoms are introduced at selected sites in the molecule without distorting the crystal structure.[136,141,142] The electron-dense heavy atom scatters the X-rays more strongly than the lighter atoms (H, C, N, O, and S) of the enzyme and will add its scattering power to the diffracted rays. By measuring the changes in intensities of the diffracted rays it is possible to extract the required information on the phases; usually, at least two heavy-atom derivatives are required for this purpose. Heavy atoms could be introduced by chemical modification of amino-acid side-chains, e.g. the reaction of cysteine side-chains with mercurial compounds, or by soaking the crystal in a solution of a heavy metal salt such as platinum chloride (usually in the form of K_2PtCl_4) or uranyl acetate ($UO_2(CH_3CO_2)_2$).[143] As more structures are being solved, there is an increasing chance that a structure similar to the one being studied has already been solved. In such cases the calculation of the phases of the unknown structure can be based on those of the known structure; clearly, the success of this method

* This phenomenon is analogous to the diffraction of visible light by a diffraction grating. The spacing between the lines on the grating is of the same order as the wavelength of the incident light.

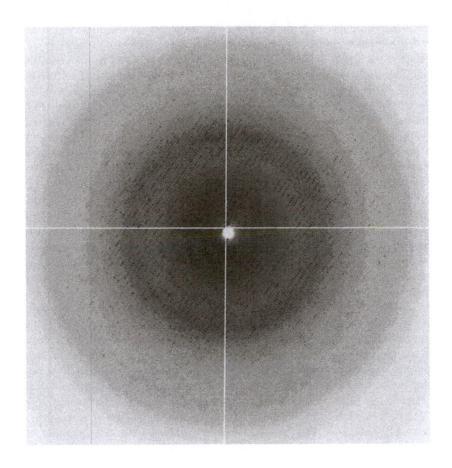

Fig. 3.19 The X-ray diffraction pattern for the type II dehydroquinase from *Streptomyces coelicolor.* This enzyme contains 12 subunits and is discussed in more detail in Chapter 5, Section 5.5.3. (The data were obtained by Dr Adrian Lapthorn at the synchrotron radiation source at Daresbury, UK, using a coupled charge device (CCD) detector.)

(known as *molecular replacement*) will depend to a high degree on the extent of similarity of the two structures.[144]

Once the phases of the diffracted rays are known, it is possible, by a process known as Fourier summation, to calculate the three-dimensional map of the electron density within the molecule.[145] By considering only a relatively limited number of spots in the diffraction pattern, i.e. those rays diffracted through the lowest angles, a map at low resolution (e.g. 0.6 nm) can be obtained. From this low-resolution map a good idea of the overall shape of the molecule can be gained. In order to obtain a map at the degree of resolution at which amino-acid side-chains can be resolved (0.3 nm or better), it is necessary to take into account the intensities of many more spots, usually several thousands, arising from rays diffracted through larger angles. The dramatic developments in computing hardware and software over the last 15 years or so have made the analysis of diffraction data and calculation of electron-density maps a relatively routine task, especially for enzymes, unless the fold is of an unusual type.

In order to interpret the electron-density map of an enzyme in terms of its atomic structure, it is necessary to know the amino-acid sequence of the molecule. We can then trace the path of the chain of α-carbon atoms and amide bonds and locate the various side-chains in the electron-density map; in fact a good quality map can be used to resolve ambiguities in the sequence (Section 3.3.2.9). The structural parameters are then refined in a cyclical process involving estimates of phases and model building.[2] The refinement process has been greatly aided by the application of the molecular dynamics technique in which the motions of atoms in the molecule are described using empirical potential energy functions as, for instance, in the program X-PLOR.[146] A full discussion of the refinement process can be found in reviews.[147,148] The correctness of the structure can be checked by calculating the so-called R-factor, which represents the difference between the diffraction intensities calculated on the basis of the proposed structure and the actual observed intensities.[2] A value of R of less than 20 per cent indicates that the proposed structure is of good quality; a value of 60 per cent would correspond to no significant agreement. When the polypeptide chain has been fitted to the electron-density map with a satisfactory R-factor, we can reasonably claim to have determined the three-dimensional structure of the enzyme.

The basic requirements for X-ray crystallography are listed in Table 3.4. Most of the steps in the process of structure determination are now much quicker than in the early 1980s. However, there can still be some frustrating obstacles to progress, particularly in the initial crystallization process, or if the enzyme crystal has unusual symmetry characteristics. The first high-resolution structure of a protein, myoglobin, was published in 1960[149] and in the subsequent 20 years about 100 structures were solved. Recent progress has, however, been spectacular. By the end of 1998, well over 7000 atomic coordinate entries had been deposited in the Protein Data Bank (see the Appendix to this chapter); this number has grown over 10-fold since 1990. In part, this growth reflects the pace of technical developments over this period, but in addition there has been a growing commitment amongst biochemists and molecular biologists to structural work, since this has been seen as essential to understanding the biological functions of proteins. Developments in recombinant DNA technology (see Chapter 2, Section 2.4.1) have played a catalytic role in the growth of this structural emphasis, since they have provided not only the required quantities of protein, but also a range of designed mutants with which to probe function. From the wealth of structural information available from X-ray crystallography, a range of useful information is available (see Section 3.4.3).

The structure of an enzyme deduced by X-ray crystallography is not entirely a static one, since atomic movements such as vibration and rotation about single

Table 3.4 The basic requirements for X-ray crystallography of enzymes

Requirement	Comments
Crystals of enzyme	Crystals should be of a suitable size, i.e. of dimensions 0.1 mm or larger. They must be reasonably stable in the X-ray beam during the course of the experiment
Isomorphous heavy-atom derivatives	The derivatives are necessary to provide information on phases. Incorporation of the heavy atoms, either by covalent modification or by soaking the crystals, must be shown not to distort the structure. If the structure of a closely related enzyme is available, the technique of molecular replacement can be used to calculate phases
Computing facilities	Facilities are required: (i) to calculate the electron density map from information on the positions, intensities, and phases of the diffracted X-rays, (ii) to carry out refinements of the structure, and (iii) to display the structure obtained
Primary structure	The sequence of amino acids is necessary to interpret the electron density map in terms of the covalent structure of the molecule

bonds can still occur even though the packing of molecules in the crystal will tend to dampen these motions.[150] A good example of atomic movements in a crystal is provided by trypsinogen, the inactive precursor of the protease trypsin (see Chapter 6, Section 6.2.1.1), of which the electron-density map has no density corresponding to four major segments of the chain: N-terminus to Gly 19, Gly 142–Pro 152, Gly 184–Gly 193, Gly 216–Asn 223.[151] It has been concluded that these segments are waggling in a flexible manner or adopting a number of conformations in the crystal leading to a 'smearing out' of electron density. In the electron-density map of trypsin these segments are well defined, presumably because the motion is now restricted. Other examples of this 'smearing out' of electron density are provided by the C-terminal regions of fructose-bisphosphate aldolase[152] and phosphoglycerate mutase.[153] As described in Section 3.4.2, the flexibility of an enzyme molecule is considerably greater in solution than in the crystal.

3.4.2 The structure of an enzyme in solution

Several lines of evidence, such as measurements of hydrogen–deuterium exchange, and NMR show that proteins in solution possess fairly flexible structures.[154,155] It is thus appropriate to think of an enzyme in solution as existing in a number of conformational states (related to each other by rotation about single bonds) of roughly equivalent energy. The structure in solution will represent a time average of these various conformational states of the enzyme.

Recent developments in NMR have now allowed the structures of enzymes of M_r up to about 25 000 in solution to be determined. As an indication of the progress made, it should be noted that over 1300 atomic coordinate structures solved by NMR had been deposited in the Protein Data Bank by the end of 1998. In order to obtain NMR signals relatively high concentrations of protein are required (about 1 mmol dm^{-3}; i.e. 20 mg cm^{-3} for a protein of M_r 20000), and it is important that the protein solutions are stable for the prolonged periods (often a few days) required for collection of data. The determination of the structure depends on measuring the strengths of interactions between different nuclei; these interactions can occur either through space or via connecting bonds. The strength of the interactions can then be used to calculate the distance between the nuclei involved. In order to obtain sufficient distance information to allow the structure to be calculated, it is necessary to examine interactions not only between hydrogen nuclei (^{1}H), but also involving carbon and nitrogen nuclei. This in turn necessitates the production of an enzyme sample which is isotopically enriched with ^{13}C and ^{15}N. The techniques involved in assigning resonances to particular nuclei and in monitoring the interactions between nuclei are beyond the scope of this book and are dealt with elsewhere.[155–158]

Based on the distances determined between pairs of nuclei, a family of structures will be generated, all of which are consistent with the distance constraints. Superposition of these structures will show that there are parts of the structure where the conformation of the protein is well defined, and other parts where the definition is much poorer. While it is probable that these latter areas correspond to highly mobile segments of the structure, it could be that the distance information in these areas is ambiguous. Additional measurements, such as those of the relaxation rates of nitrogen nuclei, would be necessary to confirm that the segments are highly mobile.[155,156] James[159] has given a good account of how the quality of NMR-derived structures can be judged.

If NMR studies cannot be carried out, more limited structural information can be obtained from the technique of circular dichroism (CD).[160,161] CD relies on the interaction of circularly polarized light with chiral (optically active) compounds; this can be used to measure the amount of regular secondary structural elements (Section 3.4.4) in an enzyme. Chemical modification of amino-acid side-chains (see Chapter 5, Section 5.4.4) can be used to divide side-chains of a particular type into those that are 'exposed', i.e. on the surface of the enzyme, and those which are 'buried', i.e. in the interior of the enzyme.[162]

From these types of studies, the overall conclusion is that the time-average structure of an enzyme in solution is generally very similar to its structure in the solid state, i.e. the crystal.[162] This is consistent with the finding that many enzymes, e.g. pancreatic ribonuclease and triosephosphate isomerase, can display catalytic activity in the crystalline state. In general, the rates of processes in the crystalline state are rather slower than those in solution, reflecting the slower diffusion of molecules through a crystal, whereas the values of equilibrium properties such as dissociation constants are similar in the two phases.[162] A number of techniques, including the use of intense synchrotron sources and of 'caged' substrates or enzymes (see Chapter 5, Section 5.4.3), have been developed to monitor structural changes which occur during catalysis.[2] These observations provide the justification for using X-ray crystallographic data as the basis for understanding the properties of the enzyme in solution.

3.4.2.1 The role of enzyme flexibility in catalysis

The finding that enzymes can possess large degrees of flexibility indicates that this may well be crucial to function. As detailed in Chapters 5 and 6, binding of substrates or other ligands can bring about conformational changes which can be monitored by X-ray, NMR, or other spectroscopic techniques such as CD or fluorescence.[163] These structural changes ('induced fits') are involved in relaying information between sites as in the case of allosteric effectors (see Chapter 6, Section 6.2.2), or play an essential role in the cycle of formation and breakdown of the catalytically active complex. In the case of phosphoglycerate kinase, for example, the binding of the two substrates causes the two domains (see Section 3.4.4.5) of the enzyme to rotate by 32° with respect to each other so as to bring the substrates into close proximity with each other.[164] This closing of the domains helps to exclude water from the active site of the enzyme, thereby discouraging the transfer of the phosphoryl group of ATP to water. A similar situation has been found in other enzymes, e.g. hexokinase, which transfer the phosphoryl group of ATP to a second substrate.[165] In the case of citrate synthase (see Section 3.4.6.2), the structural changes on binding the first substrate (oxaloacetate) lead to the formation of the binding site for the second substrate, acetyl-CoA.

The structural changes may prove to be the rate-limiting step in the reaction cycle. In the case of lactate dehydrogenase, for example, the slow step (release of the product NADH) has been shown to be limited by the rate of movement of a flexible loop (amino acids 98 to 120) which encloses the active site in the catalytically active complex (see Chapter 5, Section 5.5.4).

The importance of flexibility is also emphasized by observations on enzymes from thermophilic organisms (those which thrive at high temperatures). Many such enzymes work at temperatures above 70°C where most enzymes from mesophiles (organisms which thrive at moderate temperatures, i.e. between 20 and 40°C) would be rapidly denatured (Section 3.6.1). Although the precise basis for the additional stabilization of thermophilic enzymes is still not completely elucidated, it is clear that a number of extra interactions stabilize the tertiary structure (see Section 3.6.1). This extra stabilization is inevitably achieved at some cost to the flexibility of the enzyme. As the temperature is increased, the thermophilic enzyme will become more flexible and thus achieve more efficient catalysis. However, at the optimum temperature, its flexibility will be only roughly equivalent to that of a mesophilic enzyme at its (lower) optimum temperature. In this case, the maximum catalytic rate of both enzymes at their respective optima will be similar, i.e. the thermophilic enzyme will not gain the normal kinetic advantage observed when reactions are carried out at higher temperatures. This has been found to be the case for phosphoglycerate kinase and a number of other glycolytic enzymes.[166,167]

3.4.3 The importance of knowing the three-dimensional structure of an enzyme

The advances in our detailed understanding of the properties of enzymes over the last 20 years or so can largely be ascribed to the increase in detailed structural information provided by X-ray crystallography. These advances in understanding are discussed in the next sections (3.4.3.1 to 3.4.3.4).

3.4.3.1 To test models of macromolecular structure

The three-dimensional structure provides the experimental data with which to test theoretical models of macromolecular structure. The relative contributions of the various types of forces involved in the maintenance of the overall structure (Section 3.4.5) can be evaluated by referring to the experimentally determined structure. We can also use the experimental structure to test the success of structure prediction methods (Section 3.6.3).

3.4.3.2 To propose a mechanism of catalysis

The structure of an enzyme provides the framework within which the catalytic power of that enzyme can be understood. It is possible (see Chapter 5) to propose chemically reasonable mechanisms for a number of enzymes by combining these structural data with results from other studies such as enzyme kinetics, which tell us about the dynamic aspects of catalysis.

3.4.3.3 To explore similarities between enzymes

In Section 3.3.2.11 we described how the comparisons of amino-acid sequences have helped to explore the relationships between enzymes. This type of analysis is extended significantly if we can make suitable comparisons between the three-dimensional structures of enzymes. An early illustration of this was the finding by Rossmann and his co-workers that a common structural motif consisting of helix and sheet elements (Section 3.4.6.4) was present in a number of nucleotide binding proteins such as dehydrogenases and kinases.[168,169] This type of structural regularity would not have been deduced from comparisons of the relevant amino-acid sequences. A second type of nucleotide binding motif, based on a four helices and four strands of sheet motif, has been found in a number of other nucleotide-binding enzymes such as adenylate kinase (Section 3.4.6.4).[170]

With the great increase in the numbers of protein structures in the databases, new methods have been developed for the comparisons of three-dimensional structures.[171–173] These are considered in Section 3.4.4.6.

3.4.3.4 To assist rational drug design

The growth of the X-ray structural information on enzymes has given enormous impetus to the field of rational drug design. Many drugs have been identified by a process of random screening in order to develop so-called 'lead' compounds which then act as a focus for further modification and development. The more recent approach which is being actively exploited by major pharmaceutical companies involves identification of suitable target enzymes in appropriate metabolic pathways, determination of the structures of those enzymes, and then exploiting the powerful computational methods available to design ligands which will make complementary interactions with these enzymes. These ligands (and related compounds) can then be tested for their interaction with the enzyme and, if these trials are successful, further studies on the potential uptake of the compounds, etc. can be undertaken as a prelude to the eventual aim of clinical trials and commercial launch. Reviews of developments in this area have been given[2,174]. The development of inhibitors of angiotensin-converting enzyme for the control of hypertension is discussed in Section 3.4.6.5 and in Chapter 10, Section 10.7.3; the latter section also discusses the development of inhibitors of HIV protease for the treatment of AIDS. Given the enormous costs involved in the successful development of a single drug to the clinical use stage (estimated as around 350 million US dollars at 1998 prices), it is hard not to believe that the rational design approach will play an increasing role in the search for more effective drugs.

3.4.4 Features of structures adopted by enzymes

In this section we shall describe the principal structural elements that have been found in proteins and then proceed to outline the principles that govern the three-dimensional structures adopted by enzymes. A later section (3.4.6) will look at the detailed structures of four particular enzymes which exemplify the major types of structural classes of proteins, namely citrate synthase. α-chymotrypsin, adenylate kinase, and thermolysin.

3.4.4.1 The amide bond

The amide bond which provides the means of linking amino acids in the polypeptide chain (Section 3.3.2) is not adequately represented by the formula

$$\begin{array}{c} O \\ \parallel \\ -C-NH-, \end{array}$$

since this fails to show that there is in fact considerable resistance to rotation about the C–N bond.

Estimates of the activation energy for this rotation are of the order of 80 kJ mol^{-1}.[175] It is therefore more appropriate to depict the amide bond as a planar unit with extensive delocalization of the lone pair of electrons on the nitrogen atom imparting a partial double-bond character to the C–N bond (Fig. 3.20).

The form of the amide bond unit shown in Fig. 3.20, in which the two α-carbon atoms are *trans* to one another, is generally more stable than the *cis* form by some 12 kJ mol^{-1} because of the steric crowding in the latter.[175] However, when the amide bond involves the amino acid proline, the difference in energy between *cis* and *trans* forms is much less pronounced and the *cis* form is found occasionally in proteins.[175] As mentioned in Section 3.6.2, the *cis/trans* isomerization of amide bonds to proline can be a rate-limiting step in protein folding. Rotation is allowed about the single bonds linking the α-carbon atoms to the carbonyl carbon atom and to the nitrogen atom of the amide bond. We can thus

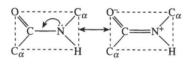

Fig. 3.20 The amide bond showing delocalization of electrons. The dotted lines represent the planar unit of two α-carbon atoms and the carbon, nitrogen, oxygen, and hydrogen atoms. The length of the amide C–N bond is 0.132 nm and the distance between the two α-carbon atoms in this *trans* configuration is approximately 0.38 nm. While the above diagram accounts for the partial double-bond character of the C–N bond, it does not provide a complete picture. Since the nitrogen atom in the amide bond does in fact have a net negative charge,[176,177] there must be extensive back-donation of σ electrons from C to N to compensate for the π electron movement.

describe the structure of a dipeptide unit in terms of the dihedral angles (ϕ, ψ), i.e. the angles of rotation about these bonds (Fig. 3.21).

Rotations about these bonds, i.e. changes in the angles ϕ and ψ, will change the distances between non-bonded atoms and thus the energy of the dipeptide unit will vary. (This is because when non-bonded atoms are brought into close proximity with each other, overlap of the electron clouds leads to repulsion and hence destabilization.) Calculations of the energy of the dipeptide unit as a function of the angles ϕ and ψ have been made,[178] and the resulting plot (known as a Ramachandran diagram) show that there are certain well-defined conformations of peptide units that are of lowest energy, i.e. of greatest stability.*

The most important of these regular secondary structures are the α-helix and β-sheet structures, which had already been proposed as the basic structural units of fibrous proteins;[180] thus, wool consists largely of an α-helical structure and silk largely of a β-sheet structure.

3.4.4.2 The α-helix

The α-helix is depicted in Fig. 3.22(a). It consists of a right-handed helix[†] (i.e. the direction of twist from the N-terminus to the C-terminus is that of a right-handed corkscrew, see Fig. 3.22(b)) in which hydrogen bonds (Section 3.4.5.1) are formed between the carbonyl group of the amide bond between amino acids n and $n + 1$ and the amino group of the amide bond between amino acids $n + 3$ and $n + 4$ (Figs 3.22(a) and 3.22(c)). A detailed analysis of the energy of the α-helix, taking into account the partial charges which occur on the main-chain atoms, has indicated that electrostatic interactions between the carbonyl group of amino acid n and those of amino acids $n + 3$ and $n + 4$ will also contribute to the stability of the helix (the carbon and oxygen atoms of the carbonyl group carry partial positive and negative charges respectively).[181] Because of the restrictions on rotation around the bonds to the α-carbon atom in proline, this amino acid cannot be incorporated into an α-helix without seriously distorting it. Proline therefore acts as a helix-breaker.

Since there are approximately 3.6 amino acids per turn of the α-helix and a total of 13 atoms in the ring closed by formation of the hydrogen bond, the α-helix is sometimes referred to as the 3.6_{13} helix. The pitch of the α-helix is 0.54 nm (Fig. 3.22(a)) which corresponds to a translation of 0.15 nm per amino acid. The ways in which α-helices can pack together in protein structures have been reviewed.[179,182] A useful way of representing the amino-acid sequence of the helical segment of a protein is known as the 'helical wheel'.[183] In this representation, amino acids are plotted at 100° intervals around a circle (note, there are 3.6 residues per turn of the helix, i.e. per 360°). When this is done, a view is obtained of the peptide chain along the axis of the helix, and the nature of the side-chains will indicate where the helix is likely to occur in the protein (Fig 3.22(c)). Thus, if the side-chains on both sides of the helix are non-polar, the helix would be very likely to be buried within the interior of a protein (see Section 3.4.4.5). A helical transmembrane segment of a protein would also have non-polar side-chains on both sides. If one side were polar and the other non-polar, this *amphipathic* helix would be likely to occur near the surface of a protein with one face pointing inwards and the other towards the solvent. Other types of helical structure are possible, such as the 3_{10} helix, but these are less stable than the α-helix because the hydrogen bonds are somewhat distorted. The 3_{10} helix occurs to a limited extent in enzymes;[184] short lengths are found, for instance, in lysozyme, carbonate dehydratase, and thermolysin (Section 3.4.6.5).

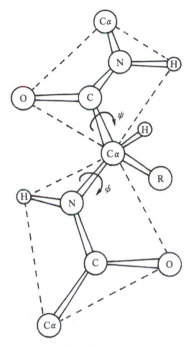

Fig. 3.21 A dipeptide unit of a polypeptide chain showing the planar amide units and the relevant angles of rotation about the bonds to the central α-carbon atom. $\phi = \psi = 180°$ corresponds to the fully extended polypeptide chain.

* This statement applies to all amino acids except glycine, where the Ramachandran plot shows a wide scatter of dihedral angles. Because glycine has such a small side-chain (–H), it can adopt a wide range of dihedral angles in proteins.[179] This allows glycine residues to be involved in unusual conformations in proteins.

† For L-amino-acids (Section 3.3.1) the right-handed helix is far more stable than the left-handed helix because of the crowding of the side-chains in the latter. The right-handed helix is found almost exclusively in proteins (but see the example of thermolysin in Section 3.4.6.5).

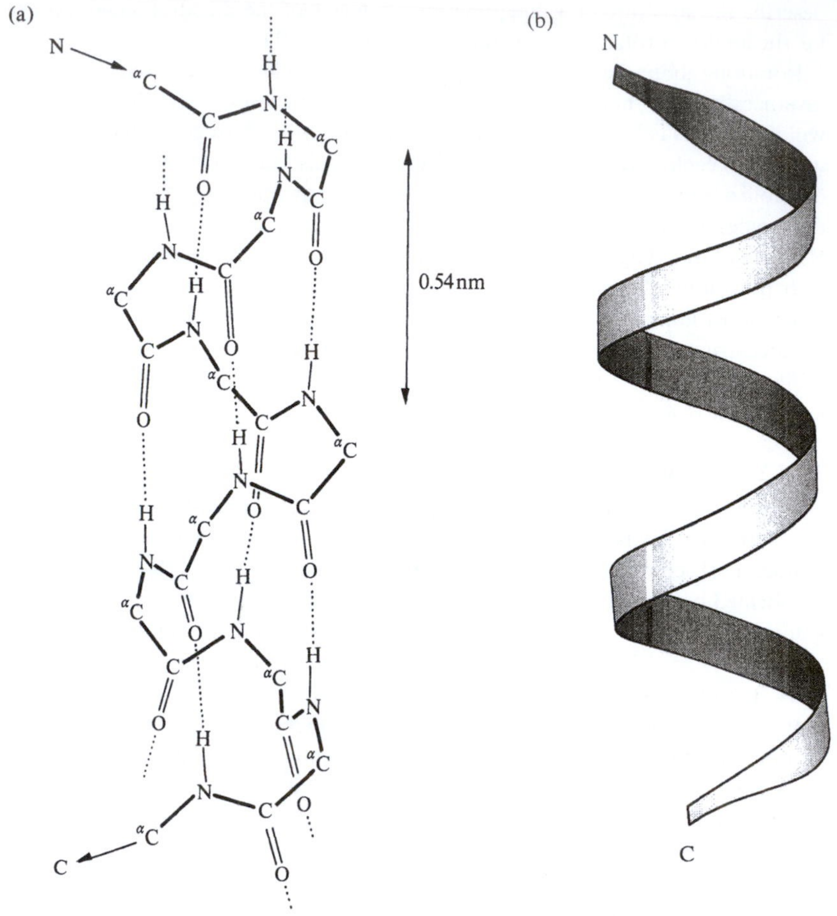

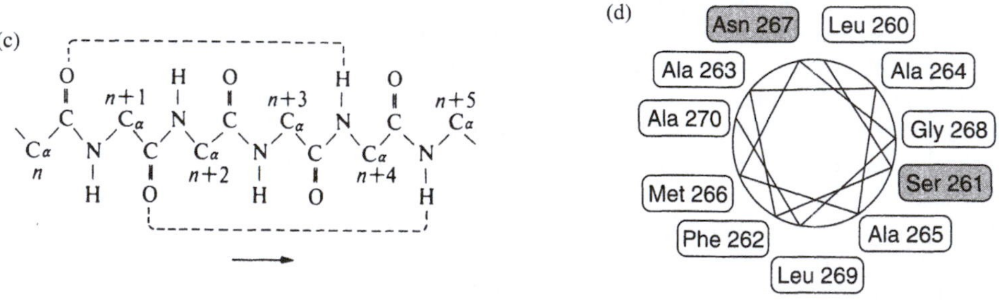

(a)

(b)

0.54 nm

Fig. 3.22 Features of the α-helix found in proteins. For the α-helix, the values of ϕ and ψ (see Fig. 3.21) are approximately $-57°$ and $-47°$, respectively. (a) Path of the polypeptide chain. The arrows show the direction of the chain (N → C). (b) A perspective view of the α-helix shown in (a). (c) The hydrogen bonding arrangement in the α-helix. (d) The 'wheel' arrangement of side-chains in a helical segment. The positions of 11 residues are plotted at 100° intervals in a clockwise direction (looking down the helix from the N-terminus). From the polarity of the side-chains the location of the helix can be predicted. For instance, in the case of citrate synthase (see Section 3.4.6.2) the sequence from Leu 260 (1) to Ala 270 (11) (within helix M), all amino acids except Ser 261 and Asn 267 are non-polar. This helix is buried in the interior of the enzyme.

(c)

(d)

Asn 267	Leu 260
Ala 263	Ala 264
Ala 270	Gly 268
Met 266	Ser 261
Phe 262	Ala 265
Leu 269	

3.4.4.3 The β-sheet

There are two types of β-sheet structures according to whether the polypeptide chains between which hydrogen bonds are formed are running in parallel or antiparallel directions. From the patterns shown in Figs 3.23(a) and 3.23(b) it can be seen that the hydrogen bonds in the parallel sheets are somewhat distorted: no such distortion occurs in the antiparallel sheet. A more detailed analysis of the interactions between the main-chain atoms in both parallel and antiparallel sheets has shown that electrostatic interactions between the carbonyl groups on

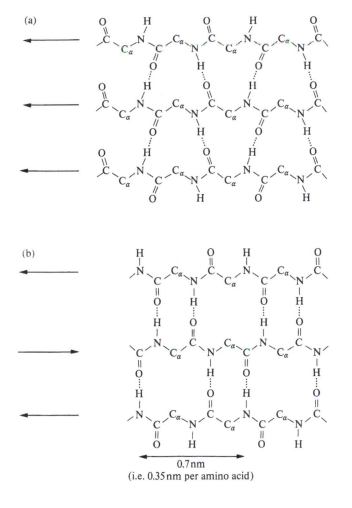

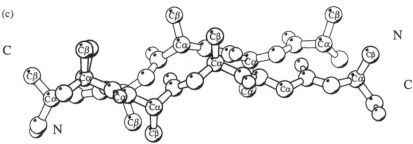

0.7 nm
(i.e. 0.35 nm per amino acid)

Fig. 3.23 Sheet structures found in proteins. (a) Hydrogen bonding in the parallel β-sheet. (b) Hydrogen bonding in the antiparallel β-sheet. For the parallel and antiparallel β-sheet structures the values of (ϕ, ψ) (see Fig. 3.21) are approximately $(-119°, +113°)$ and $(-139°, +135°)$, respectively. In three dimensions the sheet structures represented in (a) and (b) are in fact corrugated, with the side-chains of the amino acids projecting perpendicularly to the plane of sheet, and a twist between adjacent strands. These aspects are shown in (c), which depicts part of the antiparallel sheet structure within superoxide dismutase (see Fig. 3.26).

adjacent chains contribute significantly to the stability of the structure; indeed it has been estimated that the contribution of these interactions is about 80 per cent of those due to hydrogen bonds.[181] Several enzymes contain extensive networks of sheet structure: part of the arrangement found in carboxypeptidase A[185] is illustrated in Fig. 3.24. Other examples of β-sheet networks are illustrated in Branden and Tooze.[182] As originally pointed out by Chothia,[186] the β-sheets in globular proteins show a right-handed twist (as viewed along the polypeptide chain) rather than being flat as in silk. This twist is seen clearly in the structure

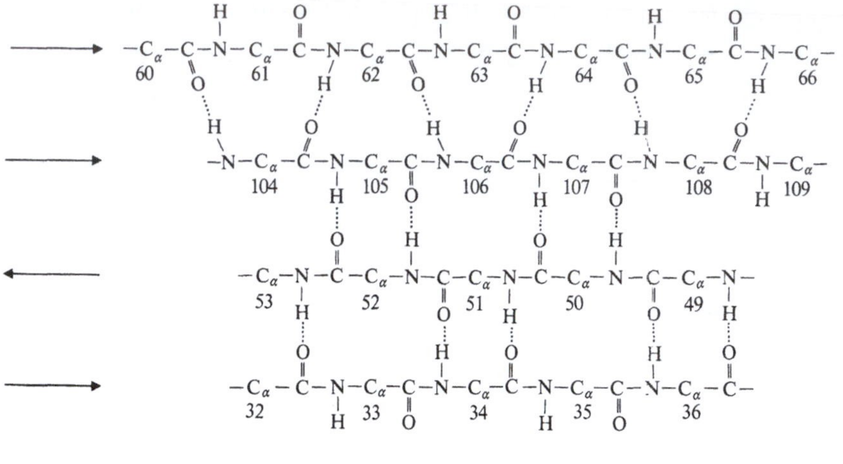

Fig. 3.24 Part of the β-sheet structure of carboxypeptidase A.[185] The directions of the strands are shown by arrows. Hydrogen bonds are represented by dotted lines.

of adenylate kinase (Fig. 3.33 in Section 3.4.6.4). The explanation for the sense of this twist appears to lie in the electrostatic interactions which can occur between the main-chain carbonyl groups of adjacent amino acids in each strand; these interactions are much more favourable in the case of the right-handed twist rather than the left-handed twist.[176]

3.4.4.4 Other structural features

Of the various other elements of regular structure in proteins, the most abundant is the β-turn. The β-turn is a sequence of amino acids in a protein, in which the polypeptide chain folds back on itself by nearly 180° (Fig. 3.25), thus giving the protein a globular rather than a linear shape.[187] Many different types of β-turn have been identified which differ in terms of the number of amino acids in the loop and the angles of rotation (ϕ, ψ in Fig. 3.21) about the various single bonds involved.[187] A detailed analysis of the protein structures shows that the most frequently occurring turns consist of between two and five amino acids.[188] A β-bulge occurs where an extra amino acid is inserted in one strand of a β-sheet, disrupting the hydrogen bonding[189].

The importance of these structural features in proteins has been reviewed.[179,190,191]

3.4.4.5 Principles governing the structures adopted by enzymes

Enzymes, and globular proteins in general, do not possess the simple regular structures that characterize fibrous proteins such as wool, silk, or collagen. However, from an analysis of the known three-dimensional structures, certain general principles can be derived that appear to govern the structures adopted by enzymes.[175,182,191,192]

1. Enzymes are generally very closely packed globular structures with only a small number of internal cavities which are normally filled by water molecules.

2. The structural elements (amide bonds, helices, sheets, etc.) generally possess similar geometries to those observed in model compounds. However, some small deviations from the planarity of amide bonds have been noted in lysozyme, and α-helices can be distorted towards 3_{10} helices.

3. Non-polar amino-acid side-chains, e.g. those of Leu, Val, Phe, etc., are generally buried in the interior of the enzymes away from the solvent water. Ionized side-chains, e.g. those of Lys, Asp, etc., tend to be on the exterior of

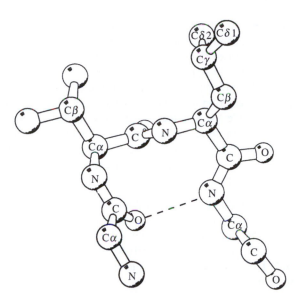

Fig. 3.25 Diagrammatic illustration of a type I β-turn found in proteins. The dotted line indicates the O and N atoms linked by a hydrogen bond.

enzymes and able to interact with the solvent. Exceptions to these generalizations usually point to side-chains that have particular functions, e.g. the buried side-chain of aspartic acid 194 in chymotrypsin that is involved in a strong electrostatic interaction with the protonated α-amino group of isoleucine 16 (see Chapter 5, Section 5.5.1.2). Non-polar side-chains on the exterior of enzymes may well be involved in binding to a non-polar substrate or to a membrane.

4. Polar groups in the interior of enzymes are normally paired in hydrogen bonds. A very large percentage of the carbonyl and amino groups of amide bonds that are placed in the interior by the folding of the polypeptide chain are paired in hydrogen bond formation or form hydrogen bonds to internal water molecules and are thus rendered essentially non-polar.[193] While these internal hydrogen bonds may not be of great significance in terms of the energy of the folded structure, they are undoubtedly important in specifying the correct folded structure (Section 3.4.5.5).

5. The polypeptide chains of most larger enzyme molecules tend to exist in structural domains, i.e. independent folded, globular, units which are connected by segments of the peptide chain.[194] The majority of domains in proteins tend to be in the M_r range from 15 000 to 20 000, i.e. correspond to a globule some 2.5 nm in diameter,[184] although these figures should only be regarded as general guidelines. The secondary structural patterns within a domain are discussed below (Section 3.4.5.6). Relatively simple examples of domain types of structure are found amongst the kinases and dehydrogenases,[165,169] which usually consist of two domains or 'lobes'. Figure 3.29 shows the structure of phosphoglycerate kinase where the two domains are distinguished by shading. One of the domains contains the binding site for the common substrate, i.e. ATP for kinases; the other domain binds the second substrate and will vary between different enzymes. Movements of the domains with respect to each other seem to be an important part of the mechanism of catalysis in such enzymes (see Section 3.4.2.1).

Some enzymes consist of a number of domains (which may well correspond to exons) linked together.[195] A number of multiple domain systems are found in enzymes of the blood-clotting system (see Chapter 6, Section 6.2.1.1).[182] Thus, for example, Factor IX consists of a calcium-binding domain, two EGF (epidermal growth factor)-type domains, and a serine proteinase domain. The group of enzymes known as matrix metalloproteinases, which are responsible for degradation of extracellular components during morphogenesis and in disease processes such as arthritis and tumour invasion, consist of multiple domains.[196,197] The occurrence of multiple domains has greatly facilitated the application of nuclear magnetic resonance to the analysis of such structures.[198]

* These limits are guidelines only. Thus the 4-oxalocrotonate tautomerase from *Pseudomonas putida* has a subunit M_r of only 7000,[199] and the isoleucine-tRNA ligase from *E. coli* is a monomer of M_r 114 000.[200]

6. In addition to the trend towards multiple domains, most large enzymes consist of multiple polypeptide chains or subunits. Thus when the M_r is less than about 30 000, monomeric (single subunit) enzymes are the norm, whereas with the M_r above about 50 000, such structures are relatively uncommon.* The significance of multiple subunits in enzymes is discussed in detail in Section 3.5.

3.4.4.6 Classification of enzyme structures

In the previous section (3.4.4.5) we described some of the basic principles that seem to govern the types of structures adopted by enzymes. A given enzyme will possess certain amounts of helix, sheet, and other elements of regular secondary structure, but this does not tell us very much about the overall structure of the enzyme. Among the various classification schemes for the structural elements either in whole proteins or in individual domains, the most widely used has been the four-class system introduced by Levitt and Chothia.[201] The four classes of protein structure are as follows.

1. All-α-proteins which have only α-helix structure. Myoglobin and citrate synthase are examples of this type of protein.

2. All-β-proteins which have mainly β-sheet structure. Chymotrypsin and the constant and variable regions of immunoglobulins belong to this class.

3. α/β-proteins, which have mixed or alternating segments of α-helix and β-sheet structure. The kinases and dehydrogenases with their nucleotide-binding domains are good examples of this class, which also includes other enzymes of the glycolytic pathway such as triosephosphate isomerase and phosphoglycerate mutase.

4. $\alpha + \beta$-proteins, which have α-helix and β-sheet structural segments that do not mix but are separated along the polypeptide chain. Thermolysin, lysozyme, and ribonuclease belong to this class of proteins. Since this class is rather small, some authors[182] have chosen to regard it as a subset of the α/β class.

A number of excellent discussions of these classes of proteins have been given.[179,182,191,202] Branden and Tooze have discussed how considerations of the topology of secondary structural elements in an enzyme can be used to predict the likely position of the active site.[182] Several computer-based procedures are now available which classify protein structures in terms of the arrangement of secondary structural elements (known as the *protein fold*). One of the most useful is scop (structural classification of proteins),[173,203] which is available via the

World Wide Web as explained in the Appendix to this chapter. The scop procedure classifies proteins at three levels. Protein fold families have similar arrangements of secondary structures, but there is no clear evidence for an evolutionary relationship at either a structural or functional level. Within a fold family there could be large insertions or deletions, so long as these do not affect the core fold arrangement. The next level of classification is protein superfamilies where there is a probable evolutionary relationship as shown by similar functions (e.g. binding of nucleotides). The third level is that of protein families where there is very strong evidence for evolutionary relationships, with a sequence identity of generally more than 30 per cent.

As more structures are being solved, the number of protein fold families has continued to increase steadily; by 1996, well over 300 types of fold had been described. From a consideration of the various stereochemical constraints involved in the packing of helix and sheet elements, it has been estimated that the total possible number of folds is probably less than 1000,[204] although other work has indicated that this may be a considerable underestimate.[172] Of course, not all these folds are populated to the same extent among the known protein structures, some folds having only a single representative. From an analysis of the protein structures available up to April 1993, it was concluded that only nine fold families contained proteins which had no sequence or functional similarity to each other.[172] These nine so-called superfolds were found to account for a significant fraction (over 30 per cent) of all known structures. The superfolds (listed in order of decreasing population) are known as: α/β doubly wound, TIM barrel, split $\alpha\beta$ sandwich, Greek key-immunoglobulin, α-up-down, globin, jelly roll, trefoil, and UB $\alpha\beta$ roll. Representatives of three of the more highly populated fold families are depicted in Fig. 3.26.

The occurrence of these protein folds in a wide variety of apparently unrelated proteins suggest that they represent especially stable structural units. It is tempting to propose that they might be important in forming nucleating centres in protein folding, thereby directing the way in which a polypeptide chain could fold to adopt its correct three-dimensional structure (Section 3.6.2).

3.4.5 The forces involved in stabilizing the folded structures of enzymes

The forces that maintain the folded three-dimensional structure of a globular protein such as an enzyme are weak, non-covalent interactions. Experiments of the type described in Section 3.6.2 have shown that in the case of enzymes that possess disulphide bonds, these bonds are formed by oxidation of pairs of cysteine side-chains only after the three-dimensional structure of an enzyme has been acquired. It would therefore seem that disulphide bonds are important in stabilizing a folded structure that has already formed rather than directing the acquisition of this structure. The bond energy of a disulphide bond (approximately 200 kJ mol^{-1}) is much greater than that of the non-covalent interactions we shall describe in Sections 3.4.5.1 to 3.4.5.4.

The non-covalent interactions can be discussed in four categories: hydrogen bonds, electrostatic forces, van der Waals forces, and hydrophobic forces. The division between these categories is somewhat arbitrary, since all attractive forces between atoms ultimately arise from favourable interactions between electrons and nuclei. Very extensive discussions of the forces involved in stabilizing the folded structures of proteins have been given.[205,206]

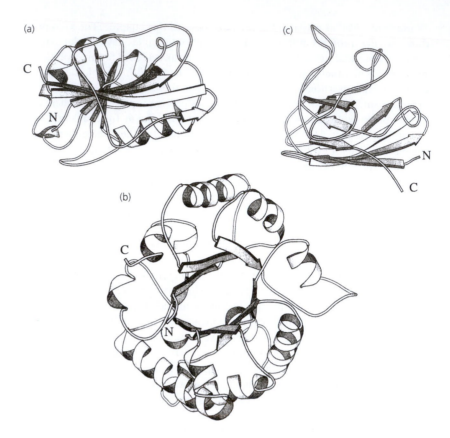

Fig. 3.26 Three of the most commonly occurring secondary fold families of proteins. Structures are shown in Molscript[216] in the cartoon format, in which α-helices and β-sheets are depicted as ribbons and arrows respectively. (a) α/β doubly wound dihydrofolate reductase (1dhf); (b) TIM barrel, triosephosphate isomerase (1ypi); (c) Greek key, superoxide dismutase (2sodB). In (b) and (c) single subunits of the enzymes are shown. The Protein Data Bank (PDB) accession codes for each protein are shown in brackets (e.g. 1dhf). The N and C termini of each polypeptide chain are indicated.

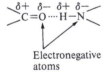

Electronegative
atoms

Fig. 3.27 The hydrogen bond involved in helix and sheet structures in enzymes and proteins.

3.4.5.1 Hydrogen bonds

The hydrogen bond arises from the attraction between a hydrogen atom attached to an electronegative atom (O and N are the most important examples in biological systems) and another electronegative atom. Figure 3.27 shows the arrangement that is important in α-helix and β-sheet structures.

The hydrogen bond is most stable when: (i) the distance between the two electronegative atoms is within closely specified limits—in the case shown in Fig. 3.27 the $O-N$ distance is within the range 0.28–0.30 nm; and (ii) the arrangement of atoms $O...H-N$ is linear. The criteria for designating hydrogen bonds in proteins have been discussed extensively.[181,207,208]

A typical value for the energy of a hydrogen bond *in vacuo* is approximately 20 kJ mol^{-1}, but the values of hydrogen bonds in proteins have been shown experimentally to vary widely depending on the nature of the acceptors and donors involved.[209] For instance, the energy of interaction between the phenolic group of a tyrosine side-chain and a positively charged amino group or a substrate corresponds to 15.5 kJ mol^{-1}, whereas the energy of interaction between two uncharged phenolic side-chains of tyrosine corresponds to only 2.2 kJ mol^{-1}. It is unlikely that the many internal hydrogen bonds in enzymes (Section 3.4.4.5) make any substantial contribution to the stabilization of the folded structure. This is because in the unfolded state the 'internal' polar groups could form alternative hydrogen bonds with the solvent water[169] and there is therefore little gain in stability when comparing the folded state with the unfolded. However, because of the relatively strict geometrical requirements referred to above, hydrogen bonds are undoubtedly important in specifying the correct folded structure of the protein (Section 3.4.5.5).

3.4.5.2 Electrostatic forces

This category consists of the attractive forces that exist between either fully charged groups in amino acids, e.g. between $-NH_3^+$ (Lys) and $-COO^-$ (Asp), or between atoms which carry partial charges. The former interactions are also known as salt bridges. An example of the latter is provided by the interactions between adjacent carbonyl groups of peptide bonds in proteins (note that the carbon and oxygen atoms of this bond are estimated to carry charges of $+0.46$ and -0.46 respectively[181]); these interactions are thought to be important in stabilizing certain secondary structural features of proteins (Section 3.4.4).

The force between two charges is inversely proportional to the dielectric constant of the medium separating them, and the lack of detailed knowledge of the value of the dielectric constant in an enzyme makes any quantitative assessment of the contribution of electrostatic forces to the stability of the folded structure a difficult task. Some progress has been made in determining the effective dielectric constant within proteins by examining the effects of changing charged amino-acid side-chains elsewhere in subtilisin on the pK_a of the active-site histidine side-chain (His 64) of the enzyme. The results suggest that the effective dielectric constant is between 30 and 60 (the values for water and benzene are 80 and 2.3, respectively). The relatively high value for the protein is thought to be due partly to the surrounding water and also to polar side-chains.[210,211] Although salt bridges are known to play a crucial role in particular enzymes, e.g. in the activation of chymotrypsinogen (see Chapter 5, Section 5.5.1.2), it appears that in general the number of salt bridges is limited and hence their contribution to the stability of enzyme structures is probably rather small. There is some evidence, however, that the enhanced thermal stability of a number of thermophilic enzymes arises from additional salt bridges (Section 3.6.1).

3.4.5.3 Van der Waals forces

Van der Waals forces are weak forces that occur when molecules or groups of atoms are in close contact with one another (less than about 0.4 nm apart in the case of atoms found in proteins).* Attractive forces result from favourable interactions between dipoles (separations of charge) that may be permanent or transient.

Transient dipoles arise by the local fluctuations in electron density in atoms and can also be induced by the presence of a neighbouring dipole. A variety of interactions are possible: dipole–dipole, dipole–induced dipole, and induced dipole–induced dipole. All these forces are weak, with energies of the order of 10 kJ mol^{-1} or less, but the large number of van der Waals contacts that occur in proteins may well mean that they make a significant contribution to the stability of protein structures.

* When atoms are brought too close to each other, the electron clouds repel one another. The balance of attractive and repulsive forces leads to an optimum distance, the van der Waals distance, between the atoms. This is the sum of the van der Waals radii of the atoms concerned.

3.4.5.4 Hydrophobic forces

It is well known that hydrocarbons, such as methane, benzene, etc., are much more soluble in liquid hydrocarbons or most other organic solvents than in water. In fact, water has an almost unique place as an extremely poor solvent for hydrocarbon solutes.[212] The probable reason for this lies in the fact that water consists of transient clusters of hydrogen-bonded molecules; when a hydrocarbon is introduced into the aqueous medium there is a reorganization of the water molecules around the hydrocarbon that makes the system more ordered. This increase in order, i.e. decrease in entropy, is unfavourable on thermodynamic grounds and hence there is a tendency to force the hydrocarbon out of contact with the water and into the organic phase, so that this decrease in entropy does not occur.

In the case of proteins, these considerations dictate that the non-polar portions of the molecule, for example the side-chains of amino acids Val, Leu, Ile, Phe, Trp, should be buried in the interior of the molecule away from the solvent water. Calculations show that this burying of the non-polar side-chains makes by far the major contribution to the stability of the folded structure compared with that of the unfolded structure.[213] For lysozyme this stabilization energy is calculated to be of the order of 1200 kJ mol^{-1}. *It should be emphasized that it is the burying of the non-polar side-chains away from water that provides this stabilization, rather than interactions between the non-polar side-chains themselves.*[214]

3.4.5.5 Some conclusions about the forces involved in the maintenance of the folded structures of enzymes

From the discussion in the previous section (3.4.5.4) it is evident that hydrophobic forces make the greatest contribution in energy terms to the stability of the folded state compared with the unfolded state. However, hydrophobic forces do not confer any geometrical specificity on interactions, since they arise essentially from exclusion from the aqueous phase,[214] and thus do not serve to specify any particular folded structure. In order to specify a structure, interactions such as hydrogen bonding that possess geometrical requirements (Section 3.4.5.1) are of great importance. The overall stability of the folded structure compared with the unfolded structure is much smaller than the values of hydrophobic stabilization energy would suggest; typical values for the overall stabilities of globular proteins are between 20 and 60 kJ mol^{-1} (Section 3.6.1). This difference can largely be accounted for by the fact that it is highly unfavourable on entropy grounds to convert a highly flexible unfolded polypeptide chain to a more ordered folded chain.[205,213]

3.4.6 Examples of enzyme structures

In this section we shall describe the three-dimensional structures of four enzymes which have been chosen as examples of the four secondary structural classes (Section 3.4.4.6), namely citrate synthase (all α), α-chymotrypsin (all β), adenylate kinase (α/β), and thermolysin ($\alpha + \beta$). However, it is first necessary to mention some of the ways of depicting these structures.

3.4.6.1 The representation of enzyme structures

Considerable problems are posed by trying to depict the complex three-dimensional structures adopted by enzymes, especially when using two-dimensional representations. Up till the mid to late 1980s the most informative representations were generally assumed to be three-dimensional scale models built directly from the coordinates of the electron-density map. An example of such a model is shown in Fig. 3.28 which depicts the structure of adenylate kinase from pig muscle at a scale of 20 cm = 1 nm. This model uses wire parts that keep the bond angles and bond lengths of the planar amide bond units constant but allow rotation about the single bonds to the α-carbon atoms. Although features such as the deep cleft at the bottom right-hand side of the model are clearly visible, it is difficult to appreciate

Fig. 3.28 The structure of adenylate kinase shown as a three-dimensional scale model built with wire parts (scale 20 cm = 1 nm). Reproduced with permission from *Principles of protein structure* by G. E. Schulz and R. H. Schirmer, p. 136. Copyright Springer, New York (1979).

the details of the structure without having the model on hand, which, given its bulkiness, is clearly impractical for most purposes. Since the late 1980s, however, the problems of representation have been very largely addressed by developments in computer programs and graphics. Thus, even with a relatively modest desktop personal computer, it is now possible to view protein structures, to rotate these structures, to zoom in on areas of interest, and to highlight particular side-chains. With more powerful workstations, full three-dimensional effects can be appreciated, giving superb insights into the structural basis of protein function. One of the most widely used graphics programs is Rasmol,[215] details of which can be found in the Appendix to this chapter. Rasmol can take the coordinates of a protein structure from the protein data bank and display the structure in a variety of formats; some of these are shown in Fig. 3.29 for the example of phosphoglycerate kinase. In this figure the graphics program Molscript[216] has been used, since it tends to produce better line drawings to illustrate perspective.

The wireframe display in Rasmol is equivalent to the three-dimensional scale model shown in Fig. 3.28. Although particular features can be highlighted in colour, it is difficult to appreciate the nature of the protein fold in such a model. In the space fill representation each atom is represented by a sphere, sometimes flattened, of radius equal to its van der Waals radius. Space-filling models are very suitable for highlighting features of the surface of an enzyme, such as a cleft

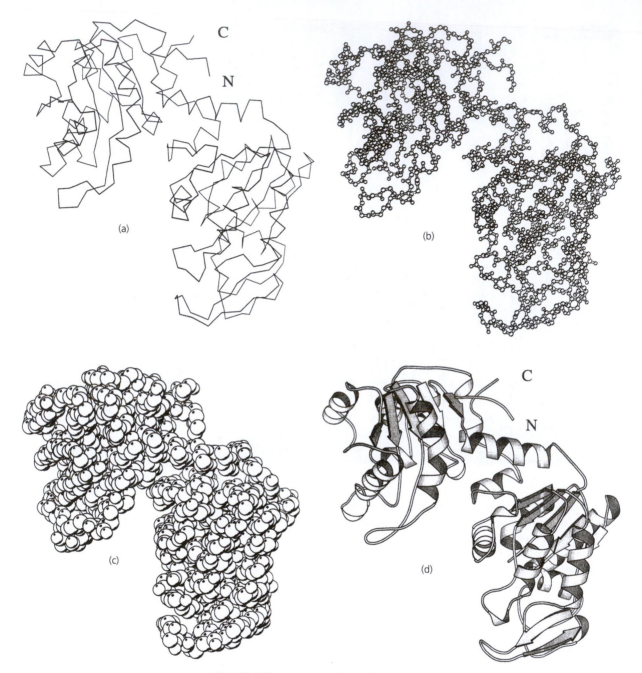

Fig. 3.29 Different representations of the structure of phosphoglycerate kinase. Structures of the enzyme from *Bacillus stearothermophilus* (1php) are drawn in Molscript.[216] (a) Polypeptide backbone, (b) ball and stick, (c) space filling, (d) cartoon.

or depression or clusters of non-polar side-chains, but are of little use in illustrating internal features such as the folding of the polypeptide chain. The path of the polypeptide chain through the molecule can be seen in the backbone representation which shows the α-carbon atoms linked by straight lines. Two limitations of the backbone models are: (i) it is not always easy to pick out elements of regular secondary structure, and (ii) the nature of the side-chains in any particular region of the enzyme such as the active (catalytic) site is not shown. In order to meet the first limitation, it has become popular to emphasize the elements of regular secondary structure in a protein by the use of stylized representations of helices and sheets in ribbon, strand, or cartoon format.[179,202] From these types of models, the nature of the fold family (Section 3.4.4.6) for the protein can be readily discerned. The second limitation can be overcome by including selected side-chains, e.g. those known to be at the active site on the simplified structure of the rest of the protein. This is shown in Figs 3.30 to 3.34 for a number of different enzymes.

Clearly, the different types of representation of protein structures have their own strengths and weaknesses and should be seen as complementary to one another. The power of the graphics programs is well illustrated by considering one of the largest enzymes whose structure has been solved, namely the 20S proteasome from *Thermoplasma acidophilum* which consists of 28 subunits, 14 of each of two types, α and β.[217] The subunits are arranged as four stacked rings $\alpha_7\beta_7\beta_7\alpha_7$. Figure 3.30(a) shows a side-view of the α-carbon backbones of the polypeptide chains of the whole molecule (a cylinder 15 nm in height and 11 nm in diameter, whose $M_r = 670\,000$). When the whole molecule is viewed from the top (Fig. 3.30 (b)), the sevenfold nature of the rings can be appreciated. By focusing on a single β subunit, the features of secondary structure can be depicted (Fig. 3.30 (c)); in this representation, the catalytically important residues (Thr 1, Glu 17 and Lys 33) in this β subunit can be highlighted.

3.4.6.2 The structure of citrate synthase (an all-α enzyme)

Citrate synthase from higher organisms consists of two identical subunits each of M_r about 50 000. The structure of the enzyme from pig heart was solved by Remington and his co-workers[218,219] and is illustrated in Fig. 3.31(a, b) in space-filling format. Figure 3.31(c) shows a single subunit of the enzyme from hyper-thermophilic bacterium *Pyrococcus furiosus* in cartoon format.

In each subunit of the muscle enzyme, there are 20 helices which pack tightly together. Over 65 per cent of the total amino acids in the enzyme are in the helical elements, some of which are kinked or bent over a large angle. Each subunit consists of two domains, the larger one containing 15 helices and the smaller five, with a substrate-binding cleft running between the two. This cleft is close to the subunit interface so that the actual binding site consists of amino acids from both subunits. The enzyme crystallizes in a number of forms which can be characterized as being of either 'open' or 'closed' type; the difference between the two is best seen in the space-filling representations in Fig. 3.31. The change between the open and closed forms amounts to a rotation of the small domain by about 20° relative to the large one. Binding of the first substrate (oxaloacetate) brings about the structural change and thereby generates a high-affinity site for the second substrate acetyl-Coenzyme A (acetyl-CoA). The catalytically important side-chains include two His and one Asp side-chains (Fig. 3.31(c)), which are all involved in proton donating and accepting steps in the proposed mechanism of the enzyme.[219]

3.4.6.3 The structure of α-chymotrypsin (an all-β enzyme)

α-Chymotrypsin consists of three polypeptide chains held together by five disulphide bonds: the origin of this three-chain structure is discussed in Chapter 5,

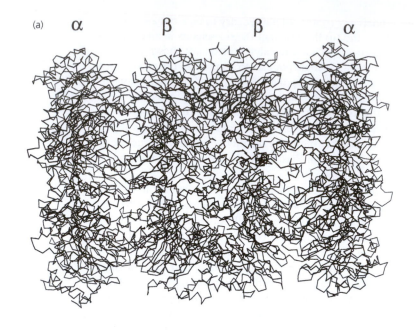

(a)   α   β   β   α

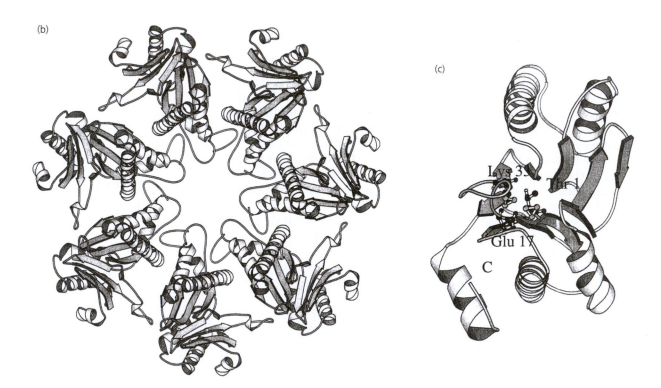

Fig. 3.30 Representations of 20S proteasome from *Thermoplasma acidophilum,* drawn in Molscript.[216] (a) A side-view showing the polypeptide backbones of the $\alpha_7\beta_7\beta_7\alpha_7$ structure (1pma); (b) a top-view showing the ring of α subunits (α_7) in cartoon format; (c) a single β-subunit shown in cartoon format with the catalytically important side-chains (Thr 1, Glu 17, and Lys 33) highlighted.

Fig. 3.31 Structure of citrate synthase, an all-α protein. (a) and (b) show the open (5csc) and closed (4cts) forms of the dimeric muscle protein in space filling format respectively. (c) shows a Molscript[216] drawing of a single subunit of the enzyme from *Pyrococcus furiosus* (1aj8) in cartoon format, with the catalytically important His 223, His 262 and Asp 312 residues highlighted. (In the muscle enzymes these residues are 274, 320 and 375 respectively, reflecting the fact that the *P. furiosus* enzyme has a 50 residue deletion at the N-terminus.)

Section 5.5.1.1. The chains run between amino acids 1–13 (chain A), 16–146 (chain B), and 149–245 (chain C). The five disulphide bonds are between cysteine side-chains at positions 1 and 122, 42 and 58, 136 and 201, 168 and 182, and 191 and 220.

The structure of the enzyme is represented in Fig. 3.32(a), which shows the positions of the three catalytically important side chains, namely His 57, Asp 102 and Ser 195 (see Chapter 5, Section 5.5.1.2). The overall shape of the molecule is that of an ellipsoid with a maximum dimension of 5.1 nm. There is a shallow depression at the active site in which the three side-chains are located.

There are only two short stretches of helix between amino acids 164 and 173 and between amino acids 235 and 245. Both helices are somewhat distorted from an α-helix towards a 3_{10} helix. There is a considerable amount of antiparallel β-sheet structure in the enzyme and in fact the structure can be described as consisting of two folded units (amino acids 27–112 and 133–230) in each of which there are six strands of antiparallel sheet forming a highly distorted hydrogen-bonded 'cylinder'. The hydrogen-bonding pattern in part of this structure is shown in Fig. 3.32(b). A fuller discussion of the structure of the enzyme has been given by Blow.[220]

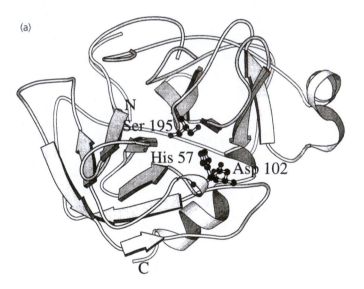

Fig. 3.32 The structure of α-chymotrypsin, an all-β protein. (a) A cartoon representation drawn in Molscript[216] of the structure of the bovine enzyme (2cha) with the residues His 57, Asp 102, and Ser 195 highlighted. (b) The arrangement of hydrogen bonds in part of the antiparallel sheet structure of the enzyme.

3.4.6.4 The structure of adenylate kinase (an α/β enzyme)

Adenylate kinase from *Escherichia coli* consists of a single polypeptide chain of 214 amino acids. Figure 3.33 shows a cartoon representation of the structure.[221] A deep cleft divides the molecule into two distinct domains. Altogether there are ten α-helices and seven strands of β-sheet in the enzyme so that a very high proportion (over two-thirds) of the amino acids in adenylate kinases from a variety of sources (including muscle)[222] are in some type of regular secondary structure. This may account for the great stability of the enzyme at low pH, which is exploited in the purification of the enzyme (see Chapter 2, Section 2.8.1).

3.4.6.5 The structure of thermolysin (an $\alpha + \beta$ enzyme)

Thermolysin is a zinc-dependent protease (see Chapter 5, Section 5.5.1.7) from the thermophile *Bacillus thermoproteolyticus*; the specificity of the enzyme has already been referred to in Section 3.3.2.2. Thermolysin consists of a single polypeptide chain of 316 amino acids. The structure of the enzyme has been solved by X-ray crystallography to high resolution[223,224] and is shown in Fig. 3.34 in a cartoon representation. Overall, the enzyme consists of two distinct domains separated by a deep cleft at the bottom of which the zinc atom is located. A stretch of α-helix (residues 137 to 150) runs parallel to the cleft through the centre of the molecule. The classification of thermolysin as an $\alpha + \beta$ protein arises from the fact that the C-terminal part of the enzyme contains 7 out of the total of 9 stretches of α-helix, and the N-terminal part contains 10 out of the total of 11 strands of β-sheet. Examination of the structure at high resolution reveals that thermolysin has a number of unusual features including a single turn of left-handed α-helix (residues 226 to 229).

Limited proteolysis of thermolysin by subtilisin leads to a derivative which retains a small amount of activity and consists of two tightly associated fragments of apparent M_r values 24 000 and 10 000.[225] The identified sites of proteolytic cleavage (between Thr 4 and Ser 5, between Thr 224 and Gln 225, and between Gln 225 and Asp 226) correspond to solvent-exposed loops in the protein, which possess high mobility.

The structures of a number of enzyme–inhibitor complexes have been solved allowing a number of features of the active site to be discerned and a mechanism

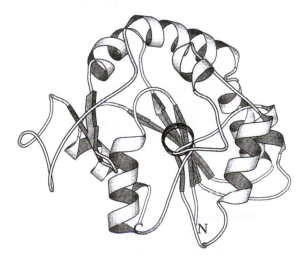

Fig. 3.33 The structure of adenylate kinase, an α/β protein. The enzyme from *Escerichi coli* (1ank) is shown in cartoon format using Molscript.[216]

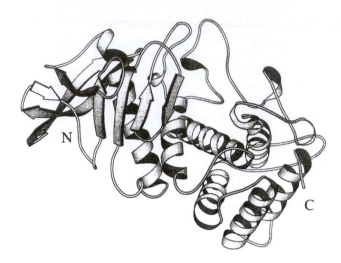

Fig. 3.34 Structure of thermolysin, an
$\alpha + \beta$ protein. The enzyme (4 tln) is
shown in cartoon form using
Molscript[216].

of action to be proposed.[226] This involves the attack of a water molecule activated by the side-chain of Glu 143 on the peptide bond to be cleaved. The essential zinc ion polarizes the carbonyl moiety of the peptide bond and may also
polarize the attacking water molecule (see Chapter 5, Section 5.5.1.7). From an
understanding of the action of a number of zinc-dependent proteases including carboxypeptidase and thermolysin, it was possible to design potential
inhibitors of another zinc-dependent protease angiotensin-converting enzyme
(ACE). ACE catalyses the removal of the C-terminal dipeptide from the inactive
decapeptide angiotensin I leading to the formation of the octapeptide angiotensin II. The latter peptide has a number of effects including the constriction
of blood vessels and thereby plays a key role in the control of blood pressure.
The inhibitors of ACE served as 'lead compounds' for the design of further
inhibitors; a number of these compounds such as lisinopril have found
extensive clinical applications and proved very valuable for the control of
hypertension.[174,226]

3.5 The determination of quaternary structure

* The definition of the term subunit
has been given in Section 3.1.

It has been found that most large enzymes consist of a number of subunits* held
together by non-covalent forces, i.e. they are *oligomeric*. In some ways the occurrence of multiple subunits in an enzyme can be considered analogous to the
occurrence of folded globular domains (Section 3.4.4.5) along a single polypeptide chain. The term quaternary structure refers to the arrangement of the
subunits in oligomeric enzymes (or proteins in general). In this section we shall
deal with the following questions that can be asked about oligomeric enzymes.
(i) How many subunits are there and of what type? (Section 3.5.1). (ii) How are
the subunits arranged? (Section 3.5.2). (iii) What forces are involved in holding
subunits together? (Section 3.5.3). (iv) What is the significance of multiple
subunits in an enzyme? (Section 3.5.4).

3.5.1 Number and type of subunits

In this section we shall discuss the various experimental approaches to determining the number and type of subunits in an oligomeric enzyme.

3.5.1.1 Studies of M_r

An indication that an enzyme consists of multiple subunits is provided by results of studies of M_r (Section 3.2) performed in the absence and presence of denaturing agents such as guanidinium chloride (Section 3.6.1). In favourable cases the number of subunits can be deduced directly from the data. For instance, alcohol dehydrogenase from yeast has an M_r of about 145 000 as determined by ultracentrifugation. However, under denaturing conditions a value of 36 000 is obtained, suggesting that the enzyme consists of four subunits that are probably identical. In this type of work it is particularly important to make sure that traces of proteases are absent from the enzyme preparations, since such proteases may well be active under the denaturing conditions and cause extensive fragmentation of the unfolded polypeptide chains of the enzyme of interest. In Chapter 7 (Sections 7.9 and 7.11.2) we shall refer to similar difficulties in the determination of the subunit structures of tryptophan synthase and of the *arom* multienzyme polypeptide protein.

3.5.1.2 Cross-linking studies

A second method for determining the number of subunits in an oligomeric enzyme relies on the use of cross-linking agents such as dimethylsuberimidate.[227] This compound reacts with pairs of lysine side-chains, which almost invariably occur on the surfaces of enzymes (Section 3.4.4.5) to form cross-linkages that are stable in the presence of denaturing agents (Fig. 3.35).

If we react an enzyme that consists of, say, four subunits with this cross-linking agent, a mixture of species will be formed in which different numbers of subunits are cross-linked (Fig. 3.36). The various species can be separated by sodium dodecylsulphate polyacrylamide gel electrophoresis (Section 3.2.3) giving a total of four bands.

The number of bands will give the number of subunits directly. It is clearly important in this work: (i) to avoid complete intersubunit cross-linking, otherwise the mixture of species shown in Fig. 3.36 will not result and only one band corresponding to an M_r of 4X will be seen on electrophoresis — this would be the case with glutaraldehyde as a cross-linking agent, for instance;[228] (ii) to avoid intermolecular cross-linking, otherwise large aggregates will be formed. These complications can be minimized by using low concentrations of cross-linking agent and of protein respectively.

3.5.1.3 Studies of ligand binding

The number of binding sites on an enzyme for a substrate or other ligand can often be used to indicate the number of subunits. For instance, the alcohol dehydrogenases from yeast and liver bind four and two moles of NADH per mole of

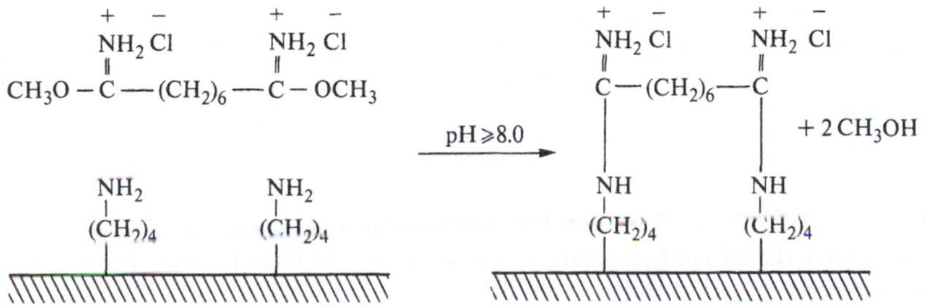

Fig. 3.35 Cross-linking of lysine side-chains by reaction with dimethylsuberimidate.

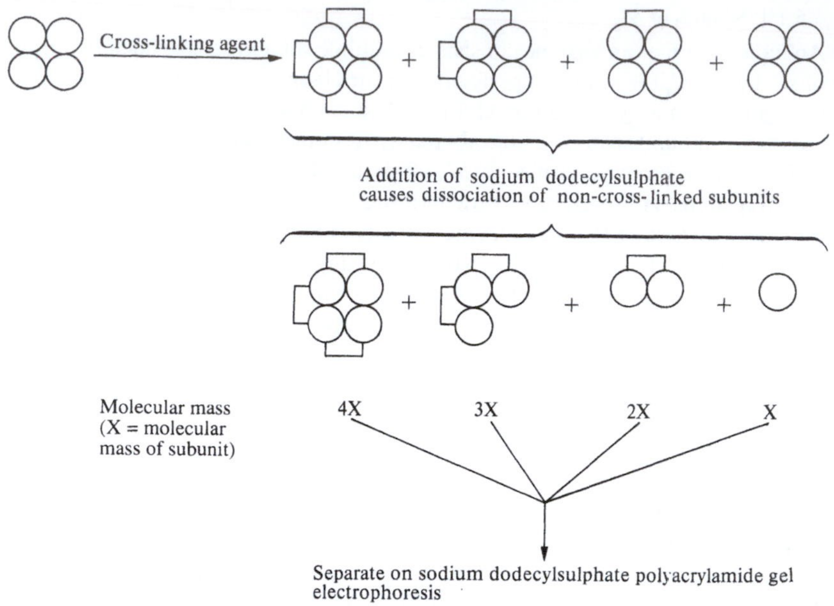

Fig. 3.36 The use of a cross-linking agent to determine the number of subunits in an oligomeric enzyme. For convenience the intrasubunit cross-linkages are not shown.

enzyme, respectively, in agreement with the tetrameric (four subunit) and dimeric (two subunit) structures of the respective enzymes.

3.5.1.4 Studies of symmetry

The type of symmetry possessed by an enzyme, deduced from X-ray crystallographic studies (Section 3.4.1), can sometimes be used to indicate the number of subunits in an enzyme or at least to exclude various proposed subunit structures. The observation of both a threefold and a twofold axis of symmetry in aspartate carbamoyltransferase from *E. coli*[229] excluded the previously proposed subunit structure of the enzyme (four catalytic plus four regulatory subunits). In conjunction with other results from studies of M_r and of the C-terminal amino acids of the subunits, these data indicated that the enzyme consisted of six catalytic and six regulatory subunits.

3.5.1.5 Identity of subunits

If, under denaturing conditions, an enzyme gives rise to a number of species that differ in M_r, it can be concluded that the enzyme contains non-identical subunits (assuming of course that the enzyme preparation is pure and that no proteolytic damage has occurred). In order to establish the relative numbers of each type of subunit, we can perform quantitative determinations of the N- and C-terminal amino acids of the various subunits (Section 3.3.2.4). The subunit composition can then be confirmed by checking that the observed M_r of the intact enzyme equals the sum of the M_r values of the component parts, e.g. the M_r of aspartate carbamoyltransferase from *E. coli* (300 000) corresponds to the sum of the M_r of six catalytic (6 × 34 000) and six regulatory (6 × 17 000) subunits. In certain cases, however, it is not difficult to 'miss' a subunit of very different size from the other subunits: thus, in the case of phosphorylase kinase (see Chapter 6, Section 6.4.2.1) a small calcium-binding subunit was only detected a number of years after the subunit structure of the enzyme had been thought to be firmly established.

When subunits appear to be identical in terms of M_r (and/or terminal amino acids), it is important to check on their identity in terms of amino-acid sequence. Because it is capable of measuring M_r values very accurately, mass spectrometry (Section 3.2.4) provides an excellent check on the identity of subunits.

Extensive compilation have been made of the subunit stoichiometries of oligomeric enzymes.[200] In the majority of such enzymes, subunits are identical with one another; some of the more notable examples of enzymes that contain non-identical subunits are listed in Table 3.5. It should also be noted that almost all oligomeric enzymes that consist of identical subunits contain even numbers of such subunits. There are, however, some well-documented examples of enzymes containing odd numbers of identical subunits, such as phospho-2-keto-3-deoxygluconate aldolase from *Pseudomonas putida*,[230] chloramphenicol acetyl-transferase from *E. coli*,[231] and ornithine carbamoyltransferase from *Bacillus subtilis*,[232] each of which is a trimeric enzyme with subunit M_r values of 24 000, 25 000, and 47 000, respectively. 4-Oxalocrotonate tautomerase from *Pseudomonas putida* is a pentamer of subunit M_r 7000[199] and heptameric rings of subunits are found in the chaperonin 60 protein[233] (see Section 3.6.2) and in the 20S proteasome complex (see Section 3.4.6.1).[217,234]

Finally, it should be remembered that some enzymes are capable of indefinite association. The examples of glutamate dehydrogenase[30] and acetyl-CoA carboxylase[32] have already been mentioned (Section 3.2.5). In such systems it is inappropriate to speak of a definite quaternary structure except perhaps under very carefully defined conditions.

3.5.2 Arrangement of subunits

The arrangement of the subunits in an oligomeric enzyme can usually be deduced from the type of symmetry possessed by the molecules (this is found by X-ray crystallography; Section 3.4.1). In general, it is found that the arrangement of the

Table 3.5 Some enzymes composed of non-identical subunits

Enzyme	Source	Subunit composition	Comments
Lactose synthase	Bovine mammary tissue	$\alpha\beta$	
Haemoglobin[a]	Human red blood cells	$\alpha_2\beta_2$	
Tryptophan synthase (see Chapter 7, Section 7.9.1)	*E. coli*	$\alpha_2\beta_2$	
cAMP-dependent protein kinase (see Chapter 6, Section 6.4.2.1)	Rabbit skeletal muscle	$\alpha_2\beta_2$	α and β represent catalytic and regulatory subunits, respectively
Aspartate carbamoyltransferase	*E. coli*	$\alpha_6\beta_6$	α and β represent catalytic and regulatory subunits, respectively. The molecule is assembled as $(\alpha_3)_2(\beta_2)_3$
Ribulosebisphosphate carboxylase	Spinach	$\alpha_8\beta_8$	
20S proteasome (see Section 3.4.6.1)	*Thermoplasma acidophilum*	$\alpha_{14}\beta_{14}$	The molecule is assembled as $\alpha_7\beta_7\beta_7\alpha_7$
RNA polymerase (see Chapter 7, Section 7.3)	*E. coli*	$\alpha_2\beta\beta'\sigma$	
Phosphorylase kinase (see Chapter 6, Section 6.4.2.1)	Rabbit skeletal muscle	$\alpha_4\beta_4\gamma_4\delta_4$	
Adenosinetriphosphatase	Beef-heart mitochondria	$\alpha_3\beta_3\gamma\delta\varepsilon$	

[a]Haemoglobin is included (although it is not an enzyme) since a great deal is known about the inter-subunit contacts in this protein (Section 3.5.3).

subunits is such as to maximize the number of intersubunit contacts, so that for a tetrameric enzyme such as lactate dehydrogenase the preferred geometrical arrangement will be tetrahedral, rather than square or linear.[235] In the case of a hexameric enzyme, the preferred arrangement will be octahedral.

The detailed arrangement of the subunits and the nature of the intersubunit contacts can only be decided when the complete three-dimensional structure of the enzyme has been obtained. The recent solutions of the structures of the 20S proteasomes from *Thermoplasma acidophilum*[217] and *Saccharomyces cerevisiae* (yeast),[234] each of which consist of 28 subunits of two main types, show how rapid progress has been in this area. However, even in systems which have not been solved by X-ray crystallography it is possible to learn something about the arrangement of the subunits by the use of cross-linking agents. This type of approach showed that the subunits of the Ca^{2+}- and Mg^{2+}-activated adenosine-triphosphatase of the inner membrane of *E. coli* were probably arranged as shown in Fig. 3.37. In this work, a reversible cross-linking agent, dithiobis(succinimidyl propionate) was used.[236] This bifunctional ester results in the cross-linking of pairs of lysine side-chains by two propionyl groups that are joined by a disulphide bond. It is thus possible to identify the components of the cross-linked species by treatment with dithiothreitol (Section 3.3.2) to break the disulphide bond, and subsequent electrophoresis in the presence of sodium dodecylsulphate. The results showed that it was possible to cross-link α and β subunits to each other or to the γ subunit. However, no α–α or β–β cross-links could be formed. Other evidence showed that the smaller δ and ε subunits were attached only peripherally (Fig. 3.37).

Support for this type of structure was subsequently obtained by the solution of the structure of the related F1 adenosinetriphosphatase from bovine mitochondria where a hexameric unit with alternating α and β subunits was arranged around a spindle consisting of the γ subunit.[237]

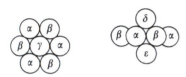

Top view   Side view

Fig. 3.37 Proposed arrangement of the Ca^{2+}- and Mg^{2+}-activated adenosinetriphosphatase from *E. coli*, as deduced by cross-linking studies.[236] The subunit stoichiometry of this enzyme is $\alpha_3\beta_3\gamma\delta\varepsilon$ (see Table 3.5).

3.5.3 Forces involved in the association between subunits

Oligomeric enzymes are usually dissociated into subunits by treatment with denaturing agents (Section 3.6.1) such as guanidinium chloride. The forces involved in the association of the subunits are thus of the weak, non-covalent type that are involved in stabilizing the folded structure of a polypeptide chain (Section 3.4.5), i.e. hydrogen bonds, electrostatic forces, van der Waals forces, and hydrophobic forces. From the results of X-ray crystallographic studies on oligomeric proteins, it is clear that surfaces involved in subunit association are predominantly (> 67 per cent) non-polar in nature. Only a small number of polar or charged atoms become inaccessible to solvent and most of these form hydrogen bonds across the subunit interface. However, formation of these hydrogen bonds is likely to make only a relatively small contribution to the energetics of the interaction, and the same is very likely to be true of the more numerous van der Waals interactions. By far the largest contribution comes from hydrophobic interactions. By considering the reduction in surface area accessible to solvent on association and using a figure based on the free energy of transfer of amino acids from a non-polar solvent to water, it is possible to show that, for instance, the non-polar interactions account for > 90 per cent of the observed energy of interaction of the subunits of the insulin dimer.[235,238,239]

Fisher has suggested that there is a strong correlation between the tendency for subunits to associate and the content of amino acids with non-polar side-chains.[240] Proteins that contain more than about 30 per cent non-polar amino acids are unable to bury the non-polar side-chains within the interior of a folded polypeptide chain. The exposed non-polar groups will then tend to interact with others on other subunits giving rise to an oligomeric system.

As mentioned previously (Section 3.4.5.5), hydrophobic forces do not confer any geometrical specificity to an interaction. Specific interactions between subunits must therefore rely on forces such as hydrogen bonding that have distinct geometrical requirements. The importance of interactions between particular side-chains in different subunits has been particularly well explored in the case of haemoglobin where a large number of naturally occurring mutants have been characterized.[241] In normal haemoglobin, the side-chain of tyrosine 35 on the β-chain forms a hydrogen bond with the side-chain of aspartic acid 126 on the α-chain. In haemoglobin Philly, the tyrosine is replaced by phenylalanine, which cannot form the hydrogen bond. This loss of a hydrogen bond leads to increased dissociation of haemoglobin Philly into subunits and is associated with clinical symptoms of mild anaemia. In tyrosyl-tRNA synthetase, which is dimeric, the replacement of phenylalanine 164, which lies near the subunit interface, by aspartic acid leads to dissociation of the subunits at pH 7.78; at this pH the aspartic acid carries a negative charge. At low pH, the side-chains are protonated and the subunits can associate.[242] In the case of the metJ repressor protein, the replacement of alanine 60 by the more bulky threonine leads to increased dissociation of the dimer.[243] It should be noted that formation of a small number of extra electrostatic interactions between the subunits of glyceraldehyde-3-phosphate dehydrogenase from *Bacillus stearothermophilus* leads to marked stabilization of the tetramer compared with the corresponding enzyme from lobster muscle.[244]

3.5.4 The significance of multiple subunits

There seem to be at least four possible reasons why the possession of multisubunit enzymes is advantageous to an organism. First, the presence of multiple subunits confers additional possibilities of regulating the catalytic activity of an enzyme. This is discussed in detail in Chapter 6, Section 6.2.2.1. Second, the assembly of different types of subunits into a large complex permits variation in catalytic properties; this is well illustrated by the examples of tryptophan synthase and RNA polymerase discussed in Chapter 7, Sections 7.9 and 7.3, respectively. Another example of this type of system is lactose synthase (Table 3.5), which consists of two different subunits. One of these subunits (β) is lactalbumin, which is a protein that is devoid of any catalytic activity but is able to alter the specificity of the other subunit (α), a galactosyltransferase, so that it can catalyse the synthesis of lactose rather than *N*-acetyllactosamine.[245]

α subunit
UDPgalactose + *N*-acetylglucosamine $\rightleftharpoons$ UDP + *N*-acetyllactosamine

αβ complex
UDPgalactose + D-glucose $\rightleftharpoons$ UDP + lactose.

Lactalbumin thus acts as a specifier protein that can alter the specificity of the α subunit so that when necessary (e.g. during lactation) lactose can be synthesized in preference to *N*-acetyllactosamine that is used in the biosynthesis of glycoproteins.

Many oligomeric enzymes do not display these properties; thus, lactate dehydrogenase does not show regulatory characteristics. In such cases we must look for another reason for the occurrence of multiple subunits. A third reason may well

* Information has been gained in this area using enzymes linked to insoluble supports, thereby permitting the preparation of matrix-bound subunits.[246] Other evidence for the existence of catalytically active subunits can be obtained from a study of the kinetics of regain of activity during refolding.[247]

be to confer stability on the enzyme. It is difficult to gain reliable data on this point because treatments that disrupt the bonds between subunits also tend to disrupt the folded structure of the subunits themselves, so that correctly folded isolated subunits are not easy to obtain.* The observation that the surfaces of subunits contain a number of non-polar side-chains (Section 3.5.3) suggests that correctly folded subunits, were they to occur, would, in general, be rather unstable.

A final reason for the occurrence of multiple subunits is that such association can generate large structures of defined geometry which may be required for particular biological functions. This can be achieved with considerable economy of genetic information if multiple copies of a single (or small number of) type(s) of subunit are involved in the association process. For instance, the association of the 14 identical subunits of the chaperonin 60 protein (see Section 3.6.2) in two rings of seven generates a central cavity of a size large enough to accommodate a partially folded polypeptide chain and thus facilitate its further correct folding.[233] The proteasome complex (see Section 3.4.6.1) consists of 14 α- and 14 β-type chains, arranged as stacked rings of 7.[217,234] The active sites are in the interior of the particle and are accessible by narrow side-entrances. The different forms of the subunits of the complex act in a coordinated way to break down proteins.

3.6 The unfolding and folding of enzymes

In Sections 3.4 and 3.5 we discussed the three-dimensional structures adopted by enzymes. It is now appropriate to consider how the three-dimensional structure of an enzyme can be perturbed (Section 3.6.1) and how the structure is acquired (Section 3.6.2). Section 3.6.3 deals with methods of predicting the three-dimensional structure of an enzyme.

3.6.1 Unfolding of enzymes

† This does not apply to enzymes isolated from extremophiles, which are usually much more stable than their counterparts. In the case of thermophilic enzymes, it appears in many cases that the additional stabilization results from an increased number of electrostatic interactions within and between subunits;[167,244] see Section 3.5.3.

‡ Chaotropic agents are ions of low charge density, e.g. trichloroacetate, that disorder water structure and facilitate the transfer of non-polar groups to the solvent water.

The compact folded form of an enzyme is generally thermodynamically more stable than the modified form by only a relatively small margin[248] of the order of 20–60 kJ mol^{-1} (Section 3.4.5.5). It is therefore not surprising that most enzymes† can be relatively easily unfolded (and dissociated in the case of multisubunit enzymes) by a variety of conditions such as extremes of pH, heating, and addition of organic solvents, detergents, chaotropic agents,‡ or high concentrations of urea or guanidinium chloride. Loss of the three-dimensional structure of the folded enzyme is known as denaturation and agents that bring about denaturation are known as denaturing agents. Denaturation of an enzyme can be monitored by the loss of catalytic activity, by changes in spectroscopic parameters such as circular dichroism (Section 3.4.2) or fluorescence.

The modes of action of denaturing agents have been extensively investigated.[249,250] Heating will disrupt the folded structure of an enzyme by increasing the vibrational and rotational motions of atoms. A change in pH will affect the state of ionization of amino-acid side-chains that may well be involved in the maintenance of the folded structure. Addition of organic solvents or detergents would be expected to unfold an enzyme by interacting with the non-polar amino-acid side-chains previously buried in the interior of the folded structure. Careful measurements have shown that both urea and guanidinium chloride also act by increasing the solubility of the non-polar amino-acid side-chains[251] while maintaining the hydrogen-bonding capacity of the aqueous solvent.[252] Guanidinium chloride has some advantages over urea as a denaturing agent, since solutions of the latter slowly decompose to yield cyanate.[253]

In many cases that have been investigated, the transition between folded and unfolded states seems to be relatively sharp, i.e. there is a small range of concentration of denaturing agent or range of temperature over which the enzyme changes more or less completely from one form to the other. In such cases it has proved satisfactory to employ a two-state model, in which only the fully folded and fully unfolded states are considered, to analyse quantitatively the thermodynamic and some of the kinetic aspects of denaturation.[248,254,255]

However, in a significant number of cases, unfolding clearly involves intermediate states which can be populated at certain concentrations of denaturants, etc. One of the most extensively investigated of these intermediate states is the so-called molten globule, first characterized by Ptitsyn and his group in studies of α-lactalbumin at low pH.[256] In this state the protein is compact and retains the majority of its native secondary structure but has lost the interactions which stabilize the tertiary structure, i.e. the side-chains are fluid or 'molten'. There is a good deal of evidence to suggest that the molten globule state may be an important intermediate in protein folding[257,258] and may play a part in dictating the specificity of certain protein–ligand interactions.[259]

3.6.2 Folding of enzymes

In the early 1960s, Anfinsen performed an important experiment that showed that a denatured enzyme could regain its correct folded structure when the denaturing agent was removed.[260] In outline, Anfinsen's experiment (Fig. 3.38) involved treating pancreatic ribonuclease with 2-mercaptoethanol to break the four disulphide bonds (Section 3.3.2) and then adding 8 mol dm^{-3} urea to produce inactive, unfolded, reduced enzyme. When the urea and reducing agent were removed by dialysis and the enzyme was oxidized to re-form the disulphide bonds, it was found that almost 100 per cent of the original enzyme activity was regained. Anfinsen argued that the complete regain of activity showed that the refolded enzyme possessed the correct disulphide bonds and therefore that the polypeptide chain had folded in such a way as to bring the correct pairs of cysteine side-chains into juxtaposition. There are 105 possible ways ($8!/2^4 \times 4!$) in which four disulphide bonds can be formed from eight cysteine side-chains,[261] so formation of the disulphide bonds on a random basis would lead to the regain of only about 1 per cent enzyme activity. The Anfinsen experiment led to the conclusion that the primary structure of an enzyme contained the information required to specify its three-dimensional structure.

Later work showed that the refolding of unfolded, reduced, ribonuclease is in fact a rather complex process. Initially a number of distinct intermediates are

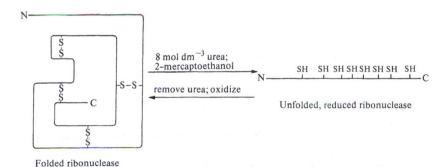

Folded ribonuclease

Fig. 3.38 The reduction and re-oxidation of pancreatic ribonuclease.

observed in which different numbers of incorrect disulphide bonds are formed; these intermediates slowly rearrange to give the final, active product containing the correct disulphide bonds.[262] The enzyme protein disulphide-isomerase catalyses the conversion of those species containing incorrect disulphide bonds to that containing the correct bonds. In the cell, this enzyme (which occurs in the lumen of the endoplasmic reticulum; see Chapter 8, Section 8.2.5) ensures that newly synthesized proteins will contain the correct disulphide bonds.[263]

Since detailed structural studies of the mechanism of protein folding during biosynthesis are technically very difficult, the refolding of denatured proteins has been widely studied as a convenient experimental model for this process. In the cases of small single-domain proteins such as myoglobin or ribonuclease, there are two principal reasons for believing that the refolding process is a satisfactory model: (a) the rate of refolding is comparable with the rate at which proteins are presumed to fold *in vivo*, i.e. a few seconds, and (b) the final product is indistinguishable from the original native protein.[264] The folding process clearly does not involve a random search of all possible conformations until the most stable one is found since this would take an impossibly long time (calculated as over 10^{60} years for a polypeptide chain of 100 amino acids, compared with the age of the earth which is only 4.6×10^9 years). The current view is that for most proteins, the folding process involves a number of intermediates.[264] Early events are likely to involve the formation of small local elements of structure such as α-helix or β-sheet (Section 3.4.4); at about the same time the polypeptide chain collapses in order to bury the hydrophobic side-chains. The resulting compact possibly resembles the molten globule state (Section 3.6.1).[257] In the next stages, the elements of secondary structure could then act as nucleation centres to direct further folding of the polypeptide chain before the tertiary interactions form and the biologically active final folded state is reached.[264-267]

The refolding of more complex proteins consisting of multiple domains (Section 3.4.4.5) or multiple subunits (Section 3.5) is much less efficient, both in terms of rate and yield.[264] Indeed, some proteins, for instance the two-domain monomeric enzyme rhodanese or the hexameric enzyme glutamate dehydrogenase, have defied all attempts to regain activity on refolding after denaturation. In these cases formation of aggregated protein is the predominant process. The formation of inclusion bodies when a number of eukaryotic proteins are expressed in *E. coli* (see Chapter 2, Section 2.4.1) also reflects the inability of certain proteins to fold efficiently. Since the late 1980s, it has become clear that the correct folding of proteins in the cell involves the intervention of a number of proteins collectively termed 'chaperones', which do not form part of the final folded structure.[268,269] The mechanism of action of chaperonin 60, closely related forms of which occur in bacteria, mitochondria, and chloroplasts, has been investigated in detail.[233,269,270] In essence, its role is to bind to hydrophobic surfaces of folding intermediates of a target protein so as to prevent improper interactions which would lead to aggregated protein. In an ATP-dependent process also involving the smaller chaperonin 10, the target protein undergoes cycles of binding to and dissociation from the chaperonin 60 as folding progresses. Folding occurs within a cavity 5 nm in diameter within the double 'doughnut' structure of the chaperonin 60 protein (see Section 3.5.4). This cavity is the location where a single polypeptide chain of the target protein can fold without encountering other copies of the polypeptide chain and has been termed the 'Anfinsen cage'.

Netzer and Hartl[271] have investigated the expression of an artificial two-domain modular polypeptide (H-Ras linked to dihydrofolate reductase) in both eukaryotic and prokaryotic host systems. In the eukaryotic host, e.g. reticulocyte lysate, the polypeptide folds efficiently by sequential folding of the domains, i.e. the folding is co-translational. When the same polypeptide was expressed in *E. coli*, the folding was post-translational, and this leads to intramolecular misfolding of concurrently folded domains. This gives an explanation of the formation of insoluble protein in the form of inclusion bodies when many eukaryotic genes are expressed in *E. coli* (see Chapter 2, Section 2.4.1).

In addition to the chaperones, two enzyme systems play roles in protein folding in the cell. Protein disulphide isomerase is involved in ensuring the correct formation of disulphide bonds.[263,272] Peptidyl prolyl isomerase (see Section 3.3.2.11) catalyses the *cis/trans* isomerization of peptide bonds to proline residues in the polypeptide chain (Section 3.4.4.1); this process may be a rate-determining step in the folding of certain proteins.[273,274]

Clearly, the folding of proteins, both *in vivo* and *in vitro*, is a complex process and has been reviewed extensively.[161,265,268,269,275] There are several reasons for the current interest in protein folding: first, the intrinsic scientific interest of the 'second half of the genetic code'; second, a number of commercially important protein products have to be recovered from inclusion bodies (see Chapter 2, Section 2.8.7); and third, a number of disease states such as cystic fibrosis, α_1-antitrypsin deficiency, and Creutzfeld–Jacob disease are linked to the formation and accumulation of misfolded proteins.[161,276,277]

3.6.3 Structure-prediction methods

The experiment outlined in Fig. 3.38 shows that the amino-acid sequence of a protein specifies its three-dimensional structure, although as indicated by the experiments outlined in Section 3.6.2 it may well be that other proteins (e.g. chaperones) are required for the protein in question to be able to achieve the folded structure. For many years, biochemists have been investigating the possibility of predicting the three-dimensional structure merely from the sequence data. For a new protein which has at least 25 per cent sequence identity to one of known three-dimensional structure, the method of homology modelling[278] can be used successfully to predict the structure of the new protein. If, however, there is no suitable homologue in the database, prediction methods must be used. The most rigorous approach would be one in which the methods of quantum mechanics are used to calculate the most stable conformation of a polypeptide chain by taking into account all possible interactions between the atoms of the chain including the non-bonded ones, as well as the interactions with the solvent. This approach is extremely laborious and useful results have been obtained only for very short peptide chains. More success has been obtained with a variety of semi-empirical approaches in which the database of known three-dimensional protein structures are used to guide the prediction methods. In the first types of such methods, the database of structures was analysed to determine the probability that a certain amino acid, or sequence of amino acids, occurred in a particular element of regular structure on a more than random basis. From this analysis a set of rules could then be drawn up and used to predict structural features in other proteins. In one of the most widely used of such methods (developed by Chou and Fasman[279]) the different amino acids are rated according to the likelihood that they occur in α-helices, β-sheets, or β-turns. For instance, glutamic acid shows a strong preference for α-helices, a very low preference for β-sheets, and a middle to low preference for β-turns; proline and glycine have very low preferences for α-helices. The empirical Chou–Fasman rule for α-helix formation is that a cluster of four 'helix-forming' amino acids out of a sequence of six will nucleate a helix; the helical segment will extend in both directions until terminated by a sequence of four 'helix-breaking' amino acids. Analogous rules for the formation of β-sheet and β-turn structures were also given.[279]

The Chou–Fasman and other prediction methods, such as the GOR (Garnier–Osguthorpe–Robson) method based on ideas of information theory,[280] have been able to predict the elements of regular secondary structure (α-helix and β-sheet) with a success rate of about 60 per cent or slightly higher.[281] There is some evidence that the use of combinations of separate prediction methods can give more reliable results.[282] The fact that the success rates of the prediction methods are only relatively modest implies that while short-range forces, extending over a few amino acids in the sequence, are important in specifying the final folded structure of a protein, the contribution of long-range forces between amino acids more remote in the sequence cannot be neglected. In an attempt to improve the success rate of prediction methods, Rost and Sander have developed a range of profile-based neural network methods[283,284] in which information from evolutionary comparisons between proteins is used. In the first step of these methods, the BLAST program (Section 3.3.2.11) is used to search the protein sequence database in order to generate a multiple sequence alignment based on the sequence of the new protein. This initial alignment is then refined using a different program. It is important for the success of the prediction procedure that the alignment contains a broad spectrum of homologous sequences. In the second step, the alignment is fed into a neural network system in which both local and global structural information is employed to give the predicted structure. A final filter is used to correct what are deemed to be obvious unrealistic predictions, for instance where a run of helical residues is interrupted by a single amino acid predicted to be in a β-sheet. Different versions of the neural network method (termed PHD) can be used to give for each amino acid in the sequence the type of secondary structure (helix, sheet, or loop) and its accessibility to solvent; it is also possible to predict transmembrane helices. The success rate of the PHD methods (72% for secondary structure; 74% for accessibility; 95% for transmembrane helices) is significantly higher than the earlier prediction methods. However, it is still true to say that we are some way from being able to predict the structure of a protein from knowledge of the amino-acid sequence alone.*

* The success of the various structure-prediction methods have been tested in the CASP (Critical Assessment of Structure Prediction) projects. In the first of these exercises the sequences of 33 proteins were given to a number of groups in the structure-prediction field *before* the results of X-ray crystallography or NMR spectroscopy on these proteins were revealed. The actual structures were then compared with the various predicted structures of these proteins in order to assess the usefulness of the prediction models.[285] Further information on the CASP experiments is given in the Appendix to this chapter.

3.7 Concluding remarks

The last two decades have witnessed an explosive growth in the amount of information concerning the structures of enzymes. This has been particularly true for the three-dimensional structures revealed by X-ray crystallography. We are now in a position to undertake detailed comparisons between families of enzymes, to understand the mechanism of enzyme catalysis in much greater detail than was previously possible, and to use this information as the basis for the rational alteration of enzyme properties or design of ligands. Many of the succeeding chapters in this book will illustrate the substantial progress that has been made in these directions.

Probably the aspect of protein structure over which the greatest question mark remains is that of the refolding of enzymes (Section 3.6.2) and the nature of the relationship between the amino-acid sequence of an enzyme and its three-dimensional structure. The improvement in structure-prediction methods when structural and sequence data are combined as in the neural network methods

(Section 3.6.3) shows how the increasing amount of information in the databases can be harnessed. It is reasonable to expect that, as this trend continues, more light will be shed on these intriguing questions.

References

1. Bairoch, A. and Boekmann, B., *Nucl. Acids Res.* **22**, 3578 (1994).

2. Wess, T. J., *Biotech. Appl. Biochem.* **26**, 127 (1997).

3. *Methods Enzymol.* **183** (1990).

4. *Methods Enzymol.* **266** (1996).

5. Byron, O., in *Proteins Labfax* (Price, N. C., ed.). Bios Scientific Publishers, Oxford (1996). See Ch. 16A.

6. Price, N. C. and Dwek, R. A., *Principles and problems in physical chemistry for biochemists* (2nd edn). Clarendon Press, Oxford (1979). See p. 73.

7. Lee, J. C. and Timasheff, S. N., *Biochemistry* **13**, 257 (1974).

8. Perkins, S. J., *Eur. J. Biochem.* **157**, 169 (1986).

9. Aune, K. C., *Methods Enzymol.* **48**, 163 (1978).

10. Harding, S. E. and Rowe, A., in *Enzymology Labfax* (Engel, P. C., ed). Bios Scientific Publishers, Oxford (1996). See p. 66.

11. Andrews, P., *Methods Biochem. Anal.* **18**, 1 (1970).

12. Scopes, R. K., *Protein purification: principles and practice* (3rd edn). Springer, New York (1994). See p. 238ff.

13. Aldaheff, J. A., *Biochem. J.* **173**, 315 (1978).

14. Nimmo, G. A. and Cohen, P., *Eur. J. Biochem.* **87**, 341 (1978).

15. Mann, K. G. and Fish, W. W., *Methods Enzymol.* **26**, 28 (1972).

16. Regnier, F. F., *Methods Enzymol.* **91**, 137 (1983).

17. Hedrick, J. L. and Smith, A. J., *Arch. Biochem. Biophys.* **126**, 155 (1968).

18. Weber, K., Pringle, J. R., and Osborn, M., *Methods Enzymol.* **26**, 3 (1972).

19. Hames, B. D. and Rickwood, D., *Gel electrophoresis of proteins: a practical approach* (2nd edn). IRL Press, Oxford (1990). See p. 117ff.

20. Laemmli, U. K., *Nature, Lond.* **227**, 680 (1970).

21. Morris, H. R., Williams, D. H., and Ambler, R. P., *Biochem. J.* **125**, 189 (1971).

22. Williams, D. H., Bradley, C. V., Santikarn, S., and Bojesen, G., *Biochem. J.* **201**, 105 (1982).

23. Seifert, W. E. Jr. and Caprioli, R. M., *Methods Enzymol.* **270**, 453 (1996).

24. Pitt, A. R., in *Proteins Labfax* (Price, N. C., ed.). Bios Scientific Publishers, Oxford (1996). See Ch. 16A.

25. Banks, J. F. Jr. and Whitehouse, C. M., *Methods Enzymol.* **270**, 486 (1996).

26. Beavis, R. C. and Chait, B. T., *Methods Enzymol.* **270**, 519 (1996).

27. Yates, J. R., *Methods Enzymol.* **271**, 351 (1996).

28. Nairn, J., Price, N. C., Kelly, S. M., Rigden, D., Fothergill-Gilmore, L. A., and Krell, T., *Biochim. Biophys. Acta* **1296**, 69 (1996).

29. Lightwahl, K. J., Schwartz, B. L., and Smith, R. D., *J. amer. chem. Soc.* **116**, 5271 (1994).

30. Reisler, E., Pouyet, J., and Fisenberg, H., *Biochemistry* **9**, 3095 (1970).

31. Khew-Goodall, Y. S., Johannsen, W., Attwood, P. V., Wallace, J.C., and Keech, D. B., *Arch. Biochem. Biophys.* **284**, 98 (1991).

32. Lane, M. D., Moss, J., and Polakis, S. F., *Curr. Top. Cell. Reg.* **8**, 139 (1974).

33. Cohen, R. and Mire, M., *Eur. J. Biochem.* **23**, 267 (1971).

34. Cohen, R. and Mire, M., *Eur. J. Biochem.* **23**, 276 (1971).

35. Rogers, K. S., Hellerman, L., and Thompson, T. E., *J. biol. Chem.* **240**, 198 (1965).

36. Price, N. C. and Jaenicke, R., *FEBS Lett.* **143**, 283 (1982).

37. Leland, S. G. and Zimmer, T.-L., *Essays Biochem.* **9**, 31 (1973).

38. IUPAC–IUB Joint Commission on Biochemical Nomenclature: Recommendations (1983). Reprinted in *Biochem. J.* **219**, 345 (1984).

39. Wess, T. J., Miller, A., and Bradshaw, J. P., *J. molec. Biol.* **213**, 1 (1990).

40. Liu, T.-Y. and Chang, Y. H., *J. biol. Chem.* **246**, 2842 (1971).

41. Hugli, T. E. and Moore, S., *J. biol. Chem.* **247**, 2828 (1972).

42. Inglis, A. S., *Methods Enzymol.* **91**, 137 (1983).

43. Spackman, D. H., Stein, W. H., and Moore, S., *Analyt. Chem.* **30**, 1190 (1958).

44. Svasti, J., *Trends Biochem. Sci.* **5**, VIII (January 1980).

45. Bottom, C. B., Hanna, S. S., and Siehr, D. J., *Biochem. Educ.* **6**, 4 (1978).

46. Udenfriend, S., Stein, S., Böhlen, P., Dairman, W., Leimgruber, W., and Weigele, M., *Science, NY* **178**, 871 (1972).

47. Benson, J. R. and Hare, P. F., *Proc. natn. Acad. Sci. USA* **72**, 619 (1975).

48. Simons, S. S. Jr. and Johnson, D. F., *J. amer. chem. Soc.* **98**, 7098 (1976).

49. Böhler, P., *Methods Enzymol.* **91**, 17 (1983).

50. Chang, J.-Y., Knecht, R., and Braun, D. G., *Methods Enzymol.* **91**, 41(1983).

51. Glazer, A. N., DeLange, R. J., and Sigman, D. S., in *Laboratory techniques in biochemistry and molecular biology*

(Work, T. S. and Work, E., eds), Vol. 4, p. 3. North-Holland, Amsterdam (1976). See p. 13ff.

52. Glazer, A. N., DeLange, R. J., and Sigman, D. S., in *Laboratory techniques in biochemistry and molecular biology* (Work, T. S. and Work, E., eds), Vol. 4, p. 3. North-Holland, Amsterdam (1976). See p. 21.

53. Tang, J., Sepulveda, P., Marciniszyn, J. Jr *et al.*, *Proc. natn. Acad. Sci. USA* **70**, 3437 (1973).

54. Hatch, F. T., *Nature, Lond.* **206**, 777 (1965).

55. Sandermann, J. Jr and Strominger, J. L., *Proc. natn. Acad. Sci. USA* **68**, 2441 (1971).

56. Price, N. C., in *Enzymology Labfax* (Engel, P. C., ed.). Bios Scientific Publishers, Oxford (1996). See p. 34.

57. Sanger, F. and Thompson, E. O. P., *Biochem. J.* **53**, 366 (1953).

58. Maxam, A. M. and Gilbert, W., *Proc. natn. Acad. Sci. USA* **74**, 560 (1977).

59. Sanger, F., Nicklen, S., and Coulson, A. R., *Proc. natn. Acad. Sci. USA* **74**, 5463 (1977).

60. Allen, G., in *Laboratory techniques in biochemistry and molecular biology* (Work, T. S. and Burdon, R. H., eds), Vol. 9. North-Holland, Amsterdam (1981).

61. *Methods Enzymol.* **152** (1987).

62. Old, R. W. and Primrose, S. B., *Principles of gene manipulation* (5th edn). Blackwell, Oxford (1994).

63. Sambrook, J., Fritsch, E. F., and Maniatis, T., *Molecular cloning: a laboratory manual* (2nd edn). Cold Spring Harbor, New York (1989).

64. Bradley, A. F., *Pure and Applied Chem.* **68**, 1907 (1996).

65. Ambler, R. P. and Scott, G. K., *Proc. natn. Acad. Sci. USA* **75**, 3732 (1978).

66. Sutcliffe, J. G., *Proc. natn. Acad. Sci. USA* **75**, 3737 (1978).

67. Blattner, F. R., Plunkett, G., Bloch, C. A., Perna, N. T., Burland, V., Riley, M. *et al.*, *Science NY* **277**, 1453 (1997).

68. Tomb, J.-F., White, O., Kerlarage, A. R., Clanton, R. A., Sulton, G. G., Fleischmann, R. D. *et al.*, *Nature, Lond.* **388**, 539 (1997).

69. Kunst, F., Ogasawara, N., Moszer, I., Albertini, A. M., Alloni, G., Azevedo, V. *et al.*, *Nature, Lond.* **390**, 249 (1997).

70. Goffeau, A., Aert, R., Agostini-Carbone, M. L., Ahmed, A., Aigle, M., Alberghina, L. *et al.*, *Nature, Lond.* **387** (Suppl.), 5 (1997).

71. Clayton, R. A., White, O., Ketchum, K. A. and Venter, J. C., *Nature, Lond.* **387**, 459 (1997).

72. Keil, B., *Enzymes* (3rd edn) **3**, 249 (1971). See p. 263.

73. Mitchell, W. M. and Harrington, W. F., *Enzymes* (3rd edn) **3**, 699 (1971). See p. 710.

74. Blow, D. M., *Enzymes* (3rd edn) **3**, 185 (1971). See p. 205.

75. Hartley, B. S. and Shotton, D. M., *Enzymes* (3rd edn) **3**, 323 (1971). See p. 333.

76. Glazer, A. N. and Smith, E. L., *Enzymes* (3rd edn) **3**, 501 (1971). See p. 519.

77. Aleclo, M. R., Dann, M. L., and Lowe, G., *Biochem. J.* **141**, 495 (1974).

78. Matsubara, H. and Feder, J., *Enzymes* (3rd edn) **3**, 721(1971). See p. 781.

79. Houmard, J. and Drapeau, G. R., *Proc. natn. Acad. Sci. USA* **69**, 3506 (1972).

80. Fruton, J. S., *Enzymes* (3rd edn) **3**, 119 (1971). See p. 140.

81. Butler, P. J. G., Harris, J. I., Hartley, B. S., and Leberman, R., *Biochem. J.* **103**, 78P (1967).

82. Dixon, H. B. F. and Perham, R. N., *Biochem. J.* **109**, 312 (1968).

83. Thomas, J. O., in *Companion to biochemistry* (Bull, A. T., Lagnado, J. R., Thomas, J. O., and Tipton, K. F., eds), Vol. 1. Longman, London (1974). See p. 101.

84. Gross, F., *Methods Enzymol.* **11**, 238 (1967).

85. Heil, A., Müller, G., Noda, L., Pinder, T., Schirmer, H., Schirmer, I. *et al.*, *Eur. J. Biochem.* **43**, 131 (1974).

86. Stark, G. R., *Methods Enzymol.* **47**, 129 (1977).

87. Mahoney, W. C., Smith, P. K., and Hermodson, M. A., *Biochemistry* **20**, 443 (1981).

88. Spande, T. F., Witkop, B., Degani, Y., and Patchornik, A., *Adv. Protein Chem.* **24**, 97 (1970).

89. Waterfield, M. D., in *Practical protein chemistry: a handbook* (Darbre, A., ed.). Wiley, Chichester (1986). See Ch. 6.

90. Gray, W. R., *Methods Enzymol.* **25**, 121 (1972).

91. Schroeder, W. A., *Methods Enzymol.* **25**. 138 (1972).

92. Ambler, R. P., *Methods Enzymol.* **25**, 143 (1972).

93. Ward, C. W., in *Practical protein chemistry: a handbook* (Darbre, A., ed.). Wiley Chichester (1986). See Ch. 18.

94. Edman, P. and Begg, G., *Eur. J. Biochem.* **1**, 80 (1967).

95. Martinez, A., Knappskog, P. M., Olafsdottir, S., Døskeland, A. P., Eiken, H. E., Svebak, R. M. *et al.*, *Biochem. J.* **306**, 589 (1995).

96. Waterfield, M. D., Scrace, G., and Totty, N., in *Practical protein chemistry: a handbook* (Darbre, A., ed.). Wiley, Chichester (1986). See Ch. 13.

97. Hunkapiller, M. W. and Hood, L. E., *Methods Enzymol.* **91**, 486 (1983).

98. Hunkapiller, M. W., Hewick, R. M., Dreyer, W. J., and Hood, L. E., *Methods Enzymol.* **91**, 399 (1983).

99. Shotton, D. M. and Hartley, B. S., *Nature, Lond.* **225**, 802 (1970).

100. Narita, K., in *Protein sequence determination* (Needleman, S. B., ed.), p. 25. Chapman and Hall, London; Springer, Berlin (1970). See p. 82.

101. Aitken, A., in *Proteins Labfax* (Price, N. C., ed.). Bios Scientific Publishers, Oxford (1996). See Ch. 23.

102. Welinder, K. G., *Eur. J. Biochem.* **96**, 483 (1979).

103. Carr, S. A., Biemann, K., Shoji, S., Parmelee, D. C., and Titani, K., *Proc. natn. Acad. Sci. USA* **79**, 6128 (1982).

104. Offord, R. E., *Nature, Lond.* **211**, 591 (1966).

105. Perkins, R. E., Conroy, S. C., Dunbar, B., Fothergill, L. A., Tuite, M. F., Dobson, M. J. *et al.*, *Biochem. J.* **211**, 199 (1983).

106. Bryant, T. N., Watson, H. C., and Wendell, P. J., *Nature, Lond.* **247**, 14 (1974).

107. Scopes, R. K., *Biochem. J.* **122**, 89 (1971).

108. Fifis, T. and Scopes, R. K., *Biochem. J.* **175**, 311 (1978).

109. Yoshida, A., *Analyt. Biochem.* **49**, 320 (1972).

110. Dobson, M. J., Tuite, M. F., Roberts, N. A., Kingsman, A. J., Kingsman, S. M., Perkins, R. E. *et al.*, *Nucl. Acids Res.* **10**, 2625 (1982).

111. Fowler, A. V. and Zabin, I., *J. biol. Chem.* **253**, 5521 (1978).

112. Campbell, J. W., Watson, H. C., and Hodgson, G. I., *Nature, Lond.* **250**, 301 (1974).

113. Winn, S. I., Watson, H. C., Fothergill, L. A., and Harkins, R. N., *Biochem. Soc. Trans.* **5**, 657 (1977).

114. Slotboom, A. J. and de Haas, G. H., *Biochemistry* **14**, 5394 (1975).

115. Blobel, G. and Dobberstein, B., *J. cell Biol.* **67**, 835 (1972).

116. Hurt, E. C. and van Loon, A. P. G. M., *Trends Biochem. Sci.* **11**, 204 (1986).

117. Engelmann, D. M., Steitz, T. A., and Goldman, A, *A. Rev. Biophys. Biophys. Chem.* **15**, 321 (1986).

118. Bairoch, A. and Bucher, P., *Nucl. Acids Res.* **22**, 3583 (1994).

119. Takahashi, N., Hayano, T., and Suzuki, M., *Nature, Lond.* **337**, 473 (1989).

120. Fischer, G., Wittmann-Liebold, B., Lang, K., Kiefhaber, T., and Schmid, F.X., *Nature, Lond.* **337**, 476 (1989).

121. Schreiber, S. L., *Science NY* **251**, 283 (1991).

122. Madden, T. L., Tatusov, R. L. and Zhang, J., *Methods Enzymol.* **266**, 131 (1996).

123. Pearson, W. R., *Methods Enzymol.* **266**, 277 (1996).

124. Sturrock, S. S. and Collins, J. F., *MPsrch version 1.3*. Biocomputing Research Unit, University of Edinburgh (1993). See also Shomer, B., Harper, R. A. L., and Cameron, G. N., *Methods Enzymol.* **266**, 3 (1996).

125. Taylor, W. R., *Methods Enzymol.* **266**, 343 (1996).

126. Dayhoff, M. O., Barker, W. C., and Hunt, L. T., *Methods Enzymol.* **91**, 524 (1983).

127. Smith, E. L., *Enzymes* (3rd edn) **1**, 267 (1970).

128. Teng, D.-F. and Doolittle, R. F., *Methods Enzymol.* **266**, 368 (1996).

129. *Methods Enzymol.* **266** (1996), see pp. 343–494 (many articles).

130. Barker, W. C., *Methods Enzymol.* **266**, 59 (1996).

131. Rawlings, N. D. and Barrett, A. J., *Methods Enzymol.* **244**, 461 (1994).

132. Irwin, D. M., *Methods Enzymol.* **224**, 552 (1993).

133. Patthy, L., *Cell* **41**, 657 (1985).

134. Markland, F. S. Jr and Smith, E. L., *Enzymes* (3rd edn) **3**, 561 (1971).

135. Kleanthous, C., Deka, R., Davis, K., Kelly, S. M., Cooper, A., Harding, S. E. *et al.*, *Biochem. J.* **282**, 687 (1992).

136. Blundell, T. L. and Johnson, L. N., *Protein crystallography*. Academic Press, New York (1976).

137. *Methods Enzymol.* **276** (1997).

138. *Methods Enzymol.* **277** (1997).

139. Glusker, J. P. and Trueblood, K. N., *Crystal structure analysis: a primer* (2nd edn). Oxford University Press (1985).

140. Drenth, L., *Principles of protein X-ray crystallography*. Springer, Berlin (1994).

141. Green D. W., Ingram, V. M., and Perutz, M. F., *Proc. R. Soc.* **A225**, 287 (1954).

142. Ke, H. M., *Methods Enzymol.* **276**, 448 (1997).

143. Rould, M. A., *Methods Enzymol.* **276**, 461 (1997).

144. Navaza, J. and Saludjian, P., *Methods Enzymol.* **276**, 581 (1997).

145. Blundell, T. L. and Johnson, L. N., *Protein crystallography*. Academic Press, New York (1976). See Ch. 12.

146. Brünger, A. T., Kuriyan, J., and Karplus, M., *Science NY* **235**, 435 (1987).

147. Kleywegt, G. J. and Jones, T. A., *Methods Enzymol.* **277**, 208 (1997).

148. Lamzin, V. S. and Wilson, K. S., *Methods Enzymol.* **277**, 269 (1997).

149. Kendrew, J. C., Dickerson, R. E., Strandberg, B. E., Hart, R. G., Davies, D. R., Phillips, D. C. *et al.*, *Nature, Lond.* **185**, 422 (1960).

150. Artymiuk, P. J., Blake, C. C. F., Grace, D. E. P., Oatley, S. J., Phillips, D. C., and Sternberg, M. J. E., *Nature, Lond.* **280**, 563 (1979).

151. Huber, R. and Bode, W., *Acc. Chem Res.* **11**, 114 (1978).

152. Sygusch, J., Beaudry, D., and Allaire, M., *Proc. natn. Acad. Sci. USA* **84**, 7846 (1987).

153. Fothergill-Gilmore, L. A. and Watson, H. C., *Adv. Enzymol. Relat. Areas Mol. Biol.* **62**, 227 (1989).

154. Bai, Y., Sosnick, T. R., Mayne, L., and Englander, S. W., *Science NY* **269**, 192 (1995).

155. Whitehead, B. and Waltho, J. P., in *Proteins Labfax* (Price, N. C., ed.). Bios Scientific Publishers, Oxford (1996). See Ch. 19.

156. Evans, J. N. S., *Biomolecular NMR spectroscopy*. Oxford University Press (1995).

157. Roberts, G. C. K. (ed.), *NMR of macromolecules: a practical approach*. IRL Press, Oxford (1993).

158. *Methods Enzymol.* **239** (1994).

159. James, T. L., *Methods Enzymol.* **239**, 416 (1994).

160. Fasman, G. D. (ed.), *Circular dichroism and the conformational analysis of biomolecules*. Plenum Press, New York (1996).

161. Kelly, S. M. and Price, N. C., *Biochim. Biophys. Acta* **1338**, 161 (1997).

162. Rupley, J. A., in *Structure and stability of biological molecules* (Timasheff, S. N. and Fasman, G. D., eds), p. 291. Marcel Dekker, New York (1969).

163. Citri, N., *Adv. Enzymol.* **37**, 397 (1973).

164. Bernstein, B. E., Michels, P. A. M., and Hol, G. W. J., *Nature, Lond.* **385**, 275 (1997).

165. Anderson, C. M., Zucker, F. H., and Steitz, T. A., *Science NY* **204**, 375 (1979).

166. Varley, P. G. and Pain, R. H., *J. molec. Biol.* **220**, 531 (1991).

167. Jaenicke, R., Schuring, H., Beaucamp, N., and Ostendorp, R., *Adv. Protein Chem.* **48**, 181 (1996).

168. Rossmann, M. G., Moras, D., and Olsen, K. W., *Nature, Lond.* **250**, 194 (1974).

169. Rossmann, M. G., Liljas, A., Bränden, C.-I., and Banaszak, L. J., *Enzymes* (3rd edn) **11**, 61 (1975).

170. Milner-White, E. J., Coggins, J. R., and Anton, I. A., *J. molec. Biol.* **221**, 751 (1991).

171. Johnson, M. S., Sali, A., and Blundell, T. L., *Methods Enzymol.* **183**, 670 (1990).

172. Orengo, C. A., Jones, D. T., and Thornton, J. M., *Nature, Lond.* **372**, 631 (1994).

173. Barton, G. J., *Trends Biochem. Sci.* **19**, 554 (1994).

174. Perutz, M. F., *Protein structure: new approaches to disease and therapy*. W. H. Freeman, New York (1992).

175. Creighton, T. E., *Prog. Biophys. molec. Biol.* **33**, 231 (1978).

176. Maccallum, P. H., Poet, R., and Milner-White, E. J., *J. molec. Biol.* **248**, 374 (1995).

177. Milner-White, E. J., *Protein Sci.* **6**, 2477 (1997).

178. Ramachandran, G. N. and Sasisekharan, V., *Adv. Protein Chem.* **23**, 283 (1968).

179. Richardson, J. S., *Adv. Protein Chem.* **34**, 167 (1981).

180. Dickerson, R. E. and Geis, I., *The structure and action of proteins*. Harper and Row, New York (1969). See Ch. 2.

181. Maccallum, P. H., Poet, R., and Milner-White, E. J., *J. molec. Biol.* **248**, 361 (1995).

182. Branden, C. and Tooze, J., *Introduction to protein structure* (2nd edn). Garland, New York (1999).

183. Schiffer, M. and Edmundson, A. B., *Biophys. J.* **7**, 121 (1967).

184. Schulz, G. E. and Schirmer, R. H., *Principles of protein structure*. Springer, New York (1979). See Ch. 5.

185. Lipscomb, W. H., Reeke, G. N. Jr, Hartsuck, J. A., Quiocho, F. A., and Bethge, P. H., *Phil. Trans. R. Soc.* **B257**, 177 (1970).

186. Chothia, C., *J. molec. Biol.* **75**, 295 (1973).

187. Chou, P. Y. and Fasman, G. D., *J. molec. Biol.* **115**, 135 (1977).

188. Sibanda, B. L. and Thornton, J. M., *Nature, Lond.* **316**, 170 (1985).

189. Richardson, J. S., Getzoff, E. D., and Richardson, D. C., *Proc. natn. Acad. Sci. USA* **75**, 2574 (1978).

190. Milner-White, E. J. and Poet, R., *Trends Biochem. Sci.* **12**, 189 (1987).

191. Chothia, C., *A. Rev. Biochem.* **53**, 537 (1984).

192. Creighton, T. E., *Proteins* (2nd edn). Freeman, New York (1993). See Ch. 6.

193. Chothia, C., *Nature, Lond.* **248**, 338 (1974).

194. Janin, J. and Chothia, C., *Methods Enzymol.* **115**, 420 (1985).

195. Baron, M., Norman, D. G., and Campbell, I. D., *Trends Biochem. Sci.* **16**, 13 (1991).

196. Dioszegi, M., Cannon, P., and van Wart, H. E., *Methods Enzymol.* **248**, 413 (1995).

197. Nagase, H., *Methods Enzymol.* **248**, 449 (1995).

198. Barlow, P. N. and Campbell, I. D., *Methods Enzymol.* **239**, 464 (1994).

199. Chen, L. H., Kenyon, G. L., Curtin, F., Harayama, S., Bembenek, N. E., Hajipour, G. *et al.*, *J. biol. Chem.* **267**, 17716 (1992).

200. Dixon, M., Webb, E. C., Thorne, C. J. R., and Tipton, K. F., *Enzymes* (3rd edn). Longman, London (1979). See pp. 550–567.

201. Levitt, M. and Chothia, C., *Nature, Lond.* **261**, 552 (1976).

202. Richardson, J. S., *Methods Enzymol.* **115**, 349 (1985).

203. Murzin, A. G., Brenner, S. E., Hubbard, T., and Chothia, C., *J. molec. Biol.* **247**, 536 (1995).

204. Chothia, C., *Nature, Lond.* **357**, 543 (1993).

205. Dill, K. A., *Biochemistry* **29**, 7133 (1990).

206. Makhatadze, G. I., *Adv. Protein Chem.* **47**, 307 (1995).

207. Kabsch, W. and Sander, C., *Biopolymers* **22**, 2577 (1983).

208. Rose, G. D. and Wolfenden, R., *A. Rev. Biophys. Biomol. Struct.* **22**, 381 (1983).

209. Fersht, A. R., Shi, J.-P., Knill-Jones, J., Lowe, D. M., Wilkinson, A. J., Blow, D. M. *et al.*, *Nature, Lond.* **314**, 235 (1985).

210. Gilson, M. K. and Honig, B. H., *Nature, Lond.* **330**, 84 (1987).

211. Sternberg, M. J. E., Hayes, F. R. F., Russell, A. J., Thomas, P. G., and Fersht, A. R., *Nature, Lond.* **330**, 86 (1987).

212. Tanford, C., *Science, NY* **200**, 1012 (1978).

213. Creighton, T. E., *Prog. Biophys. molec. Biol.* **33**, 231 (1978). See p. 248.

214. Janin, J. and Chothia, C., *J. molec. Biol.* **100**, 197 (1976).

215. Sayle, R. A. and Milner-White, E. J., *Trends Biochem. Sci.* **20**, 374 (1995).

216. Kraulis, P., *J. Appl. Crystallogr.* **24**, 946 (1991).

217. Löwe, J., Stock, D., Jap, B., Zwickl, P., Baumeister, W., and Huber, R., *Science, NY* **268**, 533 (1995).

218. Remington, S., Wiegand, G., and Huber, R., *J. molec. Biol.* **158**, 111 (1982).

219. Remington, S. J., *Curr. Top. Cell. Reg.* **33**, 209 (1992).

220. Blow, D. M., *Enzymes* (3rd edn) **3**, 185 (1971). See p. 194.

221. Berry, M. B., Meador, B., Bilderback, T., Liang, P., Glaser, M., and Phillips, G. N., *Proteins Struct. Funct. Genet.* **19**, 183 (1994).

222. Schulz, G. E., Elzinga, M., Marx, F., and Schirmer, R. H., *Nature, Lond.* **250**, 120 (1974).

223. Matthews, B. W., Weaver, L. H., and Kester, W. R., *J. biol. Chem.* **249**, 8030 (1974).

224. Holmes, M. A. and Matthews, B. W., *J. molec. Biol.* **160**, 623 (1982).

225. Vita, C., Dalzoppo, D., and Fontana, A., *Biochemistry* **24**, 1798 (1985).

226. Matthews, B. W., *Acc. Chem. Res.* **21**, 333 (1988).

227. Davies, G. E. and Stark, G. R., *Proc. natn. Acad. Sci. USA* **66**, 651 (1970).

228. Jaenicke, R. and Rudolph, R., *Methods Enzymol.* **131**, 218 (1986).

229. Wiley, D. C. and Lipscomb, W. N., *Nature, Lond.* **218**, 1119 (1968).

230. Wood, W. A., *Trends Biochem. Sci.* **2**, 223 (1977).

231. Lewendon, A., Murray, I. A., Kleanthous, C., Cullis, P. M., and Shaw, W. V., *Biochemistry* **27**, 7385 (1988).

232. Simon, J.-P. and Stalon, V., *Europ. J. Biochem.* **88**, 287 (1978).

233. Braig, K., Otwinowski, Z., Hegde, R., Boisvert, D. C., Joachimiak, A., Horwich, A L. *et al.*, *Nature, Lond.* **371**, 578 (1994).

234. Groll, M., Ditzel, L., Löwe, J., Stock, D., Bochtler, M., Bartunik, H. D. *et al.*, *Nature, Lond.* **386**, 463 (1997).

235. Schulz, G. E. and Schirmer, R. H., *Principles of protein structure.* Springer, New York (1979). See p. 100.

236. Bragg, P. D. and Hou, C., *Arch. Biochem. Biophys.* **167**, 311 (1975).

237. Abrahams, J. P., Leslie, A. G. W., Lutter, R., and Walker, J. E., *Nature, Lond.* **370**, 621 (1994).

238. Price, N. C., in *Mechanisms of protein folding* (Pain, R. H., ed.). IRL Press, Oxford (1994). See Ch. 7.

239. Chothia, C. and Janin, J., *Nature, Lond.* **256**, 705 (1975).

240. Fisher, H. F., *Proc. natn. Acad. Sci. USA* **51**, 1285 (1964).

241. Klotz, I. M., Langerman, N. R., and Darnall, D. W., *A. Rev. Biochem.* **39**, 25 (1970).

242. Jones, D. H., McMillan, A. J., Fersht, A. R., and Winter, G., *Biochemistry* **24**, 5852 (1985).

243. Rafferty, J. B., Somers, W. S., Saint-Girons, I., and Phillips, S. E. V., *Nature, Lond.* **341**, 705 (1990).

244. Walker, J. E., Wonacott, A. J., and Harris, J. I., *Eur. J. Biochem.* **108**, 581 (1980).

245. Hall, L. and Campbell, P. N., *Essays Biochem.* **22**,1 (1986).

246. Chan, W. W.-C., *Can. J. Biochem.* **54**, 521 (1976).

247. Grossman, S. H., Pyle, J., and Steiner, R. J., *Biochemistry* **20**, 6122 (1981).

248. Pace, C. N., *Trends Biotech.* **8**, 93 (1990).

249. Tanford, C., *Adv. Protein Chem.* **23**, 121 (1968).

250. Tanford, C., *Adv. Protein. Chem.* **24**, 1 (1970).

251. Nozaki, Y. and Tanford, C., *J. biol. Chem.* **238**, 4074 (1963).

252. Roseman, M. and Jencks, W. P., *J. amer. chem. Soc.* **97**, 631 (1975).

253. Pace, C. N., and Vanderburg, K. E., *Biochemistry* **18**, 288 (1979).

254. Pace, C. N., in *Proteins Labfax* (Price, N. C., ed.). Bios Scientific Publishers, Oxford (1996). See Ch. 21.

255. Pace, C. N., *Methods Enzymol.* **131**, 266 (1986).

256. Dolgikh, D. A., Gilmanshin, R. I., Brazhnikov, E. V., Bychkova, V. E., Semisotnov, G. V., Venyaminov, S. Yu. *et al.*, *FEBS Lett.* **136**, 311 (1981).

257. Ptitsyn, O. B., *Adv. Protein Chem.* **47**, 83 (1995).

258. Ptitsyn, O. B., *Trends Biochem. Sci.* **20**, 376 (1995).

259. Hornby, D. P., Whitmarsh, A., Pinabarsi, H., Kelly, S. M., Price, N. C., Shore, P. D. *et al.*, *FEBS Lett.* **355**, 57 (1994).

260. Anfinsen, C. B., *Harvey Lect.* **61**, 95 (1967).

261. Anfinsen, C. B. and Scheraga, H. A., *Adv. Protein Chem.* **29**, 205 (1975).

262. Creighton, T. E., *J. molec. Biol.* **129**, 411 (1979).

263. Freedman, R. B., Hirst, T. R., and Tuite, M. F., *Trends Biochem. Sci.* **19**, 331 (1994).

264. Jaenicke, R., *Prog. Biophys. molec. Biol.* **49**, 117 (1987).

265. Jaenicke, R., *Biochemistry* **30**, 3147 (1990).

266. Matthews, C. R., *A. Rev. Biochem.* **62**, 653 (1993).

267. Weissman J. S., *Chem. Biol.* **2**, 255 (1995).

268. Hlodan, R. and Hartl, F. U., in *Mechanisms of protein folding* (Pain, R. H., ed.). IRL Press, Oxford (1994). See Ch. 8.

269. Hartl, F. U., *Nature, Lond.* **381**, 571 (1996).

270. Mayhew, M., da Silva, A. C. R., Martin, J., Erdjument-Bromage, H., Tempst, P., and Hartl, F. U., *Nature, Lond.* **379**, 420 (1996).

271. Netzer, W. J. and Hartl, F. U., *Nature, Lond.* **388**, 343 (1997).

272. Gilbert, H. F., in *Mechanisms of protein folding* (Pain, R. H., ed.). IRL Press, Oxford (1994). See Ch. 5.

273. Nall, B. T., in *Mechanisms of protein folding* (Pain, R. H., ed.). IRL Press, Oxford (1994). See Ch. 4.

274. Schmid, F. X., Mayr, L. M., Mucke, M., and Schonbrunner, E. R., *Adv. Protein Chem.* **44**, 25 (1993).

275. Gething, M.-J. and Sambrook, J., *Nature, Lond.* **355**, 33 (1992).

276. Thomas, P. J., Qu, B.-H., and Pedersen, P. L., *Trends Biochem. Sci.* **20**, 456 (1995).

277. Sifers, R. N., *Struct. Biol.* **2**, 355 (1995).

278. Sander, C. and Schneider, R., *Nucl. Acids Res.* **22**, 3597 (1994).

279. Chou, P. Y. and Fasman, G. D., *A. Rev. Biochem.* **47**, 251 (1978).

280. Garnier, J., Osguthorpe, D., and Robson, B., *J. molec. Biol.* **120**, 97 (1978).

281. Kabsch, W. and Sander, C., *FEBS Lett.* **155**, 179 (1983).

282. Sawyer, L., Fothergill-Gilmore, L. A., and Freemont, P. S., *Biochem. J.* **249**, 789 (1988).

283. Rost, B. and Sander, C., *Proc. natn. Acad. Sci. USA* **90**, 7558 (1993).

284. Rost, B., *Methods Enzymol.* **266**, 525 (1996).

285. *Proteins: Structure, Function and Genetics* **23** (Part 3) (1995).

Appendix 3.1

Protein structures on the World Wide Web (WWW)

The rapid developments in structural techniques and in the availability of computing facilities now make it possible for anyone with access to a networked desktop computer to be able to explore (via the WWW) many of the databases of structural information on biological molecules and to analyse these structures. Since developments in this area are occurring very rapidly, it is possible to give only very limited information in this appendix. Our intention is to give some of the more important details to allow the reader to start using the databases; from this it should be possible to develop appropriate search patterns.

A very detailed account of this area has been given in a volume of *Methods in Enzymology*.[1] Two good shorter accounts have been given by Gray *et al.*[2] and by Milner-White.[3] The former article gives a very useful list of websites as well as providing a clear explanation of the relationships between the various databases and compilations. It also addresses the important aspect of the quality of data in the databases. The information in the article can be accessed via the WWW* at *http://www.ucmb.ulb.ac.be:80/StructResources*. The chapter by Milner-White[3] explains how to use many of the programs available and the significance of the information obtained. Another useful collection of websites has been given by Peitsch *et al.*[4]

* If a World Wide Web address is given at the end of a sentence, the final full stop does not form part of the address.

Primary databases

The Protein Data Bank (PDB) at Brookhaven National Laboratories contains three-dimensional structural information on proteins from either X-ray crystallography or NMR studies *http://www.pdb.bnl.gov.*

SWISS-PROT is an annotated protein sequence database set up at Geneva University Hospital and the University of Geneva and is now maintained in a collaboration between Geneva and the European Bioinformatics Institute *http://expasy.hcuge.ch/sprot/sprot-top.html.*

BioMagResBank is a database of NMR spectroscopic data on peptides, proteins, and nucleic acids *http://www.brmb.wisc.edu.*

Structural comparisons between proteins

The 3D-ali database provides alignments of protein sequences based on 3D superpositions[3,5] *http://embl-heidelberg.de/argos/ali/ali.html*

The scop (structural classification of proteins) database[6,7] has been outlined in Section 3.4.4.6. It provides a broad survey of the known folds of proteins and of the evolutionary and structural relationships between proteins *http://scop.mrc-lmb.cam.ac.uk/scop.*

Representations of protein structures

Two of the most widely used programs are Rasmol[3,8] and Molscript[9] (see Section 3.4.6.1). Rasmol can be obtained by anonymous file transfer protocol (ftp) from *ftp.dcs.ed.ac.uk.*

Compilations of resources on the WWW

The compilations provide convenient entries to the databases and the means of analysing them. One of the most widely used is the European Bioinformatics

Institute (EBI) located at Hinxton near Cambridge. The EBI is an outstation of the European Molecular Biology Laboratory (EMBL) in Heidelberg. The home page of EBI (*http://www.ebi.ac.uk*) provides a very convenient route to the major databases and analysis of them. For instance, the SWISS-PROT sequence database can be accessed and similarity searches between a specified protein sequence and the sequences in the databases can be performed using the programs FASTA and BLITZ (see Section 3.3.2.11). In addition, the EBI acts as a mirror site for the Protein Data Bank. A convenient way of exploring the PDB is via the 3DB browser (*http://www2.ebi.ac.uk/pdb-bin/pdbmain* or *http://www2.pdb.bnl.gov/pdb-bin/pdbmain*) where either keywords or the PDB code for a protein can be entered (e.g. 1ald for aldolase). Once the appropriate PDB file has been retrieved, it can be shown as a Rasmol view or saved into a suitable file for subsequent display. Further details about the EBI services have been given by Shomer *et al.*[10]

Other compilations of resources can be found at ExPASy (Expert Protein Analysis System) *http://expasy.hcuge.ch* and Pedro's home page *http://public.iastate.edu/~pedro/research_tools.html*. This latter site is mirrored in various countries, e.g. in Switzerland *http://www.fmi.ch/biology/research_tools.html*.

Protein structure prediction

Details of the CASP (Critical Assessment of techniques for protein Structure Prediction) projects can be found at *http://predictioncenter.llnl.gov*.

References for Appendix 3.1

1. *Methods Enzymol.* **266** (1996).
2. Gray, P. M. D., Kemp, G. J. L., Rawlings, C. J., Brown, N. P., Sander, C., Thornton, J. M. *et al.*, *Trends Biochem. Sci.* **21**, 251 (1996).
3. Milner-White, E. J., in *Proteins Labfax* (Price, N. C., ed.). Bios Scientific Publishers, Oxford (1996). See Ch. 22.
4. Peitsch, M. C., Wells, T. N. C., Stampf, D. R., and Sussman, J. L., *Trends Biochem. Sci.* **20**, 82 (1995).
5. Pascarella, S. and Argos, P., *Protein Engineering* **5**, 121 (1992).
6. Barton, G. J., *Trends Biochem. Sci.* **19**, 554 (1994).
7. Murzin, A. G., Brenner, S. E., Hubbard, T., and Chothia, C., *J. molec. Biol.* **247**, 536 (1995).
9. Kraulis, P., *J. Appl. Crystallography* **24**, 946 (1991).
10. Shomer, B., Harper, R. A. L., and Cameron, G. N., *Methods Enzymol.* **266**, 3 (1996).

4

An introduction to enzyme kinetics

4.1 Outline of the chapter

The subject of enzyme kinetics often generates considerable trepidation in biochemistry students. Many treatments of the topic make two assumptions: first, that the readers can confidently handle complex algebraic equations, and, second, that a detailed knowledge of kinetics is an indispensable part of a biochemist's training. In this chapter we do not make these assumptions but instead introduce some of the main ideas in enzyme kinetics that are useful in understanding the mechanism of action and control of isolated enzymes (see Chapters 5 and 6) and the role of enzymes in the cell (see Chapter 8). The emphasis will be on the information that can be gained from a study of enzyme kinetics.

The first part of the chapter (Section 4.2) describes some practical aspects of enzyme kinetics, answering the question 'How do we obtain kinetic data?' The next part (Section 4.3) deals with the analysis of the results, discussing the theoretical background, equations involved, and information obtained. In these sections we shall be dealing with 'steady-state kinetics', where an enzyme is present at very small molar concentrations (usually much less than 1 per cent) compared with the molar concentration(s) of substrate(s) acted upon. Under these conditions, the equations involved are *comparatively* straightforward and the data are usually collected over a time-scale of minutes, so that 'conventional' methods of mixing and observation can be employed. In Section 4.4 we refer in outline to experiments in which enzyme and substrate(s) are present at comparable concentrations. Under these conditions it is usually necessary to use special techniques to ensure rapid mixing and observation. This type of experiment has given detailed insights into the various steps in an overall reaction, particularly those occurring in the early, 'pre-steady-state' period (see Fig. 4.2); in fact, these studies are often termed 'pre-steady-state kinetics'. As described in Chapter 8 (Section 8.5.4) there are many examples known where, in the cell, the concentration of an enzyme is comparable with that of its substrate and in such cases the study of pre-steady-state kinetics is particularly relevant.

We end this section by stating three of the reasons why it is important for the biochemistry student to have some knowledge of enzyme kinetics. First, kinetics, *in conjunction with other techniques*, provides valuable information on the mechanism of action of an enzyme (see Chapter 5). Second, it can give an insight into the role of an enzyme under the conditions that exist in the cell and the response of an enzyme to changes in the concentrations of metabolites (see Chapter 8). Third, it can help to show how the activity of an enzyme can be controlled, which may provide a valuable pointer to mechanisms of regulation under physiological conditions (see Chapters 6 and 8).

4.2 How do we obtain kinetic data?

The aim of an experiment is to measure the rate of formation of product (or disappearance of substrate) under specified conditions. It is then possible to vary in turn certain parameters such as the concentration of substrate(s), pH, temperature, or concentration of modifying ligands, and collect data to analyse in terms of theoretical models.

The rate of a particular enzyme-catalysed reaction can often be measured in a number of ways, but there is normally one method that is more convenient than the others. This point can be illustrated by reference to the reaction catalysed by hexokinase:

$$\text{D-Glucose} + \text{ATP} \xrightarrow{\text{Mg}^{2+}} \text{D-Glucose 6-phosphate} + \text{ADP}.$$

The rate of this reaction could be monitored by removing samples from the reaction mixture at known times after addition of enzyme, stopping the reaction quickly (e.g. by addition of acid to inactivate the enzyme), and measuring the amount of product formed. Ion-exchange chromatography (see Chapter 2, Section 2.6.2.1) and high-performance liquid chromatography (h.p.l.c.) would be useful techniques for separating the products from the substrates in this case. Clearly, such a 'stop and sample' (or *discontinuous*) assay procedure involves possible sampling errors and considerable work in separation and estimation of the products.

A more convenient method would involve the *continuous* measurement of some property that changes during the course of the reaction. In the case of the reaction above, there is no convenient change in absorbance, for example, but we could bring about such a change if we *couple* the production of D-glucose 6-phosphate to the reduction of NADP$^+$ to NADPH using glucose-6-phosphate dehydrogenase:

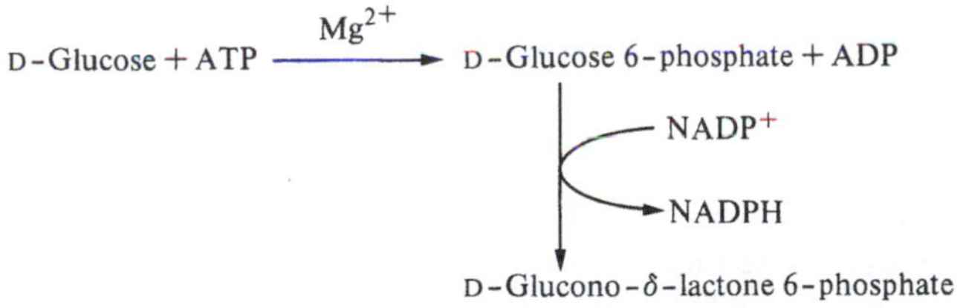

NADP$^+$ does not absorb at 340 nm, whereas NADPH does, so it is possible to monitor the production of NADPH (and hence of D-glucose 6-phosphate) continuously. If we use a *coupled assay procedure* such as this we must add sufficient coupling enzyme and substrate(s) so that the D-glucose 6-phosphate formed in the first step is 'immediately' converted to D-glucose-δ-lactone 6-phosphate, i.e. so that the coupling reaction is not rate limiting. Several detailed analyses of the kinetics of coupled assay systems have indicated the conditions under which the true rate of the reaction of interest can be measured.[1-5] In practice, a useful test is to check whether the observed rate of reaction is proportional to the amount of enzyme of interest which is added to the assay mixture. It is, of course, obvious that the coupling enzyme should be highly purified and certainly free from detectable quantities of the enzyme we are trying to assay.

In a number of cases it is not possible to monitor the reaction continuously (either directly on the reaction of interest or by using a coupled assay) so a 'stop and sample' method must be employed. For instance, in the reaction catalysed by ornithine decarboxylase,

$$NH_2-(CH_2)_3-\underset{\underset{^{14}CO_2H}{|}}{\overset{\overset{NH_2}{|}}{CH}} \longrightarrow {}^{14}CO_2 + NH_2(CH_2)_4NH_2$$

L-Ornithine ${}^{14}CO_2H$ 1,4-diaminobutane
(putrescine)

* Improvements in electrophoretic separation and fluorescence techniques have allowed the assay of single molecules of enzymes such as lactate dehydrogenase[6] and alkaline phosphatase[7]. These have shown that individual enzymes can differ markedly in terms of their activities, energies of activation, etc.

if L-[1-^{14}C]ornithine is used, the CO_2 liberated will be radioactive. This CO_2 can be trapped by a suitable base (e.g. ethanolamine dissolved in 2-methoxyethanol) and the radioactivity estimated by scintillation counting. Determination of the amount of CO_2 formed after various times of reaction allows the rate of the reaction to be calculated. Assay procedures involving substrates which are radioactively labelled or give rise to highly fluorescent products are very sensitive and are of particular value when low concentrations of substrates are being used or when the amount of enzyme activity added to the assay system is low.*

When the method of assay has been decided upon, it is important to take a number of precautions to obtain reliable data.[1,8,9]

1. The substrates, buffers, etc. should be of as high a purity as possible, since contaminants may affect the activity of enzymes. For example, commercial preparations of NAD^+ sometimes contain inhibitors of dehydrogenases[10,11] and it has been found that certain preparations of ATP contain trace amounts of vanadate (VO_4^{3-}) ions, which act as a powerful inhibitor of adenosinetriphosphatase (Na^+, K^+-activated).[12]

2. It must be ascertained that the enzyme preparation does not contain any compound (or other enzyme) that interferes with the assays. The possibility of non-enzyme-catalysed conversion of substrate to product should be tested for by performing appropriate control experiments (e.g. using heat-inactivated enzyme).

3. The enzyme should be stable (i.e. not lose any significant amount of catalytic activity) during the time taken for assay. There should be no breakdown of substrate other than by the enzyme.

4. Since the activity of an enzyme can be markedly affected by changes in pH, temperature, etc., it is important to ensure that these parameters are stabilized by use of buffers, thermostatted baths, etc.

5. It should be checked that, once the steady state has been achieved (Section 4.3.1.1), the measured rate of reaction is constant over the period of interest and is proportional to the amount of enzyme added.

6. The *initial* rate of reaction should be measured to avoid possible complications arising from product inhibition, occurrence of the reverse reaction, and depletion of substrate.[1,13]

4.3 How do we analyse kinetic data?

As in many other branches of experimental science, kinetic data are analysed in terms of theoretical models in order to test the correctness of the models and to deduce the values of constants in the equations derived from the models. In this section we shall describe for a number of situations the theoretical background to steady-state kinetics, then show how to treat data in terms of the resulting equations, and finally describe the significance of the results obtained.

4.3.1 One-substrate reactions

It is easiest to deal initially with reactions in which only one substrate is acted on by an enzyme. This would include reactions catalysed by enzymes in the following groups (see Chapter 1, Section 1.5.1): hydrolases (if H_2O is considered to be in a large excess), isomerases, and most lyases.

4.3.1.1 Theoretical background

We shall assume that catalysis occurs via rapid and reversible formation of a complex between enzyme, E, and substrate, S. (The part of the enzyme to which the substrate binds is known as the *active site* of the enzyme.) This complex then breaks down in a slow step to give the product, P, and regenerate enzyme (Scheme 4.1). In practice, this scheme is likely to be an oversimplification (see Section 4.3.1.4).

$$E + S \underset{k_{-1}}{\overset{k_1}{\rightleftharpoons}} ES$$

$$ES \overset{k_2}{\longrightarrow} E + P$$

Scheme 4.1. The conversion of substrate to product in an enzyme-catalysed reaction. k_1, k_{-1}, k_2 represent the rate constants for the individual steps.

The equation describing the kinetics of this scheme can be derived by making one of two types of assumption as follows.

(i) The equilibrium assumption
Here we assume that the $E + S \rightleftharpoons ES$ equilibrium is only slightly disturbed by the breakdown of ES to give product. This is clearly a better assumption the lower the value of k_2 relative to k_{-1}.

The equilibrium constant, K, is defined by

$$K = \frac{[E][S]}{[ES]}, \tag{4.1}$$

where [E] and [S] are the concentrations of *free* enzyme and *free* substrate, respectively. However, since the total concentration of substrate is much greater than the total concentration of enzyme (see Section 4.1), essentially all the substrate is free and we can set $[S]_{free}$ equal to $[S]_{total}$.

At any concentration of S we can evaluate the fraction, F, of enzyme present as ES as follows:

$$F = \frac{[ES]}{[E] + [ES]}.$$

Now, from (4.1)

$$[ES] = \frac{[E][S]}{K},$$

$$\therefore \quad F = \frac{[E][S]}{K} \Bigg/ \left([E] + \frac{[E][S]}{K} \right)$$

$$= \frac{[S]}{K + [S]}. \tag{4.2}$$

If the total concentration of enzyme is $[E]_0$, then $[ES] = F[E]_0$; thus, from (4.2)

$$[ES] = \frac{[E]_0[S]}{K + [S]}.$$

The rate of product formation, v, is given by

$$v = k_2[ES]$$

$$\therefore \quad v = \frac{k_2[E]_0[S]}{K + [S]} \tag{4.3}$$

Sometimes the equation (4.3) is written in the form

$$v = \frac{k_{cat}[E]_0[S]}{K + [S]}$$

where k_{cat} is a first-order rate constant equal to k_2 (k_{cat} is sometimes known as the *turnover number* of the enzyme). This formulation will be referred to later (see Section 4.3.1.3).

An equation of the type shown in (4.3) means that v will tend towards a maximum (or limiting) value as [S] increases (Fig. 4.1).

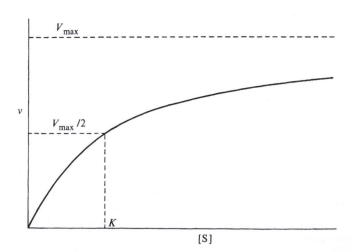

Fig. 4.1 Dependence of velocity (v) on substrate concentration ([S]) according to eqn (4.3).

The maximum rate will be observed when all the enzyme is in the form of the ES complex. Let the maximum (limiting) rate $= V_{max}$; this will be equal to $k_2[E]_0$: then eqn (4.3) can be rewritten as:

$$v = \frac{V_{max}[S]}{K + [S]}. \qquad (4.4)$$

We should note from eqn (4.4) and Fig. 4.1 that when [S] is small compared with K, the reaction is first order in [S], since $v = V_{max}[S]/K$. When [S] is large compared with K, the reaction is zero order in [S], since $v = V_{max}$. At intermediate values of [S], the reaction is of a fractional order in [S].

When $v = V_{max}/2$, $[S] = K$. Thus K corresponds to the concentration of substrate when the velocity is half-maximal. It is known as the Michaelis* constant, and is normally written as K_m. If we have used the equilibrium assumption, we can equate K_m with the dissociation constant of the ES complex (K in eqn (4.1)).

(ii) The steady-state assumption
In this approach, we abandon the assumption that the E + S $\rightleftharpoons$ ES equilibrium is not perturbed by the breakdown of ES. Instead, it is assumed that ES is in a 'steady state', i.e. that the concentration of ES remains constant because the rate of its formation equals the rate of its breakdown. If we were to examine the variation of [ES] with time in a typical experiment we would obtain a graph of the type shown in Fig. 4.2.

After an initial phase (the pre-steady-state period) the concentration of ES remains fairly constant and it is thus in order to apply the steady-state assumption to Scheme 4.1 to evaluate the fraction of enzyme in the form of the ES complex. (For the steady-state assumption to be valid, the rate of change of [ES] must be small compared with the rate of change of [S] or [P]. In experiments in which the concentration of substrate is much greater than that of enzyme, the maximum value of [ES], and hence the rate of change of [ES], will be small.)

Now, according to the steady-state assumption, the rate of production of ES ($= k_1[E][S]$) must equal the rate of its breakdown ($= k_{-1}[ES] + k_2[ES]$)

$$\therefore k_1[E][S] = k_{-1}[ES] + k_2[ES]$$
$$\therefore [ES] = \frac{k_1[E][S]}{k_{-1} + k_2}.$$

Proceeding as before,

$$F = \frac{[ES]}{[E] + [ES]}$$
$$= \frac{[S]}{\left[\dfrac{k_{-1} + k_2}{k_1}\right] + [S]}.$$

If the total concentration of enzyme is $[E]_0$, then

$$[ES] = F[E]_0$$
$$= \frac{[E]_0[S]}{\left[\dfrac{k_{-1} + k_2}{k_1}\right] + [S]}$$

* Michaelis was one of the first workers to develop the mathematical analysis of enzyme kinetics along the lines indicated here. An equation of the type (4.4) is often referred to as the Michaelis–Menten equation.

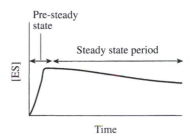

Fig. 4.2 The concentration of the ES complex as a function of time of reaction.

and the rate of product formation, v, is given by:

$$v = k_2[ES]$$

$$= \frac{k_2[E]_0[S]}{\left[\dfrac{k_{-1} + k_2}{k_1}\right] + [S]}.$$

Putting $k_2[E]_0 = V_{max}$

* This equation is sometimes referred to as the Briggs–Haldane equation because these workers (in 1925) first applied the steady-state assumption to enzyme kinetics.

$$v = \frac{V_{max}[S]}{\left[\dfrac{k_{-1} + k_2}{k_1}\right] + [S]}. \tag{4.5)*}$$

Thus using the steady-state approximation, the term K in the denominator of eqn (4.4) has been replaced by $(k_{-1} + k_2)/k_1$. Only if $k_2 \ll k_{-1}$, *when the equilibrium assumption becomes a limiting case of the steady-state assumption,* does this latter term become equal to the dissociation constant of the ES complex ($K = k_{-1}/k_1$). We should therefore find that the Michaelis constant, the concentration of substrate when the velocity is half-maximal, is not generally equal to this dissociation constant.

Using either assumption, we derive the fundamental equation (4.6) to describe the kinetics of the reaction shown in Scheme 4.1:

$$v = \frac{V_{max}[S]}{K_m + [S]}. \tag{4.6}$$

4.3.1.2 Treatment of data

It is very difficult to determine the limiting value of v (i.e. V_{max}) directly from a plot of v against [S] (Fig. 4.1) and therefore K_m cannot readily be determined in this way either. To overcome these difficulties, eqn (4.6) can be rearranged in a number of ways to give convenient graphical representations. Four of the best-known rearranged forms are given below:

(i) The Lineweaver–Burk equation[14]

This equation is obtained by taking reciprocals of the two sides of eqn (4.6):

$$\frac{1}{v} = \frac{K_m}{[S]} \cdot \frac{1}{V_{max}} + \frac{1}{V_{max}}.$$

Thus a plot of $1/v$ against $1/[S]$ gives a straight line of slope K_m/V_{max} and intercepts on the x and y axes of $-1/K_m$ and $1/V_{max}$, respectively (Fig. 4.3(a)).

(ii) The Eadie–Hofstee equation[15,16]

Equation (4.6) is rearranged to give:

$$\frac{v}{[S]} = \frac{v_{max}}{K_m} - \frac{v}{K_m}.$$

Thus a plot of $v/[S]$ against v gives a straight line of slope $-1/K_m$ and an x-axis intercept of V_{max} (Fig. 4.3(b)).

(iii) The Hanes equation[17]

Equation (4.6) is rearranged to give:

$$\frac{[S]}{v} = \frac{[S]}{V_{max}} + \frac{K_m}{V_{max}}.$$

Thus a plot of $[S]/v$ against $[S]$ is linear with a slope of $1/V_{max}$ and an x-axis intercept of $-K_m$ (Fig. 4.3(c)).

(iv) The direct linear plot[18,19,20]
This method uses a rather different approach from those described in (i)–(iii). Equation (4.6) is arranged to give:

$$V_{max} = v + \frac{v}{[S]} \cdot K_m \qquad (4.7)$$

V_{max} and K_m are now treated as variables and v and $[S]$ as constants. If the first pair of observed values of v and $[S]$ (i.e. v_1 and $[S]_1$) are plotted as shown in Fig. 4.4(a) then reference to eqn (4.7) shows that the line connecting them describes pairs of values of V_{max} and K_m that are consistent with the observed values of v and $[S]$.

If we have a second set of values of v and $[S]$ (i.e. v_2 and $[S]_2$), a new line can be drawn (Fig. 4.4(b)). The point of intersection of the lines defines uniquely the values of V_{max} and K_m that satisfy the two sets of data. This procedure can be repeated for further data points (Fig. 4.4(b)). If there were no experimental error the various lines would all intersect at a common point. In practice, a number of points of intersection are obtained (the maximum number of such points is $n(n-1)/2$, where n is the number of observations of v and $[S]$). The correct procedure is then to use the medians* of the values of V_{max} and K_m as the best-fit values.[18,19,20]

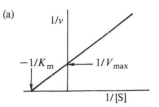
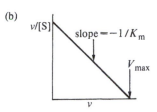
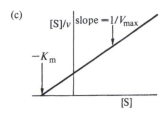

Fig. 4.3 Graphical representations of enzyme kinetic data according to the equations of (a) Lineweaver and Burk, (b) Eadie and Hofstee, and (c) Hanes. It should be remembered that the equation of a straight line is $y = mx + c$, where m is the slope and c is the intercept on the y-axis. The intercept on the x-axis equals $-c/m$.

* The median is the middle value of a set of values arranged in order of magnitude, or the mean of the middle pair if the number of values is even.

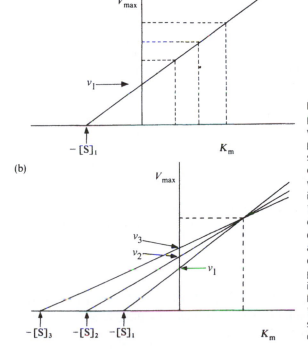

Fig. 4.4 The direct linear plot in which experimental values of v and $[S]$ are plotted directly. (a) A straight line drawn through one set of experimental values; the dashed lines indicate pairs of values of V_{max} and K_m that are consistent with the experimental values. In (b) a number of sets of experimental values are plotted. The point of intersection of the various lines gives the values of V_{max} and K_m that uniquely satisfy the experimental data.

It is not possible to give a simple answer to the question of which of the various methods of plotting data should be used. In any case it should be emphasized that the data should be as good as possible, since no graphical transformation allows sound conclusions to be drawn from poor data. The Lineweaver–Burk plot (Fig. 4.3(a)) is still the most commonly used and it has the advantage that the variables v and [S] are plotted on separate axes. However, an analysis of the errors involved in the collection of the data (and hence in the determination of the parameters K_m and V_{max}) shows that there is a highly non-uniform distribution of errors over the range of values of $1/v$ and $1/[S]$ in the Lineweaver–Burk plot.[20–23] For this reason, the use of the Eadie–Hofstee and Hanes plots has been recommended, since in these plots the distribution of errors is more uniform.

The direct linear plot has a number of advantages: (i) values of v and [S] are plotted directly so that V_{max} and K_m can be determined without the need for calculations; (ii) it is statistically sound—the use of median values of V_{max} and K_m minimizes the influence of extreme values of v and [S] on these parameters.[20,21] However, the direct linear plot has some disadvantages. It is not very suitable for the display of data in multisubstrate reactions (Section 4.3.5) because of the resulting large number of lines. In addition, it is not easy to detect departures from the basic equation (4.6).

It has now become more usual to obtain the 'best fit' values of V_{max} and K_m by use of non-linear regression computer programs in which the data are fitted directly to eqn (4.6);[24,25,26] a number of such programs are available commercially.

Finally, it should be mentioned that determinations of V_{max} and K_m can also be made from the integrated form of eqn (4.6). Substrate (or product) concentrations are monitored over a large fraction of the overall reaction, i.e. well beyond the 'initial rate' period. The values of V_{max} and K_m can be determined from a single progress curve that essentially represents a series of rate measurements at different values of substrate concentration.[13,21,25,27,28] One way in which this is done is described in Appendix 4.1. A major disadvantage of the method is that the accuracy of the estimates of V_{max} and K_m can be very sensitive to errors in the estimates of the end-point of the reaction.[13] In addition, the progress curve can be used to calculate the true initial rates of reaction in those cases where there has been a significant decline in the rate during the time taken for mixing the assay components and obtaining readings.[13,25,28]

The units of K_m and V_{max} will be those in which [S] and v are measured. Thus K_m is usually expressed in units of concentration, i.e. mol dm^{-3}. The velocity, v, can be expressed in a number of ways, depending on the information available. For a purified enzyme, v is expressed as moles substrate consumed per unit time, per weight of enzyme. In the SI system this would be in terms of katal kg^{-1}, whereas in the older system it would be in terms of units mg^{-1} (see Chapter 2, Section 2.3.1). If the M_r of the enzyme is also known, we can calculate the molar concentration of enzyme active sites in the solution and hence evaluate k_{cat} ($= V_{max}/[E]_0$), which is also known as the *turnover number* of the enzyme (see Section 4.3.1.1).

4.3.1.3 Significance of results

The parameters K_m and V_{max} (k_{cat}) are of value (i) in characterizing the specificity of an enzyme for a particular substrate, (ii) in deciding between steady-state and equilibrium mechanisms, and (iii) in indicating the role of an enzyme in metabolism.

(i) The specificity of an enzyme for a substrate can be described as follows:
From eqn (4.3) we have

$$v = k_{cat} \frac{[E]_0 [S]}{K_m + [S]}.$$

Substituting $[E]_0 = [E] + [ES]$ and noting that $[ES] = [E][S]/K_m$, we have

$$v = \frac{k_{cat}}{K_m} [E][S]. \tag{4.8}$$

Thus k_{cat}/K_m is an apparent second-order rate constant that describes the rate in terms of the concentrations of the free enzyme and free substrate.

If there are two competing substrates S_1 and S_2 for the enzyme, it follows from eqn (4.8) that the rates of reaction are

$$v_{S_1} = \left(\frac{k_{cat}}{K_m} \right)_{S_1} [E][S_1]$$

$$v_{S_2} = \left(\frac{k_{cat}}{K_m} \right)_{S_2} [E][S_2].$$

The ratio of these rates of reaction is given by

$$\frac{v_{S_1}}{v_{S_2}} = \frac{(k_{cat} / K_m)_{S_1}}{(k_{cat} / K_m)_{S_2}} \cdot \frac{[S_1]}{[S_2]}.$$

Thus, at equal concentrations of S_1 and S_2 the relative rates of reaction of the two substrates are determined by the relative values of k_{cat}/K_m. The ratio k_{cat}/K_m can thus be used as a measure of the specificity of an enzyme for a substrate. For example, in the reaction catalysed by fumarate hydratase, the ratios k_{cat}/K_m for fumarate, fluorofumarate, chlorofumarate, and bromofumarate are 1.6×10^8, 9.8×10^7, 2.0×10^5, and 2.5×10^4 s^{-1} (mol dm^{-3})$^{-1}$ respectively.[29] There is thus high specificity shown towards fumarate and fluorofumarate; substitution by the larger halogens leads to a marked decline in the reactivity of the substrate.

(ii) The ratio of k_{cat}/K_m can be used to test for the applicability of the steady-state or equilibrium mechanisms (Section 4.3.1.1). Using the steady-state assumption, we note from eqns (4.5) and (4.6) that

$$K_m = \frac{k_{-1} + k_2}{k_1}.$$

Now, if $k_2 > k_{-1}$ (which is the extreme form of the steady-state assumption), then $K_m = k_2/k_1$. Replacing k_2 by k_{cat}, it is clear that k_{cat}/K_m is equal to k_1, the rate constant for the association of enzyme with substrate. From fast-reaction studies (Section 4.4.3) it is known that the diffusion-controlled rate constant for association of enzyme with substrate is of the order of 10^9 (mol dm^{-3})$^{-1}$ s^{-1}. Thus, if k_{cat}/K_m is of this order of magnitude, it can be concluded that the steady-state mechanism operates. This is the case for fumarate hydratase, catalase, and triosephosphate isomerase, for which k_{cat}/K_m values are 1.6×10^8, 4×10^7, and 2.4×10^8 s^{-1} (mol dm^{-3})$^{-1}$ respectively.[30] However, if the value of k_{cat}/K_m is much lower than these values, it is reasonable to conclude that the equilibrium mechanism is more appropriate.

(iii) The role of an enzyme in metabolism may be judged by relating the K_m value to the prevailing concentration of substrate. This will be discussed further in Chapter 8 (Section 8.3.2) but two examples will serve to illustrate the concepts involved.

The isoenzyme IV of hexokinase (also known as glucokinase) is confined to the liver and has a high K_m for glucose (10 mmol dm^{-3}), whereas isoenzymes I–III have a much wider tissue distribution and a low K_m (40 μmol dm^{-3}). Thus, at the prevailing levels of blood glucose in a fasting subject (*ca.* 3 mmol dm^{-3}), isoenzymes I–III are working essentially at their maximum velocity, whereas isoenzyme IV is only working at about 25 per cent of its maximum velocity. After an intake of carbohydrate, the level of blood glucose rises to about 9.5 mmol dm^{-3} and the isoenzyme IV is now working at about half its maximum velocity (Fig. 4.5). Hence the liver isoenzyme (IV) can 'deal with' this extra glucose, converting it to D-glucose 6-phosphate, which is the first step in the process of storage as glycogen.

The studies on ribulose bisphosphate carboxylase (see Chapter 2, Section 2.8.2), which catalyses a reaction involving assimilation of CO_2 in photosynthetic organisms, illustrate the need to exercise care in interpreting data on K_m and V_{max}. This enzyme is present in very large amounts in such organisms and is relatively easy to purify. However, early measurements showed that the activity of the purified enzyme was far too low to account for the observed rates of photosynthetic CO_2 fixation, and that, in particular, the K_m for CO_2 was about 50-fold higher than the apparent K_m measured in intact leaves or isolated chloroplasts. The solution to this problem is thought[31] to lie in the fact that the enzyme is slowly activated by CO_2 and Mg^{2+}, and so considerable care has to be taken in performing and interpreting the assays of the purified enzyme. Activated enzyme has a K_m for CO_2 that is comparable with the apparent K_m in the intact system.

From this second example it is clear that it can be misleading to interpret the parameter K_m in anything other than purely operational terms (i.e. the concentration of substrate at which the velocity is half maximal). It is *sometimes* helpful

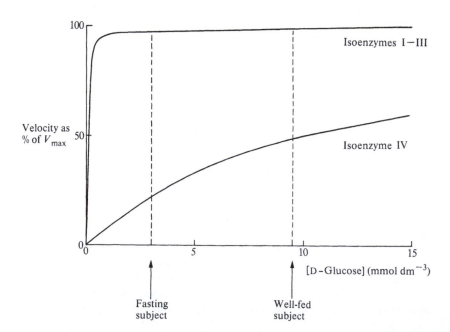

Fig. 4.5 The variation in velocity of the reactions catalysed by hexokinase isoenzymes with concentration of D-glucose.

to think of K_m as a crude measure of the affinity of an enzyme for its substrate (a high K_m implying weak affinity), but the comment made at the end of Section 4.3.1.1 must be borne in mind.

4.3.1.4 Some extensions to the simple model (Scheme 4.1)

It is worthwhile at this stage to mention some extensions to the relatively simple model (Scheme 4.1) we have considered up to now.

(i) More than one intermediate

In most cases it is probably unrealistic to suggest that only one intermediate (enzyme-containing complex) is involved in the reaction pathway (Scheme 4.1). There is likely to be a 'Michaelis complex' representing the initial (non-covalent) association of enzyme and substrate and at least one additional complex in which some bond rearrangement has taken place. In the case of the hydrolysis of ester and amide substrates catalysed by chymotrypsin, there is clear evidence for the involvement of an acyl–enzyme intermediate (Chapter 5, Section 5.4.1.5). If we extend Scheme 4.1 as follows:

$$E + S \rightleftharpoons ES \rightleftharpoons ES' \longrightarrow E + P,$$

then it can be shown by applying the steady-state assumption to the various intermediates in the pathway (ES and ES') that the rate equation is of the same form as eqn (4.6) but that the interpretation of K_m in terms of the rate constants of the individual steps will be different from that in the simple case (eqn (4.5)).[32]

(ii) Substrate inhibition

Occasionally it is found that there is a decrease in rate at high substrate concentrations. In the Lineweaver–Burk plot (Fig. 4.3(a)), for example, this would be manifested by an upward curvature at low values of $1/[S]$. For example, in the case of the hydrolysis of D-fructose 1,6-bisphosphate (FBP) catalysed by fructose-bisphosphatase, upward curvature is noted at FBP concentrations above 0.1 mmol dm^{-3} (Fig. 4.6).

This phenomenon is known as substrate inhibition and is usually interpreted in terms of the existence of two types of substrate-binding site in the enzyme. Occupation of the first, high-affinity, type at low [FBP] leads to 'normal' kinetic behaviour (the linear part of Fig. 4.6). At high [FBP], the second, low-affinity, type of site becomes occupied and this is presumed to inhibit the catalytic reaction taking place at the first type of site. Substrate inhibition is also observed in the case of 6-phosphofructokinase (see Chapter 6, Section 6.4.1.1). For a fuller discussion of this phenomenon, see references 33 and 34.

(iii) Multiple active sites

Complex kinetics are often observed when an enzyme is composed of a number of subunits and possesses more than one active site. If interactions occur between these various sites, non-linearity in the kinetic plots (Fig. 4.3) will be observed, characteristic of positive or negative cooperativity. These points are discussed in more detail in Chapter 6 (Section 6.2.2.2).

Other more detailed texts should be consulted for details of additional complications to the simple model (Scheme 4.1) such as inhibition by products[20,35] and interference from reverse reactions.[20,22,35]

4.3.2 Inhibition of one-substrate reactions

The study of the effects of inhibitors on enzyme-catalysed reactions is important not only to introduce various terms such as competitive inhibition, but also to

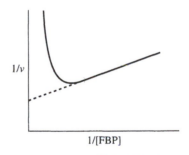

Fig. 4.6 Lineweaver–Burk plot for the reaction catalysed by fructose-bisphosphatase. The solid line represents the experimental data and shows that substrate inhibition occurs at high concentrations of FBP (above about 0.1 mmol dm^{-3}).

* The effect is reversible if it is decreased by lowering the concentration of the inhibitor (e.g. by dilution or dialysis). The distinction between reversible and irreversible inhibition is not absolute and may be difficult to make if the inhibitor binds very tightly to the enzyme and/or is released very slowly. In these circumstances the inhibitors are termed tight-binding inhibitors.[36]

give information on the active site of an enzyme (see Chapter 5, Section 5.4.1.3) and on inhibition which may be of possible physiological significance (see Chapter 6, Section 6.2.2.1). We shall confine our attention to *reversible inhibitors*, i.e. inhibitors which combine reversibly with an enzyme* rather than those which cause irreversible covalent modification. The use of the latter in the determination of enzyme mechanisms is discussed in Chapter 5, Section 5.4.4.

4.3.2.1 Theoretical background

One of the ways in which inhibition of enzyme-catalysed reactions can be discussed is in terms of a general scheme shown below (Scheme 4.2).

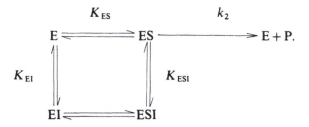

Scheme 4.2 A general scheme for inhibition of enzyme-catalysed reactions. K_{ES}, K_{ESI}, and K_{EI} represent dissociation constants. ESI is assumed to be inactive.

We shall assume that the enzyme-containing complexes are in equilibrium with each other, i.e. that the breakdown of ES to generate product does not significantly disturb the equilibrium. A general kinetic equation for this scheme can then be derived as described in Appendix 4.2 to this chapter:

$$v = V_{max} \frac{\dfrac{[S]}{K_{ES}}}{1 + \dfrac{[S]}{K_{ES}} + \dfrac{[I]}{K_{EI}} + \dfrac{[S][I]}{K_{ESI} \cdot K_{ES}}},$$

or in reciprocal form:

$$\frac{1}{v} = \frac{1}{V_{max}}\left[1 + \frac{[I]}{K_{ESI}}\right] + \frac{K_{ES}}{V_{max}}\left[1 + \frac{[I]}{K_{EI}}\right]\frac{1}{[S]}. \qquad (4.9)$$

If we apply the steady-state approximation to Scheme 4.2, a complex kinetic expression containing terms in $[I]^2$ and $[S]^2$ is derived that is difficult to test experimentally[37] and will not be discussed further here.

The general equation (4.9) is simplified if certain assumptions are made about the magnitudes of the various dissociation constants. These *limiting cases* will be discussed in terms of the Lineweaver–Burk plots (Fig. 4.3(a)) of kinetic data, but we could, of course, use the other plots shown in Figs 4.3 and 4.4. A more detailed classification of the various types of inhibition[36] and a description of the Dixon plot for the determination of inhibitor constants are given in Appendix 4.2.

4.3.2.2 Treatment of data

Case (i). Competitive inhibition

If we assume that $K_{ESI} = \infty$ (i.e. that the ES complex cannot combine with I nor the EI complex with S), then eqn (4.9) reduces to

$$\frac{1}{v} = \frac{1}{V_{max}} + \frac{K_{ES}}{V_{max}}\left[1 + \frac{[I]}{K_{EI}}\right]\frac{1}{[S]}$$

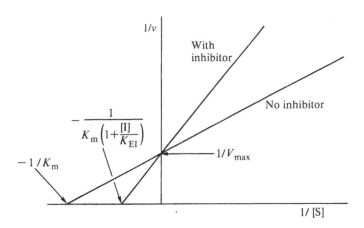

Fig. 4.7 The effect of a competitive inhibitor on a Lineweaver–Burk plot of enzyme kinetic data.

and the effect of the inhibitor on the Lineweaver–Burk plot is shown in Fig. 4.7. V_{max} is unaffected but the apparent K_m is increased by a factor $(1 + [I]/K_{EI})$. Effectively, the inhibitor pulls some of the enzyme over into the form of the EI complex. When the concentration of S is increased sufficiently, the effect on the velocity can be overcome (thus giving rise to the term 'competitive'). Competitive inhibitors have been widely used in the elucidation of enzyme mechanisms by X-ray crystallography (see Chapter 5, Section 5.4.3). Many examples of competitive inhibition are found in one-substrate reactions; thus, carbamoylcholine is a competitive inhibitor with respect to acetylcholine in the reaction catalysed by acetylcholinesterase from bovine erythrocytes:

$$(CH_3)_3\overset{+}{N}—CH_2—CH_2—O—\overset{\displaystyle O}{\overset{\|}{C}}—NH_2 \qquad \text{carbamoylcholine}$$

$$(CH_3)_3\overset{+}{N}—CH_2—CH_2—O—\overset{\displaystyle O}{\overset{\|}{C}}—CH_3 \qquad \text{acetylcholine}$$

Although in this case the structural similarity between inhibitor and substrate makes it very likely that the two molecules bind to the same site on the enzyme, it is not always a justified conclusion that a competitive inhibitor binds to the active site.[38] (Such behaviour could arise from an indirect effect, whereby the inhibitor bound at a distinct site and exerted its effect on the active site via structural changes in the enzyme; see Chapter 6, Section 6.2.2).

Case (ii). Non-competitive inhibition
If we assume that $K_{ESI} = K_{EI}$ (i.e. that the binding of S to the enzyme does not affect the binding of I), then eqn (4.9) reduces to

$$\frac{1}{v} = \frac{1}{V_{max}}\left[1 + \frac{[I]}{K_{EI}}\right] + \frac{K_{ES}}{V_{max}}\left[1 + \frac{[I]}{K_{EI}}\right]\frac{1}{[S]}$$

and the effect of the inhibitor on the Lineweaver–Burk plot is shown in Fig. 4.8.
K_m remains unaffected, but V_{max} is decreased by a factor

$$1\Big/\left(1 + \frac{[I]}{K_{EI}}\right).*$$

* Reference to Appendix 4.2 shows that this situation is more correctly termed pure non-competitive inhibition.

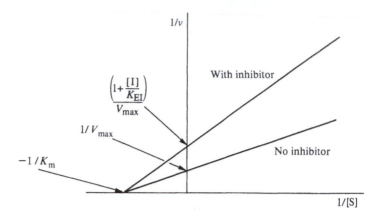

Fig. 4.8 The effect of a non-competitive inhibitor on a Lineweaver–Burk plot of enzyme kinetic data.

The inhibitor pulls both E and ES over into inactive forms (EI and ESI, respectively) but does not affect the distribution between them. Hence. increasing the concentration of S does not serve to overcome the effect of the inhibitor on the velocity. Examples of non-competitive inhibition in one-substrate reactions are less common than those of competitive inhibition. In the case of fructose-bisphosphatase, AMP acts as a non-competitive inhibitor with respect to the substrate fructose 1,6-bisphosphate. There are many examples, however, of non-competitive inhibition in multisubstrate reactions (see Section 4.3.5.5).

Case (iii). Uncompetitive inhibition
If we assume that $K_{EI} = \infty$ (i.e. that I cannot combine with E, but only with the ES complex) then eqn (4.9) reduces to

$$\frac{1}{v} = \frac{1}{V_{max}}\left[1 + \frac{[I]}{K_{ESI}}\right] + \frac{K_{ES}}{V_{max}}\frac{1}{[S]}$$

and the effect of the inhibitor on the Lineweaver–Burk plot is shown in Fig. 4.9. Both K_m and V_{max} are affected by the inhibitor giving rise to parallel lines. There are very few cases indeed of uncompetitive inhibition in one-substrate reactions (an example is the inhibition of alkaline phosphatase from rat intestine by L-phenylalanine[39]) but more examples are found in multisubstrate reactions (e.g.

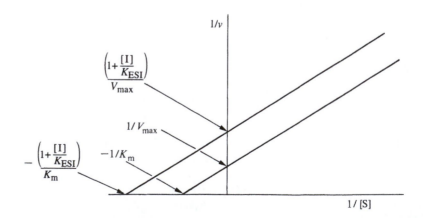

Fig. 4.9 The effect of an uncompetitive inhibitor on a Lineweaver–Burk plot of enzyme kinetic data.

S-adenosylmethionine behaves as an uncompetitive inhibitor towards ATP in the reaction catalysed by methionine adenosyltransferase from yeast.[40]

4.3.2.3 Significance of results

We have introduced the terms competitive, non-competitive, and uncompetitive inhibitors by referring to the model depicted in Scheme 4.2 for one-substrate reactions. In our view it is better to define these types of inhibitors in terms of their effects on the parameters K_m and V_{max} (Figs 4.7, 4.8, and 4.9) rather than by referring to relationships between the binding sites for S and I on the enzyme, because the latter are much more difficult to determine.[38] As we shall see later (Section 4.3.5.4), the terms competitive, non-competitive, and uncompetitive are extremely useful for classifying the effects of inhibitors in multi-substrate reactions, which can be of value in distinguishing between various types of reaction mechanism. In a number of cases (see Chapter 6, Section 6.2.2.1) the effects of an inhibitor do not fall into any of the three limiting cases depicted in Figs 4.7, 4.8, and 4.9. In such cases the inhibition is usually referred to as being of a 'mixed' type and would require different equations and assumptions from those described in this section.[33,34,41] (See the discussion in Appendix 4.2.)

4.3.3 The effect of changes in pH on enzyme-catalysed reactions

There are a number of distinct effects that a change in pH can have on enzyme-catalysed reactions, e.g. inactivation of the enzyme outside a certain pH range or a change in the ionization state of the substrate(s). A third possibility is that there could be a change in the equilibrium position if H^+ is involved in the reaction, e.g. in the reaction catalysed by creatine kinase:

$$\text{Creatine} + \text{MgATP}^{2-} \rightleftharpoons \text{phosphocreatine}^{2-} + \text{MgADP}^- + \text{H}^+.$$

In this case, increasing the pH will displace the equilibrium in favour of phosphocreatine synthesis.

However, the possibility of most interest to us is that there are changes in the ionization state of amino-acid side-chains that are essential for the catalytic activity of the enzyme. In this case we might hope to obtain information about the nature of these side-chains.

4.3.3.1 Theoretical background

The effects of pH on the velocity of an enzyme-catalysed reaction can be complex, since both K_m and V_{max} can be affected, and, in order to undertake a detailed analysis of the ionizations involved, values of both parameters should be obtained over a range of pH values. A number of detailed treatments of the effects of pH on K_m and V_{max} have been given.[42–45] We should note that it is usually easier to analyse changes in V_{max}, since this parameter generally reflects a single rate constant, whereas K_m is a function of several rate constants.*

In order to see how pK_a values can be obtained from the experimental data, we can consider the simple case of a single ionizing side-chain in the enzyme. (Similar equations would result if we considered a single ionizing side-chain in the enzyme–substrate complex.)

$$\text{EH}^+ \rightleftharpoons \text{E} + \text{H}^+$$

* Analysis of a scheme in which the free enzyme and enzyme–substrate complex each possess two ionizing side-chains shows that changes in V_{max} depend on ionizations of the enzyme–substrate complex; changes in V_{max}/K_m depend on ionizations of the free enzyme, and changes in K_m depend on ionizations of both the free enzyme and the enzyme–substrate complex. For further details of these effects and the analysis of more complex schemes, references 42–45 should be consulted.

We shall assume that EH^+ is inactive and that E is the active form. Now the acid dissociation constant, K_a, is given by

$$K_a = \frac{[E][H^+]}{[EH^+]},$$

$$\therefore [EH^+] = \frac{[E][H^+]}{K_a}.$$

The fraction, F, of enzyme in the unprotonated (active) form is given by

$$F = \frac{[E]}{[E]+[EH^+]} = \frac{K_a}{K_a +[H^+]}.$$

Let $(V_{max})_m$ equal the maximum rate when all the enzyme is in the unprotonated form. Then at any pH, the observed V_{max} is given by

$$V_{max} = (V_{max})_m \cdot F.$$

Thus

$$V_{max} = (V_{max})_m \cdot \frac{K_a}{K_a +[H^+]}. \tag{4.10}$$

If we have a second ionizing group in the enzyme, such as shown in the following scheme:

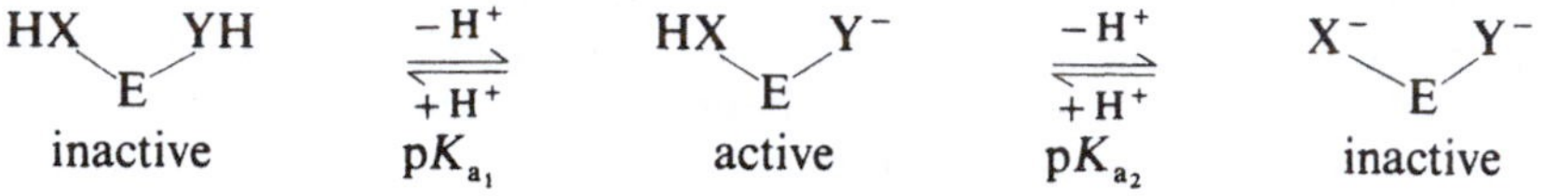

the expression for V_{max} at any pH can be shown to be

$$V_{max} = \frac{(V_{max})_m}{1 + \dfrac{[H^+]}{K_{a_1}} + \dfrac{K_{a_2}}{[H^+]}}. \tag{4.11}$$

4.3.3.2 Treatment of data

Consider eqn (4.10). At pH values well below the pK_a (i.e. when $[H^+] \gg K_a$), then

$$V_{max} = (V_{max})_m \cdot \frac{K_a}{[H^+]}.$$

Taking logarithms:

$$\log_{10} V_{max} = \log_{10} (V_{max})_m - pK_a + pH,$$

so that a plot of $\log_{10} V_{max}$ against pH will be linear with a slope of 1. (In practice this will work well up to about 1.5 pH units below the pK_a.)

At pH values well above the pK_a (i.e. when $[H^+] \ll K_a$) then

$$V_{max} = (V_{max})_m,$$

i.e. there is no variation of V_{max} with pH.

The plot of $\log_{10} V_{max}$ against pH will therefore be of the form shown in Fig. 4.10. Extrapolation of the linear portions of the plot will give the value of pK_a at the point of intersection.

A similar analysis of eqn (4.11) shows that if two ionizing groups were involved, a plot of $\log_{10} V_{max}$ against pH would have three distinct regions of slope 1, 0, and –1 as shown in Fig. 4.11. These regions correspond to the cases where $[H^+] \gg K_{a_1}$, $K_{a_1} \ll [H^+] \gg K_{a_2}$, and $K_{a_2} \gg [H^+]$ respectively.

As shown in Fig. 4.11, the values of pK_{a_1} and pK_{a_2} can be obtained from the points of intersection of the extrapolated linear portions of the plot. However, if the two pK_a values are closer than about 1.5 pH units, the ionizations will not be independent and the pK_a values derived may need correction to obtain the true values.[42,46]

4.3.3.3 Significance of results

A study of the effect of changes in pH on the velocity of an enzyme-catalysed reaction generally leads to the conclusion that there is an 'optimum pH' for that reaction. In view of the various types of effect of pH mentioned in Section 4.3.3, it is clear that such a term should be used with a certain degree of caution. There are cases known in which studies of enzyme kinetics at the optimum pH have given misleading results when considering the *in vivo* situation (see Chapter 8, Section 8.3.2).

It is often tempting to try to assign the measured pK_a values (Figs 4.10 and 4.11) to particular types of amino-acid side-chain, using the values of pK_a for free amino acids (see Chapter 3, Section 3.3.1). However, there are considerable pitfalls in this procedure, since the environment of a side-chain in an enzyme can be very different from that of the free amino acid and this can cause a large shift in pK_a. Thus, in pepsin, one aspartic acid side-chain has a pK_a of about 1.0, some 3 pH units lower than that of free aspartic acid,[47] and in triosephosphate isomerase the pK_a of His 95 is at least 2 pH units lower than that of free histidine (see Chapter 5, Section 5.5.2.4). Nevertheless, in some cases it has been possible from pH-rate studies to implicate particular side-chains in the mechanisms of certain enzymes, e.g. histidine side-chains in pancreatic ribonuclease,[48] chymotrypsin,[49] and type I dehydroquinase (see Chapter 5, Section 5.5.3.2) and carboxylic-acid side-chains in lysozyme.[50] As described in Chapter 5 (Section 5.4), these conclusions can be supported by the results of other experiments.

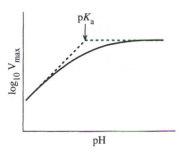

Fig. 4.10 The effect of pH on the V_{max} of an enzyme-catalysed reaction if only the unprotonated form of the enzyme is catalytically active. The solid curve represents the experimental data.

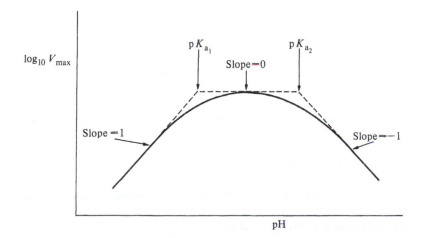

Fig. 4.11 The effect of pH on the V_{max} of an enzyme-catalysed reaction when two ionizing groups are involved. The solid curve represents the experimental data.

* A notable exception is the classic study of the chymotrypsin-catalysed hydrolysis of tryptophanamide, in which, from the temperature dependence of the reaction rate and other data, an energy profile (see Chapter 5, Fig. 5.1) for the reaction could be constructed.[51]

4.3.4 The effect of changes in temperature on enzyme-catalysed reactions

In general, the effects of changes in temperature on the rates of enzyme-catalysed reactions do not provide much useful information as far as the mechanism of catalysis is concerned.* However, these effects can be important in indicating structural changes in enzymes and in considering enzyme activity in poikilothermic organisms, in which intracellular temperatures vary considerably, so a short discussion of the effects will be given in this section. More detailed discussions can be found in references 42, 52, and 53. The effects of temperature on the stability of enzymes have been mentioned in Chapter 3, Section 3.6.1.

4.3.4.1 Theoretical background

From the transition-state theory of chemical reactions, we can derive an expression for the variation of the rate constant, k, with temperature. This is of the form of eqn (4.12), which is sometimes referred to as the Arrhenius expression (see also Chapter 5, Section 5.3),

$$k = A e^{-E_a/RT} \tag{4.12}$$

where A is known as the pre-exponential factor,
  R is the gas constant,
  T is the (absolute) temperature, and
  E_a is the activation energy for the reaction.

From eqn (4.12) it is clear that there is an exponential increase in reaction rate with temperature. We can introduce a quantity, known as the Q_{10}, which is the ratio (or quotient) of the reaction rate at $(T + 10)$ K compared with that at T K. Using eqn (4.12) it can be shown that at temperatures around 300 K (27°C), Q_{10} is approximately given by $e^{E_a/75000}$. For many chemical reactions, the values of Q_{10} are in the range 2–4, corresponding to activation energies of about 50–100 kJ mol^{-1}.

Similar considerations apply in the case of enzyme-catalysed reactions, but the values of E_a (and hence of Q_{10}) are generally lower than the corresponding values for non-enzyme-catalysed reactions when a comparison can be made. For instance, in the hydrolysis of urea catalysed by acid, the value of E_a is 100 kJ mol^{-1}, whereas the same reaction catalysed by urease has a much lower E_a (42 kJ mol^{-1}).

Above a certain temperature an enzyme will tend to lose the compact three-dimensional structure that is required for catalytic activity. Incubation of many enzymes at temperatures above about 323 K (50°C) leads to a fairly rapid loss of catalytic activity, although it should be noted that many enzymes can withstand exposure to high temperatures (see Section 4.3.4.3 and Chapter 3, Section 3.6.1). The combination of these effects leads to a maximum being observed in the graph of activity against temperature. The term 'optimum temperature', referring to the temperature at which the activity of a given enzyme is a maximum, should not be used.[42] This is because the actual value obtained will usually depend on the precise experimental conditions employed, in particular the length of time for which the enzyme is incubated at the temperatures being investigated.

*Cornish-Bowden[42] has pointed out that the analysis of Arrhenius plots is more satisfactory if the variations of K_m and V_{max} are analysed separately over a range of temperatures. At any temperature, the observed velocity will depend on both parameters.

4.3.4.2 Treatment of data

According to the Arrhenius expression (eqn (4.10)), a plot of ln(velocity) against $1/T$ gives a straight line of slope $-E_a/R$. Taking into account the loss of catalytic activity at high temperatures, we would expect the Arrhenius plot for an enzyme-catalysed reaction to resemble that shown in Fig. 4.12(a).*

Complications arise when an enzyme can arise in two (or more) interconvertible forms with different activation energies. There will then be a discontinuity in the Arrhenius plot around the temperature where the change-over between the two forms becomes significant (Fig. 4.12(b)). An example of this type of behaviour is provided by the enzyme adenosinetriphosphatase (Na⁺, K⁺-activated) for which the transition probably arises from structural changes in the tightly bound phospholipid molecules associated with the enzyme.[54] A detailed discussion of the causes of discontinuous Arrhenius plots has been given.[55]

4.3.4.3 Significance of results

As mentioned in Section 4.3.4.1 the values of Q_{10} (and hence of E_a) for enzyme-catalysed reactions are generally lower than for reactions not subject to enzyme catalysis. For most enzymes in homoiothermic species (e.g. mammals) the value of Q_{10} is approximately 2. However, in species that have to adapt to cold conditions it is important that the value of Q_{10} is lower than this so that essential metabolic reactions are not slowed down too much. It is found, for instance, that many enzymes from intertidal species such as anemones and winkles have values of Q_{10} of about 1 (i.e. the rates do not change significantly with temperature).[56] The structural alterations associated with such kinetic behaviour would be of considerable interest. In a number of species of terrestrial insects there are major alterations in metabolism to coordinate the production of cryoprotectants such as glycerol and sorbitol. In such cases there appear to be differential effects on the kinetics and regulatory properties of the glycolytic enzymes.[57]

Although, as shown in Fig. 4.12, enzymes generally lose activity at temperatures above about 323 K (50°C), it should be noted that the enzymes from thermophilic bacteria (e.g. *Thermotoga maritima*) are stable at temperatures up to 363 K (90°C). The cause of the enhanced thermal stability of these enzymes is not well understood, although it may be associated with a greater proportion of hydrophobic amino acids and increased packing density of the core of the protein.[58] In the case of glyceraldehyde-3-phosphate dehydrogenase (which contains four subunits) it has been suggested that additional inter-subunit salt bridges may also contribute to the greater thermal stability of the enzyme from thermophiles.[58,59]

4.3.5 Two-substrate reactions

Up to now our discussion has been concerned with enzyme-catalysed reactions involving a single substrate and this has allowed us to introduce terms such as K_m, V_{max}, competitive inhibition, etc. This treatment would appear to be of limited value, since oxidoreductases, transferases, and ligases (Chapter 1, Section 1.5.1) catalyse reactions which by definition involve more than one substrate. Nevertheless, many of the concepts involved in one-substrate kinetics can be useful in the analysis of the kinetics of these more complex reactions. In this section we shall explain the broad divisions of types of mechanism of two-substrate reactions and indicate how the necessary equations are derived. We shall then show how to use these equations in the analysis of experimental data, and conclude by describing how it is possible to distinguish between the various types of possible mechanism. More detailed treatments of these topics can be found in a number of textbooks.[60–65]

4.3.5.1 Theoretical background

We can divide two-substrate reactions into two main categories.

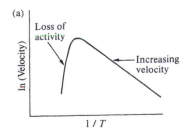

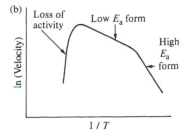

Fig. 4.12 Arrhenius plots for enzyme-catalysed reactions illustrating the loss of catalytic activity at high temperature. (a) Only one form of an enzyme involved; (b) two interconvertible forms with different activation energies involved.

(i) *Those involving a ternary complex (i.e. a complex containing enzyme and both substrates)*

In these cases the reaction

$$E + A + B \rightarrow E + P + Q$$

proceeds via ternary complexes of the type EAB and EPQ:

$$E + A + B \rightarrow EAB \rightarrow EPQ \rightarrow E + P + Q.$$

This category can be further subdivided:

(i) (a) Those reactions in which the ternary complex is formed in an *ordered* manner, i.e. the second substrate (say, B) can bind to the enzyme only after A has already bound:

$$E + A \rightarrow EA$$

$$EA + B \rightarrow EAB \qquad (\text{but } E + B \xrightarrow{\quad\times\quad} EB).$$

(i) (b) Those reactions in which the ternary complex is formed in a *random* manner (i.e. either substrate can bind first):

$$E + A \rightarrow EA \qquad\qquad E + B \rightarrow EB$$

or

$$EA + B \rightarrow EAB \qquad\qquad EB + A \rightarrow EAB.$$

(ii) *Those not involving a ternary complex*

The most important class of reactions in this category proceed by *enzyme substitution* or *ping-pong* mechanisms, i.e. a modified form of the enzyme (E′) is formed together with the first product, before the second substrate is bound:

$$E + A \rightarrow E' + P$$

$$E' + B \rightarrow E + Q.$$

A second class of enzymes in this category operates via a Theorell–Chance mechanism (named after the original investigators) in which a ternary complex is presumably formed but its breakdown to yield the first product is very fast so that the ternary complex is kinetically insignificant. This type of mechanism has been shown to apply in the oxidation of ethanol and other primary alcohols by NAD^+ catalysed by alcohol dehydrogenase from horse liver.[66,67]

An alternative way of representing enzyme-catalysed reactions has been proposed by Cleland.[68] The progress of the reaction is shown by a horizontal line (branched if necessary) with enzyme forms depicted below the line. Successive additions of substrates and release of products are depicted by vertical arrows. The rate constants can be indicated, if necessary adjacent to these arrows.

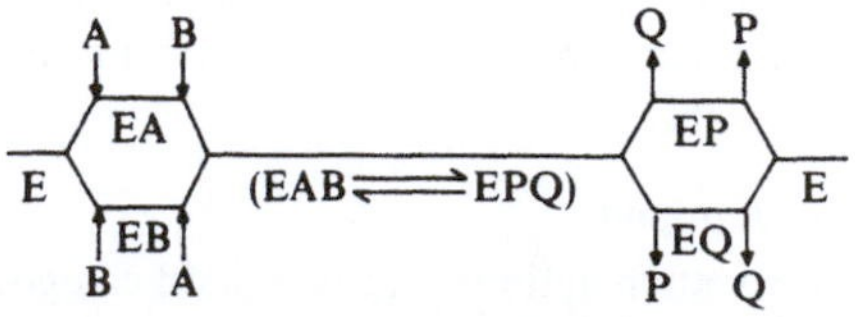

Thus, the random ternary complex mechanism (type (i) (b) above) would be and the enzyme-substitution mechanism (type (ii) above) would be

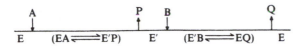

4.3.5.2 Derivation of equations for two-substrate reactions

The derivation of equations to describe the kinetics of the various types of mechanism does not involve any fundamental principles in addition to those used for one-substrate reactions (Section 4.3.1.1), although the algebra is obviously more complex. A good account of the derivations is given by Cornish-Bowden; in this book there is also an explanation of the King–Altman procedure, which simplifies the derivations considerably.[69,70] Computer-assisted methods for deriving the rate equations have also been described.[71] The basic idea is to evaluate the concentrations of the various enzyme-containing complexes in terms of the total concentration of enzyme, under the stated conditions of substrate concentrations. In order to do this we use the steady-state assumption* in the same way as for one-substrate reactions (Section 4.3.1.1). The velocity of the overall reaction is then equal to the concentration of the complex that precedes regeneration of free enzyme multiplied by the rate constant for the regeneration step, e.g. in the reaction scheme

$$E \rightleftharpoons EA \rightleftharpoons EAB \rightleftharpoons EPQ \longrightarrow EP \xrightarrow{k_i} E + P$$

the rate of the overall reaction, v, equals $k_1[EP]$. This is because we need to regenerate free enzyme (by the k_1 step above) in order to allow the reaction to continue.

The equations for v, the initial rate of the reaction, which result from these treatments are of the following forms.[†]

For the ternary complex mechanisms ((i) (a) and (i) (b) in Section 4.3.5.1):

$$v = \frac{V_{max}[A][B]}{K'_A K_B + K_B[A] + K_A[B] + [A][B]} \tag{4.13}$$

(an equation of this type is also derived for the Theorell–Chance mechanism).

For the enzyme substitution mechanisms ((ii) in Section 4.3.5.1):

$$v = \frac{V_{max}[A][B]}{K_B[A] + K_A[B] + [A][B]}. \tag{4.14}$$

4.3.5.3 Significance of the parameters in the equations

V_{max} in eqns (4.13) and (4.14) represents the maximum velocity at saturating levels of substrates A and B.

In a purely *practical* sense, the constants K_A and K_B in eqns (4.13) and (4.14) represent the Michaelis constants (for substrates A and B, respectively) in the presence of saturating concentrations of the other substrate. This is readily shown; e.g. in eqn (4.13) if we divide the numerator and denominator by [B] we obtain

$$v = \frac{V_{max}[A]}{\dfrac{K'_A K_B}{[B]} + \dfrac{K_B[A]}{[B]} + K_A + [A]},$$

* If we apply the steady-state assumption to the case of the random order ternary complex mechanism (i) (b) in Section 4.3.5.1, the equation obtained is complex, with terms containing the square of the concentrations of substrate. In this case we can apply the equilibrium assumption (Section 4.3.1.1) and assume that E, EA, EB, and EAB are all in equilibrium with one another. The resulting equation (4.13) describes many reactions of this type and to that extent the equilibrium assumption is justified.

† There are a number of different formulations of these equations. Details of some of these are given in Appendix 4.3. We have used the forms shown in eqns (4.13) and (4.14) to show the analogies with the equation for one-substrate reactions (eqn (4.4)).

and when $[B] \to \infty$, $1/[B] \to 0$, so that

$$v = \frac{V_{max}[A]}{K_A + [A]},$$

which is of the same form as eqn (4.6).

K_A' in eqn (4.13) does not have a simple practical meaning.

In terms of the *mechanisms* of the reactions, the constants K_A', K_A, and K_B represent combinations of the rate constants of individual steps in the reaction (compare eqn (4.5) for one-substrate reactions). Their precise meanings vary according to the type of mechanism under discussion. In the case of the random-order ternary complex mechanism ((i) (b) in Section 4.3.5.1), K_A', K_A, and K_B have simple meanings in terms of dissociation constants, because in this case we used the equilibrium assumption to derive eqn (4.13).

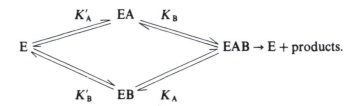

K_A', K_A, and K_B represent the dissociation constants for EA, EAB to yield EB, and EAB to yield EA respectively. (Note that $K_B' = K_A' K_B / K_A$.)

4.3.5.4 Treatment of data

The values of the parameters V_{max}, K_A', K_A, and K_B in eqn (4.13) can be derived from the experimental data by a computer fitting procedure[22] analogous to that mentioned for one-substrate reactions (Section 4.3.1.2). However, the values are more commonly obtained by graphical methods involving primary and secondary plots.

The velocity, v, is measured at various values of [A], keeping the concentration of B constant; this procedure is then repeated at other fixed values of [B].

By taking the inverse of eqn (4.13), i.e.

$$\frac{1}{v} = \left[1 + \frac{K_A}{[A]} + \frac{K_B}{[B]} + \frac{K_A' K_B}{[A][B]} \right] \frac{1}{V_{max}}, \tag{4.15}$$

we see that a *primary* plot of $1/v$ against $1/[A]$ at a fixed value of [B] will be linear (Fig. 4.13) with

$$slope = \frac{1}{V_{max}} \left[K_A + \frac{K_A' K_B}{[B]} \right] \tag{4.16}$$

and

$$intercept\ (on\ y\text{-}axis) = \frac{1}{V_{max}} \left[1 + \frac{K_B}{[B]} \right]. \tag{4.17}$$

As [B] increases, both the slope and intercept will decrease.

(The lines in the primary plot (Fig. 4.13) intersect at a point that can be above, on, or below the x-axis, depending on the relative values of K_A', K_A, and K_B. At

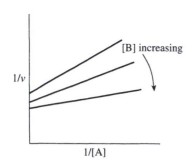

Fig. 4.13 Primary plot of enzyme kinetic data according to eqn (4.15).

any value of [B] an 'apparent' K_m for A can be derived from the intercept on the x-axis. From eqn (4.15) it can be shown that this K_m is $([B]K_A + K'_A K_B)/([B] + K_B)$; K_m will vary with [B], unless $K'_A = K_A$ when the point of intersection is on the x-axis.)

Secondary plots of the slopes and intercepts of the primary plot against 1/[B] can then be constructed (Fig. 4.14(a) and (b)).

By inspection of eqn (4.16) we see that a plot of the slope against 1/[B] (Fig. 4.14(a)) is linear with a *slope* of $K'_A K_B/V_{max}$ and an *intercept on the y-axis* of K_A/V_{max}.

From eqn (4.17) we see that a plot of the intercepts against 1/[B] (Fig. 4.14(b)) is linear with a *slope* of K_B/V_{max} and an *intercept on the y-axis* of $1/V_{max}$. Thus it is possible to determine in turn V_{max}, K_B, K_A, and K'_A from the slopes and intercepts of these secondary plots.

As mentioned in Section 4.3.5.3, the parameters K'_A, K_A, and K_B have a simple meaning in the random-order ternary complex mechanism ((i) (b) in Section 4.3.5.1). If it is known from other studies (described in Section 4.3.5.5) that this mechanism applies to a particular enzyme then we can draw some conclusions from the relative values of these parameters. Thus, for example, in the case of creatine kinase from rabbit muscle assayed in the direction of phosphocreatine synthesis, it is found that $K'_{MgATP} > K_{MgATP}$ which means that the binding of one substrate is enhanced by the binding of the other. (This is known as *substrate synergism*.) By contrast, in the oxidation of ethanol by NAD^+ catalysed by alcohol dehydrogenase from yeast, $K'_{NAD^+} = K_{NAD^+}$ and hence the binding of one substrate is independent of the binding of the other.

Turning now to eqn (4.14) for the enzyme substitution mechanism ((ii) in Section 4.3.5.1) we find that a primary plot of $1/v$ against $1/[A]$ at fixed values of [B] consists of a set of parallel lines (Fig. 4.15). This can be shown by taking the inverse of eqn (4.14):

$$\frac{1}{v} = \left[1 + \frac{K_A}{[A]} + \frac{K_B}{[B]} \right] \frac{1}{V_{max}}. \tag{4.18}$$

Hence, a plot of $1/v$ against $1/[A]$ has a *slope* of K_A/V_{max}; this slope is independent of [B], resulting in a set of parallel lines. We can derive the parameters K_B and V_{max} (and hence K_A) by performing a *secondary* plot of the y-axis intercepts of the primary plot against $1/[B]$. This secondary plot will have a *slope* of K_B/V_{max} and a *y-axis intercept* of $1/V_{max}$.

4.3.5.5 Significance of results: distinction between the various mechanisms for two-substrate reactions (Section 4.3.5.1)

From the discussion in Section 4.3.5.4 it can be seen how an enzyme substitution mechanism can be distinguished from a ternary complex mechanism, since the former gives rise to a series of parallel lines in the primary plot (Fig. 4.15). It should be emphasized that considerable care is needed to ensure that lines in a plot such as Fig. 4.15 are truly parallel, and it is probably best to make an estimate of the magnitude of the K'_A term in the general equation (4.13) and of the error of this estimate (K'_A should equal zero for the enzyme substitution mechanism). If the primary plot suggests that an enzyme substitution mechanism is operative in a particular case, confirmatory evidence can be obtained from:

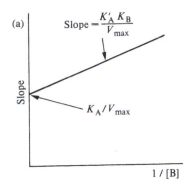

(a) Slope $= \dfrac{K'_A K_B}{V_{max}}$

K_A/V_{max}

1 / [B]

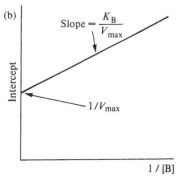

(b) Slope $= \dfrac{K_B}{V_{max}}$

$1/V_{max}$

1 / [B]

Fig. 4.14 Secondary plots of (a) slopes and (b) intercepts of primary plot (Fig. 4.13) against l/[B].

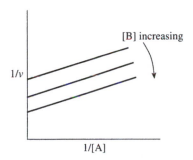

[B] increasing

$1/v$

1/[A]

Fig. 4.15 Primary plot of enzyme kinetic data according to eqn (4.18) (the enzyme substitution mechanism).

(1) demonstration of partial reactions, i.e. conversion of A to P in the absence of B:

$$E + A \rightleftharpoons E' + P$$

(2) isolation of a modified form of the enzyme (E′ in the above scheme), often using a suitable radioactive label.

Thus, in a reaction catalysed by nucleoside diphosphate kinase from erythrocytes, an enzyme substitution mechanism is indicated by the kinetic data:

$$GTP + dGDP \rightleftharpoons GDP + dGTP.$$

In this case it is possible to isolate a modified radioactive enzyme, in which a histidine side-chain is phosphorylated,[72] after incubation of the enzyme with $[\gamma\text{-}^{32}P]GTP$ (i.e. GTP labelled at the γ position of the triphosphate group).

$$E + GTP \rightleftharpoons E - \textcircled{P} + GDP,$$

which would be followed by

$$E - \textcircled{P} + dGDP \rightleftharpoons E + dGTP.$$

(E – $\textcircled{P}$ represents the phosphorylated enzyme.)

*There are in fact two types of phosphoglycerate mutases. The enzyme from plant sources is independent of the 2,3-bisphospho-D-glycerate primer, whereas the enzyme from yeast or mammalian sources requires this primer. However, both types are now classified as isomerases (EC 5.4.2.1).

Other enzymes that operate via an enzyme substitution mechanism include the aminotransferases, in which the pyridoxal phosphate cofactor (see Chapter 1, Section 1.4) attached to the enzyme undergoes a covalent change to form E′, and phosphoglycerate mutase from yeast or mammalian sources*. The reaction catalysed by the latter enzyme would appear to be a one-substrate reaction in which 2-phospho-D-glycerate (2PG) is converted into 3-phospho-D-glycerate (3PG). However, in this case a primer molecule of 2,3-bisphospho-D-glycerate (BPG) is required. We thus have two partial reactions:

$$E + BPG \rightleftharpoons E - \textcircled{P} + 2PG,$$

$$E - \textcircled{P} + 3PG \rightleftharpoons E + BPG,.$$

giving a net reaction

$$3PG \rightleftharpoons 2PG.$$

† Although the Theorell–Chance mechanism (Section 4.3.5.1) also gives an expression of the form of eqn. (4.13), it can be distinguished from the ternary complex mechanisms by comparison of the magnitudes of the various parameters in the rate equation with the rates of the forward and reverse reactions (see reference 73).

From the steady-state kinetic analysis given in Section 4.3.5.4 we are unable to distinguish between an ordered and a random ternary complex mechanism because both give rise to an expression of the form of eqn (4.13).† However, a distinction between the mechanisms can be made if other data are available, e.g. from the following types of experiment.

(i) Substrate-binding experiments

In the random-order ternary complex mechanism, each substrate, A and B, should bind separately to the enzyme, whereas in an ordered mechanism the second substrate (e.g. B) cannot bind in the absence of the first substrate, A. For instance, in the reaction catalysed by lactate dehydrogenase,

$$\text{L-Lactate} + NAD^+ \rightleftharpoons \text{pyruvate} + NADH + H^+,$$

it is found that NAD^+ binds tightly to the enzyme but there is no detectable binding of lactate.[74] This suggests that the reaction proceeds via an ordered mechanism with NAD^+ binding preceding that of lactate, a conclusion that is supported by other evidence (see Chapter 5, Section 5.5.4).

(ii) Product-inhibition patterns

The type of inhibition shown by the products P and Q towards the substrates A and B can be used to indicate a probable mechanism for the reaction (or exclude an alternative mechanism). The equations underlying these experiments are complex and full details can be found in more comprehensive textbooks and reviews.[60,61,63,75] An example of the method is to distinguish between the ordered and random ternary complex mechanisms by studying the inhibition caused by product Q. In the random mechanism, Q behaves as a competitive inhibitor with respect to substrate B, whereas in the ordered mechanism, Q behaves as a mixed-type inhibitor with respect to B (i.e. the Lineweaver–Burk plots in the presence and absence of inhibitor intersect at a point to the left of the *y*-axis). This type of approach can be used with a range of 'dead-end inhibitors' other than the products of the reaction.[76]

(iii) Isotope exchange at equilibrium[77–80]

In this type of experiment we study the enzyme-catalysed rate of exchange of isotope between substrate(s) and product(s) when the components of the reaction mixture are present at their equilibrium concentrations. For instance, in the reaction catalysed by malate dehydrogenase,

$$\text{L-Malate} + \text{NAD}^+ \rightleftharpoons \text{oxaloacetate} + \text{NADH} + \text{H}^+,$$

we set up a mixture containing substrates and products so that, at the pH in question, the ratio [oxaloacetate][NADH]/[malate][NAD$^+$] equals the previously measured equilibrium constant for the reaction. One of the substrates, say NAD$^+$, is isotopically labelled (e.g. with ^{14}C) and after addition of a catalytic amount of enzyme we study the rate at which this isotope is incorporated into NADH. (Note that there is no *net* conversion of NAD$^+$ to NADH, because the system is at equilibrium; there will, however, be forward and backward reactions occurring so that *exchange* between substrate and product can occur.) It was found that as the concentrations of malate and oxaloacetate were raised (in a constant ratio so that the equilibrium state was maintained), the rate of isotope exchange between NAD$^+$ and NADH fell to zero, whereas the rate of exchange between malate and oxaloacetate increased to a plateau value.[81] This result indicates that the enzyme obeys an ordered mechanism of the type

$$\text{E} + \text{NAD}^+ \rightleftharpoons \text{E}^{\text{NAD}^+} \rightleftharpoons \text{E}^{\text{NAD}^+}_{\text{malate}} \rightleftharpoons \text{E}^{\text{NADH}}_{\text{oxaloacetate}} \rightleftharpoons \text{E}^{\text{NADH}} \rightleftharpoons \text{E} + \text{NADH}.$$

As we raise the concentrations of malate and oxaloacetate, we increase the rate at which the binary complexes (E$^{\text{NAD}^+}$ and E^{NADH}) are converted to the ternary complexes. The enzyme thus becomes confined to the central 'box' shown in the above scheme and NAD$^+$ (or NADH) never dissociates to join the pool of free NAD$^+$ (or NADH). Since the enzyme is present only in very small amounts compared with the concentrations of substrates, this means that we will observe no exchange of isotope between free NAD$^+$ and free NADH. The results are not consistent with a random ternary complex mechanism. The scope of the isotope exchange technique has been extended by the development of computer simulation programs which allow the experimental data to be analysed in terms of complex reaction schemes.[80]

As a result of experiments of the type described in (i), (ii), and (iii) it has been possible to assign a number of enzyme-catalysed reactions to the types of mechanism discussed in Section 4.3.5.1. The reactions catalysed by alcohol dehydrogenase (from yeast), creatine kinase, and phosphorylase appear to proceed via random ternary complex mechanisms, whereas the reactions catalysed by lactate dehydrogenase and malate dehydrogenase operate via ordered ternary complex mechanisms; a compilation of some of these results has been made by Fromm.[82]

4.3.6 Reactions involving more than two substrates

A number of enzyme-catalysed reactions involve three substrates; some examples are shown below.

Glyceraldehyde 3-phosphate dehydrogenase

$$\text{D-Glyceraldehyde-3-phosphate} + \text{NAD}^+ + \text{orthophosphate} \rightleftharpoons$$
$$\text{3-phospho-D-glyceroyl phosphate} + \text{NADH}$$

Glutamate dehydrogenase

$$\text{2-Oxoglutarate} + \text{NH}_4^+ + \text{NAD(P)H} \rightleftharpoons \text{L-Glutamate} + \text{NAD(P)}^+ + \text{H}_2\text{O}$$

Isocitrate dehydrogenase

$$\text{2-Oxoglutarate} + \text{CO}_2 + \text{NADH} \rightleftharpoons \textit{threo}\text{-D}_\text{S}\text{-isocitrate} + \text{NAD}^+$$

Tyrosyl-tRNA ligase

$$\text{ATP} + \text{L-tyrosine} + \text{tRNA}^\text{Tyr} \rightleftharpoons \text{AMP} + \text{pyrophosphate} + \text{L-tyrosyl-tRNA}^\text{Tyr}$$

In these reactions there are a number of possible mechanisms, e.g. enzyme substitution or quaternary complex formation by ordered, random, or partly ordered mechanisms. The principles involved in the derivation of equations and the treatment of data are similar to those used in the analysis of two-substrate reactions (Section 4.3.5), although the algebra is considerably more complex[60,61,83] and the requirement for accurate data is more stringent. An example of the analysis of a three-substrate reaction is provided by the studies on glutamate dehydrogenase, in which it was shown that the reaction is most likely of the random type with 2-oxoglutarate, NH_4^+, and NADH binding in any order to form the quaternary complex.[84]

4.4 Pre-steady-state kinetics[85–89]

4.4.1 Background: the need for special techniques

From kinetic studies performed under steady-state conditions we can obtain only very limited information about the rates of individual steps in an enzyme-catalysed reaction. The maximum velocity (V_{max}) can be used to calculate the value of k_{cat}, or turnover number (see Section 4.3.1.1), according to the equation:

$$k_{cat} = \frac{V_{max}}{[\text{E}]_0}$$

where $[\text{E}]_0$ is the concentration of enzyme active sites.

Values of k_{cat} are known for many enzymes and range from about 10 to about 10^7 s^{-1}.[90] The value of k_{cat} gives an idea of the rate constant of the slowest step in an enzyme-catalysed reaction. In steady-state experiments we generally work with very low concentrations of enzyme (1 nmol dm^{-3} or less is not unusual) and under these conditions the reaction is slow enough to allow the use of 'conventional' methods (e.g. manual addition of enzyme to start the reaction and observation by a spectrophotometer). For instance if $[E]_0$ = 1 nmol dm^{-3} and k_{cat} = 10^3 s^{-1}, V_{max} would equal 60 μmol dm^{-3} min^{-1}, which would be quite convenient to monitor spectrophotometrically.

However, if we wish to examine processes that occur on a time-scale of less than about a few seconds (and these will include the steps other than the slowest step of the overall reaction), we need to employ faster methods of mixing and observation. The problem has been neatly solved by the introduction of 'stopped-flow' methods, which can allow observation of reactions that occur on a time scale of a few milliseconds. The basic elements of a 'stopped-flow' apparatus are shown in Fig. 4.16.

The drive barrier is pushed in, usually mechanically, and the contents of the two reactant syringes (e.g. enzyme and substrate(s)) are mixed. The flow of liquid forces the piston of the stopping syringe out until it hits the stop barrier, when the flow stops. At this point the data recorder is triggered and the changes in absorbance or fluorescence of the liquid in the observation chamber can be recorded. The kinetic trace on the data recorder can then be analysed in terms of possible kinetic schemes to assess which scheme gives the best fit to the data and to determine the values of the appropriate rate constants. For this type of apparatus there is a 'dead' time, corresponding to the time interval between the start of the reaction by mixing and the stop of the flow of liquid, of about 1 ms. In some cases, particularly where there is no convenient optical change on reaction, it can be convenient to use a 'quench-flow' approach in which the reaction is stopped shortly after mixing, e.g. by rapid cooling or by addition of an inhibitor or a denaturing agent.[87,88,91] By varying the interval between the start of the reaction and the quench, it is possible to build up a kinetic description of the process. Considerable care is needed in the choice of quenching conditions, and appropriate controls (such as the mixing of enzyme and quenching agent before addition of substrate(s)) must be performed.[88,91] It should also be remembered that, compared with 'stopped-flow', the 'quench-flow' method uses considerably more enzyme and substrate(s), since the progress of the reaction is built up in a point-by-point fashion.

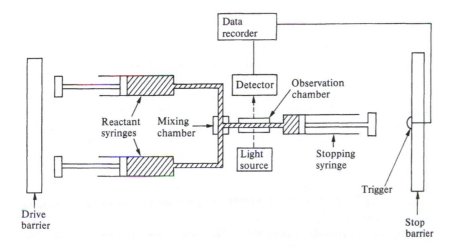

Fig. 4.16 A typical stopped-flow apparatus.

4.4.2 Applications of rapid-mixing techniques

The stopped-flow and quenched-flow techniques have been valuable in elucidating the details of enzyme-catalysed reactions. We shall describe two of the most important types of study; the determination of rate constants of individual steps in the reaction and the identification of transient species.

4.4.2.1 Determination of rate constants

An example of the determination of the rate constant of an individual step in a reaction is provided by the binding of NADH to lactate dehydrogenase (isoenzyme LDH-5) which was monitored by the increase in the fluorescence of NADH that occurs on binding.[92] Equal concentrations of enzyme active sites and substrate (8 μmol dm^{-3} of each) were mixed and the fluorescence increase was monitored over a period of about 16 ms.*

By analysing the data in terms of equations developed for reversible reactions, the rate constants could be deduced as:

$$E + NADH \underset{k_{-1}}{\overset{k_1}{\rightleftharpoons}} E.NADH$$

$$k_1 = 6.3 \times 10^7 \, (mol \, dm^{-3})^{-1} \, s^{-1}$$
$$k_{-1} = 450 \, s^{-1}$$

As will be seen in Chapter 5 (Section 5.5.4.2), the determination of the rate constant for the dissociation step (k_{-1}) is of value in deciding the nature of the slow step of the overall reaction. Further work, using stopped-flow measurements of the fluorescence changes in a mutant form of lactate dehydrogenase, showed that the slow step of the overall reaction corresponds to the movement of a flexible loop region in the enzyme. This movement causes the loop to close over the active site, with consequent changes in the fluorescence of the tryptophan side-chain in the loop (see Chapter 5, Section 5.5.4.5).

4.4.2.2 Identification of transient species

In the stopped-flow technique, the concentration of enzyme is high enough to permit the detection and identification of enzyme-containing complexes in the reaction pathway, and to determine the rate constants of the steps that involve formation and decay of these species. Using absorption spectrophotometry, concentrations of the order of 10 μmol dm^{-3} can be readily detected; using fluorescence the limit is over 10-fold lower. An early experiment by Chance showed that the oxidation of malachite green by H_2O_2 catalysed by peroxidase involved an enzyme–H_2O_2 complex; this experiment provided the first direct evidence for the participation of an enzyme–substrate complex in an enzyme-catalysed reaction.[93] In Chapter 5 (Section 5.5.4.2) we shall see that the application of the stopped-flow technique to the reaction catalysed by lactate dehydrogenase (i.e. L-lactate + NADH$^+$ $\rightleftharpoons$ pyruvate + NADH + H$^+$) has demonstrated the transient production of enzyme-bound NADH. Detailed studies of this reaction have allowed the rate constants of all the elementary steps to be determined.[94,95] Stopped-flow experiments have also been used to provide complete kinetic descriptions for the reaction catalysed by dihydrofolate reductase, namely the reduction of dihydrofolate to tetrahydrofolate by NADPH. Apart from the substrate association, hydride transfer, and product dissociation steps, the reaction involves an isomerization of two forms of the enzyme.[88,96] The quench-flow approach has been used

* For a second-order reaction in which the concentrations of reactants are equal, the half time is given by $1/k[A]$, where [A] is the initial concentration of the reactants and k is the rate constant. In this case (ignoring the reverse reaction) we can see that if $k = 6.3 \times 10^7 \, (mol \, dm^{-3})^{-1} \, s^{-1}$ and [A] = 8 μmol dm^{-3}, then $t_{1/2} \approx 2$ ms, which is within the time-scale of the stopped-flow apparatus.

to examine the individual steps of the reactions catalysed by DNA polymerase and various adenosinetriphosphatases.[91]

An extension of the stopped-flow technique has been described; this involves the use of a rapid scanning detecting device that can scan a spectrum over a 200 nm range within 1 ms.[97] By this means it is possible to record rapid changes in the absorption spectrum and hence identify transient species with more certainty. The technique has been used to show, for instance, that the NADH produced in the rapid phase of the reaction catalysed by alcohol dehydrogenase from horse liver (i.e. ethanol + NAD$^+$ $\rightleftharpoons$ acetaldehyde + NADH + H$^+$) is bound to the enzyme, because its absorption maximum (320 nm) is different from that of free NADH (340 nm). This approach has also been used to analyse the properties of the intermediates formed in the reaction between myeloperoxidase and H$_2$O$_2$.[98]

4.4.3 Methods not involving rapid mixing

Stopped-flow and quench-flow methods (Section 4.4.1) are extremely useful for the study of processes that occur over a period of several milliseconds. The lower limit is set by the time required to achieve uniform mixing of liquids and this appears to be of the order of 1 ms. We have seen in Section 4.4.2.1 that it is possible to use stopped-flow methods to study the reaction between NADH and lactate dehydrogenase provided that relatively low concentrations of substrate and enzyme are employed. However, if it is not possible to adopt this 'low concentration' procedure to bring the rate of a rapid reaction into a convenient timescale,* we can use alternative methods, which do not involve the mixing of reactants.

One approach to this problem is the use of unreactive 'caged' substrates (usually 2-nitrophenylethyl derivatives), which can be activated rapidly by laser irradiation in the near UV.[99,100] The use of 'caged' substrates in examining rapid structural changes in enzymes is discussed in Chapter 5, Section 5.4.3. The more usual approach is to use the so-called 'relaxation' methods, in which a system at equilibrium is perturbed by a sudden change in temperature or other parameter (pressure, pH, etc.) and then allowed to 'relax' to the new position of equilibrium. (By discharging a condenser between electrodes in the solution the temperature can be raised by 5–10 K in 1 μs. The position of equilibrium will change provided that the enthalpy change (ΔH^o) for the reaction is not equal to zero). The rate of the 'relaxation' is monitored by appropriate spectroscopic (or other) techniques, and can be related theoretically to the rate constants for the forward and reverse reactions occurring at equilibrium. From measurements at a variety of concentrations of reactants, the rate constants can be evaluated. More details of these procedures are given in specialized textbooks and review articles.[86–89]

The temperature-jump method has been used, for instance, to show that the rate constant for the association of NADH with malate dehydrogenase ($k = 5 \times 10^8$ (mol dm^{-3})$^{-1}$ s^{-1}) is close to the limit calculated for diffusion control of association of enzyme with substrate ($k = 10^9$ (mol dm^{-3})$^{-1}$ s^{-1}), i.e. that almost every collision or encounter between enzyme and substrate leads to formation of a complex.

In a second type of application, the reaction catalysed by alkaline phosphatase from *E. coli* was studied. This enzyme catalyses the hydrolysis of a variety of phosphate esters, R—O—Ⓟ, via formation and breakdown of a phosphorylenzyme intermediate. Because there is little dependence of the rate of reaction on the nature of the R group, it had been postulated that the hydrolysis of the phosphoryl enzyme was the common rate-limiting step for hydrolysis of the different phosphate esters. However, the rate of this step (k_4 in the scheme below) was mea-

* This could be, for instance, because of the small size of an absorbance change at low concentrations or because of the order of the process. The half time of a first-order process does not depend on the concentration of the reactant.

sured by stopped-flow techniques and found to be faster than the rate of the overall reaction.[101] The nature of the slow step was deduced from experiments in which the rate of binding of a non-hydrolysable substrate analogue, 4-nitrobenzylphosphonate, to the enzyme was studied by the temperature-jump method.[102] The results

$$NO_2 - \langle\!\!\!\bigcirc\!\!\!\rangle - CH_2 - PO_3{}^{2-}$$

indicated that a structural change in the enzyme occurred after binding the inhibitor and that the rate of this step (k_2 in the scheme below) in fact corresponded to that of the slow step of the overall reaction:

$$E + R - O - \textcircled{P} \underset{}{\overset{k_1}{\rightleftharpoons}} E.R - O - \textcircled{P} \underset{}{\overset{k_2}{\rightleftharpoons}} E^*.R - O - \textcircled{P} \overset{k_3}{\rightarrow} E - \textcircled{P}$$

slow step

$$+ ROH \qquad k_4$$

$$E \longleftarrow$$

$$+ \text{ orthophosphate}$$

4.5 Concluding remarks

In this chapter we have introduced the principles and equations that are required to give an understanding of the important features of enzyme kinetics. The application of both steady-state and pre-steady-state measurements is essential in order to deduce the sequence of enzyme-containing complexes in a reaction. We shall see in Chapter 5 how these data can be used in conjunction with other information to formulate a detailed picture of the mechanism of an enzyme-catalysed reaction in both kinetic and structural terms. As will be shown in Chapters 6 and 8, kinetic information is of value in gaining an understanding of the control of enzyme activity and of the role of an enzyme in the metabolic processes in the cell.

References

1. Cornish-Bowden, A. C., *Fundamentals of enzyme kinetics* (revised edn). Portland Press, London (1995). See Ch. 3.

2. Storer, A. C. and Cornish-Bowden, A., *Biochem. J.* **141**, 205 (1974).

3. Rudolph, F. B., Baugher, B. W., and Beissner, R. S., *Methods Enzymol.* **63**, 22 (1979).

4. Easterby, J. S., *Biochem. J.* **219**, 843 (1984).

5. Yang, S-Y. and Schulz, H., *Biochemistry* **26**, 5579 (1987).

6. Xue, Q. and Yeung, E. S., *Nature* **373**, 681 (1995).

7. Craig, D. B., Arriaga, E. A., Wong, J. C. Y., Lu, H. and Dovichi, N. J., *J. amer. chem. Soc.* **118**, 5245 (1996).

8. Allison, R. D. and Purich, D. L., *Methods Enzymol.* **63**, 3 (1979).

9. Engel, P.C., in *Enzymology Labfax* (Engel, P. C., ed.). Bios Scientific Publishers, Oxford (1996). See Ch. 3A.

10. Dalziel, K., *J. biol. Chem.* **238**, 1538 (1963).

11. Winer, A. D., *J. biol. Chem.* **239**, PC 3598 (1964).

12. Cantley, L. C. Jr, Josephson, L., Warner, R., Yanagisawa, M., Lechene, C., and Guidotti, G., *J. biol. Chem.* **252**, 7421 (1977).

13. Wharton, C. W., *Biochem. Soc. Trans.* **11**, 817 (1983).

14. Lineweaver, H. and Burk, D., *J. amer. chem. Soc.* **56**, 658 (1934).

15. Eadie, G. S., *J. biol. Chem.* **146**, 85 (1942).

16. Hofstee, B. H. J., *J. biol. Chem.* **199**, 357 (1952).

17. Hanes, C. S., *Biochem. J.* **26**, 1406 (1932).

18. Eisenthal, R. and Cornish-Bowden, A., *Biochem. J.* **139**, 715 (1974).

19. Cornish-Bowden, A. and Eisenthal, R., *Biochem. J.* **139**, 721 (1974).

20. Cornish-Bowden, A. C., *Fundamentals of enzyme kinetics* (revised edn). Portland Press, London (1995). See Ch. 2.

21. Cornish-Bowden, A. C., in *Enzymology Labfax* (Engel, P. C., ed.). Bios Scientific Publishers, Oxford (1996). See Ch. 3C.

22. Cornish-Bowden, A. C. and Wharton, C. W., *Enzyme kinetics*. IRL Press, Oxford (1988). See Ch. 1.

23. Schulz, A. R., *Enzyme kinetics from diastase to multienzyme systems*. Cambridge University Press (1994). See Ch. 1.

24. Cleland, W. W., *Methods Enzymol.* **63**, 103 (1979).

25. Atkins, G. L. and Nimmo, I. A., *Analyt. Biochem.* **104**, 1 (1980).

26. Cornish-Bowden, A. C., *Analysis of enzyme kinetic data*. Oxford University Press (1995).

27. Duggleby, R. G., *Methods Enzymol.* **249**, 61 (1995).

28. Moreno, J., *Biochem. Educ.* **13**, 64 (1985).

29. Teipel, J. W., Hass, G. M., and Hill, R. L., *J. biol. Chem.* **243**, 5684 (1968).

30. Fersht, A. R., *Structure and mechanism in protein science*. Freeman, New York (1999). See p. 166.

31. Jensen, R. G. and Bahr, J. T., *A. Rev. Plant Physiol.* **28**, 379 (1977).

32. Fersht, A. R., *Structure and mechanism in protein science*. Freeman, New York (1999).

33. Cornish-Bowden, A. C., *Fundamentals of enzyme kinetics* (revised edition). Portland Press, London (1995). See Ch. 5.

34. Schulz, A. R., *Enzyme kinetics from diastase to multienzyme systems*. Cambridge University Press (1994). See Ch. 3.

35. Schulz, A. R., *Enzyme kinetics from diastase to multienzyme systems*. Cambridge University Press (1994). See Ch. 5.

36. *Symbolism and terminology in enzyme kinetics*. Recommendations (1981) of the Nomenclature Committee of the International Union of Biochemistry (now the International Union of Biochemistry and Molecular Biology). Reprinted in *Eur. J. Biochem.* **128**, 281 (1982).

37. Engel, P. C., *Enzyme kinetics* (2nd edn). Chapman and Hall, London (1981). See p. 34.

38. Price, N. C., *Trends Biochem. Sci.* **4**, N272 (1979).

39. Ghosh, N. K. and Fishman, W. H., *J. biol. Chem.* **241**, 2516 (1966).

40. Greene, R. C., *Biochemistry* **8**, 2255 (1969).

41. Todhunter, J. A., *Methods Enzymol.* **63**, 383 (1979).

42. Cornish-Bowden, A. C., *Fundamentals of enzyme kinetics* (revised edn). Portland Press, London (1995). See Ch. 8.

43. Dixon, M., Webb, F. C., Thorne, C. J. R., and Tipton, K. F., *Enzymes* (3rd edn). Longman, London (1979). See pp. 138–64.

44. Schulz, A. R., *Enzyme kinetics from diastase to multienzyme systems*. Cambridge University Press (1994). See Ch. 10.

45. Tipton, K. F. and Dixon, H. B. F., *Methods Enzymol.* **63**, 183 (1979).

46. Engel, P. C., *Enzyme kinetics* (2nd edn). Chapman and Hall, London (1981). See p. 41.

47. Fersht, A. R., *Structure and mechanism in protein science*. Freeman, New York (1999). See p. 487.

48. Findlay, D., Mathias, A. P., and Rabin, B. R., *Biochem. J.* **85**, 139 (1962).

49. Hammond, B. R. and Gutfreund, H., *Biochem. J.* **61**, 187 (1955).

50. Parsons, S. M. and Raftery, M. A., *Biochemistry* **11**, 1623 (1972).

51. Bender, M. L., Kézdy, F. J., and Gunter, C. R., *J. amer. Chem. Soc.* **86**, 3714 (1964).

52. Dixon, M., Webb, E. C., Thorne, C. J. R., and Tipton, K. F., *Enzymes* (3rd edn). Longman, London (1979). See pp. 164–82.

53. Laidler, K. J. and Peterman, B. F., *Methods Enzymol.* **63**, 234 (1979).

54. Dahl, J. L. and Hokin, L. E., *A. Rev. Biochem.* **43**, 327 (1974). See p. 343.

55. Londesborough, J., *Eur. J. Biochem.* **105**, 211 (1980).

56. Hazel, J. R. and Prosser, C. L., *Physiol. Rev.* **54**, 620 (1974). See p. 621.

57. Storey, K. B. and Storey, J. M., *Trends Biochem. Sci.* **8**, 242 (1983).

58. Jaenicke, R., Schurig, H., Beaucamp, N., and Ostendorp, R., *Adv. Prot. Chem.* **48**, 181 (1996).

59. Walker, J. F., Wonacott, A. J., and Harris, J. I., *Eur. J. Biochem.* **108**, 581 (1980).

60. Cornish-Bowden, A. C., *Fundamentals of enzyme kinetics* (revised edition). Portland Press, London (1995). See Ch. 6.

61. Dickinson, F. M., in *Enzymology Labfax* (Engel, P. C., ed.). Bios Scientific Publishers, Oxford (1996). See Ch. 3B.

62. Dixon, M., Webb, F. C., Thorne, C. J. R., and Tipton, K. F., *Enzymes* (3rd edn). Longman, London (1979). See pp. 82–119.

63. Schulz, A. R., *Enzyme kinetics from diastase to multienzyme systems*. Cambridge University Press (1994). See Chs 5 and 6.

64. Engel, P. C., *Enzyme kinetics* (2nd edn). Chapman and Hall, London (1981). See Ch. 5.

65. Cornish-Bowden, A. C. and Wharton, C. W., *Enzyme kinetics*. IRL Press, Oxford (1988). See Ch. 3.

66. Theorell, H. and Chance, B., *Acta chem. Scand.* **5**, 1127 (1951).

67. Dalziel, K. and Dickinson, F. M., *Biochem. J.* **100**, 34 (1966).

68. Cleland, W. W., *Enzymes* (3rd edn) **2**, 3 (1970). See p. 5.

69. Cornish-Bowden, A. C., *Fundamentals of enzyme kinetics* (revised edn). Portland Press, London (1995). See Ch. 4.

70. Huang, C. Y., *Methods Enzymol.* **63**, 54 (1979).

71. Fromm, H. J., *Methods Enzymol.* **63**, 84 (1979).

72. Parks, R. E. Jr. and Agarwal, R. P., *Enzymes* (3rd edn) **8**, 307 (1973). See p. 315.

73. Engel, P. C., *Enzyme kinetics* (2nd edn). Chapman and Hall, London (1981). See p. 59.

74. Holbrook, J. J., Liljas, A., Steindel, S. J., and Rossmann, M. G., *Enzymes* (3rd edn) **11**, 191 (1975). See p. 281.

75. Cooper, B. F. and Rudolph, F. B., *Methods Enzymol.* **249**, 188 (1995).

76. Fromm, H. J., *Methods Enzymol.* **249**, 123 (1995).

77. Purich, D. L. and Allison, R. D., *Methods Enzymol.* **64**, 3 (1980).

78. Cornish-Bowden, A. C., *Fundamentals of enzyme kinetics* (revised edn). Portland Press, London (1995). See Ch. 7.

79. Schulz, A. R., *Enzyme kinetics from diastase to multi-enzyme systems*. Cambridge University Press (1994). See Ch. 8.

80. Welder, F. C., *Methods Enzymol.* **249**, 443 (1995).

81. Silverstein, E. and Sulebele, G., *Biochemistry* **8**, 2543 (1969).

82. Fromm, H. J., *Methods Enzymol.* **63**, 42 (1979).

83. Dalziel, K., *Biochem. J.* **114**, 547 (1969).

84. Engel, P. C. and Dalziel, K., *Biochem. J.* **118**, 409 (1970).

85. Hammes, G. G. and Schimmel, P. R., *Enzymes* (3rd edn) **2**, 67 (1970).

86. Fersht, A. R., *Structure and mechanism in protein science*. Freeman, New York (1999). See Ch. 4.

87. Cornish-Bowden, A. C., *Fundamentals of enzyme kinetics* (revised edn). Portland Press, London (1995). See Ch. 11.

88. Fierke, C. A. and Hammes, G. G., *Methods Enzymol.* **249**, 3 (1995).

89. Gutfreund, H., *Kinetics for the life sciences: receptors, transmitters and catalysts*. Cambridge University Press (1995). See Chs 5 and 6.

90. Fersht, A. R., *Structure and mechanism in protein science*. Freeman, New York (1999). See p. 166.

91. Johnson, K. A., *Methods Enzymol.* **249**, 38 (1995).

92. Stinson, R. A. and Gutfreund, H., *Biochem. J.* **121**, 235 (1971).

93. Chance, B., *J. biol. Chem.* **151**, 553 (1943).

94. Südi, J., *Biochem. J.* **139**, 251 (1974).

95. Südi, J., *Biochem. J.* **139**, 261 (1974).

96. Fierke, C. A., Johnson, K. A., and Benkovic, S. J., *Biochemistry* **26**, 4085 (1987).

97. Holloway, M. R. and White, H. A., *Biochem. J.* **149**, 221 (1975).

98. Marquez, L. A., Huang, J. T., and Dunford, H. B., *Biochemistry* **33**, 1447 (1994).

99. McCray, J. A. and Trentham, D. R., *A. Rev. Biophys. Biophys. Chem.* **18**, 239 (1989).

100. Gutfreund, H., *Kinetics for the life sciences: receptors, transmitters and catalysts*. Cambridge University Press, Cambridge (1995). See Ch. 8.

101. Trentham, D. R. and Gutfreund, H., *Biochem. J.* **106**, 455 (1968).

102. Halford, S. E., Bennett, N. G., Trentham, D. R., and Gutfreund, H., *Biochem. J.* **114**, 243 (1969).

Appendix 4.1

The integrated form of the Michaelis–Menten equation[1,2]

Equation (4.6) states that

$$v = \frac{V_{max}[S]}{K_m + [S]}.$$

The velocity, v, can be expressed as $-d[S]/dt$, therefore

$$-\frac{d[S]}{dt} = \frac{V_{max}[S]}{K_m + [S]}.$$

Separating variables and integrating between limits at time zero when the substrate concentration $= [S]_0$ and time t where the substrate concentration $= [S]_t$:

$$\int_{[S]_0}^{[S]} d[S] + K_m \int_{[S]_0}^{[S]} \frac{d[S]}{[S]} = -V_{max} \int_0^t dt,$$

therefore $[S] - [S]_0 + K_m \cdot \ln \frac{[S]}{[S]_0} = -V_{max} \cdot t.$

Rearranging

$$\frac{[S]_0 - [S]}{t} = -K_m \cdot \frac{1}{t} \cdot \ln \frac{[S]_0}{[S]} + V_{max}.$$

Thus a plot of $([S]_0 - [S])/t$ against $1/t \cdot \ln([S]_0/[S])$ (Fig. A4.1) gives a straight line of slope $- K_m$ and intercept V_{max}.

Appendix 4.2

The interaction of an enzyme with substrate (S) and inhibitor (I)

(a) General considerations

Consider the scheme below, where E, S, and I represent enzyme, substrate, and inhibitor, respectively:

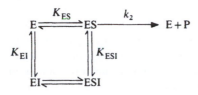

ESI is assumed to be inactive and K_{ES}, K_{EI}, and K_{ESI} represent dissociation constants. Now

$$[ES] = \frac{[E][S]}{K_{ES}}, \quad [EI] = \frac{[E][I]}{K_{EI}},$$

and

$$[ESI] = \frac{[ES][I]}{K_{ESI}} = \frac{[E][S][I]}{K_{ESI} \cdot K_{ES}}.$$

The fraction (F) of enzyme in the form of the ES complex is given by

$$F = \frac{[ES]}{[E] + [ES] + [EI] + [ESI]}$$

$$= \frac{\dfrac{[S]}{K_{ES}}}{1 + \dfrac{[S]}{K_{ES}} + \dfrac{[I]}{K_{EI}} + \dfrac{[S][I]}{K_{ESI} \cdot K_{ES}}}.$$

Now the observed velocity, v, is related to the maximum velocity, V_{max}, by

$$v = V_{max} \cdot F$$

Hence,

$$v = V_{max} \frac{\dfrac{[S]}{K_{ES}}}{1 + \dfrac{[S]}{K_{ES}} + \dfrac{[I]}{K_{EI}} + \dfrac{[S][I]}{K_{ESI} \cdot K_{ES}}}.$$

(b) Classification of inhibition

Because of inconsistencies in the usage of various terms relating to enzyme inhibition, particularly relating to the terms *mixed* and *non-competitive inhibition*, the International Union of Biochemistry and Molecular Biology (formerly the

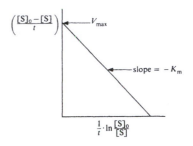

Fig. A4.1 Analysis of the progress curve of an enzyme-catalysed reaction using the integrated form of the Michaelis–Menten equation.

International Union of Biochemistry) has produced a number of recommendations.[3] The classification of types of inhibition should be in terms of the effects on V_{max} and V_{max}/K_m.

If the apparent value of V_{max}/K_m is decreased by the inhibitor. the inhibition is said to have a *competitive* component; if the inhibitor has no effect on the apparent value of V_{max} the inhibition is said to be *competitive*.

If there is an effect on the apparent value of V_{max}, the inhibition has an *uncompetitive component;* if the inhibitor has no effect on the apparent value of V_{max}/K_m, the inhibition is said to *be uncompetitive*.

If both competitive and uncompetitive components are present, the inhibition is said to be *mixed.* The case in which the effects on V_{max}/K_m are greater than on V_{max} is termed *predominantly competitive inhibition;* that in which the effects on V_{max} are greater than on V_{max}/K_m is termed *predominantly uncompetitive inhibition.* Where the effects on V_{max}/K_m and V_{max} are the same (i.e. there is no change in K_m) this may be called *pure non-competitive inhibition* (see Section 4.3.2.2). However, usage of the term non-competitive inhibition is to be discouraged because of its use for various types of mixed inhibition.

(c) The Dixon plot[4]

The general equation (4.9) can be rearranged as

$$\frac{1}{v} = \frac{1}{V_{max}} + \frac{K_{ES}}{V_{max}} \cdot \frac{1}{[S]} + [I]\left(\frac{1}{V_{max} \cdot K_{ESI}} + \frac{1}{V_{max}} \cdot \frac{K_{ES}}{K_{EI}} \cdot \frac{1}{[S]}\right).$$

Thus, a plot of $1/v$ against $[I]$ at constant $[S]$ will be a straight line. If two such lines are drawn (from measurements at two different values of $[S]$), the values of K_{EI} can be found from the point of intersection (see Fig. A4.2). (This can be shown by setting the values of $1/v$ equal at two values of $[S]$.) The plot is therefore particularly useful if v is measured at a large number of inhibitor concentrations but only a limited number of substrate concentrations (compare Figs 4.7, 4.8, and 4.9). In the case of competitive inhibition, the lines intersect above the abscissa (Fig. A4.2(a)), whereas in the case of pure non-competitive inhibition, the point of intersection is on the abscissa (Fig. A4.2(b)). There is no point of intersection in the case of uncompetitive inhibition, since the lines are parallel (Fig. A4.2(c)). The Dixon plot cannot be used to deduce the value of K_{ESI}.

Appendix 4.3

Formulation of equations for two-substrate kinetics

The International Union of Biochemistry and Molecular Biology[3] proposes the following formulation for the initial rate, v, of a two-substrate reaction obeying Michaelis–Menten kinetics:

$$v = \frac{[E]_0}{\dfrac{1}{k_0} + \dfrac{1}{k_A[A]} + \dfrac{1}{k_B[B]} + \dfrac{1}{k_{AB}[A][B]}},$$

where $[E]_0$ is the total concentration of enzyme and the ks are individual rate constants. In terms of eqn (4.13),

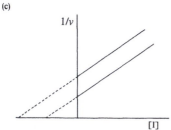

Fig. A4.2 The Dixon plot for determining inhibitor constants: (a), (b), and (c) show the plots for the cases of competitive, pure non-competitive, and uncompetitive inhibition, respectively. Each line in the plots represents data obtained at a constant [S].

$$k_0 = \frac{V_{max}}{[E]_0}, \quad k_A = \frac{k_0}{K_A},$$

$$k_B = \frac{k_0}{K_B} \quad \text{and} \quad k_{AB} = \frac{k_0}{K_B \cdot K_A'}.$$

The formulation of Dalziel[5] for such a reaction is

$$\frac{[E]_0}{v} = \phi_0 + \frac{\phi_1}{[A]} + \frac{\phi_2}{[B]} + \frac{\phi_{12}}{[A][B]}$$

where ϕ_0, etc. are known as Dalziel coefficients. These coefficients allow the equation to be written more concisely, since they represent combinations of constants.

In terms of eqn (4.13),

$$\phi_0 = \frac{[E]_0}{V_{max}}, \quad \phi_1 = \frac{[E]_0 K_A}{V_{max}},$$

$$\phi_2 = \frac{[E]_0 K_B}{V_{max}} \quad \text{and} \quad \phi_{12} = \frac{[E]_0 K_A' K_B}{V_{max}}.$$

For further details of the various formulations of these equations, see references 6–8.

References for Appendices

1. Moreno, J., *Biochem. Educ.* **13**, 64 (1985).

2. Duggleby, R. G., *Methods Enzymol.* **249**, 61 (1995).

3. Symbolism and Terminology in Enzyme Kinetics. Recommendations (1981) of the Nomenclature Committee of the International Union of Biochemistry (now the International Union of Biochemistry and Molecular Biology). Reprinted in *Europ. J. Biochem.* **128**, 281(1982).

4. Dixon, M., *Biochem. J.* **55**, 170 (1953).

5. Dalziel, K., *Acta chem. Scand.* **11**, 1706 (1957).

6. Rudolph, F. B. and Fromm, H. J., *Methods Enzymol.* **63**, 138 (1979).

7. Cornish-Bowden, A. C., *Fundamentals of enzyme kinetics* (revised edn). Portland Press, London (1995). See Ch. 6.

8. Dickinson, F. M., in *Enzymology Labfax* (Engel, P. C., ed.). Bios Scientific Publishers, Oxford (1996). See Ch. 3B.

5

The mechanism of enzyme action

* This sequence constitutes the *elementary steps* of the overall reaction.

5.1 Introduction

In this chapter, we shall consider the mechanism of action of enzymes and gain some insight into two of the remarkable properties of enzymes referred to in Chapter 1, Section 1.3; namely catalytic power and specificity. In Section 5.3 we introduce some of the basic principles of catalysis and show how the study of the mechanisms of reactions in organic chemistry can provide a framework in which to interpret catalysis by enzymes. The main experimental approaches available to determine the mechanism of enzyme action are described in Section 5.4. Section 5.5 shows how these different approaches have been combined to provide coherent pictures of the mechanism of five particular enzymes, namely chymotrypsin, triosephosphate isomerase, dehydroquinase, lactate dehydrogenase, and DNA methyltransferase. In the concluding section (5.6), some more general aspects of enzyme catalysis are discussed.

The enormous progress which has been made in the understanding of enzyme mechanisms in recent years has been largely due to the wealth of structural information available from X-ray crystallographic and NMR studies (see Chapter 3, Section 3.4). However, although chemically plausible mechanisms have been proposed for many enzymes, it has not yet been possible to account precisely from quantum mechanical calculations for the catalytic power displayed by any one enzyme.

5.2 Definition of the mechanism of an enzyme-catalysed reaction

We can justifiably claim to have deduced the mechanism of an enzyme-catalysed reaction if we have determined: (i) the *sequence* of the enzyme-containing complexes* as substrate(s) is (are) converted to product(s); (ii) the *rates* at which these complexes are interconverted; and (iii) the *structures* of these complexes. This definition implies that we require both kinetic and structural information and it is therefore not surprising that kinetic and structural techniques have made major contributions to our understanding of enzyme mechanisms. Valuable information has also been gained from studies involving detection of intermediates (Section 5.4.2), chemical modification of amino-acid side-chains (Section 5.4.4) and site-directed mutagenesis (Section 5.4.5).

5.3 **Background to catalysis**

Studies of mechanisms in organic chemistry have highlighted various factors that contribute to the enhancement of the rate of reactions. These factors are described (Sections 5.3.1 to 5.3.6) so that their importance in the specific examples of enzyme catalysis discussed in Section 5.5 can be assessed. However, not all the factors are necessarily important in every enzyme-catalysed reaction.

Inspection of the 'energy profile' for a typical reaction (Fig. 5.1(a)) shows that in order to proceed from reactant(s) to product(s) an energy barrier ($\Delta G^{\ddagger}$) must be surmounted. The highest point of the energy profile is designated the *transition state* of the reaction. From the transition-state theory of reaction rates,[1] an equation can be derived that expresses the relationship between the rate constant, k', and the free energy of activation ($\Delta G^{\ddagger}$):

$$k' = \frac{kT}{h} e^{-\Delta G^{\ddagger}/RT} = \frac{kT}{h} e^{-\Delta H^{\ddagger}/RT} \cdot e^{\Delta S^{\ddagger}/R}$$

where k is the Boltzmann constant,

h is the Planck constant,

T is the temperature,

R is the gas constant,

and $\Delta H^{\ddagger}$ and $\Delta S^{\ddagger}$ are the enthalpy and entropy of activation, respectively.

In order to speed up the rate of a reaction at constant temperature, the value of $\Delta G^{\ddagger}$ must be reduced. Therefore, the function of a catalyst is to provide an alternative reaction pathway so that a lowered energy barrier has to be surmounted (see, for example, the dashed line in Fig. 5.1(a)). One particular way in which a catalyst might function is via the rapid formation and decomposition of an *intermediate*, which is represented by the local minimum in the dotted line in Fig. 5.1(a). This is known as covalent catalysis. In all cases, the catalyst does not alter the magnitude of free-energy change (ΔG in Fig. 5.1(a)) and therefore does not cause a shift in the equilibrium between reactant(s) and product(s);* it merely increases the rate at which that equilibrium is attained.

These general considerations also apply to catalysis by enzymes; however, we should note that the energy profile corresponding to even the simplest type of enzyme mechanism is more complex (see Fig. 5.1(b)). The energy profiles of a number of enzyme-catalysed reactions have been determined, e.g. triosephosphate isomerase (Fig. 5.29), proline racemase,[2] and tyrosyl-tRNA synthetase.[3]

* This is of course true for catalysis brought about by concentrations of enzymes which are small by comparison with the concentrations of substrates or products. However, if the concentration of an enzyme is comparable with that of the substrates and products there can be an *apparent* shift in the equilibrium, see Chapter 8, Section 8.5.4. The latter situation can often apply in the cell.

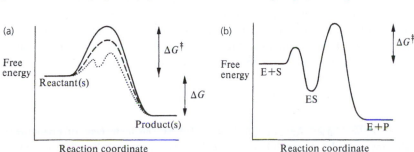

(a)

Free energy

Reactant(s)

$\Delta G^{\ddagger}$

ΔG

Product(s)

Reaction coordinate (progress of reaction)

(b)

Free energy

E+S

ES

E+P

$\Delta G^{\ddagger}$

Reaction coordinate

Fig. 5.1 Energy profiles for reactions. (a) Profiles for an uncatalysed reaction (solid line), a catalysed reaction (dashed line), and a catalysed reaction involving intermediate formation (dotted line). (b) Profile for an enzyme-catalysed reaction E + S $\rightleftharpoons$ ES $\longrightarrow$ E + P. Note that comparisons should not be made between the vertical scales in (a) and (b).

Reacting groups
separated

$N(CH_3)_3$
+
$CH_3-\overset{O}{\underset{\|}{C}}-O-$⟨benzene ring⟩$-NO_2$

$k = 4.3 \ (mol \ dm^{-3})^{-1} \ min^{-1}$

Reacting groups
combined

⟨structure: CH_2, CH_2, CH_2 chain with $N(CH_3)_2$; $\overset{O}{\underset{\|}{C}}-O-$⟨benzene ring⟩$-NO_2$⟩

$k = 21\,500 \ min^{-1}$

⟨structure (I): benzene ring with OH and side chain CH_2 CH_2 CO_2H⟩

(I)

⟨structure (II): benzene ring with OH, side chain CH_2 $C(CH_3)_2$ CO_2H, and two CH_3 groups on ring⟩

(II)

Various factors leading to rate enhancements that may be relevant to enzyme-catalysed reactions are proximity and orientation effects, acid–base catalysis, covalent catalysis, changes in environment, and strain or distortion. These will now be discussed in turn.

5.3.1 Proximity and orientation effects

It seems intuitively obvious that an enzyme could increase the rate of a reaction involving more than one substrate by binding the substrates at adjacent sites and therefore bringing them into close proximity with each other. The reaction would then occur more readily than if it depended on chance encounters between the reacting molecules in solution. Many examples from organic chemistry suggest that reactions can be greatly accelerated when the reacting groups are combined within a single molecule. For example, different rate constants were obtained for the amine-catalysed hydrolysis of 4-nitrophenyl esters[4] (a reaction involving nucleophilic attack by the amine of the carbonyl group of the ester), depending on whether the reacting groups are separated or combined.

Difficulties arise in such situations when we try to establish the correct basis for comparison; thus, in this example we are comparing the magnitudes of rate constants that are of different orders and hence of different dimensions. There could also be some differences between the detailed mechanisms of the two reactions. Despite these difficulties, Page and Jencks[5,6] have given a detailed thermodynamic treatment of proximity effects in enzyme-catalysed reactions and their conclusions are mentioned at the end of this section after the importance of orientation effects has been discussed.

There is no doubt that the orientation of the reacting molecules with respect to each other can greatly influence the rate of reaction. Catalysis could thus be achieved by ensuring that the reactants are in the correct orientation as they approach each other. A good example of rate enhancement by this means is provided by the rates of lactone (internal ester) formation in compounds (I) and (II).[7]

The presence of the two methyl groups on the carbon atom adjacent to the aromatic ring in (II) increases the rate of lactone formation by a factor of 4×10^5. The restrictions placed on rotation about single bonds in the carboxylate side-chain of (II) by the bulky substituents ensure that the preferred orientation of the reacting groups closely resembles that of the transition state of the reaction. A simple way of looking at this is to say that the molecule will spend a greater proportion of the time in the conformation that leads to reaction, and therefore less rotational entropy (fewer degrees of freedom) will be lost as the reaction proceeds towards the transition state. The smaller negative entropy of activation will lead to an increase in the rate of reaction.

Estimates of the effects of orientation and proximity on the entropy of activation, and hence on the rate of reaction, have been given by Page and Jencks.[5,6] Their calculations suggest that each factor can contribute a rate enhancement of at least 10^4-fold for a bimolecular reaction, giving a possible total rate enhancement of at least 10^8-fold from these two factors combined.

5.3.2 Acid–base catalysis

Many reactions of the type catalysed by enzymes are known to be catalysed by acids and/or bases. An example is provided by the hydrolysis of phenyl-β-D-glucopyranoside[8] shown in Fig. 5.2. Acid catalysis proceeds (Fig. 5.2(a)) via protonation of the glycosidic oxygen followed by dissociation of phenol to yield a

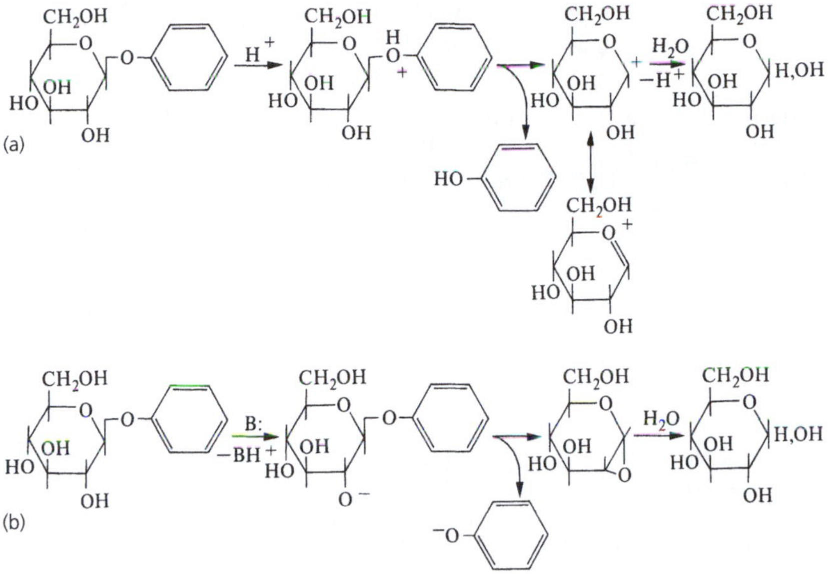

Fig. 5.2 Hydrolysis of phenyl-β-D-glucopyranoside subject to catalysis by acid (a) or base (b).

stabilized carbocation which is then attacked by water to form glucose. On the other hand, base catalysis proceeds (Fig. 5.2(b)) via abstraction of a proton from the C2 OH group followed by nucleophilic attack of the –O$^-$ group on C1 to form an oxirane (epoxide). The oxirane is then attacked by water to form glucose.

Since enzymes contain a number of amino-acid side-chains that are capable of acting as proton donors or acceptors (see Chapter 3, Section 3.3.1), it is reasonable to suppose that acid–base catalysis would be important in enzyme-catalysed reactions. In simple organic reactions, acid catalysis is divided into *specific acid catalysis* (where the rate expression contains contributions from H$^+$ only) and *general acid catalysis* (where the rate expression includes contributions from H$^+$ and from other potential proton donors in the solution, e.g. CH_3CO_2H). A similar division is made for base catalysis. As far as enzyme-catalysed reactions are concerned, only *general* acid or base catalysis can occur, because enzymes do not possess any mechanism for concentrating H$^+$ or OH$^-$. A good example of general acid catalysis occurs in the hydrolysis of oligosaccharides catalysed by lysozyme.[9] In this enzyme the side-chain of Glu 35 is in a highly non-polar environment, and this has the effect of raising its pK_a so that the carboxyl group remains protonated at pH values up to about 6 (the pK_a of the side-chain of free glutamic acid is 4.3). The carboxyl group donates a proton to the glycosidic oxygen of the substrate, facilitating cleavage of the C$-$O bond, and leading to the formation of a carbocation that is stabilized by the neighbouring ionized side-chain of Asp 52 (Fig. 5.3). The analogies with the acid-catalysed hydrolysis of the glucoside in Fig. 5.2(a) should be noted.

Histidine side-chains often play important roles as acid–base catalysts in enzyme action (see, for example, the cases of chymotrypsin, triosephosphate isomerase, dehydroquinase, and lactate dehydrogenase discussed in Section 5.5). This is because the histidine side-chain has a pK_a close to 7, so that at pH values near neutrality there is a reasonable balance between the proton-donating (protonated) and proton-accepting (deprotonated) forms.

It is conceivable that suitably positioned proton-donating and proton-accepting side-chains in an enzyme could act in a concerted fashion, so that electrons are both 'pulled' and 'pushed'. This type of effect has been demonstrated in a model

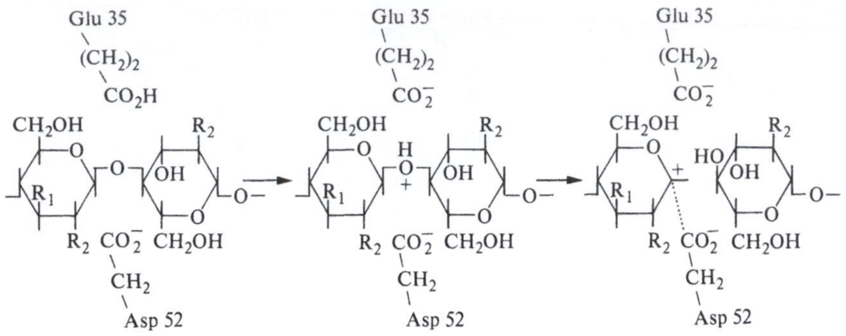

Fig. 5.3 Part of the proposed mechanism of action of lysozyme.[9] The process is completed by reaction of the carbocation with water, with overall retention of configuration at C1.

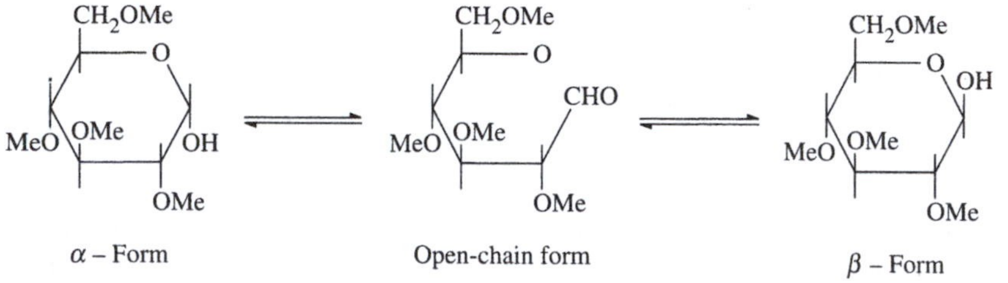

Fig. 5.4 The mutarotation of tetra-*O*-methyl glucose in which α- and β-forms are interconverted via the open-chain form. 2-Hydroxypyridine, in which acid and base groups are combined in one species, acts as a much more effective catalyst than a mixture of pyridine and phenol, indicating that concerted acid–base catalysis occurs.[10]

system, namely the mutarotation of tetra-*O*-methylglucose (Fig. 5.4), but the importance of the effect in enzyme-catalysed reactions is less clear.[10]

A fuller discussion of acid–base catalysis and its importance in enzyme-catalysed reactions has been given by Fersht.[11]

5.3.3 Covalent catalysis (intermediate formation)

It has been recognized for many years that reactions can be speeded up by the formation of intermediates, provided that such intermediates are both rapidly formed and rapidly broken down (see the dotted line in Fig. 5.1(a)). A classic example is provided by the decarboxylation of acetoacetate, which is catalysed by primary amines via the rapid formation and breakdown of an imine (Schiff base),[12] see Fig. 5.5.

The advantage conferred by intermediate formation in this case is that the Schiff base is readily protonated, thus providing greater electron-withdrawing power to aid the loss of CO_2 than would be provided by the carbonyl group itself.

Many of the examples of covalent catalysis in enzyme-catalysed reactions involve attack by a nucleophilic side-chain at an electron-deficient centre in the

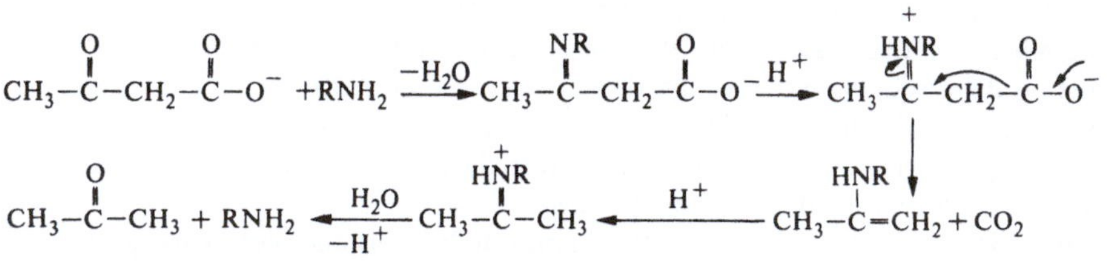

Fig. 5.5 The decarboxylation of acetoacetate catalysed by primary amines.

substrate; such attack is termed nucleophilic catalysis. Examples of nucleophilic catalysis are provided by chymotrypsin (Section 5.5.1), dehydroquinase (Section 5.5.3), and DNA methylase (Section 5.5.5), which involve the nucleophilic side-chains of serine, lysine, and cysteine, respectively. Among the other types of side-chain that have been found to participate in nucleophilic catalysis is histidine, as in the examples of phosphoglycerate mutase and nucleosidediphosphate kinase.

5.3.4 Changes in environment

The rates of many organic reactions are highly sensitive to the nature of the solvents in which they occur. In particular, dipolar aprotic solvents such as dimethyl sulphoxide and dimethylformamide, which are not capable of solvating anions, are extremely good solvents for nucleophilic displacement reactions. The reaction below occurs over 12 000 times faster in dimethyl sulphoxide than in water, reflecting the fact that in water the solvation shell of the anion must be removed to allow reaction to occur.

$N_3^- +$ [benzene ring with F at top and NO_2 at bottom] $\longrightarrow$ $F^- +$ [benzene ring with N_3 at top and NO_2 at bottom]

X-ray crystallographic studies show that enzymes are capable of providing unusual environments for reactions. For instance, in the case of lysozyme, the cleft in the molecule that forms the substrate-binding site is lined by a number of non-polar amino-acid side-chains; these clearly provide an environment markedly different from that of the solvent water. In addition, within the active site there is an appropriately positioned negative charge, the ionized side-chain of Asp 52, which helps to stabilize the positive charge on the carbocation in the transition state of the reaction (Fig. 5.3). Calculations[13] suggest that this electrostatic stabilization, which would be strong in a medium of low dielectric constant, may be the most important factor in this case, contributing a rate enhancement of 3×10^6-fold.

5.3.5 Strain or distortion

The bonds in a substrate may be distorted on binding to the appropriate enzyme. This would speed up the reaction if the distortion (or strain) lowered the free energy of activation by making the geometry and electronic structure of the substrate more closely resemble that of the postulated transition state.[14] Examples of this effect can be found in model systems, e.g. the very rapid hydrolysis of strained cyclic phosphate esters compared with the open-chain forms[15] (Fig. 5.6).

The evidence is more difficult to interpret in enzyme-catalysed reactions. It has been proposed that the glucose ring at the point of cleavage of an oligosaccharide substrate is distorted on binding to lysozyme, so that the geometry approaches that in the postulated transition state, which involves a carbocation[9] (see Figs 5.3 and 5.7); this would be expected to lead to an enhancement of the rate of reaction.* Other examples of substrate distortion include the distortion of dihydroxyacetone phosphate on binding to triosephosphate isomerase[16] (Section 5.5.2.5) and the distortion of DNA on binding to DNA methyltransferase (Section 5.5.5.4).

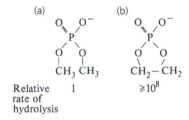

(a)	(b)
[phosphate ester structure with CH₃ CH₃]	[phosphate ester structure with CH₂—CH₂]

Relative rate of hydrolysis   1 $\geqslant 10^8$

Fig. 5.6 Structures of phosphate esters. The faster rate of hydrolysis of (b) is attributed to the strain in the molecule, which is relieved on hydrolysis.

* Detailed calculations of the energy of the lysozyme–substrate complex suggest, however, that this distortion does not in fact contribute greatly to the rate enhancement[13] (see Section 5.3.5).

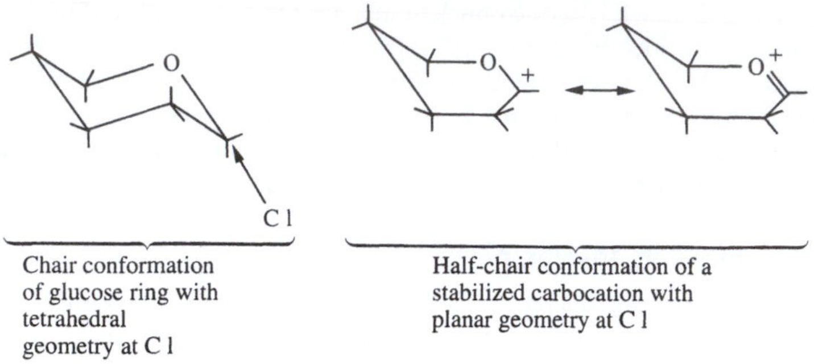

Fig. 5.7 Distortion of glucose ring on formation of a carbocation.

Chair conformation of glucose ring with tetrahedral geometry at C 1

Half-chair conformation of a stabilized carbocation with planar geometry at C 1

5.3.6 Transition-state analogues

An alternative way of regarding strain is to think in terms of stabilization by the enzyme of the transition state of the reaction rather than in terms of destabilization of the substrate, i.e. that an enzyme is more suited to making favourable interactions with the transition state for the reaction than with the substrate(s) in its (their) normal conformation. From this it would follow that a compound that resembles the transition state in geometry and electronic structure will bind more tightly to an enzyme than will the substrate(s). There has been considerable interest[17–19] in the search for such 'transition-state analogues', since these would help to confirm proposals about the mechanisms of reactions, as well as providing potentially useful enzyme inhibitors. A detailed discussion has been given of the design and application of transition-state analogues for enzymes in each of the six main classes (see Chapter 1, Section 1.5.1).[19] Transition-state analogues for triosephosphate isomerase are described in Section 5.5.2.3. Some of the analogues have been found to bind very tightly to enzymes; for example 3,4-dihydro-uridine binds to cytidine deaminase 2×10^9-fold more tightly than uridine (the product of the reaction). Analogues for a number of two-substrate reactions have also been described; these usually combine the binding determinants of both substrates, and thereby possess large entropic advantages over the binding of the separate substrates. The analogue N-phosphonoacetyl-L-aspartate (Fig. 5.8) acts as a powerful inhibitor of aspartate carbamoyltransferase, with a K_d of approximately 25 nmol dm^{-3}.[20] Intriguingly, although the enzyme–analogue complexes almost always have very slow rates of dissociation, in a number of cases the rates of association are surprisingly slow.[21] The slow rates may well reflect the need for the interaction between enzyme and analogue to occur in a number of stages with relatively slow adjustments of the enzyme structure required.

5.3.7 Discussion of the factors likely to be involved in enzyme catalysis

In Sections 5.3.1 to 5.3.5 we have discussed various factors that are generally accepted to be important in bringing about rate enhancement in enzyme-catalysed reactions. We should also remember that the structural changes in enzymes which accompany binding of substrates can play crucial roles in the catalytic process (see Chapter 3, Section 3.4.2.1). The specific examples of enzyme mechanisms described in Section 5.5 will show how these various factors operate in particular situations. Additional factors may be involved in certain cases; for example, in a metalloenzyme the metal ion can display powerful electron-withdrawing ability thereby polarizing a chemical bond and making it more susceptible to nucleo-

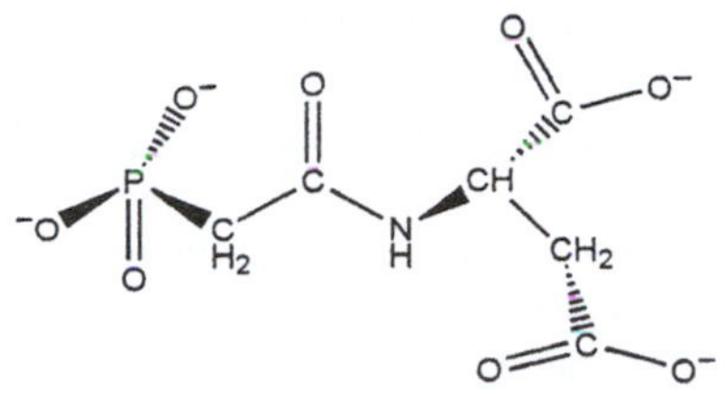

N-phosphonoacetyl-L-aspartate (PALA)

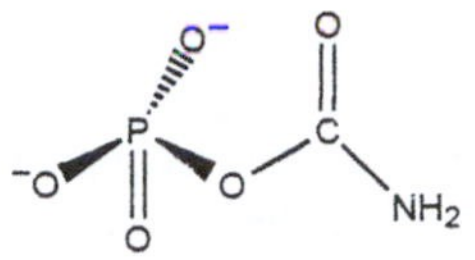

Carbamyl phosphate 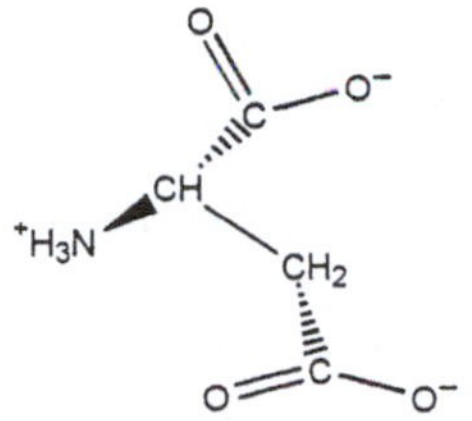   L-Aspartate

Fig. 5.8 Structures of N-phosphonoacetyl-L-aspartate, carbamoyl phosphate, and L-aspartate.

philic attack. An instance of this type of effect is provided by carboxypeptidase, where a zinc ion serves to polarize the carbonyl group of an amide substrate (see Section 5.5.1.7).

Although the causes of rate enhancement are reasonably well understood in qualitative terms, we are still some way from a *quantitative* understanding of their importance in enzyme-catalysed reactions. However, progress is being made in this direction (see, for example, the work on triosephosphate isomerase, Section 5.5.2.5). It has been claimed that the 10^{16}-fold rate enhancement of DNA hydrolysis shown by staphylococcal nuclease can be accounted for by a combination of transition-state stabilization, metal-ion catalysis, general base catalysis, and catalysis by proximity of attacking water.[22]

5.3.8 Design of novel enzymes

It is possible to make practical applications of the knowledge we have concerning rate enhancements to designing synthetic enzymes that are capable of catalysing new types of reactions or display enhanced stability compared with normal enzymes. One approach involves incorporating functional groups of enzymes into a non-protein support. As an example of this approach an 'artificial' chymotrypsin has been made, in which a catalytic site containing imidazole, carboxyl, and hydroxyl groups (see Section 5.5.1.2) was incorporated into an oligosaccharide (cyclodextrin), which provides the binding site. The artificial enzyme had an M_r only about 5 per cent that of chymotrypsin but showed comparable kinetic features in the esterase reaction and enhanced stability at elevated temperatures and high pH.[23] A number of reviews of the design and applications of synthetic enzymes have been given.[23-26]

A second approach has involved the production of catalytic antibodies or 'abzymes' (see Chapter 1, Section 1.2). This involves the raising of antibodies towards stable transition-state analogues for reactions, and then screening the resulting range of antibodies in order to identify those which may possess catalytic properties in the desired reaction. This approach has now been successfully applied to a wide range of types of chemical reaction.[27,28] Detailed analysis of the structures of a limited number of abzymes has shown that the catalytic mechanisms have similarities with those of enzymes catalysing related reactions. As an illustration, an antibody raised against a norleucine phosphonate derivative (presumed on the basis of geometry and charge distribution to resemble the transition state for ester hydrolysis) was found to catalyse the hydrolysis of norleucine and methionine phenyl esters. X-ray analysis of the structure of the antibody showed that the active site contained serine and histidine side-chains, a lysine side-chain to stabilize the developing negative charge on the oxyanion, and a hydrophobic binding pocket to bind the side-chains of the amino-acid ester substrates.[29] All these features are analogous to those seen in serine proteinases (Section 5.5.1.7).

A third approach is to graft features of one enzyme on to the structural framework provided by another, in order to generate novel activities or specificities. This is illustrated by the application of the site-directed mutagenesis technique (Section 5.4.5) to mutate three selected amino acids in cyclophilin (a protein which catalyses the interconversion of *cis* and *trans* peptide bonds to proline) close to the peptide-binding cleft to form a charge relay system similar to that found in the serine proteases (see Section 5.5.1.2). These changes gave rise to a new enzyme (which was termed cyproase I) with the ability to hydrolyse Xaa–Pro peptide bonds in both model peptides and proteins.[30] The redesign of the active site of lactate dehydrogenase to accommodate a range of substrates is discussed in Section 5.5.4.

5.4 Experimental approaches to the determination of enzyme mechanisms

Having discussed the theoretical basis of rate enhancements, we now describe the main experimental approaches that can be used to elucidate the mechanism of action of an enzyme. These approaches will be dealt with under five main headings: kinetic studies (Section 5.4.1), detection of intermediates (Section 5.4.2), X-ray crystallography (Section 5.4.3), chemical modification of amino-acid side-chains (Section 5.4.4), and site-directed mutagenesis (Section 5.4.5). These approaches are to be regarded as complementary; the information gained from one approach should be interpreted in the light of results from the others in order to build up the overall picture of the mechanism.

5.4.1 Kinetic studies

In Chapter 4, we described how the catalytic activity of an enzyme could be assayed and then proceeded to introduce some of the principal concepts and equations of enzyme kinetics. Our object in this section is to show how information about enzyme mechanisms can be gained from these studies. The types of information available from kinetic studies are summarized in Table 5.1 and the various techniques are discussed in turn (Sections 5.4.1.1 to 5.4.1.5).

5.4.1.1 Variation of substrate concentration(s)

Steady-state kinetic studies support the proposal that one-substrate reactions proceed via the formation and decay of one or more enzyme–substrate complexes

Table 5.1 The types of information on enzyme mechanisms available from kinetic studies

Experiment	Information available
Variation of substrate concentration(s)	Sequence of complexes in reaction Distinction between possible mechanisms
Variation of substrate structure	Structural features responsible for binding and catalytic activity (Mapping the active site)
Reversible inhibition (see Chapter 4, Section 4.3.2)	Competitive inhibitors can help to define the active site
Variation of pH (see Chapter 4, Section 4.3.3)	The pK_a values of side-chains involved in catalytic activity. Assignment of the pK_a values to particular amino-acid side-chains may be possible
Pre-steady-state kinetics (see Chapter 4, Section 4.4)	Detection of enzyme-containing complexes. Rates of elementary steps in the reaction

(see Chapter 4, Section 4.3.1), but cannot by themselves give any indication of the sequence of such complexes. In the case of two-substrate reactions, steady-state kinetic studies can distinguish between ternary complex and enzyme substitution mechanisms (see Chapter 4, Section 4.3.5.1). With additional information from product inhibition, substrate binding, or isotope-exchange experiments, it is possible to distinguish between mechanisms in which a ternary complex is formed in an ordered or a random fashion (see Chapter 4, Section 4.3.5.5).

5.4.1.2 Variation of substrate structure

A good deal has been learnt about the general features of enzyme active sites by correlating the rates of reactions with the structures of the substrates. For instance, by comparing the rates of hydrolysis of a large number of amide derivatives of amino acids it has been demonstrated that chymotrypsin has a strong preference for substrates containing aromatic or bulky hydrophobic R groups, whereas elastase has a strong preference for substrates containing small hydrophobic R groups. Clearly the substrate-binding sites of these enzymes must contain features that account for the observed specificities (see Section 5.5.1.2).

$$\begin{array}{c} \text{O site of hydrolysis} \\ \parallel \ \downarrow \\ \text{X—NH—CH—C—NHY} \\ | \\ \text{R} \end{array}$$

It is possible to develop this approach and map active sites in greater detail. The classic example of this was the extensive studies carried out by Berger and Schechter on the specificity displayed by papain towards synthetic peptide substrates.[31] These studies showed that there are seven 'subsites' on the enzyme (see Fig. 5.9), and that subsite S_2 interacts specifically with an L-phenylalanine side-chain. Later work showed that subsite S_1' is stereospecific for L-amino acids with a preference for the hydrophobic side-chains of leucine and tryptophan.[32] The interactions between enzyme and substrate in this case have been studied in detail by X-ray crystallography, and the locations of a number of the subsites (e.g. S_1', S_1, and S_2) in the enzyme have been deduced.[31,33] The nomenclature introduced by Berger and Schechter has been widely used to describe the binding subsites of a number of hydrolytic enzymes when they act on complex substrates.[34]

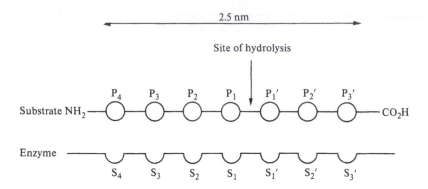

Fig. 5.9 The 'subsites' of papain as revealed by active-site mapping.

5.4.1.3 Reversible inhibition

The study of inhibition of enzyme-catalysed reactions (see Chapter 4, Section 4.3.2) can give information on the structures of active sites. One of the most likely explanations for competitive inhibition is that substrate and inhibitor bind to the same site on an enzyme. By comparison of the structures of the substrate and the competitive inhibitor it is possible to define the essential structural features of these molecules that are involved in their binding to the active site. A detailed study was carried out in the case of papain;[31] thus the tripeptide Ala-Phe-Arg acted as a powerful competitive inhibitor of the enzyme, presumably since it occupied the subsites S_3, S_2, and S_1 (Fig. 5.9) and could not therefore undergo hydrolysis. Competitive inhibitors are also of value in X-ray crystallographic work, where it is usually difficult to study the enzyme–substrate complex directly (Section 5.4.3). A special class of inhibitors are 'transition-state analogues' (Section 5.3.6) which usually bind very tightly to enzymes, and can provide important confirmation about proposed mechanisms.

5.4.1.4 Variation of pH

The catalytic activity of many enzymes is markedly dependent on pH. As described in Chapter 4, Section 4.3.3, there are a number of explanations for this phenomenon, but the one of most immediate concern here is the ionization of amino-acid side-chains that are involved in the catalytic mechanism. By suitable analysis of a plot of reaction rate against pH,[35] it is usually possible to deduce the pK_a values of these ionizing side-chains and the identity of the side-chains can be guessed at by comparison with the pK_a values of side-chains of free amino acids (see Chapter 3, Section 3.3.1) or of small peptides. However, it should be noted that the environment of a side-chain in an enzyme can shift the pK_a of the side-chain by up to four units from the value for the free amino acid. More definite deductions about the identities of the side-chains can be made if other information, such as the effect of solvent polarity on the pK_a, is available.*

This approach was used, for instance, to implicate two histidine side-chains as being involved in the catalytic mechanism of ribonuclease,[36] a conclusion later confirmed by X-ray crystallography, NMR, and other studies.[37]

5.4.1.5 Pre-steady-state kinetics

In Chapter 4, Section 4.4, we described how studies of pre-steady-state kinetics could be used to detect enzyme-containing complexes in a reaction and to determine the rates of formation and decay of such complexes. In these experiments, the concentration of enzyme more closely approaches that of the substrates than

* If the dissociating group is neutral, e.g. $-CO_2H \rightleftharpoons -CO_2^- + H^+$, dissociation is accompanied by charge separation. A decrease in solvent polarity will discourage dissociation, i.e. raise the pK_a. However, if the dissociating group is cationic, e.g. $-ImH^+ \rightleftharpoons -Im + H^+$, there is no separation of charge and the pK_a will be much less sensitive to changes in solvent polarity. (Im = imidazole.)

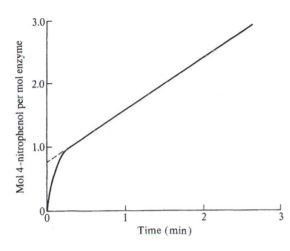

Fig. 5.10 The hydrolysis of 4-nitrophenyl acetate catalysed by chymotrypsin. The solid line represents the experimental data.

is the case with steady-state studies, and usually special techniques are required to achieve rapid mixing and rapid detection of the changes occurring.

A simple illustration of the use of pre-steady-state kinetics is afforded by the chymotrypsin-catalysed hydrolysis of 4-nitrophenyl acetate. The production of 4-nitrophenol (Fig. 5.10) shows a 'burst' phase[38] (the size of which corresponds to approximately 1 mol per mol enzyme) followed by a slower steady-state rate. This observation is consistent with the type of mechanism described in Section 5.5.1.3 for this enzyme, in which a fast step, corresponding to the formation of acyl enzyme and release of 4-nitrophenol, is followed by a slow step, corresponding to the rate of hydrolysis of the acyl enzyme to regenerate enzyme:

$$E-CH_2-OH+NO_2-\text{⟨⟩}-O-\overset{\displaystyle O}{\overset{\|}{C}}-CH_3 \xrightarrow{\text{fast}} E-CH_2-O-\overset{\displaystyle O}{\overset{\|}{C}}-CH_3+NO_2-\text{⟨⟩}-OH$$

$$\text{slow}$$

$$CH_3CO_2H \qquad H_2O$$

The steady-state rate in this chymotrypsin-catalysed reaction is sufficiently slow that the burst phase can be observed without the need for specialized apparatus. However, more information on the mechanism, such as the rate of formation of the acyl enzyme, would be available from measurements of the early phase of the reaction using fast-reaction techniques. The use of fast-reaction techniques is well illustrated by experiments carried out on the reaction catalysed by alcohol dehydrogenase from horse liver.[39] If the enzyme is mixed rapidly with saturating concentrations of the substrates ethanol and NAD^+, there is rapid production of 2 mol NADH per mol of the dimeric enzyme, followed by a slower steady-state rate (Fig. 5.11). If the production of NADH is monitored by fluorescence, when free and enzyme-bound NADH can be readily distinguished, it is found that the NADH produced in the rapid phase is enzyme-bound. These studies show that the release of NADH from the enzyme is the rate-determining (i.e. slowest) step in the overall reaction. Values of the rate constants of some of the individual steps in the reaction could be deduced from more detailed measurements.[39] A further example of pre-steady-state kinetic

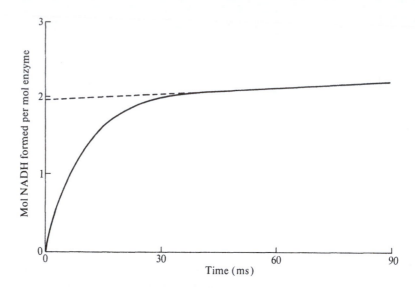

Fig. 5.11 Stopped-flow study of the reaction between ethanol and NAD⁺ catalysed by alcohol dehydrogenase from horse liver (a dimeric enzyme).[39]

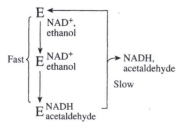

Acetaldehyde is also known as ethanal.

studies is provided by the work on the reaction catalysed by lactate dehydrogenase (Section 5.5.4.2).

5.4.2 Detection of intermediates

One of the most direct methods for obtaining information about the pathway of a reaction is to detect any intermediates that may be involved in the reaction. In some cases an intermediate is sufficiently stable to be isolated and characterized (the stability will depend on the depth of the minimum in the energy profile shown in Fig. 5.1(a)). In other cases, such as in the reaction catalysed by alcohol dehydrogenase from horse liver, it is possible to infer the existence of an intermediate by spectroscopic means. Whatever the method of detection, it must be shown that the rates of formation and decay of any presumed intermediate are consistent with the overall rate of the reaction. Thus, a postulated intermediate that is so stable that it breaks down more slowly than the overall rate of the reaction cannot be involved in that reaction. Intermediates can provide useful information for the design of transition-state analogues (Section 5.3.6).

We have already noted (Section 5.4.1.5) that pre-steady-state kinetic studies suggest that an acyl enzyme intermediate is involved in the chymotrypsin-catalysed hydrolysis of esters. The rate of breakdown of the acyl enzyme is very low at acid pH, so that the intermediate can be isolated (and crystallized) if the enzyme and ester substrate are mixed and the pH is then rapidly lowered. X-ray crystallographic work has shown that it is the side-chain of Ser 195 that becomes acylated.[40] As we shall see in Section 5.5.1.3, the chymotrypsin-catalysed hydrolysis of amide substrates is also thought to follow a pathway involving formation and breakdown of an acyl enzyme. However, in this case the breakdown step is faster than the formation step, so that the acyl enzyme does not accumulate. It is nevertheless possible to infer its participation in the overall reaction by 'trapping' experiments, as described in Section 5.5.1.3.

Another example of 'trapping' intermediates is provided by the work on dehydroquinase, where it is possible to trap the imine (Schiff base) intermediate by reduction with sodium borohydride (Section 5.5.3.2).

$$\text{\Large$>$}\!C=NH\!\!-\quad \xrightarrow{\text{NaBH}_4}\quad \text{\Large$>$}\!CH\!-\!NH_2\!\!-$$

Protonated Schiff   Protonated
base (unstable)   amine (stable)

It may also be possible to stabilize intermediates at low temperatures, where the rate of breakdown will be decreased.[41,42] In such experiments it is usually necessary to work in mixed-solvent systems such as dimethyl sulphoxide–water to prevent freezing, although these systems can cause problems by altering reaction mechanisms. In some cases, the use of concentrated solutions of salts such as ammonium acetate may be preferable.[43] Considerable success has been achieved with this cryoenzymological approach; thus, in the case of ribonuclease, a series of enzyme-containing complexes on the reaction pathway has been studied by X-ray crystallography.[41] In studies of the elastase-catalysed hydrolysis of a 4-nitroanilide substrate, the structure of a tetrahedral intermediate (Section 5.5.1.3) has been deduced.[42]

5.4.3 X-ray crystallographic studies

In Chapter 3 (Section 3.4) we saw that the use of X-ray crystallography has given a wealth of detail about the structures of enzymes. Clearly, we need to be able to locate the active site in an enzyme and then examine the mode of binding of substrate(s) in order to pinpoint the functional amino-acid side-chains that are involved in the catalytic mechanism. Up to fairly recently it has not proved generally possible to study the structure of the catalytically active complex of enzyme and substrate(s) directly, because the collection of diffraction data takes several hours, by which time conversion of substrate(s) to product(s) will have occurred. In the case of an enzyme that catalyses a reaction involving two or more substrates, it may be possible to examine the complex formed between the enzyme and one of the substrates; see the example of lactate dehydrogenase in Section 5.5.4.3. A number of alternative approaches can be adopted to gain information on the structure of the catalytically active complex.[44]

First, it may be possible to examine the active complex in the case of a one-substrate reaction when the equilibrium lies very much to one side. Thus, in the case of triosephosphate isomerase, which catalyses the reaction below, it has been possible to examine the crystal structure of the enzyme–dihydroxyacetone phosphate (glycerone phosphate)* complex (Section 5.5.2.2):

Dihydroxyacetone phosphate $\rightleftharpoons$ D-glyceraldehyde 3-phosphate.

* Dihydroxy acetone phosphate is also known as glycerone phosphate, see p. 186.

Second, an enzyme can be studied in the presence of a very poor substrate or a competitive inhibitor. These molecules are likely to bind to the active site in a manner similar to the substrate but will remain essentially unchanged over the time-course of the experiment. By inspection of a three-dimensional model of the enzyme it is then possible to work out how the normal substrate binds to the enzyme. This type of approach is illustrated in the cases of chymotrypsin (Section 5.5.1.2) and lactate dehydrogenase (Section 5.5.4.3). Variations on this approach include the use of altered pH and of mutant enzymes of low catalytic activity in order to slow down the reaction. In the case of DNA methyltransferase (Section 5.5.5.4), it was possible to examine the structure of a complex in which the enzyme had reacted with a 'mechanism-based inhibitor' (Section 5.4.4.3).

Third, it may be possible to examine unstable complexes at low temperatures, where the rate of decomposition of such complexes will be decreased. This type of approach has been mentioned in connection with work on ribonuclease and elastase (Section 5.4.2).

Fourth, the time-scale of the X-ray experiment can be shortened by the application of synchrotron radiation (the radiation produced during the acceleration of charged particles). The high intensity of the radiation compared to that of conventional X-ray sources means that the time required for data collection is dramatically reduced and in principle it is possible to obtain information on structural changes or catalytic steps that occur on the millisecond time-scale.[44,45] In a pioneering experiment, the catalysis by phosphorylase b in the presence of AMP (see Chapter 6, Section 6.4.2.1) of the reaction between the poor substrate heptenitol and orthophosphate to yield heptulose 2-phosphate was studied. The structure of the ternary complex (which accumulates because the rate of its conversion to product is slow) was determined.[45] One of the major difficulties in this type of work is the need to ensure that the reaction between enzyme and substrate occurs synchronously, since the diffusion of substrate throughout the crystal can take a significant time. One solution to this problem involves the use of 'caged' substrates, which are unreactive until a blocking group is removed. For example, the 2-nitrophenylethyl blocking group can be removed from caged GTP by flash photolysis with a UV source. This approach was used to study the hydrolysis of GTP by p21, a gene product of the *ras* oncogene family.[46] (It should be noted that the p21 used in this work lacked the C-terminal 23 amino acids of the cellular protein; the complexes involving this shorter p21 were sufficiently long-lived for the X-ray studies.) After hydrolytic removal of the γ-phosphate group of GTP, structural changes in the protein were seen in the effector loop which is implicated in the binding of p21 to the GTPase-activating protein. A further example is provided by the use of a pH-jump to initiate a reaction. The acyl–enzyme intermediate (Sections 5.4.1.5 and 5.5.1.3) formed by reaction of trypsin with 4-nitrophenyl guanidinobenzoate is stable for several days at pH 5.5, but at pH 8.0 has a half-life of the order of one hour. When crystals of the modified trypsin prepared at low pH were subjected to a pH-jump (to pH 8.5), it was possible to monitor the deacylation process and to observe a water molecule positioning itself to attack the acyl group.[47]

Finally, it is possible to explore the interaction between enzyme and substrate(s) or other ligand(s) using a variety of computer programs which are now available. In these programs, the chemical and stereochemical features of ligands can be modelled into the three-dimensional structure of the enzyme, and the likely modes of interaction established. This type of approach is now widely used in the search for 'lead' compounds in drug design projects (see Chapter 3, Section 3.4.3.4).

The detailed information available from X-ray crystallography is now regarded as essential in providing a framework in which to interpret the results of other experiments. Apart from locating the active site and providing information on the nature of side-chains likely to be involved in the catalytic mechanism, it is also possible to assess the extent of any structural (or conformational*) changes in the enzyme that are associated with the binding of substrate(s), and in some cases that accompany the catalytic process (Section 5.6). However, it should be remembered that even such a powerful technique as X-ray crystallography does have its limitations. Sometimes, attempts to prepare an enzyme–substrate complex by soaking crystals of enzyme in a solution of the substrate lead to cracking of the crystals, thus hampering further crystallographic analysis. The cracking is presumably a result of conformational changes within the enzyme that disrupt the

* Conformations of molecules are related to each other by rotations about single bonds. Conformational changes refer to interconversions between different conformations of an enzyme molecule.

crystal lattice. In the studies on phosphorylase *a*, the problem posed by the cracking of the crystals on addition of substrates was overcome by prior treatment of the crystals with a cross-linking agent.[48] A second type of difficulty is presented by the rather artificial conditions under which crystals are studied. It is possible that in concentrated salt solutions, e.g. 3 mol dm^{-3} ammonium sulphate, the interaction between an enzyme and its substrate may be rather different from that which occurs in solutions of lower ionic strength. Thus, there could be no binding at all or weaker binding to additional sites on the enzyme. These problems have led to considerable difficulties involved in locating the active sites of enzymes such as phosphorylase *a*[49] and adenylate kinase.[50]

5.4.4 Chemical modification of amino-acid side-chains

The principle underlying the application of chemical modification techniques to the study of enzyme mechanisms is very simple. If an amino-acid side-chain involved in the catalytic activity is chemically modified, the enzyme will be inactivated. Provided that the identity of the modified side-chain can be established by standard structural techniques, e.g. isolating and sequencing a modified peptide (see Chapter 3, Section 3.3), then it should be possible to determine which side-chain is involved in the catalytic mechanism. Recently, it has been demonstrated that the application of high-resolution mass spectrometry (see Chapter 3, Section 3.2.4) to the analysis of chemical modification mixtures in the early stages of reaction can give extremely valuable insights into complex modification reactions (Section 5.5.3.2).

However, chemical modification techniques pose problems both of design and of interpretation.[51-53] On the experimental side it is highly desirable that the modification be specific. Thus it should be checked that only one type of side-chain is modified, and further that only one of this type has reacted (e.g. only one of, say, 30 Lys side-chains in the enzyme). The requirement for specificity presents considerable problems when it is remembered that many amino-acid side-chains are nucleophilic, so that an electrophile such as 1-fluoro-2,4-dinitrobenzene could be attacked by a variety of side-chains, e.g. those of cysteine, lysine, histidine, tyrosine, etc. In Sections 5.4.4.1 to 5.4.4.3 we shall mention some approaches to improve the specificity of modification reactions; Section 5.4.4.4 will deal with problems encountered in the interpretation of chemical modification experiments.

5.4.4.1 Application of chemical principles

It is well known that mercury has a very strong affinity for sulphur (cf. the very low solubility product of HgS), so it would be expected that mercurial reagents should bring about highly specific modification of cysteine side-chains in enzymes:

$$E-CH_2-SH + Cl-Hg-\underset{\text{4-Chloromercuribenzoate}}{\boxed{}}-CO_2^- \longrightarrow E-CH_2-S-Hg-\boxed{}-CO_2^- + Cl^- + H^+$$

We would also expect that the activated aromatic ring in a tyrosine side-chain would be especially susceptible to electrophilic substitution, for example by I_2 or by tetranitromethane, $C(NO_2)_4$:

$$E-CH_2-\boxed{}-OH + C(NO_2)_4 \longrightarrow E-CH_2-\overset{NO_2}{\boxed{}}-OH + H^+ + C(NO_2)_3^-$$

(It has been suggested[54] that the reaction between tetranitromethane and proteins takes place via initial formation of a charge-transfer complex between the reagent and the phenolate ion, followed by electron transfer.)

It is also useful to take note of the different orders of reactivity observed for the side-chains of free amino acids with various reagents:

acylating reagents   Cys > Tyr > His > Lys
   (e.g. iodoacetamide, iodoacetate)

arylating reagents   Cys > Lys > Tyr > His
   (e.g. 2,4,6-trinitrobenzene sulphonate,
     1-fluoro-2,4-dinitrobenzene)

and apply this information in enzyme-modification experiments. Thus, arylating agents such as 2,4,6-trinitrobenzenesulphonate should be used in preference to acylating agents to modify lysine side-chains.

Using these principles it is possible to draw up a list of reagents that should display reasonable specificity for modifying particular types of amino-acid side-chain (Table 5.2). However, the specificity of any modification reaction should always be checked, e.g. by analysis of the modified enzyme, since the reactivity of a side-chain can be considerably influenced by its local environment within an enzyme.

Changes in the selectivity of certain reagents can be brought about by changing the pH. For instance, at pH values above about 7, the side-chain of cysteine ($pK_a \approx 8$) is extremely reactive towards iodoacetate because there is a significant proportion of the ionized form, $-CH_2-S^-$, which is the reactive nucleophile. At pH values below 6, the fraction of the ionized form of the cysteine side-chain is much less and the side-chain of methionine ($-CH_2-CH_2-S-CH_3$) is a more reactive nucleophile. By working at pH 5.6 it was possible to react a methionine side-chain in pig heart isocitrate dehydrogenase with iodoacetate without any modification of cysteine side-chains.[55]

Table 5.2 Some chemical modification procedures that show reasonable specificity for amino-acid side-chains in enzymes

Side-chain	Reagent(s) used
Cysteine	Mercurials, e.g. 4-chloromercuribenzoate
	Disulphides, e.g. 5,5'-dithiobis-(2-nitrobenzoic acid)
	Iodoacetamide
	Maleimide derivatives, e.g. N-ethylmaleimide
Lysine	2,4,6-Trinitrobenzenesulphonate
	Pyridoxal phosphate (± reducing agent such as NaBH$_4$)
Histidine	Diethylpyrocarbonate
	Photo-oxidation
Arginine	Phenylglyoxal
	2,3-Butanedione
Tyrosine	Tetranitromethane
	N-Acetylimidazole
	Iodine
Tryptophan	N-Bromosuccinimide
Aspartic acid or Glutamic acid	Water-soluble carbodiimide plus nucleophile, e.g. glycine methylester

Details of these procedures are given in references 51 and 53.

5.4.4.2 Super-reactive side-chains

Sometimes one particular amino-acid side-chain in an enzyme is especially reactive because of its unique environment. In such cases, specific modifications can be achieved in an 'accidental' manner. A good example of such a super-reactive group is provided by chymotrypsin, where it is found that diisopropylfluorophosphate modifies only the side-chain of Ser 195 and does not react with any of the other 27 serine side-chains in the enzyme or with free serine. The reason for the super-reactivity of the side-chain of Ser 195 has been found from X-ray crystallographic studies (Section 5.5.1.2). Another example of a super-reactive group is provided by glutamate dehydrogenase from beef liver, where the side-chain of one lysine (Lys 126) out of a total of over 30 lysine side-chains in each subunit is especially reactive towards 2,4,6-trinitrobenzenesulphonate.[56] In chymotrypsin and glutamate dehydrogenase these super-reactive groups are involved in the catalytic mechanisms, but this is not always the case. For instance, the highly reactive thiol group in creatine kinase does not appear to be essential for the catalytic activity of the enzyme (Section 5.4.4.4). Conversely, the side-chain of Glu 35 in lysozyme (which is essential for the catalytic activity) has proved impossible to modify without prior denaturation of the enzyme.[57]

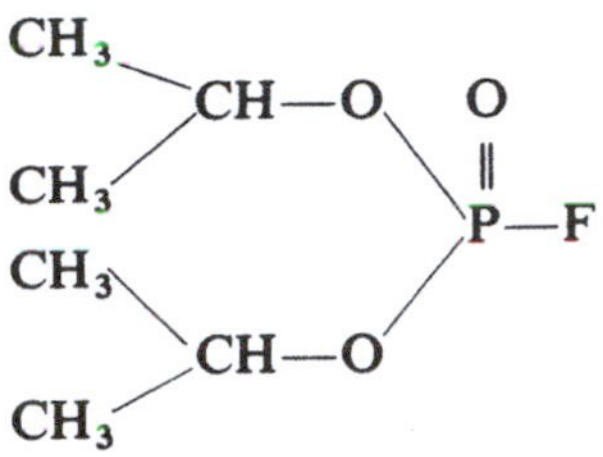

5.4.4.3 Affinity labelling

One way of improving the specificity of a modifying reagent is to incorporate within it some structural feature that will 'direct' it to a certain site on the enzyme, such as the active site. The reactive part of the reagent will then react with an amino-acid side-chain in the vicinity of that site. In the case of triosephosphate isomerase, bromohydroxyacetone phosphate acts as an affinity label;[58] the resemblance between the affinity label and the substrate for the enzyme is clearly seen in the structures shown.

The affinity label binds to the active site of the enzyme and the labile Br atom (activated by the adjacent carbonyl group) can be displaced by a suitably positioned nucleophilic amino-acid side-chain (Section 5.5.2.4). Affinity labelling of chymotrypsin and of trypsin is described in Section 5.5.1.4.

The ideas underlying affinity labelling have been extended to the design of mechanism-based enzyme inhibitors ('suicide substrates') for enzymes.[59] When acted upon by the appropriate enzyme such a substrate is converted to a product that essentially irreversibly inactivates the enzyme, usually by covalent modification (see the scheme below).

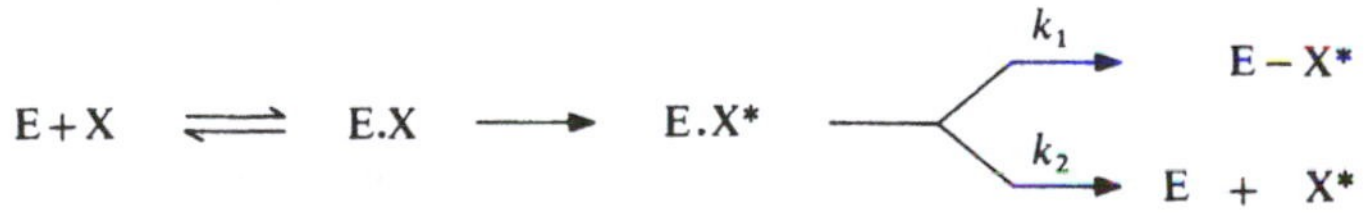

Diisopropylfluorophosphate

Substrate
CH_2OH
|
$C=O$
|
$CH_2OPO_3^{2-}$

Dihydroxyacetone phosphate

Affinity label
CH_2Br
|
$C=O$
|
$CH_2OPO_3^{2-}$

Bromohydroxyacetone phosphate

Reaction of an enzyme, E, with a suicide substrate, X. After binding to the enzyme, the substrate is converted by the enzyme to a reactive form (X*). The E·X* complex can either form E–X*, an irreversibly inactivated complex, or can dissociate to form E and X*. The balance between these two processes will depend on the relative values of the rate constants k_1 and k_2.

A number of mechanism-based inhibitors (e.g. clavulanic acid and sulbactam) have been designed for penicillinase (the enzyme that confers resistance towards penicillin), and it is clear that this is an area of considerable pharmaceutical importance.[59] A detailed review of the design of such inhibitors for other enzymes including monoamine oxidase has been given by Silverman.[60]

5.4.4.4 Interpretation of chemical modification experiments

Considerable caution is required in interpreting the results of chemical modification experiments. If modification of a particular amino-acid side-chain leads to inactivation of an enzyme, it does not necessarily follow that the side-chain is directly involved in the catalytic mechanism. It could happen, for instance, that the reacting side-chain is some way from the active site, but that modification causes a conformational change in the enzyme leading to loss of activity. There is obviously a whole range of possibilities for the positioning of such a side-chain relative to the active site. In many cases it may be possible to make an assessment of the extent of involvement of a side-chain in the catalytic mechanism by comparing the effects of modification by groups of different sizes. Thus, in the case of creatine kinase from rabbit muscle, modification of a single cysteine side-chain by iodoacetamide leads to complete loss of activity (a); however, the smaller perturbation (b) leads to only a 30 per cent loss of activity.[61]

(a) $E–CH_2–SH \rightarrow E–CH_2–S–CH_2–CO–NH_2$

(b) $E–CH_2–SH \rightarrow E–CH_2–S–CN$

The cysteine side-chain cannot therefore be 'directly' involved (i.e. in a covalent fashion) in the catalytic mechanism of the enzyme. Side-chains that can be ascribed an 'indirect' role may still play an extremely important part in binding of substrates or in conformational changes in the enzymes that are involved in catalysis. Any modification of a side-chain that was directly involved in the catalytic mechanism, such as that of Ser 195 of chymotrypsin which takes part in the formation of the acyl enzyme, would lead to essentially complete inactivation; the modification of 'indirectly' involved amino acids would have varying effects depending on the severity of the perturbation.

In a chemical modification experiment there are two criteria that must be fulfilled before it can be stated with any degree of confidence that a modified side-chain is at the active site of an enzyme.[62]

First, there must be a stoichiometric relationship between the extent of inactivation and the extent of modification, so that complete inactivation corresponds to the modification of one side-chain per active site (Fig. 5.12(a)). In those cases where side-chains other than an essential one are modified, the analysis can be complex; an example is given in Section 5.5.3.2. Second, the addition of substrate or competitive inhibitor must protect against inactivation (Fig. 5.12(b)), since the side-chain is no longer accessible to the modifying reagent. The need for proper analysis of such protection experiments has been emphasized.[63]

These criteria should be regarded only as minimal and it is possible that they can be fulfilled even when modification is occurring at a side-chain remote from the active site; the example of the modification of the side-chain of Cys 165 in lactate dehydrogenase is mentioned in Section 5.5.4.4. Despite these problems, chemical modification techniques have been used very extensively to help elucidate details of enzyme mechanisms. We shall refer to chemical modification data in the examples of mechanisms discussed in Section 5.5.

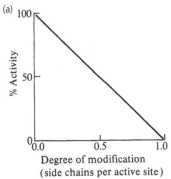

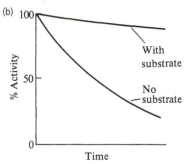

Fig. 5.12 Criteria for modification at the active site of an enzyme. (a) Stoichiometric relationship between extent of inactivation and extent of modification. (b) Protection against inactivation by addition of substrate (or competitive inhibitor).

5.4.5 Site-directed mutagenesis

Since the early 1980s the development of recombinant-DNA technology (see Chapter 2, Section 2.4.1) has led to the possibility of introducing amino-acid replacements at specified positions in a protein, provided that (a) the gene coding for the protein is available, and (b) there is a suitable system for expression of the gene. The details of this technique, known as site-directed mutagenesis or oligonucleotide-directed mutagenesis, are beyond the scope of this book, but can be found in a number of articles and reviews.[64–67] The principal methods which have been used are based on primer extension or on the polymerase chain reaction (PCR). A mutant enzyme in which, for example, a Ser at position 195 had been replaced by Ala would be denoted as Ser195Ala (S195A in the one-letter code; see Chapter 3, Section 3.3.1).

In the primer extension method (Fig. 5.13), the gene of interest is cloned into a vector such as the filamentous phage M13. The DNA of this phage, which is in the form of a covalent circle of single-stranded DNA (ssDNA), goes through a covalent double-stranded form (duplex) during replication in the *E. coli* host. One strand of the DNA is isolated. An oligonucleotide (typically 10–20 nucleotides in length) is synthesized that is complementary to the region of the gene to be mutated except for a single base (or double base) change that changes the codon from that (those) of the target amino acid(s) to that (those) of the desired replacement amino acid(s). (For example, the change 5′-TTG-3′ to 5′-TTC-3′ changes the codon from that for Gln to that for Glu.) This oligonucleotide is annealed to the DNA strand of the mutagenesis vector and serves as the primer for DNA synthesis by either the Klenow fragment of DNA polymerase I (this fragment contains the 5′ → 3′ polymerase and 3′ → 5′ exonuclease activities, but not the 5′ → 3′ exonuclease activity of the intact enzyme) or the DNA polymerase from phage T7. After joining (by DNA ligase) the ends of the new strand of DNA, the heteroduplex is used to transform a host (e.g. the JM strain of *E. coli*). Clones containing the mutant DNA can be selected by their ability to hybridize (form hydrogen bonds) to the original oligonucleotide under more forcing conditions of temperature and ionic strength than clones containing the wild-type DNA. The mutant enzyme can then be produced in large quantities in the transformed cells.

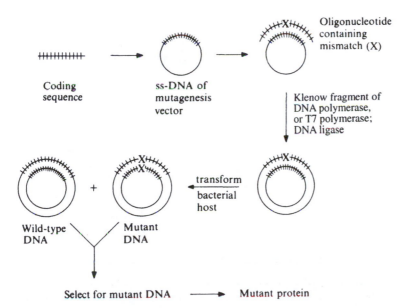

Fig. 5.13 Typical procedure for site-directed mutagenesis using the primer extension method. The mismatch is depicted as (X) within a complementary (i.e. base-pairing region).

A number of techniques are available to increase the frequency of mutants in the transformed cells. One of these involves the incorporation of sulphur-containing nucleotide analogues (phosphothioates) during the *in vitro* synthesis of DNA by the Klenow fragment, which then protects the mutant strand against 'nicking' by certain restriction endonucleases.[68,69] This allows the amount of non-mutant strand (which is subject to 'nicking') to be reduced by digestion with exonuclease III.

PCR (polymerase chain reaction) methods involve the amplification of a portion of DNA by the use of primers designed to be complementary to the regions of sequence flanking this portion.[70] The reaction cycle has three phases, strand separation, primer annealing, and chain extension, which take place at different temperatures. Thermostable DNA polymerases such as *Taq* polymerase from *Thermus aquatus* are used in PCR, since these survive the temperatures (typically around 90–95°C) used for strand separation. One widely used method[67,71,72] uses three primers to bring about site-directed mutagenesis (Fig. 5.14). Of the first two primers, one includes the mutation to be introduced, the other defines one end of the DNA encoding the protein of interest. The product of this first round of PCR amplification is then used as a primer for later rounds of PCR in conjunction with the third primer which defines the other end of the DNA coding region. By this means it is possible to produce the desired mutation with 100 per cent efficiency. However, because the thermostable DNA polymerases are prone to introduce occasional errors into the copied DNA, it is always important to check the sequence of the mutated gene to make sure that only the desired mutation has been introduced.

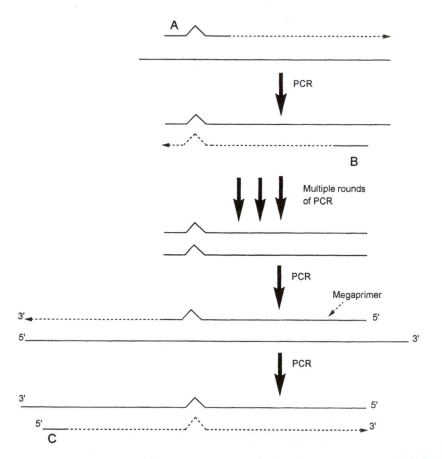

Fig. 5.14 A typical PCR-based method of site-directed mutagenesis using three primers. In this method, primers A and B are used in the first round of PCR, to generate a megaprimer used in conjunction with primer C in the second round of PCR. The dashed lines represent newly synthesized DNA.

5.4.5.1 Strategies in choice of mutations

Site-directed mutagenesis is clearly a very powerful method for assessing the importance of particular amino-acid side-chains in an enzyme. Given that in principle any amino acid in a protein can be changed to any other amino acid, it is clear that for even a moderate-sized protein, an almost astronomical number of mutants could be made and examined. (For a protein of 140 amino acids, there are 2660 single mutants, over 7 million double mutants, and over 1.8×10^{10} triple mutants.) It is therefore necessary to consider how a rational choice of mutants can be made. This question has two parts: (a) which amino acids in the wild-type protein should be mutated, and (b) which mutations of these amino acids should be made? A detailed discussion of these questions has been presented by Plapp.[73]

The amino acids to be mutated are usually selected to answer specific questions about the mechanism of a reaction or to bring about specific changes in the properties of the enzyme. It is now regarded as almost essential to have a detailed three-dimensional structure of the enzyme available or at least the structure of a closely related enzyme on which the enzyme of interest can be modelled (see Chapter 3, Section 3.4.1). On the basis of the structure, certain amino acids may be proposed as being important for substrate recognition or may play roles in catalysis, e.g. by donating or accepting protons. These amino acids would be the likely targets to be mutated in a systematic fashion. Other information which would allow amino acids to be chosen would include that available from chemical modification experiments (Section 5.4.4) and from comparisons of amino-acid sequences; highly conserved amino acids are presumed to be likely to play important structural or functional roles in an enzyme. The latter points are well illustrated by the work on the mechanism of dehydroquinases (Section 5.5.3).

Having selected the amino acids to mutate, the choice of amino-acid replacements must be considered. If the importance of a particular amino acid in an enzyme mechanism is to be assessed, the replacement should be similar in terms of overall size and polarity, but differing in the important chemical characteristic, e.g. replacement of Asp by Asn would remove the negative charge. In general, replacement of an amino acid by a smaller one, thereby creating a void, causes less disturbance to the overall structure than replacement with a larger amino acid, which could lead to steric interference. If the importance of flexibility or a particular part of the polypeptide chain is to be examined, the effects of removal or addition of proline can be examined, since this amino acid imposes special requirements on the local geometry of the chain (see Chapter 3, Section 3.4.4.2). Taking these various factors and the pattern of naturally occurring mutations in proteins into account, it is possible to produce a table of suggested changes (Table 5.3), but it should be emphasized that this is only an overall guide and the effects of mutations need to be examined in each case, since these can often be subtle (and sometimes unexpected).[73]

For some purposes, the replacement of amino acids will differ from the pattern shown above. For instance, the purpose of mutagenesis may be to try to improve the stability of a protein by the introduction of one or more disulphide bonds; in this case amino acids which would be in the correct position to form such a bond are identified and both are mutated to Cys; under oxidative conditions a disulphide bond should be formed, assuming that the mutations have not affected the geometry of the enzyme. In the case of subtilisin a disulphide bond was introduced by the replacement of both Thr 22 and Ser 87 by Cys; the mutated enzyme

Table 5.3 Some common amino acid replacements in proteins

Amino acid	Replacement
Ala	Ser, Gly, Thr
Arg	Lys, His, Gln
Asn	Asp, Gln, Glu, Ser, His, Lys
Asp	Asn, Gln, Glu, His
Cys	Ser
His	Asn, Asp, Gln, Glu, Arg
Leu	Met, Ile, Val, Phe
Lys	Arg, Gln, Asn
Ser	Ala, Thr, Asn, Gly
Tyr	Phe, His, Trp
Trp	Phe, Tyr

It should be noted that this Table is not exhaustive; further details can be found in reference 73.

was somewhat more stable towards thermal denaturation than the wild-type enzyme.[74] In other cases, the purpose of the mutation may be to change the specificity of the enzyme; in these cases the changes may be designed to alter the size or polarity of the active site. This type of approach is discussed in the case of lactate dehydrogenase in Section 5.5.4.5.

It should be mentioned that the technique of site-directed mutagenesis can be adapted to allow the incorporation of non-natural amino acids into proteins. In this case, the codon specifying the amino acid to be mutated is replaced by the stop codon AUG (this requires the sequence TAG in the DNA). Certain bacterial strains (amber suppressor mutants) produce tRNAs which recognize the AUG codon and insert an amino acid. By synthesizing chemically tRNA molecules which carry a non-natural amino acid, it is possible to incorporate this amino acid into the polypeptide chain. This approach has been used, for example, to introduce non-natural amino acids into β-lactamase[75] and lysozyme.[76] It would also be possible to incorporate amino acids Cys and Met with selenium replacing sulphur into proteins by this route; the heavier Se atom can be useful in the determination of structures by X-ray crystallography (see Chapter 3, Section 3.4.1). However, seleno-amino acids can be fairly readily incorporated in place of their sulphur analogues under normal growth conditions.[77] A selenoMet derivative of T5 5′-exonuclease was prepared by using a mutant strain of *E. coli* (which requires methionine) for growth as the expression system; after overnight starvation of Met, selenoMet was added to the culture as expression of the required enzyme was induced. The availability of the selenoMet derivative allowed the X-ray structure of the enzyme to be solved; previous attempts at introducing heavy atoms had failed.[78]

5.4.5.2 Site-directed mutagenesis compared with chemical modification

The techniques of site-directed mutagenesis (Section 5.4.5) and chemical modification (Section 5.4.4) both involve the production of chemically altered proteins; they should be regarded as complementary to each other. In both cases it is necessary to check that the overall structural features of the enzyme have not been altered, since otherwise it would be difficult to ensure that any change in the properties is solely a consequence of the alteration introduced. Spectroscopic techniques such as circular dichroism (see Chapter 3, Section 3.4.2) are useful in this regard, since the overall structural features of an enzyme can be ascertained quickly,[79] in contrast to the laborious process of structure determination by X-ray crystallography.

Site-directed mutagenesis requires the gene encoding the enzyme of interest and a suitable expression system to be available. Chemical modification of amino-acid side-chains is useful in those cases where these conditions are not fulfilled and to gain an idea of the types of side-chains that may be important for the function of an enzyme. (Of course it may well be worthwhile investing the effort required to obtain the gene and expression system in order to overexpress the enzyme.) In one sense, chemical modification can be regarded as more versatile since it is possible to introduce a wider range of functionalities into a protein. Thus, heavy atoms can be introduced for X-ray crystallography (see Chapter 3, Section 3.4.1) and spin labels or fluorescent groups can be introduced for spectroscopic studies.[80] However, site-directed mutagenesis allows the chemical nature of an amino-acid side-chain to be varied in a more systematic fashion, and also allows even buried amino acids which are not readily accessible to chemical reagents to be mutated. The interplay of the two techniques on studies of enzyme mechanisms is illustrated in the examples discussed in Section 5.5.

5.5 Examples of enzyme mechanisms

In this section we shall discuss current ideas about the mechanisms of action of five enzymes; chymotrypsin, triosephosphate isomerase, dehydroquinase, lactate dehydrogenase, and DNA methyltransferase. These enzymes catalyse different types of reactions, i.e. they belong to different classes in the classification scheme described in Chapter 1, Section 1.5.1. In each example it should become apparent how various experimental approaches have been used in a complementary fashion to formulate a reasonably detailed picture of the mechanism. Comments on the mechanisms of related enzymes will be made where appropriate.

5.5.1 Chymotrypsin

Chymotrypsin is a protein-hydrolysing enzyme. It is synthesized in the acinar cells of the pancreas as an inactive precursor (zymogen) known as chymotrypsinogen that is activated in the small intestine by the action of proteases (see Chapter 6, Section 6.2.1.1). The mechanism of chymotrypsin has been intensively studied[81] and is probably better understood than any other. We shall discuss in turn the information gained from studies of X-ray crystallography, site-directed mutagenesis, detection of intermediates, chemical modification, and kinetics, before outlining the proposed mechanism of action.

5.5.1.1 Background information on the structure and action of the enzyme

Chymotrypsin has an M_r of about 25 000 and consists of three polypeptide chains held together by disulphide bonds. This unusual structure arises as a result of the activation of the zymogen by proteolytic cleavage (Fig. 5.15). The activation of chymotrypsinogen occurs in stages.[81] There is an initial cleavage, catalysed by trypsin, of the Arg 15–Ile 16 bond that yields a fully active, two-chain, enzyme known as π-chymotrypsin. Subsequent cleavage, catalysed by chymotrypsin, then takes place to yield δ-chymotrypsin (by cleavage of the Leu 13–Ser 14 bond), and finally α-chymotrypsin (by cleavage of the Tyr 146–Thr 147 and Asn 148–Ala 149 bonds). α-Chymotrypsin is the most commonly studied form of the enzyme.

Chymotrypsin catalyses the hydrolysis not only of proteins but also of small substrates and in the latter case will act on both ester and amide bonds. The enzyme shows relatively little specificity when acting on ester substrates: a finding that can be correlated with the inherently higher reactivity of esters. In the case of amide bonds (which are intrinsically less easily hydrolysed) the only substrates that are acted on by chymotrypsin are those with large hydrophobic side-chains, principally those of Trp, Tyr, and Phe, and, to a lesser extent, those of Leu and Met. Specificity is shown for the nature of the R group attached to the carbon

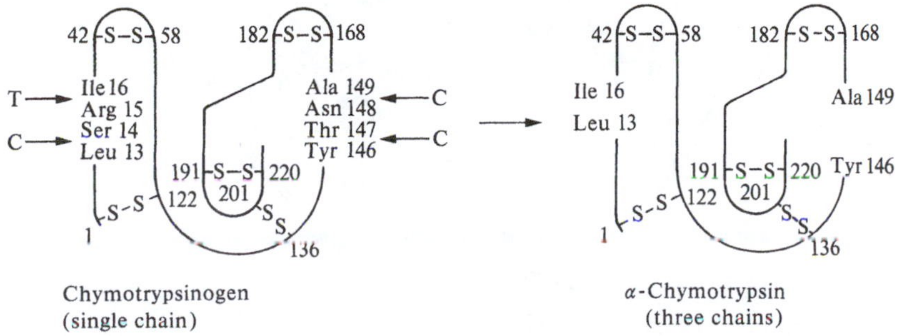

Chymotrypsinogen
(single chain)

α-Chymotrypsin
(three chains)

Fig. 5.15 Activation of chymotrypsinogen. T and C represent the actions of trypsin and chymotrypsin, respectively.

$$\text{X—NH—CH—}\overset{\overset{\displaystyle O}{\|} {\downarrow}}{\text{C}}\text{—NH—Y}$$
$$|$$
$$R$$

Cleavage of peptide bond catalysed by chymotrypsin R is a bulky hydrophobic side-chain.

* The numbering of amino acids in chymotrypsin refers to the sequence of the intact polypeptide chain of chymotrypsinogen.

atom on the amino side of the amide bond. Despite the differences in specificity as far as amide and ester substrates are concerned, there are good grounds for believing that the mechanism of hydrolysis is the same for both types of substrates (Section 5.5.1.3).

5.5.1.2 X-ray crystallographic studies

The structure of α-chymotrypsin has been discussed briefly in Chapter 3, Section 3.4.6.3; the molecule consists of two folded domains (amino acids 27–112 and 133–230*) in each of which there are six strands of antiparallel β-sheet, forming a distorted hydrogen-bonded cylinder (Fig. 5.16). There is a shallow depression at the active site between the two domains, and also a distinct pocket that plays an important part in the binding of specific substrates.[82] Of special note is the finding that certain charged groups, e.g. the α-amino group of Ile 16 and the carboxylate side-chains of Asp 102 and Asp 194, are buried in the interior of the molecule.[82] The significance of these groups is discussed later in this section.

X-ray crystallographic studies have provided answers to a number of questions about chymotrypsin:

(i)   Why is the side-chain of Ser 195 especially reactive?

(ii) How do substrates bind to the enzyme?

(iii) How do the differences in specificity between chymotrypsin and related proteases arise?

(iv) How is the zymogen activated?

(i) Why is the side-chain of Ser 195 especially reactive?

Chymotrypsin is one of a number of enzymes (trypsin, elastase, and thrombin are other examples) that are classified as serine proteases, on the basis that modification of a single serine side-chain in each molecule by a reagent such as diisopropylfluorophosphate (see Section 5.4.4.2) leads to inactivation. In chymotrypsin this reactive serine has been identified as Ser 195. The reason for the high reactivity of this side-chain (the other 27 serine side-chains in chymotrypsin do not react with diisopropylfluorophosphate, nor does free serine) was revealed by X-ray crys-

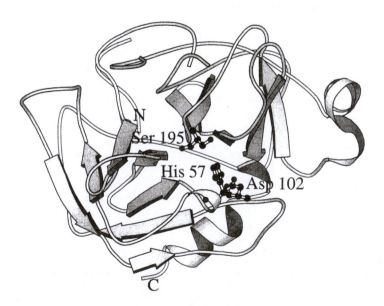

Fig. 5.16 Structure of chymotrypsin (2cha) drawn in Molscript,[195] with key side-chains highlighted.

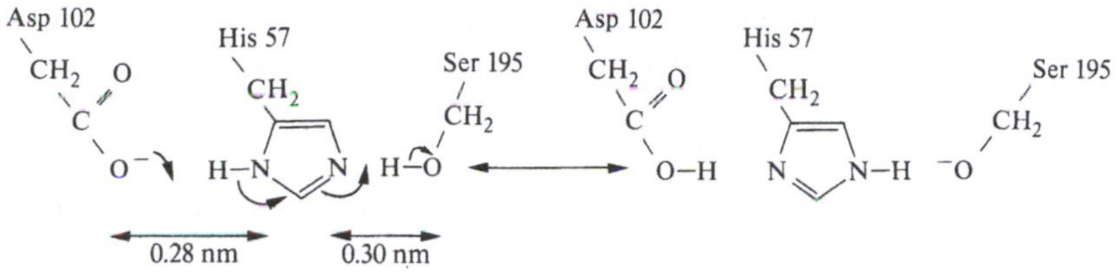

Fig. 5.17 The charge-relay system in chymotrypsin.[83]

tallography which showed that there was a precisely arranged group of three side-chains in the enzyme.[83] This arrangement allows the possibility of partial ionization of the side-chain of Ser 195 by a 'charge relay' mechanism (Fig. 5.17).

Normally, ionization of serine side-chains is insignificant because of the high pK_a (≈ 14) of the $-CH_2-OH$ group. The partial negative charge on the oxygen atom increases the nucleophilicity of the serine side-chain enormously; this is exploited in the catalytic mechanism that proceeds via acyl enzyme formation. All other serine proteases so far examined possess this catalytic triad or charge-relay arrangement of Asp ... His ... Ser side-chains. The extensive similarities of amino-acid sequences around these amino acids in various serine proteases have already been noted (see Chapter 3, Section 3.3.2.11). The details of the charge-relay systems in these enzymes have been extensively investigated. Using NMR[84,85] and neutron diffraction[86] techniques it has been shown that His 57 is the ionizing group of pK_a around 7 and that the buried Asp 102 is of low pK_a. Site-directed mutagenesis has also been used to assess the importance of Asp 102 in the charge-relay system.[87] The mutant trypsin Asp 102 → Asn showed a k_{cat} value some 5000-fold lower than that of the native enzyme, showing that Asp 102 is important but not absolutely required for catalysis. Using affinity labels that modified different side-chains at the active site, it was shown that the Asp 102 → Asn mutation had considerably greater effects on the reactivity of Ser 195 than on that of His 57. Other studies using site-directed mutagenesis have shown that the three side-chains in the charge-relay system act synergistically to cause rate enhancement.[88]

(ii) How do substrates bind to the enzyme?

A number of crystallographic studies have been made of the binding of poor substrates or competitive inhibitors to the enzyme. The results of such studies[89] on the binding of the inhibitor N-formyl tryptophan (amides of which are good substrates for chymotrypsin) are shown in Fig. 5.18. The aromatic side-chain of the inhibitor binds in the pocket, which is lined with non-polar side-chains. The

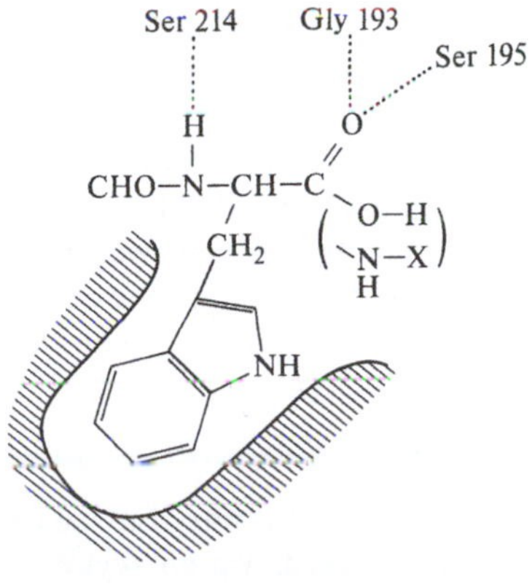

Fig. 5.18 Binding of the inhibitor N-formyl tryptophan to chymotrypsin. The structure of the potential substrate is indicated in brackets.

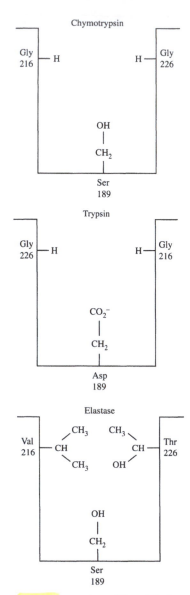

Fig. 5.19 Comparison of the substrate binding pockets of the serine proteases chymotrypsin, trypsin, and elastase.

* Some careful work has shown that chymotrypsinogen does show very slight activity in catalysing the hydrolysis of esters; the activity is some 106-fold less than that of chymotrypsin. The mechanism of the zymogen-catalysed reaction appears to be similar to that of the active enzyme, proceeding via an acyl enzyme.[93]

–NH– group forms a hydrogen bond with the peptide-chain carbonyl group of Ser 214 and the carboxyl group (and hence, by implication, the amide bond of a substrate) is close to the side-chain of Ser 195. The carbonyl group can also form hydrogen bonds with the peptide chain –NH– group of Ser 195 and/or Gly 193.

(iii) How do the differences in specificity between chymotrypsin and related proteases arise?

The serine proteases chymotrypsin, trypsin, and elastase possess very similar three-dimensional structures, but display quite different specificities for substrates. Chymotrypsin is specific for amides with aromatic or other large hydrophobic side-chains, trypsin is specific for amides with positively charged side-chains (Lys or Arg), and elastase has a somewhat broader specificity, showing a preference for amides with small hydrophobic side-chains, e.g. Ala.[90] The results of X-ray crystallography have shown that these differences in specificity can be correlated with differences in the substrate-binding pockets of the three enzymes (Fig. 5.19). In the case of chymotrypsin, the hydrophobic side-chain of the substrates makes a number of favourable contacts with the non-polar side-chains lining the pocket. At the bottom of the pocket is the uncharged side-chain of Ser 189. In trypsin, this serine is replaced by aspartic acid and the negatively charged side-chain is then exactly placed to bind electrostatically to the positively charged side-chain of a substrate. Elastase has a pocket similar to that of chymotrypsin; however, access to the pocket is obstructed by the bulky side-chains of Val 216 and Thr 226. This gives rise to a shallow hydrophobic depression which can accommodate small hydrophobic side-chains of potential substrates.[90] Site-directed mutagenesis has been used to investigate the specificity of trypsin;[87] in these experiments the side-chains of Gly 216 or Gly 226 or both have been replaced by Ala. The mutant enzymes show differing specificities towards lysine and arginine substrates; thus, the mutation Gly 226 → Ala increases the preference of the enzyme for lysine rather than arginine. A review of the specificities of these proteases and attempts to manipulate them has been given.[91]

(iv) How is the zymogen activated?

A comparison of the three-dimensional structures of chymotrypsinogen and chymotrypsin allows us to identify the differences between the two molecules and to explain the activation process in structural terms. One perhaps surprising finding is that the Asp … His … Ser charge-relay system is already present in the zymogen, with the three side-chains occupying very similar or identical positions to those found in the active enzyme.[92] The zymogen, however, is inactive* because the substrate-binding pocket is not properly formed, and thus a substrate cannot be positioned precisely to take advantage of the reactive serine side-chain. Formation of the substrate-binding pocket during the activation process proceeds as follows.[92] Cleavage of the Arg 15–Ile 16 bond by trypsin creates a new positive charge at the α-amino group of Ile 16. A strong electrostatic force between this positive charge and that of the side-chain of Asp 194 helps to move other parts of the molecule, such as the side-chains of Arg 145 and Met 192, and to form the substrate-binding pocket. The electrostatic interaction between Ile 16 and Asp 194 is strong because it occurs in a region of low dielectric constant in the interior of the enzyme (see Chapter 3, Section 3.4.5.2).

5.5.1.3 Detection of intermediates

We have already mentioned (Section 5.4.1.5) that the chymotrypsin-catalysed hydrolysis of esters such as 4-nitrophenyl acetate proceeds via an acyl enzyme intermediate. The rate of the deacylation step is pH-dependent and can be slowed down to such an extent that at low pH the acyl enzyme can be isolated; the acyl group is linked to the side-chain of Ser 195.[40]

In the hydrolysis of amides, it is found that the deacylation step is more rapid than formation of acyl enzyme; this can be correlated with the lower reactivity of amides compared with esters. There is, therefore, no accumulation of the acyl enzyme and its participation in the reaction can only be inferred indirectly by 'trapping' experiments. A suitable 'trapping' experiment involves conducting the enzyme-catalysed hydrolysis of a substrate in the presence of a high concentration of a nucleophile such as alaninamide or hydroxylamine that can compete with water in the breakdown of the acyl enzyme:

The ratio of products (a)/(b) was determined using an ester substrate for which the involvement of an acyl enzyme intermediate had been established directly. It was then found that, at the same concentration of nucleophile, an identical product ratio was observed using an amide substrate.[94] This provides very good evidence for the participation of an acyl enzyme intermediate in the hydrolysis of amides.

By analogy with hydrolytic reactions in organic chemistry, formation and decomposition of the acyl enzyme are presumed to proceed via tetrahedral intermediates (Fig. 5.20), in which developing negative charge on the oxygen is stabilized by hydrogen-bonding interactions with the peptide –NH– groups of Gly 193 and Ser 195, in what is termed the 'oxanion hole'. In the case of the serine protease subtilisin, site-directed mutagenesis has shown that disruption of the hydrogen bonds which stabilize the tetrahedral intermediate reduces the value of k_{cat} by up to 1000-fold.[95]

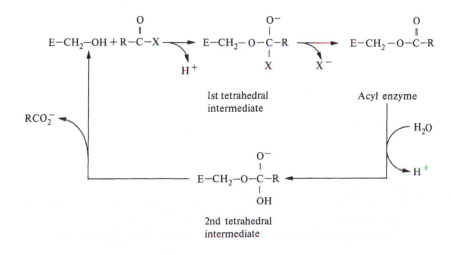

Fig. 5.20 The mechanism of action of chymotrypsin showing the tetrahedral intermediates involved in the formation and breakdown of acyl enzyme.

Direct evidence for the existence of tetrahedral intermediates has been gathered from a number of experiments, mainly on enzymes related to chymotrypsin. First, stopped-flow studies (Chapter 4, Section 4.4) on the elastase-catalysed hydrolysis of an amide substrate, a 4-nitroanilide derivative of a tripeptide. have shown that the decomposition of the first tetrahedral intermediate to the acyl enzyme is the rate-limiting step of the overall hydrolytic reaction.[96] Second. X-ray crystallographic studies have been made of the very strong complexes formed by trypsin with small protein inhibitor molecules such as bovine pancreatic trypsin inhibitor and soybean trypsin inhibitor.[97,98] These studies have shown that the strong interactions arise at least in part from the fact that in these complexes the crucial amide bonds of the inhibitors are distorted so as to resemble a tetrahedral intermediate. Third, as we have already mentioned (Section 5.4.2), it has proved possible by working at low temperatures to stabilize a tetrahedral intermediate in the elastase-catalysed hydrolysis of a 4-nitroanilide substrate.[42]

5.5.1.4 Chemical modification of amino-acid side-chains

There have been a large number of chemical-modification studies on chymotrypsin, but we shall mention only some of those that have confirmed the roles of side-chains implicated by X-ray crystallographic work.

The importance of the side-chain of Ser 195 in the catalytic mechanism was originally deduced from chemical-modification studies using diisopropylfluorophosphate (Section 5.4.4.2). This side-chain can also be modified by phenylmethanesulphonyl fluoride which, being a solid, has certain advantages in handling over diisopropylfluorophosphate, which is a liquid.*

Phenylmethanesulphonyl fluoride is widely used in enzyme-purification procedures to minimize degradation by proteases (see the isolation of the *arom* multienzyme protein discussed in Chapter 2, Section 2.8.4).

Some elegant affinity-labelling experiments (Section 5.4.4.3) have shown that the side-chain of His 57 is involved in the catalytic activity of the enzyme. The close structural relationship between the reagent *N*-4-toluenesulphonyl-L-phenylalanine chloromethylketone (TPCK) and the ester substrate is shown in Fig. 5.21.

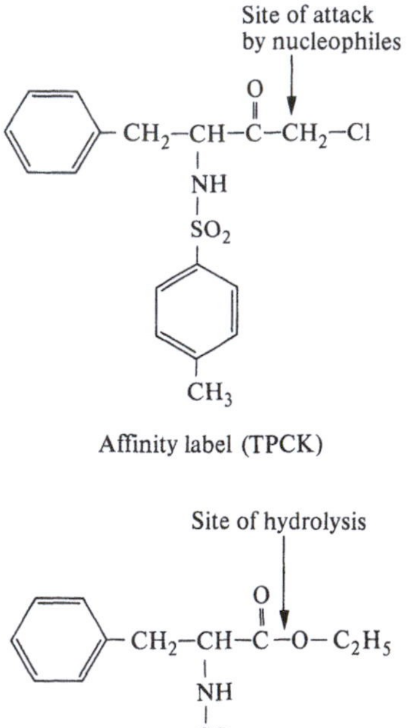

TPCK inactivates chymotrypsin and fulfils the usual criteria for modification at the active site (Section 5.4.4.4). The site of modification is the side-chain of His 57.[81] That the affinity label does indeed recognize a particular site in chymotrypsin is nicely demonstrated by the finding that TPCK does not inactivate trypsin, which possesses a different substrate-binding pocket (Fig. 5.19). Conversely, the reagent *N*-4-toluenesulphonyl-L-lysine chloromethylketone (TLCK) inactivates trypsin but has no effect on chymotrypsin.[99] The two reagents find considerable application in protein-sequencing studies (Chapter 3, Section 3.3.2.2), since in the preparation of peptide fragments it is essential to ensure that the chymotrypsin used is not contaminated by trypsin or *vice versa*. (Thus it is desirable in such studies to use TPCK-treated trypsin and TLCK-treated chymotrypsin to guarantee the specificity.)

The importance of the α-amino group of Ile 16 in chymotrypsin is suggested by chemical modification studies.[81] It would normally be extremely difficult to modify this amino group selectively in view of the presence of the other amino groups in the molecule. This problem was overcome by a 'double-labelling' experiment as outlined in Fig. 5.22, in which chymotrypsinogen was first reacted with acetic anhydride (Ac$_2$O) to acetylate all the amino groups (α- and ε-) in the molecule. This acetylated chymotrypsinogen could then be cleaved by trypsin to yield

* Diisopropylfluorophosphate is an extremely toxic substance, since it inactivates acetylcholinesterase by reaction with a highly reactive serine side-chain, thereby blocking the neurotransmission function of acetylcholine.

Site of attack by nucleophiles

Affinity label (TPCK)

Site of hydrolysis

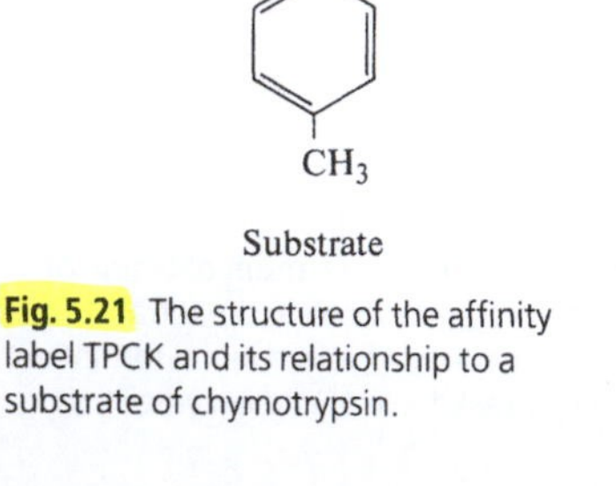

Substrate

Fig. 5.21 The structure of the affinity label TPCK and its relationship to a substrate of chymotrypsin.

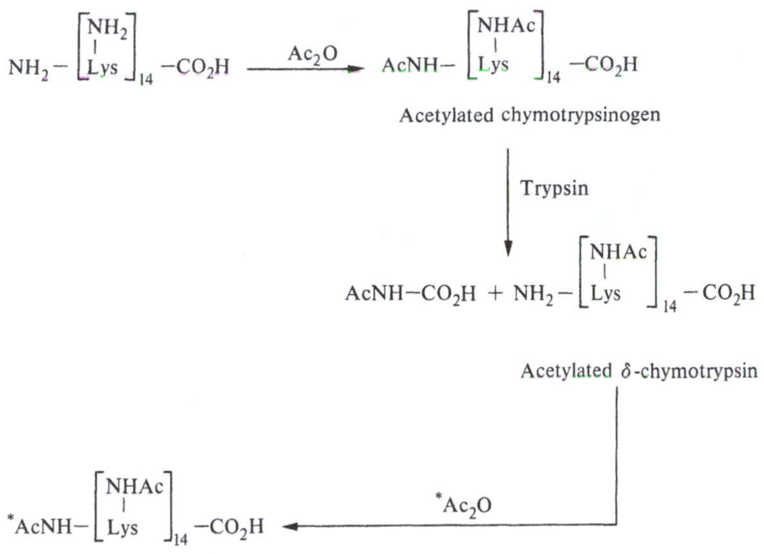

$$NH_2 - \left[\begin{array}{c} NH_2 \\ | \\ Lys \end{array} \right]_{14} -CO_2H \quad \xrightarrow{Ac_2O} \quad AcNH- \left[\begin{array}{c} NHAc \\ | \\ Lys \end{array} \right]_{14} -CO_2H$$

Acetylated chymotrypsinogen

Trypsin

$$AcNH-CO_2H \; + \; NH_2 - \left[\begin{array}{c} NHAc \\ | \\ Lys \end{array} \right]_{14} -CO_2H$$

Acetylated δ-chymotrypsin

$$ {}^*AcNH- \left[\begin{array}{c} NHAc \\ | \\ Lys \end{array} \right]_{14} -CO_2H \quad \xleftarrow{\;{}^*Ac_2O\;}$$

Fig. 5.22 Acetylation of the α-amino group of Ile 16 in chymotrypsin by a double-labelling experiment. Ac_2O is acetic anhydride.

a fully active acetylated δ-chymotrypsin which has a free α-amino group at Ile 16. On treatment with ^{14}C-labelled acetic anhydride it is found that activity is lost in proportion to the amount of radioactivity incorporated into the α-amino group of Ile 16, thus confirming the importance of this group in the activity of the enzyme.[81]

5.5.1.5 Kinetic studies

The kinetics of the reactions catalysed by chymotrypsin have been very intensively investigated. The results of these experiments are more easily interpreted if small synthetic ester or amide substrates are employed; many of these substrates (e.g. 4-nitrophenyl esters or 4-nitroanilide derivatives) possess convenient spectroscopic properties that allow the hydrolysis reactions to be monitored easily. The results of steady-state and pre-steady-state kinetic studies have allowed conclusions to be drawn regarding (i) the reaction pathway and the involvement of the acyl enzyme, (ii) the specificity of the enzyme and the nature of the substrate-binding site, and (iii) the magnitudes of the rate constants of a number of the individual steps in Fig. 5.18. For a summary of this work, reference 100 should be consulted.

5.5.1.6 The proposed catalytic mechanism of chymotrypsin

From the various types of experiments described in Sections 5.5.1.2 to 5.5.1.5, a mechanism for the chymotrypsin-catalysed hydrolysis of amide (and ester) substrates has been proposed.[83] The mechanism is outlined in Fig. 5.23, where for the sake of clarity details of the substrate-binding site and the movements of substrate and enzyme have been omitted. A fuller account of these aspects is given in reference 101. In the acyl enzyme, the charge-relay system is disrupted, since no proton is available to link the side-chains of His 57 and Ser 195. However, in the deacylation step, the $R'NH_2$ leaving group is replaced by H_2O and the charge-relay system can be re-established. As mentioned in Section 5.4.3, time-resolved X-ray crystallography has been used to demonstrate directly the attack of water on the acyl enzyme intermediate.[47]

Asp 102

CH$_2$
C O
O
H

His 57—CH$_2$ N
N
H   O
R'—N—C—R
H   O

Ser 195
CH$_2$

→

Asp 102

CH$_2$
C O
O$^-$
H

His 57—CH$_2$ N
N....H........O
R'—N   C—R
H   O

Ser 195
CH$_2$

→ R' NH$_2$

← H$_2$O

Asp 102

CH$_2$
C O
O
H

His 57—CH$_2$ N
N
H   O
O—C—R
H   O

Ser 195
CH$_2$

Ser 195 ← His 57—CH$_2$ N
N....H........O
O   C—R
H   O

Ser 195
CH$_2$

Fig. 5.23 The proposed mechanism of action of chymotrypsin on amide substrates.[83]

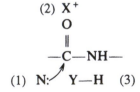

(2) **X$^+$**

O
‖
—C—NH—

(1) **N:**   **Y—H**   (3)

5.5.1.7 The relationship between the mechanism of action of chymotrypsin and those of other proteases

If we consider the hydrolysis of an amide bond, we can formulate the essential chemical requirements of a catalyst as the possession of the following:

(1) a nucleophilic group, N:, to attack the carbonyl group, leading to the formation of a tetrahedral intermediate;

(2) some positively charged species, X$^+$, in the neighbourhood of the oxygen atom of the carbonyl group. This would not only increase the susceptibility of the carbonyl group to nucleophilic attack but would also stabilize the tetrahedral intermediate;

(3) a proton donor, Y–H, to make the NH– moiety a better leaving group.

We have already seen how these requirements are met in the case of chymotrypsin and related serine proteases. The nucleophilic group is the side-chain of Ser 195 activated by the charge-relay mechanism. Polarization of the carbonyl group is enhanced by hydrogen-bonding interactions between the oxygen atom and the peptide chain –NH– groups of Gly 193 and Ser 195. (As noted in Section 5.5.1.3, these interactions stabilize the tetrehedral intermediates in the 'oxyanion hole'.) The proton-donating species is the side-chain of His 57. A number of variations on the theme of the charge-relay system have been observed in related enzymes. Thus in acetylcholine esterase, there is a Ser ... His ... Glu system,[102] and in the 20S proteasome the important nucleophile is the N-terminal Thr of the β subunits which forms a system Thr ... Glu ... Lys (see also Chapter 9, Section 9.6.2.2).[103]

At first sight, other types of proteases possess very different mechanisms, but it is possible to interpret these mechanisms in terms of the requirements outlined

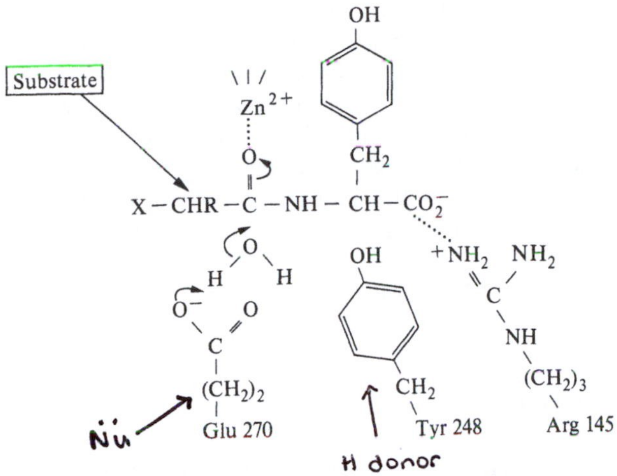

Fig. 5.24 Part of the proposed mechanism of action of carboxypeptidase.[101,104,105]

above. For instance, in carboxypeptidase (Fig. 5.24), the nucleophilic group is a water molecule activated by the side-chain of Glu 270 and the zinc ion acts as the polarizing influence on the carbonyl group. [101,104]

On the basis of earlier chemical-modification and X-ray data, it had been proposed that the phenolic –OH group of the side-chain of a Tyrosine (probably Tyr 248) acted as a proton donor to the leaving group. However, this possibility was ruled out by site-directed mutagenesis experiments[105] in which Tyr 248 was replaced by Phe. In the mutant the k_{cat} was unchanged although the K_m for peptide and ester substrates is raised. These results suggest that Tyr 248 plays a role in the binding of substrates. Similar results have been found for the mutant in which Tyr 198 (also close to the active site) has been replaced by Phe and in the double mutant (Tyr 248 and Tyr 198 both replaced by Phe).[106]

In thiol proteases such as papain, the nucleophilic group is a cysteine side-chain that is partially ionized as a result of the presence of a neighbouring histidine side-chain. The polarization of the carbonyl group of the amide bond occurs by hydrogen bonding as in the case of chymotrypsin, and the proton-donating group is a histidine side-chain.[101,107–109] In acid (aspartic) proteases such as pepsin and penicillopepsin, the nucleophilic group appears to be a water molecule activated by an aspartic acid side-chain (Asp 32 in pepsin). The diad of Asp 32 and Asp 215 acts as a proton donor to the –NH leaving group.[110] Interestingly, in the case of the protease encoded by human immunodeficiency virus (HIV), the two Asp side-chains are contributed by the two polypeptide chains in the dimeric enzyme (Asp 25 and Asp 25'). The unusual symmetry displayed in this enzyme has provided a useful opportunity for the design of symmetrical inhibitors of HIV protease (see Chapter 10, Section 10.7.3). The design of these and other inhibitors has been discussed in detail.[111,112]

From this discussion it can be appreciated that the various types of proteases have adopted different solutions to the problems associated with the hydrolysis of amide bonds, although all the different solutions can be rationalized in terms of chemical principles. The differences in specificity between the various proteases can be understood in terms of the details of the interactions between enzyme and substrate (see the examples of chymotrypsin, trypsin, and elastase in Section 5.5.1.2). The specificity displayed by carboxypeptidase is for the position of an amide bond in a polypeptide chain, since this enzyme removes the C-terminal amino acid (see Chapter 3, Section 3.3.2.4). By referring to the structure of the active site of the enzyme (Fig. 5.24) we can see that the positively charged side-chain of Arg 145 is in an appropriate position to bind to the C-terminal carboxylate group of the substrate. There are in fact two types of carboxypeptidases, A and B, which are very similar in overall structure. However, carboxypeptidase B, which is specific for the removal of C-terminal Lys or Arg, has an appropriately positioned negatively charged side-chain (that of Asp 255) to bind the positively charged side-chains of the substrate.[113] Carboxypeptidase A will remove all C-terminal amino acids other than Lys or Arg from a polypeptide chain.

5.5.2 Triosephosphate isomerase

Triosephosphate isomerase catalyses the following reaction:

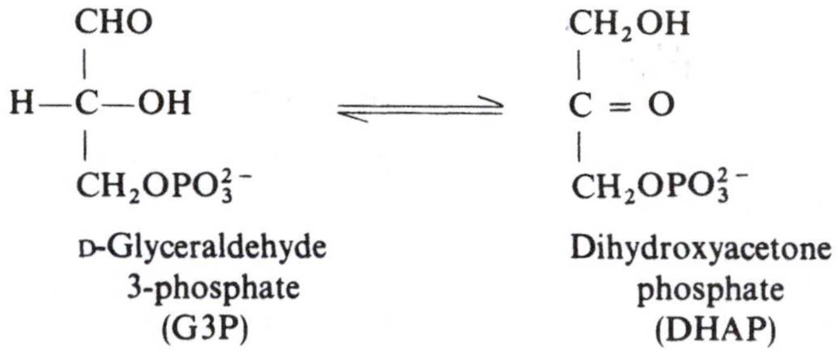

$$
\begin{array}{ccc}
\text{CHO} & & \text{CH}_2\text{OH} \\
| & & | \\
\text{H}-\text{C}-\text{OH} & \rightleftharpoons & \text{C}=\text{O} \\
| & & | \\
\text{CH}_2\text{OPO}_3^{2-} & & \text{CH}_2\text{OPO}_3^{2-} \\
\text{D-Glyceraldehyde} & & \text{Dihydroxyacetone} \\
\text{3-phosphate} & & \text{phosphate} \\
\text{(G3P)} & & \text{(DHAP)}
\end{array}
$$

The physiological importance of this reaction is that it allows interconversion of the two three-carbon units produced by the cleavage of D-fructose 1,6-bisphosphate catalysed by fructose-bisphosphate aldolase (see Chapter 6, Section 6.4.1). D-Glyceraldehyde 3-phosphate is an intermediate in the glycolytic pathway leading from glucose to pyruvate. Dihydroxyacetone phosphate (also known as glycerone phosphate) can be reduced to glycerol 1-phosphate, which acts as a precursor in the synthesis of various lipids.

The equilibrium of the triosephosphate isomerase-catalysed reaction is strongly in favour of dihydroxyacetone phosphate:

$$K = \frac{[\text{DHAP}]}{[\text{G3P}]} = 367 * \text{ at 298 K (25°C)}.$$

* Earlier values of K were quoted as being around 20. However, it is recognized that both G3P and DHAP can exist as hydrated forms in aqueous solution.[114] These hydrated (diol) forms do not act as substrates for the enzyme. When the corrections for the amounts of true (unhydrated) substrate and product are made, K becomes much larger.

A good understanding of the mechanism of the enzyme has been achieved by the use of X-ray crystallography, affinity labelling, site-directed mutagenesis, and experiments involving the isotopic labelling of substrates. Attention will be focused on the triosephosphate isomerases from chicken breast muscle and yeast, but it is clear that the enzyme from a wide variety of other sources has very similar properties.

5.5.2.1 Background information on the structure and action of the enzyme

Triosephosphate isomerase has an M_r of 53 000 and is composed of two subunits of identical sequence. It is of special note that the enzyme is an extremely efficient catalyst, with a turnover number, expressed per active site, of about 250 000 min⁻¹ in the direction of conversion of D-glyceraldehyde 3-phosphate to dihydroxyacetone phosphate. As will be mentioned later (Section 5.5.2.5), there are good grounds for believing that triosephosphate isomerase is an almost perfectly evolved enzyme, i.e. that further evolution to produce a more efficient catalyst is not possible.[115]

5.5.2.2 X-ray crystallographic studies

Crystallographic studies have shown that each subunit of the enzyme is roughly spherical with a diameter of approximately 3.5 nm. There is a regular manner of chain folding so that each subunit consists of an inner cylinder or 'barrel' of eight strands of parallel-pleated sheet. The adjacent strands are linked mainly by helical segments, which thus form the outer face of each 'barrel' (Fig. 5.25).[116] Triosephosphate isomerase is thus an excellent example of the α/β type of structure described by Levitt and Chothia[117]); indeed the so-called 'TIM-barrel' structure represents one of the most commonly occurring 'superfolds' in proteins (see Chapter 3, Section 3.4.4.6). In the contact area between the subunits there is a significant number of amino acids with polar side-chains.

One important aspect of the X-ray crystallographic work is that it is possible to study a true enzyme–substrate complex, because the equilibrium in this one substrate reaction lies so much over to one side. An analysis of the complex of enzyme with dihydroxyacetone phosphate showed that the substrate is bound to each subunit in a 'pocket' that includes a few side-chains from the adjacent subunit, i.e. the active site seems to contain contributions from both subunits. Indeed, the importance of subunit interactions in the catalytic properties of the enzyme has been emphasized by site-directed mutagenesis experiments in which side-chains near the subunit interface, e.g. Asn 14 and Asn 78, were mutated with consequent effects on k_{cat}.[118]

A good deal of additional information concerning the active site and mechanism of the enzyme has come from X-ray studies of the complex formed between the yeast enzyme and a transition state analogue, phosphoglycolohydroxamate (see Section 5.5.2.3).[119] From this work, it can be seen that the principal amino-acid side-chains that are within about 0.4 nm of the substrate include Glu 165, His 95, and Lys 12.*

* The numbering refers to the yeast enzyme; the corresponding residue in the muscle enzyme is Lys 13.

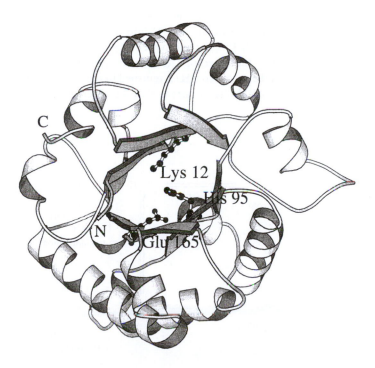

Fig. 5.25 The structure of one subunit of yeast triosephosphate isomerase (1ypi) drawn in Molscript,[195] with residues Lys 12, His 95, and Glu 165 highlighted.

By comparison of the structures of the enzyme in the presence and absence of the transition-state analogue, it can be seen that a mobile loop (residues 166–176) undergoes major structural changes so as to enclose the substrate. The α-carbon of Thr 172 moves 0.71 nm in this process, which results in the main chain –NH– of Gly 171 forming a hydrogen bond with the phosphate of the analogue.[120] Deletion of residues 170–173 in the loop region leads to a nearly 100 000-fold loss in activity.[121]

5.5.2.3 Isotopic labelling of substrates

The effects of isotopic substitution on the rate of a reaction can give valuable information about the elementary steps of that reaction (for reviews, see references 122 and 123). However, we shall confine our attention to experiments in which isotopes are used as labels to follow the fate of a particular atom or group in a reaction.

Early work established that, in the presence of enzyme, up to one atom of tritium (^{3}H) can be incorporated into dihydroxyacetone phosphate from the solvent ^{3}H$_2$O. This atom of ^{3}H is incorporated in a stereospecific manner.[124]

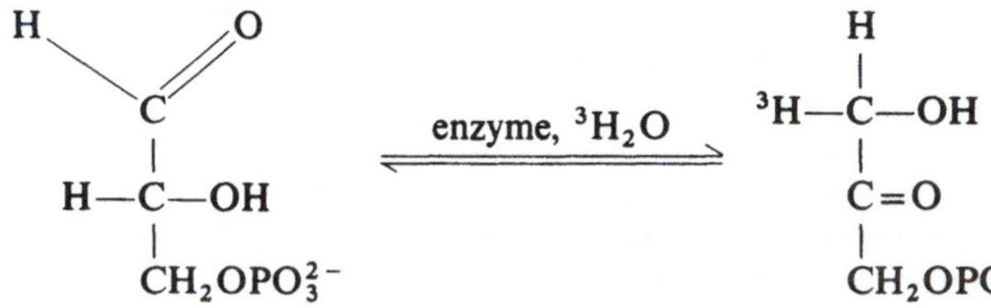

This finding suggests that the reaction proceeds via abstraction of a proton from the substrate by a base (B), followed by an exchange with the solvent. On chemical grounds it seems reasonable that a *cis*-enediol intermediate is involved (Fig. 5.26). The extent of incorporation of ^{3}H into the product will depend on how the rate of exchange of BH$^+$ with the solvent compares with the rate at which the proton is returned to the *cis*-enediol to give the product. The *cis*-enediol mechanism implies that the proton (or ^{3}H) is returned to the same face of the diol from which it is abstracted; this is consistent with the observed stereochemistry of the reaction.

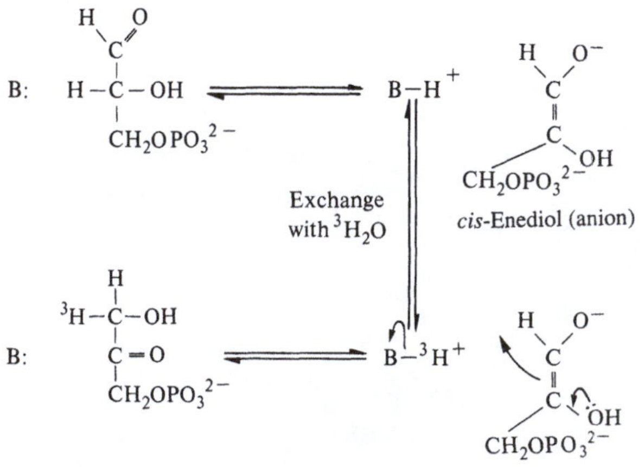

Fig. 5.26 Incorporation of ^{3}H into dihydroxyacetone phosphate via a *cis*-enediol intermediate in the reaction catalysed by triosephosphate isomerase.

Further support for the idea of a *cis*-enediol intermediate in the enzyme-catalysed reaction was obtained from studies of enzyme inhibition. 2-Phosphoglycolate is a powerful competitive inhibitor of the enzyme, with a dissociation constant of 6 μmol dm^{-3}. The fact that it shows a marked similarity in terms of electron distribution and stereochemistry to the *cis*-enediol intermediate has led to the proposal[17] that 2-phosphoglycolate acts as a transition-state analogue (Section 5.3.6). Another powerful competitive inhibitor of the enzyme which has been used in X-ray studies[119] is phosphoglycolohydroxamate, also an analogue of the *cis*-enediol intermediate.[125]

5.5.2.4 Affinity labelling

Some elegant affinity-labelling experiments have helped to identify the base involved in the proton-transfer reactions (Fig. 5.26). Two different affinity labels have been used, bromohydroxyacetone phosphate (I),[58] and glycidol phosphate (II).[126]

Both inhibitors conformed to the usual criteria for modification at the active site of the enzyme (Section 5.4.4.4), i.e. stoichiometric inactivation and protection by competitive inhibitors. Inhibitor I clearly resembles dihydroxyacetone phosphate and has an activated Br atom that can be displaced by a nucleophilic group. Inhibitor II resembles the *cis*-enediol intermediate because of the steric requirements of the oxirane ring; this ring can be opened following attack by a suitable nucleophile.

In each case the modified, inactivated, enzyme was subjected to proteolytic digestion and the modified peptide isolated. By sequencing the peptides it was shown that in each case the site of modification was the side-chain of Glu 165, which is also implicated by crystallographic studies as being at the active site (Section 5.5.2.2). In the case of enzyme modified by inhibitor I, it was found that the attached label could subsequently migrate to the adjacent side-chain, that of Tyr 164; this migration could be prevented by reduction of the modified enzyme with sodium borohydride.[58] This example serves to emphasize that in some chemical-modification experiments there can be a possibility of migration of the modifying group, so that the site of attachment deduced by the normal methods of structural characterization (Chapter 3, Section 3.3) may not correspond to the initial site of modification where the observed effect, e.g. on enzyme activity, occurred.

The affinity-labelling experiments show that the side-chain of Glu 165 is indeed at the active site of the enzyme, and the fact that the side-chain can act as a nucleophile makes it very likely that it acts as the general base involved in proton-transfer reactions (Fig. 5.26).

The important role played by Glu 165 has been confirmed by site-directed mutagenesis experiments in which Glu 165 was replaced by Asp.[127] The mutant enzyme had a k_{cat} some 10^3-fold lower than that of the native enzyme, with an altered K_m. Determination of the rate of the individual steps in the catalytic process showed that the mutation had decreased the rates of the enolization steps (ES $\rightleftharpoons$ EZ and EZ $\rightleftharpoons$ EP in Fig. 5.29) rather than those of the binding steps (E + S $\rightleftharpoons$ ES and E + P $\rightleftharpoons$ EP). X-ray studies of the Glu165Asp mutant enzyme and its complex with phosphoglycolohydroxamate have shown that although the overall structure of the enzyme is unchanged compared with the wild-type enzyme, there are subtle changes in the substrate-binding region. Thus, replacement of Glu by the shorter Asp increases the distance between the carboxylate side-chain and the substrate by 0.1 nm; in addition, the orientation of

2-Phosphoglycolate

Phosphoglycolohydroxamate

cis-Enediol intermediate

$$CH_2Br$$
$$|$$
$$C=O$$
$$|$$
$$CH_2OPO_3^{2-}$$

(I)

$$CH_2$$
$$| \quad \diagdown O$$
$$CH \diagup$$
$$|$$
$$CH_2OPO_3^{2-}$$

(II)

Affinity labels for triosephosphate isomerase.

the carboxylate group with respect to the substrate is altered in the mutant enzyme. It has been proposed that a combination of these two factors is responsible for the observed decrease in k_{cat} in the mutant.[128]

The roles of His 95 and Lys 12, which have been mentioned as being close to the substrate (Section 5.5.2.2) and which are conserved in all triosephosphate isomerases, have been explored in detail using a combination of site-directed mutagenesis, kinetic and structural techniques. Replacement of His 95 by Gln or Asn leads to approximately 1000-fold losses in activity, suggesting that it plays an important role in catalysis.[129] Interestingly, His 95 appears to have a very low pK_a (< 4.5), since ^{15}N NMR studies have shown that it remains uncharged over the pH range from 5 to 9.9.[130] It has therefore been proposed that this uncharged histidine acts (in a highly unusual fashion) as an acid, donating a proton to, and thereby polarizing, the carbonyl group of the substrate.[130,131] The presence of a polarizing group is not only desirable on chemical grounds, since it helps to make the hydrogen attached to the adjacent carbon atom more acidic, but is also indicated by the fact that dihydroxyacetone phosphate bound to the enzyme is much more readily reduced by hydride ion than is free dihydroxyacetone phosphate[132] (Fig. 5.27). Additional evidence for the polarization of the carbonyl group of the substrate when it is bound to the enzyme has also been obtained from IR spectroscopic studies that have provided direct evidence for distortion of the bound substrate.[16]

The role of Lys 12 (yeast enzyme numbering) has been explored in detail.[133,134] Replacement of this Lys by Arg reduces the k_{cat} by a factor of about 200-fold and weakens the binding of substrate. When replaced by His the resulting enzyme is essentially inactive at neutral pH, but shows a small amount of activity (approximately 0.1 per cent of that of the wild-type) at pH 6.1, where it is presumably protonated. These results demonstrate the importance of a positively charged group at this position, presumably to neutralize the negative charge on the phosphate group of the substrate, as well as possibly assisting proton transfer reactions. As mentioned in Section 5.5.2.2, residues within the mobile loop region may also interact with the phosphate group of the substrate.

5.5.2.5 The mechanism of triosephosphate isomerase

From the evidence discussed above, the outline mechanism of action of the enzyme can be proposed in which Glu 165 and His 95 act as a base and an acid respectively (Fig. 5.28). Lys 13 (Lys12 in yeast) interacts with the phosphate group of the substrate.

Very detailed measurements have been made of the rates of exchange of isotope between substrate and solvent during the enzyme-catalysed reaction. These results have been used to calculate the rate constants of the individual steps of the reaction and have enabled a free-energy profile of the reaction pathway to be constructed,[115,135] (Fig. 5.29). The reaction occurs very nearly at a rate that is limited by the rate at which substrate can encounter the enzyme by the normal process of diffusion in solution,[115,136] i.e. no further 'evolutionary improvement' in catalytic power is possible. (It is, of course, possible that further evolution might lead

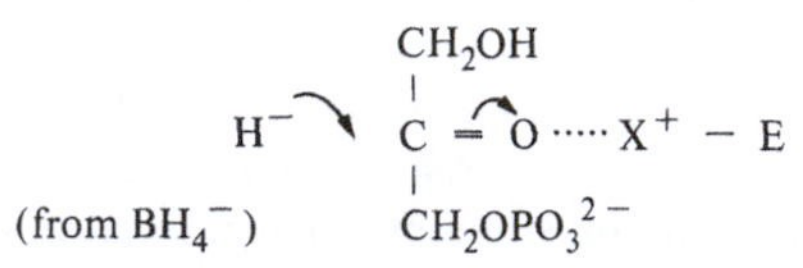

Fig. 5.27 Polarization of the carbonyl group of dihydroxyacetone phosphate bound to triosephosphate isomerase. This would facilitate the reduction of the substrate by borohydride.[132]

$$\text{H}^- \searrow \quad \begin{array}{c} \text{CH}_2\text{OH} \\ | \\ \text{C} = \text{O} \cdots \text{X}^+ - \text{E} \\ | \\ \text{CH}_2\text{OPO}_3^{2-} \end{array}$$

(from BH_4^-)

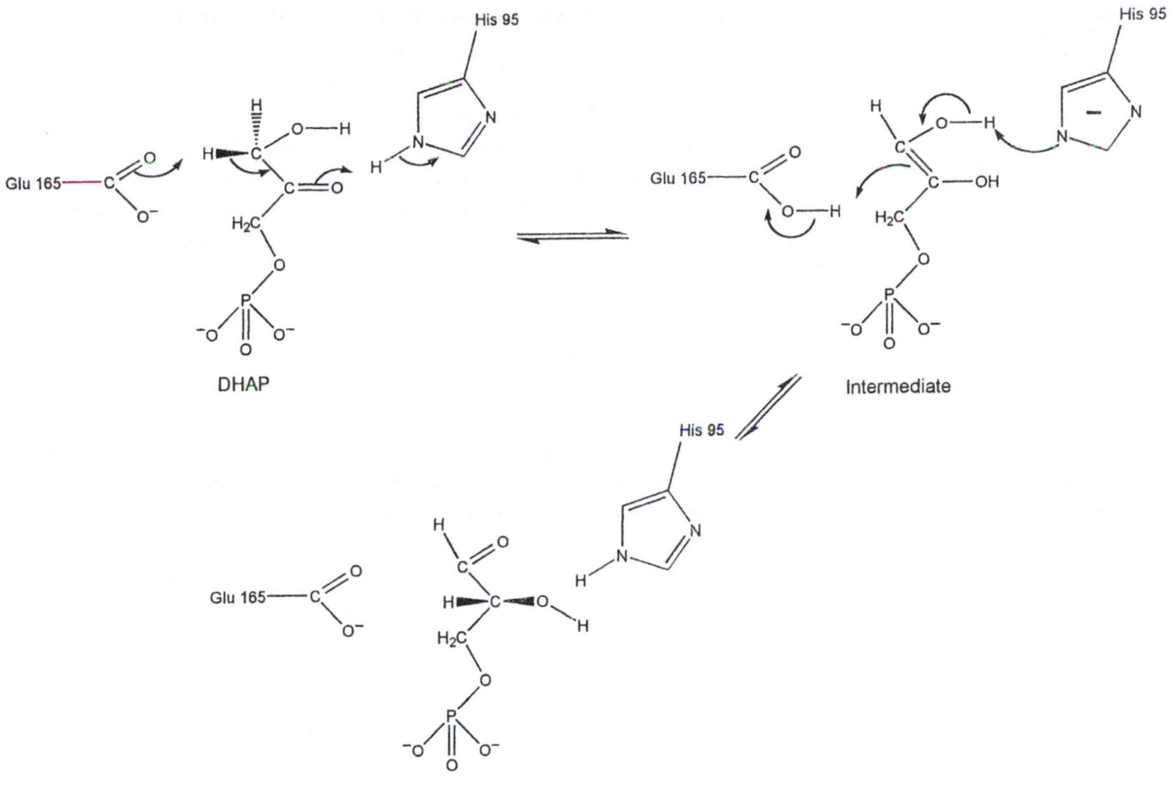

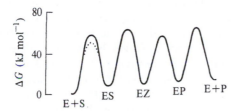

Fig. 5.28 Outline of the mechanism of the reaction catalysed by triosephosphate isomerase. Details of the proton-transfer reactions in the second step have been omitted.

Fig. 5.29 Free-energy profile for the reaction catalysed by triosephosphate isomerase.[135] S is dihydroxyacetone phosphate, P is D-glyceraldehyde 3-phosphate, and Z is the *cis*-enediol intermediate. The dashed line between E + S and ES represents the diffusion-controlled encounter rate. Free energies are referred to a standard state of 40 μmol dm^{-3}, which is the concentration of dihydroxyacetone phosphate *in vivo*.

to 'desirable' changes in the regulatory properties of the enzyme.) Several key features of the energy profile of the reaction pathway have been simulated using an approach based on quantum mechanical calculations.[137]

The mechanism of the triosephosphate isomerase-catalysed reaction is probably analogous to the mechanisms of a number of other aldose–ketose isomerases. For instance, the mechanism of glucosephosphate isomerase, which catalyses the interconversion of D-glucose 6-phosphate and D-fructose 6-phosphate, is also thought to proceed via a *cis*-enediol intermediate.[138]

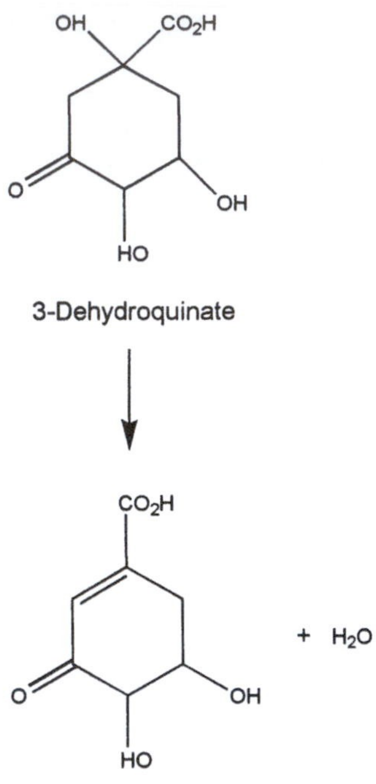

OH   CO$_2$H

O

HO   OH

3-Dehydroquinate

↓

CO$_2$H

+ H$_2$O

O   OH

HO

3-Dehydroshikimate

5.5.3 Dehydroquinase (3-dehydroquinate dehydratase)

Dehydroquinase (3-dehydroquinate dehydratase) catalyses the removal of water from 3-dehydroquinate so as to generate 3-dehydroshikimate.

This reaction represents a central step in the pathway of biosynthesis of aromatic amino acids, and the enzyme is widely distributed in bacteria, plants, and lower eukaryotes such as fungi. However, the reaction also forms part of a catabolic pathway in which quinate is degraded. The enzyme has a number of additional features of interest which relate to its occurrence in multienzyme systems (see Chapter 7, Section 7.11.2) and to the fact that there are two mechanistically distinct types of enzyme, as discussed below.

5.5.3.1 Background information on the structure of the enzymes and the reaction catalysed

As explained in the previous section, the conversion of 3-dehydroquinate to 3-dehydroshikimate occurs in two distinct metabolic pathways. In a number of fungi including *Neurospora crassa*, genetic studies have shown that there were two distinct enzymes involved, termed biosynthetic and catabolic (bDHQase and cDHQase respectively). Thus the two enzymes were encoded by two distinct genetic loci, subject to different control mechanisms. In *Aspergillus nidulans* cDHQase is induced by the presence of quinate and is encoded by the QUTE gene, whereas bDHQase is constitutively produced and is encoded by the AROM gene which encodes a pentafunctional polypeptide chain that catalyses five adjacent reactions in the biosynthetic pathway (see Chapter 7, Section 7.11.2). In bacteria these five steps are catalysed by five distinct enzymes; in plants dehydroquinase occurs in a bifunctional polypeptide chain with shikimate dehydrogenase.[139] The amino-acid sequence information on the multienzyme polypeptides suggests that they arose from a number of gene fusion events.

When the individual enzymes were purified and characterized,[140–142] it became clear that there were major differences between representative examples of bDHQase and cDHQase in terms of subunit M_r, quaternary structure, and susceptibility towards certain chemical-modification reagents. In addition, there is virtually no significant sequence identity between the two types. Taken together, these lines of evidence suggested that the two types had not arisen from a common precursor by divergent evolution, but instead represented a case of convergent (or parallel) evolution (see Chapter 3, Section 3.3.2.11). Further work, however, suggested that the distinction between bDHQase and cDHQase might be difficult to make in certain organisms. Thus, for instance, in *Streptomyces coelicolor*, there is a single type of DHQase which had the physical properties of the cDHQase found in other organisms, but was clearly involved in the biosynthetic pathway. From this and other evidence, it was proposed that a more appropriate classification of the dehydroquinases could be based on the physical and mechanistic properties of the enzymes, i.e. type I (biosynthetic-like) and type II (catabolic-like) DHQases.

Detailed work on the dehydroquinases required the cloning and overexpression of the genes encoding the enzymes.[143,144] When they occur separately (e.g. in bacteria), type I enzymes are dimeric with a subunit M_r of about 27 000. The enzymes are fairly susceptible to denaturation by heat or agents such as guanidinium chloride (Gdm Cl) (see Chapter 3, Section 3.6.1); for instance, the denaturation temperature of the *E. coli* enzyme is 57°C and the enzyme is 50 per cent unfolded in 1.3 M Gdm Cl. By contrast, the type II enzymes are dodecameric (i.e. consist of 12 subunits) with a subunit M_r of about 16 000 and are much more stable. The type II enzyme from *A. nidulans*, for example, shows two thermal transitions in the range 85–95°C, and the concentration of Gdm Cl required to bring about 50 per cent unfolding is about 5 M.[142]

Detailed analysis of the substrates and products has shown that the type I and type II enzymes also differ in terms of the stereochemistry of the reaction catalysed. The type I enzymes catalyse a *cis* (*syn*) elimination reaction in which –H and –OH are removed from the same side of the substrate, whereas the type II enzymes catalyse a *trans* (*anti*) elimination (Fig. 5.30).[145]

5.5.3.2 Chemical-modification and site-directed mutagenesis experiments

Type I enzymes

Lysine A number of chemical-modification experiments have been performed on the type I enzyme from *E. coli*. The involvement of a lysine side-chain in the mechanism was indicated by the formation of a Schiff base (imine) when the enzyme was incubated with dehydroquinate; the imine could be trapped by addition of sodium borohydride (see Section 5.4.2), leading to inactivation of the enzyme. By using the radioactive NaB^3H_4 to reduce the imine, it was possible to identify the modified peptide after digestion of the inactivated enzyme by CNBr and trypsin (see Chapter 3, Section 3.3.2.2). Sequencing of the modified peptide allowed the Lys side-chain to be identified as Lys 170, a residue found to be conserved in a range of type I enzymes. An analogous modification could be carried out of the type I domain of the pentafunctional enzyme of *Neurospora crassa*.[146]

Interestingly, formation of the modified (reduced imine) enzyme leads to a marked increase in stability towards denaturation by GdnHCl or heat; this is correlated with a significant decrease in the flexibility of the enzyme revealed by fluorescence quenching.[147]

The importance of Lys 170 in the type I enzyme has been confirmed by site-directed mutagenesis. The mutant Lys170Ala showed a reduction in catalytic activity of approximately one millionfold, although the overall structure of the active site was largely unaffected since the binding of substrate or product was only about threefold weaker than for the wild-type enzyme.[148]

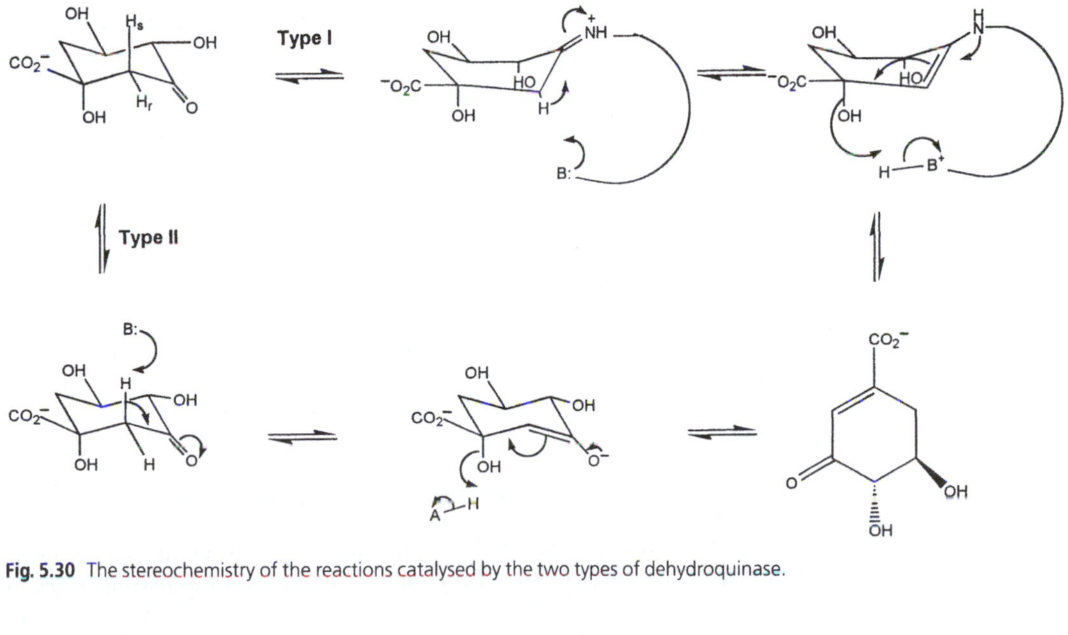

Fig. 5.30 The stereochemistry of the reactions catalysed by the two types of dehydroquinase.

Histidine Reaction of the type I enzyme with diethylpyrocarbonate (see Section 5.4.4.1) was found to lead to inactivation, suggesting that one or more His side-chains may be involved in the mechanism of the enzyme.[149] Additional evidence for the involvement of His came from the dependence of the V_{max} for the enzyme on pH, which showed a single ionization ascending limb behaviour (see Section 5.4.1.4), with a pK_a of 6.1. The modification of the type I enzyme by diethylpyrocarbonate is discussed in some detail as it illustrates a number of important points about the execution and interpretation of chemical modification experiments.

Inclusion of substrate (dehydroquinate)* led to protection against inactivation, indicating that modification occurred within the active site. The modified enzyme showed an increased absorbance at 240 nm, characteristic of the formation of carbethoxy-histidine; from the increase in absorbance it was concluded that complete inactivation was associated with modification of six His per polypeptide chain. Further evidence for the involvement of His was afforded by the fact that addition of hydroxylamine (which was known to reverse the modification of His side-chains) led to the regain of most of the activity and a corresponding decrease in the absorbance at 240 nm. In order to determine the number of 'essential' His side-chains in the enzyme, the data correlating the fraction of activity remaining with the number of modified side-chains were analysed by a statistical method developed by Tsou.[150] The data were only consistent with a model in which one of the six His modified was essential for activity. In order to identify the essential His, the modification was carried out in the presence and absence of the substrate/product mixture, and the chymotryptic peptide profiles obtained on reverse-phase high-performance liquid chromatography (see Chapter 3, Section 3.3.2.9) compared. This showed that a single peptide differed in the two mixtures; the sequence of this corresponded to residues 141–158. The peptide contains two His side-chains (His 143 and His 146). Evidence from sequence alignments indicated that only the His 143 was conserved, so this was proposed to be the essential His side-chain.

These conclusions have been elegantly confirmed by site-directed mutagenesis.[148] Each of the two histidines in the peptide (141–158) was replaced in turn by Ala. The His146Ala mutant retained full activity, whereas the His143Ala mutant had an activity about one millionfold lower than the wild-type enzyme.

Other residues The reaction of type I dehydroquinase from *E. coli* with the arginine-specific reagent phenylglyoxal (Section 5.4.4.1) leads to loss in activity. Using mass spectrometry to analyse the tryptic peptides produced by digestion of enzyme which had been modified to a low (20 per cent) extent, it was shown that the most reactive Arg was Arg 213.[144] Comparison of a number of sequences of type I enzymes showed that this Arg was conserved in all species, suggesting that it could well play an important role in the catalytic mechanism, possibly by binding to the carboxylate group of the substrate. Reaction of the enzyme with iodoacetate leads to loss of activity and modification of two highly reactive methionine side-chains, identified as Met 23 and Met 205.[151] However, site-directed mutagenesis studies in which either of these was replaced by Leu had no effect on activity showing that they are not essential for activity.[148] Thus the inactivation caused by reaction with iodoacetate must reflect a secondary effect of the introduction of the charged grouping in the proximity of the active site.

Type II enzymes
Although the studies on the type II enzymes have been less extensive than those on the type I, a number of side-chains have been implicated as playing a role in

* The addition of substrate leads to the formation of an equilibrium mixture of substrate and product; the ratio of dehydroshikimate to dehydroquinate is about 15.

catalysis. It should be emphasized that there is no evidence for the involvement of any Lys side-chain in the mechanism of the type II enzymes. Indeed, the enzymes from *Streptomyces coelicolor* and *Mycobacterium tuberculosis* each have only one Lys side-chain per subunit which is not at a conserved position in the sequence.

Histidine Reaction of the type II enzyme from *A. nidulans* with diethylpyro-carbonate leads to loss in activity; this loss is prevented if the reaction is carried out in the presence of the substrate/product mixture, suggesting that one or more His side-chains may be involved in the catalytic mechanism.[142] However, the number or identity of these side-chains has not been determined. Sequence alignment show that there are two conserved His side-chains in the type II enzymes, which may represent the sites of modification.

Arginine The type II enzymes from *S. coelicolor* and *A. nidulans* are both inactivated by reaction with phenylglyoxal (PGO). Mass spectrometry proved very useful in identifying the most reactive Arg residue and in establishing the nature of the product.[152] Enzyme samples inactivated to varying extents were analysed by electrospray mass spectrometry, and it could be concluded that the principal product formed in the early stages of reaction corresponded to the condensation product formed by reaction of one molecule of PGO per subunit with the loss of water, i.e. giving a mass increase of 116 units. Further reaction led to the formation of this mono-condensation derivative at other Arg side-chains, as well as formation of a product involving two PGO moieties per Arg side-chain; this gives a mass increase of 250 units. From an analysis of the masses of chymotryptic peptides derived from enzyme modified to moderate extents, it was possible to show that inactivation was correlated with reaction of Arg 23 (*S. coelicolor*) or the corresponding Arg 19 (*A. nidulans*), which represent conserved residues in all type II enzymes. The role of this Arg has been confirmed by site-directed mutagenesis;[144] replacement of Arg 23 in the *S. coelicolor* enzyme by Lys, Gln, or Ala led to a 3000–30 000-fold loss in activity (k_{cat}), although the binding of substrate as indicated by K_m and direct-binding experiments was somewhat strengthened in the mutants.

Tyrosine Reaction of the type II enzyme from *S. coelicolor* with the tyrosine-specific reagent tetranitromethane (Section 5.4.4.1) led to inactivation. Analysis of the product by mass spectrometry confirmed the nature of the product as the mono-nitro derivative, and examination of tryptic peptides of the modified enzyme showed that inactivation was due to modification of Tyr 28, a residue conserved in type II enzymes.[144]

5.5.3.3 X-ray structures of dehydroquinases
The structures of both type I and type II enzymes have now been solved by X-ray crystallography, allowing much of the detailed work described above to be interpreted and mechanisms proposed.

Type I enzymes
The X-ray structure of the type I enzyme from *Salmonella typhi* has been solved to a resolution of 0.21 nm.[153] The overall structure (Fig. 5.31) is that of an eight-stranded β-barrel surrounded by α-helices, an arrangement first found in triosephosphate isomerase (Fig. 5.25) and subsequently in a number of other enzymes and proteins (see Chapter 3, Section 3.4.4.6). Interestingly, a number of

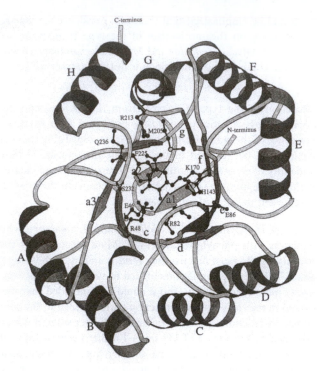

Fig. 5.31 The structure of one subunit of type I dehydroquinase (1qfe), drawn in Molscript.[195] The 8 parallel β-strands forming the β-barrel are viewed approximately end on, with the hairpin loop comprising strands a1 and a2 blocking off the far end of the barrel. The borohydride-reduced product is shown attached to the side chain of Lys 170 (K170 in the one-letter code).

these enzymes, e.g. Class I fructose bisphosphate aldolase,[154] function via imine mechanisms analogous to that of type I dehydroquinase. In addition to the eight strands forming the barrel, there are two small antiparallel strands which effectively block one end of the barrel. Helices F, G, and H in the β-barrel form the interface between the dimers, with the corresponding helices in the other subunit running in an antiparallel fashion. The side-chain of Arg 213, which has been identified by chemical modification experiments as important in catalysis, forms an interaction with the carboxylate of the bound substrate and is also involved in interactions between the subunits. It would appear that the dimeric structure is important for maintaining the structure of the active site. The binding of substrate to the enzyme involves a complex network of hydrogen bonds in which two water molecules also play a part (Fig. 5.32). Lysine 170 (identified as the side-chain forming the imine intermediate) is located towards the centre of the barrel on strand f. Other side-chains involved in binding of substrate include Arg 48 and Glu 46, which form hydrogen bonds to the C5 hydroxyl group, and Glu 46 and Arg 82, involved in hydrogen bonding to the C4 hydroxyl group. The C5 and C4 hydroxyls and the imine-linked C3 lie against a basic region of the enzyme consisting of Arg 48, Arg 82, and His 143. This His is suitably positioned to play the role of a general base; it may well be oriented by interaction with the neighbouring Glu 86, conserved in all type I enzymes. The two Met side-chains modified by iodoacetate (Section 5.5.3.2) are close to the active site, but make no direct contact with the substrate.

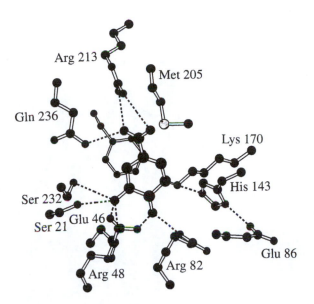

Fig. 5.32 The hydrogen-bonding network involved in binding of substrate to type I dehydroquinase as seen in the borohydride-reduced complex. For the sake of clarity, two water molecules which form hydrogen bonds to the substrate have been omitted.

Type II enzymes

From the structures of the type II enzymes from *M. tuberculosis* and *S. coelicolor*, it is found that the 12 subunits of the enzymes are arranged as a tetramer of trimeric units. This is a relatively unusual arrangement for a dodecameric enzyme; e.g. in the case of glutamine synthetase the 12 subunits are arranged as two eclipsed rings of six.[155] Each subunit consists of a five-stranded parallel β-sheet core flanked by four α-helices, with the order of the strands 21345 (Fig. 5.33), a fold similar to that seen in the flavin-containing redox protein flavodoxin.[156] Within each trimer, there are strong interactions between the subunits consisting of electrostatic bonds (salt bridges) and hydrogen bonds; there are three and one salt bridges per interface in the *M. tuberculosis* and *S. coelicolor* enzymes respectively. By contrast, the forces between the trimeric units are relatively weak, consisting mainly of main-chain hydrogen bonding involving the C-terminal β-strands. This distinction between the forces between and within the trimeric units is consistent with the finding that low concentrations of GdnHCl (0.5 M) dissociate the dodecamer into trimeric units with little change in the structure of the subunits or in the enzyme activity.

The location of the active sites within each subunit has been deduced mainly by taking the evidence from the chemical-modification and site-directed mutagenesis work, which has shown the importance of Arg 23 and Tyr 28 (*S. coelicolor* numbering), and the pattern of conserved residues. These have shown that the active site consists of a cavity close to the subunit interfaces within the trimeric units (Fig. 5.34). Some of the conserved side-chains could be assigned roles in the proposed mechanism of the enzyme, as described in the next section.

5.5.3.4 The mechanisms of the two types of dehydroquinases

Taking all the evidence into account, it is possible to formulate reasonably detailed proposals for the mechanisms of the two enzymes.

(a)

(b)

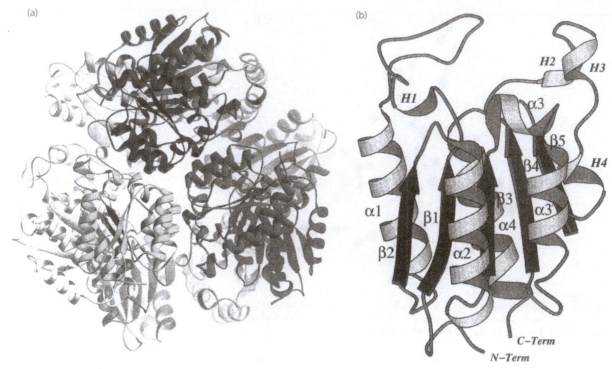

Fig. 5.33 The structure of type II dehydroquinase (1doi) drawn using Molscript[195] and Ribbons[196] showing (a) the dodecamer and (b) an individual subunit. In (a) the dodecamer is viewed down the three-fold axis of a trimeric unit; each monomer is individually shaded. The overall molecule is roughly spherical in shape with a diameter of 10 nm, and possesses tetrahedral symmetry with four trimers packing together. In (b) the strands of β-sheet are numbered, with the four α-helices flanking the sheet labelled α1–4. Helix α3 is split into two parts (α3' and α3''). There are four 3_{10} helices labelled H1–4. A colour version of (a) is shown in Plate 3 at the back of this book.

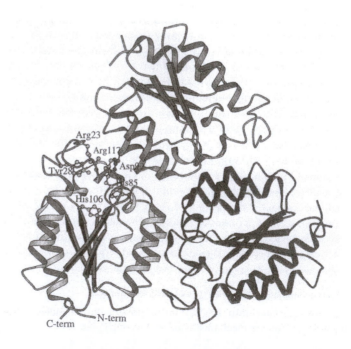

Fig. 5.34 The trimeric unit of type II dehydroquinase (1doi) viewed down the 3-fold axis drawn using Molscript[195] and Ribbons[196]; the active site residues implicated in the catalytic mechanism are highlighted in one of the subunits. A colour version of this figure is shown in Plate 4 at the back of this book.

For the type I enzymes, the key residues which are involved in catalysis are Lys 170 and His 143. Lys 170 forms an imine with the carbonyl group of the substrate; protonation of this imine generates the electrophilic centre which serves to polarize the C–H bond on the adjacent C2. The side-chain of His 143 acts as a general base which abstracts a proton from C2 to give a carbanion intermediate. Loss of a hydroxyl group (as OH⁻) from C1 leads to the formation of the double bond, completing the elimination reaction. Finally, the imine dissociates to generate the dehydroshikimate product (Fig. 5.35). The role of Arg 213 and other side-chains in forming a network of substrate-binding interactions has been referred to in Section 5.5.3.3.

Type II dehydroquinases clearly follow a different mechanism. Studies of the variation of V_{max} and K_m with pH point to the involvement of an enolate intermediate.[157] The reactive Arg identified by chemical modification and site-directed mutagenesis (Arg 23 in the *S. coelicolor* enzyme) is likely to be involved in stabilization of the enolate intermediate. Arg 113 appears to be in a suitable position to play a role in substrate binding (to the carboxylate group). The conserved His side-chain (His 106) could be involved in removal of the axial hydrogen from C2 (Fig. 5.36); the basicity of this His is enhanced by the neighbouring Glu 104. The identity of the side-chain involved in abstracting the OH group from C1 in this mechanism is not yet known.

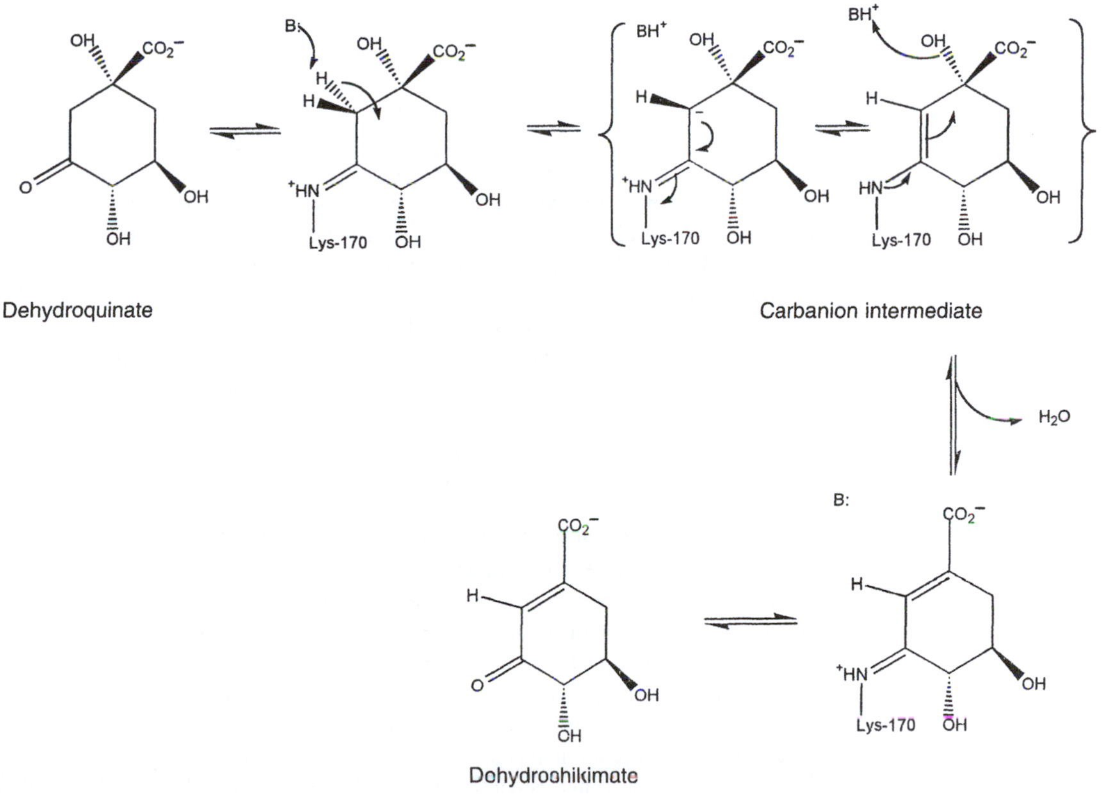

Fig. 5.35 Mechanism of the type I dehydroquinase reaction.[148] His 143 is likely to act as the general base B̈. The stereochemical aspects of the reaction are also illustrated in Fig. 5.30.

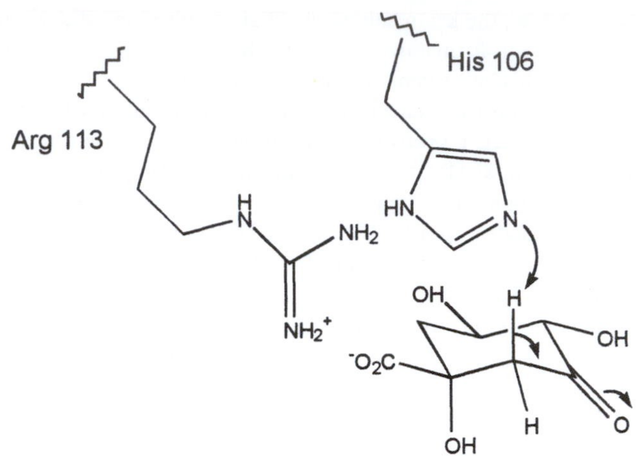

Fig. 5.36 Partial mechanism of the type II dehydroquinase reaction.[157] The stereochemical aspects of the reaction are also illustrated in Fig. 5.30.

5.5.4 Lactate dehydrogenase

The enzyme lactate dehydrogenase catalyses the reaction:

$$H - \underset{\underset{CO_2^-}{|}}{\overset{\overset{CH_3}{|}}{C}} - OH \;\; + NAD^+ \; \rightleftharpoons \; \underset{\underset{CO_2^-}{|}}{\overset{\overset{CH_3}{|}}{C}} = O + NADH + H^+$$

S-**Lactate**   **Pyruvate**

S-Lactate is also referred to as L-Lactate

In the reverse direction, this reaction represents the last step in the process of anaerobic glycolysis and provides a means of regeneration of NAD^+ required for the reaction catalysed by glyceraldehyde-3-phosphate dehydrogenase. The enzyme is present in the cytosol in sufficiently high concentrations that the reaction is close to equilibrium; this is discussed further in Chapter 8, Sections 8.3.1.4 and 8.5.4. Lactate dehydrogenase has been isolated from many sources, but most of the detailed work relating to the mechanism has been performed on the enzymes isolated from dogfish and pig (for a comprehensive review of this work, see reference 158). We shall discuss the results of kinetic studies (Section 5.5.4.2), X-ray crystallographic work (Section 5.5.4.3), experiments involving chemical modification of amino-acid side-chains (Section 5.5.4.4), and site-directed mutagenesis (Section 5.5.4.5). In Section 5.5.4.6 we shall discuss the mechanism of lactate dehydrogenase in relation to that of other dehydrogenases.

5.5.4.1 Background information on the structure of the enzyme

Lactate dehydrogenase is a tetramer of M_r 140 000. The enzyme provides a good example of the occurrence of isoenzymes (see Chapter 1, Section 1.5.2); in most tissues there are five forms of the enzyme, which can be separated by electrophoresis. The different forms arise from the five possible ways of assembling a tetramer from two types of subunit (α_4, $\alpha_3\beta$, $\alpha_2\beta_2$, $\alpha\beta_3$, and β_4). It is found that LDH-1 predominates in heart muscle, and is often referred to as the H_4 form, whereas LDH-5 predominates in skeletal muscle, and is often referred to as the M_4 form. Various other isoenzymes (e.g. the C, E, and F forms) are also known,[158] but they are much less widely distributed than the heart and muscle forms and will not be discussed further here.

5.5.4.2 Kinetic studies

Detailed steady-state kinetic studies on lactate dehydrogenase have shown that the enzyme follows an ordered ternary complex mechanism (see Chapter 4, Section 4.3.5.1) with NAD⁺ (or NADH) binding preceding that of lactate (or pyruvate). This is shown in the following scheme, where the proton involved in the reaction has been omitted:

$$E + NAD^+ \rightleftharpoons E^{NAD+} \underset{-\text{lactate}}{\overset{+\text{lactate}}{\rightleftharpoons}} E^{NAD+}_{\text{lactate}} \overset{\text{catalyst}}{\rightleftharpoons}$$

$$E^{NADH}_{\text{pyruvate}} \underset{+\text{pyruvate}}{\overset{-\text{pyruvate}}{\rightleftharpoons}} E^{NADH} \rightleftharpoons E + NADH.$$

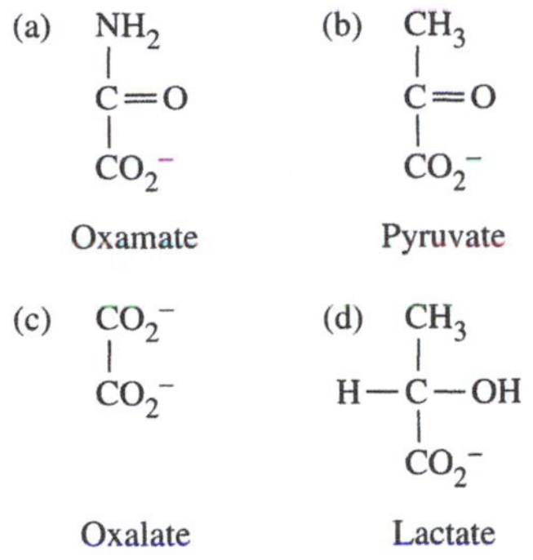

The ordered ternary complex mechanism was deduced from studies of product inhibition and of substrate binding. For instance, the enzyme will not bind lactate or pyruvate in the absence of the dinucleotide substrate.[159]

Steady-state kinetic studies have also shown that oxamate (Fig. 5.37(a)) acts as a competitive inhibitor with respect to pyruvate (Fig. 5.37(b)) (note that the two molecules are isoelectronic) and that oxalate (Fig. 5.37(c)) is competitive with respect to lactate.[160] These inhibitors are very likely to bind to the active site of the enzyme and have found considerable value as substrate analogues in X-ray crystallographic studies on lactate dehydrogenase, allowing conclusions to be drawn regarding the structure of the catalytically active complex.

A more detailed analysis of the rates of the individual steps of the reaction can be undertaken using stopped-flow techniques (see Chapter 4, Section 4.4). The enzyme (LDH-l from pig was used in these experiments) was mixed rapidly with saturating concentrations of the substrates NAD⁺ and lactate, and production of NADH was monitored spectrophotometrically[161] (Fig. 5.38).

There was an initial burst of NADH production within the 'dead time' of the instrument, followed by a slow steady-state production of NADH. The size of the burst was equal to 1 mol NADH per mol of active sites. From these data, it can be concluded that all the steps in the reaction up to formation of enzyme-bound NADH are very rapid. Similar results were obtained with LDH-5 from pig,[162] but the size of the burst was only equal to 1 mol NADH per mol active sites at pH values of 8.0 or above.*

Since NADH binds very tightly to the enzyme it might be thought that the dissociation of the E^NADH complex would be the slowest (rate-determining) step of the overall reaction. However, this is not the case, since the rate constant for the dissociation (measured in a separate experiment as 450 s⁻¹) is considerably greater than the rate constant for oxidation of lactate in the steady state (80 s⁻¹).[162] It was therefore concluded that there is an extra step in the overall reaction that occurs after the interconversion of the ternary complexes that produces enzyme-bound NADH, but which precedes the dissociation of NADH. This extra step appears to represent a slow conformational change in the enzyme (see Section 5.5.4.5). The use of detailed pre-steady-state kinetic measurements has allowed the rate constants of all the individual steps in the reaction catalysed by LDH-l from pig to be determined.[163,164]

5.5.4.3 X-ray crystallographic studies

The first detailed X-ray crystallographic studies were performed on LDH-5 from dogfish muscle,[158] but the structures of the enzyme from several other sources including pig muscle and *Bacillus stearothermophilus* are now available.[165] All the

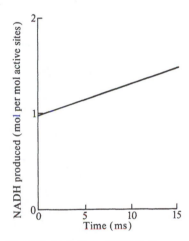

Fig. 5.37 Structures of some substrates and inhibitors of lactate dehydrogenase.

Fig. 5.38 Production of NADH in the reaction catalysed by lactate dehydrogenase.[161]

* At lower pH values, the size of the burst was less; e.g. at pH 6.8 it was 0.5 mol per mol of active sites.

enzymes have very similar overall three-dimensional structures (Fig. 5.39). There is a considerable amount of secondary structure (see Chapter 3, Section 3.4.4) in the molecule; ==approximately 40 per cent of the amino acids occur in α-helices and a further 23 per cent in various forms of β-structure.==

Of particular interest is the occurrence of supersecondary structures in the molecule (see Chapter 3, Section 3.4.4.6). In the N-terminal half of the molecule there is a ==six-stranded parallel sheet,== and in the C-terminal half there are two three-stranded antiparallel sheet structures.[158] The six-stranded sheet in the N-terminal part of the enzyme is involved in the binding of NAD[+] and seems to be a general feature of all the dehydrogenases whose structures have been examined so far, and has led to the development of ideas[166] about the evolutionary relationships between these and other enzymes such as kinases that act on mononucleotide substrates (e.g. ATP).*

* The tertiary structures of lactate dehydrogenase and malate dehydrogenase are very similar indeed; by changing three amino acids in the active-site region of the enzyme it is possible to change the specificity of lactate dehydrogenase to that of malate dehydrogenase (see Section 5.5.4.5).

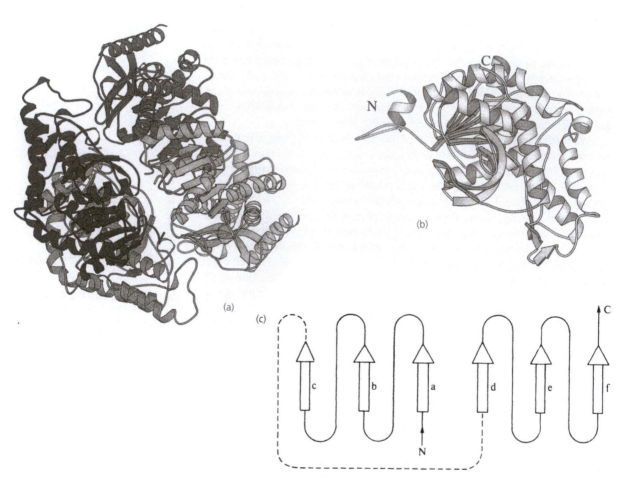

Fig. 5.39 Structure of lactate dehydrogenase; (a) and (b) show cartoon representations of the tetramer (1ldm) and a single subunit, respectively, drawn in Molscript.[195] The arrangement of the tetramer as a dimer of dimers is clearly brought out in (a). (c) The pattern of the six-stranded sheet which forms the NAD[+]-binding region. The arrows represent strands of sheet and the connecting lines represent helices (between strands c and d there is a non-helical stretch of peptide chain.) The nicotinamide ring of NAD[+] binds between strands d and e and the adenine ring between strands a and b. A colour version of (a) is shown as Plate 5.

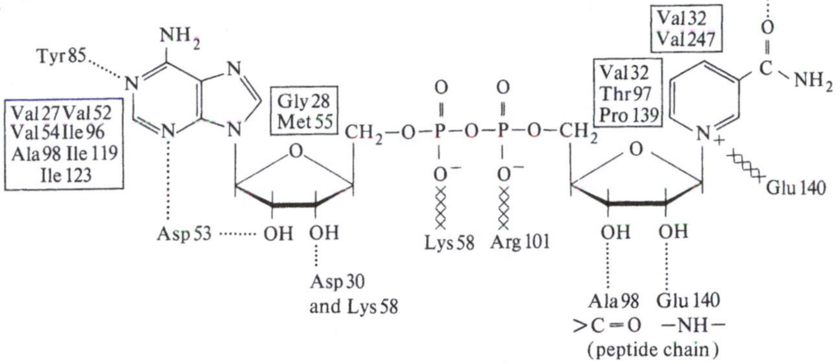

[handwritten annotations:]
Adenine binds in HP pocket. H-bond w/ Try 85 and asp 53 gives specificity

gives specificity for A'-side (LDH, MDH, ADH)

Both ribose are involved in H-bonds. Interaction at 2'-OH explains why NADP(H) is not a substrate

Fig. 5.40 Some of the interactions involved in the binding of NAD⁺ to lactate dehydrogenase.[158] The dotted lines represent hydrogen bonds, the crosses electrostatic interactions, and the boxes hydrophobic interactions.

The binding of NAD⁺ to the enzyme has been examined in great detail.[120] When we consider the large number of amino-acid side-chains involved in the binding, it is not difficult to appreciate how the remarkable specificity displayed by this and other enzymes arises. Some of the important features of the binding of NAD⁺ to lactate dehydrogenase are shown in Fig. 5.40.

The adenine ring binds in a hydrophobic pocket provided by side-chains of Val, Ile, and Ala. Hydrogen bonds to the side-chains of Tyr 85 and Asp 53 (Fig. 5.40) impart specificity to the orientation of the bound adenine ring. Both ribose rings are involved in hydrogen bonds, as shown in Fig. 5.40; the specific interaction with the 2'-OH group of the adenine ribose helps to explain why NADP⁺, which has a phosphate group at this position, is not a substrate for the enzyme.†

As far as the nicotinamide ring is concerned, we should note that there is a very important hydrogen-bonding interaction between the carbonyl oxygen of the side-chain and the side-chain of Lys 250. This interaction serves to orientate the nicotinamide ring in such a way as to expose the 'A'-side of the ring to the other substrate, lactate, thereby accounting for the 'A'-side specificity of the enzyme (Fig. 5.41). In the complex formed between NAD⁺ and glyceraldehyde-3-phosphate dehydrogenase, there is a specific hydrogen bond between the carbonyl oxygen of the side-chain of the nicotinamide ring and the side-chain of Asn 313 that serves to expose the 'B'-side of the ring to the other substrate, accounting for the 'B'-side specificity of this enzyme (Fig. 5.41).[167,168]

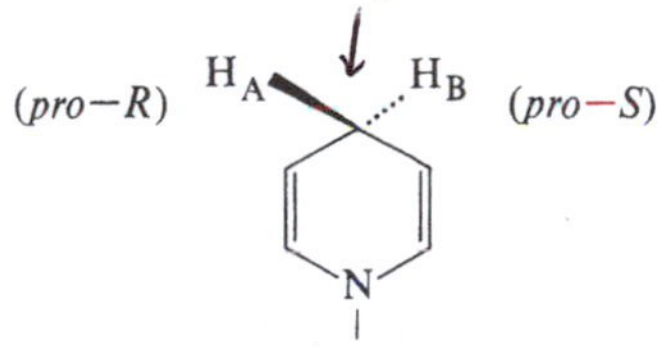

The structure of the catalytically active complex of lactate dehydrogenase has been deduced from an examination of the structures of complexes containing inhibitors $E_{pyruvate}^{NAD^+}$, $E_{oxalate}^{NAD^+}$, and $E_{oxamate}^{NADH}$ and of the structure of a complex containing an NAD–pyruvate adduct.[169] These structures indicate that the site for lactate is most probably located between the nicotinamide ring and the side-chain of His 195, as shown in Fig. 5.42.

The side-chains of Arg 109 and Arg 171 can bind electrostatically to the carboxylate group of lactate. The side-chain of His 195 can form a hydrogen bond to the hydroxyl group of lactate, permitting a movement of electrons in the catalytic process as shown in Fig. 5.42.

Conformational changes in the enzyme appear to be of crucial importance in the formation of the catalytically active complex.[158] These changes are particularly marked on binding the second substrate to the binary E^NAD⁺ complex. There is a flexible loop (between amino acids 98 to 120), which in the absence of substrates extends into the solvent, but in the ternary complex moves down to enclose the

† In general, a dehydrogenase acts on *either* NAD⁺ or NADP⁺. Glutamate dehydrogenase is almost unique in that it can act on both (see Chapter 8, Section 8.3.1.4).

[handwritten:] prochirality

Fig. 5.41 The specificity of NADH-linked dehydrogenases. 'A'-side dehydrogenases (e.g. malate, lactate, and alcohol dehydrogenases) select the H_A hydrogen from NADH. 'B'-side dehydrogenases (e.g. glutamate and glyceraldehyde-3-phosphate dehydrogenases) select the H_B hydrogen from NADH. (See also Fig. 1.4.)

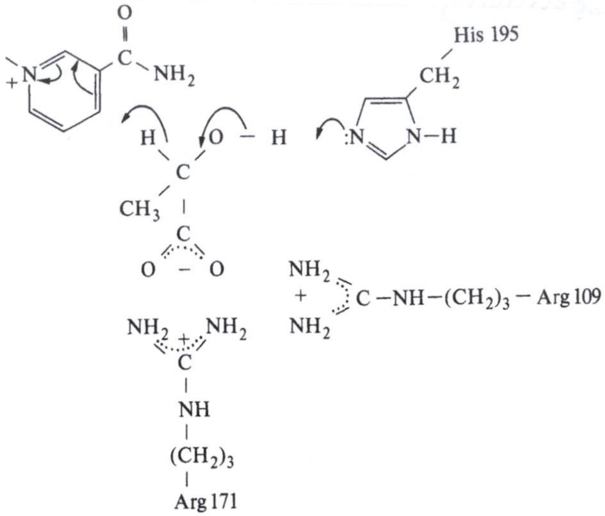

Fig. 5.42 Movement of electrons in the reaction between NAD⁺ and lactate catalysed by lactate dehydrogenase.

substrates and thereby exclude water from the active site of the enzyme (see Section 5.5.4.5). For instance, the side-chain of Arg 109 moves 1.4 nm to be close to the substrate (Fig. 5.42). A smaller (0.1–0.2 nm)[169] but highly important change occurs so as to bring the side-chain of His 195 into contact with the substrate and permit the movement of electrons as shown (Fig. 5.42).

5.5.4.4 Chemical modification of amino-acid side-chains

In this section we shall discuss experiments involving the modification of histidine and arginine side-chains. The results of these studies complement the information provided by X-ray crystallography.

(i) Histidine

3-Bromoacetylpyridine

Bromopyruvate

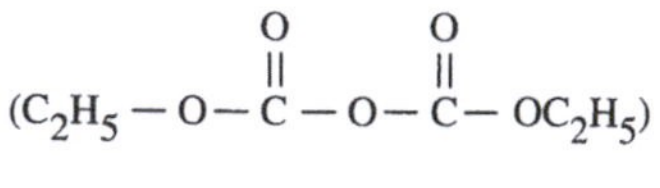

The idea that there is a histidine side-chain at the active site of lactate dehydrogenase was suggested by the results of experiments using the affinity labels 3-bromoacetylpyridine and bromopyruvate.[158] (The latter acts as an affinity label in the presence of NAD⁺.) In each case it was shown that the side-chain of His 195 was modified. It is also possible to modify the side-chain of His 195 selectively using diethylpyrocarbonate because of the much higher reactivity of this side-chain compared with the other His side-chains in the enzyme.[170] In all three cases modification led to inactivation of the enzyme, in keeping with the proposed role of the side-chain of His 195 (Fig. 5.42).

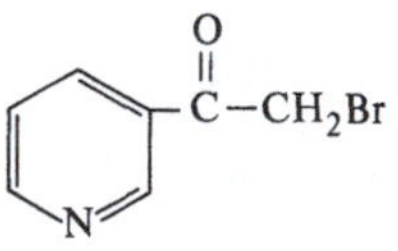

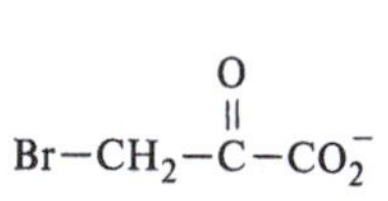

(ii) Arginine

Diethyl pyrocarbonate

Phenylglyoxal, which is fairly specific for modification of arginine side-chains (Section 5.4.4.1), inactivates lactate dehydrogenase. The findings that only one arginine side-chain per subunit was modified and that formation of the ternary complex provided protection against inactivation suggested that there is an arginine side-chain at the active site of the enzyme.[158] The X-ray crystallographic data (Fig. 5.42) show that two arginine side-chains are likely to be involved in the binding of substrate.

5.5.4.5 Site-directed mutagenesis experiments

Because of the wealth of structural and kinetic data available on the enzyme, lactate dehydrogenase has been subjected to extensive studies by site-directed mutagenesis.[171,172] One set of experiments has been aimed at understanding the roles of certain amino acids in the catalytic activity of the enzyme. A second set has been aimed at altering the specificity of the enzyme, in order to allow chiral reduction of α-keto acids to α-hydroxy acids for a wide range of synthetic applications, particularly in the pharmaceutical industry. Some examples of these set of experiments will be discussed in turn. The enzyme used for most of these studies is that from the thermophilic bacterium *Bacillus stearothermophilus* for which the appropriate gene has been isolated. The amino-acid sequence and overall three-dimensional structure of this enzyme, as well as its catalytic properties (when measured in the presence of the activator FBP), are similar to those of eukaryotic lactate dehydrogenase.[165]

(i) Several experiments have contributed to our understanding of the roles played by amino acids in the flexible loop region of the enzyme (Section 5.5.4.3). In the first of these,[173] Arg 109 in the flexible loop was replaced by Gln, thereby changing a positively charged side-chain into a neutral one. The mutation had no effect on the binding of NADH to the enzyme but decreased k_{cat} by some 400-fold. The results suggest that the side-chain of Arg 109 enhances the polarization of the carbonyl group of pyruvate (Fig. 5.42) and stabilizes the transition state of the reaction.

In a second experiment,[174] the rate of the movement of the loop region that is associated with formation of the catalytically active complex was measured. For this experiment Gly 106 in the loop was replaced by Trp and the other Trp side-chains (Trp 80, Trp 150, and Trp 203) were replaced by Tyr. (It should be noted that in proteins, Trp side-chains generally show much more intense fluorescence than Tyr side-chains because of the quenching of the excited state of the latter.[80]) The quadruple mutant enzyme thus has effectively a single fluorescent side-chain (Trp 106) within the loop region. The k_{cat} value for the mutant enzyme was approximately 60 per cent of that of the native enzyme and the K_m for pyruvate was unchanged. On mixing the mutant E^{NADH} complex with oxamate to form the ternary complex (Section 5.5.4.3), it was found that the fluorescence of Trp 106 decreased, reflecting movement of the loop. By stopped-flow measurements (see Chapter 4, Section 4.4.1) the rate of this movement was found to be 125 s^{-1}, i.e. very similar to the overall turnover rate of the enzyme as measured by k_{cat}. It is thus clear that the rate-determining step in the overall reaction corresponds to the movement of the loop to close over the active site (Section 5.5.4.3).

In a different type of experiment in this group, it has proved possible to design a mutant form of lactate dehydrogenase which alters the behaviour of the enzyme at high concentrations of pyruvate. Most lactate dehydrogenases are inhibited by high concentrations of pyruvate, probably by the formation of an abortive $E^{NAD^+}_{pyruvate}$, complex in which a covalent adduct is formed between the NAD$^+$ and pyruvate. By examining the X-ray structure of the mammalian enzyme and comparing the sequence of this enzyme with that from the malarial parasite *Plasmodium falciparum* (which is not subject to inhibition), it was proposed that replacement of Ser 163 in the mammalian enzyme by Leu should remove the inhibition. This in fact proved to be the case, although the effects of the mutation were somewhat more complex than anticipated.[175]

(ii) One of the first experiments aimed at altering the specificity of lactate dehydrogenase was designed to allow a mutant enzyme to accept malate as a substrate, i.e. to function as a malate dehydrogenase.[176] To take account of the increased size and additional negative charge of the new substrate, the following changes were introduced into the lactate dehydrogenase from *Bacillus stearothermophilus*: (i) the volume of the active site was increased (Thr 246Gly); (ii) an acidic side-chain was replaced by a neutral one (Asp 197Asn); (iii) a basic side-chain was introduced (Gln 102Arg). These changes led to a new 'malate' dehydrogenase that catalysed the reduction of oxaloacetate by NADH 500 times faster than that of pyruvate. (The corresponding ratio for the original lactate dehydrogenase is 0.001.) This experiment confirms that the overall structures of the two enzymes are sufficiently similar that relatively small changes at the active site can cause marked changes in specificity.[176]

Later experiments involving multiple changes at positions 102–105 and 236–237 (which were implicated in substrate recognition) were designed to allow the active site to accommodate large hydrophobic substrate side-chains. Mutations in these regions led to an enzyme with a broader specificity for branched-chain keto acids, although the k_{cat} values were less than 10 per cent of the value for the wild-type enzyme acting on pyruvate.[165]

5.5.4.6 The mechanism of lactate dehydrogenase and some other dehydrogenases

The various approaches outlined in Sections 5.5.4.2 to 5.5.4.5 have given detailed insights into the mechanism of the reaction catalysed by lactate dehydrogenase. The structure of the active complex and the movement of electrons within it have been described (Fig. 5.42) and to this must be added details of the binding sites (e.g. Fig. 5.40) and of the rates of the various elementary steps in the reaction.

The mechanisms of a number of other dehydrogenases (e.g. alcohol dehydrogenase, malate dehydrogenase, and glyceraldehyde-3-phosphate dehydrogenase) are also known and certain generalizations can be made.[101] All dehydrogenases contain a recognizable nucleotide binding domain in each subunit (see Section 5.5.4.3), with a second (catalytic) domain that can be variable in structure. We have already noted that the mode of binding of the nicotinamide ring of NAD^+ to the various enzymes can explain the specificity of hydrogen transfer (Fig. 5.41). The mode of binding of NAD^+ to lactate dehydrogenase explains why $NADP^+$ does not function as a substrate for this enzyme. (In the case of glutathione reductase from *E. coli*, a successful attempt was made to alter the specificity from that of $NADP^+$ to NAD^+, but this involved mutating no less than seven amino acids.)[177] In all the dehydrogenases mentioned, except alcohol dehydrogenase, it appears that a histidine side-chain acts as a general base catalyst by abstracting a proton from the –OH group of the substrate,* thereby facilitating transfer of the hydride ion to NAD^+ (see Fig. 5.42). In the case of alcohol dehydrogenase, a zinc ion is involved in the catalytic activity and it is suggested that this zinc ion, or an ionized water molecule (OH^-) attached to the zinc ion, could play the part of the basic catalyst.[178]

5.5.5 DNA methyltransferase

DNA methyltransferases are found in almost all organisms; they catalyse the transfer of a methyl group from the donor molecule (*S*-adenosyl-L-methionine; AdoMet) to a base (cytosine or adenine) within a specific target DNA sequence. The products are the methylated DNA and *S*-adenosylhomocysteine (AdoHcy).

In prokaryotes the functions of the methylation reaction are well understood.[179–181] Methylation is used as a means of protecting the DNA of the host organism against the action of restriction endonucleases which digest unmethylated 'foreign' DNA. It can also be used to correct errors in DNA replication by the mismatch repair system; the methylation of the parent DNA strand distin-

* In the case of glyceraldehyde-3-phosphate dehydrogenase, the –OH group is that of a thiohemiacetal attached to the side-chain of Cys 149. The thiohemiacetal is formed by the attack of this side-chain on the carbonyl group of the aldehyde substrate.

guishes it from the new strand which requires repair. In eukaryotes, the function of DNA methylation is not so well defined. It is thought to be involved in the control of gene expression and in developmental regulation, and has been shown to be essential for normal development of the mouse embryo.[179,182]

One type of methylase acts to add a methyl group to the amino group substituent of adenine (to give N6-methyladenine) or cytosine (to give N4-methylcytosine). A second type adds a methyl group to a ring carbon atom of cytosine (to give C5-methylcytosine).

A variety of techniques including kinetics, chemical modification, site-directed mutagenesis, and X-ray crystallography have given insights into the mechanism of action of the methyltransferases and shown some remarkable features about the interactions between enzyme and substrate during the catalytic cycle. The enzyme which has been most intensively investigated at the structural level is the *Hha*I enzyme from *Haemophilus haemolyticus* which methylates at the C5 position the internal (highlighted) cytosine in the sequence 5′-GCGC-3′.

5.5.5.1 Background information on the structure of the enzyme

*Hha*I methyltransferase is one of the smallest DNA methyltransferases known, consisting of a single subunit of 327 amino acids with a molecular mass of 37 000. The gene encoding the enzyme has been cloned and overexpressed in *E. coli*, permitting large quantities of material to be prepared for structural studies. Comparison of the sequences of a large number of enzymes which catalyse the transfer of a methyl group to the C5 of cytosine shows a pattern of conserved sequence motifs.[183,184] Of the 10 conserved motifs (designated I–X), six are very strongly conserved (I, IV, VI, VIII, IX, and X) with the others less so. Motif IV contains a conserved Pro-Cys dipeptide sequence which is involved in the catalytic mechanism of the enzyme (see Section 5.5.5.3) and which is also conserved in thymidylate synthase, which catalyses a related one-carbon transfer reaction.[184] Between motifs VIII and IX there is a region which is highly variable among the DNA methyltransferases both in terms of size and sequence; in the HhaI enzyme this region encompasses amino acids 171–271. Within this variable region, a small element (the Target Recognition Domain) is thought to be involved in binding to the target DNA sequence; in the HhaI enzyme this element is made up of amino acids 231–253.

5.5.5.2 Kinetic studies of the enzyme

Detailed studies of the kinetics of the *Hha*I enzyme were undertaken by Wu and Santi,[185] using poly(dG-dC) (2000–10 000 base pairs in length) and AdoMet as substrates. At least 95 per cent of the cytosine residues in the substrate could be methylated if the reaction was allowed to go to completion. The reaction was found to follow an ordered ternary complex mechanism (see Chapter 4, Section 4.3.5.1) with DNA binding first and methylated DNA dissociating last. From the variation of rate with substrate concentrations, the kinetic parameters were determined. The K_m values are 2.3 and 15 nmol dm^{-3} for poly(dG-dC) and AdoMet respectively and the value of k_{cat}, while quite low (1.3 min^{-1}), is similar to those reported for other DNA methyltransferases. The k_{cat}/K_m values for substrates are in the range 10^6–10^7 s^{-1} (mol dm^{-3})$^{-1}$.

As the reaction progresses, there is a decline in rate, which is due to the formation of a tight complex between E^{DNA} and the product AdoHcy. AdoHcy acts as a competitive inhibitor with respect to AdoMet with a K_i equal to 2 nmol dm^{-3}.

5.5.5.3 **Chemical studies of the enzyme mechanism**

The transfer of a methyl group to the C5 of cytosine is analogous to a number of other one-carbon transfer reactions (e.g. thymidylate synthase, dUMP, and dCMP hydroxymethyltransferases). As far as the mechanism is concerned, it is important to note that the C5 atom of cytosine (or pyrimidines in general) is not a sufficiently powerful nucleophile to attack the methyl donor AdoMet. However, following attack by a nucleophilic side-chain on the enzyme, the pyrimidine can be activated by formation of a carbanion intermediate for subsequent attack on the methyl donor.

Wu and Santi showed that the *Hha*I enzyme was inactivated by *N*-ethyl-maleimide, a chemical which is reasonably specific for modification of Cys side-chains. Addition of the poly(dG-dC) substrate provided protection against inactivation; however AdoMet did not provide protection. These results suggested that the nucleophile involved in the mechanism was Cys. and it was proposed that this residue was the Cys in the conserved Pro-Cys sequence.[185]

The kinetic studies provided the groundwork for the development of a mechanism-based inhibitor (see Section 5.4.4.3) of the enzyme.[186] 5-Fluorodeoxycytidine was incorporated into a synthetic DNA polymer which contained the recognition sequence of *Hha*I methylase, i.e. –GCGC–. In the absence of AdoMet, poly(FdC-dG) binds in a fashion competitive with respect to poly(dG-dC). However, in the presence of AdoMet, there was a time-dependent, first-order inactivation of the enzyme, in which the enzyme first bound to the DNA analogue, followed by AdoMet, leading to the formation of a dead-end complex. This complex is very stable with no dissociation over 3 days, and stable to 95°C in the presence of SDS. The rationale for the formation of the stable complex is that the transfer of the methyl group gives an intermediate, which cannot break down since that would involve the loss of the highly unstable F^+.

The mechanism-based inhibitor was used to identify the active site nucleophile in the enzyme directly.[187] In this work, the *Hae*III methyltransferase from *Haemophilus aegyptius* (which is highly homologous to the *Hha*I enzyme) was inactivated by the inhibitor. Analysis of the complex contained one molecule of the FdC-duplex, one molecule of enzyme, and one methyl group derived from AdoMet. Proteolytic digestion of the complex allowed the isolation of a DNA-containing 20 amino-acid peptide, containing the conserved Pro-Cys sequence. In this peptide, the only amino acid which had been modified by the inhibitor was the Cys in the Pro-Cys sequence.

Further support for the role of this Cys was provided by site-directed mutagenesis. Substitution of the Cys by a number of other amino acids led to over 1000-fold loss of acitivity.[188] The mutation Cys → Gly proved cytotoxic to *E. coli* cells expressing this mutant enzyme, due to very tight binding of DNA which interfered with replication and/or transcription.

5.5.5.4 X-ray crystallographic studies of the enzyme

Cheng and colleagues[184] determined the X-ray structure of the *Hha*I methyltransferase and its complex with AdoMet. The structure of the molecule is shown in Fig. 5.43.

The enzyme has dimensions $4 \times 5 \times 6$ nm and is folded into three parts, a small domain, a hinge region which is involved in forming a cleft which can accommodate DNA, and a large domain. The large domain (1–193 and 304–327) is a mixed α/β structure, with a twisted six-stranded β-sheet core. The small domain (194–275) consists of seven strands of β-sheet; five of these are arranged in a circular formation. The hinge region (276–303) is made up of an α–β–α structure which connects the two domains at the base of the cleft.

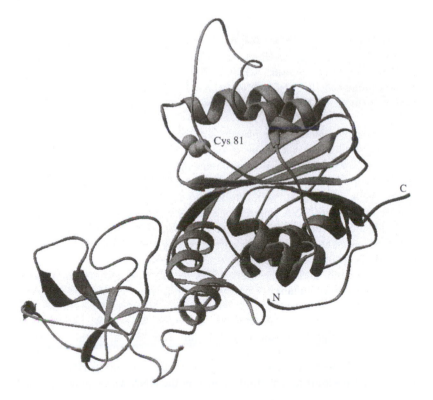

Fig. 5.43 Structure of the *Hha*I DNA methyltransferase enzyme (hmy), drawn using the graphics program Ribbons.[196] The large domain on the right contains the binding site for S-adenosylmethionine and the catalytically essential Cys 81. The binding site for DNA lies between the domains. A colour version of this structure is shown as Plate 6 at the back of this book).

On the large-domain side of the cleft, there is a cavity in which AdoMet binds. The adenosine part of the substrate is inserted into a pocket within the cleft, which the methionine part extends into the cleft. The binding involves hydrophobic, electrostatic, and hydrogen-bond interactions. Many of the structural elements involved in binding appear to be conserved between AdoMet binding enzymes. The catalytically important Cys (Cys 81) is located close to the AdoMet binding pocket.

It was proposed that the cleft between the two domains was the DNA binding region;[184] the cleft is large enough to accommodate double-stranded DNA. The target recognition domain is on the side of the cleft on the small domain.

In later work,[189] the X-ray structure of a chemically trapped complex between the *Hha*I enzyme, the product AdoHcy, and a double-stranded synthetic 13 base-pair oligonucleotide containing 5-fluorocytosine at its target base was determined. The presence of the fluorine at the C5 position does not allow the covalent adduct to break down, as discussed in Section 5.5.5.3. The solution of the structure of this complex allowed details of the DNA binding and of the structural changes undergone by the enzyme and substrate to be determined.

Overall, the structure of the enzyme in the complex is very similar to that observed in the complex with AdoMet.[184] The major difference is in the active-site loop region (amino acids 80–99) in the large domain. The loop contains six conserved residues including the active-site nucleophile, Cys 81. Upon binding of DNA, this loop undergoes a large conformational change and moves towards the DNA binding cleft with three amino acids (Ser 85, Lys 89, and Arg 97) forming interactions with the sugar–phosphate backbone in the minor groove of DNA. The effect of these changes is to bring the side-chain of the active site Cys 81 into close proximity with the target cytosine. There is also a structural change in the small domain, which leads to a smaller structural change towards the DNA binding cleft (Fig. 5.44).

The structure of the DNA substrate bound to the enzyme is that of standard B form DNA with Watson–Crick base pairing apart from the G–C base pair involving the target cytosine, which is disrupted as a result of the cytosine flipping completely out of the DNA helix (Fig. 5.44). There are accompanying local distortions of the phosphodiester backbone on the strand containing this cytosine.

The target cytosine is held in place by a number of interactions with conserved residues, including the side-chains of Glu 119 and Arg 165 and the main-chain carbonyl oxygen of Phe 79. Two side-chains of the enzyme (Gln 237 and Ser 87) occupy the position vacated by the target cytosine which has flipped out of the double helix, forming a hydrogen-bonded arrangement which restores the stacked structure of DNA. Interestingly, these two amino acids which provide the 'infill' are not conserved in other methyltransferases, although it is suggested that other side-chains could form corresponding interactions with each other and with the guanine complementary to the cytosine.

Three regions of the enzyme interact with the DNA. In addition to the active-site loop (80–99) containing the catalytic cysteine from the large domain, there are two glycine-rich sequences (233–240 and 250–257) which form surface loops between β-strands in the small domain. These two loops make almost all the base-specific contacts with the DNA and are thus known as the recognition loops.

In summary, the X-ray studies show how the interaction of the DNA substrate with the enzyme leads to structural changes in the DNA which allows the target

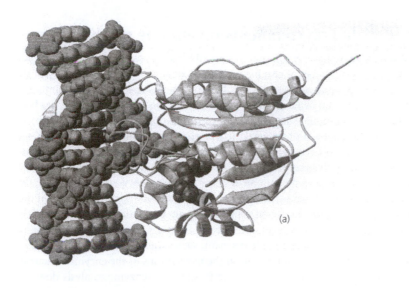

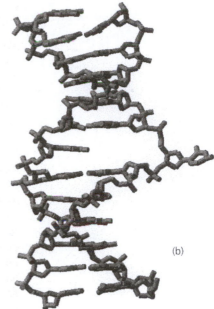

Fig. 5.44 The structure of the DNA methyltransferase–DNA complex (1mht), showing the movement of the active-site loop and distortion of DNA so as to flip out the target cytosine base. (a) The DNA is shown in space-filling form and the enzyme in a cartoon format using the program Ribbons.[196] The location of the S-adenosylmethionine (space-filling) binding site in the enzyme is indicated. (b) The distortion of the DNA in the enzyme–DNA complex is shown in ball-and-stick format. A colour version of (a) is shown as Plate 7 at the back of this book.

cytosine to be flipped out of the double helix so that it is accessible for methyl transfer. The structural changes in the active-site loop of the enzyme serve to bring the catalytic Cys side-chain into the correct orientation for attack on the target cytosine. Despite a good deal of experimental work, a suitable candidate for the base proposed in the mechanism has not yet been identified. This will await the refinement of the structure to higher resolution.

5.5.5.5 The mechanism of DNA methyltransferase and relationship to other enzymes

The results obtained with the *Hha*I DNA methyltransferase provide the basis for understanding the mechanisms of many other enzymes. For instance, we have already noted that other DNA methyltransferases and other one-carbon transferases have conserved Pro-Cys sequences which provide the nucleophilic side-chain for attack on the target base. In terms of the structural organization of the enzymes, it appears that all DNA methyltransferases have a two-domain structure, dictated by the conserved motif pattern (see Section 5.5.5.2). Although X-ray structural evidence for the base-flipping mechanism has not been obtained for other enzymes, other types of experiment have shown that distortion of DNA occurs on binding. Band mobility shift analysis was used to study the *Eco*RV DNA methyltransferase, which transfers a methyl group to the 6-amino group of (the highlighted) adenine in the sequence 5′-GATATC-3′.[190,191] It was shown that binding of DNA leads to a 60° bending of the bound DNA. Use of analogues of the target sequence modified at the target A and its complementary T suggest that base flipping occurs in this enzyme. The results obtained are analogous to those obtained with the cytosine methyltransferases, suggesting that base-flipping is a common mechanism amongst these types of enzyme.

5.6 Concluding comments on enzyme mechanisms

In this chapter we have shown that it is possible to propose fairly detailed mechanisms for a large number of enzyme-catalysed reactions. The understanding of the catalytic processes has allowed the catalytic power and specificity of a number of enzymes to be manipulated in a rational fashion (see Section 5.5.4.5), and has been invaluable in the design of artificial enzymes (Section 5.3.8). Undoubtedly, the increase in the amount of detailed structural information available from X-ray crystallography has been the most profound influence on the rate of the progress in this area, but in this chapter we have tried to show that other techniques, including site-directed mutagenesis, have had a complementary role to play in this process. Significant progress has been made in the development of techniques which allow detailed structural information to be gathered during the course of the catalytic events (Section 5.4.3). Many of the factors suggested by model studies (Section 5.3) as leading to rate enhancement, such as acid–base catalysis and intermediate formation, are seen to occur in the mechanisms of the enzymes discussed in Section 5.5. It is therefore reasonable to state that enzyme catalysis does not involve any new principles in addition to those involved in non-enzyme-catalysed reactions; enzymes merely achieve highly effective combinations of the various established factors leading to rate enhancement. However, such a statement cannot be made with certainty until the observed rate enhancements in a number of enzyme-catalysed reactions can be accounted for quantitatively which, as yet, is not the case, although progress is being made in this area (Section 5.5.2.5).

In the cases of lactate dehydrogenase (Section 5.5.4) and DNA methyltransferase (Section 5.5.5), it is clear that binding of substrates is accompanied by large conformational changes in the enzyme. These types of structural changes have been shown to occur in a very large number of other enzymes by X-ray crystallographic or other less-direct spectroscopic methods such as circular dichroism and fluorescence.[192] These observations provide support for Koshland's 'induced-fit' hypothesis,[193] which proposed that binding of substrate to enzyme induces a structural change in the enzyme so that the important amino-acid side-chains are brought into the correct spatial relationship for catalysis to occur. This 'induced fit' has been viewed as playing a crucial role in enzyme catalysis and specificity (i.e. different substrates would be distinguished from one another by the extents to which they would induce the active conformation of the enzyme). However, the quantitative significance of 'induced fit' has been called into question,[194] and it has been suggested that specificity arises because of the differences in binding of the transition states of different substrates to the enzyme, i.e. the transition states of better substrates bind more effectively, see Section 5.3.6. There is no doubt that in many reactions, perhaps the majority, significant conformational changes in the enzyme accompany the conversion of substrate(s) to product(s). In at least some of these cases, the structural changes could constitute the rate-limiting step of the catalytic cycle (see Section 5.5.4.5, and Chapter 3, Section 3.4.2.1).

References

1. Price, N. C. and Dwek, R. A., *Principles and problems in physical chemistry for biochemists* (2nd edn). Clarendon Press, Oxford (1979). See p. 154.

2. Albery, W. J. and Knowles, J. R., *Biochemistry* **25**, 2572 (1986).

3. Fersht, A. R., Leatherbarrow, R. J., and Wells, T. N. C., *Phil. Trans. R. Soc. Lond.* **A317**, 305 (1986).

4. Bruice, T. C. and Benkovic, S. J., *J. amer. chem. Soc.* **85**, 1 (1963).

5. Page, M. I. and Jencks, W. P., *Proc. natn. Acad. Sci. USA* **68**, 1678 (1971).

6. Jencks, W. P. and Page, M. I., *Biochem. Biophys. Res. Commun.* **57**, 887 (1974).

7. Caswell, M. and Schmir, G. L., *J. amer. chem. Soc.* **102**, 4815 (1980).

8. Dahlquist, F. W., Rand-Meir, T., and Raftery, M. A., *Proc. natn. Acad. Sci. USA* **61**, 1194 (1968).

9. Imoto, T., Johnson, L. N., North, A. C. T., Phillips, D. C., and Rupley, J. A., *Enzymes* (3rd edn) **7**, 665 (1972).

10. Swain, C. G. and Brown, J. F. Jr, *J. amer. chem. Soc.* **74**, 2538 (1952).

11. Fersht, A. R., *Structure and mechanism in protein science*. Freeman, New York (1999). See Ch. 2.

12. Westheimer, F. H., *Proc. chem. Soc.* **253** (1963).

13. Warshel, A. and Levitt, M., *J. molec. Biol.* **103**, 227 (1976).

14. Jencks, W. P., *Adv. Enzymol.* **43**, 219 (1975). See p. 362.

15. Jencks, W. P., *Catalysis in chemistry and enzymology*. McGraw-Hill, New York (1969). See p. 305.

16. Belasco, J. G. and Knowles, J. R., *Biochemistry* **19**, 472 (1980).

17. Wolfenden, R., *Acc. Chem. Res.* **5**, 10 (1972).

18. Wolfenden, R., *A. Rev. biophys. Bioeng.* **5**, 271 (1976).

19. Radzicka, A. and Wolfenden, R., *Methods Enzymol.* **249**, 284 (1995).

20. Collins, K. D. and Stark, G. R., *J. biol. Chem.* **246**, 6599 (1971).

21. Schloss, J. V., *Acc. Chem. Res.* **21**, 348 (1988).

22. Serpersu, F. H., Shortle, D., and Mildvan, A. S., *Biochemistry* **26**, 1289 (1987).

23. D'Souza, V. T. and Bender, M. L., *Acc. Chem. Res.* **20**, 146 (1987).

24. Wharton, C. W., *Int. J. biol. Macromolec.* **1**, 3 (1979).

25. Breslow, R., *Adv. Enzymol.* **58**, 1 (1986).

26. Anderson, S., Anderson, H. L., and Sanders, J. K. M., *Acc. Chem. Res.* **26**, 469 (1993).

27. Lerner, R. A., Benkovic, S. J., and Schultz, P. G., *Science* **252**, 659 (1991).

28. Wagner, J., Lerner, R. A., and Barbas, C. F. III, *Science* **270**, 1797 (1995).

29. Zhou, G. W., Guo, J., Huang, W., Fletterick, R. J., and Scanlan, T. S., *Science* **265**, 1059 (1994).

30. Quéméneur, E., Moutiez, M., Charbonnier, J.-B., and Ménez, A., *Nature* **391**, 301 (1998).

31. Berger, A. and Schechter, I., *Phil. Trans. R. Soc.* **B257**, 249 (1970).

32. Alecio, M. R., Dann, M. L., and Lowe, G., *Biochem. J.* **141**, 495 (1974).

33. Lowe, G. and Yuthavong, Y., *Biochem. J.* **124**, 107 (1971).

34. Perona, J. J. and Craik, C. S., *J. biol. Chem.* **272**, 29987 (1997).

35. Cornish-Bowden, A., *Fundamentals of enzyme kinetics* (revised edn). Portland Press, London (1995). See Ch. 8.

36. Findlay, D., Mathias, A. P., and Rabin, B. R., *Biochem. J.* **85**, 139 (1962).

37. Roberts, G. C. K., Dennis, F. A., Meadows, D. H., Cohen, J. S., and Jardetzky, O., *Proc. natn. Acad. Sci. USA* **62**, 1151 (1969).

38. Hartley, B. S. and Kilby, B. A., *Biochem. J.* **56**, 288 (1954).

39. Shore, J. D. and Gutfreund, H., *Biochemistry* **9**, 4655 (1970).

40. Henderson, R., *J. molec. Biol.* **54**, 341 (1970).

41. Douzou, P. and Petsko, G. A., *Adv. Protein Chem.* **36**, 246 (1984).

42. Fink, A. L. and Petsko, G. A., *Adv. Enzymol.* **52**, 177 (1981).

43. Cartwright, S. J. and Waley, S. G., *Biochemistry* **26**, 5329 (1987).

44. Wess, T. J., *Biotechnol. Appl. Biochem.* **26**, 127 (1997).

45. Hajdu, J., Acharya, K. R., Stuart, D. I., Barford, D., and Johnson, L. N., *Trends Biochem. Sci.* **13**, 104 (1988).

46. Schlichting, I., Almo, S. C., Rapp, G., Wilson, K., Petratos, K., Lentfer, A. *et al.*, *Nature* **345**, 309 (1990).

47. Singer, P. T., Smalås, A., Carty, R. P., Mangel, W. F., and Sweet, R. M., *Science* **259**, 669 (1993).

48. Madsen, N. B., Kasvinsky, P. J., and Fletterick, R. J., *J. biol. Chem.* **253**, 9097 (1978).

49. Sygusch, J., Madsen, N. B., Kasvinsky, P. J., and Fletterick, R. J., *Proc. natn. Acad. Sci. USA* **74**, 4757 (1977).

50. Fry, D. C., Kuby, S. A., and Mildvan, A. S., *Biochemistry* **26**, 1645 (1987).

51. Means, G. E. and Feeney, R. E., *Chemical modification of proteins*. Holden-Day, San Francisco (1971).

52. Cohen, L. A., *Enzymes* (3rd edn) **1**, 147 (1970).

53. Lundblad, R. L., *Techniques in protein modification*. CRC Press, Boca Raton, Florida (1995).

54. Bruice, T. C., Gregory, M. J., and Walters, S. L., *J. amer. chem. Soc.* **90**, 1612 (1968).

55. Colman, R. F., *J. biol. Chem.* **243**, 2454 (1968).

56. Smith, E. L., Austen, B. M., Blumenthal, K. M., and Nyc, J. F., *Enzymes* (3rd edn) **11**, 293 (1975).

57. Malcolm, B. A., Rosenberg, S., Corey, M. J., Allen, J. S., Debaetselier, A., and Kirsch, J. F., *Proc. natn. Acad. Sci. USA* **86**, 133 (1989).

58. De La Mare, S., Coulson, A. F. W., Knowles, J. R., Priddle, J. D., and Offord, R. E., *Biochem. J.* **129**, 321 (1972).

59. Walsh, C. T., *Trends Biochem. Sci.* **8**, 254 (1983).

60. Silverman, R. B., *Methods Enzymol.* **249**, 240 (1995).

61. Der Terrossian, E. and Kassab, R., *Eur. J. Biochem.* **70**, 623 (1976).

62. Singer, S. J., *Adv. Protein Chem.* **22**, 1 (1967).

63. Rakitzis, E. T., *Biochem. J.* **217**, 341 (1984).

64. Winter, G., Fersht, A. R., Wilkinson, A. J., Zoller, M., and Smith, M., *Nature* **299**, 756 (1982).

65. Rossi, J. and Zoller, M., *Protein engineering*. Alan R. Liss, New York (1987). See Ch. 4.

66. Fersht, A. R., *Structure and mechanism in protein science*. Freeman, New York (1999). See Ch. 14.

67. Old, R. W. and Primrose, S. B., *Principles of gene manipulation* (5th edn). Blackwell, Oxford (1994). See Ch. 11.

68. Taylor, J. W., Schmidt, W., Cosstick, R., Okruszek, A., and Eckstein, F., *Nucl. Acids Res.* **13**, 8749 (1985).

69. Taylor, J. W., Ott, J., and Eckstein, F. J., *Nucl. Acids Res.* 13, 8765 (1985).

70. Old, R. W. and Primrose, S. B., *Principles of gene manipulation* (5th edn). Blackwell, Oxford (1994). See Ch. 10.

71. Sarkar, G. and Sommer, S. S., *Biotechniques* **8**, 404 (1990).

72. Silver, J., Limjoco, T., and Feinstone, S., in *Site-specific mutagenesis using the polymerase chain reaction in PCR strategies* (Innis, J. J., Gelfand, D. H., and Sninsky, J. J., eds). Academic Press, New York (1995). See p. 179.

73. Plapp, B. V., *Methods Enzymol.* **249**, 91 (1995).

74. Pantoliano, M. W., Ladner, R. C., Bryan, P. N., Rollence, M. L., Wood, J. F., and Poulos, T. L., *Biochemistry* **26**, 2077 (1987).

75. Noren, C. J., Anthony-Cahill, S. J., Griffith, M. C., and Schultz, P. G., *Science* **244**, 182 (1989).

76. Mendel, D., Ellman, J. A., Chang, Z., Veenstra, D. L., Kollman, P. A., and Schultz, P. G., *Science* **256**, 1798 (1992).

77. Stadtman, T. C., *Adv. Enzymol.* **48**, 1 (1979).

78. Ceska, T. A., Sayers, J. R., Stier, G., and Suck, D., *Nature* **382**, 90 (1996).

79. Kelly, S. M. and Price, N. C., *Biochim. Biophys. Acta* **1338**, 161 (1997).

80. Campbell, I. D. and Dwek, R. A., *Biological spectroscopy.* Benjamin/Cummings, Menlo Park, California (1984). See Chapters 5 and 7.

81. Hess, G. P., *Enzymes* (3rd edn) **3**, 213 (1971).

82. Blow, D. M., *Enzymes* (3rd edn) **3**, 185 (1971).

83. Blow, D. M., Birktoft, J. J., and Hartley, B. S., *Nature* **221**, 337 (1969).

84. Bachovchin, W. W., Kaiser, R., Richards, J. H., and Roberts, J. D., *Proc. natn. Acad. Sci. USA* **78**, 7323 (1981).

85. Porubcan, M. A., Westler, W. M., Ibañez, I. B., and Markley, J. L., *Biochemistry* **18**, 4108 (1979).

86. Kossiakoff, A. A. and Spencer, S. A., *Biochemistry* **20**, 6462 (1981).

87. Rutter, W. J., Gardell, S. J., Roczniak, S., Hilvert, D., Sprang, S., Fletterick, R. J. *et al.*, *Protein engineering.* Alan R. Liss, New York (1987). See Ch. 23.

88. Carter, P. and Wells, J. A., *Nature* **332**, 564 (1988).

89. Steitz, T. A., Henderson, R., and Blow, D. M., *J. molec. Biol.* **46**, 337 (1969).

90. Hartley, B. S. and Shotton, D. M., *Enzymes* (3rd edn) **3**, 323 (1971).

91. Perona, J. J. and Craik, C. S., *J. biol. Chem.* **272**, 29987 (1997).

92. Kraut, J., *Enzymes* (3rd edn) **3**, 165 (1971).

93. Gertler, A., Walsh, K. A., and Neurath, H., *Biochemistry* **13**, 1302 (1974).

94. Fastrez, J. and Fersht, A. R., *Biochemistry* **12**, 2025 (1973).

95. Wells, J. A. and Estell, D. A., *Trends Biochem. Sci.* **13**, 291 (1988).

96. Hunkapiller, M. W., Forgac, M. D., and Richards, J. H., *Biochemistry* **15**, 5581 (1976).

97. Rühlmann, A., Kukla, D., Schwager, P., Bartels, K., and Huber, R., *J. molec. Biol.* **77**, 417 (1973).

98. Blow, D. M., Janin, J., and Sweet, R. M., *Nature* **249**, 54 (1974).

99. Keil, B., *Enzymes* (3rd edn) **3**, 249 (1971).

100. Fersht, A. R., *Structure and mechanism in protein science.* Freeman, New York (1999). See Ch. 7.

101. Fersht, A. R., *Structure and mechanism in protein science.* Freeman, New York (1999). See Ch. 16.

102. Maelicke, A., *Trends Biochem. Sci.* **16**, 355 (1991).

103. Groll, M., Ditzel, L., Löwe, J., Stock, D., Bochtler, M., Bartunik, H. D. *et al.*, *Nature* **386**, 463 (1997).

104. Christianson, D. W., David, P. R., and Lipscomb, W. N., *Proc. natn. Acad. Sci. USA* **84**, 1512 (1987).

105. Gardell, S. J., Craik, C. S., Hilvert, D., Urdea, M. S., and Rutter, W. J., *Nature* **317**, 551 (1985).

106. Gardell, S. J., Hilvert, D., Barnett, J., Kaiser, E. T., and Rutter, W. J., *J. biol. Chem.* **262**, 576 (1987).

107. Drenth, J., Jansonius, J. N., Koekoek, R., and Wolthers, B. G., *Adv. Protein Chem.* **25**, 79 (1971).

108. Angelides, K. J. and Fink, A. L., *Biochemistry* **18**, 2355 (1979).

109. Polgar, L. and Halasz, P., *Biochem. J.* **207**, 1 (1982).

110. Polgar, L., *FEBS Lett.* **219**, 1 (1987).

111. Ringe, D., *Methods Enzymol.* **241**, 157 (1994).

112. Perutz, M. F., *Protein structure: new approaches to disease and therapy.* Freeman, New York (1992). See Ch. 5.

113. Schmid, M. F. and Herriott, J. R., *J. molec. Biol.* **103**, 175 (1976).

114. Reynolds, S. J., Yates, D. W., and Pogson, C. I., *Biochem. J.* **122**, 285 (1971).

115. Knowles, J. R. and Albery, W. J., *Acc. chem. Res.* **10**, 105 (1977).

116. Banner, D. W., Bloomer, A. C., Petsko, G. A., Phillips, D. C., Pogson, C. I., Wilson, I. A. *et al.*, *Nature* **255**, 609 (1975).

117. Levitt, M. and Chothia, C., *Nature* **261**, 552 (1976).

118. Casal, J. I., Ahern, T. J., Davenport, R. C., Petsko, G. A., and Klibanov, A. M., *Biochemistry* **26**, 1258 (1987).

119. Davenport, R. C., Bash, P. A., Seaton, B. A., Karplus, M., Petsko, G. A., and Ringe, D., *Biochemistry* **30**, 5821 (1991).

120. Joseph, D., Petsko, G. A., and Karplus, M., *Science* **249**, 1425 (1990).

121. Pompliano, D. L., Peyman, A., and Knowles, J. R., *Biochemistry* **29**, 3186 (1990).

122. Fersht, A. R., *Structure and mechanism in protein science.* Freeman, New York (1999). See pp. 96–9.

123. Cornish-Bowden, A., *Fundamentals of enzyme kinetics* (revised edn). Portland Press, London (1995). See Ch. 7.

124. Rieder, S. V. and Rose, I. A., *J. biol. Chem.* **234**, 1007 (1959).

125. Collins, K. D., *J. biol. Chem.* **249**, 136 (1974).

126. Miller, J. C. and Waley, S. G., *Biochem. J.* **123**, 163 (1971).

127. Raines, R. T., Sutton, E. L., Straus, D. R., Gilbert, W., and Knowles, J. R., *Biochemistry* **25**, 7142 (1986).

128. Joseph-McCarthy, D., Rost, L. E., Komives, E. A., and Petsko, G. A., *Biochemistry* **33**, 2824 (1994).

129. Nickbarg, E. B., Davenport, R. C., Petsko, G. A., and Knowles, J. R., *Biochemistry* **27**, 5948 (1988).

130. Lodi, P. J. and Knowles, J. R., *Biochemistry* **30**, 6948 (1991).

131. Knowles, J. R., *Nature* **350**, 121 (1991).

132. Webb, M. R. and Knowles, J. R., *Biochem. J.* **141**, 589 (1974).

133. Lodi, P. J., Chang, L. C., Knowles, J. R., and Komives, E. A., *Biochemistry* **33**, 2809 (1994).

134. Joseph-McCarthy, D., Lolis, E., Komives, E. A., and Petsko, G. A., *Biochemistry* **33**, 2815 (1994).

135. Albery, W. J. and Knowles, J. R., *Biochemistry* **15**, 5627 (1976).

136. Blacklow, S. C., Raines, R. T., Lim, W. A., Zamore, P. D., and Knowles, J. R., *Biochemistry* **27**, 1158 (1988).

137. Bash, P. A., Field, M. J., Davenport, R. C., Petsko, G. A., Ringe, D., and Karplus, M., *Biochemistry* **30**, 5826 (1991).

138. Rose, I. A., *Adv. Enzymol.* **43**, 491 (1975).

139. Mousdale, D. M., Campbell, M. S., and Coggins, J. R., *Phytochemistry* **26**, 2665 (1987).

140. Chaudhuri, S., Lambert, J. M., McColl, L. A., and Coggins, J. R., *Biochem. J.* **239**, 699 (1986).

141. White, P. J., Young, J., Hunter, I. S., Nimmo, H. G., and Coggins, J. R., *Biochem. J.* **265**, 735 (1990).

142. Kleanthous, C., Deka, R., Davis, K., Kelly, S. M., Cooper, A., Harding, S. E. *et al.*, *Biochem. J.* **282**, 687 (1992).

143. Duncan, K., Chaudhuri, S., Campbell, M. S., and Coggins, J. R., *Biochem. J.* **238**, 475 (1986).

144. Krell, T., Horsburgh, M. J., Cooper, A., Kelly, S. M., and Coggins, J. R., *J. biol. Chem.* **271**, 24492 (1996).

145. Harris, J., Kleanthous, C., Coggins, J. R., Hawkins, A. R., and Abell, C. A., *J. Chem. Soc. Chem. Commun.* 1080 (1993).

146. Chaudhuri, S., Duncan, K., Graham, L. D., and Coggins, J. R., *Biochem. J.* **275**, 1 (1991).

147. Moore, J. D., Hawkins, A. R., Charles, I. G., Deka, R., Coggins, J. R., Cooper, A. *et al.*, *Biochem. J.* **295**, 277 (1993).

148. Leech, A. P., James, R., Coggins, J. R., and Kleanthous, C., *J. biol. Chem.* **270**, 25827 (1995).

149. Deka, R. K., Kleanthous, C., and Coggins, J. R., *J. biol. Chem.* **267**, 22237 (1992).

150. Tsou, C. L., *Scientia Sinica* **11**, 1535 (1962).

151. Kleanthous, C. and Coggins, J. R., *J. biol. Chem.* **265**, 10935 (1990).

152. Krell, T., Pitt, A. R., and Coggins, J. R., *FEBS Lett.* **360**, 93 (1995).

153. Gourley, D. G., Shrive, A. K., Polikarpov, I., Krell, T., Coggins, J. R., Hawkins, A. R. *et al.*, *Nature Struct. Biol.* **6**, 521 (1999).

154. Sygusch, J., Beaudry, D., and Allaire, M., *Proc. natn. Acad. Sci. USA* **84**, 7846 (1987).

155. Almassy, R. J., Janson, C. A. , Hamlin, R., Xuong, N.-H., and Eisenberg, D., *Nature* **323**, 304 (1986).

156. Burnett, R. M., Darling, G. D., Kenall, D. S., Le Quesene, M. E., Mayhew, S. G., Smith, W. W. *et al.*, *J. biol. Chem.* **249**, 4383 (1974).

157. Harris, J. M., Gonzalez-Bello, C., Kleanthous, C., Hawkins, A. R., Coggins, J. R., and Abell, C., *Biochem. J.* **319**, 333 (1996).

158. Holbrook, J. J., Liljas, A., Steindel, S. J., and Rossmann, M. G., *Enzymes* (3rd edn) **11**, 191 (1975).

159. Takenaka, Y. and Schwert, G. W., *J. biol. Chem.* **223**, 157 (1956).

160. Novoa, W. B., Winer, A. D., Glaid, A. J., and Schwert, G. W., *J. biol. Chem.* **234**, 1143 (1959).

161. Heck, H. d'A., McMurray, C. H., and Gutfreund, H., *Biochem. J.* **108**, 793 (1968).

162. Stinson, R. A. and Gutfreund, H., *Biochem. J.* **121**, 235 (1971).

163. Südi, J., *Biochem. J.* **139**, 251 (1974).

164. Südi, J., *Biochem. J.* **139**, 261 (1974).

165. Wilks, H. M., Halsall, D. J., Atkinson, T., Chia, W. N., Clarke, A. R., and Holbrook, J. J., *Biochemistry* **29**, 8587 (1990).

166. Rossman, M. G., Liljas, A., Bränden, C.-I., and Banaszek, L. J., *Enzymes* (3rd edn) **11**, 61 (1975).

167. Biesecker, G., Harris, J. I., Thierry, J. S., Walker, J. E., and Wonacott, A. J., *Nature* **266**, 328 (1977).

168. Moras, D., Olsen, K. W., Sabesan, M. N., Buehner, M., Ford, G. C., and Rossmann, M. G., *J. biol. Chem.* **250**, 9137 (1975).

169. Adams, M. J., Buehner, M., Chandrasekhar, K., Ford, G. C., Hackert, M. L., Liljas, A. *et al.*, *Proc. natn. Acad. Sci. USA* **70**, 1968 (1973).

170. Holbrook, J. J. and Ingram, V. A., *Biochem. J.* **131**, 729 (1973).

171. Clarke, A. R., Atkinson, T., and Holbrook, J. J., *Trends Biochem. Sci.* **14**, 101 (1989).

172. Clarke, A. R., Atkinson, T., and Holbrook, J. J., *Trends Biochem. Sci.* **14**, 145 (1989).

173. Clarke, A. R., Wigley, D. B., Chia, W. N., Barstow, D., Atkinson, T., and Holbrook, J. J., *Nature* **324**, 699 (1986).

174. Waldman, A. D. B., Hart, K. W., Clarke, A. R., Wigley, D. B., Barstow, D. A., Atkinson, T. *et al.*, *Biochem. Biophys. Res. Commun.* **150**, 752 (1988).

175. Eszes, C. M., Sessions, R. B., Clarke, A. R., Moreton, K. M., and Holbrook, J. J., *FEBS Lett.* **399**, 193 (1996).

176. Clarke, A. R., Smith, C. J., Hart, K. W., Wilks, H. M., Chia, W. N., Lee, T. V. *et al.*, *Biochem. Biophys. Res. Commun.* **148**, 15 (1987).

177. Scrutton, N. S., Berry, A., and Perham, R. N., *Nature* **343**, 38 (1990).

178. Branden, C. and Tooze, J., *Introduction to protein structure* (2nd edn). Garland, New York (1999). See Ch. 1.

179. Adams, R. L. P., *Biochem. J.* **265**, 309 (1990).

180. Wilson, G. G. and Murray, N. E., *A. Rev. Genet.* **25**, 585 (1991).

181. Modrich, P., *A. Rev. Genet.* **25**, 229 (1991).

182. Li, E., Bestor, T. H., and Jaenisch, R., *Cell* **69**, 915 (1992).

183. Posfai, J., Bhagwat, A. S., Posfai, G., and Roberts, R. J., *Nucl. Acids Res.* **17**, 2421 (1989).

184. Cheng, X., Kumar, S., Posfai, J., Pflugrath, J. W., and Roberts, R. J., *Cell* **74**, 299 (1993).

185. Wu, J. C. and Santi, D. V., *J. biol. Chem.* **262**, 4778 (1987).

186. Osterman, D. G., DePillis, G. D., Wu, J. C., Matsuda, A. and Santi, D. V., *Biochemistry* **27**, 5204 (1988).

187. Chen, L., MacMillan, A. M., Chang, W., Ezaz-Nikpay, K., Lane, W. S., and Verdine, G. L., *Biochemistry* **30**, 11018 (1991).

188. Mi, S. and Roberts, R. J., *Nucl. Acids Res.* **21**, 2459 (1993).

189. Klimasauskas, S., Kumar, S., Roberts, R. J., and Cheng, X., *Cell* **76**, 357 (1994).

190. Cal, S. and Connolly, B. A., *J. biol. Chem.* **271**, 1008 (1996).

191. Cal, S. and Connolly, B. A., *J. biol. Chem.* **272**, 490 (1997).

192. Citri, N., *Adv. Enzymol.* **37**, 397 (1973).

193. Koshland, D. E. Jr, *Proc. natn. Acad. Sci. USA* **44**, 98 (1958).

194. Fersht, A. R., *Structure and mechanism in protein science.* Freeman, New York (1999). See pp. 369. 372.

195. Kraulis, P., *J. Appl. Crstallogr.* **24**, 946 (1991).

196. Carson, M., *J. Appl. Crstallogr.* **24**, 958 (1991).

6

The control of enzyme activity

* This would be the case in an
oxidizing atmosphere, i.e. under
aerobic conditions. Under anaerobic
conditions, glucose would be
converted to pyruvate and thence
lactate (see Fig. 6.21).

*Regulation needed
for control and
organization.*

6.1 Introduction

In this chapter we shall consider the third of the remarkable properties of
enzymes mentioned in Chapter 1, Section 1.3, namely the control of enzyme
activity. It should be self-evident why enzyme activity needs to be controlled in
living organisms, since if no such control existed all metabolic processes would
tend towards states in which they were at equilibrium with their surroundings.
Thus, for instance, storage of glycogen as an energy reserve in muscle would be
impossible, since there is an enzyme (phosphorylase) catalysing the breakdown of
glycogen into glucose units, which would then be acted upon to yield ultimately
carbon dioxide and water.*

Clearly, the existence of glycogen as an energy store implies that the activity of
phosphorylase can be controlled in such a way as to allow the store to be called
on as the situation demands. We shall see later (Section 6.4.2) that there are a
number of elaborate mechanisms to effect this control in muscle, thereby allowing the rate of glycogen breakdown to vary by several orders of magnitude in
different circumstances.

Regulation of enzyme activities can be brought about in two ways. First,
the amount of an enzyme can be altered as a result of changes in the rate of
enzyme synthesis or degradation (or both). This type of regulation is suitable
for long-term regulation, e.g. in response to changes in nutrients, and is discussed in more detail in Chapter 9. Second, the activities of enzymes already
present in a cell or organism can be altered, permitting a rapid response to
changes in conditions. This chapter will be confined to this second means of
regulation.

We shall first outline in Section 6.2 some of the mechanisms that have been
shown to be important in controlling the activity of single enzymes, and then
proceed in Sections 6.3 and 6.4 to consider the control of a sequence of enzyme-
catalysed reactions that constitute a metabolic pathway. In the cell, there are
usually a number of interconnected metabolic pathways and the control of these
presents more complex features (see Chapter 8, Section 8.3).

6.2 Control of the activities of single enzymes

The principal mechanisms that exist for control of enzyme activity can be broadly classified into two categories: (i) those mechanisms that involve a change in the covalent structure of an enzyme, and (ii) those mechanisms that involve conformational changes in an enzyme caused by the reversible binding of 'regulator' molecules. These categories are discussed in Sections 6.2.1 and 6.2.2, respectively.

There are various other mechanisms by which enzyme activity could be controlled, including the following.

(i) Specific inhibitor macromolecules

The best studied cases of such macromolecules act as inhibitors of proteases. We have already mentioned how bovine pancreatic trypsin inhibitor and soybean trypsin inhibitor bind at the active site of trypsin by mimicking the structure of the tetrahedral intermediate that occurs in the enzyme-catalysed reaction (see Chapter 5, Section 5.5.1.3).* The function of this type of inhibitor molecule will be mentioned in Section 6.2.1.1.

In most other cases the precise details of the binding of inhibitor to enzyme are less clear. However, the interactions can be of great physiological significance, as is illustrated by the role of the phosphatase inhibitor in the regulation of glycogen metabolism (Section 6.4.2.4). A compilation of proteins that act as inhibitors of intracellular enzymes has been made. It has been suggested that these proteins be termed 'regulatory subunits' rather than 'antizymes'.[3] An example of such a protein is discussed in Chapter 9, Section 9.6.2.4.

(ii) Availability of substrate or cofactor

The intracellular concentrations of the substrates for many enzymes are significantly lower than the values of the Michaelis constant (K_m). For example, the concentration of D-glyceraldehyde 3-phosphate in muscle (approximately 3 μmol dm^{-3}) is well below the K_m of this substrate for glyceraldehyde-3-phosphate dehydrogenase (approximately 70 μmol dm^{-3}).[4] Under these conditions the rate of the enzyme-catalysed reaction will depend on the concentration of substrate (see Chapter 4, Fig. 4.1). If the prevailing concentration of substrate is much greater than the K_m, changes in the concentration of substrate will have little effect on the rate of reaction. This would be the case for fructosebisphosphate aldolase in mouse brain, where the concentration of fructose bisphosphate (approximately 200 μmol dm^{-3}) is well above its K_m (12 μmol dm^{-3}).[5]

When data on the intracellular concentrations of substrates are being used to decide whether a certain enzyme is essentially saturated, it is important to bear in mind that cells usually contain a number of distinct organelles, such as nuclei, mitochondria, lysosomes, etc. These organelles are bounded by membranes which often display selective permeability and thus enable an uneven distribution of a metabolite to exist within the various compartments of a cell. Many methods of determining the concentrations of metabolites give only an average intracellular concentration, rather than the concentration in a particular compartment where the relevant enzyme may be located. The consequences of this compartmentation within cells are discussed more fully in Chapter 8, Section 8.3.

(iii) Product inhibition

The product of an enzyme-catalysed reaction may act as an inhibitor of that reaction, so that under circumstances in which product accumulates. the enzyme is inhibited and the rate of product formation is decreased. (See for example the

* Some other examples of protease inhibitors include α_1-antichymotrypsin, α_1-antiprotease, and α_2-macroglobulin, which are all found in blood plasma.[1] There are also several protease inhibitors of bacterial origin, e.g. leupeptin and antipain, which are of relatively low M_r.[2]

inhibition of hexokinase caused by D-glucose 6-phosphate discussed in Section 6.4.1.2.)

(iv) Non-enzyme-catalysed reactions

A fourth possibility is that in certain cases the rate of non-enzyme-catalysed reactions may be rate limiting. Several enzymes involved in carbohydrate metabolism display anomeric specificity, i.e. they act on only one anomer of a sugar.[6] 6-phosphofructokinase acts on the β-anomer of D-fructose 6-phosphate (F6P), whereas fructose-bisphosphatase acts on the α-anomer of D-fructose 1,6-bis-phosphate (FBP) (see Fig. 6.1).

The rates of anomerization (i.e. interconversion of α- and β-forms) of F6P and FBP are known. Calculations suggest that the non-enzyme-catalysed anomerization α-F6P $\rightarrow$ β-F6P, would not be rate limiting in glycolysis (glucose breakdown), but that the anomerization β-FBP $\rightarrow$ α-FBP might be so in gluconeogenesis (glucose synthesis).[6] These types of non-enzyme-catalysed reactions might well be rate limiting in other metabolic processes, especially when there are relatively high concentrations of the relevant enzymes, as is often the case in cells (see Chapter 8, Section 8.5).

Having briefly described these additional mechanisms of control, we shall now turn our attention to the principal means of regulating the activity of single enzymes.

6.2.1 Control of activity by changes in the covalent structures of enzymes

This category of control can be subdivided into those cases in which the change is essentially irreversible (Section 6.2.1.1) and those cases in which the change can be reversed provided that a source of energy, such as ATP, is available (Section 6.2.1.2). There are relatively few examples of irreversible covalent modification of enzymes and these are largely restricted to extracellular processes (Table 6.1). By contrast, there is a very large number of examples of reversible covalent modification and these are all intracellular processes. Covalent modification with ubiquitin is an intracellular process by which proteins become tagged prior to degradation (see Chapter 9, Section 9.6.1). It might be regarded as irreversible, although when deubiquitination occurs it is generally only after the tagged protein has become degraded.

6.2.1.1 Control of activity by essentially irreversible changes in covalent structure

It has been recognized for many years that a number of proteases are synthesized in the form of inactive precursors or zymogens, which can then be activated by the action of other proteases. The activation of chymotrypsinogen was discussed in detail in Chapter 5, Section 5.5.1.2. Some examples of enzymes that are activated in this way are given in Table 6.1. The first four enzymes mentioned in the table

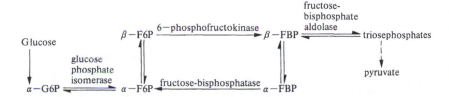

Fig. 6.1 Part of the pathways of glycolysis and gluconeogenesis, illustrating anomeric specificities of certain enzymes.[6] G6P, F6P, and FBP represent D-glucose 6-phosphate, D-fructose 6-phosphate, and D-fructose 1,6-bisphosphate, respectively.

Table 6.1 Some enzymes activated by proteolytic action

Enzyme	Precursor	Function
Trypsin Chymotrypsin Elastase Carboxypeptidase	Trypsinogen Chymotrypsinogen Proelastase Procarboxypeptidase	Pancreatic secretion (see Section 6.2.1.1)
Phospholipase A_2[7]	Prophospholipase A_2	Pancreatic secretion
Pepsin	Pepsinogen	Secreted into gastric juice: most active in pH range 1–5
Thrombin	Prothrombin	Part of the blood coagulation system (see Section 6.2.1.1)
C$\bar{1}$r	C1r	Part of the first component of the complement system (see Section 6.2.1.1)
Chitin synthase[8]	Zymogen	Involved in the formation of the septum during budding and cell division in yeast

are synthesized as zymogens in the acinar cells of the pancreas and stored in zymogen granules. The release of the zymogens from these granules into the duodenum is under the control of hormones, primarily cholecystokinin-pancreozymin and secretin. Digestion of proteins requires the coordinated action of these various enzymes with their different specificities (see Chapter 5, Section 5.5.1.2); coordination is achieved by the common factor that the zymogens are all activated by the action of trypsin (Fig. 6.2.). In each case, the peptide bond cleaved during the activation of the zymogen lies on the C-terminal side of a lysine or arginine side-chain, in accordance with the known specificity of trypsin. For example, the activation of trypsinogen involves the removal of a hexapeptide Val-Asp-Asp-Asp-Asp-Lys from the N-terminus of the polypeptide chain. In order to account for the role of trypsin in initiating the activation of the zymogens, we must consider the involvement of another protease, enteropeptidase (previously referred to as enterokinase). Enteropeptidase, which is synthesized in the brush border of the epithelial cells of the small intestine, is an enzyme of high specificity catalysing the activation of trypsinogen as it enters the duodenum. The trypsin produced then catalyses the activation of the zymogens (Fig. 6.2). The scheme illustrates a number of important control features. It is essential that there is no premature activation of the zymogens, since this would result in considerable damage to the pancreas.*

* In the condition known as acute pancreatitis, which can be lethal in severe cases, there is premature activation of the precursors of proteases and lipases synthesized in the pancreas.

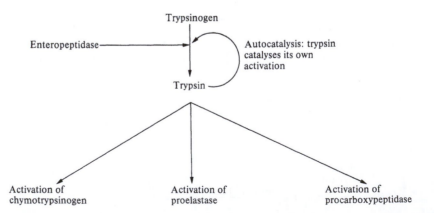

Fig. 6.2 Coordinated activation of the pancreatic proteases.

In order to achieve this, the initial trigger provided by enteropeptidase is located at a site distinct from the site of production of the zymogens. The amount of active proteases produced is controlled by the amount of zymogens secreted into the duodenum, which is in turn under hormonal control. Eventually, the proteases digest themselves into amino acids and small peptides, which are re-absorbed from the intestinal tract and used for protein synthesis. The average daily production of the short-lived hydrolytic enzymes by the human pancreas is of the order of 10 g.[9]

An additional mechanism to prevent premature activation of trypsinogen is provided by the presence of a trypsin inhibitor protein in the pancreatic secretion. This protein, which is present to the extent of about 2 per cent of the protein content of the secretion (very much less in molar terms than the concentration of trypsinogen), binds very tightly to any active trypsin that might be present in the zymogen granules, and thus ensures that the activation of trypsinogen occurs only at the physiologically appropriate time and place, i.e. when it encounters enteropeptidase in the small intestine. The control of the production of enteropeptidase is less well understood and remains a problem for future work.

The scheme shown in Fig. 6.2 illustrates the principle of *signal amplification*, which will provide a recurrent theme in this chapter. A small amount of enteropeptidase can produce rapidly a larger amount of trypsin, since one mole-cule of enzyme can act on many molecules of substrate in a short time. The trypsin, in turn, gives rise to an even larger quantity of active proteases.

This feature of amplification is also well illustrated by the systems involved in blood coagulation and complement activation. In the blood coagulation systems,[10] a series of enzymes and other protein factors act on each other in a sequential fashion, known as a *cascade*, eventually leading to the activation of the zymogen prothrombin to yield thrombin. Thrombin then catalyses the hydrolysis of a limited number of Arg-Gly bonds in fibrinogen to produce fibrin, which spontaneously forms an insoluble clot, strengthened by cross-linking between fibrin molecules. There are two pathways of activation of blood coagulation; extrinsic, which is triggered by a lipoprotein released from injured tissue, and intrinsic, which is triggered by contact of one of the factors with a surface such as the fibrous protein collagen that surrounds the damaged blood vessel. In both pathways, the sequence of enzyme-catalysed reactions provides a large amplification of the initial signal to ensure that sufficient fibrin is formed rapidly at the site of injury to restrict blood loss.

The complement system consists of a series of blood plasma proteins that cause damage to, and eventually lysis of, an invading organism, such as a bac-terium.[11] The trigger for the system is usually provided by the aggregation of antibody molecules bound to the surface of the invading organism. This aggre-gation activates the components of the complement system by a sequence of activations of zymogens, leading to the rupture of the membrane of the invading organism.

In the various cases discussed, it is clear that the sequences of enzyme-catalysed reactions provide a *rapid, amplified response* to a small initial signal. Once the zymogens have been activated and the resulting proteases have performed their particular functions, the active enzymes are degraded rather than being converted back to zymogens. It is obviously important that the systems are 'switched off' at an appropriate time, so that for instance fibrin is formed only at the site of an injury and not in the entire blood system. The 'switching off' aspect has so far received less attention than the 'switching on'.

6.2.1.2 **Control of activity by reversible changes in covalent structure**

It is now clear that a large number of enzymes exist in two forms (of different catalytic properties), which can be interconverted by the action of other enzymes. The classic example of this type of enzyme is phosphorylase, which catalyses the reaction:

$$(\text{Glycogen})_n + \text{orthophosphate} \rightleftharpoons (\text{glycogen})_{n-1} + \alpha\text{-D-glucose 1-phosphate}.$$

The enzyme exists in two forms: phosphorylase b, which requires AMP (or a few other ligands) for activity, and phosphorylase a, which is active in the absence of AMP, although AMP does activate it to a small extent. The only difference in covalent structure between the a and b forms of phosphorylase is that the side-chain of Ser 14 is phosphorylated in the a form but not in the b form.[12] Comparison of the three-dimensional structures of the two forms shows that the conformations of the first 19 amino acids from the N-terminus are markedly different. This segment is flexible in the b form, but well ordered in the a form, with the phosphorylated side-chain of Ser 14 interacting with the positively charged side-chain of Arg 69.[12]

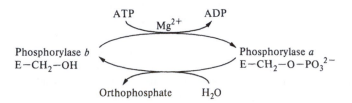

In the direction $b \rightarrow a$, the reaction requires ATP and Mg^{2+} and is catalysed by phosphorylase kinase; in the direction $a \rightarrow b$, the reaction is catalysed by phosphorylase phosphatase and liberates orthophosphate. If the reaction proceeds in a cyclical fashion, the net result would be the hydrolysis of ATP to ADP and orthophosphate.

In 1986 there were over 140 well-documented examples[13] of enzymes and other proteins whose biological activity can be controlled by reversible covalent modification, but it is known that as many as one-third of all proteins in mammalian cells contains covalently bound phosphate,[14] so the total number must be at least an order of magnitude higher; some examples of these are listed in Table 6.2. In the vast majority of cases phosphorylation is the form of covalent modification that occurs, but adenylylation and ADP-ribosylation are also used to regulate the activity of certain enzymes (Table 6.2). In addition, other types of protein modification include methylation, acetylation, myristoylation, palmitoylation, farnesylation, geranylgeranylation, tyrosinolation, and sulphation, but these have not been shown to have an important role in regulating enzyme activity.[13,18]

Reversible covalent modification by a phosphorylation–dephosphorylation cycle occurs at either a serine, threonine, or tyrosine side-chain. The enzymes catalysing phosphorylation and dephosphorylation are known as protein kinases and protein phosphatases, respectively. The importance of protein kinases in regulating cellular activities is evident from the large number of protein kinase genes that present in eukaryote genomes. It is estimated that humans have as many as 2000 protein kinase genes and may have in the region of 1000 protein phosphatase genes.[19] Many of these protein kinases act on a number of different protein substrates, and so it would be generally unsatisfactory to classify them on

A lot of enzymes draw E from ATP.

Table 6.2 Some enzymes and proteins whose activity can be controlled by reversible covalent modification

Enzyme or protein	Modification	Biological function
Glycogen phosphorylase	Phosphorylation	Glycogen metabolism (see Section 6.4.2)
Glycogen synthase	Phosphorylation	
Phosphorylase kinase	Phosphorylation	
Phosphatase inhibitor protein	Phosphorylation	
Fructose 2,6-bisphosphatase/6-phosphofructo-2-kinase	Phosphorylation	Regulation of glycolysis (see Section 6.4.1.1)
Pyruvate dehydrogenase complex (mammalian)	Phosphorylation	Entry of pyruvate into tricarboxylic-acid cycle (see chapter 7, Section 7.7.5)
Branched chain 2-oxoacid dehydrogenase complex	Phosphorylation	Breakdown of leucine, isoleucine, and valine
Acetyl-CoA carboxylase	Phosphorylation	Synthesis of fatty acids
Troponin-1	Phosphorylation	Muscular contraction
Myosin light chain	Phosphorylation	
cdc kinase	Phosphorylation	Regulation of mitosis[15] (see Chapter 9, Section 9.7.4)
Glutamine synthetase (*E. coli*)	Adenylylation	Glutamine acts as N donor in a wide range of biosynthetic reactions
Glutamine synthetase (mammalian)	ADP-ribosylation	
RNA polymerase (*E. coli*)	ADP-ribosylation	On infection by T4 phage, an Arg side chain in the α subunit becomes modified. This shuts off transcription of the host genes
G-protein	ADP-ribosylation	G-proteins can act as transducing agents relaying the effect of hormone binding to the activation of adenylate cyclase (see Chapter 8, Section 8.4.5)
Nitrogenase	ADP-ribosylation	Regulation in response to ammonia in nitrogen-fixing bacteria[16]
Fructose 2,6-bisphosphatase/6-phosphofructo-2-kinase	ADP-ribosylation	Possible regulation of glycolysis[17] (see Section 6.4.1.2)

For further details, see reference 13 and 15–17.

the basis of the reaction catalysed (see Chapter, 1 Section 1.5.1), since they all catalyse the reaction:

$$\text{Protein} + \text{ATP} \rightleftharpoons \text{phosphoprotein} + \text{ADP}.$$

They can be initially divided on the basis of the amino-acid residue phosphorylated: (i) serine/threonine or (ii) tyrosine. There are also a few kinases that phosphorylate histidine, lysine, arginine, glutamate, or aspartate side-chains. Phosphorylation of proteins at either serine or threonine residues is generally associated with metabolic control mechanisms, whereas phosphorylation of tyrosine residues is associated with cell growth and differentiation. A further subdivision is made on the basis of the nature of the intracellular regulator, since many kinases, although not all, are controlled by a signal transduction system. A simplified classification of the main groups is given in Table 6.3. For a more comprehensive classification see reference 20.

The best documented example of covalent modification by adenylylation is that of glutamine synthase. In this case, a tyrosine side-chain (Tyr 397) in each of the 12 subunits of the enzyme can be adenylylated.[22] When fully adenylylated it shows very low catalytic activity.

$$\text{E}-\text{CH}_2-\langle\ \rangle-\text{OH} + \text{ATP} \xrightarrow{\text{Mg}^{2+}} \text{E}-\text{CH}_2-\langle\ \rangle-\text{O}-\text{AMP} + \text{Pyrophosphate}$$

Table 6.3 Classification of protein kinases

Enzyme family	Amino-acid acceptor	Regulator	Processes regulated
cAMP-dependent PK (cAPK)	Ser/Thr	cAMP	glycolysis, gluconeogenesis, triglyceride and cholestrol metabolism, catecholamine metabolism
cGMP-dependent PK (cGPK)	Ser/Thr	cGMP	similar to cAPK but restricted distribution, e g. smooth muscle
protein kinase C	Ser/Thr	diacylglycerol Ca^{2+}	exact role unknown, control of intracellular $[Ca^{2+}]$, phosphorylation of EGF receptor
Ca^{2+}/calmodulin PK	Ser/Thr	Ca^{2+}	phosphodiesterase, Ca^{2+}/Mg^{2+} ATPase, myosin LC kinase, phosphorylase kinase
Cyclin-dependent kinase	Ser/Thr	cyclins	cell cycle regulation by phosphorylation of lamins, vimentin, histone H1, and others
Mitogen activated protein kinases (MAP kinases)	Ser/Thr	growth factors cytokines, pheromones	translocation to nucleus to activate transcriptional factors
Receptor tyrosine kinases	Tyr	growth factors	activation of enzymes including phosphatidylinositol 3-kinase, GTPase activation protein, MAP kinases
Cytosolic tyrosine kinases	Tyr	cytokines	transcriptional activation

For further details see references 20 and 21. (PK = protein kinase)

The enzyme nitrogenase from the nitrogen-fixing bacteria *Rhodospirillum rubrum* is regulated by covalent modification. Arginine 101 residue becomes ADP-ribosylated by NAD^+ in response to increasing concentrations of ammonium ions and this effects a reduction in its catalytic activity:[16]

$$\text{Enz-arg} + NAD^+ \rightleftharpoons \text{Enz-arg-ADP} + \text{nicotinamide.}$$

The roles of a number of the kinases and phosphatase enzymes and proteins will be discussed in connection with the regulation of glycogen metabolism (Section 6.4.2). The significance of reversible covalent modification of enzymes has been discussed in a number of reviews.[23-26] Two of the more noteworthy aspects of these systems are the following.

1. Because the conversion of an enzyme from one form to another is enzyme catalysed, there can be a *rapid change* in the amount of active enzyme present, and a *large amplification* of an initial signal. The degrees of amplification that can be achieved in various circumstances are discussed in references 13 and 24.

2. Reversible modification permits much more controlled responses to different metabolic circumstances than is the case with irreversible covalent modification. In the former case, a system can be viewed as being 'poised' for response with continual activation and inactivation of the enzyme.

For this type of system to be effective the enzymes catalysing the opposing steps must be nearly in balance. This has been found to be the case with Ser/Thr protein kinases and Ser/Thr protein phosphatases where the intrinsic catalytic activities of the two groups are approximately the same.[19] It is not the case with the protein tyrosine kinases and phosphatases where the activity of the latter is often as much as two orders of magnitude greater.[27] This explains not only the very low level of tyrosine phosphorylation of cellular proteins (usually not more than 0.05 per cent of the available amino acid residues) but also the transient nature of responses to protein tyrosine kinase activation.

Such 'poising' obviously consumes a certain amount of energy, usually in the form of ATP, but some calculations suggest that this represents only a small fraction of the total cellular turnover of ATP.[13,28] However, this conclusion has been challenged by Goldbeter and Koshland,[29] who maintained that a significant fraction of the total energy expenditure is required for the large number of reactions that involve reversible covalent modification of proteins. Whatever the actual fraction, however, it is clear that this is the price that the organism pays in order to support its sophisticated control mechanism. Under the influence of an appropriate stimulus, a system can rapidly be activated to produce almost exclusively the active form of the required enzyme. When the stimulus is removed, the system can be converted back to its resting state (almost exclusively the inactive form of the enzyme) and it is also possible to generate a whole range of intermediate levels of response. By contrast, in the cases of irreversible covalent modification, such as blood coagulation and complement activation, there is a rapid amplified response to some emergency (e.g. injury or infection) but, after use, the whole cascade systems must be replenished, since the steps in the cascades operate in one direction only.

6.2.2 Control of activity by ligand-induced conformational changes in enzymes

We shall first describe (Section 6.2.2.1) some of the observations that led to the formulation of ideas about the role of conformational changes in controlling the activities of enzymes, and then proceed (Section 6.2.2.2) to outline some of the theoretical models that have been used to describe such control systems. A final section (6.2.2.3) will discuss the biological significance of this type of control.

6.2.2.1 Experimental observations on regulatory enzymes*

Many of the ideas about the role of conformational changes in controlling the activities of enzymes were developed as a result of work on biosynthetic pathways in microorganisms. In the mid-1950s, for example, it was found that threonine dehydratase, the first enzyme in the pathway of isoleucine biosynthesis in *E. coli*, was strongly inhibited by the end-product, isoleucine, which bears only a limited structural resemblance to the substrate or the product of the reaction[30] (Fig. 6.3). Only the first enzyme in the pathway shows this inhibition.

* In one sense most, if not all, enzymes are regulatory, since their activities can be changed to some extent by, for example, changes in the concentration of substrates, etc. However, in this context, regulatory enzymes are those whose activity is controlled by conformational changes that accompany the binding of ligands. This definition includes those enzymes that exhibit sigmoid kinetics (see Fig. 6.4).

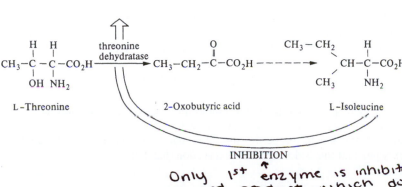

Fig. 6.3 Inhibition of threonine dehydratase, the first enzyme in the pathway of isoleucine biosynthesis in *E. coli*.

A more striking example was provided by aspartate carbamoyltransferase, the enzyme catalysing the first step in pyrimidine biosynthesis in *E. coli*, which is inhibited by the end-products of the pathway, especially CTP.[31] It was also found that the purine nucleotide ATP could activate aspartate carbamoyltransferase, thus providing a mechanism for achieving a balance between the production of pyrimidine and purine nucleotides, which is required for the synthesis of nucleic acids.

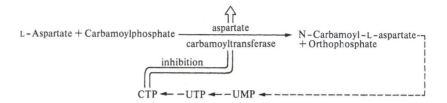

By the early 1960s a number of examples had been described of this type of regulation, in which only the first enzyme in a pathway was subject to *feedback inhibition* by the end-product of the pathway. In an incisive review[32] of these systems, Monod, Changeux, and Jacob noted several common features.

1. The ligands that were involved in regulation of enzyme activity (referred to as *regulator molecules* or *effectors*) were usually structurally distinct from the substrates or products of the relevant enzyme-catalysed reactions. It was therefore unlikely that an effector would bind at the active site of an enzyme.

2. Many of the enzymes whose activity was controlled in this way did not show the normal type of kinetic behaviour, i.e. the plots of velocity against substrate concentration were of a sigmoidal (Fig. 6.4) rather than a hyperbolic (see Chapter 4, Fig. 4.1) nature.

3. It was often possible to distinguish between the binding of effector and the binding of substrate to a regulatory enzyme. Various physical or chemical treatments could lead to desensitization of the enzyme, i.e. loss of response to regulator molecules with no loss of catalytic activity. In the case of aspartate carbamoyltransferase, mild heat treatment yielded an active enzyme that was not inhibited by CTP. The desensitized enzyme showed normal, hyperbolic kinetic behaviour (Fig. 6.5).

4. In general, the regulatory enzymes were found to be oligomeric, i.e. they consisted of multiple subunits held together by non-covalent forces as described in Chapter 3, Section 3.5.

Monod, Changeux, and Jacob[32] proposed that the regulator molecule (or effector) bound to a site distinct from the active site, so that the relationship between regulator molecule and substrate was an indirect or *allosteric* one (the term allosteric is derived from the Greek meaning 'different solid'). The binding of the regulator molecule to the regulator site induces a reversible conformational change in the enzyme that causes an alteration of the structure of the active site and a consequent change in the kinetic properties of the enzyme (Fig. 6.6). Monod, Wyman and Changeux proposed a model[33] to explain the relationship between the kinetic properties and the conformational changes in allosteric enzymes, and this is described in next section (6.2.2.2) .

(a)

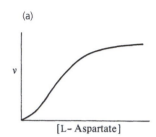

(b)

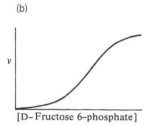

Fig. 6.4 Sigmoidal plots of velocity against substrate concentration for: (a) aspartate carbamoyltransferase from *E. coli*; (b) 6-phosphofructokinase from rabbit skeletal muscle (assayed at high concentrations of ATP; see Fig. 6.22).

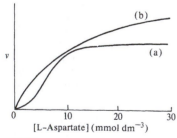

Fig. 6.5 Kinetic behaviour of aspartate carbamoyltransferase from *E. coli*. Curve (a) was obtained with enzyme assayed in the absence of CTP. Curve (b) was obtained with desensitized enzyme assayed in the absence or presence of CTP (0.2 mmol dm^{-3}).

6.2.2.2 Theoretical models which account for the behaviour of regulatory enzymes

A number of theoretical models have been proposed to try to account quantitatively for the behaviour of regulatory enzymes. In these models the major emphasis is placed on the interactions between the active sites on different subunits that can lead to sigmoidal kinetic plots of the type shown in Fig. 6.4. The models can then be extended to include the effects of regulator molecules on the kinetics. We shall describe four different models, those due to (i) Hill; (ii) Adair; (iii) Monod, Wyman, and Changeux; and (iv) Koshland, Némethy, and Filmer. Models (i) and (ii) are purely mathematical and make little reference to the behaviour of the enzyme or protein concerned, whereas models (iii) and (iv) make substantial reference to the behaviour of the enzyme or protein. For a more detailed account of these and other models see references 34 and 35.

Before considering these models it should be remembered that deviations from hyperbolic saturation curves had been detected in the early 1900s, when a study was made of the binding of oxygen to haemoglobin (Hb) and to myoglobin (Mb) (Fig. 6.7). It was later shown that myoglobin consisted of a single subunit with one oxygen-binding site, whereas haemoglobin consisted of four subunits each with an oxygen-binding site. The saturation curve for haemoglobin indicates that the binding of oxygen is *cooperative*, i.e. that binding of the first molecule of oxygen facilitates the binding of subsequent molecules.

The hyperbolic saturation curve for myoglobin (note the analogy with the hyperbolic kinetic plot shown in Chapter 4, Fig. 4.1) can be accounted for as follows.

Consider the binding of a ligand, A, to a protein, P:

$$P + A \rightleftharpoons PA.$$

The dissociation constant, K, is given by

$$K = \frac{[P][A]}{[PA]},$$

$$\therefore \quad [PA] = \frac{[P][A]}{K}. \tag{6.1}$$

The fractional saturation, Y (i.e. the concentration of sites on the protein that are actually occupied, divided by the total concentration of such sites), is given by

$$Y = \frac{[PA]}{[P]+[PA]}.$$

Substituting from (6.1),

$$Y = \frac{[A]}{K+[A]}. \tag{6.2}$$

The plot of Y against [A] is shown in Fig. 6.8.

(i) The Hill equation

In 1910, Hill[36] suggested the use of an equation to account for oxygen binding to haemoglobin that resembles (6.2) but has the concentration of free ligand, [A], raised to the power h:

$$Y = \frac{[A]^h}{K+[A]^h}. \tag{6.3}$$

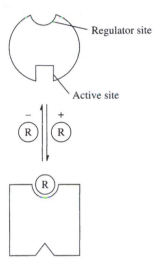

Fig. 6.6 Alteration of the structure of the active site of an enzyme by reversible binding of a regulator molecule R to a regulator site.

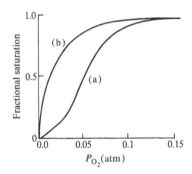

Fig. 6.7 Oxygen saturation curves for (a) haemoglobin and (b) myoglobin.

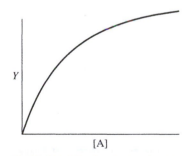

Fig. 6.8 Plot of fractional saturation, Y, against concentration of free ligand, [A], according to eqn (6.2).

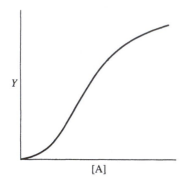

Fig. 6.9 Plot of fractional saturation, Y, against concentration of free ligand, [A], according to eqn (6.3) when $h = 3$.

An equation of this type can give rise to a sigmoidal plot of Y against [A] (see Fig. 6.9 for the case where $h = 3$), similar to the observed behaviour of the haemoglobin–oxygen system (Fig. 6.7). This suggests that there may be some physical basis for the assumption underlying the derivation of the equation.

An equation of the type given in (6.3) can be derived by assuming that a protein, P, has n binding sites for a ligand, A, and that the binding of A is *extremely cooperative*. This means that, as soon as one site on P is occupied, the remaining sites are immediately occupied, so that essentially we can consider an equilibrium between only two species P and PA_n, with no significant contribution from intermediate complexes (e.g. PA_{n-1}):

$$P + nA \rightleftharpoons PA_n,$$

$$K = \frac{[P][A]^n}{[PA_n]}. \tag{6.4}$$

The fractional saturation, Y, is given by

$$Y = \frac{[PA_n]}{[P] + [PA_n]}.$$

Substituting from (6.4), this becomes

$$Y = \frac{[A]^n}{K + [A]^n} \quad \text{(which is of the same form as eqn (6.3))}$$

In eqn (6.3), h is known as the Hill coefficient. The value of h can be derived from experimental measurements of Y as a function of [A]. Rearranging eqn (6.3):

$$\frac{Y}{1-Y} = \frac{[A]^h}{K},$$

$$\therefore \quad \log\left[\frac{Y}{1-Y}\right] = h \log[A] - \log K. \tag{6.5}$$

Thus a plot of $\log[Y/(1 - Y)]$ against $\log$ [A] is linear with a slope equal to h (Fig. 6.10).

For the binding of oxygen to haemoglobin, a straight line is obtained in this type of plot over most of the range of saturation of binding sites (at very low and very high degrees of saturation, deviations from linearity are observed because of a breakdown of the assumption underlying the derivation of eqn (6.3)). From the slope of the linear part of the plot, a value of $h = 2.6$ can be derived. The value of h can be taken as a measure of the strength of the cooperativity in ligand binding. If h is found to be equal to the number of binding sites, n (four in the case of haemoglobin), this would imply extreme cooperativity in ligand binding, i.e. that intermediate species of the type PA_{n-1} make no significant contribution. In most cases, h is found to be less than the number of binding sites and is not necessarily integral. A value of h greater than 1 but less than the number of binding sites implies that the binding of ligand is cooperative (but not extremely cooperative). When h is equal to 1, there is no cooperativity in ligand binding and the saturation curve is hyperbolic (see eqn (6.2) and Fig. 6.8). If h is found to be less than 1, the binding could be described as *negatively cooperative*, i.e. the

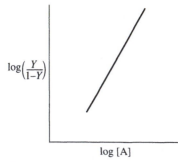

Fig. 6.10 Plot of $\log[Y/(1 - Y)]$ against $\log$[A] according to eqn (6.5). The slope of the line gives the value of h, the Hill coefficient.

binding of the first ligand molecule discourages the binding of subsequent molecules of ligand.

An equation analogous to the Hill equation can be used to analyse enzyme kinetic data. In this case a plot of $\log[v/(V_{max} - v)]$ against $\log[\text{substrate}]$ gives a straight line, the slope of which equals h.

In summary, it can be stated that although the assumption underlying the derivation of eqn (6.3), i.e. that there is extreme cooperativity in ligand binding, is highly questionable in most cases, because intermediate complexes do make a significant contribution, the Hill coefficient, h, is nevertheless widely used as an empirical measure of the strength of cooperativity of ligand binding. For example, h for the enzyme aspartate carbamoyltransferase (see Section 6.2.2.1 and Fig. 6.12) in the absence of effectors is 1.8–2.0, but in the presence of the allosteric inhibitor CTP it is 2.3, and in the presence of the activator ATP it is 1.4.[35] The limitations of the Hill equations are further discussed in reference 37.

(ii) The Adair equation

In 1925, Adair[38] suggested that a more realistic description of the binding of oxygen to haemoglobin would include the various intermediate complexes. There would therefore be four equilibria, each with a characteristic dissociation constant:

$$Hb + O_2 \rightleftharpoons HbO_2 \qquad K_1$$
$$HbO_2 + O_2 \rightleftharpoons Hb(O_2)_2 \qquad K_2$$
$$Hb(O_2)_2 + O_2 \rightleftharpoons Hb(O_2)_3 \qquad K_3$$
$$Hb(O_2)_3 + O_2 \rightleftharpoons Hb(O_2)_4 \qquad K_4$$

The fractional saturation of sites (Y) can be expressed in terms of the concentration of free ligand ([A]) and the four dissociation constants (eqn (6.6)). For a derivation of this equation, see Appendix 6.1.

$$Y = \cfrac{\cfrac{[A]}{K_1} + \cfrac{2[A]^2}{K_1 K_2} + \cfrac{3[A]^3}{K_1 K_2 K_3} + \cfrac{4[A]^4}{K_1 K_2 K_3 K_4}}{4\left[1 + \cfrac{[A]}{K_1} + \cfrac{[A]^2}{K_1 K_2} + \cfrac{[A]^3}{K_1 K_2 K_3} + \cfrac{[A]^4}{K_1 K_2 K_3 K_4}\right]}. \qquad (6.6)$$

From the experimental values of Y as a function of [A], it is possible to deduce the 'best fit' values of the four dissociation constants by curve-fitting procedures. The relative values of these dissociation constants (strictly speaking, of the *intrinsic* dissociation constants)* give a clue to the cooperativity of ligand binding. For a further discussion of the Adair equation, references 39 and 40 should be consulted.

(iii) The model of Monod, Wyman, and Changeux

This model[33] was proposed to account for the behaviour of allosteric proteins and enzymes, i.e. those that showed allosteric, or indirect, interactions between ligand-binding sites.[†]

Most allosteric proteins are oligomeric, that is they consist of more than one subunit. The model makes four principal statements about the symmetry and the conformation of an allosteric protein.

* In considering the binding of ligand to more than one site on a protein, statistical factors have to be taken into account. The dissociation constants (K_1, etc.) are related to intrinsic dissociation constants (K'_1, etc.) for a protein containing four sites, as follows:
$K_1 = K'_1/4, \quad K_2 = 2K'_2/3,$
$K_3 = 3K'_3/2, \quad K_4 = 4K'_4$
(see Appendix 6.1).
If the *intrinsic* dissociation constants followed a sequence $K'_1 > K'_2 > K'_3 > K'_4$, the binding would be said to show (positive) cooperativity at each stage.

† There are the two classes of allosteric effects.[33] *Homotropic* effects are interactions between identical ligands, e.g. between substrate molecules bound to different subunits in an enzyme. *Heterotropic* effects are interactions between different ligands, e.g. between substrate and regulator molecule.

(a) The subunits occupy equivalent positions, so that within the oligomer, there is at least one axis of symmetry. (It now appears that the dimeric enzyme hexokinase from yeast may be an exception to this statemen:.[41])

(b) The conformation of each subunit depends on its interaction with the other subunits.

(c) There are two conformational states available to the oligomer. These are designated R (relaxed) and T (tense) and they differ in affinity for a given ligand.

(d) When the conformation of the protein changes from R to T (or *vice versa*), the symmetry of the oligomer is conserved.

If the conformations of the subunits in the T and R states are denoted by circles and squares, respectively, then statements (c) and (d) would mean that for a dimeric protein there is an equilibrium:*

T   R

* The equilibrium between the R and T states of a dimeric protein.

There would be no 'hybrid' species ($\square\bigcirc$), since in such species the symmetry of the oligomer would be lost. In other words, all the subunits change conformation in a 'concerted' fashion—this gives rise to the alternative name 'concerted' for this model. It is also sometimes known as the symmetry model.[34]

The various equilibria involved in binding of a ligand, A, to a dimeric protein are illustrated in Fig. 6.11.

Two parameters ($\bar{Y}$ and $\bar{R}$) are used to describe the state of the allosteric system.[33] $\bar{Y}$ is the fractional saturation of ligand sites on the protein, and $\bar{R}$ is the fraction of protein that is in the R state. The values of both $\bar{Y}$ and $\bar{R}$ are in the range from 0 to 1.

Using the equilibrium expressions, $\bar{Y}$ and $\bar{R}$ were evaluated in terms of the number of ligand-binding sites, n, the concentration of free ligand, α,[†] and two constants L and c that could be adjusted to give the best fit to the experimental data. L is the ratio of the concentration of protein in the T state to that in the R state in the absence of ligand, i.e. $L = [T_0]/[R_0]$ in Fig. 6.11, and c is the ratio of the intrinsic dissociation constants for the ligand from the R and T states, respectively, i.e. $c = K_R/K_T$ in Fig. 6.11.

The expressions derived[33] were

[†] For convenience, the concentration of free ligand [A] is divided by the dissociation constant of the ligand from the R state, K_R, to give a dimensionless ratio, α.

$$\bar{Y} = \frac{L c \alpha (1 + c \alpha)^{n-1} + \alpha (1 + \alpha)^{n-1}}{L(1 + c \alpha)^n + (1 + \alpha)^n}, \tag{6.7}$$

$$\bar{R} = \frac{(1 + \alpha)^n}{L(1 + c \alpha)^n + (1 + \alpha)^n}. \tag{6.8}$$

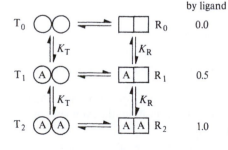

Fractional saturation by ligand

T_0 ... R_0 0.0

T_1 ... R_1 0.5

T_2 ... R_2 1.0

Fig. 6.11 Equilibria involved in binding of a ligand, A, to a dimeric protein according to the model of Monod, Wyman, and Changeux.[33] K_T and K_R are the intrinsic dissociation constants for ligand from the T and R states, respectively. Circles and squares represent the conformations of subunits in the T and R states, respectively.

It is important to appreciate that the successive intrinsic dissociation constants for the T state are the same ($= K_T$) and that the successive intrinsic dissociation constants for the R state are the same ($= K_R$). *According to the model of Monod, Wyman, and Changeux, cooperativity in ligand binding arises because the protein is initially predominantly in one state (say T), but the ligand binds preferentially to the other state (R). As more ligand is added to the system, the protein is gradually swung over into the tighter-binding R state.* For the haemoglobin–oxygen system, the Monod, Wyman, and Changeux model gives a good fit to the experimental saturation curve with $L = 9050$ and $c = 0.014$,[33] i.e. the protein is initially more than 99.9 per cent in the T state, but oxygen binds over 70 times more tightly to the R state. An analysis of eqn (6.7) shows that cooperativity in ligand binding will be more pronounced when L is large and c is small.

Up to this point we have considered the Monod, Wyman, and Changeux model in terms of ligand saturation functions, which are readily applicable to a binding protein such as haemoglobin. However, in order to analyse kinetic data for an enzyme, where the observed velocity reflects both a Michaelis constant and a maximum velocity, some modifications to the basic equations are necessary. These modifications, in which dissociation constants are replaced by Michaelis constants and provision is made for the case that the two forms of the enzyme have different maximum velocities, are described by Dalziel.[42]

Effectors of allosteric enzymes are considered to exert their effects on the R $\rightleftharpoons$ T equilibrium in one of two ways. In a K system, both the substrate and the effector have different affinities for the T and R states of the enzyme. The presence of the effector modifies the affinity of the enzyme for the substrate, i.e. it affects K_m. Assuming that the substrate binds more tightly to the R state, it would follow that an effector that binds more tightly to the T state would displace the R $\rightleftharpoons$ T equilibrium towards the T state, thereby raising the K_m and decreasing the velocity at a given concentration of substrate. By contrast, an effector that binds more tightly to the R state would increase the velocity at a given concentration of substrate. Using these arguments, Monod, Wyman, and Changeux[33] were able to account for the behaviour of a K system such as aspartate carbamoyltransferase in the presence of activators (e.g. ATP) and inhibitors (e.g. CTP) (Fig. 6.12).

In a V system, the substrate has the same affinity for the R and T states of the enzyme. However, the two states differ in their catalytic activities, so that the

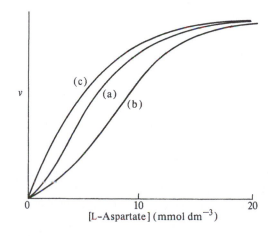

Fig. 6.12 Kinetic behaviour of aspartate carbamoyltransferase from *E. coli.* Curve (a) was obtained in the absence of effectors, curve (b) in the presence of 0.5 mmol dm^{-3} CTP, and curve (c) in the presence of 2 mmol dm^{-3} ATP.

effector, by displacing the R $\rightleftharpoons$ T equilibrium, affects V_{max} rather than K_m. An example of a V system is phosphorylase b, which is activated by AMP.

(iv) The model of Koshland, Némethy, and Filmer

This model[43] which accounts for the ligand-binding properties of oligomeric proteins and enzymes can be seen as an extension of Koshland's 'induced-fit' hypothesis (see Chapter 5, Section 5.6) and as an attempt to define relationships between the various constants in the Adair equation (eqn (6.6)). In the 'induced-fit' hypothesis, it is postulated that the binding of substrate to an enzyme induces the appropriate conformation of the enzyme to allow catalysis to occur. Similarly, the binding of a ligand to one subunit in an oligomeric protein changes the conformation of that subunit, thereby altering the interaction of that subunit with its neighbours. In diagrammatic terms the binding of a ligand A to a dimeric protein can be represented by the following equilibria:

$$\text{OO} + A \rightleftharpoons \text{O}\boxed{A}$$
$$\text{O}\boxed{A} + A \rightleftharpoons \boxed{A}\boxed{A}$$

Features of the model are as follows.

(a) In the absence of ligand, the protein exists in one conformational state rather than as an equilibrium mixture of two conformational states as in the model of Monod, Wyman, and Changeux.

(b) The conformational change in the protein is *sequential*, i.e. the subunits change conformation sequentially as ligands bind, rather than in a concerted fashion as in the model of Monod, Wyman, and Changeux.

(c) The interactions between the subunits can be of a positive or a negative type, so that the binding of the second (and later) molecule of ligand can be more favourable (positive cooperativity) or less favourable (negative cooperativity) than the binding of the first molecule. By contrast, the model of Monod, Wyman, and Changeux allows only positive cooperativity in ligand binding, since the binding of ligand brings about a concerted transition of all the subunits to the form that has higher affinity for the ligand.

An expression for the fractional saturation of sites by ligand ($\bar{Y}$) is derived by considering the various equilibria with their characteristic dissociation constants. Each dissociation constant is considered[43] to have three components:

(a) a dissociation constant characterizing the binding of ligand by the more favoured subunit conformation ($\square$ in this scheme);

(b) an equilibrium constant for the conformational change of the subunit (i.e. $\bigcirc \rightleftharpoons \square$). This constant will be related to the difference in free energy between the two conformations;

(c) one or more interaction constants that depend on the degree of stability of the complex between the subunits relative to the stability of the isolated subunits. These constants depend on the geometrical arrangement of the oligomer, since this will determine the number and strength of the interactions between the subunits. For a tetrameric protein the possible arrangements are shown opposite; the form of the interaction constants would differ in each case.

Square

Linear

Tetrahedral

Assuming a particular arrangement of subunits in the oligomer, an equation can be derived for the $\bar{Y}$ in terms of the free ligand concentration and the various constants referred to above. The equations derived from the model of Koshland, Némethy, and Filmer are more complex than those derived from the model of Monod, Wyman, and Changeux and elaborate curve-fitting procedures are usually necessary to derive the values of the various constants that give the best fit to the experimental data. For detailed accounts of the model of Koshland, Némethy, and Filmer, references 34, 35, 40, 43, and 44 should be consulted.

(v) Relationships between the model of Monod, Wyman, and Changeux and the model of Koshland, Némethy, and Filmer

The concerted model of Monod, Wyman, and Changeux and the sequential model of Koshland, Némethy, and Filmer can be viewed as limiting cases of a more general model involving all possible conformational and liganded states of an oligomeric protein[45] (Fig. 6.13). These more general models have been analysed (see, for instance, references 46 and 47), but the resulting equations are extremely complex and for our purposes it is more useful to discuss only the limiting cases (i.e. the concerted and the sequential models).

In some cases one of the models clearly provides a better description of the behaviour of a particular protein or enzyme, but with others this is not so. There are at least three types of observation in which a distinction between the models can be made.

(a) Negative cooperativity in ligand binding. As mentioned in (iv) above, the concerted model cannot account for negative cooperativity in ligand binding, whereas the sequential model can account for either positive or negative cooperativity. Thus, the observation of negative cooperativity in ligand binding can be used to exclude the use of the concerted model in a description of that particular system. Of the known examples of negative cooperativity,[45] the best-investigated is that of the binding of NAD^+ to the tetrameric enzyme glyceraldehyde-3-phosphate dehydrogenase from rabbit muscle, where the successive dissociation constants have been determined as:

$$K_1 = 10^{-11} \text{ mol dm}^{-3}, \quad K_2 = 10^{-9} \text{ mol dm}^{-3}, \quad K_3 = 10^{-6} \text{ mol dm}^{-3},$$

$$K_4 = 2 \times 10^{-5} \text{ mol dm}^{-3}.$$

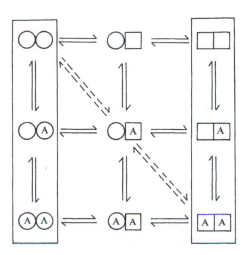

Fig. 6.13 A general model for the equilibria involved in the binding of a ligand, A, to a dimeric protein. The two vertical columns outlined show the species considered in the model of Monod, Wyman, and Changeux.[33] The dashed lines show the species considered in the model of Koshland, Némethy, and Filmer.[43]

It should be pointed out that it is very difficult to distinguish between negative cooperativity and non-identical binding sites. Thus, if in a dimeric protein there were one tight-binding site and one weak-binding site for a ligand, we would appear to observe negative cooperativity in the binding of the ligand. Presumably, to prove that the binding showed negative cooperativity we would need to establish that the sites were identical in the absence of ligands—a not inconsiderable task.

In its most extreme form, negative cooperativity can be manifested as 'half-of-the-sites reactivity', where, for instance, only half of the active sites of an oligomeric enzyme are able to catalyse the reaction at the same time.[48-50] For example, the dimeric tyrosyl tRNA ligase can bind only one molecule of tyrosine per molecule of enzyme.[49]

(b) *Measurements of conformational changes and ligand binding.* The conformational changes in enzymes and proteins that occur on binding of ligands can be monitored by a variety of means such as spectroscopic probes, changes in sedimentation coefficient, or changes in the reactivity of amino-acid side-chains.[51] Using such methods, the value of $\bar{R}$ (i.e. the fraction of protein in the R state) can be evaluated as increasing concentrations of ligand are added to the protein. The values of $\bar{Y}$ (i.e. the fraction of sites saturated) can be measured by direct-binding studies.[52] The shape of the plot of $\bar{R}$ against $\bar{Y}$ can be used to distinguish between the concerted and sequential models (Fig. 6.14).

In the sequential model of Koshland, Némethy, and Filmer the conformational change in a subunit occurs only when a ligand binds to that subunit, and hence we would expect the plot of $\bar{R}$ against $\bar{Y}$ to be linear* (see (a) in Fig. 6.14). However, in the concerted model of Monod, Wyman, and Changeux, we would expect a non-linear plot of $\bar{R}$ against $\bar{Y}$ (see, for example, (b) in Fig. 6.14) because of the concerted nature of the R $\rightleftharpoons$ T transition. By examining the experimental data for $\bar{R}$ and $\bar{Y}$, it has been possible to classify the binding of AMP to phosphorylase *a* as being of the concerted type[53] and the binding of NAD⁺ to glyceraldehyde-3-phosphate dehydrogenase from rabbit muscle as being of the sequential type.[54]

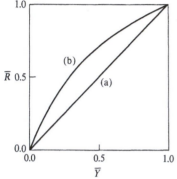

Fig. 6.14 Plots of $\bar{R}$ (the fraction of protein in the R state) against $\bar{Y}$ (the fraction of sites saturated) for a protein that conforms to: (a) the sequential model of Koshland, Némethy, and Filmer; and (b) the concerted model of Monod, Wyman, and Changeux.

* It should be noted that a linear plot would only be expected in the simplest form of the sequential model, in which the ligand-induced conformational change is restricted to one subunit. If the conformational change extended across a subunit interface, deviations from linearity would be expected.

(c) *Demonstration of isomerization reactions.* The binding of NAD⁺ to glyceraldehyde-3-phosphate dehydrogenase from yeast has been studied in great detail by stopped-flow[55] and temperature-jump[56] methods (see Chapter 4, Section 4.4). The results were found to be in accord with predictions of the concerted model, but not with those of the sequential model, since the latter model cannot account for the observed isomerization reaction that occurs in the absence of ligand. (This isomerization reaction is denoted by the $T_0 \rightleftharpoons R_0$ equilibrium in the concerted model, see Fig. 6.11.) Other evidence for the existence of the $T_0 \rightleftharpoons R_0$ equilibrium comes from the transmembrane protein, nicotinic acetylcholine receptor. This is a ligand-gated cation channel, which in the presence of the ligand, acetyl choline, allows Na⁺ and K⁺ to pass through the membrane. In this case it is possible to study a single ion receptor molecule by a technique known as patch clamp, and detect the membrane current when cations pass through. It is found that in the presence of acetyl choline the gate is open, with brief flickers of closure, but in the absence of acetyl choline the gate is closed with brief flickers in the open state. This is interpreted as evidence of the $T_0 \rightleftharpoons R_0$ equilibrium.[57,58]

(vi) *Usefulness of the models*

The various models described in (i)–(iv) can give only a limited insight into the nature of subunit interactions and cooperativity in ligand binding in regulatory enzymes. It is necessary to have X-ray crystallographic data to understand the structural basis for these properties. Structural information is now available on ligand-induced conformational changes in a number of enzymes, including phosphorylase a,[12] lactate dehydrogenase (see Chapter 5, Section 5.5.4), glyceraldehyde-3-phosphate dehydrogenase,[59] fructose 1,6-bisphospatase,[60] and aspartate carbamoyltransferase,[61,62] but probably the best understood example is the haemoglobin–oxygen system. The elegant studies of Perutz[63,64] have shown how the binding of oxygen at the haem group on one subunit of deoxyhaemoglobin leads to a small movement (≈ 0.07 nm) of the iron into the plane of the haem; this movement triggers a series of structural alterations in the protein that can be sensed at the other haem groups in the molecule that are between 2.5 and 3.7 nm distant. The structural differences between the R and T forms have been solved by X-ray crystallography for aspartate carbamoyltransferase, phosphofructokinase, and fructose 1,6-bisphosphatase. Aspartate carbamoyltransferase crystallizes in the T form in the presence of the allosteric inhibitor CTP and in the R form in the presence of the bisubstrate analogue N-phosphono-L-aspartate.[61,62] Phosphofructokinase crystallizes in the T state in the presence of the inhibitor PEP, and in the R state in the presence of the activator MgADP.[65] Fructose 1,6-bisphosphatase crystallizes in the T form in the presence of fructose-6-phosphate, AMP, and Mg^{2+}, and in the presence of AMP alone, and it crystallizes in the R form in the presence of either fructose-6-phosphate, fructose 2,6-bisphosphate, or fructose 1,6-bisphosphate.[60]

The concerted model of Monod, Wyman, and Changeux and the sequential model of Koshland, Némethy, and Filmer have provided a relatively simple conceptual basis for understanding the cooperative binding of ligands to proteins. A number of allosteric proteins appear to fit the concerted model reasonably well, e.g. glycogen phosphorylase, pyruvate kinase, fructose bisphosphatase, and aspartate carbamoyltransferase.[35] The models often act as a good starting framework, but can only account for some of the properties of these systems. For example, phosphofructokinase partly fits the concerted model, but other phenomena such as association–dissociation and slow conformational changes are also considered to be involved.[35,66] Fluorescence spectroscopic methods also suggest that the free enzyme has a different conformation from either of those with the activator MgATP bound or the inhibitor PEP bound.[67] It also appears that although haemoglobin has four subunits ($\alpha_2\beta_2$) it only undergoes the T $\rightarrow$ R switch after each $\alpha\beta$ dimer has one molecule of ligand bound.[68]

There are some enzymes showing cooperative kinetics which have been explained by other factors. For example, glucokinase (hexokinase IV) shows sigmoid kinetics with respect to glucose ($h = 1.7$); it is a monomer thought to exist in two different conformations which show a slow conformational change. The contribution that the two enzyme conformations make to the overall rate of catalysis depends on the glucose concentration.[34,69] Similar processes may be involved with octopine dehydrogenase and fumarate hydratase.[35]

Some of the experimental approaches used to study regulatory enzymes are summarized in Table 6.4.

6.2.2.3 The significance of allosteric and cooperative behaviour in enzymes

Allosteric interactions between regulator molecules and substrates provide a versatile means of regulation, since they allow the activity of an enzyme to be controlled by changes in the concentrations of species other than the substrate and product of that enzyme-catalysed reaction. We have already mentioned this point

Table 6.4 Some methods of study of regulatory enzymes

Type of information sought	Methods of study
Is cooperative behaviour observed in enzyme kinetic studies?	Perform assays of enzyme activity over a wide range of substrate concentrations. Examine the kinetic plots (see Chapter 4, Fig. 4.3) for non-linearity. Use the Hill plot (Fig. 6.10) to give measure of cooperativity
Is cooperative behaviour observed in ligand binding?	Use direct-binding methods,[52] e.g. equilibrium dialysis or ultracentrifugation (these methods separate free ligand from bound ligand). Test for cooperativity using, for example, the Hill plot (Fig. 6.10) or the Scatchard plot (Fig. 6.15)
Do regulator molecules bind to a site distinct from the active site?	Test effects of regulator molecules on the kinetic plots. Are these effects consistent with binding to distinct sites? (See for example Chapter 4, Section 4.3.2.3) A distinction between substrate and effector sites can be made if the enzyme can be desensitized to the effects of the regulator molecule while retaining activity (Section 6.2.2.1)
Do conformational changes occur on binding of ligand?	X-ray crystallographic data give the most detailed information, but less-direct methods, involving spectroscopic probes or measurements of the reactivity of amino-acid side-chains, can also be useful[51]
Does the enzyme consist of subunits?	The number, type, and arrangement of subunits in an enzyme can be determined from determinations of M_r, symmetry, and cross-linking patterns. These methods are discussed in Chapter 3, Section 3.5

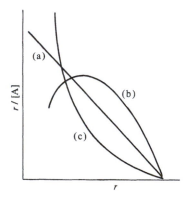

Fig. 6.15 Scatchard plots of ligand-binding data (see Appendix 6.1): r represents the number of molecules of A bound per molecule of P and [A] equals the concentration of free ligand. Case (a) corresponds to binding sites that are equivalent and independent; case (b) corresponds to positive cooperativity between ligand-binding sites; case (c) corresponds to negative cooperativity between ligand-binding sites (or possibly non-identical binding sites).

in connection with the feedback inhibition of early enzymes in biosynthetic pathways (Section 6.2.2.1) and shall return to it in Section 6.3.1.

Enzymes that show cooperativity in kinetics have significantly different responses to changes in the concentrations of substrates from those that exhibit hyperbolic kinetics. The difference in response can be expressed in quantitative terms by calculating the change in substrate concentration required to increase the velocity from 10 per cent of V_{max} to 90 per cent of V_{max} as a function of the Hill coefficient, h, in eqn (6.9), which is analogous to the Hill equation (eqn 6.3):

$$v = \frac{V_{max}[S]^h}{K_m + [S]^h} \qquad (6.9)$$

Value of h in eqn (6.9)	Required change in [S] to increase velocity from 10 per cent of V_{max} to 90 per cent of V_{max}
1	81-fold
2	9-fold
3	4.33-fold
4	3-fold
0.5	6561-fold

For an enzyme displaying hyperbolic kinetics ($h = 1$ in eqn (6.9)) an 81-fold change in substrate concentration is required; this change is reduced to just over fourfold for an enzyme showing cooperativity in kinetics with a Hill coefficient of 3. The enhanced response could well be important in enabling an enzyme to adjust to changes in conditions, but it should be stressed that large variations in the concentrations of substrates are unlikely to occur in cells. Additional means of amplifying the response of enzymes to changes in conditions are discussed in Section 6.3.2 in connection with the regulation of metabolic pathways.

It is also evident that an enzyme that shows negatively cooperative kinetics ($h < 1$ in eqn (6.9)) has a diminished response to changes in substrate concentration (over certain ranges of concentration) compared with an enzyme that displays hyperbolic kinetics. Negative cooperativity could thus serve to insulate an enzyme from the effects of changes in substrate concentration.[48] In at least one case (the binding of NAD^+ to glyceraldehyde-3-phosphate dehydrogenase from yeast) it appears that an enzyme can show a mixture of positive and negative cooperativity in ligand binding,[48,70] permitting further variations in the response of the enzyme to changes in substrate concentration.

The best understood example of cooperativity in ligand binding is the haemoglobin–oxygen system. In this case the physiological significance of the sigmoidal saturation curve (Fig. 6.7) is clear.[71] Haemoglobin is almost fully saturated with oxygen in the lungs, where the partial pressure of oxygen is about 0.13 atm, and yet unloads the oxygen in the tissue capillaries, where the partial pressure of oxygen is about 0.01 atm, much more completely than if the binding curve were hyperbolic with the four binding sites acting independently. The equilibrium between oxyhaemoglobin and deoxyhaemoglobin is also influenced by other factors, including the concentration of 2,3-bisphospho-D-glycerate, which acts as an allosteric regulator.[71] The significance of cooperative and allosteric behaviour in regulatory enzymes is not so well understood, since the concentrations of substrates and effectors *in vivo* are not known with the same degree of accuracy. However the significance can be appreciated in qualitative terms (see the example of 6-phosphofructokinase in Section 6.4.1.1).

A final word should be added about the rate at which ligand-induced conformational changes occur in regulatory enzymes. Several lines of evidence have indicated that proteins in solution can have highly flexible structures (see Chapter 3, Section 3.4.2). The time-scale of conformational changes in enzymes appears to be generally in the range 10^{-2} to 10^{-4} s,[72] which is roughly equivalent to the time-scale of catalytic events (see Chapter 4, Section 4.4). However, a number of cases of slow conformational changes in enzymes are now known.[40,72–74] An example is the inhibition of hexokinase isoenzyme II from ascites tumour cells by D-glucose 6-phosphate, which has a half-time of over 10 s.[75] The physiological function of slow conformational changes is not well understood; some possibilities have been discussed.[73,76]

6.3 Control of metabolic pathways

In Section 6.2 we were largely concerned with the more important ways in which the activities of single enzymes can be controlled. We now turn to considering the regulation of metabolic pathways that consist of sequences of enzyme-catalysed reactions. It is convenient to distinguish between two types of control that can be exerted on metabolism. *Intrinsic* (or *internal*) *control* refers to the regulation of enzyme activity by the concentrations of metabolites, e.g. substrates, products, end-products of pathways, or adenine nucleotides (see the example of 6-phosphofructokinase in Section 6.4.1.1). This type of control is predominant in unicellular organisms. In multicellular organisms, the metabolism within any given type of cell must be related to the requirements of the whole organism. In such cases an extra level of control, *extrinsic* (or *external*) *control* is brought about by means such as hormones or nervous stimulation (see Section 6.4.2). These signals generally act via *second messengers* to affect the activities of target enzymes.[77] Examples of second messengers include 3':5'-cyclic AMP (cAMP),[78] 3':5'-cyclic GMP (cGMP),[79] cADPribose,[80,81] Ca^{2+},[82] inositol 1,4,5-trisphosphate,[83] and diacylglycerol[83] (Fig. 6.16).

The first messengers interact with receptors on the extracellular surface of the target cells of tissues. This usually causes the activation of an enzyme associated with the cell membrane, which in turn catalyses the formation of a second messenger. A generalized scheme is shown in Fig. 6.17. Many of the second messengers activate protein kinases and they in turn catalyse phosphorylation of a number of target enzymes. The most thoroughly understood system is that in which adenylate cyclase becomes activated, and this leads to the activation of cAMP-dependent protein kinase. The interaction between hormone receptor, transducer, and adenylate cyclase is discussed in Chapter 8, Section 8.4.5, and the activation of cAMP-dependent protein kinase in Section 6.4.2.1. This type of mechanism, illustrated on Fig. 6.17, has a number of physiological advantages.

(i)   Each stage in the mechanism involves amplification of the signal, so that the circulating concentration of hormones is typically in the nanomolar range, whereas the change in intracellular concentration of cAMP may be in the millimolar range.

(ii) The targets cells are specified by the presence or absence of appropriate receptors.

(iii) The number of receptors per cell will determine the cells sensitivity to a particular hormone. For example, there are $\approx$ 200 000 insulin receptors per cell of adipocyte* and hepatocyte, whereas erythrocytes, which are relatively insensitive to insulin, have $\approx$ 40 per cell.[84]

(iv) The protein kinases activated by the different second messengers can phosphorylate several different enzymes. Each kinase, e.g. cAMP-dependent protein kinase, protein kinase C, calmodulin kinase, and multiprotein kinase, recognizes a specific consensus sequence on its substrate proteins.[85,86] For example, cAMP-dependent protein kinase catalyses the phosphorylation of phosphorylase kinase, glycogen synthase, pyruvate kinase, 6-phosphofructo-2-kinase, hormone-sensitive lipase, acetyl-CoA carboxylase, phenylalanine hydroxylase, tyrosine hydroxylase, cholesterol ester hydrolase, and cardiac phospholamban.[77] This arrangement ensures coordinated metabolic responses within each cell.

* Adipocytes are the principal cells that make up adipose tissue. They are roughly spherical in shape and $\approx$ 70% of the cell volume comprises lipid droplets. Hepatocytes are the principal cells of the liver, sometimes referred to as parenchymal cells, and they make up $\approx$ 70% of liver cells.

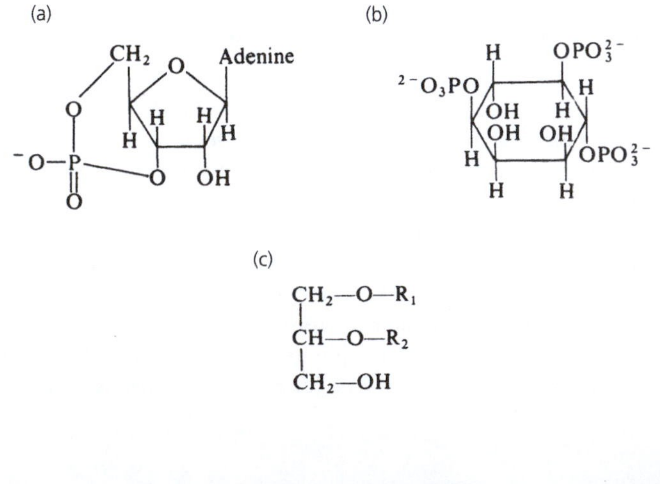

(a)   (b)

Fig. 6.16 Structures of some 'second messengers': (a) 3':5'-cyclic AMP (cAMP); (b) inositol 1,4,5-trisphosphate; (c) diacylglycerol, where R_1 and R_2 represent fatty-acid chains, commonly stearic and arachidonic acids, respectively. Details of the action of these and other messengers can be found in reference 77.

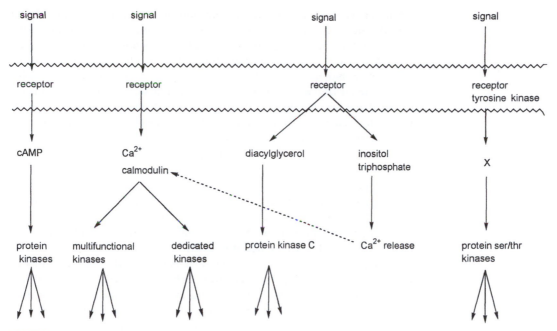

Fig. 6.17 Examples of different signal transduction systems used by eukaryote cells.

6.3.1 **General considerations**

We shall consider a metabolic pathway in which A is converted to F in a sequence of steps catalysed by enzymes E_1 to E_5:

$$\begin{array}{ccccc} E_1 & E_2 & E_3 & E_4 & E_5 \\ A \rightarrow B \rightarrow C \rightarrow D \rightarrow E \rightarrow F. \end{array}$$

It would be most economical to regulate the flux through the pathway, i.e. the rate at which F is formed from A, by regulating the catalytic activity of an early enzyme in the pathway such as E_1. This would then avoid the wasteful accumulation of intermediates. If E_1 were subject to feedback inhibition (Section 6.2.2.1) by F, then if F accumulates or is supplied exogenously, the production of F would be shut off. On the other hand, if the concentration of F falls, the production of F would be resumed.

It might also be expected that if a pathway were branched (see opposite), there would be a control point near the branch point. If an intermediate C is required for the synthesis of two products E and G, then it would seem reasonable that there should be some control mechanism to allow the rate of production of C to be altered to deal with various situations; for example, one in which E might be in excess but G not in excess. Such control could be achieved by a variety of interlocking controls near the branch point (for a review see reference 87). A good example is provided by the enzyme carbamoyl-phosphate synthase in *E. coli*. This enzyme catalyses the reaction:

Glutamine + 2ATP + bicarbonate → carbamoyl phosphate + 2ADP

+ orthophosphate + glutamate.

Carbamoyl phosphate is used in the production of pyrimidines and of arginine. As shown in Fig. 6.18, there is a variety of control mechanisms to ensure that an appropriate rate of production of carbamoyl phosphate is maintained in a variety of circumstances. Carbamoyl-phosphate synthase is subject to strong feedback inhibition by pyrimidine nucleotides but not by arginine. Thus, when pyrimidine nucleotides are in excess, the synthesis of carbamoyl phosphate is inhibited and may become too slow to support adequate synthesis of arginine. Under these circumstances, ornithine will accumulate and, at sufficiently high concentrations, will antagonize the effects of pyrimidine nucleotides and thus restore the activity of carbamoyl-phosphate synthase. When this happens the carbamoyl phosphate produced will be used exclusively for the production of arginine since the enzyme catalysing the production of N-carbamoyl-L-aspartate (aspartate carbamoyltransferase) is feedback-inhibited by the high concentrations of pyrimidine nucleotides. If arginine is present in excess, the enzyme catalysing the production of N-acetyl-L-glutamate (amino-acid acetyltransferase) will be feedback-inhibited (Fig. 6.18).

Other elaborate control mechanisms that have been demonstrated to exist include sequential inhibition, cumulative inhibition, and enzyme multiplicity.[87] An example of enzyme multiplicity is found in the biosynthesis of the aromatic amino acids in *E. coli*.[88] The first step in the pathway, i.e. the production of 7-phospho-2-keto-3-D-*arabino*-heptanoate from phosphoenolpyruvate and D-erythrose 4-phosphate is catalysed by three distinct enzymes, each of which is subject to control by a different end-product. The first enzyme is subject to feedback inhibition by phenylalanine, the second to feedback inhibition by tyrosine, and synthesis of the third enzyme is repressed by tryptophan.

A final point that should be made is that within a cell there could be compartmentation of metabolites and enzymes. The proposed mechanisms of control must therefore take into account any membrane-permeability barriers that may exist (see Chapter 8, Sections 8.2 and 8.3).

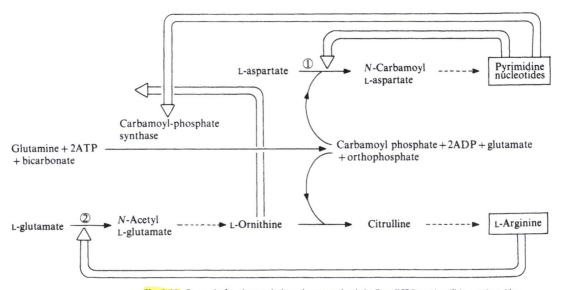

Fig. 6.18 Control of carbamoyl phosphate synthesis in *E. coli*.[87] Reaction (l) is catalysed by aspartate carbamoyltransferase. Reaction (2) is catalysed by amino-acid acetyltransferase.

6.3.2 **Amplification of signals**

One of the features of a metabolic pathway (when compared with a single enzyme) is that considerable amplification of an initial signal is possible. The initial signal is usually a change in the concentration of a substrate or other ligand. Two particular mechanisms are considered to be important in signal amplification: these are substrate cycles (Section 6.3.2.1) and interconvertible enzyme cycles (Section 6.3.2.2).

6.3.2.1 **Substrate cycles**
Consider the following segment of a metabolic pathway:

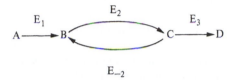

in which there are separate enzymes to catalyse the conversion of B to C (E_2) and of C to B (E_{-2}). If E_2 and E_{-2} are active simultaneously, a *substrate cycle* will occur in which B and C are interconverted. E_2 and E_{-2} both catalyse exergonic reactions. An example of a substrate cycle occurs in the glycolytic pathway (Fig. 6.19).[89]

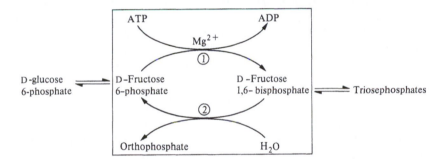

Fig. 6.19 The substrate cycle between D-fructose 6-phosphate and D-fructose 1,6-bisphosphate. Reaction (1) is catalysed by 6-phosphofructokinase and reaction (2) by fructose-bisphosphatase. Note the anomeric specificities of these enzymes (see Fig. 6.1).

Reaction (1) is catalysed by 6-phosphofructokinase, reaction (2) by fructose-bisphosphatase. If both enzymes are active simultaneously, the net result is the hydrolysis of ATP to yield ADP and orthophosphate. The importance of the substrate cycle in this case is that the activities of the two enzymes involved can be separately controlled (Section 6.4.1.1) and hence the net flux through this step, i.e. the rate of reaction (1) minus the rate of reaction (2), can be controlled much more precisely than if only a single enzyme were involved. The activities of the two enzymes are controlled in a reciprocal fashion, e.g. AMP activates 6-phosphofructokinase but inhibits fructose-bisphosphatase, so the effect of small changes in the concentrations of AMP on the net flux can be greatly amplified (Section 6.4.1.1).

A substrate cycle may also operate between D-glucose and D-glucose 6-phosphate in the liver.[90,91]

→ Price : ATP consumption

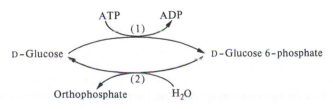

Reaction (1) is catalysed by hexokinase type IV (glucokinase) and reaction (2) by glucose 6-phosphatase. There is evidence that this substrate cycle may be important in the regulation of glucose metabolism in the liver.[90,91]

Other examples where there is evidence for substrate cycles are in the cycling of fatty acids and triacylglycerol in adipose tissue[92] and in the interconversion of phospho*enol*pyruvate and pyruvate in glycolysis and gluconeogenesis.[93] In the case of fatty acid/triacylglycerol the cycling rate is estimated to be between 10 and 20 times the net rate of lipolysis in adipose tissue.[92]

Substrate cycles confer on the cell the possibility of more sensitive control of metabolic pathways. The price paid for this is consumption of energy in terms of ATP hydrolysis (Section 6.2.1.2). In addition, they may control the direction of flow at branch points and in bidirectional pathways, they may also have a role in control of thermogenesis in brown adipose tissue and in flight muscles of bumble-bees, and they may act to buffer metabolite concentrations.[93] We shall discuss the evidence that substrate cycling does occur *in vivo* in Section 6.4.1.1 and further details are also given in reference 93.

6.3.2.2 Interconvertible enzyme cycles

In Table 6.2 we listed some examples of enzymes whose activity can be controlled by reversible covalent modification. An interconvertible enzyme cycle can provide a large amplification of an initial signal provided that the activities of the enzymes catalysing the interconversions are carefully controlled.[13,28,94] In such systems, a 0.5 per cent change in modifier concentration can result in the fraction of covalently modified enzyme increasing from 10 to 90 per cent with a corresponding increase in enzyme activity.[95] However, for maximum sensitivity of response by covalent modification certain conditions must prevail. (1) The modifiers must act at near saturation concentrations. (2) They must affect opposing reactions in opposite directions e.g. activate a protein kinase and inhibit a protein phosphatase. (3) The enzyme subject to inhibition must respond to a lower modifier concentration than the enzyme subject to activation.[95]

For example, the activity of the pyruvate dehydrogenase multienzyme complex from mammalian sources can be altered by phosphorylation (see Chapter 7, Section 7.7.4):

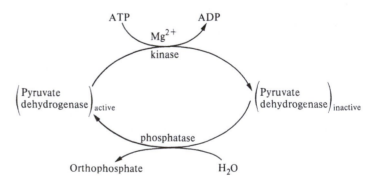

The kinase is active when the ratios of [NADH]/[NAD+] and/or [acetylCoA]/[CoA] are high, so that when the concentration of NADH and/or acetylCoA is high, pyruvate dehydrogenase will be inactivated and the entry of pyruvate into the tricarboxylic acid cycle will be blocked.[23,96] Some reciprocal effects are observed on the phosphatase; for example, the phosphatase is inhibited by a high

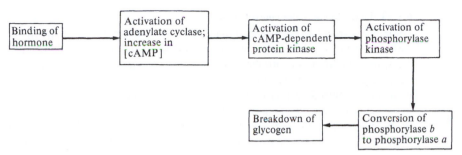

Fig. 6.20 Sequence of events involved in initiation of the breakdown of glycogen following binding of a hormone to a receptor on the cell surface.

ratio of [NADH]/[NAD$^+$], so that the balance between the active and inactive forms of pyruvate dehydrogenase can be very sensitive to changes in the concentrations of these metabolites.[23]

In the case of phosphorylase, the conversion of the inactive b form to the active a form is catalysed by phosphorylase kinase, which is itself subject to enzyme-catalysed activation and inactivation. A sequence of events can be traced back to the binding of a hormone to a receptor on the cell surface (Fig. 6.20).

The enzyme-catalysed activation of phosphorylase and of phosphorylase kinase can be reversed by the action of phosphatases that are themselves subject to regulation. There is also an elaborate mechanism to ensure that when phosphorylase is activated, the enzyme catalysing the synthesis of glycogen (glycogen synthase) is inactivated (Section 6.4.2.5).

Figure 6.20 provides an excellent example of the type of signal amplification mentioned previously in Section 6.2.1.2. Experimentally it has been shown that there is only a small change in the concentration of cAMP on administration of a hormone such as adrenalin;[97] however, each molecule of protein kinase that is activated can rapidly activate many molecules of phosphorylase kinase and hence a great number of molecules of phosphorylase, thus leading to rapid breakdown of glycogen. It has been estimated that 50 per cent of the phosphorylase b could be converted to the active (a) form with only a 1 per cent increase in the concentration of cAMP,[98] the process occurring within about 2 s of the administration of hormone.[99]

6.3.3 Formulation of theories for the control of metabolic pathways

The development of a theory for the control of a particular metabolic pathway usually involves the following steps:

(1) detection of the major control points in the pathway;

(2) a study of the properties of the control enzymes;

(3) formulation of an initial theory;

(4) testing the correctness of the theory by obtaining and assessing experimental data (preferably from *in vivo* experiments); making any necessary modifications to the original theory.

We shall discuss these various steps in turn (Sections 6.3.3.1 to 6.3.3.4) and then illustrate them by discussing the present-day theories for the control of glycolysis (Section 6.4.1) and of glycogen metabolism (Section 6.4.2) in muscle.

6.3.3.1 Detection of major control points

For many years it had been general practice to divide the enzymes catalysing the various steps in a metabolic pathway into controlling and non-controlling types. Thus, the activity of a controlling enzyme would be changed by the regulatory signal, e.g. a change in the concentration of some metabolite, whereas the activity of a non-controlling enzyme would only change as a consequence of a change in the activity of a controlling enzyme. Gradually it has become appreciated that the activities of all the enzymes in a pathway may have some influence on the overall flux through the pathway. This represents a change from an 'all-or-none' approach to one in which all enzymes have some degree of influence on the overall flux through a pathway. In order to define metabolic control more precisely, and to provide experimentally testable models of control, a number of detailed mathematical analyses have been undertaken.[100–104] One of the best-known, referred to as 'metabolic control analysis', is that due to Kacser and Burns[100,101] and Heinrich and Rapoport.[105] In this analysis, each enzyme in a pathway is assigned a flux control coefficient, which is a quantitative measure of the extent to which a particular enzyme can affect the overall flux through a pathway. Values generally range between 0 and 1, the highest values having greatest influence on the overall flux. Since 'metabolic control analysis' is a more recent development, and has built on the 'classical methods' of studying metabolic control, it will be considered separately in Section 6.3.3.2.

For our present purposes we can effectively speak of major controlling and minor controlling enzymes. In the terminology of metabolic control analysis, these would be enzymes having high and low flux control coefficients, respectively. A major controlling enzyme should catalyse the slowest (or rate-limiting) step in the pathway. A reaction that is not at equilibrium *in vivo* is likely to be a rate-limiting step, since the reason that equilibrium is not achieved is presumably that there is insufficient active enzyme present to allow equilibration to occur. The initial identification of controlling enzymes therefore involves a search for those enzymes that (i) are present at low activities and (ii) catalyse reactions that are not at equilibrium. Additional methods involve searches for cross-over points and mathematical modelling techniques (see (iii) and (iv) below, respectively).

(i) The activity of each enzyme in a metabolic pathway is measured *in vitro* by performing an appropriate assay (see Chapter 4, Section 4.2) in a homogenate of the tissue concerned. It is important to measure the maximum velocity of each enzyme under conditions of pH, ionic strength, etc. that are similar to those existing in the cell. This may involve a partial purification of the homogenate to remove an inhibitor, for example. The results of the assays usually show that a reasonably clear distinction can be drawn between enzymes present in the cell at 'high activities' and those present at 'low activities'. In the glycolytic pathway (Fig. 6.21) hexokinase and 6-phosphofructokinase are 'low-activity' enzymes, whereas triosephosphate isomerase and pyruvate kinase are 'high-activity' enzymes.[5,106]

(ii) In order to decide whether a given reaction in a pathway is at equilibrium *in vivo*, we must be able to measure the intracellular concentrations of the particular metabolites involved. The equilibrium constant for the reaction under conditions of pH, ionic strength, etc. that are similar to those that are thought to obtain in the cell can be determined by separate experiments using the isolated enzyme.

A widely used method for determining the concentrations of metabolites in a given tissue involves the technique of 'freeze clamping'.[89] A sample of tissue is very rapidly frozen by compression between aluminium plates that are cooled to the temperature of liquid nitrogen, i.e. 77 K ($-196°C$). This rapid freezing effectively stops any processes that might tend to change the concentrations of metabolites. Before analysis of the tissue, it is essential to inhibit any enzyme-catalysed reactions; this is achieved by treating the frozen tissue with frozen perchloric acid, which will inactivate enzymes as the preparation is allowed to thaw. After centrifugation to remove protein and other precipitated material, the supernatant is neutralized and analysed for the metabolites of interest by appropriate chemical or enzymatic methods (see Chapter 10, Section 10.7). The content of a metabolite, determined as micromoles per gram of tissue can be expressed as a concentration, provided that the intracellular water content is known.*

Results obtained on a rat heart perfused with glucose were:[108]

Metabolic	Intracellular concentration
ATP	11.5 mmol dm^{-3}
ADP	1.3 mmol dm^{-3}
D-Fructose 6-phosphate (F6P)	0.09 mmol dm^{-3}
D-Fructose 1,6-bisphosphate (FBP)	0.02 mmol dm^{-3}
AMP	0.17 mmol dm^{-3}

Thus, for the reaction catalysed by 6-phosphofructokinase,

$$F6P + ATP \xrightleftharpoons{Mg^{2+}} FBP + ADP,$$

the ratio of intracellular concentrations, known as the mass action ratio, Γ, is given by

$$\Gamma = \frac{[FBP][ADP]}{[F6P][ATP]}$$
$$= 0.025.$$

Since the equilibrium constant for this reaction is about 1200, it is clear that this reaction is far removed from equilibrium *in vivo*.

By contrast, for the reaction catalysed by adenylate kinase,

$$ATP + AMP \xrightleftharpoons{Mg^{2+}} 2ADP,$$

the mass action ratio is 0.85 which is close to the equilibrium constant for this reaction (0.44). Therefore, adenylate kinase catalyses a reaction that is close to equilibrium *in vivo*. Rolleston[109] suggests that if the ratio of Γ/K (K is the equilibrium constant) for a reaction is less than 0.05, the reaction can be considered as being non-equilibrium. ^{31}P nuclear magnetic resonance (NMR) can be used to measure the intracellular concentrations of phosphorus-containing metabolites such as ATP, ADP, AMP, phosphocreatine, etc.[110–112] It is possible to measure the concentration of these metabolites in a sample of tissue, an organ, or even a whole organism inserted into the wide-bore magnet of the NMR spectrometer, and to detect changes in the concentrations caused by treatments such as anoxia, or addition of metabolites or hormones. The results obtained have confirmed the general conclusions reached using the freeze-clamping method, but it is evident that the non-destructive NMR technique allows a more detailed insight into the balance of metabolites in cells than was previously possible.

* The total water content of a tissue can be determined from the decrease in weight after drying. The contribution of extracellular water to this total can be assessed by measuring the concentration of some added compound (e.g. sorbitol) which is known not to penetrate the cells. The balance is the intracellular water.[107]

One difficulty in this work is that only the total concentrations of metabolites are measured, and no account is taken of the possible effects of compartmentation within cells (see Chapter 8, Section 8.2). It appears, for instance, that a high proportion of the ADP in muscle is bound to the protein actin[113] and so the concentration of free ADP is likely to be much lower than the measured total concentration. For the purposes of calculating mass action ratios, the concentrations of free metabolites should be used. We should therefore make deductions about whether particular reactions are at equilibrium *in vivo* with a certain degree of caution.

(iii) A third method of locating control points in a metabolic pathway is based on 'cross-over' experiments.[105,109] In this type of experiment, a system is perturbed, e.g. by addition of some inhibitor or activator or by anoxia, and the changes in the concentrations of the various metabolites in a metabolic pathway are measured. The simple 'cross-over' theorem states that 'following a perturbation of a metabolic steady state, the variations in the concentrations of the metabolites before and after a controlling enzyme have different signs'. For instance, in the case of mouse brain, it was found that anoxia causes a decrease in the concentrations of D-glucose 6-phosphate and D-fructose 6-phosphate, but an increase in the concentrations of D-fructose, 1,6-bisphosphate, dihydroxyacetone phosphate, and D-glyceraldehyde 3-phosphate.[114] There is thus a 'cross-over' between D-fructose 6-phosphate and D-fructose 1,6-bisphosphate and the perturbation (anoxia) has clearly activated 6-phosphofructokinase so that the concentrations of its substrates are lowered and of its products raised. This clearly indicates that 6-phosphofructokinase is a controlling enzyme whose activity can respond to the perturbation. For a detailed appraisal of the 'cross-over' theorem, reference 105 should be consulted.

6.3.3.2 Metabolic control analysis

Metabolic control analysis is an approach developed by Kacser and Burns[100,101] and by Heinrich and Rapoport[105] to understand how the flux through a metabolic pathway is controlled by its component enzymes, and particularly the sensitivity of different control points. For a metabolic sequence in a steady state the rate of catalysis by each enzyme step will be the same. (A steady state is when the concentration of intermediates in the pathway remains constant.) However, certain enzymes will be acting well below their maximum velocities, whilst others will be acting at near maximum velocity. For any particular metabolic pathway or segment of a pathway this will depend on the amount of each enzyme present and their catalytic properties. For each enzyme in a particular system there will be a flux control coefficient, C, which is a measure of the effect of a change in the activity of that enzyme on the flux through the pathway.

Flux control coefficients usually range between 0 and 1. A flux control coefficient of 1.0 would mean that the flux through a pathway would increase in the same proportion as the increase in enzyme activity, e.g. if the amount of enzyme doubled, then the flux through the pathway would also double. In the glycolytic pathway, hexokinase and 6-phosphofructokinase have large values of C (0.7 and 0.3, respectively, in erythrocytes; 0.47 and 0.53, respectively, in *S. cerevisiae*[115]). The other enzymes in the pathway have very low C values and thus make a negligible contribution to the control of the flux. The sum of all the flux control coefficients in a system is equal to 1.

$$C_1^J + C_2^J \dots C_n^J = 1 \tag{6.11}$$

Mathematically, the flux control coefficient (C_1) for an enzyme E_1 is defined as

$$C_1 = \frac{\left(\dfrac{dJ}{J}\right)}{\left(\dfrac{dE_1}{E_1}\right)}$$

$$= \frac{d \ln J}{d \ln E_1} \tag{6.10}$$

where J is the flux through the pathway and E_1 is the catalytic activity of enzyme E_1.

The flux control coefficient can also be negative, e.g. in the case of a branched pathway where a metabolite can be transformed into two different end-products. Here, an increase in the enzyme activity in one branch of the pathway might lead to increased flux down that branch, but decrease the flow down the second branch giving a negative flux control coefficient. It is also theoretically possible for the coefficient to be greater than one although no such experimental values have yet been found. For a fuller discussion of the interpretation of flux control coefficients see reference 116.

The flux control coefficient is a property of an enzyme in a particular system and cannot be considered in isolation since it also depends on the amounts of other enzymes in the metabolic path and is related to the properties of the enzyme itself. If the activity of an enzyme is changed, then it will have an effect on the steady state concentration of its substrate and its product. In the metabolic sequence

$$A \xrightarrow{E_1} B \xrightarrow{E_2} C \xrightarrow{E_3} D$$

increasing the activity of E_2 will have the effect of lowering the steady state concentration of B and raising that of C. The extent to which changing the enzyme activity will affect the steady state concentration of the substrates is defined by the term elasticity, E, where

$$E_S^v = \frac{d \ln|v|}{d \ln S}$$

S is the substrate concentration and v the velocity, regardless of its direction.

Kacser and Burns also derived the link between the flux control coefficients and the elasticities for the enzymes in a metabolic pathway known as the connectivity theorem.[100,101]

For a pathway in which the enzymes are labelled i, j, and k, and their respective flux control coefficients and elasticities are C_i^J, C_j^J, C_k^J, E_S^i, E_S^j, and E_S^k, the following relationship holds:

$$C_i^J E_S^i + C_j^J E_S^j + C_k^J E_S^k = 0$$

The mathematical statement of this theorem is:

$$\sum_{i=1}^{n} C_i^J E_S^i = 0 \tag{6.12}$$

Having developed these theoretical concepts we now need to see how they work out in practice. It should be possible to replace the rate-limiting steps in a pathway by a set of flux control coefficients which give some idea of the relative importance of different steps in the overall control, and eventually to use computer modelling to predict the effects of changes in activities of particular enzymes on the overall flux through a pathway. For further details on computer modelling see references 117–120.

Next we briefly review some of the results obtained so far. Various methods have been used to change the enzyme activities in particular pathways. These include (i) increasing the amount of an enzyme by increasing the gene dosage or by other genetic manipulation, (ii) the addition of a purified enzyme to a cell-free system such as a homogenate, and (iii) reducing the amount of an enzyme by addition of a specific enzyme inhibitor. Four examples will be given to illustrate this.

1. Kacser's group[121] studied control in the pathway leading to the formation of arginine from glutamate in *Neurospora crassa*.

glutamate → ornithine → citrulline → arginino-succinate → arginine

The mycelia of *N. crassa* are multinuclear, each nucleus being haploid. It is possible to form heterokaryons which have varying proportions of wild-type and mutant nuclei in the multinuclear mycelia and in this way to vary the gene dosage. They studied the effect of reducing the level of ornithine carbamoyltransferase (step 2) and of argininosuccinate lyase (step 4) on the flux through the pathway to arginine. When the ornithine carbamoyltransferase activity was reduced, the flux control coefficient for that enzyme changed from 0.02 to 0.31, and when argininosuccinate lyase activity was reduced, its flux control coefficient increased from 0.07 to 0.42. The flux control coefficients of acetyl-ornithine aminotransferase (step 1) and argininosuccinate synthase (step 3) in the wild type are about 0.06 and 0.2, respectively. Two conclusions can be drawn: (i) none of the four enzymes studied exercise major control over arginine biosynthesis in the wild type, and (ii) the influences of the enzymes are not all-or-nothing, but vary depending on the activity of each enzyme.

2. The enzyme tryptophan 2,3-dioxygenase catalyses the first step in tryptophan breakdown and higher amounts of the enzyme can be induced in rat liver after feeding on a high-protein diet. (For details of the mechanism see Chapter 9, Section 9.7.3.) When the higher levels of tryptophan 2,3-dioxygenase have been induced, the flux control coefficient falls from 0.75 on a normal diet to 0.25 after a high-protein diet.[122] It then has less influence on the overall flux.

3. The upper part of the glycolytic pathway was studied in rat liver homogenates using an excess of fructose bisphosphate aldolase and glycerol phosphate dehydrogenase. This traps all the fructose 1,6-bisphosphate produced and reduces it to glycerol-3-phosphate. The accompanying NADH oxidation can be used to measure the flux through the pathway. The flux control coefficients for the three enzymes studied were 0.79 for hexokinase, 0.0 for glucose-6-phosphate isomerase, and 0.21 for phosphofructokinase.[123] Under these conditions hexokinase exerts the major controlling influence. The control of glycolysis is discussed further in Section 6.4.1.

4. Using the connectivity theorem (eqn (6.12)) it is possible to determine the flux control coefficients from the elasticities by solving a number of simultaneous equations. This method was used by Groen *et al.*[124] to determine the flux control coefficients and hence the rate-limiting step in gluconeogenesis between lactate and phosphoenolpyruvate. At the time two possible candidates were pyruvate carboxylase and phosphoenolpyruvate carboxykinase (PEPCK). The flux control coefficients were determined in rat hepatocytes incubated with and without glucagon. Glucagon promotes gluconeogenesis particularly in liver. The results are summarized in Table 6.5.

It can be seen that the pyruvate carboxylase reaction is the rate-limiting step; an increase in activity after glucagon stimulation leads to a decrease in its flux control coefficient.

For more detailed accounts of metabolic control analysis see references 104 and 126.

6.3.3.3 **Properties of the controlling enzymes**

From the various approaches described in Sections 6.3.3.1 and 6.3.3.2, it is possible to locate the probable sites of control of a metabolic pathway. The next step

Table 6.5 Flux control coefficients in gluconeogenesis from lactate in rat hepatocytes

Step(s)	Flux control coefficient	
	+ Glucagon	– Glucagon
1. Pyruvate transport	0.01	0.00
2. Pyruvate carboxylase	0.83	0.51
3. Oxaloacetate transport	0.04	0.02
4. Phosphoenolpyruvate carboxykinase	0.08	0.05
5. Pyruvate kinase	0.00	– 0.17
6. Enolase to phosphoglycerate kinase	0.00	0.29
7. Triose phosphate isomerase to fructose bisphosphatase	0.03	0.27
8. Phosphoglucose isomerase to glucose 6-phosphatase	0.00	0.02

Table summarized from Reference 125.

is to study the catalytic properties of the controlling enzymes, especially those properties that might be involved in regulatory mechanisms. Of particular interest are the following questions.

1. What type of kinetic behaviour is observed (hyperbolic, sigmoidal, etc.)? How does the activity of the enzyme vary with changes in the concentration of substrate, particularly in the range of substrate (and enzyme) concentrations that occur *in vivo*?

2. Is the activity of the enzyme affected by any regulator molecules at concentrations that are known to occur in the cell? What is the nature of the regulation (allosteric, competitive inhibition, etc.)?

3. Is the activity of the enzyme subject to regulation by changes in its covalent structure, e.g. by phosphorylation or dephosphorylation? How are the enzymes catalysing these changes themselves regulated?

6.3.3.4 Formulation of an initial theory for the control of a metabolic pathway

From the studies of the controlling enzymes (Section 6.3.3.3), it should be possible to decide which means of regulation is most likely to be involved in the control of a metabolic pathway. On the basis of the initial theory for the controls it is possible to make predictions that can be then tested.

6.3.3.5 Test of the correctness of a theory for the control of a metabolic pathway

In order to test the correctness of a theory, it is necessary to gather data about the activities of enzymes, the concentrations of substrates and effectors, and the state of covalent modification, etc. in a variety of metabolic circumstances. It is important to determine these parameters under conditions that reflect the *in vivo* situation as closely as possible; non-destructive techniques such as ^{31}P NMR[110–112] (Section 6.3.3.1) have great potential in such studies. If it emerges that, for instance, the variations in the concentrations of substrates and/or effectors are too small to account for the observed changes in enzyme activity or in the flux through the pathway, then the initial theory must be modified or discarded altogether and a new theory formulated and tested.

6.4 Examples of control of metabolic pathways

We shall now discuss how present-day ideas about the control of metabolic pathways have been developed by considering two particular pathways, namely glycolysis (Section 6.4.1) and glycogen metabolism (Section 6.4.2). These pathways illustrate most of the points made in Section 6.3.3. The experiments described refer mainly to the regulation of the pathways in muscle that have been particularly well studied because muscle is a highly specialized tissue in which the fluxes through the pathways vary dramatically in different metabolic circumstances. These metabolic pathways in muscle must therefore be under a high degree of control.

6.4.1 Regulation of glycolysis

The glycolytic pathway is the sequence of reactions in which glucose is converted to pyruvate, and is shown in Fig. 6.21. This pathway appears to be of almost universal importance in energy production.

Using the various approaches described in Section 6.3.3.1, it has been shown that the major controlling points for the pathway are the steps catalysed by 6-phosphofructokinase and by hexokinase, since these enzymes are present at low

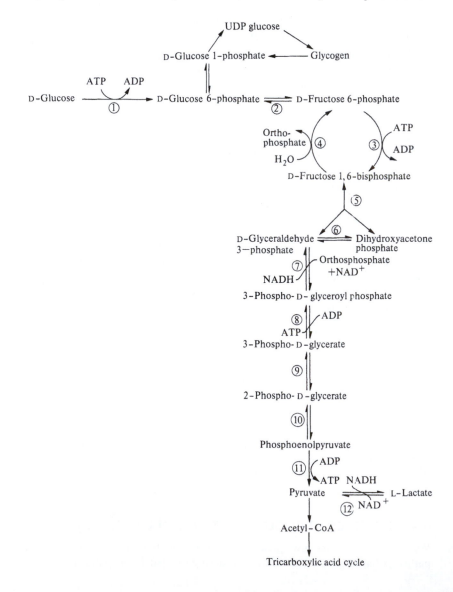

Fig. 6.21 The glycolytic pathway in muscle. The various steps are catalysed by the following enzymes: (1), hexokinase; (2), glucose-6-phosphate isomerase; (3), 6-phosphofructokinase; (4), fructose-bisphosphatase; (5), fructose-bisphosphate aldolase; (6), triosephosphate isomerase; (7), glyceraldehyde-3-phosphate dehydrogenase; (8), phosphoglycerate kinase; (9), phosphoglycerate mutase; (10), enolase; (11), pyruvate kinase; (12), lactate dehydrogenase. The magnesium ions associated with the kinase-catalysed reactions have been omitted.

activities and catalyse reactions which, *in vivo*, are far removed from equilibrium.[89,106] Pyruvate kinase is not at equilibrium *in vivo*, but since the activity of this enzyme is relatively high in muscle, its role in the control of the pathway is somewhat open to question.[98,106] The importance of 6-phosphofructokinase and hexokinase as controlling enzymes in the glycolytic pathway in a variety of tissues has also been confirmed by 'cross-over' experiments.[114] Metabolic control analysis (Section 6.3.3.2) suggests a somewhat different emphasis with a major controlling roles for glucose transport, hexokinase, enolase, and pyruvate kinase, and a lesser role for phosphofructokinase. In cardiac muscle, flux control coefficients for the first steps in glycolysis were estimated to be as follows: hexokinase 0.82, glucose 6-phosphate isomerase 0.01, 6-phosphofructokinase 0.17, and fructose 1,6-bisphosphate aldolase 0.01,[127] giving hexokinase a much greater controlling role than 6-phosphofructokinase.

6.4.1.1 Properties of 6-phosphofructokinases

6-Phosphofructokinase is an allosteric enzyme having both catalytic and regulatory sites (see Section 6.2.2.2). In most bacteria it exists as a homotetramer, each subunit having an M_r of $\approx 33\,000$. The X-ray structure has been solved for the enzyme from *B. stearothermophilus*.[65] Each subunit has both the catalytic domain and effector domain. The *E. coli* enzyme shows sigmoid kinetics with respect to the substrate F6P but not ATP and is allosterically activated by ADP or GDP and allosterically inhibited by PEP.[65]

In yeasts the enzyme is an octamer (A_4B_4) in which A has the effector binding site and B the active site. In mammals it exists as a tetramer with larger subunits ($M_r \approx 85\,000$) having the catalytic domain at the N-terminal half and the effector domain at the C-terminal half. It is thought to have arisen by a process of gene duplication as there are sequence similarities in the N-terminal and C-terminal halves. In the protozoan parasite *Trypanosoma brucei* 6-phosphofructokinase exists as a homotetramer, with subunits of $M_r \approx 50\,000$, but unlike the other enzymes it is not sensitive to effectors. It is present in the glycosomes (see Chapter 8, Section 8.2.5) together with seven other glycolytic enzymes which catalyse the conversion of glucose to 3-phosphoglycerate.[127]

The free energy change for the reaction catalysed by 6-phosphofructokinase (EC 2.7.1.11) suggests that it is effectively irreversible under physiological conditions.[128]

$$\text{ATP} + \text{F6P} \rightarrow \text{FBP} + \text{ADP} \qquad \Delta G^{o'} = -18.4 \text{ kJ mol}^{-1}$$

This is what would be expected of a rate-determining step in a pathway. However, increasing the amount of 6-phosphofructokinase does not seem to affect the flux through the glycolytic pathway appreciably. For example, overexpressing 6-phosphofructokinase by three- to fivefold over the wild-type level in *Aspergillus niger* or yeast does not increase the rate of flux through the glycolytic pathway.[129,130] Similar results have been obtained in transgenic potato tissue in which there is massive overexpression of 6-phosphofructokinase, but no significant increase in flux.[131]

In some unicellular eukaryotes and higher plants anaerobic glycolysis relies on a diphosphate-fructose-6-phosphate 1-phosphotransferase (EC 2.7.1.90) which catalyses the following reaction:

$$\text{PP}_i + \text{F6P} \rightleftharpoons \text{FBP} + \text{P}_i \qquad \Delta G^{o'} = -8.7 \text{ kJ mol}^{-1}$$

PP_i = pyrophosphate
P_i = orthophosphate

In this case the much lower free energy change means that the reaction is reversible under physiological conditions and may serve in both glycolysis and gluconeogenesis.[128]

(a)

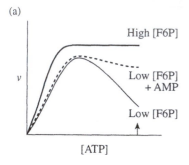

(b)

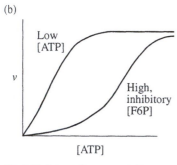

Fig. 6.22 Schematic representation of the kinetic behaviour of 6-phosphofructokinase.[132,134] Variation of velocity with concentration of (a) ATP (the arrow on the abscissa indicates the prevailing physiological concentration of ATP) and (b) F6P (D-fructose 6-phosphate).

---

* The dependence of the activity of 6-phosphofructokinase on the concentrations of the adenine nucleotides has been explored by Atkinson[137,138] who introduced the term 'energy charge'. This is defined as 'one half of the phosphate anhydride bonds per adenosine' and is equal to

$$\frac{[ATP] + \frac{1}{2}[ADP]}{[ATP] + [ADP] + [AMP]}.$$

The 'energy charge' varies from 0, when all the adenosine in a cell is present as AMP, to 1, when all the adenosine is present as ATP. Enzymes that are energy-utilizing, such as ATP citrate (*pro-3S*)-*lyase*, have a different dependence on energy change from those that are energy-regenerating, such as 6-phosphofructokinase.

6.4.1.2 Regulation of the activity of 6-phosphofructokinase

The kinetic behaviour of 6-phosphofructokinase *in vitro* is highly complex.[132–134] As shown schematically in Fig. 6.22, there are mutual interactions between the two substrates D-fructose 6-phosphate (F6P) and ATP. At low concentrations (in the physiological range) of F6P, it is found that high concentrations of ATP (in the physiological range) cause inhibition (Fig. 6.22(a)). This inhibition is due to binding of ATP to a site distinct from the active site (see Chapter 4, Section 4.3.1.4) as is shown by the fact that the enzyme can be desensitized (Section 6.2.2.1) to ATP inhibition by photo-oxidation in the presence of methylene blue.[135] The inhibition at high concentrations of ATP is accentuated by citrate but can be counteracted by various metabolites including AMP (see Fig. 6.22(a)), ADP, orthophosphate, and D-fructose 1,6-bisphosphate. These latter compounds tend to accumulate when ATP is degraded, for example during anoxia. At high concentrations (in the physiological range) of ATP, the plot of velocity against concentration of F6P is markedly sigmoidal with a very low velocity at low concentrations of F6P (Fig. 6.22(b)).

An initial theory, based on the data shown in Fig. 6.22(a), is that the concentration of ATP controls the activity of 6-phosphofructokinase and hence the flux through the glycolytic pathway. Thus, when the concentration of ATP is high, as in resting muscle, 6-phosphofructokinase is inhibited and hence the rate of production of ATP falls. (Note that for each molecule of glucose converted to two molecules of pyruvate there is a net production of two molecules of ATP.) If this theory is correct, it would be expected that significant changes in the concentration of ATP would occur *in vivo* to account for the changes in the activity of 6-phosphofructokinase (Fig. 6.22(a)). However, measurements on the flight muscles of blowflies showed that during flight there was only a small (10 per cent) decrease in the concentration of ATP compared with resting muscle, even though the flux through the glycolytic pathway had increased 100-fold.[136] Reference to the data shown in Fig. 6.22(a) shows that this change in the concentration of ATP is nowhere near large enough to account for the change in the flux through the pathway.

It has been proposed that the effects of ATP on the activity of 6-phosphofructokinase can be amplified by other metabolites, especially ADP and AMP, which relieve the inhibition caused by high concentrations of ATP. The adenine nucleotides are linked in the reaction catalysed by adenylate kinase,

$$ATP + AMP \overset{Mg^{2+}}{\rightleftharpoons} 2ADP,$$

which is at equilibrium in muscle (Section 6.3.3.1). Since the concentrations of nucleotides in muscle are [ATP] > [ADP], [AMP], a relatively small change in the concentration of ATP brings about much larger changes in the concentrations of ADP and AMP.[89,134] In the blowfly flight muscle, for instance, it was found that during flight the concentration of AMP increased by about 2.5-fold compared with resting muscle[120] (cf. a 10 per cent decrease in the concentration of ATP). The revised proposal is therefore that the activity of 6-phosphofructokinase is controlled by a combination of changes in the concentrations of ATP, AMP, ADP,* and other metabolites such as orthophosphate.

However, from detailed considerations of the kinetic behaviour of 6-phosphofructokinase (Fig. 6.22), it is considered unlikely that the combination of the changes in the concentrations of adenine nucleotides is large enough to account for the observed changes in the flux through the glycolytic pathway.[89,134] Various suggestions have been made as to how the activity of 6-phosphofructokinase might be made even more sensitive to changes in the concentration of metabolites. One of these proposals[89,134] is based on a substrate cycle between D-fructose 6-phosphate (F6P) and D-fructose 1,6-bisphosphate (FBP) (Fig. 6.19 and Section 6.3.2.1). Fructose-bisphosphatase, which is strongly inhibited by AMP, is found in significant amounts (approximately 10 per cent of the maximal activity of 6-phosphofructokinase) in various types of muscle where no significant gluconeogenesis occurs. The proposal has therefore been made that in resting muscle, in which the concentration of AMP is low, the low rate of conversion of F6P to FBP catalysed by 6-phosphofructokinase is opposed by conversion of FBP to F6P catalysed by fructose-bisphosphatase, giving a very low net flux through this step in the glycolytic pathway. However, when the muscle is made to do work, the concentration of ATP falls and that of AMP rises; 6-phosphofructokinase is now activated (i.e. no longer inhibited) and fructosebisphosphatase is inhibited, giving a large net flux through this step. Calculations[139] suggest that the observed changes in net flux can be accounted for by changes in the concentration of AMP that are of the same order of magnitude as those detected by freeze-clamping measurements.[136] (Thus, a 250-fold change in net flux could be accounted for by a 2.5-fold change in the concentration of AMP.[139]) The hypothesis of substrate cycling accounts not only for the sensitivity of net flux to changes in the concentration of AMP, but also for the presence and distribution of fructose-bisphosphatase in various types of muscle. Fructose-bisphosphatase is not present in muscles such as heart muscle, where the energy demands are fairly constant and hence the need for regulation of glycolysis is much less than in the case of skeletal muscle. Lardy and his co-workers have also obtained direct evidence for substrate cycling in various tissues such as bumble-bee flight muscle[140] and rat liver[141] using isotopically labelled phosphorylated monosaccharides (see Fig. 6.23).

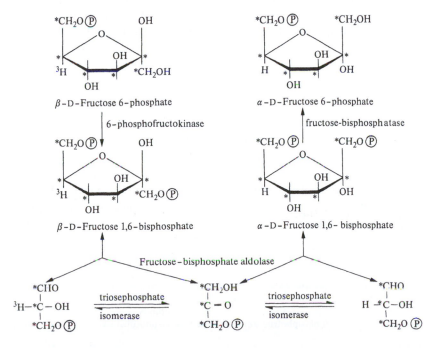

β-D-Fructose 6-phosphate

α-D-Fructose 6-phosphate

6-phosphofructokinase

fructose-bisphosphatase

β-D-Fructose 1,6-bisphosphate

α-D-Fructose 1,6-bisphosphate

Fructose-bisphosphate aldolase

triosephosphate isomerase

triosephosphate isomerase

Fig. 6.23 System used to demonstrate substrate cycling. An asterisk represents ^{14}C and Ⓟ a phosphoryl group. On addition of [5-^{3}H, U-^{14}C]F6P to the system, the action of 6-phosphofructokinase and fructose-bisphosphate aldolase produces isotopically labelled D-glyceraldehyde 3-phosphate and dihydroxyacetone phosphate. The action of triosephosphate isomerase allows ^{3}H at C-2 of D-glyceraldehyde 3-phosphate to be exchanged for H from solvent water (see Chapter 5, Section 5.5.2.3). The action of fructose bisphosphate aldolase and fructose-bisphosphatase then generates F6P in which ^{3}H is lost. If the remaining F6P is isolated at various times after the addition of [5-^{3}H, U-^{14}C]F6P and analysed for the content of ^{3}H and ^{14}C, the ratio of $^{3}H/^{14}C$ will fall if substrate cycling occurs. From the rate of decrease of this ratio, the rate of cycling can be deduced.[142]

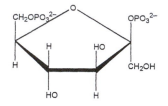

β-D-Fructose-2,6-bisphosphate

However, the assumptions underlying the interpretation of the experimental data in these experiments have been questioned, particularly in the case of rat liver.[91] Also, mathematical modelling of the situation in rat liver suggests that under physiologically realistic conditions, when the compartmentation and the states of protonation of adenine nucleotides are taken into account, the rate of substrate cycling is negligible.[143]

Fructose 2,6-bisphosphate was discovered in 1980 during investigations into the mechanism whereby glucagon, via its second messenger cAMP, stimulates gluconeogenesis and inhibits glycolysis in liver.[17] It is particularly important in liver where it is the most potent known activator of liver 6-phosphofructokinase, with effects being shown in the 0.1 μmol dm^{-3} concentration range. The activation of 6-phosphofructokinase is synergistic with AMP. F26BP is also a powerful inhibitor of liver fructose-bisphophatase, again acting with AMP. The synthesis and degradation of F26BP are catalysed by a kinase (6-phosphofructo-2-kinase) and a phosphatase (fructose-2,6-bisphosphatase) whose activities are subject to control by reversible phosphorylation (Fig. 6.24). The latter kinase and phosphatase activities are located on a homodimeric bifunctional enzyme. The X-ray structure of this enzyme has been solved; the PFK-2 activity resides in the 250-residue N-terminal domain, and the FBPase-2 activity in the 218-residue C-terminal domain.[144]

All the available evidence shows that, in liver, F26BP plays a major part in the hormonal and nutritional control of 6-phosphofructokinase and fructosebisphosphatase activity and in determining the balance between glycolysis and gluconeogenesis.[145] Thus, the glycolytic flux in liver can be correlated with the level of F26BP but not with the level of ATP, AMP, or citrate. The presence of F26BP is the signal that the level of glucose is high and that gluconeogenesis can be switched off. The presence of glucagon, however, causes the level of F26BP to fall (see Fig. 6.24) and hence gluconeogenesis is favoured.

The importance of F26BP in the regulation of glycolysis in other tissues is less firmly established. The concentration of F26BP in muscle is lower than that in the liver and this, coupled with the smaller effect of F26BP on muscle 6-phosphofructokinase compared with the liver enzyme, means that F26BP is unlikely to

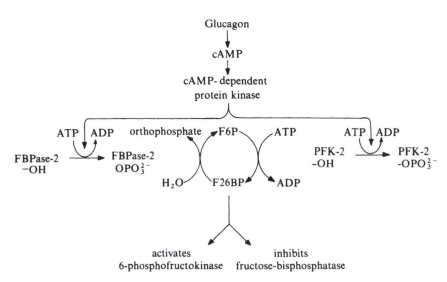

Fig. 6.24 Control of the synthesis and degradation of fructose 2,6-bisphosphate (F26BP).[145] In this scheme, FBPase-2 and PFK-2 represent fructose-2,6-bisphosphatase and 6-phosphofructo-2-kinase, respectively.

have a major effect on the glycolytic flux in muscle. There are a number of different isoforms of PFK-2/FBPase-2 which exist in different tissues, and they differ from one another both in their affinities for the substrates and whether they are regulated by cAMP-dependent protein kinase.[17] PFK-2/FBPase-2 in skeletal muscle lacks a phosphorylation site and is thus not sensitive to cAMP-dependent protein kinase. In cardiac muscle not only does phosphorylation of PFK-2/FBPase-2 occur at a site different from that of the liver enzyme, but hormones that cause an increase in cAMP also effect an increase in the intracellular concentration of F26P, thereby stimulating glycolysis. This is in contrast to liver where glycolysis is inhibited. These observations fit in with the different physiological functions of the two tissues. In liver, when glycogen breakdown is required, it is associated with increased glucose release into the blood and thus inhibition of glycolysis. In cardiac muscle glycogenolysis is associated with glycolysis, since the glucose from glycogen breakdown is required in cardiac tissue.

Among other factors that may be important in the regulation of the activity of 6-phosphofructokinase in muscle are the following.

1. 6-Phosphofructokinase and fructose-bisphosphatase show different anomeric specificities (see Figs. 6.1 and 6.23), acting on the β-anomer of F6P and the α-anomer of FBP, and being activated by the α-anomer of FBP and the β-anomer of F6P, respectively. These specificities, taken together with the slow, spontaneous $\alpha \rightarrow \beta$ anomerizations of F6P and FBP (rate constants approximately 1 s^{-1}, reference 6) create non-equilibrium intracellular distributions of the α- and β-anomers of F6P and FBP. These non-equilibrium distributions could then amplify the effects of AMP and other metabolites on the enzymes.[146]

2. The calcium-binding protein calmodulin has been shown to interact with 6-phosphofructokinase from muscle.[147,148] Calmodulin acts as a Ca^{2+}-dependent inhibitor of the enzyme and appears to compete with ATP for binding to the regulatory site of the enzyme.

3. 6-Phosphofructokinase from muscle can be reversibly phosphorylated. There is evidence that this affects its state of oligomerization and the way in which allosteric inhibitors such as citrate interact with the enzyme.[149,150] However, it does not appear to be a major regulatory mechanism.

It should be noted that the factors (1)–(3) have been studied *in vitro* and their possible importance *in vivo* has not been clarified.

In summary, it can be stated that although the importance of the step D-fructose 6-phosphate $\rightarrow$ D-fructose 1,6-bisphosphate as a control point in glycolysis is very widely accepted, the different analytical approaches, e.g. metabolic control analysis versus mass action ratios, differ in the degree of importance they attach to this step.

6.4.1.3 Regulation of the activity of hexokinase

Up to now we have concentrated attention on the regulation exerted at the D-fructose 6-phosphate $\rightarrow$ D-fructose 1,6-bisphosphate step in glycolysis. However, it is obvious that when 6-phosphofructokinase is inhibited, there would be an accumulation of the phosphorylated hexoses D-glucose 6-phosphate and D-fructose 6-phosphate unless some means existed to control the activity of hexokinase that catalyses the conversion of D-glucose to D-glucose 6-phosphate at the expense of ATP (see Fig. 6.21). Evidence from metabolic control analysis suggests that in the metabolic pathway from D-glucose to pyruvate, hexokinase[151] has the highest flux control coefficient (see Section 6.3.3.2). When comparisons are made between the flux rates and the maximum flux capacity (V_{max}) for

* Hexokinase isoenzymes I, II, and III have low K_ms for glucose (7–150 μM), are inhibited by glucose 6-phosphate, and have M_r values of $\approx 100\,000$. Hexokinase IV (glucokinase) has a $K_m \approx 10$ mM, is not inhibited by glucose 6-phosphate (see Chapter 4, Section 4.3.1.3), and has an $M_r \approx 50\,000$. Evidence from sequence comparisons suggests that hexokinases I, II, and III have evolved by gene duplication, and that the glucose 6-phosphate inhibitory site evolved in the duplicated forms but not in hexokinase IV. Isoenzyme IV (also known as glucokinase) is not inhibited by D-glucose 6-phosphate.[153]

hexokinases, it is found that in most mammalian muscles the enzyme is operating at between 6 and 28 per cent of V_{max}, whereas in the flight muscles of the honey-bee and hummingbird the levels are between 74 and 98 per cent of V_{max}.[152] It is also found that hexokinases from a variety of mammalian tissues are inhibited by D-glucose 6-phosphate;* the inhibition is non-competitive (see Chapter 4, Section 4.3.2.2) with respect to D-glucose.[75] The inhibition of hexokinase would thus serve to regulate the rate of production of D-glucose 6-phosphate and hence D-fructose 6-phosphate.

There is, however, an additional factor to be borne in mind. As indicated in Fig. 6.20, hexokinase occurs near a branch point in the metabolism of monosaccharides, since D-glucose 6-phosphate is not only used to furnish D-fructose 6-phosphate for the glycolytic pathway but can also be used to produce D-glucose 1-phosphate via the action of phosphoglucomutase for the synthesis of glycogen. Under conditions in which the concentration of D-glucose is high, e.g. after an intake of carbohydrate in the food, it is desirable to build up reserves of glycogen. This would not be possible in a tissue such as muscle lacking glucokinase (hexokinase IV) if the only hexokinase present were inhibited by high prevailing concentrations of D-glucose 6-phosphate. The hormone insulin, secreted in response to high circulating concentrations of D-glucose, not only facilitates the transport of D-glucose into the cell, but also stimulates the synthesis of glycogen (Section 6.4.2.5). Thus when D-glucose concentrations are high, the excess D-glucose can be channelled via D-glucose 6-phosphate, D-glucose 1-phosphate, and UDP glucose to glycogen (Fig. 6.21).

In liver, where glucokinase and hexokinase I are both present, it is D-glucose 6-phosphate produced by glucokinase that activates hepatic glycogen synthase to promote glycogen synthesis. The evidence that supports this is that when hepatocytes are induced to overexpress glucokinase using an adenovirus-mediated overexpression system, this greatly increased the rate of glycogen synthesis. When a similar experiment was carried out but hexokinase I was overexpressed, it had no effect.[154]

The dual control system afforded by regulation of the activities of hexokinase† and of 6-phosphofructokinase that occur before and after the branch point, respectively, thus allows the cell to respond to a variety of metabolic situations and provides more flexible regulation of the glycolytic pathway than could be achieved with a single control point.

6.4.2 Regulation of glycogen metabolism

The second example of control of a metabolic pathway that we shall discuss is the regulation of glycogen metabolism in muscle. Glycogen is a polysaccharide made up of D-glucopyranose units linked by α-1,4 glycosidic bonds with branches formed by 1,6 bonds about every 12–18 groups (Fig. 6.25), i.e. there is no unique molecular structure for glycogen. The M_r of glycogen is in the range from about 10^6 to 10^8 (consisting of between 6000 and 600 000 D-glucose units). In a variety of organisms, e.g. E. coli[155] and rat liver,[156] a protein molecule is involved as a 'primer' backbone in the biosynthesis of glycogen.

The reactions involved in glycogen metabolism are relatively few in number (see Fig. 6.26) but the study of these reactions has revealed some very complex control mechanisms.

In both muscle and liver, glycogen has the function of being a reserve of D-glucose units. Liver glycogen can be used to help replenish D-glucose in the blood between meals and during exercise. In muscle, however, the enzyme glucose

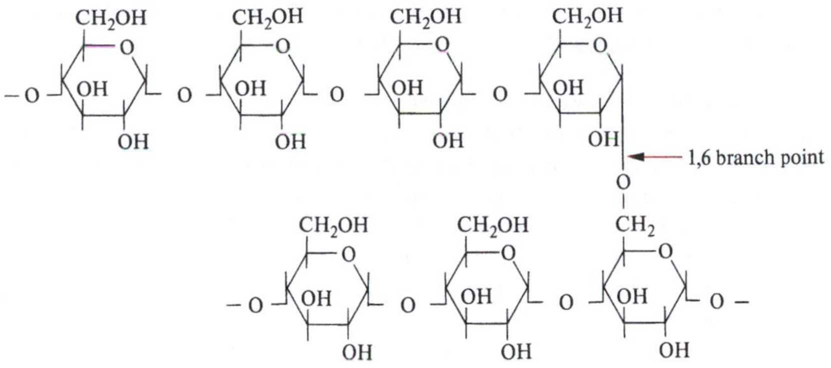

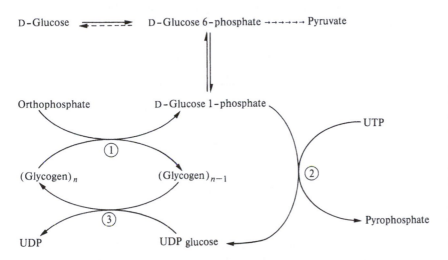

Fig. 6.25 The structure of glycogen showing the α-1,4 glycosidic bonds linking D-glucopyranose units and a 1,6 branch point.

Fig. 6.26 Reactions involved in glycogen metabolism. The dashed line represents a reaction that occurs in liver but not in muscle. Reaction (1) is catalysed by phosphorylase plus debranching enzyme (which possesses two activities: amylo-l,6-glucosidase and 4-α-D-glucanotransferase[157]). Reaction (2) is catalysed by glucose 1-phosphate uridylyltransferase. Reaction (3) is catalysed by glycogen synthase plus branching enzyme (l,4-α-glucan branching enzyme[158]).

6-phosphatase is lacking and hence the D-glucose 1-phosphate produced by the breakdown of glycogen is metabolized via the glycolytic pathway to provide ATP that acts as the fuel for muscular contraction. The amount of ATP that can be generated from the muscle glycogen is sufficient to sustain contraction for only a short time (of the order of one minute). Prolonged exercise requires the utilization of other sources of energy, e.g. triglycerides.[159]

Using the criteria for identification of controlling enzymes (Section 6.3.3.1), it has been established that the control points for glycogen breakdown and synthesis are the reactions catalysed by phosphorylase and glycogen synthase respectively.[89,134,160]

The main features of phosphorylase that implicate it as a controlling enzyme are that it is present at low activity (similar to the activity of 6-phosphofructokinase) and that it catalyses a reaction that, *in vivo*, is not at equilibrium*.

A 'cross-over' type of experiment showed that when the breakdown of glycogen is stimulated by adrenalin, the catalytic activity of phosphorylase increases even though the concentration of its substrate decreases, i.e. phosphorylase is the controlling enzyme.[160] In the case of glycogen synthase, the main evidence that it is a controlling enzyme comes from the finding that it is present at low activity. 'Cross-over' experiments have helped to confirm this conclusion.[160] Comparisons between the rate of glycogenolysis *in vivo* and the V_{max} for glycogen phosphorylase

* See notes overleaf.

* The equilibrium constant for the reaction catalysed by phosphorylase is given by

$$K = \frac{[(\text{Glycogen})_{n-1}]}{[(\text{Glycogen})_n]} \times \frac{[\text{G1P}]}{[\text{Orthophosphate}]}$$

but since it is impossible to determine the first term because of the variety of sizes of glycogen molecules, we write

$$K = \frac{[\text{G1P}]}{[\text{Orthophosphate}]}$$

At equilibrium this ratio is 0.3, but in muscle the ratio is approximately 100-fold lower. This is because in muscle the concentration of orthophosphate is high, and the concentration of G1P is low (the equilibrium between G1P and G6P lies 95 per cent to the side of G6P). Hence, in the muscle phosphorylase acts to break down glycogen. Glycogen synthesis proceeds via a different route, as shown in Fig. 6.26. Patients suffering from McArdle's disease lack phosphorylase but their ability to synthesize glycogen is not impaired.[161]

in vitro suggest that in exercising mammalian muscle glycogen phosphorylase is operating at between 6 and 28 per cent of its maximum rate.[152]

6.4.2.1 Regulation of the activity of phosphorylase

In 1936, phosphorylase purified from muscle was shown to require AMP for activity and so may be considered to be the earliest example of allostery.[162] By 1941 there were indications that it existed in a second form which contained a phosphate residue and so it could also be considered the first example of regulation by covalent modification.[163] Since these initial studies glycogen phosphorylase has been studied intensively and much detail is known about its structure, function, and regulation.[163–166]

The two forms of phosphorylase that can be isolated from muscle are phosphorylase *b*, which is inactive in the absence of certain ligands, notably AMP, and phosphorylase *a*, which is active in the absence of these added ligands. The interconversions of the two forms of phosphorylase are catalysed by a kinase and a phosphatase that can be isolated from muscle extracts.

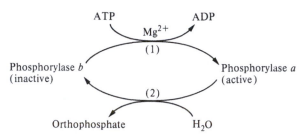

(1) is catalysed by phosphorylase kinase
(2) is catalysed by phosphorylase phosphatase

As mentioned earlier (Section 6.2.1.2) the conversion of phosphorylase *b* to phosphorylase *a* involves phosphorylation of a single side-chain (Ser 14) in each subunit. The conversion is accompanied by a change in the quaternary structure of the enzyme (the *b* form is a dimer, the *a* form a tetramer) but this change is not an integral part of the activation process.[167]

Thus there appear to be two major means of regulating the activity of phosphorylase.

1. By ligand-induced conformational changes in the enzyme. AMP (or IMP) will activate phosphorylase *b;* this effect is counteracted by certain other ligands, notably ATP, ADP, UDPglucose, and D-glucose 6-phosphate.[163] Hence, under conditions in which the concentration of AMP is high and the concentrations of ATP and D-glucose 6-phosphate, in particular, are low, phosphorylase *b* will be active. For phosphorylase *b* to be active, two molecules of AMP and two molecules of D-glucose 1-phosphate have to be bound to the effector and catalytic sites of the dimer, respectively.[168] This provides a 'crude' control mechanism for the enzyme.

2. By a change in the covalent structure of the enzyme, i.e. conversion of phosphorylase *b* to phosphorylase *a*, catalysed by phosphorylase kinase. Phosphorylase kinase is itself subject to regulation and can be activated in two main ways: by phosphorylation catalysed by a cAMP-dependent protein kinase and by Ca^{2+}-mediated stimulation.

The X-ray structure has been determined for phosphorylase a, phosphorylase b, and phosphorylase b with AMP bound,[165,169,170] and from these structures it is apparent that the same conformational changes are induced by phosphorylation of phosphorylase b to phosphorylase a as by binding AMP to phosphorylase b, although the regulatory signals function through different local structural mechanisms.

The sequence of events in which phosphorylase kinase is activated by cAMP-dependent protein kinase is shown in Fig. 6.27.

Binding of the hormone to the cell surface receptor activates adenylate cyclase, which is a membrane-bound enzyme[171] (see Chapter 8, Section 8.4.5). The cAMP produced binds to the protein kinase causing dissociation of the inactive R_2C_2 complex to yield active catalytic subunits and a dimer of regulatory subunits.[172,173]

The active catalytic subunits catalyse the activation of phosphorylase kinase and hence of phosphorylase. This cascade mechanism allows for considerable amplification of the initial signal (Section 6.3.2.2). The administration of adrenalin to various tissues causes only small increases (less than fivefold) in the concentration of cAMP,[160] and indeed in the case of rabbit gracilis muscle it has been shown that the administration of 4 pmol of isoproterenol, an analogue of adrenalin, did not change the concentration of cAMP within experimental error, although the amount of phosphorylase a increased threefold.[174]

Ca^{2+} ions in the concentration range of 0.1 to 1 μmol dm^{-3} can activate phosphorylase kinase in the presence of Mg^{2+} and ATP. This activation occurs by a process of autophosphorylation in which the enzyme catalyses its own phosphorylation,[175] at sites that are distinct from those that are phosphorylated by the action of cAMP-dependent protein kinase.[176,177] Ca^{2+} ions in this concentration range are released from the sarcoplasmic reticulum in muscle in response to a nerve impulse and can then act as a trigger for muscular contraction by allowing the components of the contractile apparatus (actin and myosin) to interact. Thus there is a clear link between the stimulation of muscular contraction, which requires ATP as a fuel, and activation of glycogen breakdown, which can furnish ATP via the glycolytic pathway (Fig. 6.21).

There are a number of protein kinases[178] that are regulated by Ca^{2+}. For most of these, intracellular Ca^{2+} ions interact with the calcium-binding protein,

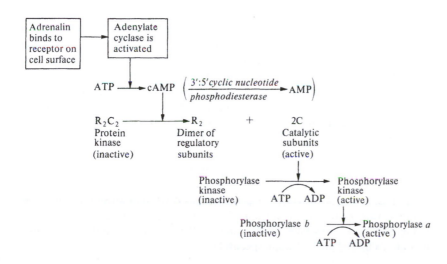

Fig. 6.27 The cascade mechanism for hormone-triggered activation of phosphorylase.

calmodulin. These include phosphorylase kinase, myosin light chain kinase, calmodulin kinases I–IV, and calmodulin kinase kinase.[178] Most of these are multifunctional kinases phosphorylating a number of different intracellular proteins.[179] However, a few, such as phosphorylase kinase, are dedicated kinases acting on only one protein, in this case phosphorylase.[179] Phosphorylase kinase is a large protein, M_r 1.3×10^6, with a quaternary structure $(\alpha\beta\gamma\delta)_4$.[180] The γ subunit (M_r 45 000) contains the catalytic domain, the α (M_r 145 000) and β (M_r 128 000) are regulatory subunits containing the sites of phosphorylation, and the δ subunit (M_r 16 000) is calmodulin. The X-ray structure of the catalytic domain of the γ subunit is known.[181] Activation of phosphorylase kinase by Ca^{2+} can proceed via Ca^{2+} binding to the δ subunit, which triggers a conformational change in the γ subunit converting it to an active form.[180] A further degree of stimulation can be achieved by the binding of troponin-C (the Ca^{2+}-binding protein involved in the regulation of muscular contraction). Troponin-C interacts only with the β subunit of phosphorylase kinase and only in the presence of Ca^{2+} ions.[177] Since the concentration of troponin-C in muscle is much higher than that of calmodulin, it may well be that activation of phosphorylase kinase via the binding of troponin-C is important in the synchronization of glycogen breakdown with muscular contraction.[177,180]

In summary, we have seen how the conversion of phosphorylase b to phosphorylase a can be brought about by the action of cAMP-dependent protein kinase or by Ca^{2+}/calmodulin which forms an integral part of phosphorylase kinase. Certain hormones such as adrenalin, when binding to β-receptors,* and glucagon activate adenylate cyclase, causing activation of cAMP-dependent protein kinase. Increased intracellular concentrations of cAMP have been detected in intact cells and tissues following hormone treatment. Other hormones, such as vasopressin, angiotensin, and adrenalin, when binding to α-receptors also effect the conversion of phosphorylase b to phosphorylase a, but with no detectable increase in intracellular cAMP. Evidence suggests that these act through increasing intracellular Ca^{2+} and operate through the Ca^{2+}/calmodulin route. Their actions can be suppressed by the presence of EGTA, a Ca^{2+}-chelating agent.[184]

Nerve impulses also act via Ca^{2+} ions. Collectively, these provide a versatile amplified response to a variety of initial signals. It is now becoming increasingly recognized that the hormone-stimulated and nerve-impulse-stimulated pathways are interwoven.[177,185,186] For instance, Ca^{2+} ions acting via calmodulin are known to affect the activities of adenylate cyclase and 3′:5′-cyclic-nucleotide phosphodiesterase, which catalyse the synthesis and hydrolysis, respectively, of cAMP. The Ca^{2+}-calmodulin system is also responsible for activating a multi-protein kinase of broad specificity that phosphorylates a number of proteins including glycogen synthase and tyrosine hydroxylase (which is involved in the synthesis of catecholamines).[185] In addition, cAMP-dependent protein kinase can catalyse the phosphorylation of troponin-I which may serve to regulate the interaction of actin and myosin and affect the force of muscular contraction.[187]

A number of experimental tests must be made in order to assess whether the various mechanisms for regulation of the activity of phosphorylase are operative *in vivo*. Two particular questions that need to be answered are: what are the concentrations *in vivo* of the various ligands such as AMP that affect the activity of phosphorylase b, and can the various events of the cascade mechanism depicted in Fig. 6.27 be demonstrated to occur *in vivo*? We shall deal with these questions in Sections 6.4.2.2 and 6.4.2.3, respectively.

* Catecholamines such as adrenaline and noradrenaline bind to two types of receptors known as α- and β-adrenergic receptors. They can be distinguished by their relative affinities for adrenaline, noradrenaline, and a variety of agonists and antagonists. α- and β-receptors show different tissue distribution and different responses when activated. α-Receptors promote phospholipase C activation, which causes an increase in inositol phosphate release and hence Ca^{2+}. β-Receptors activate adenylate cyclase and hence cAMP production.[182,183]

6.4.2.2 Concentrations of ligands *in vivo*

The concentrations of a number of metabolites in muscle have been determined by the freeze-clamping technique (Section 6.3.3.1). These studies have shown that the concentration of AMP in resting muscle (approximately 100 μmol dm^{-3}) is greater than the concentration required to cause 50 per cent of the maximal activation of phosphorylase b (approximately 50 μmol dm^{-3}). The negligible activity of phosphorylase in resting muscle therefore implies that *in vivo* the effect of AMP is counteracted by high concentrations of ligands such as ATP, ADP, and D-glucose 6-phosphate; a conclusion confirmed by measurements of the concentrations of these ligands in resting muscle.[188] Ligand-induced activation of phosphorylase b can therefore occur *in vivo* only when the concentrations of activating ligands rise and/or those of counteracting ligands fall. Such a situation has been shown to arise in the perfused rat heart under anaerobic conditions,[189] and perhaps more convincingly in the muscles of I-strain mice that are deficient in phosphorylase kinase and cannot therefore activate phosphorylase by the $b \to a$ conversion.[190] A detailed study of the concentrations of various ligands in the muscles of I-strain mice before and after electrical stimulation of the muscle has led to the conclusion that, in this case, the increase in the activity of phosphorylase is largely due to a rise in the concentration of IMP rather than of AMP.[191]

Finally, mention should be made of the fact that *in vivo* a substantial proportion of the phosphorylase in muscle is probably bound to glycogen in the form of the 'glycogen particle' (see Chapter 7, Section 7.13).

6.4.2.3 Demonstration of the steps in the cascade mechanism (Fig. 6.27)

The various steps in the cascade mechanism for the activation of phosphorylase have all been shown to occur *in vivo* in response to the administration of adrenalin.

(i) Conversion of phosphorylase b to phosphorylase a*

The phosphorylation of phosphorylase b has been demonstrated *in vivo* in rabbit skeletal muscle.[192] In order to demonstrate this it is necessary to prevent the hydrolysis of phosphorylase a during the isolation procedure by homogenizing the tissue in the presence of fluoride ions, which act as powerful inhibitors of phosphorylase phosphatase.

(ii) Phosphorylation of phosphorylase kinase

Both the α- and the β-subunits of rabbit muscle phosphorylase kinase are phosphorylated in response to an intravenous injection of adrenalin.[193,194] and the sites phosphorylated were the same as those phosphorylated *in vitro*.

(iii) Activation of cAMP-dependent protein kinase

It has proved more difficult to demonstrate convincingly that the cAMP-dependent protein kinase is activated *in vivo* in response to adrenalin. In theory it is possible to seek evidence for the cAMP-induced dissociation of the enzyme (Fig. 6.27) or for the action of the kinase. Reasonably clear-cut demonstrations of the activation of cAMP-dependent protein kinase have been afforded by the work on phosphorylase kinase ((ii) above) and by the observation that, in rat liver, histone H1 could be phosphorylated in response to hormones at the site that is labelled by cAMP-dependent protein kinase *in vitro*.[195] A discussion of the problems involved in this work is given in reference 196.

(iv) Increase in concentration of cAMP

Administration of adrenalin and other hormones causes an increase in the concentration of cAMP, although the changes are relatively small (see Section 6.4.2.1). The reason that the activation of adenylate cyclase does not lead to large

* The phosphorylation of α- and β-subunits increases the activity of phosphorylase kinase some 15–20-fold, but this activity is still completely dependent on Ca^{2+}.[177] At the concentration of Ca^{2+} in resting muscle (0.1 μmol dm^{-3}), the activity is likely to be very low. The ability of adrenalin to bring about the conversion of phosphorylase b to phosphorylase a is thought to be due in large part to the inhibition of protein phosphatase-1 (see Section 6.4.2.4).[177]

* There is a large family of phosphodiesterases in the cell with differing K_m values for cAMP and cGMP, but the one generally believed to play a major role in controlling the level of cAMP has a high K_m.[197]

increases in the concentration of cAMP is that cAMP is hydrolysed to AMP in a reaction catalysed by 3′:5′-cyclic-nucleotide phosphodiesterase* (Fig. 6.27). The K_m for cAMP is higher than its prevailing concentration, so that as the concentration of cAMP is raised by activation of adenylate cyclase, the activity of the phosphodiesterase is also raised.[160]

It is therefore essential that the relatively small changes in the concentration of cAMP lead to an amplified physiological response; this amplification is achieved by the system of interconvertible enzymes shown in Fig. 6.27.

Thus, in conclusion, it can be seen that the principal events in the proposed cascade system (Fig. 6.27) have been demonstrated to occur not only *in vitro*, but also *in vivo*. Before ending this discussion of the regulation of glycogen metabolism, two further aspects should be mentioned: first, the regulation of the enzymes that reverse the actions of the enzymes in the cascade system (Fig. 6.27), and second, the regulation of glycogen synthase, the controlling enzyme in the synthesis of glycogen (Fig. 6.26). These aspects are discussed in Sections 6.4.2.4 and 6.4.2.5, respectively.

6.4.2.4 How is phosphorylase switched off?

It appears inevitable that an elaborate control system of the type shown in Fig. 6.27 should possess mechanisms able to switch off phosphorylase and so prevent unnecessary depletion of glycogen reserves. As described in Section 6.4.2.3, cAMP is hydrolysed to AMP in a reaction catalysed by 3′:5′-cyclic-nucleotide phosphodiesterase. The action of the kinases (cAMP-dependent protein kinase, Ca^{2+}-calmodulin multiprotein kinase, and phosphorylase kinase) could be reversed by the action of one or more phosphatases (Section 6.4.2.1), thereby allowing much finer control of enzyme activity (Section 6.3.2.2) at the expense of consumption of a small amount of energy. Eukaryotic protein phosphatases dephosphorylate a wide range of phosphorylated proteins. They each belong to one of three gene families known as *PPP*, *PPM*, and *PTP*. The major families are *PPP* and *PPM* which dephosphorylate phosphoserine and phosphothreonine residues; the *PTP* dephosphorylates phosphotyrosine residues (see reference 198 for a review of the *PTP* which is not considered further here).

From information gained from the nematode (*Caenorhabditis elegans*) genome sequencing project it is estimated that in humans as many as ≈ 500 genes encode serine/threonine protein phosphatases.[199] The protein phosphatases comprise catalytic and regulatory subunits and the latter have an important role in determining the specificity. Since each catalytic subunit can associate with more than one different regulatory subunit, the number of potential protein phosphatases is very large. Much of the early work on protein phosphatases was related to the control of glycogen metabolism, and this has been an important factor in the initial classification. The most abundant groups of protein phosphatases are classified as PP1, PP2A, PP2B, and PP2C.[200-202] The main properties that have been used to distinguish the four groups are (i) whether they preferentially dephosphorylate the α- or β-subunits of phosphorylase kinase, (ii) their requirement for divalent metal ions, and (iii) whether they are inhibited by certain small inhibitor proteins, e.g. inhibitor-1 (Inh-1) and inhibitor 2 (Inh-2). Inhibitor-1 is a remarkably stable[†] protein of M_r 20 000 that is present in muscle at concentrations (1.5 μmol dm^{-3}) comparable with the concentrations of PP1, phosphorylase kinase, and glycogen synthase. More recently, a phylogenetic classification has become possible based on their gene sequences; it is now clear that PP1 is more closely related to PP2A and PP2B than is PP2C.[202] The main properties of these four groups are summarized on Table 6.6. Additional groups of protein

† The purification of this protein involves heating partially purified muscle extracts at 363 K (90°C) to precipitate unwanted proteins.

Table 6.6 Properties of protein phosphatases

Property	PP1	PP2A	PP2B	PP2C
Preference for α or β subunit of phosphorylase kinase	β	α	α	α
Divalent metal ion requirement	No	No	Ca^{2+}	Mg^{2+}
Inhibition by Inh-1 and Inh-2	Yes	No	No	No
Gene family	PPP	PPP	PPP	PPM
Possible physiological roles	glycogen metabolism, Ca^{2+} transport, muscle contraction, protein synthesis, cell cycle regulation	cell cycle regulation, intracellular localization of transcription factors	T-cell activation Ca^{2+} induced gene expression	osmoregulation via the MAP kinase cascade in yeast

See references 199, 200, and 202.

phosphatases known as PP4, PP5, and PP6 have also been described; for details of their properties see reference 201.

The range of functions of the different protein phosphatases is only beginning to be understood. Their role in the regulation of glycogen metabolism has been most extensively studied. The catalytic subunit of PP1 associates with a regulatory subunit, G. The latter can target PP1 to either the glycogen particle (see Chapter 7, Section 7.13) or the sarcoplasmic reticulum. When attached to the glycogen particle it is able to dephosphorylate glycogen phosphorylase and glycogen synthase, leaving the former in its most inactive state and the latter active. The inhibitor Inh-1 also has an important role. It can only inhibit PP1 after it has been phosphorylated on Thr-35. After an injection of adrenalin, cAMP-dependent protein kinase becomes activated in skeletal muscle. This promotes phosphorylation of Inh-1, which is then able to inhibit PP1 (Fig. 6.28). Glucagon administration can effect a similar change in rabbit liver where it also activates cAMP-dependent protein kinase.[202] PP1, PP2A, and PP2C account for the vast majority of protein phosphatase activity in muscle and act on a variety of enzymes involved in major metabolic pathways, including glycogen metabolism, glycolysis, fatty-acid synthesis, and amino-acid breakdown.[202]

6.4.2.5 The regulation of glycogen synthase

As mentioned in Section 6.4.2, glycogen synthase is the controlling enzyme in the synthesis of glycogen. Extensive work has shown that glycogen synthase is subject to a variety of controlling mechanisms that nicely complement those that exist for phosphorylase, thus permitting a coordinated control of glycogen metabolism.

Glycogen synthase can be converted from an active form (known as *a*, or I because its activity is **I**ndependent of D-glucose 6-phosphate) to an inactive form (known as *b*, or D because its activity is **D**ependent on D-glucose 6-phosphate)

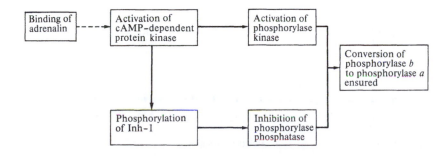

Fig. 6.28 Coordinated control of the activities of phosphorylase kinase and phosphorylase phosphatase following binding of adrenalin to the cell surface.

by the action of cAMP-dependent protein kinase.[203] Thus, as phosphorylase is switched on, glycogen synthase is switched off (Fig. 6.29).

Glycogen synthase is a tetramer (subunit M_r 86 000) and its sequence has been determined from a number of different sources.[204–206] It was the first example of regulation by 'multisite' phosphorylation,[207] unlike glycogen phosphorylase (Section 6.4.2.1) which is regulated by phosphorylation at a single site. Nine different serine residues are phosphorylated *in vivo* (Fig. 6.30) by at least six different protein kinases.[14] The cAMP-dependent protein kinase phosphorylates two serine side-chains near the C-terminus (residues 697 and 710) and a serine side-chain seven amino-acid residues from the N-terminus (N7); the latter can also be phosphorylated by phosphorylase kinase and by the Ca^{2+}-calmodulin multiprotein kinase. Sites for phosphorylation by other protein kinases are shown in Fig. 6.30. The various modifications produce forms of glycogen synthase with distinctly different catalytic properties, thereby increasing the versatility of the response to signals such as cAMP and Ca^{2+}. Phosphorylations that decrease glycogen synthase activity are those at residues 7, 10, 640, 644, and 648. Adrenalin inhibits glycogen synthase by increasing the extent of phosphorylation at all five of these sites, whereas insulin activates glycogen synthase by decreasing phosphorylation at residues 640, 644, and 648.[14]

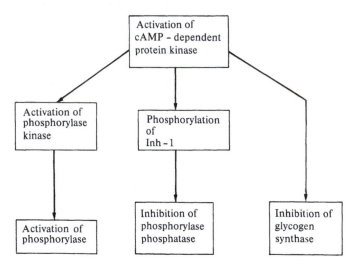

Fig. 6.29 Multiple effects of activation of cAMP-dependent protein kinase.

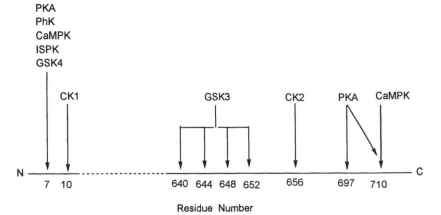

Fig. 6.30 The sites of phosphorylation of glycogen synthase.[14] PKA is a cAMP-dependent kinase; PhK is phosphorylase kinase; GSK3 and 4 are glycogen synthase kinases 3 and 4; CaMPK is calmodulin-dependent multiprotein kinase; ISPK is insulin-dependent protein kinase; and CK1 and 2 are casein kinases 1 and 2. The polypeptide chain is not shown to scale.

Insulin has multiple effects on metabolism and mitogenesis and the details of its signalling mechanism is still not completely understood.[208] It has been known for many years[209] that insulin stimulates the activity of glycogen synthase, thus allowing D-glucose to be converted to glycogen under conditions in which glycolysis is inhibited (Section 6.4.1.2). The hormone has little effect on the concentration of cAMP or on the activity of cAMP-dependent protein kinase. It does decrease phosphorylation of glycogen synthase at residues 640, 644 and 648 (Fig. 6.30). The major enzyme that phosphorylates residues 640, 644, and 648 of glycogen synthase is glycogen synthase kinase 3 (GSK3).[205] Protein phosphatase 1 (PP1) is able to dephosphorylate and hence activate glycogen synthase. The regulatory subunit (PP1-G) of PP1 (see Section 6.4.2.4) controls its binding to glycogen particles. Phosphorylation of the G subunit at a single site increases the activity of PP1 enabling it to dephosphorylate and thus activate glycogen synthase in skeletal muscle.[211] Insulin binds to its receptor, present on the cell membrane of insulin-sensitive cells. The structure of the receptor and the mechanism of activation of its protein tyrosine kinase are well understood.[208] Insulin may stimulate the activation of glycogen synthase, either by promoting its dephosphorylation involving PP1, or by promoting the inactivation of GSK3. In 1990 a cascade mechanism for activation of PP1 by insulin involving a group of kinases (insulin-sensitive protein kinase, mitogen-activated protein kinase (MAPkinase) and mitogen-activated protein kinase kinase was proposed.[212] However, more recently a second cascade involving the second messengers phosphatidylinositide-3,4,5-triphosphate (PtdIns[3,4,5]P$_3$) and phosphatidylinositide-3,4-diphosphate (PtdIns[3,4]P$_2$) has been proposed.[211] Specific drugs have been developed that can be used to inhibit either cascade. When the phosphatidylinositide cascade is inhibited, insulin no longer inactivates GSK3; by contrast when the mitogen-activated protein kinase cascade is inhibited, GSK3 activity is unaffected.[211] The insulin signal transduction mechanism that is considered to account for the stimulation by insulin of glycogen synthesis in skeletal muscle is given below.

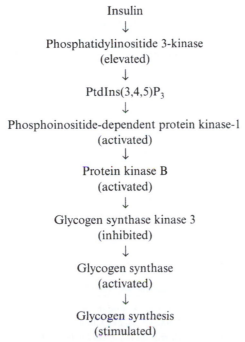

Insulin
↓
Phosphatidylinositide 3-kinase
(elevated)
↓
PtdIns(3,4,5)P$_3$
↓
Phosphoinositide-dependent protein kinase-1
(activated)
↓
Protein kinase B
(activated)
↓
Glycogen synthase kinase 3
(inhibited)
↓
Glycogen synthase
(activated)
↓
Glycogen synthesis
(stimulated)

6.4.2.6 Concluding remarks on the regulation of glycogen metabolism

The preceding Sections (6.4.2.1 to 6.4.2.5) have illustrated the variety of mechanisms by which glycogen metabolism can be regulated, e.g. ligand-induced conformational changes in enzymes, hormonal action, and nerve impulses acting via Ca^{2+} ions. These mechanisms allow a variety of responses in different metabolic circumstances. It might appear that the recent work mentioned has only served to confuse the issue, since it has now been found that a bewildering variety of enzymes and other proteins are involved in regulation. However, some of the recent work has demonstrated that considerable economy is exercised on the part of an organism, since some of the enzymes and proteins have multiple effects. For instance, cAMP-dependent protein kinase and Ca^{2+}-calmodulin multiprotein kinase have effects on enzymes involved in a number of metabolic pathways such as glycolysis, glycogen and lipid metabolism, and catecholamine synthesis.[210] Protein kinases thus appear to play a central role in transmitting the effects of intracellular signals such as cAMP and Ca^{2+} to the target enzymes, which then bring about the metabolic changes in the cell. It is also clear that the various phosphatases each have multiple activities (Section 6.4.2.4).

The regulation of glycogen metabolism is still under very active investigation and it is reasonable to expect that new features of the system will emerge; nevertheless, the account given here should give some impression of the current ideas on the subject.

6.5 Concluding remarks

In this chapter we have discussed the various means by which the activities of enzymes are controlled. The studies on single enzymes (Section 6.2) have provided the conceptual basis on which theories of the control of metabolic pathways have been formulated (Section 6.3). At present, we are in the position that the control of a number of single enzymes can be understood in quantitative terms, and some progress has been made towards a quantitative understanding of the control of metabolic pathways. However, when the cellular level of organization is considered (see Chapter 8), the situation becomes even more complicated, since a number of new factors such as compartmentation are introduced. Much of the future work will be directed towards a detailed understanding of these new factors.

References

1. Heimburger, N., in *Proteases and biological control* (Reich, E., Rilkin, D. B., and Shaw, F., eds), pp. 367–386. Cold Spring Harbor Laboratory (1975).

2. Aoyagi, T. and Umezawa, H., in reference 1, pp. 429–54.

3. Small, C. and Traut, T., *Trends Biochem. Sci.* **9**, 49 (1984).

4. Oguchi, M., Gerth, F., and Park, J. H., *J. biol. Chem.* **248**, 5571 (1973).

5. Lowry, O. H. and Passonneau, J. V., *J. biol. Chem.* **239**, 31 (1964).

6. Benkovic, S. J. and Schray, K. J., *Adv. Enzymol.* **44**, 139 (1976).

7. Dijkstra, B. W., Drenth, J., and Kalk, K. H., *Nature, Lond.* **289**, 604 (1981).

8. Cabib, E., *Trends Biochem. Sci.* **1**, 275 (1976).

9. Banks, P., Bartley, W., and Birt, L. M., *The biochemistry of the tissues* (2nd edn). Wiley, London (1976). See p. 159.

10. Furie, B. and Furie, B. C., *Cell* **53**, 505 (1988).

11. Reid, K. B. M., *Essays Biochem.* **22**, 27 (1986).

12. Fletterick, R. J. and Madsen, N. B., *A. Rev. Biochem.* **49**, 31 (1980).

13. Shacter, F., Chock, P. B., Rhee, S. G., and Stadtman, E. R., *Enzymes* (3rd edn) **17**, 21 (1986).

14. Cohen, P., *Biochem. Soc. Trans.* **21**, 555 (1993).

15. Murray, A. and Hunt, T., *The cell cycle*. Freeman, New York (1993). See Ch. 4.

16. Burris, R. H., *J. Biol. Chem.* **266**, 9339 (1991).

17. Pilkis, S. J., Claus, T. H., Kurland, I. J., and Lange, A., *A. Rev. Biochem.* **64**, 799 (1995).

18. Han, K.-K. and Martinage, A., *Int. J. Biochem.* **24**, 19 (1992).

19. Hunter, T., *Cell* **80**, 225 (1995).

20. Hunter, T., *Methods Enzymol.* **200**, 3 (1991).

21. Hancock, J. T., *Cell signalling.* Longmans, Harlow (1997). See Ch. 4.

22. Rhee, S. G., Chock, P. B., and Stadtman, E. R., *Adv. Enzymol.* **62**, 37 (1989).

23. Stadtman, E. R. and Chock, P. B., *Curr. Top. cell. Reg.* **13**, 53 (1978).

24. Chock, P. B. and Stadtman, E. R., *Methods Enzymol.* **64**, 297 (1980).

25. *Curr. Top. cell. Reg.* **27** (1985).

26. Cardenas, M. L. and Cornish-Bowden, A., *Biochem. J.* **257**, 339 (1989).

27. Sun, H. and Tonks, N. K., *Trends. Biochem. Sci.* **19**, 480 (1994).

28. Chock, P. B., Stadtman, E., Jurgensen, S. R., and Rhee, S. G., *Curr. Top. cell. Reg.* **27**, 3 (1985).

29. Goldbeter, A. and Koshland, D. F. Jr, *J. biol. Chem.* **262**, 4460 (1987).

30. Umbarger, H. E. and Brown, B., *J. biol. Chem.* **233**, 415 (1958).

31. Gerhart, J. C. and Pardee, A. B., *J. biol. Chem.* **237**, 891 (1962).

32. Monod, J., Changeux, J.- P., and Jacob, F., *J. molec. Biol.* **6**, 306 (1963).

33. Monod, J., Wyman, J., and Changeux, J.-P., *J. molec. Biol.* **12**, 88 (1965).

34. Cornish-Bowden, A., *Fundamentals of enzyme kinetics* (revised edn). Portland Press, London (1995). See Ch. 9.

35. Neet, K. E., *Methods Enzymol.* **249**, 519 (1995).

36. Hill, A. V., *J. Physiol. Lond.* **40**, iv (1910).

37. Forsén, S. and Linse, S., *Trends. Biochem. Sci.* **20**, 495 (1995).

38. Adair, G. S., *J. biol. Chem.* **63**, 529 (1925).

39. Cornish-Bowden, A., *Fundamentals of enzyme kinetics* (revised edn). Portland Press, London (1995). See p. 211ff.

40. Ricard, J. and Cornish-Bowden, A., *Eur. J. Biochem.* **166**, 255 (1987).

41. Steitz, T. A., Fletterick, R. J., Anderson, W. F., and Anderson, C. M., *J. molec. Biol.* **104**, 197 (1976).

42. Dalziel, K., *FEBS Lett.* **1**, 346 (1968).

43. Koshland, D. E. Jr, Némethy, G., and Filmer, D., *Biochemistry* **5**, 365 (1966).

44. Rubin, M. M. and Changeux, J.-P., *J. molec. Biol.* **21**, 265 (1966).

45. Eigen, M., *Q. Rev. Biophys.* **1**, 3 (1968).

46. Herzfeld, J. and Stanley, H. E., *J. molec. Biol.* **82**, 231 (1974).

47. Whitehead, E., *Prog. Biophys. molec. Biol.* **21**, 323 (1970).

48. Levitzki, A. and Koshland, D. E. Jr, *Curr. Top. cell. Reg.* **10**, 1 (1976).

49. Fersht, A. R., *Structure and mechanism in protein science.* Freeman, New York (1999). See p. 296.

50. Suter, P. and Rosenbusch, J. P., *J. biol. Chem.* **251**, 5986 (1976).

51. Citri, N., *Adv. Enzymol.* **37**, 397 (1973).

52. Price, N. C. and Dwek, R. A., *Principles and problems in physical chemistry for biochemists* (2nd edn). Clarendon Press, Oxford (1979). See Ch. 4.

53. Griffiths, J. R., Price, N. C., and Radda, G. K., *Biochim. biophys. Acta* **358**, 275 (1974).

54. Fuller Noel, J. K. and Schumaker, V. N., *J. molec. Biol.* **68**, 523 (1972).

55. Kirschner, K., *J. molec. Biol.* **58**, 51 (1971).

56. Kirschner, K., Gallego, E., Schuster, I., and Goodall, D., *J. molec. Biol.* **58**, 29 (1971).

57. Jackson, M. B., *Trends Biochem. Sci.* **19**, 396 (1994).

58. Changeux, J.-P. and Edelstein, S. J., *Trends Biochem. Sci.* **19**, 399 (1994).

59. Biesecker, G., Harris, J. I., Thierry, J. C., Walker, J. E., and Wonacott, A. J., *Nature, Lond.* **266**, 328 (1977).

60. Liang, J.-Y., Zhang, Y., Huang, S., and Lipscomb, W. N., *Proc. natn. Acad. Sci. USA* **90**, 2132 (1993).

61. Kantrowitz, E. R. and Lipscombe, W. N., *Science* **241**, 669 (1988).

62. Gouaux, J. E. and Lipscombe, W. N., *Biochemistry* **29**, 389 (1990).

63. Perutz, M. F., *Nature, Lond.* **228**, 726. (1970).

64. Perutz, M. F., *Q. Rev. Biophys.* **22**, 139 (1989).

65. Schirmir, T. and Evans, P. R., *Nature* **343**, 140 (1990).

66. Deville-Bonne, D., Bougain, F., and Garel, J.-R., *Biochemistry* **30**, 5750 (1991).

67. Johnson, J. L. and Reinhart, G. D., *Biochemistry* **36**, 12814 (1997).

68. Ackers, G. K. and Hazzard, J. H., *Trends. Biochem. Sci.* **18**, 385 (1993).

69. Pilkis, S. J., Weber, I. T., Harrison, R. W., and Bell, G. I., *J. Biol. Chem.* **269**, 21925 (1994).

70. Cornish-Bowden, A., and Koshland, D. E. Jr, *J. molec. Biol.* **95**, 201 (1975).

71. Voet, D. and Voet, J. G., *Biochemistry* (2nd edn). Wiley, New York (1995). See Ch. 9.

72. Citri, N., *Adv. Enzymol.* **37**, 397 (1973). See p. 557.

73. Frieden, C., *A. Rev. Biochem.* **48**, 471 (1979).

74. Bisswanger, H., *J. biol. Chem.* **259**, 2457 (1984).

75. Colowick, S. P., *Enzymes* (3rd edn) **9**, 1 (1973). See p. 45.

76. Neet, K. E. and Ainslie, G. R. Jr, *Methods Enzymol.* **64**, 192 (1980).

77. Hardie, D. G., *Biochemical messengers.* Chapman and Hall, London (1991). See Ch. 1.

78. Hardie, D. G., *Biochemical messengers.* Chapman and Hall, London (1991). See Ch. 7.

79. Garbers, D. L. and Lowe, D. G., *J. Biol. Chem.* **269**, 30741 (1994).

80. Galione, A., *Science* **259**, 325 (1993).

81. Guse, A. H., Da Silva, C. P., Weber, K., Ashamu, G. A., Potter, B. V. L., and Mayr, G. W., *J. Biol. Chem.* **271**, 23946 (1996).

82. Clapham, D. E., *Cell* **80**, 259 (1995).

83. Divecha, N. and Irvine, R. F., *Cell* **80**, 269 (1995).

84. White, M. F. and Kahn, C. R., *J. biol. Chem.* **269**, 1 (1994).

85. Kennelly, P. J. and Krebs, E. G., *J. biol. Chem.* **266**, 15555 (1991).

86. Pinna, L. A. and Ruzzene, M., *Biochim. Biophys. Acta* **1314**, 191 (1996).

87. Stadtman, E. R., *Enzymes* (3rd edn) **1**, 397 (1970).

88. Pittard, J. and Gibson, F., *Curr. Top. cell. Reg.* **2**, 29 (1970).

89. Newsholme, E. A. and Leech, A. R., *Biochemistry for the medical sciences*. Wiley, Chichester (1983). See Ch. 7.

90. Newsholme, E. A. and Leech, A. R., *Biochemistry for the medical sciences*. Wiley, Chichester (1983). See Ch. 11.

91. Katz, J. and Rognstad, R., *Curr. Top. cell. Reg.* **10**, 237 (1976).

92. Steinberg, D., *Biochem. Soc. Symp.* **24**, 111 (1963).

93. Fell, D., *Understanding the control of metabolism*. Portland Press, London (1997). See Ch. 7, Section 7.3.

94. Shader, E., Chock, P. B., Rhee, S. G., and Stadtman, E. R., *Enzymes* (3rd edn) **17A**, Ch. 2 (1986).

95. Cardenas, M. L. and Cornish-Bowden, A., *Biochem. J.* **257**, 339 (1989).

96. Kerbey, A. L., Randle, P. J., Cooper, R. H., Whitehouse, S., Pask, H. T., and Denton, R. M., *Biochem. J.* **154**, 327 (1976).

97. Newsholme, E. A. and Start, C., *Regulation in metabolism*. Wiley, London (1973). See p. 168.

98. Nimmo, H. G. and Cohen, P., *Adv. cyclic Nucleotide Res.* **8**, 145 (1977). See p. 193.

99. Cohen, P. and Antoniw, J. F., *FEBS Lett.* **34**, 43 (1973).

100. Kacser, H. and Burns, J. A., *Symp. Soc. exp. Biol.* **32**, 65 (1973).

101. Kacser, H. and Burns, J. A., *Biochem. Soc. Trans.* **7**, 1149 (1979).

102. Rapoport, T. A., Heinrich, R., Jacobasch, G., and Rapoport, S., *Eur. J. Biochem.* **42**, 107 (1974).

103. Crabtree, B. and Newsholme, E. A., *Curr. Top. cell Reg.* **25**, 21 (1985).

104. Fell, D., *Understanding the control of metabolism*. Portland Press, London (1997). Chapters 5 and 6.

105. Heinrich, R. and Rapoport, T. A., *Eur. J. Biochem.* **42**, 97 (1974).

106. Martin, B. R., *Metabolic regulation: a molecular approach*. Blackwell, Oxford (1987). See Ch. 1.

107. Newsholme, E. A. and Start, C., *Regulation in metabolism*. Wiley, London (1973). See pp. 142–4.

108. Williamson, J. R., *J. biol. Chem.* **240**, 2308 (1965).

109. Rolleston, F. S., *Curr. Top. cell. Reg.* **5**, 47 (1972).

110. Cohen, S. M., *Methods. Enzymol.* **177**, 417 (1989).

111. Evans, J. N. S., *Biomolecular NMR spectroscopy*. Oxford University Press (1995) See Section 6.7.

112. London, R. E., Gabel, S. A. and Perlman, M. E., in *NMR applications in biopolymers* (Finley, J. W., Schmidt, S. J., and Serianni, A. S., eds). Plenum, New York (1990). See pp. 349–60.

113. Seraydarian, K., Mommaerts, W. F. H. M., and Waliner, A., *Biochim. biophys. Acta* **65**, 443 (1962).

114. Lowry, O. H., Passonneau, J. V., Hasselberger, F. X., and Schultz, D. W., *J. biol. Chem.* **239**, 18 (1964).

115. Fell, D. A., *Trends Biochem. Sci.* **9**, 515 (1984).

116. Fell, D., *Understanding the control of metabolism*. Portland Press, London (1997). p. 105ff.

117. Cornish-Bowden, A. and Hofmeyr, J.-H. S., *Comp. Appl. Biosci.* **7**, 89 (1991).

118. Mendes, P., *Comp. Appl. Biosci.* **9**, 563 (1993).

119. El-Mansi, E. M. T., Dawson, G. C., and Bryce, C. F. A., *Comp. Appl. Biosci.* **10**, 295 (1994).

120. Sauro, H. M., *Comp. Appl. Biosci.* **9**, 441 (1993).

121. Flint, H. J., Tateson, R. W., Bartelmess, I. B., Porteous, D. J., Donachie, W. D., and Kacser, H., *Biochem. J.* **200**, 231 (1981).

122. Salter, M., Knowles, R. G., and Pogson, C. I., *Biochem. J.* **234**, 635 (1986).

123. Small, J. R. and Kacser, H., *Eur. J. Biochem.* **213**, 613 (1993).

124. Groen, A. K., van Roermund, C. W. T., Vervoorn, R. C., and Tager, J. M., *Biochem. J.* **237**, 379 (1986).

125. Fell, D., *Understanding the control of metabolism*. Portland Press, London (1997). p. 169.

126. Cornish-Bowden, A., *Fundamentals of enzyme kinetics* (revised edn). Portland Press, London (1995). See Ch. 10.

127. Michels, P. A. M., Chevalier, N., Opperdoes, F. R., Rider, M. H., and Rigden, D. J., *Eur. J. Biochem* **250**, 698 (1997).

128. Merteus, E., *FEBS Lett.* **285**, 1 (1991).

129. Ruijter, G., Panneman, H., and Visser, J., *Biochim. biophys. Acta* **1334**, 317 (1997).

130. Heinisch, J., *Mol. Gen. Genet.* **202**, 75 (1986).

131. Thomas, S., Mooney, P. J. F., Burrell, M. M., and Fell, D. A., *Biochem. J.* **322**, 119 (1997).

132. Stadtman, E. R., *Adv. Enzymol.* **28**, 41 (1966).

133. Bloxham, D. P. and Lardy, H. A., *Enzymes* (3rd edn) **8**, 239 (1973).

134. Martin, B. R., *Metabolic regulation: a molecular approach*. Blackwell, Oxford (1987). See Ch. 14.

135. Ahlfors, C. E. and Mansour, T. E., *J. biol. Chem.* **244**, 1247 (1969).

136. Sacktor, B. and Hurlbut, E. C., *J. biol. Chem.* **241**, 632 (1966).

137. Atkinson, D. E., *Biochemistry* **7**, 4030 (1968).

138. Shen, L. C., Fall, L., Walton, G. M., and Atkinson, D. E., *Biochemistry* **7**, 4041 (1968).

139. Newsholme, E. A. and Crabtree, B., *FEBS Lett.* **7**, 195 (1970).

140. Clark, M. G., Bloxham, D. P., Holland, P. C., and Lardy, H. A., *Biochem. J.* **134**, 589 (1973).

141. Clark, M. G., Bloxham, D. P., Holland, P. C., and Lardy, H. A., *J. biol. Chem.* **249**, 279 (1974).

142. Bloxham, D. P., Clark, M. G., Holland, P.C., and Lardy, H. A., *Biochem. J.* **134**, 581 (1973).

143. Garfinkel, L., Kohn, M. C., and Garfinkel, D., *Eur. J. Biochem.* **96**, 183 (1979).

144. Haseman, C. A., Istvan, E. S., Uyeda, K., and Deisenhofer, J., *Structure* **4**, 1017 (1996).

145. van Schaftingen, E., *Adv. Enzymol.* **59**, 315 (1987).

146. Koerner, T. A. W. Jr, Voll, R. J., and Younathan, E. S., *FEBS Lett.* **84**, 207 (1977).

147. Mayr, G. W., *Eur. J. Biochem.* **143**, 513 (1984).

148. Mayr, G. W., *Eur. J. Biochem.* **143**, 521 (1984).

149. Pilkis, S, J., Claus, T. H., Kountz, P. D., and El-Maghrabi, M. R., *Enzymes* (3rd edn) **18**, 3 (1987).

150. Cai, G.-Z., Calluci, T. P., Luther, M. A., and Lee, J. C., *Biophys. Chem.* **64**, 199 (1997).

151. Puigjaner, J., Rais, B., Burgos, M., Comin, B., Ovadi, B., and Cascante, M., *FEBS Lett.* **418**, 47 (1997).

152. Suarez, R. K., Staples, J. F., Lighton, J. R. B., and West, T. G., *Proc. natn. Acad. Sci. USA* **94**, 7065 (1997).

153. Iynedjian, P. B., *Biochem. J.* **293**, 1 (1993).

154. Seoane, J., Gomez-Foix, A. M., O'Doherty, R. M., Gomez-Ara, C., Newgard, C. B., and Guinovart, J. J., *J. biol. Chem.* **271**, 23760 (1996).

155. Barengo, R., Flawia, M., and Krisman, C. R., *FEBS Lett.* **53**, 274 (1975).

156. Krisman, C. R. and Barengo, R., *Eur. J. Biochem.* **52**, 117 (1975).

157. Taylor, C., Cox, A. J., Kernohan, J. C., and Cohen, P., *Eur. J. Biochem.* **51**, 105 (1975).

158. Caudwell, F. B. and Cohen, P., *Eur. J. Biochem.* **109**, 391 (1980).

159. Newsholme, E. A. and Leech, A. R., *Biochemistry for the medical sciences.* Wiley, Chichester (1983). See Ch. 9.

160. Newsholme, E. A. and Start, C., *Regulation in metabolism.* Wiley, London (1973). See Ch. 4.

161. Banks, P., Bartley, W., and Birt, L. M., *The biochemistry of the tissues* (2nd edn). Wiley, London (1976). See p. 208.

162. Cori, C. F. and Cori, G. T., *Proc. Soc. Exp. Biol. Med.* **34**, 702 (1936).

163. Fletterick, R. J. and Madsen, N. B., *A. Rev. Biochem.* **49**, 31 (1980).

164. Madsen, N. B., *Enzymes* (3rd edn) **17**, 365 (1986).

165. Johnson, L. N. and Barford, D., *J. biol. Chem.* **265**, 2409 (1990).

166. Oikonomakos, N. G., Acharya, K. R., and Johnson, L. N., *Posttranslational modification of proteins* (Harding, J. J. and Crabbe, M. J. C., eds). CRC Press, Boca Raton, Florida (1992). pp. 81–151.

167. Birkett, D. J., Radda, G. K., and Salmon, A. G., *FEBS Lett.* **11**, 295 (1970).

168. Sergienko, E. A. and Srivastava, D. K., *Biochem. J.* **328**, 83 (1997).

169. Barford, D. and Johnson, L. N., *Nature* **340**, 609 (1989).

170. Sprang, S. R., Withers, S. G., Goldsmith, E. J., Fletterick, R. J., and Madsen, N. B., *Science* **254**, 1367 (1991).

171. Taussig, R. and Gilman, A. G., *J. biol. Chem.* **270**, 1 (1995).

172. Beebe, S. J. and Corbin, J. D., *Enzymes* (3rd edn) **17A**, 44 (1986).

173. Gibson, R. M. and Taylor, S. S., *J. biol. Chem.* **272**, 31998 (1997).

174. Nimmo, H. G. and Cohen, P., *Adv. cyclic Nucleotide Res.* **8**, 145 (1977). See p. 192.

175. Nimmo, H. G. and Cohen, P., *Adv. cyclic Nucleotide Res.* **8**, 145 (1977). See p. 183.

176. Nimmo, H. G. and Cohen, P., *Adv. cyclic Nucleotide Res.* **8**, 145 (1977). See p. 185.

177. Cohen, P., *Biochem. Soc. Trans.* **15**, 999 (1987).

178. Soderling, T. R., *Biochim. biophys. Acta* **1297**, 131 (1996).

179. Braun, A. P. and Schulman, H., *A. Rev. Physiol.* **57**, 417 (1995).

180. Pickett-Giess, C. A. and Walsh, D. A., *Enzymes* (3rd edn) **17**, 396 (1986).

181. Lowe, E. D., Noble, M. E. M., Skamnaki, V. T., Oikonomakos, N. G., Owen, D. J., and Johnson, L. N., *EMBO J.* **16**, 6646 (1997).

182. Lefkowitz, R. J., Stadel, J. M., and Caron, M. G., *A. Rev. Biochem.* **52**, 159 (1983).

183. Dohlman, H. G., Thorner, J., Caron, M. G., and Lefkowitz, R. J., *A. Rev. Biochem.* **60**, 653 (1991).

184. Pickett-Geis, C. A. and Walsh, D. A. *Enzymes* (3rd edn) **17A**, Section 7 (1986).

185. Cohen, P., *Eur. J. Biochem.* **151**, 439 (1985).

186. Cohen, P., *Nature, Lond.* **296**, 613 (1982).

187. Hardie, D. G., *Biochemical messengers.* Chapman and Hall, London (1991). See p. 221.

188. Morgan, H. E. and Parmeggiani, A., *J. biol. Chem.* **239**, 2440 (1964).

189. Morgan, H. E. and Parmeggiani, A., *J. biol. Chem.* **239**, 2335 (1964).

190. Lyon, J. B. Jr. and Porter, J., *J. biol. Chem.* **238**, 1 (1963).

191. Rahim, Z. H. A., Perrett, D., Lutaya, G., and Griffiths, J. R., *Biochem. J.* **186**, 331 (1980).

192. Mayer, S. E. and Krebs, E. G., *J. biol. Chem.* **245**, 3153 (1970).

193. Pickett-Gies, C. A. and Walsh, D. A., *Enzymes* (3rd edn) **17A**, 408 (1986).

194. Pickett-Gies, C. A. and Walsh, D. A., *Enzymes* (3rd edn) **17A**, Section VIII (1986).

195. Langan, T. A., *Proc. natn. Acad. Sci. USA* **64**, 1276 (1969).

196. Nimmo, H. G. and Cohen, P., *Adv. cyclic Nucleotide Res.* **8**, 145 (1977). See p. 173.

197. Charbonneau, H., in *Cyclic nucleotide phosphodiesterases: structure, regulation and drug action* (Beavo, J. and Houslay, M. eds). Wiley, New York. (1990), pp. 267–96.

198. Fauman, E. B. and Saper, M. A., *Trends. Biochem. Sci.* **21**, 413 (1996).

199. Zolnierowicz, S. and Hemmings, B. A., in *Signal transduction* (Heldin, C.-H. and Purton, M., eds). Chapman and Hall, London (1996). See Ch. 18.

200. Barford, D., *Trends. Biochem. Sci.* **21**, 407 (1996).

201. Cohen , P. T. W., *Trends. Biochem. Sci.* **22**, 245 (1996).

202. Wera, S. and Hemmings, B. A., *Biochem. J.* **311**, 17 (1995).

203. Soderling, T. R., Hickenbottom, J. P., Reimann, E. M., Hunkeler, F. L., Walsh, D. A., and Krebs, E. G., *J. biol. Chem.* **245**, 6317 (1970).

204. Zhang, W., Browner, M. F., Fletterick, R. J., Depaoli-Roach, A. A., and Roach, P. J., *FASEB J.* **3**, 2532 (1989).

205. Bai, G., Zhang, Z., Werner, R., Nuttal, F. Q., Tan, A. W. H., and Lee, E. Y. C., *J. biol. Chem.* **265**, 7843 (1990).

206. Nuttal, F. Q., Gannon, M., Bai, G., and Lee, E. Y. C., *Arch. Biochem. Biophys.* **311**, 443 (1994).

207. Cohen, P., *Trends Biochem. Sci.* **1**, 38 (1976).

208. White, M. F. and Kahn, C. R., *J. biol. Chem.* **269**, 1 (1994).

209. Hardie, D. G., *Biochemical messengers.* Chapman and Hall (1991). See Ch. 2.

210. Hardie, D. G., *Biochemical messengers.* Chapman and Hall (1991). See Ch. 8.

211. Cohen, P., Alessi, D. R. and Cross, D. A. E., *FEBS Lett.* **410**, 3 (1997).

212. Dent, P., Lavoinne, A.., Nakielny, S., Caudwell, F. B., Watt, P. and Cohen, P. *Nature (London)* **348**, 302 (1990).

Appendix 6.1

The Adair equation and the Scatchard equation for a protein containing multiple ligand-binding sites

Consider a system in which one molecule of protein, P, can bind up to n molecules of ligand, A.

	Sites occupied	Sites unoccupied
$P + A \rightleftharpoons PA$	1	$n - 1$
$PA + A \rightleftharpoons PA_2$	2	$n - 2$
$PA_2 + A \rightleftharpoons PA_3$	3	$n - 3$
$PA_{n-1} + A \rightleftharpoons PA_n$	n	0

Let the successive dissociation constants for PA, PA_2 ... be K_1, K_2, etc.

$$\text{i.e.} \quad K_1 = \frac{[P][A]}{[PA]}, \quad K_2 = \frac{[PA][A]}{[PA_2]}, \quad \text{etc.}$$

A.6.1. The Adair equation

The fractional saturation of ligand sites (Y) is given by

$$Y = \frac{\text{total concentration of A bound}}{\text{total concentration of sites for A}}$$

$$= \frac{[PA] + 2[PA_2] + 3[PA_3]...*}{n([P] + [PA] + [PA_2] + [PA_3]...)}$$

$$= \frac{\dfrac{[P][A]}{K_1} + \dfrac{2[P][A]^2}{K_1 K_2} + \dfrac{3[P][A]^3}{K_1 K_2 K_3} + ...}{n\left[[P] + \dfrac{[P][A]}{K_1} + \dfrac{[P][A]^2}{K_1 K_2} + \dfrac{[P][A]^3}{K_1 K_2 K_3} + ...\right]},$$

* Note the factor n in the denominator. This arises because each molecule of P has n sites for A. The factors 1, 2, 3, etc. in the numerator arise because each molecule of PA_2 contains two molecules of A, and so on.

$$\therefore Y = \cfrac{\dfrac{[A]}{K_1} + \dfrac{2[A]^2}{K_1K_2} + \dfrac{3[A]^3}{K_1K_2K_3} + \ldots}{n\left[1 + \dfrac{[A]}{K_1} + \dfrac{[A]^2}{K_1K_2} + \dfrac{[A]^3}{K_1K_2K_3} + \ldots\right]}. \qquad (A.6.1)$$

This is the Adair equation.

For the special case of a tetrameric protein with four sites for A, $n = 4$. Then eqn (A.6.1) becomes

$$Y = \cfrac{\dfrac{[A]}{K_1} + \dfrac{2[A]^2}{K_1K_2} + \dfrac{3[A]^3}{K_1K_2K_3} + \dfrac{4[A]^4}{K_1K_2K_3K_4}}{4\left[1 + \dfrac{[A]}{K_1} + \dfrac{[A]^2}{K_1K_2} + \dfrac{[A]^3}{K_1K_2K_3} + \dfrac{[A]^4}{K_1K_2K_3K_4}\right]}. \qquad (A.6.2)$$

This is the Adair equation for a tetramer (see eqn (6.6)).

A.6.2 The Scatchard equation

We introduce a new parameter, r, which is the average number of molecules of A bound per molecule of P.

By definition, $r = nY$ where n equals the number of ligand-binding sites and Y equals the fractional saturation of ligand-binding sites. From eqn (A.6.1),

$$r = \cfrac{\dfrac{[A]}{K_1} + \dfrac{2[A]^2}{K_1K_2} + \dfrac{3[A]^3}{K_1K_2K_3} + \ldots}{1 + \dfrac{[A]}{K_1} + \dfrac{[A]^2}{K_1K_2} + \dfrac{[A]^3}{K_1K_2K_3} + \ldots}.$$

In order to simplify this quite general expression, we have to drive a relationship between the successive K_s. Now if it is assumed that the sites are independent and equivalent (i.e. that the free energy of interaction of the ligand with each site is the same), then the K_s are related to each other by statistical factors, i.e. A can dissociate from the PA_2 complex in two ways, but A can associate with the $(n-1)$ vacant sites in the PA complex in $(n-1)$ ways. The general relationship is that the ith dissociation constant (K_i) is given by

$$K_i = \left[\frac{i}{n-i+1}\right]K,$$

where K is an intrinsic dissociation constant (i.e. one that takes into account these statistical factors).*

* K is actually the geometric mean of all the dissociation constants, i.e. $K = (K_1K_2K_3\ldots K_n)^{1/n}$.

Thus
$$K_1 = \frac{K}{n}, \qquad K_2 = \frac{2K}{n-1}, \qquad K_3 = \frac{3K}{n-2}, \quad \text{etc.}$$

Our expression for r now becomes

$$r = \cfrac{[A]\dfrac{(n)}{K} + \dfrac{2[A]^2(n)(n-1)}{2K^2} + \dfrac{3[A]^2(n)(n-1)(n-2)}{(2)(3)K^3} + \ldots}{1 + \dfrac{[A](n)}{K} + \dfrac{[A]^2(n)(n-1)}{2K^2} + \dfrac{[A]^3(n)(n-1)(n-2)}{(2)(3)K^3} + \ldots}$$

$$= \frac{\dfrac{[A](n)}{K}\left[1 + \dfrac{[A](n-1)}{K} + \dfrac{[A]^2(n-1)(n-2)}{2K^2} + \ldots\right]}{1 + \dfrac{[A](n)}{K} + \dfrac{[A]^2(n)(n-1)}{2K^2} + \dfrac{[A]^3(n)(n-1)(n-2)}{6K^3} + \ldots}$$

The expression in the numerator and denominator in this equation are both binominal expansions. Thus the expression can be simplified to give

$$r = \frac{\dfrac{[A](n)}{K}\left[1 + \dfrac{[A]}{K}\right]^{n-1}}{\left[1 + \dfrac{[A]}{K}\right]^{n}}$$

$$= \frac{\dfrac{[A](n)}{K}}{\left[1 + \dfrac{[A]}{K}\right]}$$

$$\therefore \quad r = \frac{n[A]}{K + [A]}.$$

This equation can be rearranged to give the Scatchard equation:

$$\frac{r}{[A]} = \frac{n}{K} - \frac{r}{K}. \tag{A.6.3}$$

Thus, a plot of $r/[A]$ against r gives a straight line of slope $-1/K$ and intercept on x-axis of n, *provided that the sites are equivalent and independent* (see Fig. 6.15).

7

Enzymes in organized systems

7.1 Introduction

In the previous chapters we have discussed results obtained from the study of highly purified enzymes. By examining single purified enzymes it is possible to study the structure (Chapter 3), the kinetics (Chapter 4), and mechanism of action (Chapter 5) of a single enzyme in great detail and without interference from competing reactions, and this is essential for an understanding of the molecular basis of enzyme catalysis. However, in the intact cell, enzymes do not act as separated, isolated catalysts in dilute aqueous buffers with kinetics approximating to those described by the Michaelis and Menten equation (see Chapter 4, eqn (4.4)). In many subcellular compartments the protein concentrations are high (e.g. in the mitochondrial matrix, 500 mg protein cm^{-3}),[1] and many enzymes compete with one another for substrates and effectors and they may be physically associated with other enzymes to varying extents. In the next two chapters we consider how enzymes may behave in the intact cell, first by examining the types of organized enzyme systems that exist in the cell, and second by examining the nature of the environment in which enzymes operate *in vivo* and how this differs from the conditions under which most enzyme assays are performed *in vitro*.

7.2 Organized enzyme systems

The organization of enzymes can be seen as a progression of increasing complexity extending from those enzymes that exist as single separate polypeptide chains, through to oligomeric enzymes, which may show allosteric interactions between the subunits, and ultimately to proteins that catalyse more than one reaction. The enzyme nomenclature devised by the Enzyme Commission (see Chapter 1, Section 1.5.1, and reference 2) is primarily concerned with classifying enzymes on the basis of the reactions they catalyse rather than their physical organization. Proteins having more then one catalytic activity are referred to as enzyme systems, e.g. the pyruvate dehydrogenase system. However, when considering the physical organization of such a system it is useful to have a more discriminating nomenclature. The term multienzyme protein is used to describe all proteins with multiple catalytic domains (see Chapter 1, Section 1.5.3, and references 2 and 3). They are subdivided into those that are covalently linked (multienzyme polypeptides) and those that are non-covalently linked (multienzyme complexes).

In the previous four chapters, we have already considered enzymes having single polypeptide chains and oligomeric enzymes and thus in the next sections we examine the following types of organization: (i) enzymes which although catalysing a single reaction, also interact in a specific manner with a macromolecular template and therefore require a more complex structure; (ii) multienzyme proteins with well-defined structures, either multienzyme complexes or multienzyme polypeptides, which have more than one distinct catalytic activity; and (iii) loose associations of enzymes in which the stoichiometry may be variable. DNA-directed RNA polymerase (abbreviated to RNA polymerase in this chapter) is the best example of the first type, since, in addition to catalysing nucleotide transfers, the enzyme has to recognize specific regions on the DNA template in order to initiate and to terminate the synthesis of particular species of RNA. This is discussed in Section 7.3. The proteasome, which plays an important role in protein turnover is another example and is described in Chapter 9, Section 9.6.2. There are several examples of the second and third types (Table 7.1), e.g. pyruvate dehydrogenase (Section 7.7), tryptophan synthase (Section 7.9), and fatty-acid synthase (Section 7.11.1), in which more than one step is catalysed by the system concerned. The third type is represented by the glycogen particle (Section 7.13).

7.3   RNA polymerase from *E. coli*

RNA polymerase from *E. coli* represents an intermediate stage in the complexity of its organization between that of a typical allosteric enzyme and a large multienzyme protein. It comprises four different polypeptide chains, each having distinct functions in the complex, but at the same time each helping to coordinate the synthesis of RNA on a DNA template. The RNA polymerase from *E. coli* is discussed specifically here for the following reasons: (i) it was the first to be obtained in a highly purified state, and now using overexpression systems (see Chapter 2, Section 2.8.3) 10 mg quantities of each subunit can be readily obtained,[27] and (ii) it is responsible for catalysing the synthesis of all the RNA (mRNA, rRNA, and tRNA) present in uninfected *E. coli* cells, in contrast to the situation in eukaryotes where there is a multiplicity of RNA polymerases serving different functions. The enzyme catalyses the complete transcription cycle (Fig. 7.1), which includes the steps of (i) recognition of the specific promoter regions on the genome, (ii) transcribing the appropriate gene or group of genes controlled by a particular operator, and (iii) terminating transcription at the appropriate region of the genome.

The RNA polymerase is thus concerned with the synthesis of a large number of different mRNAs, each subject to different controls, and also with the synthesis of tRNA and rRNA. (For reviews on transcription and RNA polymerases, see references 28–31.)

7.3.1   Structure of RNA polymerase

The purified holoenzyme from *E. coli* has an M_r value of $\approx 449\,000$ and is composed of four different polypeptide chains, α (M_r 36 512), β (M_r 150 619), β' (M_r 155 162), and σ (M_r 70 236) and has the composition $\alpha_2\beta\beta'\sigma$. The genes for the polypeptides (*rpoA*, *rpoB*, *rpoC*, and *rpoD*) have been cloned and sequenced. The enzyme contains two Zn atoms, which are not removed by dialysis against

Table 7.1 Examples of multienzyme proteins

System	Function	No. of different catalytic centres	Structure-multienzyme complex or polypeptide	References
Pyruvate dehydrogenase	Pyruvate → acetyl CoA	3	Complex, three different polypeptides in prokaryotes	4–6
2-Oxoglutarate dehydrogenase	2-Oxoglutarate → succinyl CoA	3	Complex, three different polypeptides	4–6
Branch-chained oxoacid dehydrogenase	2-Oxoacid → acyl CoA	3	Complex, three different polypeptides	7
Aspartokinase homoserine dehydrogenase	Amino acid biosynthesis	2	Polypeptide	8
γ-Glutamyl kinase-glumatic acid semialdehyde dehydrogenase	Two steps in proline biosynthesis	2	Polypeptide	9
Phosphoribosyl anthranilate isomerase-indolyl glycerol phosphate synthase	Intermediate stages of tryptophan biosynthesis	2 in Gram negative bacteria 3 (linked to anthranilate synthase) in fungi	Polypeptide	10, 11
Tryptophan synthase	Final steps in tryptophan biosynthesis	2	Complex in *E. coli*, polypeptide in lower eukaryotes	12
arom complex	Aromatic amino-acid biosynthesis	5	Polypeptide (fungi)	13
CAD (see Section 7.10)	First three steps in pyrimidine biosynthesis	3	Polypeptide	13
Orotate phosphoribosyl transferase-orotidine 5′-phosphate decarboxylase	Final steps in uridylic acid synthesis orotate to UMP	2	Polypeptide	14
Aminoimidazole ribonucleotide carboxylase-SAICAR synthetase	Two steps in purine biosynthesis	2	Polypeptide	15
Enoyl CoA hydratase-enoyl CoA isomerase in *E. coli*	Fatty acid oxidation	2	Polypeptide	16
β-Oxidation complex (pig heart mitochondria)	Long-chain enoyl CoA hydratase and 3-hydroxyacyl CoA dehydrogenase	2	Polypeptide	17
F6P ⇌ Man-6P; Man-1P + GTP ⇌ GDP-Man + PP$_i$	Alginate (polysaccharide) biosynthesis	2	Polypeptide in *Pseudomonas*	18
CO$_2$ fixation complex	CO$_2$ incorporation into the Benson–Calvin cycle	5	Complex	19
Fatty-acid synthase	Malonyl CoA to long-chain Acyl CoA	7	Polypeptide	20
6-Deoxyerythronolide synthase (*Streptomyces*)	Polyketide biosynthesis	?	Polypeptide	13
Peptide synthetase	Gramicidin S biosynthesis	3	Complex	21
Glycine decarboxylase	2 glycine → serine + NH$_3$ + CO$_2$	2 catalytic + 2 binding proteins	Complex	22, 23
6-Phosphofructokinase/fructose 2,6-bisphosphatase	Interconversion of F6P and F2,6BP-regulation of glycolysis	2	Polypeptide	24
DNA polymerase III	DNA synthesis	3	Complex	25
UDP-*N*-acetylglucosamine 2-epimerase/*N*-acetylmannosamine kinase	Biosynthesis of *N*-acetylneuraminic acid	2	Polypeptide associates as A$_2$ and A$_6$	26

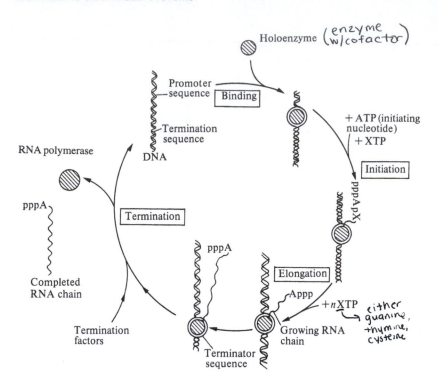

Holoenzyme *(enzyme w/cofactor)*

+*n*XTP *either guanine, +thymine, cysteine*

Fig. 7.1 The cycle of events catalysed by RNA polymerase from *E. coli*. The release of the σ subunit following initiation is not shown.

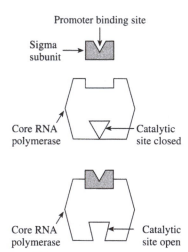

Fig. 7.2 Model showing the possible change in conformation when the RNA polymerase core enzyme binds the σ.[33]

chelating agents, although they are released when the enzyme is dissociated. They are associated with a zinc finger motif on the β' subunit which interacts with the DNA template.[32] The σ subunit can dissociate from the holoenzyme $(\alpha_2\beta\beta'\sigma)$ to yield a core enzyme $(\alpha_2\beta\beta')$. Both the holoenzyme and the core enzyme will catalyse RNA synthesis on a DNA template *in vitro* (see Fig. 7.1 under binding, initiation, and elongation), but whereas the holoenzyme only initiates RNA synthesis on one strand of the DNA, commencing at the promoter sites, the core enzyme catalyses transcription on both strands starting at random positions. A number of approaches have been used to elucidate the tertiary and quaternary structure of the enzyme. Low-resolution structures of both holo- and core enzymes have been obtained.[33] These show the overall shape of the enzyme and differences between the two forms. The holoenzyme has a thumb-like projection surrounding a deep open groove, but in the core enzyme the thumb forms a ring that surrounds the groove (Fig. 7.2). This is discussed further in the next section.

Cross-linking reagents have been used to study the topographical relationships between the subunits. The most recent of these has used a photoaffinity cross-linker, *N*-(5-azido-2-nitrobenzoyloxy)succinimide.[34] On activation of the azido group by UV radiation the reagent linked the σ subunit to the core enzyme. Using gel electrophoresis and immunodetection it was shown that the σ subunit was linked to all three subunits (α, β, and β'). RNA polymerase can be dissociated into its subunits and then reassembled *in vitro*.[35] The sequence of reassembly is as follows:

$$2\alpha \rightarrow \alpha_2 \rightarrow \alpha_2\beta \rightarrow \alpha_2\beta\beta' \rightarrow \alpha_2\beta\beta'\sigma$$

These results taken together with low-angle neutron diffraction studies support the arrangement of the subunits proposed by Hillel and Wu.[36]

There is one primary σ factor, often referred to as σ^{70} because of its M_r, that directs the bulk of transcription during exponential growth. There are other σ factors for specific groups of genes that are active under certain physiological and developmental states.[37] The three-dimensional structure of a fragment of the σ^{70} (residues 114–448) has been determined and from this a model for its binding to the promoter region of DNA and to the core polymerase has been proposed (see Section 7.3.2 and reference 37). Apart from the polypeptides that make up the holoenzyme, others are involved in regulation of certain stages of transcription. For instance, *nus* A replaces σ^{70} after initiation at the start of elongation. It can regulate the rate of transcription and has a role in termination.[30] Termination may also require an additional ρ protein.[38]

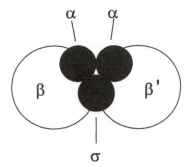

Model of RNA polymerase proposed by Hillel and Wu.[36]

7.3.2 Binding of RNA polymerase to DNA and initiation of transcription

We discuss in the next three sections initiation, elongation, and termination of transcription and the roles of the RNA polymerase subunits in these processes. The main events in the sequence of transcription are: (i) the location of the promoter by RNA polymerase and the formation of a closed complex (in which the DNA duplex is intact), (ii) the local 'melting' of the DNA duplex to form an open complex, (iii) RNA chain initiation, (iv) clearance of the promoter by RNA polymerase and release of σ^{70}, (v) RNA chain elongation, (vi) RNA chain termination and dissociation of nascent RNA and the core polymerase from the DNA duplex.

Transcription is initiated by recognition of a promoter region on the DNA template. It is estimated that the complete chromosome of *E. coli* contains in the region of 2000 promoters,[39] associated with different operons. The dissociation constants for the interaction between the core enzyme and DNA and between the holoenzyme and DNA are 0.5×10^{-11} mol dm^{-3} and 10^{-7} mol dm^{-3}, respectively. The σ subunit thus reduces the *general* binding affinity of RNA polymerase by about 20 000 times, and it is thought that this is important in enabling the holoenzyme to locate a particular region of DNA, namely the promoter site(s). The dissociation constant for the interaction of the holoenzyme with a promoter region is about 10^{-14} mol dm^{-3}. There is thus very tight binding between a promoter region and the holoenzyme and this has enabled the binary complex to be isolated on nitrocellulose filters. The different promoters will compete for RNA polymerase molecules and their effectiveness in competition will determine their frequency of usage. This is generally referred to as promoter strength, strong promoters being most effective.

[handwritten margin note: lower dissociation means higher affinity. Thus, core enzyme (10^{-11}) has higher affinity.]

The promoter regions are identified by nuclease digestion of the binary complex between DNA and the RNA polymerase, since the latter protects the promoter region. Over 100 promoter regions in *E. coli* have been sequenced[39] and by comparing these a consensus sequence can be formulated. There are two regions around positions −35 and −10 that are highly conserved. (The numbering indicates the number of nucleotides before the position at which transcription is initiated.) The consensus sequence at −35 is 5′-TTGACA-3′ and that at −10 is 5′-TATAAT-3′. There is a good correlation between promoter strength and the degree to which the −35 and −10 elements agree with the consensus sequences.[40] However, there are additional factors that regulate the rate of transcription. The rates at which genes are transcribed span a range of at least one thousand-fold. This is partly due to different rates of initiation, but also to differences in the rates of elongation. In exponentially growing *E. coli* 60 per cent of the total RNA synthesis is ribosomal RNA. The promoters of ribosomal RNA genes bind RNA polymerase very strongly, and this is due to an addition element at ≈ -55 and is AT-rich which is also recognized by RNA polymerase.[40] It is known as the UP element (upstream). There are additional proteins that regulate transcription; for example, the catabolite gene activator protein (CAP) activates transcription of the *lac* promoter. This binds both to a DNA site centred around −61 and −62 and to the RNA polymerase, thereby strengthening the DNA:RNA polymerase interaction.

The positioning of the holoenzyme at the start sequence for transcription is accurate to within ± 1 nucleotide, e.g. at the *lac* promoter in *E. coli*, transcription is initiated with the ribonucleotide sequence pppApApU, etc. or pppGpApApU, etc. An important question is how the holoenzyme locates the promoters within the double-stranded DNA of the chromosome. It is possible by using the values of estimated diffusion constants for RNA polymerase and the promoter to calculate the maximum rate at which the RNA polymerase could locate a promoter by a diffusion-controlled process.[41] However, when the calculated maximum rate is compared with experimentally determined rates for the strongest promoters, the latter are approximately 100-fold greater. In a detailed discussion von Hippel *et al.*[41] suggest that the holoenzyme locates non-specific sites on the DNA initially, thus providing an 'expanded target', and then translocates along the DNA to the promoter region. A similar process of one-dimensional diffusion is thought to occur with restriction endonucleases. The endonucleases bind non-specifically to DNA and then scan the DNA by one-dimensional diffusion until they find their recognition sequence.[42]

The details of the molecular interactions between the promoter sequences and the subunits of RNA polymerase have been studied using DNA footprinting, protease digestion of complexes, and X-ray crystallography; in addition, mutant proteins in which amino-acid residues have been either deleted or substituted have been studied. Figure 7.3 is a model showing the possible interactions involved.

The σ^{70} subunit is a V-shaped structure with promoter recognition occurring at one face and binding the RNA polymerase core with the other.[37] A helix-turn-helix DNA binding motif (residues 572–592) on σ^{70} interacts with the −35 consensus region binding motif and a second helical region (residues 425–441) interacts with the −10 consensus region.[37,40] In this latter region there are four conserved aromatic amino-acid residues which are thought to be involved in the melting of the double helix which starts at the −10 consensus region.[37] The two α subunits are in contact with σ^{70}, β, and β' (Fig. 7.3 and references 34 and 36).

*helps make complex structures into organized patterns that are easier to interpret

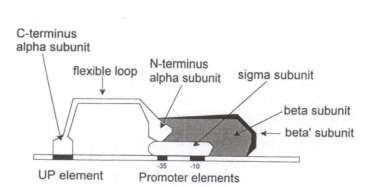

Fig. 7.3 Diagram of an initiation complex between a promoter and an RNA polymerase holoenzyme.[40]

The α subunits have 329 amino-acid residues comprising two structural domains linked by a spacer;[43] the N-terminal domain of one is in contact with the β subunit (residues 30–75) and the other with the β' (residues 175–210), but both C-terminal domains interact in tandem with the UP element of the promoter.[44-47] The N-terminal domain of β' is involved in the interaction with α.[47] The β' subunit is the most basic of the subunits; it is able to bind directly to DNA and is undoubtedly involved in template binding. The β subunit participates in binding the substrates, is involved in initiation and elongation, and it also carries the catalytic site for phosphodiester bond formation. The substrate-binding site was detected by affinity labelling (see Chapter 5, Section 5.4.4.3) using uridine triphosphate analogues. RNA polymerase from *E. coli* is strongly inhibited by the antibiotics rifamycin and streptolydigin and this is due to interaction with the β subunit. Proof of this is provided by experiments involving resistant strains of *E. coli* which have a modified β subunit; and that if RNA polymerase is reconstituted using the β subunit from a resistant strain and the other subunits from a sensitive strain, then the resultant enzyme is insensitive to the antibiotics.

7.3.3 Elongation of transcription

Elongation of the nascent RNA chain is catalysed by the core enzyme and the process continues, new nucleotides being added in a sequence complementary to that of one strand (the template strand) of the DNA. The mechanism is considered to be as shown in Fig. 7.4.

It has been estimated that errors in transcription are between 1 in 10^3 and 1 in 10^5. This is a much higher rate than that for DNA polymerase (see Chapter 1, Section 1.3.2). There is no clear evidence of a proofreading mechanism for RNA polymerase such as exists for DNA polymerase.[48] The process begins with the release of the σ factor, and the first translocation in which the RNA polymerase is displaced from the promoter. The ternary complex involving the RNA polymerase, DNA template, and nascent RNA is relatively stable and can be isolated. One reason for this stability is that the polymerase is thought to have two distinct binding sites for both the DNA and RNA.[48] A non-ionic interaction between a zinc finger on the β' subunit of RNA polymerase and DNA occurs ahead of the growing RNA chain and involves seven to nine base pairs, and an ionic interaction involving the C-terminal catalytic domain of the β subunit and about six nucleotides on the non-template strand of DNA occurs behind the growing chain.[32] Elongation does not occur at a uniform rate; there may be pauses or arrests. An arrest is a stoppage in which an accessory factor is required to restart elongation, whereas a pause is a temporary stop for a finite period. These processes are important in the regulation of transcription, e.g. operons for amino

Fig. 7.4 Proposed mechanism of transcription. A limited region of the DNA is unwound and a DNA:RNA hybrid of about 12 base pairs is formed during transcription. The axis of the DNA template rotates relative to that of the enzyme. Unwinding of the DNA:RNA hybrid occurs during the rotation. The model is supported by photochemical crosslinking, chemical modification, and nuclease protection experiments.[41]

acid biosynthesis such as those for histidine and tryptophan. Pause sites often occur where the nascent RNA is able to form a hairpin secondary structure. Pauses also occur before termination.

7.3.4 Termination of transcription

The process of termination involves three steps: (i) termination in the growth of the nascent RNA chain, (ii) release of nascent RNA, and (iii) release of the core enzyme from the DNA template. The mechanism of termination is not fully understood but there are two distinct mechanisms: an intrinsic mechanism in which termination occurs in response to certain limited sequences on the template strand of DNA, and a mechanism requiring the protein factor ρ in addition to certain sequences on the template DNA strand. For the intrinsic mechanism the termination sites have (i) a series of 4 to 10 consecutive AT base pairs in which the As are on the template strand, and (ii) a G + C-rich palindromic sequence immediately preceding the AT base pairs. This mechanism is believed to operate as follows. Once the G + C-rich sequence has been transcribed, the nascent RNA forms a hairpin loop using the palindromic sequence. This causes a pause in transcription, and also dissociation of the hairpin loop from the DNA template (Fig. 7.5).

The sequence of transcribed Us on the adjoining region forms unstable bonds with the sequence of As (AU and AT base pairs are less stable than GC pairs, and also a sequence of Us has little tendency to form a regular secondary structure) and allows the release of the nascent RNA. The ρ-dependent termination occurs at specific sequences but these are less constrained than those for intrinsic termination. Rho (ρ) is a single polypeptide having 419 amino acid residues, which forms a hexamer in the form of a ring. The full three-dimensional structure has not yet been solved, but the polypeptide consists of an N-terminal RNA-binding domain which has RNA:DNA helicase activity (it is able to dissociate RNA:DNA hybrids) and this is connected through a linker region to an ATP

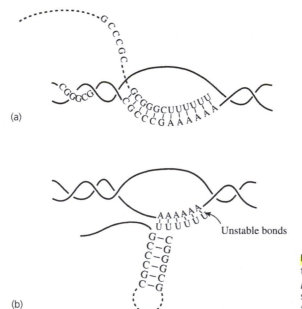

(a)

(b)

Unstable bonds

Fig. 7.5 Transcriptional termination independent of ρ: at point (a) elongation stops and is followed by the formation of a hairpin (b).[28]

binding domain. Rho acts by first binding to a site on the nascent RNA, and subsequently by using ATP hydrolysis as a source of energy to effect the dissociation of the nascent RNA from the RNA polymerase and DNA template.[38] The detailed mechanism is not yet known. The ATP binding domain has several sequence motifs that are characteristic of ATPases.

Thus, the holoenzyme is necessary for correct initiation of RNA synthesis, after which the σ factor is no longer necessary, and the additional proteins such as *nus* A and ρ may be required for elongation and termination.

7.3.5 RNA polymerases from bacteriophages

RNA polymerase from *E. coli* has to recognize a variety of promoter sites on the chromosome and transcription from these sites is subject to a variety of different controls and it is perhaps partly for this reason that the enzyme is a multisubunit protein. By contrast, the bacteriophage RNA polymerases have only to recognize a few promoters and this is reflected in a simple enzyme structure. The phage polymerases are also able to catalyse a more rapid rate of transcription, e.g. 200–400 nucleotides s^{-1} compared with 50–100 nucleotides s^{-1} for bacterial polymerases.[48] The RNA polymerases coded for by genes on the bacteriophages T_3 and T_7 comprise single polypeptide chains of M_r 110 000. These RNA polymerases are highly specific in their recognition of DNA. The T_7 enzyme interacts only weakly with the promoters of T_3 and vice versa. Thus, the complexity of any given RNA polymerases is partly a function of the range of species of RNA it is required to synthesize. The overall mechanism of catalysis by *E. coli* and by T_7 polymerases is similar, involving the recognition of promoter-specific sequences, unwinding of double-stranded DNA, and termination at G + C-rich hairpin sequences followed by a U-rich sequence. The three-dimensional structure of T_7 RNA polymerase[49] solved at 0.33 nm resolution shows that it closely resembles polymerases such as the Klenow fragment of DNA polymerase I (see Section 7.11.2) and HIV reverse transcriptase in having a polymerase domain resembling a cupped right hand, but it also has a second distinct N-terminal domain that appears to carry out functions specific to RNA synthesis. It shows much less resemblance to *E. coli* polymerase although the latter has a thumb subdomain similar to that within the cupped right hand domain of T_7 RNA polymerase.[49–51]

7.4 The occurrence and isolation of multienzyme proteins

In a metabolic pathway a number of enzymes catalyse a sequence of reactions linked so that the product of one enzyme-catalysed reaction becomes the substrate for the next enzyme. Such a sequence may be represented as

$$A \xrightarrow{E_1} B \xrightarrow{E_2} C \xrightarrow{E_3} D \xrightarrow{E_4} E \xrightarrow{E_5} F$$

in which A, B, C, D, and E are substrates for the enzymes E_1, E_2, E_3, E_4, and E_5. Both the efficiency of these reactions, measured as the yield of the end-product F from the starting substrate A, and the overall rate of conversion of A to F will depend partly on the extent of coordination occurring between the five enzymes. For some metabolic pathways it has been shown that certain of the enzymes are physically associated with one another to form multienzyme proteins, e.g. the fatty-acid synthase complex (Section 7.11.1) and enzymes of the tryptophan biosynthetic pathway (Sections 7.9 and 7.11.2).The multienzyme systems that have been most studied are probably amongst the more stable ones. There may be many others in which the associations are weak and are thus more difficult to detect as multienzymes. The isolation and characterization of a multienzyme protein usually presents more technical difficulties than isolation of a single enzyme. This is well illustrated by two examples discussed in more detail later in the chapter, namely, the pyruvate dehydrogenase complex (Section 7.7) and the yeast fatty-acid synthase complex (Section 7.11.2). In the case of the pyruvate dehydrogenase complex from *E. coli*, although it is clear that the multienzyme complex contains three different enzyme activities, the precise number of each of the polypeptide chains present in the complex has been difficult to determine. This may be because (i) different isolation procedures yield complexes of slightly different composition, (ii) some dissociation may occur during isolation, or (iii) in the intact cell complexes of slightly variable composition do exist. In a multi-enzyme complex having a large number of subunits, e.g. 50–100, a small percentage error in the determination of the M_r of the complex will significantly affect the proposed number of subunits.

Yeast fatty-acid synthase posed a different type of problem. It proved difficult to separate the polypeptide chains corresponding to the seven catalytic activities of the fatty-acid synthase. It was then shown that the whole complex comprises two multifunctional polypeptide chains and that the smaller fragments observed in earlier experiments were the result of limited proteolysis during isolation.

Evidence for the existence of a multienzyme complex usually appears when a component enzyme of the complex is being isolated and it is found to copurify with another enzyme from the same metabolic pathway. If the ratio of the enzyme activities remains constant during the isolation procedure, then this is further evidence for the existence of a definite multienzyme complex, e.g. carbamoyl phosphate synthase, aspartate carbamoyltransferase, dihydro-orotase abbreviated to CAD (see Section 7.10). Two enzyme activities that were initially shown to copurify from rat liver and were subsequently found to be two catalytic activities associated with a single polypeptide chain are fructose 2,6-bisphosphatase and 6-phosphofructo-2-kinase. The enzyme catalysing both these activities has been purified from a number of tissues and it has been shown that the two activities reside in separate parts of the polypeptide chain.[52,53] The importance of this enzyme is discussed in Chapter 6, Section 6.4.1.1. Another example of phosphatase and kinase activities residing in the same polypeptide chain is that of isocitrate dehydrogenase kinase and isocitrate dehydrogenase phosphatase. The two activities copurify and are coded for by a single gene.[54,55]

7.5 Phylogenetic distribution of multienzyme proteins

Although only a limited number of multienzyme complexes have so far been studied, it does seem that, in general, multienzyme complexes and multienzyme polypeptides are both more common and more complex in eukaryotes than in prokaryotes; this is borne out by the examples of pyruvate dehydrogenase, fatty-acid synthase, and carbamoyl phosphate synthase, aspartate carbamoyltrans-ferase, dihydro-orotase, and the enzymes of the tryptophan biosynthetic pathway discussed in this chapter. A typical prokaryote cell, e.g. *E. coli*, has a very much

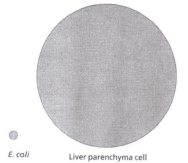

E. coli   Liver parenchyma cell

Comparative sizes of eukaryote and prokaryote cells

smaller volume than a typical eukaryote cell, e.g. liver parenchyma (1 μm^3 as compared to 6000 μm^3).

Even making allowance for the intracellular compartmentation in eukaryotes brought about by the various intracellular membranes, the volume in which free diffusion of substrates and enzymes can occur is much greater than in a prokaryote cell. Thus, if the intracellular concentrations of substrates and catalytic centres are similar in prokaryotes and eukaryotes, then the diffusion time is more likely to be rate limiting in the latter. The comparative biochemical information on carbamoyl phosphate synthase, aspartate carbamoyltransferase, dihydroorotase, the tryptophan biosynthetic pathway, and fatty-acid synthase suggests that evolution has resulted in multienzyme polypeptides and more tightly associated multienzyme complexes. In addition, genetic evidence suggests that this is accompanied by a clustering or even fusion of the genes coding for these proteins.

7.6 Properties of multienzyme proteins

What are the advantages of enzymes being physically associated with one another rather than occurring as separate entities within cells? The answers may be found by comparing the properties of multienzyme proteins with those of separated enzymes. First, the transit time, i.e. the time required for a product of one enzyme to diffuse to the catalytic site of a second enzyme, will be less in a physically associated multienzyme protein than when the enzymes are not associated. Whether the transit time becomes the rate-limiting step in a metabolic pathway will depend on the catalytic activities of the enzymes. Two instances in which the transit times may become important are (a) when an intermediate has a short half-life and (b) when an intermediate has a high M_r and thus would diffuse more slowly. There is as yet no well-documented example of the first situation, although carbamoyl phosphate probably has a short half-life in the cytosol, since there are enzymes capable of rapidly degrading it. The association of carbamoyl phosphate synthase with aspartate carbamoyltransferase, which channels carbamoyl phosphate within the multienzyme protein, probably prevents its degradation.[56] An example of the second situation is that of the pyruvate dehydrogenase system, in which the second step in the reaction sequence involves the two enzyme-bound intermediates hydroxyethyl-thiamindiphosphate-pyruvate dehydrogenase and dihydrolipoamide acetyltransferase.

Second, in addition to the transit time being reduced in a multienzyme protein, so also would be the transient time, that is the time required to change from one steady-state rate to another. Thus in a system with two enzymes

$$A \xrightarrow{E_1} B \xrightarrow{E_2} C$$

in a steady-state in a cell, if the concentration of A is increased then the time required to reach the new steady-state rate (transient time) will be much less in the case of a multienzyme protein.[57]

A third possibility afforded by multienzyme proteins is that of channelling or compartmentation. Once an intermediate has been formed in a multienzyme protein, it is not generally available to be acted upon by other enzymes outside the complex. This would also mean that the concentrations of such metabolites within a cell would be lower. Because there are relatively high concentrations of solutes, particularly macromolecules, in cell compartments (see Section 7.1), the solvent capacity of cell water is limited. Lowering the concentrations of metabolites would spare this capacity. Channelling would also reduce the loss of pathway flux by leakage or instability of free intermediates.[13] There is still much debate on (a) the extent to which channelling occurs *in vivo* and (b) whether it has a physiological role.[58,59] The arguments are based on a combination of theoretical treatments and measurements of the flux of metabolites through multienzyme proteins.[13,58–61] For example, carbamoyl phosphate formed in the multienzyme complex for pyrimidine biosynthesis in *Neurospora crassa* (see Section 7.10) is not available for the biosynthesis of arginine, and ammonia generated by glutaminase in liver mitochondria is preferentially used by carbamoyl phosphate synthase.[62] There is both kinetic and structural evidence for channelling in the biosynthesis of riboflavin.[63] The last two steps of riboflavin biosynthesis in *B. subtilis* are catalysed by the lumazine synthase/riboflavin synthase complex. This complex ($\alpha_3\beta_{60}$) comprises an icosahedral capsid (β_{60}) surrounding three α subunits. The β subunit catalyses the synthesis of 6,7-dimethyl-8-ribityl-lumazine, the conversion of which to riboflavin is catalysed by the α subunit. At low substrate concentration kinetic evidence suggests that 6,7-dimethyl-8-ribityllumazine is not released into the medium but is probably retained within the icosahedron diffusing from the β to the α subunits. At high substrate concentrations the icosahedron appears to leak some of the intermediate, but this is thought to occur only when the substrate is present above normal physiological concentrations.

There is also evidence for coordinate activation of a whole multienzyme protein. One domain, or polypeptide, may exert an allosteric effect on an adjacent domain or polypeptide, e.g. in the *arom* complex (or AROM enzyme) discussed later (Section 7.11.2), where the substrate of the first enzyme is able to activate all five catalytic activities of the multienzyme protein. Another example is ribulose bisphosphate carboxylase-oxygenase. This enzyme acts together with four other enzymes (phosphoribulokinase, phosphoribose isomerase, phosphoglycerate kinase, and glyceraldehyde phosphate dehydrogenase) to catalyse five consecutive reactions of the Benson–Calvin cycle. The free enzyme is a hexadecamer made up of small (S) and large (L) subunits having the structure L_8S_8, but when present in the five-enzyme complex has an L_2S_4 structure. There are changes in the kinetic parameters when the enzyme is associated in the complex, and its specific activity, expressed as k_{cat}/K_m (see Chapter 4, Section 4.3.1.3), is about 10-fold higher in the complex when compared with the free enzyme.[19]

The examples of multienzymes discussed in the subsequent sections of this chapter have been chosen to illustrate the various properties of multienzyme proteins. Pyruvate dehydrogenase and related oxo-acid dehydrogenases together with the tryptophan synthase systems are the multi-enzyme complexes about which most structural information is available. They also illustrate how the intermediate substrates remain attached to the multienzyme complexes. Fatty-acid synthase and the *arom* complex are examples of multienzyme polypeptides. The multienzyme complex of carbamoyl phosphate synthase and aspartate carbamoyltransferase is used to illustrate the importance of channelling an intermediate (carbamoyl phosphate) into one of two competing pathways. The 'glycogen particle' is much less well defined than the other multienzyme proteins and presumably represents a looser association between glycogen and the glycolytic enzymes.

7.7 Pyruvate dehydrogenase multienzyme complex and related systems

$P_Y \rightarrow A_{CCOA}$

In the early 1950s it was realized that the oxidation of pyruvate to acetyl coenzyme A (acetyl CoA) was catalysed by a large homogeneous enzyme preparation having an M_r value of about 4×10^6 and that the reaction involved more than one catalytic step.[64] The pyruvate dehydrogenase multienzyme system was the first multienzyme to be purified. At the time it was realized that 2-oxoglutarate dehydrogenase was a closely related system having the same cofactor requirements and also catalysing an oxidative decarboxylation. A third multienzyme system capable of oxidizing branched-chain ketoacids derived from the amino acids valine, leucine, and isoleucine was identified in the mid-1970s.[7] All three multienzymes, pyruvate dehydrogenase, 2-oxoglutarate dehydrogenase, and branched-chain oxoacid dehydrogenase, constitute a family having the following features in common: (i) they have three catalytic centres catalysing a decarboxylation, a transacylation, and a dehydrogenation as shown in the general scheme below; (ii) they use the same five cofactors, thiamin diphosphate (TDP), lipoate, coenzyme A, FAD, and NAD[+]; (iii) they show structural and mechanistic similarities, in each case having a transacylase at the core of the complex surrounded by the decarboxylase and dehydrogenase on the periphery; and (iv) all three multienzyme complexes share the same dihydrolipoamide dehydrogenase component, except in the case of *Pseudomonas putida* where three dehydrolipoamide dehydrogenases exist, each servicing different multienzyme complexes.[65]

$$E_1\text{-TDP} + CH_3COCOO^- \rightleftharpoons E_1\text{-TDP}\text{—}CHOHCH_3 + CO_2 \qquad 1$$

$$E_1\text{-TDP}\text{—}CHOHCH_3 + E_2\text{-lip}\big\langle\begin{smallmatrix}S\\|\\S\end{smallmatrix} \rightleftharpoons E_1\text{-TDP} + E_2\text{lip}\big\langle\begin{smallmatrix}SH\\\\SCOCH_3\end{smallmatrix} \qquad 2$$

$$E_2\text{-lip}\big\langle\begin{smallmatrix}SH\\\\SCOCH_3\end{smallmatrix} + CoASH \rightleftharpoons E_2\text{-lip}\big\langle\begin{smallmatrix}SH\\\\SH\end{smallmatrix} + CH_3COSCoA \qquad 3$$

$$E_2\text{-lip}\big\langle\begin{smallmatrix}SH\\\\SH\end{smallmatrix} + E_3\text{-FAD} \rightleftharpoons E_2\text{-lip}\big\langle\begin{smallmatrix}S\\|\\S\end{smallmatrix} + E_3\text{-FAD reduced} \qquad 4$$

$$E_3\text{-FAD reduced} + NAD^+ \rightleftharpoons E_3\text{-FAD} + NAD + H^+ \qquad 5$$

E_1 = pyruvate decarboxylase   TDP = thiamin diphosphate

E_2 = dihydrolipoamide acetyl transferase $\text{lip}\big\langle\begin{smallmatrix}S\\|\\S\end{smallmatrix}$ = oxidized lipoamide

E_3 = dihydrolipoamide dehydogenase $\text{lip}\big\langle\begin{smallmatrix}SH\\\\SH\end{smallmatrix}$ = reduced lipoamide

The features where they differ are the following: (i) some are based on octahedral symmetry, others on icosahedral; (ii) the decarboxylase component may comprise homodimers (α_2) of a single polypeptide of $M_r \approx 100\,000$, or heterotetramers ($\alpha_2\beta_2$) of two polypeptides of $M_r \approx 41\,000$ and $36\,000$ respectively;

(iii) mammals and yeast and probably other eukaryotes have an additional polypeptide component X which is essential for the proper assembly and functioning of the complex; and (iv) the catalytic activities of some eukaryote complexes are regulated by phosphorylation of the decarboxylase (E_1). The pyruvate dehydrogenase system catalyses an important step regulating the flow of acetyl groups into the tricarboxylic acid cycle, the 2-oxoglutarate dehydrogenase catalyses a step in the tricarboxylic acid cycle, and the branched-chain oxoacid dehydrogenase catalyses a rate-limiting step in the degradation of the three essential amino acids valine, leucine, and isoleucine. The large size of the multienzyme complexes is due to the multiple copies of each of the three component enzymes present in the complex. Table 7.2 summarizes the key features of these multienzyme systems.

The pyruvate dehydrogenase multienzyme complex from *E. coli* is one of the most thoroughly investigated multienzyme proteins. Other organisms and tissues from which the dehydrogenase complexes have been studied include: *B. stearothermophilus*, *Azotobacter vinelandii*, *Pseudomonas spp*, *Enterococcus faecalis*, *Saccharomyces cerevisiae*, *Neurospora crassa*, *Arabidopsis*, mammalian heart, kidney, and liver, and avian tissues. Because of their importance in a number of genetic deficiencies, the genes for all the components of human pyruvate dehydrogenase complex have been cloned and sequenced. In a genetically inherited disease, maple syrup urine disease (see Chapter 10, Section 10.6), there is a mutation in the branched-chain oxoacid decarboxylase (E_1) which leads to a 100-fold increase in its K_m and as a result there may be as much as a 70-fold increase in the plasma concentration of oxoacids and their parent amino acids, which can cause brain damage.[7] The oxoacid dehydrogenases have been purified from a number of bacterial,[71-73] fungal, plant, and animal sources.[70,74] The isolation and overall structure of the oxo-acid dehydrogenases are discussed in Section 7.7.1, the structure of the components in Section 7.7.2, the catalytic mechanism in Section 7.7.3, the regulation of *E. coli* pyruvate dehydrogenase system in Section 7.7.4, and regulation by covalent modification in eukaryotes in Section 7.7.5.

Table 7.2 Properties of the oxoacid dehydrogenase multienzyme complexes

System	Species or taxa	Symmetry	Approximate composition ($E_1:E_2:E_3$)	Regulation by phosphorylation	References
Pyruvate dehydrogenase	*E. coli*	Octahedral	24:24:12	No	66, 67
	Azotobacter vinelandii	Octahedral	24:24:12	No	4, 5
	Enterococcus faecalis	Icosahedral			4, 5
	B. stearothermophilus	Icosahedral	60:60:30	No	4, 5
	Saccharomyces cerevisiae	Icosahedral	60:60:30 + protein X	No	4–6
	Neurospora crassa	Icosahedral	60:60:30	Yes	68
	Avian	Icosahedral		Yes	6, 69
	Mammalian	Icosahedral	30 copies of $\alpha_2\beta_2E_1$, 60 E_2, 6–12 α_2E_3 + 12 of X	Yes	6, 69
	Plant (arabidopsis)	Icosahedral			70
2-Oxoglutarate dehydrogenase	Mammalian	Octahedral	24:24:12	No	4–6
	E. coli	Octahedral	24:24:12	No	
	B. stearothermophils	Icosahedral	60:60:30	No	
Branched-chain oxoacid dehydrogenase	Mammalian	Octahedral	24:24:12	Yes	7

7.7.1 Isolation and overall structure of the pyruvate dehydrogenase multienzyme complex

Pyruvate dehydrogenase can be described as a 'housekeeping enzyme' (an enzyme catalysing a central metabolic pathway operating in almost all cell types) which is therefore present in reasonable quantities in most cells and tissues. Most protein purification procedures used for the pyruvate dehydrogenase complex are based on that originally devised for the *E. coli* complex (see Fig. 7.6) or a procedure of poly(ethylene glycol) fractionation used for mammalian pyruvate dehydrogenase complex followed by isoelectric precipitation and gel filtration.[76]

Larger amounts of the *E. coli* complex have been obtained by using strains carrying a plasmid that encodes the complex.[77] The genes for *Saccharomyces* and human pyruvate dehydrogenase have been cloned and expressed in *E. coli* [66,78] which has enabled large quantities of the complexes to be purified for structural studies. The isolated multienzyme complex can be dissociated into its constituent subunits by use of 4 mol dm^{-3} urea, high pH, and calcium phosphate gel (Fig. 7.6), or alternatively by using a high salt concentration, e.g. 2 mol dm^{-3} NaCl.[79] The

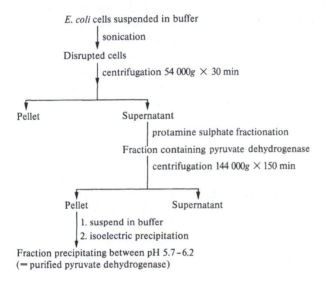

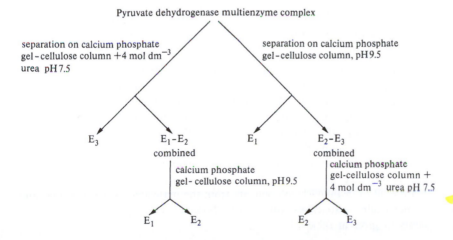

Fig. 7.6 Pyruvate dehydrogenase complex: isolation and separation of the subunits (see reference 75).

multienzyme complex may also be reconstituted from the subunits and full enzyme activity restored. It is clear from both the dissociation and reconstitution studies that E_2 has binding domains for both E_1 and E_3, but that E_1 and E_3 do not bind in the absence of E_2.

In order to determine the stoichiometry of the complexes, the M_r values of the complex, the constituent subcomplexes, and the individual polypeptides have to be determined. The M_r values of the polypeptide chains can be determined by conventional means (see Chapter 3, Section 3.2) and in many cases have been confirmed from those inferred by the DNA sequence.[80,81] The M_r of the complex presented a more difficult problem because of its large size and in the early determinations for *E. coli* pyruvate dehydrogenase complex values[82,83] ranged from 3×10^6 to 6.1×10^6. Three factors contributed to this variation: (i) methods for determination of very high M_r values are less accurate, (ii) the complex as it exists *in vivo* may not be homogeneous, and (iii) some subunits may become detached during the isolation procedure.

From studying a range of different oxo-acid dehydrogenase complexes from different species, it is clear that all have a dihydrolipoyl acyltransferase (E_2) surrounded by the decarboxylase (E_1) and dihydrolipamide dehydrogenase (E_3). Those complexes with octahedral symmetry have 24 copies of E_2, whereas those with icosahedral symmetry have 60 copies. For completely symmetrical complexes the ideal compositions of E_1:E_2:E_3 would be 24:24:12 or 60:60:30 respectively. For the *E. coli* pyruvate dehydrogenase complex this would correspond to an M_r value of 4.57×10^6. Many isolated complexes appear deficient in certain components necessary to achieve a completely symmetrical complex (Table 7.2). Both E_1 and E_3 bind strongly to E_2 and compete with each other for the subunit binding domain on E_2 and this may account for strict symmetry not being maintained.[84] An additional polypeptide X is present in pyruvate dehydrogenase complexes from yeast and mammalian sources. It remains attached to the E_2 core after E_1 and E_3 have been dissociated, and it requires chaotropic agents such as 5 mol dm^{-3} urea in order to dissociate this E_2–X subcomplex.[79] Polypeptide X is structurally similar to the N-terminal region of E_2, but differs in the C-terminal region. The probable structure of *E. coli* pyruvate dehydrogenase multienzyme complex and others based on octahedral symmetry is shown in Fig. 7.7. The 24 E_2 polypeptide chains are arranged in the form of a cube constituting the core of the complex with the E_1 and E_3 on the outside.

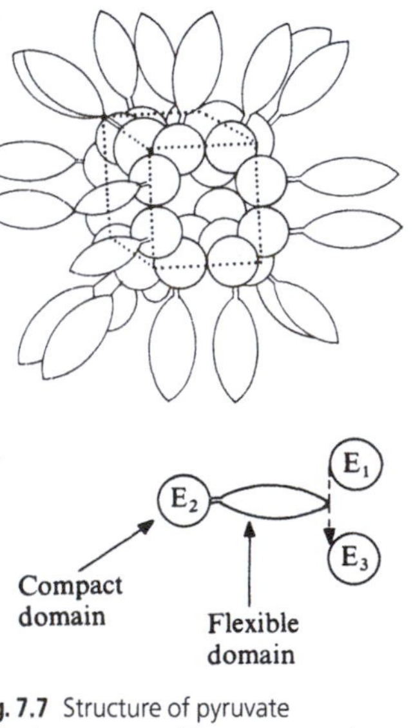

Fig. 7.7 Structure of pyruvate dehydrogenase multienzyme complex. (a) The probable arrangement of dihydrolipoamide acetyltransferase (E_2) forming the core of the pyruvate dehydrogenase multienzyme complex. Each E_2 polypeptide is thought to be made up of two domains, a central core containing the region of the active site and a flexible domain protruding out from the core. The latter contains the lipoate residues. The 24 E_2 polypeptide chains are arranged in the form of a cube (a), and the E_1 and E_3 enzymes form a larger cube encasing the E_2 cube such that the flexible domain of E_2 can move between E_1 and E_3, as shown in (b). For details, see references 4, 70, and 85.

7.7.2 Structure of the components of oxo-acid dehydrogenase multienzyme complexes

The oxo-acid dehydrogenases have M_r values of 5×10^6 to 10×10^6 and are seen as particles of 30–40 nm in diameter in the electron microscope. This is roughly the same size as a 70s ribosome (M_r 2.3×10^6). The complete structure of this complex has not yet been solved. However, the structures of individual components and of subcomplexes have been solved. First we consider the components. When comparing the enzyme components of the three different dehydrogenase complexes, it is found that although there are distinct decarboxylases (E_1) and transacylases (E_2), the dehydrogenase (E_3) is common to all three complexes; the only known exception to this is that of *Pseudomonas putida*. The letters p, o, and b are used to distinguish components from the pyruvate-, 2-oxoglutarate-, and branch-chain 2-oxoacid-dehydrogenases. Some of the properties of the components are given in Table 7.3.

Table 7.3 Properties of components of oxoacid dehydrogenase multienzyme complexes

Component	System	M_r	Structure	Activity
E_1	Gram +ve PDC, all OGDC	100 000	α_2 homodimer with TDP binding domain	Decarboxylation
$E_1 \alpha + \beta$	Gram +ve PDC, all eukaryote PDC, all BCDHC	$\alpha \approx 41\,000$, $\beta \approx 36\,000$	$\alpha_2\beta_2$ heterotetramer, TDP binding domain on α, E_2 binding domain on β	Decarboxylation
E_2	All complexes	50 000–70 000	1 catalytic domain ($\approx$ 250 residues), 1 E_3 binding domain ($\approx$ 50 residues), 1–3 lipoyl domains ($\approx$ 80 residues per domain), each linked by flexible linkers	Transacylation, binding E_1 and E_3
E_3	All complexes	50 000–55 000	4 domains: FAD-binding, NAD-binding, central and catalytic	Dihydrolipoyl dehydrogenation
X	Mammalian and yeast PDC	42 000–50 000	1 catalytic domain ($\approx$ 160 residues, 1 E_3 binding domain ($\approx$ 50 residues), 1 lipoyl domain ($\approx$ 80 residues)	Similar to E_2, but with shortened catalytic domain lacking activity. Probable function—binding E_3 for assembly of complex
Kinase	Eukaryote PDC and BCDC	Isoenzymes, 45 000–48 000	Bound to the inner lipoyl domain of E_2	Phosphorylation of $E_1\alpha$. Capable of phosphorylating 20–30 E_1s per complex
Phosphatase	Eukaryote PDC and BCDC	$\alpha = 52\,600$ $\beta = 95\,600$	$\alpha\beta$, α = regulatory, β = catalytic	Dephosphorylation of phosphorylated $E_1\alpha$

References 4–6, 66, 84, 85 and 94.

Of the three major components of the complexes, the most complete structural study has been carried out on dihydrolipoamide dehydrogenase (E_3). It belongs to the family of flavin-dependent disulphide oxidoreductases and exists as a homodimer having a structure resembling that of a pair of butterfly wings (Fig. 7.8).

It has four domains: FAD-, NAD+-, central- and interface-domains. The catalytic centre lies at the interface between the two chains, both of which are necessary for catalytic activity. Its mechanism of catalysing the oxidation of the dihydrolipoamide moiety is similar to that of other flavoprotein dehydrogenases.[86]

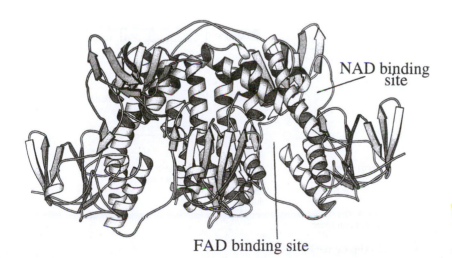

NAD binding site

FAD binding site

Fig. 7.8 The structure of dihydrolipoamide dehydrogenase dimer showing the NAD- and FAD-binding sites (see reference 5).

The acyltransferase component (E$_2$) has a somewhat unusual structure. It has lipoic acid covalently attached through an amide bond to a lysine residue.

$$\text{CH}_2\text{—CH—(CH}_2)_4\text{—CO NH(CH}_2)_4\text{—C—H}$$

Lipoate — Lysyl group

1.4 nm

The enzyme is susceptible to proteolysis, and this is due to its rather extended structure in which the lipoate is present on an exposed arm. This exposed arm is flexible and accounts for a lack of success in obtaining crystals of the complete E$_2$ polypeptide chain. The E$_2$ polypeptides from different sources are made up of between three and five domains (see Fig. 7.9) which comprise one to three lipoyl domains, one peripheral E$_3$ binding domain, and one catalytic domain. Between each of the domains are flexible regions of between 25 and 40 amino-acid residues rich in alanine and proline.

Although a complete solution of the transacylase (E$_2$) structure has not been achieved, it has been possible to use recombinant DNA technology (see Chapter 2, Section 2.4.1) to express large amounts of individual domains, and three-dimensional structures for the lipoyl domains (Fig. 7.10) and the E$_2$p catalytic domain have been solved using NMR spectroscopy and X-ray diffraction.[87] It is clear that the lipoyl residue is exposed on the end of its domain.

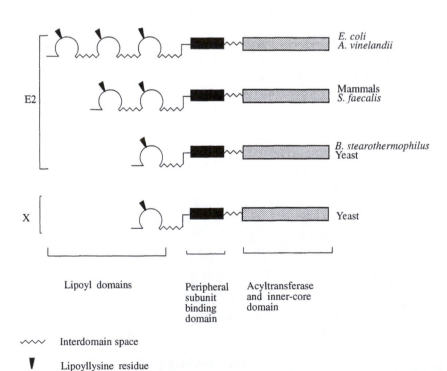

Fig. 7.9 The domain structure of E$_2$ and protein X subunits of 2-oxoglutarate dehydrogenase multienzyme complexes. The lipoyl residue is shown by an arrow and the interdomain segments by wavy lines.[4]

The E_1 component exists as a homodimer (α_2) or heterotetramer ($\alpha_2\beta_2$) in octa-hedral and icosahedral complexes, respectively. Its three-dimensional structure has not been solved, although it is evident from sequence comparisons that it contains a structural motif common to other proteins that bind TDP, e.g. transketolase.[5]

The additional protein X (also known as E_3 binding protein) has many simi-larities with E_2 (see Table 7.3 and Fig. 7.9), although its C-terminal domain has no catalytic activity or sequence similarity to the catalytic domain of E_2.[5] It is present in yeast and mammalian PDC and appears to be required to anchor E_3 to the E_2 core. Cryo-electron microscopic studies show that protein X fits within cavities present in the E_2 core.[66]

The function of the multiple lipoyl domains in some of the dehydrogenase complexes is not clear. By using recombinant-DNA techniques to remove some of the lipoyl domains, or site-directed mutagenesis to replace lipoyllysine with glutamine, it has shown that one lipoyl domain is sufficient for the complex to be active.[6] However, when isogenic strains of *E. coli* containing pyruvate dehydro-genases having one, two, or three lipoyl domains are grown in minimal medium there is a direct correlation between the maximum growth rate of the strains and the number of lipoyl domains,[4] suggesting that the extra lipoyl domains do confer some advantage.

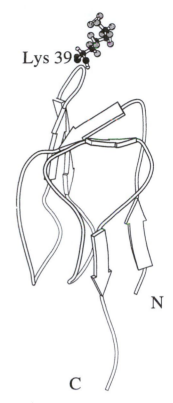

Fig. 7.10 The structure of the lipoyl domain of E_2 showing Lys 39 which forms an amide bond with lipoic acid.[87]

7.7.3 **Mechanism**

The initial decarboxylation occurs when pyruvate combines with TDP on the decarboxylase probably by the mechanism shown below.

where $R_1 = CH_3$ pyrimidine $-CH_2-$ and $R_2 = O-P-O-P-OCH_2CH_2-$

The second step involves the oxidative transfer of the hydroxyethyl group bound to E_1 on to the disulphide group of lipoamide on E_2, forming a thioester (reaction 2 in Section 7.7). For this step the two enzymes have to be in close contact. The acetyl group so formed is then transferred to CoA in the third step. The fourth and fifth steps are concerned with the regeneration of oxidized lipoamide by transfer of 2H from dihydrolipoamide on E_2 via FAD of the dehy-drogenase (E_3) to NAD^+. The fourth step also involves the interaction between two enzyme-bound intermediates of E_2 and E_3 (reaction 4, Section 7.7).

It has been suggested[70] that the transfer of the acetyl group and of the two hydrogen atoms from dihydrolipoamide involves the flexible arm of the lipoamide domain of E_2 swinging between E_1 and E_3. Free lipoamide is a very poor substrate for acylation by E_1. The value of k_{cat}/K_m (see Chapter 4, Section 4.3.1.3) is raised 10 000-fold when the lipoamide is attached to the lipoyl domain compared with that of free lipoamide. This suggests that the catalytic domain of E_1 interacts, not just with the lipoamide itself, but with a large part of the lipoyl domain.[6] The lysyllipoamide arm is approximately 1.4 nm and therefore could swing through a diameter of 2.8 nm. Estimates have been made[88,89] of the distances separating E_1 and E_3 to test this hypothesis. By using an analogue of TDP, thiochrome diphosphate, bound at E_1, it is possible to estimate the distance to the FAD on the active site of E_3 by fluorescence energy transfer. The distance appears to be about 4.5 nm, which is too great for the flexible lysyllipoamide arm alone to be the explanation, but the lysyllipoamide is on the tip of the lipoyl domain (Fig. 7.10) which is a highly flexible structure and this enables it to span the distance.

A question that arises with a multienzyme complex of this nature in which there are several copies of each of the enzymes is whether, in a given catalytic cycle, one molecule of pyruvate is oxidatively decarboxylated by a single specific cluster of E_1–E_2–E_3 within the complex. Several experiments have been aimed at answering this question. In the multienzyme complex, pyruvate decarboxylase (E_1) catalyses the rate-limiting step. This has been demonstrated by using an active-site-directed inhibitor of E_1, thiamin thiazolone diphosphate, which inhibits pyruvate dehydrogenase activity, in parallel with the decrease in the rate of conversion of pyruvate to acetyl CoA.[90] In contrast, the acetyltransferase (E_2) can be partially inhibited by reacting with N-ethylmaleimide before any reduction in the rate of conversion of pyruvate to acetyl CoA occurs. Also, it is possible to remove nearly all the lipoyl moieties using the enzyme lipoamidase, without significant loss of activity.[91]

It is possible to reconstitute a pyruvate dehydrogenase complex that is deficient in pyruvate decarboxylase (E_1) and to use this to demonstrate that pyruvate decarboxylase (E_1) is capable of transferring hydroxyethyl groups to several acetyltransferases.[92] This is done by allowing the reaction to occur in the absence of CoA and measuring the extent to which the acetyltransferase (E_2) becomes acetylated. From these results it seems probable that there is a communicating network of acetyltransferases (E_2) in which the acetyl groups may be transferred from one E_2 to another, thus enabling a transfer of acetyl groups and the associated reduction of lipoamide residues from an E_1 site to a distant E_3 site within the complex,[93] i.e. a process of random multiple coupling.

7.7.4 Control of the pyruvate and oxoglutarate dehydrogenase systems in *E. coli*

The pyruvate dehydrogenase complex occupies a key position linking the glycolytic pathway to the tricarboxylic acid cycle, and several factors control its activity in different organisms.[65] The decarboxylation step is effectively irreversible under physiological conditions and thus controls the entry of pyruvate into the tricarboxylic acid cycle. Pyruvate dehydrogenase activity is inhibited by two of the products of its activity, namely, acetyl CoA, which is a competitive inhibitor with respect to CoA, and NADH, which is competitive with respect to NAD^+ (Fig. 7.11). The combination of acetyl CoA and NADH may also inhibit by

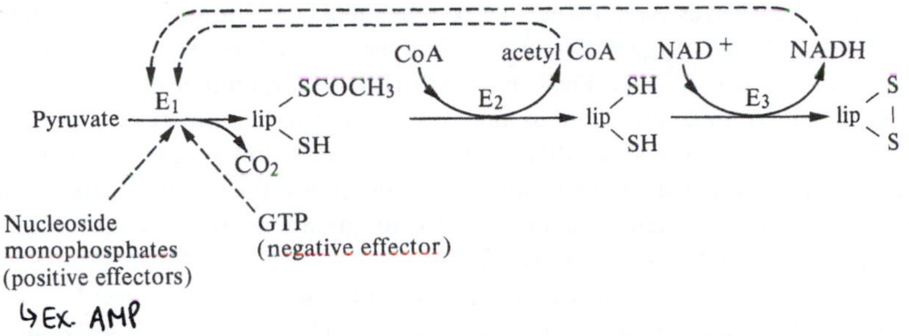

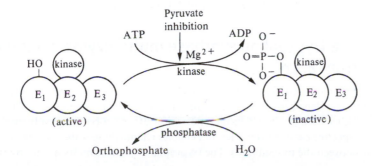

Pyruvate — E_1 → lip $\overset{SCOCH_3}{\underset{SH}{\diagdown}}$ $\xrightarrow{E_2}$ lip $\overset{SH}{\underset{SH}{\diagdown}}$ $\xrightarrow{E_3}$ lip $\overset{S}{\underset{S}{\diagdown}}$

CoA   acetyl CoA NAD^+ NADH

CO_2

Nucleoside
monophosphates
(positive effectors)

GTP
(negative effector)

↳ Ex. AMP

Fig. 7.11 Regulation of the pyruvate dehydrogenase complex in *E. coli*.

acetylation of the bound lipoamide. The activity of the pyruvate dehydrogenase complex is stimulated by nucleoside monophosphates and inhibited by GTP. The latter effect is reversed by GDP. Like the pyruvate dehydrogenase complex, oxoglutarate dehydrogenase activity is also regulated by the ratio of the concentrations of substrates and products ([succinyl CoA]/[CoA] and [NADH]/[NAD$^+$]), the products inhibiting activity. Activity is stimulated by AMP, which appears to promote tighter binding of the substrate, 2-oxoglutarate, and the cofactor TDP. ATP competes with AMP in reversing the stimulation.

7.7.5 Organization and control of the pyruvate dehydrogenase system in mammalian tissues

The pyruvate dehydrogenase complexes from mammalian and avian sources, and from *Neurospora*, interact with two additional enzymes,[65,85] namely, [pyruvate dehydrogenase (lipoamide)] kinase and [pyruvate dehydrogenase (lipoamide)] phosphatase; the former is tightly bound to the complex, whereas the latter is very easily dissociated from the complex. The kinase is bound to the acetyltransferase (E_2) and sub-complexes containing E_2-kinase have been isolated. The substrate of the kinase is pyruvate-decarboxylase (E_1), which becomes phosphorylated on one of its serine side-chains when the intracellular ratio of [ATP]/[ADP] is high; this results in inactivation of the multienzyme complex, specifically inhibiting the decarboxylation step (Fig. 7.12). A single kinase can phosphorylate 20–30 E_1 tetramers within the complex.

The phosphatase consists of a catalytic subunit (M_r 52 600) and an FAD-containing regulatory subunit (M_r 95 600) and is a member of the protein phosphatase 2C family (see Chapter 6, Section 6.4.2.4, and reference 94). It catalyses the dephosphorylation of E_1, which converts the inactive multi-enzyme complex back into the active form (further details of this mechanism are given in reference 95). Pyruvate inhibits the phosphorylation step, thus maintaining the complex in an active state. The ratio of phosphorylated to dephosphorylated

Fig. 7.12 Regulation of mammalian pyruvate dehydrogenase complexes by phosphorylation and dephosphorylation.

pyruvate dehydrogenase multienzyme complex has been shown to vary with the physiological state of the mammal, e.g. whether it is well fed or starved (see also Chapter 6, Section 6.3.2.2). The pyruvate dehydrogenase multienzyme complexes from the lower eukaryotes *Neurospora crassa* and *Saccharomyces lactis* are also subject to similar covalent modification.[96] The branched-chain oxoacid dehydrogenases are also regulated by phosphorylation/dephosphorylation but there is no evidence in mammalian oxoglutarate dehydrogenase complexes for covalent modification by kinases and phosphatases.

In eukaryotes the pyruvate dehydrogenase multienzyme complex resides in the matrix within the inner mitochondrial membrane. The three enzyme components are coded for by nuclear genes. It has been shown that E_1, E_2, and E_3 are synthesized as cytosolic precursors, each having a transit peptide, which together with chaperone proteins (see Chapter 3, Section 3.6.2) direct them into the mitochondrial matrix, where the peptide is removed by proteolysis and the multienzyme complex is assembled.[97]

7.8 Glycine decarboxylase multienzyme complex

The glycine decarboxylase system is important as a major glycine degradation pathway in mammals and is important in photorespiration in plants. The glycine decarboxylase multienzyme system (EC 2.1.12.10) has a number of features in common with the oxo-acid dehydrogenase systems (Table 7.1). Together with serine hydroxymethyltransferase (reaction 2) the enzyme catalyses the transformation of glycine into serine in photorespiration in plants.

1. Glycine + NAD^+ + THF $\rightleftharpoons$ methyleneTHF + CO_2 + NH_3 + NADH

2. Glycine + methyleneTHF + H_2O $\rightleftharpoons$ serine + THF

THF = tetrahydrofolate

Glycine decarboxylase (reaction 1) oxidatively decarboxylases glycine to yield a –CH= group which then becomes transferred to a second glycine molecule to form serine. Glycine decarboxylase comprises four different polypeptides: amino acid decarboxylase (P), a polypeptide having a bound lipoamide (H), tetrahydrofolate transferase (T), and lipoamide dehydrogenase (L) with a probable stoichiometry 4P:27H: 9T:2L.[22] Polypeptide L is identical to with the lipoamide dehydrogenase in the pyruvate dehydrogenase system[23] and is also involved in transferring hydrogen to NAD^+ from a lipoamide moiety but in this case on polypeptide H.

There are other multienzyme complexes that show structural and functional relationships with the oxo-acid dehydrogenase complexes, and which have flexible arms in which a lysine residue is either linked to lipoic acid, e.g. acetoin dehydrogenase, or to biotin, e.g. acetyl CoA carboxylase and fatty-acid synthase, and is able to swing between two components of the system.[4]

7.9 The tryptophan synthase multienzyme complex from *E. coli*

The tryptophan synthase system is much smaller than either the pyruvate or oxoglutarate dehydrogenase systems, having an M_r of approximately 150 000. It is a striking example in which the interaction of two subunits greatly modifies their respective catalytic activities and enables channelling of the intermediate to occur between the two subunits. The tryptophan synthase multienzyme complex catalyses the final steps in the biosynthesis of tryptophan (Fig. 7.13). The tryptophan synthase most studied is that from *E. coli* but it has also been investigated in a number of other bacteria, in fungi, and in plants. Mammals lack tryptophan synthase and thus require a dietary source of tryptophan. Tryptophan synthase from *E. coli* is coded for by two structural genes, which, together with three other structural genes, form an operon encoding the enzymes required for the synthesis of tryptophan from chorismate (Fig. 7.14). Tryptophan synthase comprises two different polypeptide chains α and β and has the structure $\alpha_2\beta_2$. The α and β_2 units catalyse reactions 1 and 2, respectively, whereas the $\alpha_2\beta_2$ complex catalyses reaction 3 but will also catalyse reactions 1 and 2.

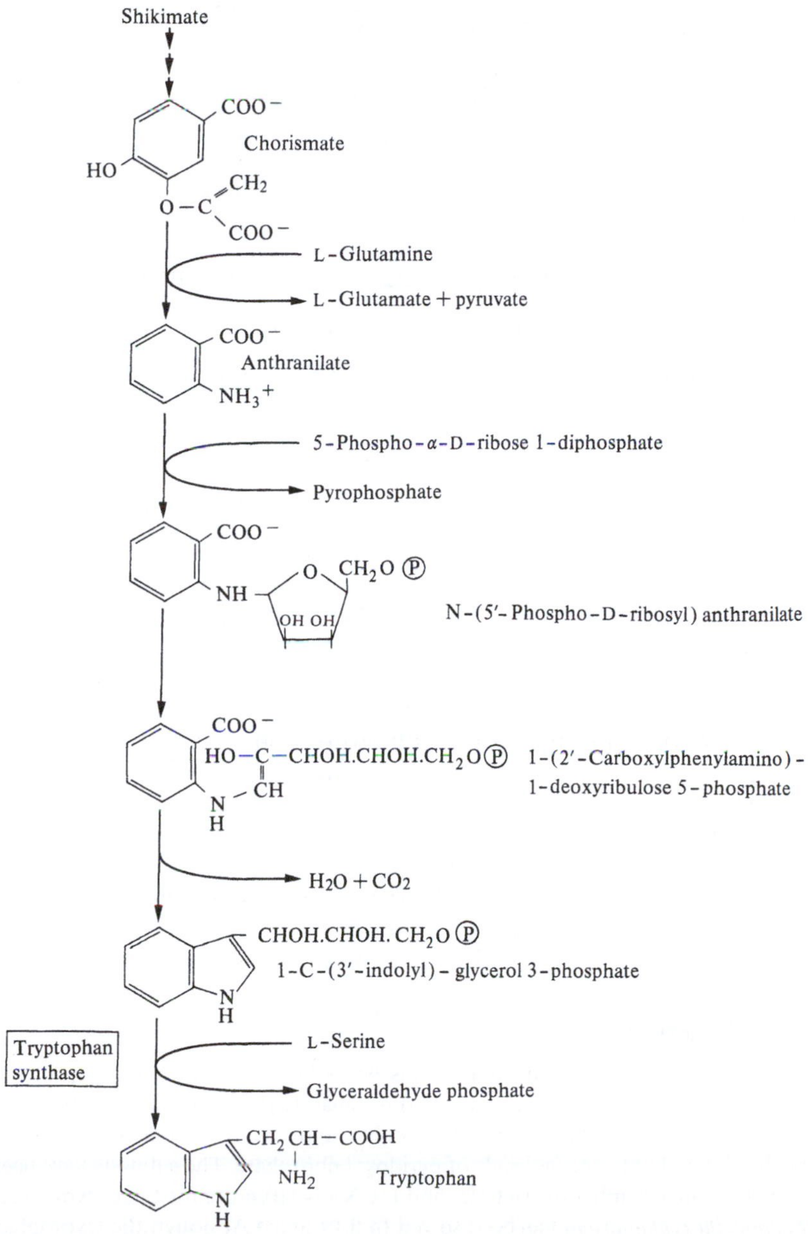

Fig. 7.13 The pathway for biosynthesis of tryptophan. $\textcircled{P}$ = phosphoryl group.

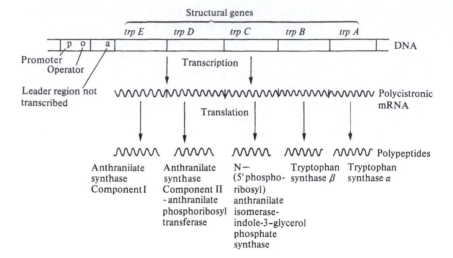

Fig. 7.14 The arrangement of the genes which constitute the tryptophan operon in *E. coli.*

Reaction 1

Indoleglycerol phosphate   Indole   Glyceraldehyde 3–phosphate

Reaction 2

Indole   L–Serine   L–Tryptophan

Reaction 3

7.9.1 Structure

The enzyme from *E. coli* can readily be dissociated into separate α and β_2 proteins, but more drastic conditions are required to separate β_2 into β; this results in loss of activity. The activity is, however, restored upon reassociation. Each β polypeptide chain binds one molecule of pyridoxal phosphate. The subunits have been sequenced in a number of bacteria, and the X-ray structure for the enzyme from *Salmonella typhimurium* has been solved to 0.19 nm.[98] Although the tryptophan

Table 7.4 Structures of tryptophan synthase from different species

Source	Structure	M_r of complex	M_r of subunits	Comments
E. coli	$\alpha_2\beta_2$	146 000	2 × 29 000 2 × 44 000	Ready dissociation to $2\alpha + \beta_2$
Neurospora crassa	A_2	150 000	2 × 75 000	
Saccharomyces cerevisiae	A_2	154 000–167 000	2 × 80 000	
Euglena	A_2?	325 000	2 × 160 000	Complex contains four catalytic activities for E_9–E_{12} on Fig. 7.25

References 99–100.

synthase system in plants is also a two-component heteropolymer, the enzymes from *Euglena*, *Neurospora crassa*, and *Saccharomyces cerevisiae* are homopolymers (Table 7.4). The enzymes from *Neurospora crassa* and *Saccharomyces cerevisiae* were initially thought to be composed of separate α and β subunits but this was due to proteolysis during isolation and both have been shown to be homopolymers.[99,100] The homopolymeric nature of these enzymes is also supported by the genetic evidence of a single gene locus. This is believed to be an example where, during the course of evolution, gene fusion has occurred. The DNA sequence of the genes in both *E. coli* and *Saccharomyces cerevisiae* have been determined and they show a high degree of similarity.[101]

The multienzyme polypeptide from *Euglena gracilis* is a homodimer capable of catalysing the five steps from anthranilate to tryptophan (see Fig. 7.25). Each polypeptide has an M_r of 136 000. A comparison of the sequence with that from *E. coli* tryptophan synthase subunits does not show the degree of similarity that would be expected had gene fusion occurred.[102]

7.9.2 **Mechanism**

The studies on the mechanism refer to the enzyme system from *E. coli*. The multienzyme complex, in addition to catalysing the complete reaction (reaction 3), catalyses the formation of indole (reaction 1) and the conversion of indole to tryptophan (reaction 2). The separate α and β_2 subunits catalyse reactions 1 and 2, but at only 3 and 1 per cent of the rate catalysed by $\alpha_2\beta_2$. For this higher rate of reaction the α and β proteins have to be in physical contact. (They cannot, for example, be separated by a dialysis membrane.) Detailed kinetic studies, both rapid and steady state, have been made on reactions 1 and 2 catalysed by α and β_2, respectively, and these have been compared with similar studies using $\alpha_2\beta_2$. The general conclusion is that the mechanism is unchanged whether the single enzyme or the multienzyme complex is used, e.g. reaction 1 when studied in the direction of indole glycerol phosphate formation proceeds via an ordered ternary complex mechanism (see Chapter 4, Section 4.3.5.1) in which glyceraldehyde 3-phosphate binds first whether it is catalysed by α or by $\alpha_2\beta_2$. Although the mechanism appears to be the same, the rate constants are appreciably changed when a subunit is used compared to the multienzyme complex.

These increases in the rates of the partial reactions (reactions 1 and 2) that occur when either α or β_2 are combined may also be brought about by combining one defective protein with a catalytically active one. It is possible to obtain defective α or β_2 proteins either from suitable mutants of *E. coli* or by chemical modification of catalytically active subunits. These defective proteins can then be used to assemble hybrid $\alpha_2 \beta_2$ proteins. A subunit does not have to be catalytically active in order to bring about a rate enhancement in the other reaction, the key requirement for the effect is binding. The explanation for the changes in rates of catalysis when the subunits are combined is evident from the three-dimensional structure of the enzyme, determined by X-ray crystallography, and the conformational changes detected using the fluorescent probe 8-anilino-1 naphthalene sulphonate.[12]

If the complete reaction (reaction 3) is carried out in the presence of [^{14}C]indole, it is found that no radioactivity is present in the tryptophan formed. This suggests that, when the complete reaction occurs, either indole is not an intermediate or indole formed in the first reaction remains enzyme bound, i.e. the indole is effectively channelled to the second catalytic site without equilibrating with indole in the surrounding medium. This second possibility has now been clearly demonstrated to be the case. The three-dimensional structure of the enzyme from *Salmonella typhimurium* shows that the α subunit has an (α helix/β sheet)$_8$ barrel folded motif analogous to that in triose phosphate isomerase (see Chapter 5, Section 5.5.2) and that a 2.5 nm tunnel (Fig. 7.15) connects the active sites of the α and β subunits. The indole formed at the active site of the α subunit passes through the tunnel to the active site of the β subunit where it interacts with the α-aminoacrylate intermediate (E(A-A)) to form tryptophan. For efficient

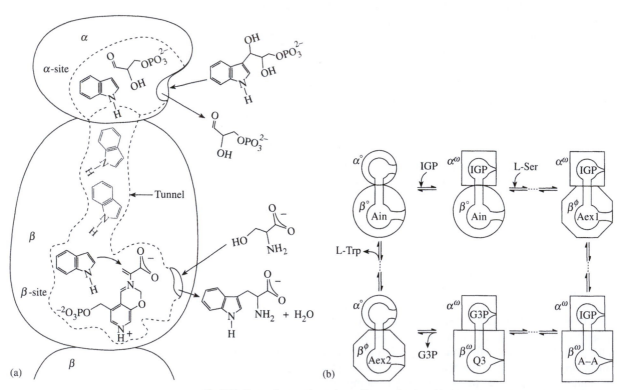

Fig. 7.15 Tryptophan synthase channelling mechanism, (a) showing the diffusion of indole from the α subunit to the β subunit , and (b) the conformational changes occurring during different stages of the reaction. The 'circular' conformations are open, the 'square' conformations closed, and the 'octagonal' conformation for the β subunit is partially open. Ain = the β subunit with pyridoxal phosphate attached. Other abbreviations are those given in the reaction scheme on p. 299.[12]

channelling between the two active sites, catalysis at both α and β subunits must be in phase and the indole intermediate must be unable to escape. Both α and β subunits are able to undergo allosteric conformational changes to coordinate this phasing. Evidence for the conformational changes is from a combination of kinetic studies and the use of substrate analogues and the conformational probe 8-anilino-1-naphthalene sulphonate.[12] Serine binds to the active site of the β subunit and is converted to the α-aminoacrylate intermediate. The formation of this intermediate induces a 25- to 30-fold activation of the α subunit as the result of conformational changes. The subsequent stage of the reaction in which $E(Q_3)$ is converted to $E(Aex_2)$ inactivates the α subunit by reversing the conformational change. In addition, there are polypeptide loops that are able to flip over the active sites to prevent access and escape of substrates and products at the appropriate stages as shown in Fig. 7.15.

7.10 Carbamoyl phosphate synthase and the associated enzymes of the pyrimidine and arginine biosynthetic pathways in *E. coli*, fungi, and mammalian cells

Carbamoyl phosphate is the metabolic precursor of both pyrimidines and arginine, and carbamoyl phosphate synthase catalyses a step that is at a branch point in the metabolic pathways (Fig. 7.16; see also Chapter 6, Section 6.3.1). The carbamoyl moiety may become transferred either to aspartate or to ornithine. In certain tissues and organisms, carbamoyl phosphate synthase has been shown to be physically associated with aspartate carbamoyltransferase and in some cases also dihydro-orotase.

The multienzymes involved illustrate two important features, namely (i) the channelling of a metabolite in a multienzyme protein and (ii) the evolutionary changes in the organization of a multienzyme protein.[103] We compare the organization of the pathways for the biosynthesis of pyrimidines and arginine in *E. coli*, *Saccharomyces cerevisiae*, and *Neurospora crassa*, and examine the structure and regulation of the multienzyme complexes involved. The main features of the systems in *E. coli*, *S. cerevisiae*, and *N. crassa* are illustrated in Fig. 7.17.

7.10.1 *E. coli*

It can be seen that in *E. coli* there is a single carbamoyl phosphate synthase that catalyses the formation of carbamoyl phosphate for both arginine and pyrimidine biosynthesis. The enzyme comprises two subunits having glutaminase and carbamoyl phosphate synthetase activities respectively. There are two genes encoding

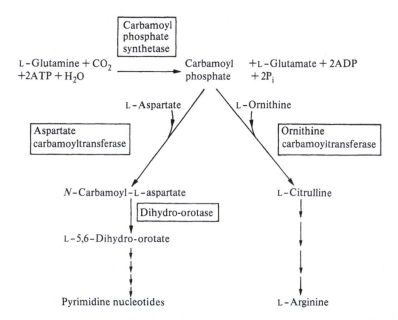

Fig. 7.16 The position of carbamoyl phosphate synthase on the metabolic pathways for arginine and pyrimidine biosynthesis. P_i = orthophosphate.

the subunits, namely *car*A and *car*B. Mutations in either can cause auxotrophy for both arginine and pyrimidines (i.e. both arginine and pyrimidine have to be supplied in the nutrient medium in order for growth to occur). Regulation of the balance in the supply of carbamoyl phosphate between pyrimidine and arginine biosynthesis is by the activator ornithine and the inhibitor UMP (Fig. 7.17) and is more fully described in Chapter 6, Section 6.3.1. There is no evidence for any physical association of the enzymes in *E. coli*.

7.10.2 *Saccharomyces cerevisiae* (yeast)

In contrast to the situation in *E. coli*, the two fungi *Saccharomyces* and *Neurospora*[104] each have two distinct carbamoyl phosphate synthetases, although there are differences between the two in the extent to which channelling of carbamoyl phosphate occurs. In *Saccharomyces* there are two distinct carbamoyl phosphate synthetases (abbreviated to CPS_{Py} and CPS_{Arg}). One of these, CPS_{Py}, is physically associated with aspartate carbamoyltransferase, but carbamoyl phosphate produced by either CPS_{Arg} or CPS_{Py} may be utilized in either biosynthetic pathway. The evidence for this is that mutants lacking either CPS_{Arg} or CPS_{Py} will grow on a minimal medium (i.e. without arginine or pyrimidine

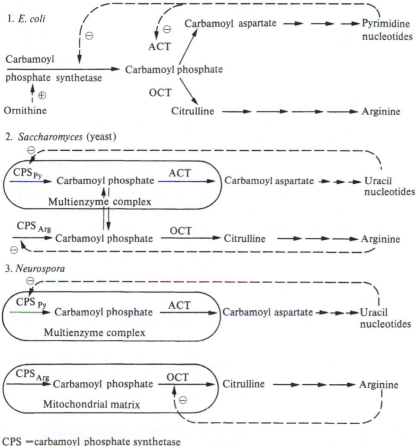

CPS =carbamoyl phosphate synthetase
ACT=aspartate carbamoyltransferase
OCT=ornithine carbamoyltransferase

Fig. 7.17 Comparison of arginine and pyrimidine biosynthesis in *E. coli*, *Saccharomyces*, and *Neurospora*. ⊕ = activation, ⊖ = inhibition.

supplementation). However, mutants lacking CPS_{Py} will not grow on arginine-supplemented media and mutants lacking CPS_{Arg} will not grow on uracil-supplemented media. This is as expected from the feedback-inhibition patterns (Fig. 7.17), e.g. uracil inhibits CPS_{Py} and thus in a mutant lacking CPS_{Arg} no carbamoyl phosphate can be formed and growth is inhibited.

The fact that mutants lacking either CPS_{Arg} or CPS_{Py} will grow on a minimal medium shows that carbamoyl phosphate formed may be exchanged between the pathways. However, the growth rate of these mutants on minimal medium is appreciably slower than that of the wild type, which suggests there is not a completely free exchange of carbamoyl phosphate between the pathways, i.e. some channelling occurs.

7.10.3 Neurospora

In *Neurospora*[105] there is efficient channelling of carbamoyl phosphate synthesized by the two enzymes. The channelling is achieved by means of a bifunctional enzyme protein containing CPS_{Py} and aspartate carbamoyltransferase and by different subcellular locations of the enzymes (Fig. 7.17). CPS_{Arg} and ornithine carbamoyltransferase are located in the mitochondria, whereas the CPS_{Py}:aspartate carbamoyltransferase multienzyme protein is present in the cytosol. Consistent with this is the finding that mutants lacking CPS_{Py} or CPS_{Arg} are both auxotrophs (i.e. they require supplementation with uracil or arginine, respectively, for growth to occur). Further evidence is that carbamoyl phosphate formed by CPS_{Py} *in vitro* does not exchange with added radioactively labelled carbamoyl phosphate. CPS_{Arg} in *Neurospora* is not feedback-inhibited by arginine as is the corresponding enzyme in *Saccharomyces*.

7.10.4 Structure and regulation of the carbamoyl phosphate synthetase:aspartate carbamoyltransferase: dihydro-orotase (CAD) complex

The organization of the enzymes catalysing the conversion of bicarbonate to L-5,6-dihydro-orotate (E_1, E_2, and E_3 in Fig. 7.18) differs according to species. There are three classes of carbamoyl phosphate synthetase known as CPS I, CPS II, and CPS III[106] catalysing the first step. In most prokaryotes there is a single CPS II comprising two polypeptide chains, one of which catalyses the conversion of glutamine into an amino group (glutaminase) and the other catalyses the formation of carbamoyl phosphate (carbamoyl phosphate synthetase). CPS II is used in both arginine and pyrimidine biosynthesis. In fungi there are two separate CPS IIs, one used for arginine and one for pyrimidine biosynthesis. The CPS II required for arginine biosynthesis, located in the mitochondria or cytosol, has two polypeptides which act in the same way as that in prokaryotes, but the CPS II used for pyrimidine biosynthesis, located in the nucleus, is part of a multienzyme polypeptide having glutaminase, carbamoyl synthetase, and aspartate carbamyoltransferase (E_2) activity. In mammals the arginine biosynthetic pathway uses CPS I which is a single polypeptide chain having a carbamoyl phosphate synthetase site and an inactive glutaminase site. Mammals use ammonia rather than glutamine as the amino donor for synthesis of carbamoyl phosphate. However, pyrimidine biosynthesis is catalysed by a multifunctional polypeptide which includes glutaminase, carbamoyl phosphate synthetase (CPS II), aspartate carbamoyltransferase (E_2), and dihydro-orotase (E_3) (Fig. 7.19). Glutamine is used as the amino donor. CPS III is present in invertebrates and fishes. It has glutaminase

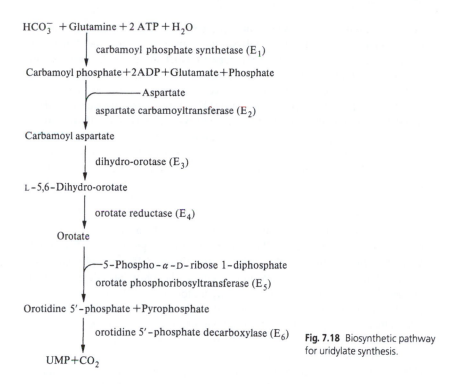

$$HCO_3^- + Glutamine + 2\,ATP + H_2O$$

carbamoyl phosphate synthetase (E_1)

Carbamoyl phosphate + 2 ADP + Glutamate + Phosphate

Aspartate

aspartate carbamoyltransferase (E_2)

Carbamoyl aspartate

dihydro-orotase (E_3)

L-5,6-Dihydro-orotate

orotate reductase (E_4)

Orotate

5-Phospho-α-D-ribose 1-diphosphate

orotate phosphoribosyltransferase (E_5)

Orotidine 5′-phosphate + Pyrophosphate

orotidine 5′-phosphate decarboxylase (E_6)

$$UMP + CO_2$$

Fig. 7.18 Biosynthetic pathway for uridylate synthesis.

and carbamoyl phosphate synthetase domains fused, and uses glutamine as the amino donor. The only evidence for any association between CPS and ornithine carbamoyltransferase is in the hyperthermophilic archaebacteria *Pyrococcus furiosus*. In this organism, which can grow at 100°C, there is evidence that carbamoyl phosphate formed by CPS is channelled directly to ornithine carbamoyltransferase to form citrulline. The half-life of carbamoyl phosphate at 100°C is < 2 s.[107]

The arrangement of the four polypeptide domains involved in the synthesis of dihydro-orotate from bicarbonate is given in Fig. 7.19.

The mammalian multienzyme CAD, catalysing the conversion of bicarbonate to dihydro-orotate, is a single polypeptide of M_r 243 000 which associates to form hexamers. Initially, there was some surprise to find that the M_r of the multienzyme from both *Saccharomyces cerevisiae* and *Schizosaccharomyces pombe* (fission yeast) was also about 240 000, since both lacked dihydro-orotase activity. The explanation is that both have a dihydro-orotase region which is catalytically inactive.[109,110] The structure of the domains of CAD has been worked out by a combination of DNA sequence determination and limited proteolysis of CAD

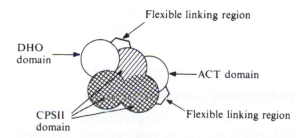

Flexible linking region

DHO domain

ACT domain

CPSII domain

Flexible linking region

Fig. 7.19 A model of the domain structure of the multifunctional polypeptide CAD.[108] Abbreviations are explained in Fig. 7.20.

overexpressed in *E. coli* cells.[13] Limited proteolysis is often used to study the separate domains of multienzyme polypeptides, since in general the domains are compact but are separated by flexible regions which are generally more sensitive to proteolysis (see also Sections 7.11.1.1 and 7.11.2). When the mammalian CAD multienzyme has been subjected to proteolysis using elastase and trypsin, it is possible to separate the dihydro-orotase (DHO), the aspartate carbamoyltransferase (ACT), and the carbamoyl phosphate synthase II (CPS II) domains.[108] There is a large linking region of 109 amino-acid residues connecting dihydro-orotase to aspartate carbamoyltransferase, but the CPS II region connects to the dihydro-orotase without an apparent linker. Mammalian aspartate carbamoyltransferase is similar to that from *E. coli* in forming trimers, but it lacks the regulatory subunits that are required for cooperative binding of aspartate in the *E. coli* enzyme (see Chapter 6, Section 6.2.2.2).

CPS catalyses the rate-limiting step of the multienzyme polypeptide; the maximum catalytic activity of ACT is about 50 times greater than that of CPS. UTP, the end-product of the pathway, inactivates CPS and phosphoribosyl pyrophosphate is an activator. A protein kinase phosphorylates the multienzyme polypeptide activating CPS and abolishing inhibition by UTP. Carrey[111] has proposed that CPS can exist in two conformational states, P and U. The P state is stabilized by PRPP or phosphorylation and has a high V_{max} and low K_m for ATP and Mg^{2+}. The U state is stabilized by UTP and has a low V_{max} and high K_m. In addition, the binding of aspartate to the catalytic site of ACT induces a change in the CPS catalytic site which prevents carbamoyl phosphate from causing product inhibition. The substrates for CPS affect the catalytic site of ACT by increasing its affinity for carbamoyl phosphate and aspartate. These combined effects, referred to as reciprocal allostery, together with the close proximity of the two catalytic sites are thought to ensure that carbamoyl phosphate formed by CPS II is efficiently used in the second step of pyrimidine biosynthesis[109] (Fig. 7.20).

There are two steps catalysed by CPS II; glutaminase activity (1), followed by formation of carbamoyl phosphate (2).

(1) glutamine + H_2O $\rightleftharpoons$ glutamate + NH_3

(2) NH_3 + HCO_3^- + 2ATP $\rightleftharpoons$ H_2N–CO·OPO_3^{2-} + 2ADP + P_i.

It is interesting to compare the organization of the structural genes for the biosynthesis of UMP from carbon dioxide and glutamine or ammonia in different species. There are altogether six enzyme-catalysed reactions in the pathway, as shown in Fig. 7.18. In *E. coli* there are six gene loci corresponding to the six structural genes in the pathway. In *Saccharomyces* and *Neurospora* there are only five structural genes, since a single gene codes for the multienzyme polypeptide that contains E_1 and E_2 activities. In mammalian systems there appear to be only three structural genes coding for E_1–E_2–E_3, E_4, and E_5–E_6.[112] There is some evidence that E_5–E_6 in mammals exists as a multienzyme protein. Thus, during the course of evolution there is clustering on the chromosomes of genes that have related functions and in some cases there is gene fusion leading to the production of multienzyme polypeptides.

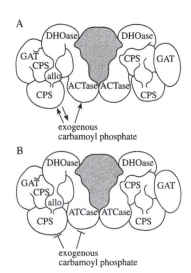

Fig. 7.20 Model of the interactions of carbamoyl phosphate synthase II (CPS II) and aspartate carbamoyltransferase (ACTase) domains of CAD. The catalytic domains of CPS II and ACT are in close proximity. Aspartate is considered to induce a conformational change from A to B bringing the two catalytic sites into closer proximity allowing carbamoyl phosphate to be transferred efficiently from one site to the other. The other catalytic domains are glutamine amidotransferase (GAT) and dihydro-orotase (DHOase).[109]

7.11 Multienzyme polypeptides: fatty-acid synthase and the *arom* complex (AROM enzyme)

If enzymes are to be isolated with their complete polypeptide chains intact it is extremely important to avoid proteolysis occurring during isolation (see Chapter

2, Section 2.8.4). Many cells and tissues contain proteases that may be present in a latent form, e.g. within lysosomes or vacuoles, and these are often released during isolation. On a simple statistical basis, proteolysis is more likely to cause a single break in a large polypeptide chain than in a small one. Thus a number of multienzyme proteins originally considered as multienzyme complexes have turned out to be multienzyme polypeptides. Although the effects of unwanted proteolysis led to early problems in isolating intact multienzyme polypeptides, controlled proteolysis is now used as a tool to provide structural information about multidomain proteins. The sensitivity of the interdomain regions of such proteins to proteolytic attack can often be turned to advantage in identifying and isolating such domains. The examples of tryptophan synthase in *Neurospora* and *Saccharomyces* were mentioned in Section 7.9.1 and CAD in Section 7.10.4, but the most striking examples of multienzyme polypeptides are the fatty-acid synthases and the complex of enzymes involved in aromatic amino-acid biosynthesis known as the *arom* complex; these are discussed in this section.

7.11.1 Fatty-acid synthase

The ability to synthesize long-chain fatty acids is an attribute of all organisms and those in which fatty-acid synthesis has been examined in any detail show the same basic metabolic pathway, though with certain minor modifications, as illustrated in Fig. 7.21. Fatty acids are synthesized from acetyl CoA and malonyl CoA and the overall process of extending an acyl chain by two methylene groups entails transacylations, two reductions, a dehydration, and a condensation and thus involves six catalytic activities as shown in Fig. 7.21. An additional enzyme, acetyl CoA carboxylase, is necessary to catalyse the formation of malonyl CoA but this is not part of the complex.

7.11.1.1 Organization of the fatty-acid synthase system

There are differences between organisms both in the predominant chain length of fatty acids they synthesize and in the mechanism of release of the long-chain acyl carboxylate, but the sequence of enzyme reactions is similar in all cases. In spite of this similarity, the organization of the enzyme systems may differ.[113] In *E. coli* six separate enzymes (ACP acetyltransferase, ACP malonyltransferase, 3-oxoacyl ACP synthase, 3-oxoacyl ACP reductase, crotonyl ACP hydratase, and enoyl ACP reductase), together with the acyl carrier protein (ACP), have been isolated and purified. These enzymes together catalyse fatty-acid synthesis and there is no evidence that they are physically associated with one another. This non-aggregated type of fatty-acid synthase system (referred to as type II) has also been found in other bacteria, in the blue-green alga *Phormidium lunidum*, in *Euglena*, *Chiamydomonas*, Avocardo mesocarp, and in the chloroplasts from lettuce and spinach. However, although the fatty-acid synthase enzymes from chloroplasts are made up of discrete soluble enzymes, there is evidence to suggest that in the chloroplasts they exist as a multienzyme complex associated with the thylakoids. Evidence from permeabilized chloroplasts shows they are capable of synthesizing long-chain fatty acids from acetate but do not use exogenous acetyl CoA or malonyl CoA. The lack of competition from exogenous intermediates together with the kinetics of fatty-acid synthesis has been explained on the basis that acetyl CoA synthetase and acetyl CoA carboxylase channel acetate directly into the fatty-acid synthase multienzyme complex.[114]

By contrast, in many animal tissues, avian tissues, yeast, *Neurospora*, and in some prokaryotes, the fatty-acid synthase system is found to exist as an aggregated

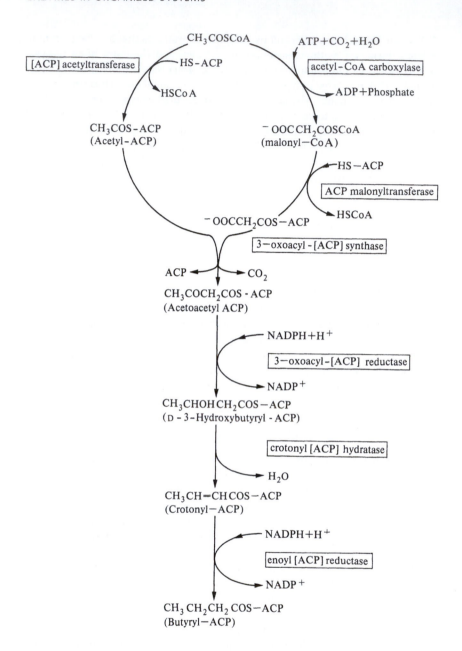

Fig. 7.21 The fatty-acid biosynthetic pathway. All reactions except the carboxylation of acetyl CoA are catalysed by the fatty-acid synthase multienzyme system.

multienzyme protein (type I). It was originally thought that these multienzyme proteins comprised several tightly associated polypeptide chains each having a single catalytic activity, but as a result of purifications, in which rigorous measures have been taken to exclude proteolysis, several of these have been shown to exist as multienzyme polypeptides. The type I fatty-acid synthases may be subdivided on the basis of their size into type IA, which have values of M_r of about $500\,000$ and subunit structures α_2, and type IB, showing a higher degree of aggregation and values of M_r from 1×10^6 to 2.4×10^6 and subunit structures α_6–α_8 or $\alpha_6\beta_6$.[113] Type IA are found in animal tissues, whereas Type IB occur in fungi, algae, and certain bacteria such as *Mycobacteria*, *Corynebacteria*, and *Streptomyces*. The subunits of both types IA and IB have M_r values of between $185\,000$ and $250\,000$.

The yeast fatty-acid synthase system has been most thoroughly investigated both from a biochemical and genetic standpoint. By methods of genetic mapping[115] it has been shown that there are two unlinked genetic loci on the chromosome designated *fas 1* and *fas 2*, which code for the whole fatty-acid synthase system. The two different subunits and the assignment of the catalytic activities to particular subunits is based on biochemical[116] and genetic[117] evidence. Fatty-acid synthase can be dissociated reversibly after modification of its free amino groups using maleic anhydride* (see Chapter 3, Section 3.3.2.2) and then the catalytic activities can be assigned to the separated polypeptide chains. The arrangement of catalytic centres on the α and β polypeptide chains is shown in Fig. 7.22, together with a possible model for the three-dimensional arrangement of the α and β subunits. The α chains are arranged in the form of a disc, and the β chains are looped across pairs of α chains above and below the plane of the disc.

In the smaller type IA fatty-acid synthases found in animal tissues, those that have been examined in detail have been found to be homodimers.[118-120] In the yeast fatty-acid synthase, the activities of ACP malonyltransferase, which is necessary to initiate fatty-acid chain growth (see Fig. 7.21, and of ACP palmityltransferase, which is necessary to terminate chain growth, are in identical regions of the polypeptide chain.[121,122] In the mammalian multienzymes each polypeptide of the homodimer (α_2) has seven component activities. Evidence from proteolytic mapping indicates that these domains are in the following order starting from the N-terminus: 3-oxoacyl ACP synthase, ACP acetyl- and malonyl-transacylases, crotonyl ACP hydratase, enoyl ACP reductase, 3-oxoacyl ACP reductase, acyl carrier protein and thioesterase.[123] Thioesterase cleaves the final acyl ACP to release the free fatty acid. The ACP malonyl and acetyl transferase activities are accounted for by a single domain having an essential serine and histidine residue at the catalytic site.[124] Arginine-606 is essential for malonyl transacylase but not acetyl transacylase activity. It is thought that the positively charged guanidinium group on arginine is necessary to bind the free carboxyl of malonyl ACP to position it at the active site.[125] The fatty-acid synthases from chicken liver,[118] pigeon liver,[119] and rabbit mammary gland[120] each contain all seven catalytic activities within a single polypeptide chain and the multienzyme appears to be a homodimer arranged in head-to-tail fashion. A model for the organization of the domains has been proposed (see Fig. 7.23). It can be seen that the growing chain is attached to the thiol group on the end of the pantotheine moiety; it has to interact with the thiol on the synthase on the opposite α chain. The evidence for this stems from the observation that when the dimers are dissociated, $\alpha_2 \rightarrow 2\alpha$, the only activity lost is 3-oxoacyl ACP synthase.[20]

A common feature of both type I and type II synthases is that the growing acyl chain is covalently attached to a protein through a phosphopantotheine

$$\text{Protein (Ser)OPO}_3^-\text{CH}_2\cdot\text{C(CH}_3)_2\cdot\text{CHOH}\cdot\text{CONH(CH}_2)_2\text{CONH(CH}_2)_2\underbrace{\text{SCOR}}.$$

acyl group

prosthetic group. The acyl carrier protein has been isolated from several prokaryotes and sequenced. Its size varies between species, ranging from M_r values of 8600 to 16 000.[126] In eukaryotes it is clear that the equivalent of an acyl carrier protein exists, since phosphopantotheine becomes covalently bound to the fatty-acid synthase and thus the 'acyl-carrier protein' is a section of the multienzyme polypeptide.

* Modification of free amino groups facilitates dissociation of the complex.[118]

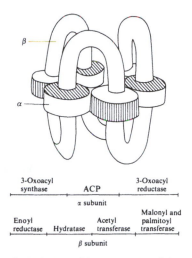

3-Oxoacyl synthase	ACP	3-Oxoacyl reductase

α subunit

Enoyl reductase	Hydratase	Acetyl transferase	Malonyl and palmitoyl transferase

β subunit

Fig. 7.22 A possible arrangement of the (α) and (β) subunits of yeast fatty-acid synthase ($\alpha_6\beta_6$), together with the location of enzyme activities on the polypeptide chains.

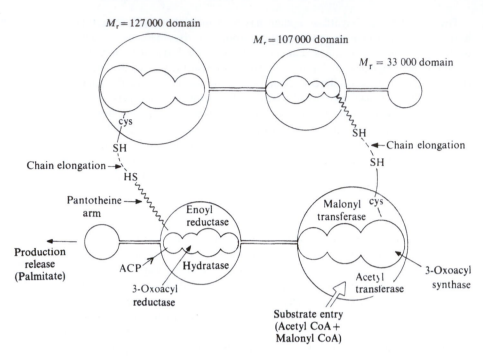

Fig. 7.23 The proposed domain structure of animal fatty-acid synthase, showing the two α subunits aligned head-to-tail.

7.11.1.2 Control of the chain length of synthesized fatty-acids

The process of fatty-acid synthesis is often described as a spiral because a sequence of reactions occurs that leads to the addition of $-CH_2CH_2-$ to a growing acyl chain and then the sequence is repeated. During this growth the acyl group is attached to either the acyl carrier protein or part of the multienzyme polypeptide. What determines the length of chain to be synthesized and how is it released from the enzyme? There is a fairly well-defined pattern for both the chain length and the degree of saturation of fatty acids synthesized by different species. In *E. coli* the predominant fatty acids synthesized are palmitic acid, palmitoleic acid, and *cis*-vaccenic acid and the ACP derivatives of these acids transfer their acyl groups directly to glycerol-l-phosphate during phospholipid biosynthesis. The type of fatty acid synthesized in *E. coli* seems to be determined largely by the specificity of 3-oxoacyl ACP synthase, which catalyses the following two partial reactions:

Palmitic acid $CH_3(CH_2)_{14}COOH$

Palmitoleic acid
$CH_3(CH_2)_5CH=CH(CH_2)_7COOH$

cis-Vaccenic acid
$CH_3(CH_2)_5CH=CH(CH_2)_9COOH$

$$\text{R-CO-S-ACP} + \text{HS-enzyme} \rightleftharpoons \text{R-CO-S-enzyme} + \text{ACP-SH}$$

$$\text{R-CO-S-enzyme} + \text{malonyl-S-ACP} \rightleftharpoons \text{R-COCH}_2\text{CO-S-ACP} + \text{HS-enzyme} + \text{CO}_2$$

where R = saturated or unsaturated alkyl group.

The acyl group undergoing elongation is transferred to a cysteine side-chain on 3-oxoacyl ACP synthase.

When the maximum velocities of this synthase are compared using acyl ACPs having different chain lengths of saturated hydrocarbons as substrates, there is very little difference comparing C_2 to C_{10} acyl ACPs, but the velocity decreases markedly for C_{14} acyl ACP and the enzyme is inactive towards C_{16} acyl ACP. When a similar comparison is made using substrates having unsaturated chains, the activity is low when hexadecenyl ACP is used and inactive with *cis*-vaccenyl ACP. Thus, the specificity of the synthase towards different acyl ACPs sets an upper limit to the number of C_2 units that may be added to the growing chain.[127]

In yeast, the chain lengths of the fatty acids synthesized are predominantly C_{16} and C_{18} and control is probably similar to that in *E. coli*, namely, the specificity of 3-oxoacyl ACP synthase. The final step in the sequence is the transfer of palmityl ACP or stearyl ACP to coenzyme A catalysed by an acyltransferase.

$$\text{acyl-S-ACP} + \text{CoA} \rightleftharpoons \text{acyl-S-CoA} + \text{ACP.}$$

In yeast, palmityltransferase and malonyltransferase activities are associated with the same catalytic site on the multienzyme polypeptide, whereas in the mammalian fatty-acid synthases the two catalytic sites are distinct. A further difference is that in the mammalian system the sequence terminates with the release of free fatty acid, whereas in yeast, acyl CoA is released. This will greatly affect the rate of release of fatty-acid chains, since free fatty acids are readily released from the multienzyme complex, whereas acyl CoAs, particularly of long-chain acids, are only slowly released. (For a review, see reference 128.)

The sequence in yeast and in mammalian and avian systems[129] is as follows.

(i) Yeast system

1. E_1-Pantotheine-S-acyl + E_2-OH $\rightleftharpoons$ E_1-Pantotheine-SH + E_2-O-acyl.

2. E_2-O-acyl + CoA-SH $\rightleftharpoons$ Acyl-S-CoA + E_2-OH.

(ii) Mammalian and avian systems

1. E_1-Pantotheine-S-acyl + E_2-OH $\rightleftharpoons$ E_1-Pantotheine-SH + E_2-O-acyl

2. E_2-O-acyl + H_2O $\rightleftharpoons$ E_2-OH + acyl-OH

where E_1 and E_2 are different catalytic centres on the multienzyme polypeptides. Phosphopantotheine is covalently attached to E_1. The acyl group is transferred from the sulphydryl on pantotheine to a serine hydroxyl group on E_2.

7.11.1.3 Control of fatty-acid synthesis in *Mycobacterium smegmatis*

The control of fatty-acid synthesis in *Mycobacterium smegmatis* differs from that of *E. coli*, yeast, or mammalian systems and will be described here briefly. (For a detailed review see reference 130.) Fatty acids synthesized by *E. coli* are used primarily for incorporation into the phospholipids of the cell membrane. In mammals, they are, in addition, incorporated into triglycerides to be stored as an energy reserve. *Mycobacterium smegmatis*, which is a complex prokaryote, synthesizes fatty acids both for incorporation into the cell membrane and into the cell wall. C_{16} and C_{18} fatty acids are used for membrane phospholipids, whereas C_{24} and C_{26} fatty acids are incorporated into the cell walls. Both C_{16}–C_{18} and C_{24}–C_{26} fatty acids are synthesized using a type I fatty-acid synthase system. The fatty acids are synthesized whilst attached to the multienzyme protein via a phosphopantotheine residue and they are eventually transferred to CoA as in the yeast system (Fig. 7.24). Release of C_{16} and C_{18} acyl CoAs from the enzyme is fairly slow (compare with yeast fatty-acid synthase, Section 7.11.1.2), but release of the C_{24} and C_{26} acyl CoAs is very slow indeed, diffusion taking of the order of minutes.

Two factors regulate the balance between synthesis of C_{16} and C_{18} fatty acids and of C_{24} and C_{26} fatty acids: (i) high concentrations of acetyl CoA stimulate the

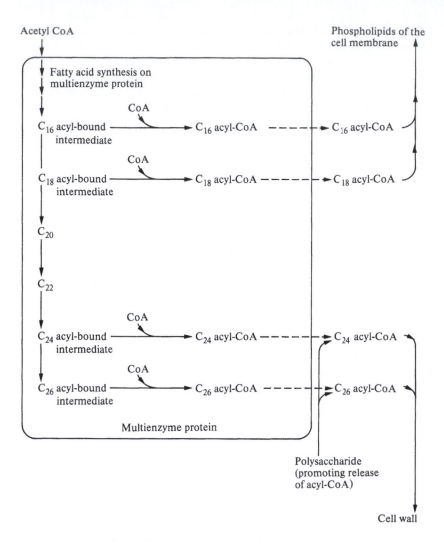

Fig. 7.24 Synthesis of fatty acids in *Mycobacterium smegmatis.*

formation of C_{16}–C_{18} fatty acids; (ii) two classes of polysaccharides that contain decasaccharide segments of 6-*O*-methylglucose and 3-*O*-methylmannose residues promote the release of C_{24}–C_{26} acyl CoAs from the fatty-acid synthase. Thus, in the presence of the polysaccharides, C_{24}–C_{26} fatty-acid synthesis is favoured, but when the polysaccharide concentration is low, slow release of C_{24}–C_{26} acyl CoAs from the enzyme retards fatty-acid synthesis. This causes a build up of acetyl CoA that in turn stimulates the synthesis of C_{16}–C_{18} acyl CoAs.

7.11.2 Multienzyme proteins involved in the biosynthesis of aromatic amino acids

The aromatic amino acids tryptophan, tyrosine, and phenylalanine are synthesized *de novo* in most bacteria, fungi, and plants, whereas they are dietary requirements in many mammals. The biochemistry and genetics of the enzymes involved in this pathway have been studied in detail, particularly in certain prokaryotes and fungi. The same pathway operates in all the organisms studied (as illustrated for *Neurospora crassa* in Fig. 7.25), but there is considerable variation between species in the organization of the genes, and hence for the enzymes for which they code,[131] and also in the pathway regulation.[132] In plants

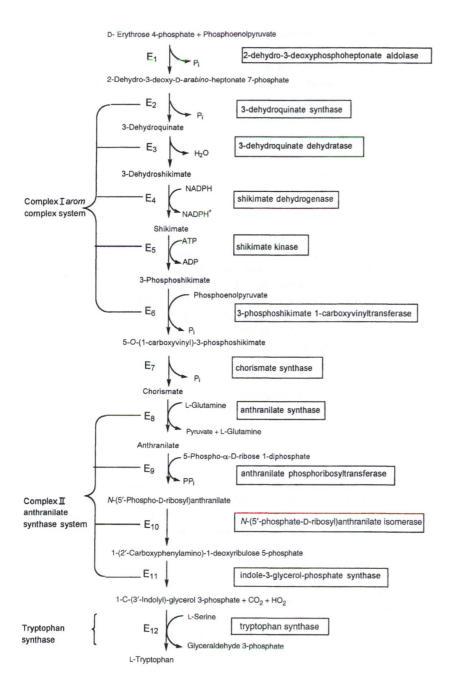

D- Erythrose 4-phosphate + Phosphoenolpyruvate

E_1 — P_i 2-dehydro-3-deoxyphosphoheptonate aldolase

2-Dehydro-3-deoxy-D-*arabino*-heptonate 7-phosphate

E_2 — P_i 3-dehydroquinate synthase

3-Dehydroquinate

E_3 — H_2O 3-dehydroquinate dehydratase

3-Dehydroshikimate

E_4 — NADPH / NADPH⁺   shikimate dehydrogenase

Shikimate

E_5 — ATP / ADP   shikimate kinase

3-Phosphoshikimate

E_6 — Phosphoenolpyruvate / P_i 3-phosphoshikimate 1-carboxyvinyltransferase

Complex I *arom* complex system

5-*O*-(1-carboxyvinyl)-3-phosphoshikimate

E_7 — P_i chorismate synthase

Chorismate

E_8 — L-Glutamine / Pyruvate + L-Glutamine   anthranilate synthase

Anthranilate

E_9 — 5-Phospho-α-D-ribose 1-diphosphate / PP_i anthranilate phosphoribosyltransferase

N-(5′-Phospho-D-ribosyl)anthranilate

E_{10} *N*-(5′-phosphate-D-ribosyl)anthranilate isomerase

1-(2′-Carboxyphenylamino)-1-deoxyribulose 5-phosphate

E_{11} indole-3-glycerol-phosphate synthase

Complex II anthranilate synthase system

1-C-(3′-Indolyl)-glycerol 3-phosphate + CO_2 + HO_2

E_{12} — L-Serine / Glyceraldehyde 3-phosphate   tryptophan synthase

Tryptophan synthase

L-Tryptophan

Fig. 7.25 Pathway for the biosynthesis of tryptophan in *Neurospora crassa*, showing the multienzyme proteins involved. P_i = orthophosphate; PP_i = pyrophosphate.

the enzymes are also involved in the biosynthesis of auxins, such as indole-3-acetic acid, antimicrobials agents such as phytoalexins, and alkaloids. When considering the conversion of chorismate to tryptophan, which is catalysed by enzymes E_8–E_{12} (Fig. 7.25), it is possible to identify seven different catalytic domains that can be completely separated in certain organisms. These are designated by the following letter code.

Domain E	catalyses the anthranilate synthase reaction with ammonia as amino group donor.
Domain G	interacts with the E domain and provides the glutamine amido-transferase activity for the glutamine-dependent anthranilate synthase reaction.
Domain D	is the catalytic site of E_9.
Domain F	is the catalytic site of E_{10}.
Domain C	is the catalytic site of E_{11}.
Domain A	catalyses the conversion of indole glycerol phosphate to indole.
Domain B	catalyses the conversion of indole to tryptophan.

The way in which these domains are linked shows considerable variation when different species are considered.[133] Table 7.5 shows the linkage of polypeptide chains and subunit organization for the best-studied cases. In *Neurospora crassa* the biosynthetic pathway from erythrose 4-phosphate to tryptophan contains three multienzyme complexes (Fig. 7.25). There are also indications that weaker interactions may hold all three multienzyme proteins together with the remaining enzymes of the pathway in a 'super-multienzyme' protein.[134] In this section we discuss the properties of the AROM enzyme, since certain features have been demonstrated with this multienzyme protein which have not been demonstrated in other multienzyme proteins. The AROM enzyme is a pentadomain enzyme catalysing the conversion of 2-dehydro-3-deoxy-D-*arabino*-heptonate 7-phosphate to 5-*O*-(1-carboxyvinyl)-3-phosphoshikimate (E_2–E_6 in Fig. 7.25) encoded by the *arom* gene cluster in filamentous fungi. It comprises two identical polypeptide chains each having an M_r of 165 000.[135] From a study of the peptides obtained from hydrolysing the multienzyme protein, it is clear that each polypeptide chain possesses the five catalytic activities (Fig. 7.25). The genes for all five catalytic activities are located in a cluster and may be transcribed as a polycistronic message, since the catalytic activities are transcribed and translated in an ordered sequence. The domain order in the AROM enzyme (E_2 E_6 E_5 E_3 E_4) is not the same as the sequence of reactions they catalyse.[13] In the case of the AROM enzyme from *N. crassa*, limited proteolysis has been used to separate two frag-

Table 7.5 Patterns of organization of the domains for the conversion of chorismate to tryptophan

Domain of organization					Organisms
Prokaryotes					
E,G–D	F–C			A,B ($\alpha_2\beta_2$)	*Escherichia coli, Salmonella, Enterobacter, Klebsiella, Citrobactor, Vibrio*
E,G($\alpha\beta$)	D	F	C	A,B ($\alpha_2\beta_2$)	*Bacillus subtilis, Pesudomonas putida, Acinebacter calcoaceticus*
E,G ($\alpha_2\beta_2$)	D	F–C		A,B ($\alpha_2\beta_2$)	*Serretia marcescens*
Eukaryotes					
E,G–F–C	D			A–B (α_2)	*Neurospora crassa, Claviceps purpurea, Aspergillus, Penicillium, Phycomyces*
E,G–C	F	D		A–B (α_2)	*Saccharomyces cerevisiae*
E–G				A–B (α_2)	*Euglena gracilis*
E,G	C	F	D	A,B?	Plants

'–' signifies a covalent link, ',' signifies a non-covalent link.
Characterization in some organisms is incomplete.

ments, one containing E_3/E_4 activity (Fig. 7.25) and the other E_6 activity. A model for the possible organization of the domains is shown in Fig. 7.26.

In *E. coli* the catalytic activities appear to be on separate polypeptide chains and are encoded in five separate genes[136] in contrast to the pentadomain enzyme found in yeast, filamentous fungi, and flagellates. The *arom* gene in filamentous fungi is considered to have arisen by multiple gene fusions.

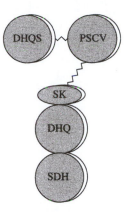

Fig. 7.26 A possible domain map for the shikimate pathway AROM protein: DHQS, dehydroquinate synthase; PSCV, 3-phosphoshikimate 1-carboxyvinyl transferase; SK, shikimate kinase; DHQ, 3-dehydroquinase; SDH, shikimate dehydrogenase.[13]

HO COO⁻ → 3-dehydroquinate dehydratase → 3-Dehydroquinate / 3-Dehydroshikimate + H_2O

A number of kinetic studies have been made to elucidate the working of the AROM enzyme. It is thought that this complex may bring about a channelling of the intermediates so as to effectively separate the biosynthetic from the degradative pathways for aromatic amino acids. There is one enzyme activity, 3-dehydroquinate dehydratase, which is common to both biosynthetic and degradative pathways, and thus without a multienzyme polypeptide, which channels bound 3-dehydroshikimate on to the next catalytic domain, there would be direct competition between the two pathways for these intermediates. A noteworthy finding is that the repressor protein that controls the expression of 3-dehydroquinate dehydratase is homologous with the C-terminal half of the AROM enzyme.[137] Calculations have been made to determine the expected transient time (see Section 7.6) for the reaction sequence from 2-dehydro-3-deoxy-D-*arabino*-heptonate 7-phosphate (DDAP) to 5-*O*-(1-carboxyvinyl)-3-phosphoshikimate (OCPS) (Fig. 7.25) on the assumption that each step was catalysed by a separate enzyme and that the product of each step was released from each catalytic centre into the medium and was subsequently taken up by the next enzyme. When this calculated transient time is compared with the experimentally observed transient time for this multienzyme polypeptide, the latter is found to be approximately 7–10 per cent of the former[138] (Fig. 7.27). Thus, when catalytic centres are in close proximity the metabolic rate can change much more rapidly to a new steady-state rate, and this is probably due to the fact

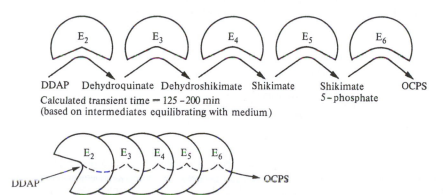

Fig. 7.27 A comparison of the calculated and experimentally observed transient times for the enzyme-catalysed conversion of DDAP to OCPS (see reference 138).

that the intermediates do not appreciably diffuse out into the medium but move directly from one catalytic centre to another.

Another related property of this multienzyme polypeptide is that of coordinate activation. It is found that the AROM enzyme can convert DDAP to OCPS (this requires E_2, E_3, E_4, E_5, and E_6; see Fig. 7.25) at approximately ten times the rate at which it can convert shikimate to OCPS (which requires E_5 and E_6 only). This is believed to be due to a process of coordinate activation of catalytic centres of the multienzyme polypeptide, with DDAP possibly bringing about a conformational change in the whole multienzyme protein. There is evidence that DDAP causes a reduction in K_m values of E_3, E_4, and E_6 for their respective substrates.[131] Also it is found that when DDAP is bound to the AROM enzyme it gives all five catalytic centres some protection against attack by proteases, suggesting possible conformational changes.

The organization of the anthranilate synthase complex (domains E, G–F–C) has been studied in *Neurospora crassa* using the technique of limited proteolysis to identify domains.[139,140] It comprises four polypeptide chains $\alpha_2\beta_2$. The α chain contains the catalytic site for anthranilate synthase capable of using ammonia as amino donor (domain E) and the β chain the catalytic site for phosphoribose anthranilate isomerase (domain F), indole glycerol phosphate synthase (domain C), and glutamine aminotransferase (G). In the genera *Bacillus*, *Acinetobacter*, and *Pseudomonas* the β subunit also serves as the glutamine aminotransferase for *p*-aminobenzoate synthase.[140] When the multienzyme complex ($\alpha_2\beta_2$) is incubated with elastase, it cleaves the β chain between residues 237 and 238, splitting the multienzyme into two fragments. One of these fragments (M_r 98 000), comprising the complete α chain (M_r 70 000) and a fragment (M_r 29 000) of the β chain, contains the anthranilate synthase activity together with its glutamine aminotransferase domain (G). A model for the probable arrangement is shown in Fig. 7.28.[139]

In *E. coli*, the phosphoribosyl anthranilate isomerase (domain F) and indole glycerol phosphate synthase (domain C) exist as a bifunctional monomeric enzyme. X-ray diffraction studies have been carried out on this enzyme using an iodinated analogue of l-(2′-carboxyphenylamino)-l-deoxyribulose 5-phosphate, the substrate of E_{11} (domain C) and the product of E_{10} (domain F).[141] This provided not only the three-dimensional structure of the enzyme but also a clear indication of the positions of the two catalytic sites. It is particularly interesting that the two catalytic sites are located back-to-back (Fig. 7.29) and this would appear to preclude the possibility of channelling of the reactants between the two catalytic sites. This is in contrast to another enzyme having two catalytic sites,

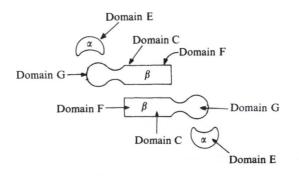

Fig. 7.28 A model for the arrangement of the anthranilate synthase complex ($\alpha_2\beta_2$) from *N. crassa*.

namely the DNA polymerase fragment known as the Klenow fragment[142] (see Chapter 1, Section 1.3.2). This fragment contains both DNA polymerase activity and also 3'–5' exonuclease activity. The latter serves an editing or proofreading function (see Chapter 1, Section 1.3.2) in excising mismatched deoxyribonucleotides. In this case the two catalytic sites are juxtaposed so that the nascent chain from DNA polymerase activity passes directly through the second catalytic site for editing.[143]

7.12 Enzymes involved in DNA synthesis

There is evidence that the enzymes required for DNA synthesis are organized into multienzyme complexes. A number of different complexes in different organisms are known. In *E. coli* infected with T4 bacteriophage a complex of $M_r \approx 1.5 \times 10^6$ known as deoxyribonucleoside triphosphate synthetase has been detected. Ten enzymes have so far been reported as components of this complex: they are dihydrofolate reductase, dCTPase–dUTPase, deoxyribonucleoside monophosphokinase, ribonucleotide reductase, thymidylate synthetase, dCMP hydroxymethylase, dCMP deaminase, nucleoside diphosphokinase, thymidine kinase, and adenylate kinase. The complex contains the enzymes necessary for the production of the four deoxyribonucleotide triphosphates and is believed to be located near the replication fork. It generates the four deoxynucleoside triphosphates at the site of DNA synthesis.[144] There is evidence for specific channelling of the nucleoside diphosphates into nascent DNA through the reactions catalysed by the sequence of enzymes from ribonucleotide reductase to DNA polymerase.[145] A 21s complex has been extracted from the cytosol of HeLa cells that contains DNA polymerase, α-primase, a 3',5' exonuclease, DNA ligase 1, RNAse H, and topoisomerase 1.[146] A 40s complex containing DNA polymerase, RNA polymerase, primase, 3',5' exonuclease and ATPase has been isolated from yeast mitochondria.[147] There is also evidence that several of the aminoacyl t-RNA synthetases from eukaryotes exist as a multienzyme complex.[148]

7.13 The glycogen particle

The evidence for the existence of a subcellular particle containing glycogen together with the enzymes concerned with glycogen metabolism comes from studies of the isolation of glycogen from skeletal muscle and from liver. Several years ago histological studies showed that glycogen is present in skeletal muscle in the form of distinct particles (Fig. 7.30).

If glycogen is isolated from skeletal muscle homogenates by procedures that do not cause protein denaturation, e.g. precipitation from the homogenate at pH 6.1 followed by high-speed centrifugation, then the particles containing glycogen also contain most of the glycolytic enzymes and they are of a similar size to those observed by histological studies. The evidence that these particles are not artefacts generated during the isolation procedure is supported by the following observations: (i) a high concentration of glycolytic enzymes is present after washing the particles; (ii) when three different methods are used to isolate the particles their composition is similar in each case.

The isolated glycogen particles have a reasonably well-defined composition, although not approaching the stoichiometric composition of the other multienzyme complexes considered in this chapter. A typical glycogen particle has a diameter of 40 nm, contains about 55 000 glucose residues and has about 20–25

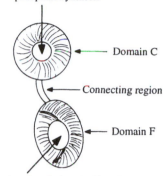

Active site of indole glycerol phosphate synthesis

Domain C

Connecting region

Domain F

Active site of phosphoribosyl anthranilate isomerase

Fig. 7.29 Diagram showing the two catalytic sites in the bifunctional enzyme phosphoribosyl anthranilite isomerase and indole glycerol phosphate synthase from *E. coli*.

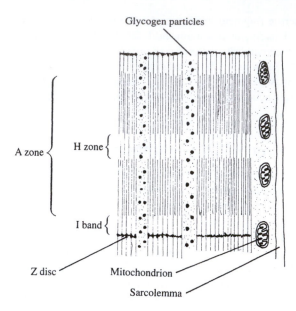

Fig. 7.30 The structure of a muscle fibre

Table 7.6 Typical enzyme content of glycogen particles

Glycogen phosphorylase
Phosphorylase kinase
Protein phosphatase 1_G (PP1$_G$)
Glycogen synthase
Protein kinase
Adenosinetriphosphatase
AMP deaminase
Adenylate kinase
Creatine kinase
Phosphoglucomutase

References 150 and 151.

* PP1$_G$ is the form of PP1 (see Chapter 6, Table 6.6) bound to glycogen.

phosphorylase molecules attached.[149] The enzymes present in glycogen particles are shown in Table 7.6. When the enzymes present in the particles are separated by polyacrylamide gel electrophoresis it is found that over 95 per cent of the total enzyme protein is accounted for by glycogen phosphorylase, glycogen synthase, debranching enzyme (see Chapter 6, Fig. 6.25), and phosphorylase kinase. Although protein phosphatase 1_G (PP1$_G$)* is tightly associated with the complex, it represents only 0.2 per cent of the protein of the complex. Approximately 60 per cent of PP1$_G$ in rabbit skeletal muscle is associated with glycogen particles. PP1$_G$ is a heterodimer made up of a catalytic subunit (M_r 37 000) and a regulatory subunit G ($M_r \approx 160\,000$). It is the G subunit that binds to glycogen with high affinity ($K_d \approx 6 \times 10^{-9}$ mol dm^{-3}), and also to the sarcoplasmic reticulum. Protein phosphatases generally have broad specificity, and it is thought that binding PP1$_G$ to glycogen is a means of targeting it more specifically towards glycogen phosphorylase. Phosphorylation of the G subunit causes the release of the catalytic subunit, but the G subunit remains attached to glycogen.[150,152]

Isolated washed glycogen particles are capable of catalysing the conversion of glycogen to lactate at a reasonable rate, e.g. 1.3 μmol per min per 100 mg protein compared to 40 μmol lactate per min per 100 mg protein in skeletal muscle *in vivo*.

The glycogen particles contain many of the enzymes that control glycogen synthase and glycogen phosphorylase. They also contain AMP deaminase, which may control AMP levels.

$$AMP + H_2O \underset{}{\overset{\text{AMP} \atop \text{deaminase}}{\rightleftharpoons}} IMP + NH_3$$

Much of the interest in the glycogen particle has been in the study of control of glycogen metabolism. In a number of situations the responses to ligands observed with the glycogen particle are similar to those observed in intact tissue. For example, the concentrations of Ca^{2+} and ATP required for phosphorylase kinase to phosphorylate phosphorylase, and the concentration of AMP required to inhibit purified phosphorylase phosphatase, in the glycogen particle are similar to those observed *in vivo*. Glycogen particles are polydisperse, varying in M_r from 6×10^6 to 1.6×10^9. Turnover of glucose moieties occurs more rapidly on the outside of the particles than on the inside. There is a different enzyme distribution in the particles depending on their size, e.g. glycogen phosphorylase activity is highest in small particles, whereas glycogen synthase activity is highest in large particles.[153] The control of glycogen metabolism is discussed further in Chapter 6, Section 6.4.2, and also in references 151 and 154.

7.14 Conclusions

In this chapter we have described a number of different multienzyme proteins, less well-defined aggregates of enzymes, and an enzyme in which the different polypeptide chains operate in a concerted manner to catalyse a complex enzyme reaction (RNA polymerase). Multienzyme proteins are generally regarded as providing a more efficient method of catalysing a sequence of reactions than can be achieved by separate enzymes. The increase in efficiency is thought to stem from (i) reduction in transit time, (ii) reduction in transient time, (iii) channelling of intermediates so as to reduce competition from other pathways, (iv) coordinate activation of a group of catalytic centres, and (v) protection of unstable intermediates from non-enzymatic degradation. (These factors were discussed in Section 7.6.) Examples in support of points (i) to (iv) have been given throughout this chapter. As yet there appears to be no well-documented example on an unstable intermediate that would be degraded spontaneously if it were not efficiently transferred from one catalytic centre to the next.

However, although it is not difficult to see that multienzyme proteins may improve efficiency, it is not clear how important or essential this is *in vivo*. For example, it has been shown that the transient time for the *arom* complex may be reduced by 10–15-fold as a result of being organized as a multienzyme protein (see Section 7.11.3), and thus the system can respond more rapidly to a change in substrate concentration. Whether or not this is important in the growth of the organism and its ability to respond to changing conditions remains to be established. The organization of the fatty-acid synthase as a multienzyme protein in yeast and mammalian systems means that a growing acyl group on the enzyme complex is free from competition from other growing chains. Only one fatty acid can be synthesized at a time per fatty-acid synthase. There is no evidence to suggest that fatty-acid synthesis is less efficient, either in terms of the maximum rates achievable or the efficiency with which the initial precursor is converted to the final products, in the organisms in which the enzymes appear to be completely separate, e.g. *E. coli* and plants, than in the multienzyme polypeptides from yeast or mammalian tissues.

References

1. Srere, P. A., *Trends Biochem. Sci.* **5**, 120 (1980).

2. *Enzyme nomenclature*, Recommendations (1992) of the Nomenclature Committee of the International Union of Biochemistry and Molecular Biology, pp. 562–3. Academic Press, Orlando (1992).

3. Karlson, P. and Dixon, H. B. F., *Trends Biochem. Sci.* **4**, N275 (1980).

4. Berg, A and de Kok, A., *Biol. Chem.* **378**, 617 (1997).

5. Mattevi, A., de Kok, A., and Perham, R. N., *Curr. Biol.* **2**, 877 (1992).

6. Perham, R. N., *Biochemistry* **30**, 8501 (1991).

7. Randle, P. J., Patston, P. A., and Espinal, J., *Enzymes* (3rd edn) **18B**, 97 (1987).

8. Cohen, G. N. and Dautry-Varsat, A., in *Multifunctional proteins* (Bisswanger, H. and Schmincke-Oft, E., eds), pp. 47–121. Wiley, New York (1980).

9. Zhang, C., Lu, Q., and Verma, D. P. S., *J. biol. Chem.* **270**, 20491 (1995).

10. Eberhard, M., Tsai-Pflugerfelder, M., Boleheska, K., Hommel, U., and Kirschner, K., *Biochemistry* **34**, 5419 (1995)

11. Li, J., Chen, S., Zhu, L., and Last, R. L., *Plant Physiol.* **108**, 877 (1995).

12. Pan, P. and Dunn, M. F., *Biochemistry* **35**, 5002 (1996).

13. Hawkins, A. R. and Lamb, H. K., *Eur. J. Biochem.* **232**, 7 (1994).

14. Yablonski, M. J., Pasek, D. A., Han, B.-D., Jones, M. E., and Traut, T. W., *J. biol. Chem.* **271**, 10704 (1996).

15. Firestone, S. M. and Davisson, V. J., *Biochemistry* **33**, 11917 (1994).

16. He, X.-Y., Deng, H., and Yang, S.-Y., *Biochemistry* **36**, 261 (1997).

17. Yang, S. Y., *Comp. Biochem. Physiol.* **109**, 557 (1994).

18. May, T. B., Shinabarger, D., Boyd, A., and Chakrabarty, A. M., *J. biol. Chem.* **269**, 4872 (1994).

19. Ricard, J., Guidici-Orticoni, M.-T., and Gontero, B., *Eur. J. Biochem.* **226**, 993 (1994).

20. Wakil, S. J. and Stoops, J. K., *Enzymes* (3rd edn) **16**, 3 (1983).

21. Stachelhaus, T. and Marahiel, M. A., *J. biol. Chem.* **270**, 6163 (1995).

22. Srinivasan, R. and Oliver, D. J., *Plant Physiol.* **109**, 161 (1995).

23. Bourguigon, J., Merand, V., Rawsthorne, S., Forest, E., and Douce, R., *Biochem. J.* **313**, 229 (1996).

24. Pilkis, S. J., Claus, T. H., Kurland, I. J., and Lange, A. J., *A. Rev. Biochem.* **64**, 799 (1995).

25. Kelman, Z. and O'Donnell, M., *A. Rev. Biochem.* **64**, 171 (1995).

26. Hinderlich, S., Stasehe, R., Zeitler, R., and Reutter, W., *J. biol. Chem.* **272**, 24313 (1997).

27. Fujita, N. and Ishihama, A., *Methods Enzymol.* **273**, 121 (1996).

28. Adams, R. L. P., Knowler, J. T., and Leader, D. P., *The biochemistry of nucleic acids* (11th edn). Chapman and Hall, London (1992).

29. Lewin, B., *Genes VI.* Oxford University Press (1997).

30. Travers, A., *DNA–protein interactions*, Chapter 4. Chapman and Hall, London (1993).

31. Das, A., *A. Rev. Biochem.* **62**, 893 (1993).

32. Nudler, E., Avetissova, E., Markovtsov, V., and Goldfarb, A., *Science* **273**, 211 (1996).

33. Polyakov, A., Severinova, E., and Darst, S. A., *Cell* **83**, 365 (1995).

34. McMahan, S. and Burgess, R. R., *Biochemistry* **33**, 12092 (1994).

35. Yura, T. and Ishihama, A., *A. Rev. Genet.* **13**, 59 (1979).

36. Hillel, Z. and Wu, C.-W., *Biochemistry* **16**, 3334 (1977).

37. Malhotra, A., Severinova, E., and Darst, S. A., *Cell* **87**, 127 (1996).

38. Richardson, J. P., *J. biol. Chem.* **271**, 1251 (1996).

39. McClure, W. R., *A. Rev. Biochem.* **54**, 171 (1985).

40. Busby, S. and Ebright, R. H., *Cell* **79**, 743 (1994).

41. von Hippel, P. H., Bear, D. G., Morgan, W. D., and McSwiggen, J. A., *A. Rev. Biochem.* **53**, 389 (1984).

42. Pingoud, A. and Jeltsch, A., *Eur. J. Biochem.* **246**, 1 (1997).

43. Blatter, E. E., Ross, W., Tang, H., Gourse, R. L., and Ebright, R. H., *Cell* **78**, 889 (1994).

44. Heyduk, T., Heyduk, E., Severinov, E., Tang, H., and Ebright, R. H., *Proc. natn. Acad. Sci. USA* **93**, 10162 (1996).

45. Murakami, K., Kimura, K., Owens, J. T., Meares, C. F., and Ishihami, A., *Proc. natn. Acad. Sci. USA* **94**, 1709 (1997).

46. Negishi, T., Fujita, N., and Ishihami, A.. *J. molec. Biol.* **248**, 723 (1995).

47. Luo, J., Sharif, K. A., Jin, R., Fujita, N., Ishihami, A., and Krakow, J. S., *Genes to Cells* **1**, 819 (1996).

48. Uptain, S. M., Kane, C. M., and Chamberlain, M. J., *A. Rev. Biochem.* **66**, 117 (1997).

49. Sousa, R., Chung, Y. J., Rose, J. P., and Wang, B.-C., *Nature* **364**, 593 (1993).

50. Bonner, G., Lafer, E. M., and Sousa, R.. *J. biol. Chem.* **269**, 25129 (1994).

51. Joyce, C. M. and Steitz, T. A., *A. Rev. Biochem.* **63**, 777 (1994).

52. Gilles, R. J., *Trends Biochem. Sci.* **8**, 301 (1985).

53. Van Schaftingen, E., *Adv. Enzymol.* **59**, 316 (1987).

54. Laporte, D. C. and Chung, T., *J. biol. Chem.* **260**, 15291 (1985).

55. Nimmo, H. G., in *Molecular aspects of cellular regulation* (Cohen, P., ed.), Vol. 3, p. 155. Elsevier, Amsterdam (1984).

56. Mori, M. and Tatibana, M., *Methods Enzymol.* **51**, 111 (1978).

57. Gaertner, F. H., *Trends Biochem. Sci.* **3**, 63 (1978).

58. Cornish-Bowden, A., in *Channelling in intermediary metabolism* (Agius, L. and Sherratt, H. S. A., eds), Chapter 4. Portland Press, London (1997).

59. Easterby, J. S., in *Channelling in intermediary metabolism* (Agius, L. and Sherratt, H. S. A., eds), Chapter 5. Portland Press, London (1997).

60. Welch, G. R. and Easterby, J. S., *Trends Biochem. Sci.* **19**, 193 (1994).

61. Cornish-Bowden, A. and Cardenas, M. L., *Eur. J. Biochem.* **213**, 87 (1993).

62. Meijer, A. J., *FEBS Lett.* **191**, 249 (1985).

63. Kis, K. and Bacher, A., *J. biol. Chem.* **270**, 16788 (1995).

64. Slater, E. C., *A. Rev. Biochem.* **22**, 17 (1953).

65. Roche, T. E. and Cox, D. J., in *Channelling in intermediary metabolism* (Agius, L. and Sherratt, H. S. A., eds), Chapter 7. Portland Press, London (1997).

66. Stoops, J. K., Cheng, R. H., Yazdi, M. A., Maeng, C.-Y., Schroeter, J. P., Klueppelberg, U. *et al.*, *J. biol. Chem.* **272**, 5757 (1997).

67. Izasrd, T., Sarfaty, S., Westphal, A., De Kok, A., and Hol, W. G. J., *Protein Sci.* **6**, 913 (1997).

68. Bessam, H., Mareck, A. M., and Foucher, B., *Biochim. biophys. Acta* **990**, 66 (1989).

69. McCartney, R. G., Sanderson, S. J., and Lindsay, G., *Biochemistry* **36**, 6819 (1997).

70. Reed, L. J., *Acc. chem. Res.* **7**, 40 (1974).

71. Bresters, T. W., De Abreu, R. A., Dekok, A., Visser, J., and Veeger, C., *Eur. J. Biochem.* **59**, 335 (1975).

72. Henderson, C. E. and Perham, R. N., *Biochem. J.* **189**, 161 (1980).

73. Jeyaseelan, K., Guest, J. R., and Visser, J., *J. gen. Microbiol.* **120**, 393 (1980).

74. Dixon, M., Webb, F. C., Thorne, C. J. R., and Tipton, K. F., *Enzymes* (3rd edn). Longman, London (1979). See p. 482.

75. Reed, L. J. and Mukherjee, B. B., *Methods Enzymol.* **13**, 55 (1969).

76. Stanley, C. J. and Perham, R. N., *Biochem. J.* **191**, 147 (1980).

77. Yi, J., Nemeria, N., McNally, A., Jordan, F., Machado, R. S., and Guest, J. R., *J. biol. Chem.* **271**, 33192 (1996).

78. Yang, D., Song, J., Wagenknecht, T., and Roche, T. E., *J. biol. Chem.* **272**, 6361 (1997).

79. Sanderson, S. J., Miller, C., and Lindsay, J. G., *Eur. J. Biochem.* **236**, 68 (1996).

80. Stephens, P. F., Darlison, M. G., Lewis, H. M., and Guest, J. R., *Eur. J. Biochem.* **133**, 155 (1983).

81. Stephens, P. F., Darlison, M. G., Lewis, H. M., and Guest, J. R., *Eur. J. Biochem.* **133**, 481 (1983).

82. Dennert, G. and Hoglund, S., *Eur. J. Biochem.* **12**, 502 (1970).

83. Danson, M. J., Hale, G., Johnson, P., Perham, R. N., Smith, J., and Spragg, P., *J. molec. Biol.* **129**, 603 (1979).

84. Lessard, I. A. D., Fuller, C ., and Perham, R. N., *Biochemistry* **35**, 16863 (1996).

85. Reed, L. J. and Yeaman, S. J., *Enzymes* (3rd edn.) **18B**, 77 (1987).

86. Mathews, F. S., *Curr. Struct. Biol.* **1**, 955 (1991).

87. Berg, A., Vervoot, J., and de Kok, A., *Eur. J. Biochem.* **244**, 352 (1997).

88. Shepherd, G. B. and Hammes, G. G., *Biochemistry* **15**, 311 (1976).

89. Scouten, W. H., Visser, A. J. W. G., Grande, H. J., DeKok, A., and Graaf-Hess, A. C., *Eur. J. Biochem.* **112**, 9 (1980).

90. Angelides, K. J. and Hammes, G. G., *Proc. natn. Acad. Sci. USA* **75**, 4877 (1978).

91. Stepp, L. R., Bleile, D. M., McRorie, D. K., Petit, F. H., and Reed, L. J., *Biochemistry* **20**, 4555 (1981).

92. Bates, P. L., Danson, M. J., Hale, G., Hooper, F. A., and Perham, R. N., *Nature, Lond.* **268**, 313 (1977).

93. Akiyama, S. K. and Hammes, G. G., *Biochemistry* **19**, 4208 (1980).

94. Lawson, J. E., Park, S. H., Mattison, A. R., Yan, J., and Reed, L. J., *J. biol. Chem.* **272**, 31625 (1997).

95. Randle, P. J., *Trends Biochem. Sci.* **3**, 217 (1978).

96. Wieland, O. H., Hartmann, U., and Siess, F. A., *FEBS Lett.* **27**, 240 (1972).

97. DeMarcucci, O. G. L., Gibb, G. M., Dick, J., and Lindsay, J. G., *Biochem. J.* **251**, 817 (1988).

98. Rhee, S., Parris, K. D., Hyde, C. C., Ahmed, S. A., Miles, E. W., and Davis, D. R., *Biochemistry* **36**, 7664 (1997).

99. Bartolomes, P., Boker, H., and Jaenicke, R., *Eur. J. Biochem.* **102**, 167 (1979).

100. Dettwiler, M. and Kirschner, K., *Eur. J. Biochem.* **102**, 159 (1979).

101. Zalkin, H. and Yanofsky, C., *J. biol. Chem.* **257**, 1491 (1982).

102. Schwarz, T., Uthoff, K., Klinger, C., Meyer, H. E., Bartholmas, P., and Kaufmann, M., *J. biol. Chem.* **272**, 10616 (1997).

103. Makoll, A. J. and Radford, A., *Microbiol. Rev.* **42**, 307 (1978).

104. Pierard, A. and Schroter, B., *J. Bacteriol.* **134**, 167 (1978).

105. Davis, R. H., in *Organizational biosynthesis* (Vogel, H. J., Lampen, J. O., and Bryson, V., eds), pp. 303–32. Academic Press, New York (1967).

106. Hong, J., Salo, W. L., Lusty, C. J., and Anderson, P. M., *J. Molec. Biol.* **243**, 131 (1994).

107. Legrain, C., Demarez, M., Glansdorff, N. and Pierard, A., *Microbiology* **141**, 1093 (1995).

108. Carrey, E. A., *Biochem. J.* **236**, 327 (1986).

109. Irvine, H. S., Shaw, S. M., Paton, A., and Carrey, E. A., *Eur. J. Biochem.* **247**, 1063 (1997).

110. Lollier, M., Jacquet, L., Nedeva, T., Lacroute, F., Potier, S., and Souciet, J.-L., *Curr. Genet.* **28**, 138 (1995).

111. Carrey, E. A., *Biochem. Soc. Trans.* **23**, 899 (1995).

112. Jones, M. E., *A. Rev. Biochem.* **49**, 253 (1980).

113. Smith, S., in *Channelling in intermediary metabolism* (Agius, L. and Sherratt, H. S. A., eds), Chapter 8. Portland Press, London (1997).

114. Roughan, P. G., *Biochem. J.* **327**, 267 (1997).

115. Schweitzer, R., Kniep, B., Castorph, H., and Holzner, U., *Eur. J. Biochem.* **39**, 353 (1973).

116. Wieland, F., Renner, L., Verfirth, C., and Lynen, F., *Eur. J. Biochem.* **94**, 189 (1979).

117. Dixon, H. B. F. and Perham, R. N., *Biochem. J.* **109**, 312 (1968).

118. Stoops, J. K. and Wakil, S. J., *J. biol. Chem.* **256**, 5128 (1981).

119. Katiyar, S. S. and Porter, J. W., *Eur. J. Biochem.* **130**, 177 (1983).

120. McCarthy, A. D. and Hardie, D. G., *Eur. J. Biochem.* **130**, 185 (1983).

121. Engeser, H., Hubner, K., Straub, J., and Lynen, F., *Eur. J. Biochem.* **101**, 407 (1979).

122. Engeser, H., Hubner, K., Straub, J., and Lynen, F., *Eur. J. Biochem.* **101**, 413 (1979).

123. Chiral, S. S., Huang, W.-Y., Jayakumar, A., Sakai, K., and Wakil, S. J., *Proc. natn. Acad. Sci. USA* **94**, 5588 (1997).

124. Rangan, V. S. and Smith, S., *J. biol. Chem.* **271**, 31749 (1996).

125. Rangan, V. S. and Smith, S., *J. Biol. Chem.* **272**, 11975 (1997).

126. Volpe, J. J. and Vagelos, P. R., *A. Rev. Biochem.* **42**, 21 (1973).

127. Greenspan, M. D., Birge, C. H., Powell, G. L., Hancock, W. S., and Vagelos, P. R., *Science* **170**, 1203 (1970).

128. Bloch, K. and Vance, D., *A. Rev. Biochem.* **46**, 263 (1977).

129. Kumar, S., *J. biol. Chem.* **250**, 5150 (1975).

130. Bloch, K., *Adv. Enzymol.* **45**, 1 (1977).

131. Welch, U. R. and Gaertner, F. H., *Arch. Biochem. Biophys.* **172**, 476 (1976).

132. Crawford, I. P., *A. Rev. Microbiol.* **43**, 567 (1989).

133. Lamb, H. K., Wheeler, K. A., and Hawkins, A. R., in *Channelling in intermediary metabolism* (Agius, L. and Sherratt, H. S. A., eds), Chapter 9. Portland Press, London (1997).

134. Welch, U. R. and DeMoss, J. A., in *Microenvironment and metabolic compartmentation* (Srere, P. and Easterbrook, R. W., eds), p. 345. Academic Press, New York (1978).

135. Lumsden, J. and Coggins, J. R., *Biochem. J.* **161**, 599 (1977).

136. Coggins, J. R., Boocock, M. R., Campbell, M. S., Chaudhuri, S., Lambert, J. M., Lewendon., A. *et al.*, *Biochem. Soc. Trans.* **13**, 299 (1985).

137. Lamb, H. K., Moore, J. D., Lakey, J. H., Levett, L. J., Wheeler, K. A., Lago, H. *et al.*, *Biochem. J.* **313**, 941 (1996).

138. Gaertner, F. H., in *Microenvironment and metabolic compartmentation* (Srere, P. and Easterbrook, R. W., eds), p. 345. Academic Press, New York (1978).

139. Walker, M. S. and DeMoss, J. A., *J. biol. Chem.* **261**, 16073 (1986).

140. Romero, R. M., Roberts, M. F., and Phillipson, J. D., *Phytochemistry* **39**, 263 (1995).

141. Priestle, J. P., Grutter, M. G., White, J. L., Vincent, M. G., Kaina, M., Wilson, E. *et al.*, *Proc. natn. Acad. Sci. USA.* **84**, 5690 (1987).

142. Joyce, C. M. and Steitz, T. A., *Trends. Biochem. Sci.* **12**, 288 (1987).

143. Ollis, D. L., Brick, P., Hamlin, R., Xnong, N. U., and Steitz, T. A., *Nature, Lond.* **313**, 762 (1985).

144. Wheeler, L. J., Ray, N. B., Ungermann, C., Hendricks, S. P., Bernard, M. A., Hanson, E. S. *et al.*, *J. biol. Chem.* **271**, 11156 (1996).

145. Christopherson, R. I. and Szabados, E., in *Channelling in intermediary metabolism* (Agius, L. and Sherratt, H. S. A., eds), Chapter 16. Portland Press, London (1997).

146. Li, C., Cao, L.-G., Wang, Y.-L., and Baril, E. F., *J. Cell. Biochem.* **53**, 405 (1993).

147. Murthy, V. and Pasupathy, K., *Mol. Biol. Rep.* **20**, 135 (1995).

148. Dang, C. V. and Dang, C. V., *Biochem. J.* **239**, 249 (1986).

149. Melendez-Hevia, E., Guinovart, J. J., and Cascante, M., in *Channelling in intermediary metabolism* (Agius, L. and Sherratt, H. S. A., eds), Chapter 14. Portland Press, London (1997).

150. Hubbard, M. J. and Cohen, P., *Trends Biochem. Sci.* **18**, 172 (1993).

151. Busby, S. J. W. and Radda, G. K., *Curr. Top. Cell. Reg.* **10**, 89 (1976).

152. Hubbard, M. J. and Cohen, P., *Eur. J. Biochem.* **186**, 701 (1989).

153. Srere, P., *A. Rev. Biochem.* **56**, 89 (1987).

154. Ottaway, J. H. and Mowbray, J., *Curr. Top. Cell. Reg.* **12**, 108 (1977).

8

Enzymes in the cell

8.1 Introduction

In order to make measurements of enzyme activity, whether of single enzymes or a sequence of enzymes, the organism or tissue concerned is disrupted and the enzyme activity measured in crude homogenates or partially purified enzyme preparations. Disruption is necessary in order that the enzymes have free access to the added substrates. In an intact cell the membrane often either prevents the substrate from entering the cell or limits the rate at which it enters. In whole tissues, only the cells on the surface would have unrestricted access to the substrates. Much or all of the subcellular organization is lost in this process of disruption and the assays for enzyme activity are usually performed under conditions which differ considerably from those *in vivo*. In this chapter we consider how the process of cell disruption, and the changed environment *in vitro* compared with that *in vivo*, may affect single enzymes, and also their interactions with other enzymes and cell structures, since this is important in assessing how far results obtained from enzyme measurements *in vitro* (cf. Chapters 4–6) may be used to understand enzyme action *in vivo*.

The principal differences between enzyme reactions measured *in vitro* and those occurring *in vivo* are as follows.

(i) Loss of organization and compartmentation

In an intact cell, enzymes, substrates, cofactors, or effectors are not distributed uniformly throughout the cell. There are often considerable differences between the concentrations of any of these in subcellular organelles; there is even evidence that concentration gradients exist within the soluble phase of the cytoplasm, i.e. the cytosol (see Section 8.2.7), but the differences may not exist once a cell or its constituent organelles have been disrupted.

(ii) The dilution factor

During the course of preparation of either a crude homogenate or a purified enzyme, the suspending medium used causes dilution of the solutes present within the cell. Protein concentrations within cell compartments are at the lower end of the range of protein concentrations that exist in protein crystals.[1] For example, muscle cells have 23 per cent protein by weight, red blood cells 35 per cent, and most dividing cells 17–26 per cent, whereas protein crystals range between 20 and 90 per cent protein. In contrast, the protein concentration in an enzyme assay using a liver homogenate might be 5 mg cm^{-3}, and with a highly purified enzyme the protein concentration might be of the order of 5 μg cm^{-3}. The dilution of the cell constituents may affect a number of interactions between macromolecules and between macromolecules and small molecules and ions.

(iii) Relative concentrations of enzyme and substrate

studied very little

In enzyme assays carried out under steady-state conditions (see Chapter 4, Section 4.1) there is a vast excess of substrate compared with that of enzyme. Typical concentrations might be in the region of 10^{-3} mol dm^{-3} for substrates and 10^{-6} mol dm^{-3} or below for enzymes. Measurements *in vivo* suggest that there are many instances in which enzymes are present in concentrations comparable with or greater than their substrates[2] and thus the kinetics observed under steady-state conditions may be somewhat different from those *in vivo*.

(iv) Closed and open systems

studied extensively

This aspect is related to the previous one. The enzyme assay carried out *in vitro* is essentially a closed system. The reactants are provided at the beginning of the assay and no replacement usually occurs during assay. The living cell, on the other hand, is an open system in which metabolites are continually entering and leaving. Thus, the concentrations of both the substrate and product of an enzyme reaction may be unchanged for a considerable period of time *in vivo* but nevertheless there may be considerable flux through the enzyme pathway (see Chapter 6, Section 6.3.1).

These four differences between enzymes in the cell and enzymes in the cuvette or test-tube will now be discussed in more detail. It will become apparent that certain aspects of this situation, for example compartmentation, have been explored extensively, whereas other areas, for example the high concentrations of proteins within organelles and their effects on reactions, have hardly been touched.

8.2 Intracellular compartmentation

Cells vary considerably in size, shape, and the amount of subcellular organization contained within them. A 'typical' eukaryotic cell illustrating subcellular organization is given in Fig. 8.1. Much of the subcellular organization can be discerned from light and electron microscopic observations, but in addition to this there are many examples of metabolic 'pools', where for a given metabolite there is kinetic evidence for compartmentation but for which no morphological counterpart can be discerned. A typical prokaryote cell has a volume in the range 0.5–500 μm^3 and well-defined subcellular structures have not been identified and isolated. A eukaryote cell is generally considerably larger,[3] in the range 200 μm^3 to 15 000 μm^3, and many well-defined subcellular organelles have been characterized. The larger size probably necessitates a greater degree of subcellular compartmentation. In the following section we briefly describe the isolation and properties of the principal subcellular organelles of eukaryote cells. The main emphasis will be to give an indication of the size of each organelle, the selectivity of its surrounding membrane where appropriate, and the nature of the substances comprising the matrix.

Since the liver parenchyma cell is perhaps the most studied eukaryotic cell, the information given refers to the liver cell unless otherwise stated.

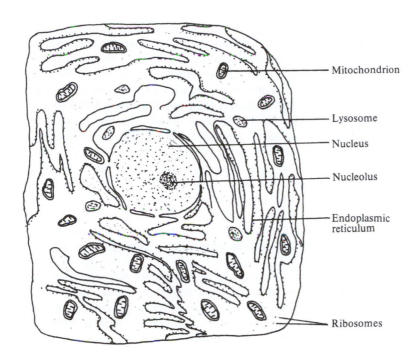

Fig. 8.1 A typical eukaryote cell showing the intracellular structure (subcellular organelles not drawn to scale).

8.2.1 Isolation of subcellular organelles

The aim of subcellular fractionation procedures is to isolate subcellular organelles as free from contaminants as possible and with as little disruption as possible. The procedure involves two principal stages: (i) disruption of the tissue or cells to release the subcellular organelles and (ii) separation of the organelles from one another. It is necessary also to have methods for assessing the integrity of the organelles and the extent of contamination by other components. There is no single procedure suitable for all cells and tissues, and methods have to be adapted for the tissue concerned.

In the first stage, cells and tissues are disrupted by a variety of different methods according to the tissue involved (see Chapter 2, Section 2.5). The second step in the procedure is to separate components from the homogenate usually by differential centrifugation. The rate of sedimentation in a centrifugal field increases with the size of a particle, the difference in density between the particle and the solvent, and the centrifugal field. There are small differences in the densities of the subcellular organelles, nuclei (buoyant density 1.3–1.4 g cm^{-3}) being denser than mitochondria (buoyant density 1.17–1.21 g cm^{-3}), but the main factor influencing the separations is the difference in size of the organelles. The organelles are usually separated by centrifugation at increasing speeds, as indicated in Fig. 8.2. In order to obtain fairly pure preparations, the sedimented fractions are usually resuspended in buffer and then resedimented once or twice more. This procedure gives a fairly good separation of the main organelles: nuclei, mitochondria, and microsomes (disrupted endoplasmic reticulum). It is difficult to separate mitochondria (buoyant density 1.17–1.21 g cm^{-3}) from lysosomes (buoyant density 1.20–1.26 g cm^{-3}) because they have similar sizes and densities. An enriched lysosomal pellet may be prepared if a number of resuspensions and resedimentations are carried out, but this is unlikely even then to contain more than 10 per cent lysosomes. A better resolution is achieved if the resuspended pellet is centrifuged through a sucrose density gradient (1.1 to 2.1 M sucrose) at $95\,000g$ for 2–5 h. Although there is some overlap of the two bands it is possible to select the fractions which contain almost entirely lysosomes. To obtain a complete separation of rat liver lysosomes from mitochondria, the rats may be injected prior to sacrifice with Triton WR1339 (an anionic detergent, the sodium salt of an alkyl aryl polyether sulphate). The Triton WR 1339 is phagocytosed by the lysosomes, causing decreases in their densities. They may then be separated from mitochondria.[4] Peroxisomes also sediment with mitochondria and lysosomes during differential centrifugation (Fig. 8.2) but they cannot be separated from lysosomes using sucrose density gradients since they have a very similar buoyant density in sucrose. However, they can be separated using a metrizamide gradient in which lysosomes have a buoyant density of 1.140 g cm^{-3}, whereas peroxisomes have a buoyant density of 1.231 g cm^{-3}. The differences in buoyant densities depend on whether the media (sucrose or metrizamide) exert osmotic effects on the organelle membrane at the concentrations at which they are used.

Further refinements to subcellular fractionations may be carried out; for example, smooth microsomes (disrupted endoplasmic reticulum lacking ribosomes) can be separated from rough microsomes (endoplasmic reticulum containing ribosomes) by centrifugation through layers of sucrose solution having different densities. The denser sucrose solutions retard the smooth microsomes

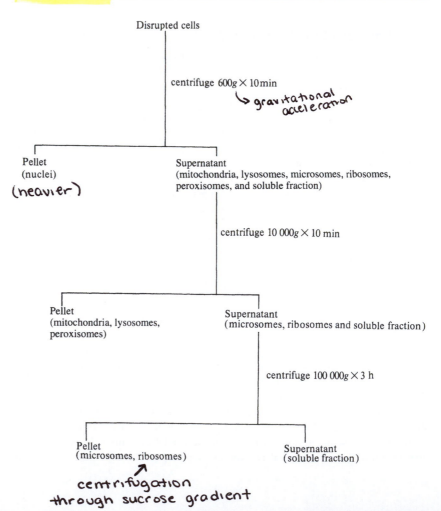

Disrupted cells

centrifuge $600g \times 10$ min

↳ gravitational acceleration

Pellet (nuclei)

(heavier)

Supernatant (mitochondria, lysosomes, microsomes, ribosomes, peroxisomes, and soluble fraction)

centrifuge $10\,000g \times 10$ min

Pellet (mitochondria, lysosomes, peroxisomes)

Supernatant (microsomes, ribosomes and soluble fraction)

centrifuge $100\,000g \times 3$ h

Pellet (microsomes, ribosomes)

Supernatant (soluble fraction)

centrifugation through sucrose gradient

Fig. 8.2 Procedure for subcellular fractionation by differential centrifugation.

(buoyant density ≈ 1.13 g cm^{-3}), which are less dense than rough microsomes (buoyant density ≈ 1.2 g cm^{-3}), because the former lack the more dense RNA.[5] For further details of subcellular fractionation, see references 4 and 6.

The purity of subcellular organelles can be assessed by biochemical and histological methods. Certain enzymes that have been found to be present either exclusively or predominantly in one subcellular organelle are used as marker enzymes to assess the extent of contamination of a particular subcellular fraction. For example, succinate dehydrogenase is a mitochondrial enzyme, and thus by measuring succinate dehydrogenase activity in a preparation of nuclei, the extent of contamination by mitochondria can be assessed. A list of marker enzymes for different subcellular fractions is given in Table 8.1. Besides the biochemical methods, it is useful to examine a preparation by light or electron microscopy in order to check that the organelles are undamaged. Histochemical reagents can also be used, e.g. histochemical assay for acid phosphatase, or Feulgen's reagent for DNA (a reaction which depends upon the fact that partial hydrolysis products of DNA restore the colour of basic fuchsin that has been decolorized by sulphurous acid).

8.2.2 The nucleus

The nucleus is the largest subcellular organelle. Its size does not vary substantially in a given type of cell, although there is a wide variation in the sizes of nuclei from different cell types and this variation is not related to the volume of cytoplasm. For example, nuclei from certain fungi may be only 4 μm^3 in volume and occupy less than 1 per cent of the volume of the cytoplasm, whereas the nuclei of cells from the thymus gland are 300–400 μm^3 and occupy 70 per cent of the cell volume. The nuclei from liver cells have a volume of 300 μm^3 which is about 5 per cent of the total cell volume. The volume of plant cell nuclei are generally larger, often $> 10\,000$ μm^3, and this seems to reflect their larger DNA content.[7] The nucleus is bounded by two membranes which show some properties similar to those of the endoplasmic reticulum, e.g. all enzymes found on the nuclear membrane except cytochrome c oxidase are also found in the endoplasmic reticulum. However, they differ from the endoplasmic reticulum in buoyant

Table 8.1 Marker enzymes used in subcellular fractionation

Enzyme	Subcellular fraction
DNA polymerase	Nuclei
Nicotinamide-nucleotide adenyltransferase	Nuclei
Succinate dehydrogenase	Mitochondria
Cytochrome c oxidase	Mitochondria
Glucose 6-phosphatase	Endoplasmic reticulum
Acid phosphatase	Lysosomes
Ribonuclease	Lysosomes
Catalase	Peroxisomes
Urate oxidase	Peroxisomes
5'-Nucelotidase	Plasma membrane
Glucose 6-phosphate dehydrogenase	Cytosol
Lactate dehydrogenase	Cytosol
6-Phosphofructokinase	Cytosol

density.[8,9] There is a polymeric protein network lining the nucleoplasmic surface of the nuclear membrane.[10]

A characteristic feature of the nuclear membrane is the presence of pores. They vary in size and number according to the type of nucleus. In many mammalian nuclei the pores occupy 10–20 per cent of the surface area of the nucleus and the pore diameter may be as large as 23 nm. Many small molecules and ions and also macromolecules, including protein and RNA, are able to enter the nucleus. The nuclear pore restricts the diffusion of macromolecules by acting as a sieve. It also carries out selective transport of many proteins between the nucleus and the cytoplasm. Proteins having M_r larger than $\approx 40\,000$ require a nuclear localization sequence in their primary structure to enable them to bind to the cytoplasmic surface of the nuclear pore prior to translocation. They are taken up against a concentration gradient in a process which requires ATP. It is not clear whether this is by active transport or by facilitated diffusion.[11,12] There are differences in the concentrations of Na^+ and K^+ in the nucleus compared with outside the nucleus but these are probably due to different proportions of ions bound to macromolecules. The space within the nucleus into which sucrose and other small molecules cannot penetrate is estimated at less than 10 per cent of the total volume.

The nucleus is the densest subcellular organelle (density approximately 1.3–1.4 g cm³) and this reflects the high concentrations of macromolecules within it (approximately 34 g protein, 9.5 g DNA, and 1.5 g RNA per 100 g nuclei).[13] The large amount of DNA in the nucleus, which can be as much as 10 per cent of the cell mass (including water), necessitates a high degree of condensation from that of the double helix by as much as 50 000-fold in order to fit within the dimensions of the nucleus. The high concentration of macromolecules in the nucleus restricts their movement within the nucleus. Two analogies which have been used to describe this situation are; traffic in a big city, only in three dimensions rather than two; and a plateful of spaghetti in thick tomato sauce.[14] The latter is very different from the standard assay conditions for an enzyme incubation mixture in a cuvette.

The enzymes of the nucleus can be grouped according to their location (Table 8.2). The enzyme content of the soluble compartment of the nucleus is similar to that of the cytosol, especially in regard to that of the enzymes of the glycolytic pathway. The second and third groups (Table 8.2) of enzymes are bound to insoluble components of the nucleus which are not readily washed out of the preparations of isolated nuclei, but they may be extracted using concentrated salt solutions, e.g. 1 mol dm⁻³ NaCl. They comprise mainly the enzymes concerned with DNA and RNA synthesis and are attached to chromatin. The fourth group is only extracted by the use of detergents and includes membrane-bound enzymes such as glucose 6-phosphatase and acid phosphatase. It is possible to isolate a nuclear matrix that comprises an insoluble skeletal framework connected to the nuclear membrane, and which has an appearance similar to that of the nucleus under the electron microscope. It comprises about 10 per cent of the nuclear proteins, and processes such as replication and transcription are believed to occur when the enzymes concerned are associated with this framework.[17] The role of the nucleolus is principally the biogenesis of ribosomes and it develops and degenerates with each cell cycle.[15]

8.2.3 The mitochondrion

The mitochondrion is typically a sausage-shaped organelle, 0.7–1 μm length and having a volume of about 1 μm³. Although it is a relatively small organelle, there

Table 8.2 Enzymes present in the nucleus

1. Enzymes in the soluble space
Glycolytic enzymes
Pentose phosphate pathway enzymes
Lactate dehydrogenase
Malate dehydrogenase
Isocitrate dehydrogenase
Arginase

2. Enzymes bound to chromatin
RNA polymerase II
RNA polymerase III
Nucleoside triphosphatase
DNA polymerase
Nicotinamide-nucleotide adenylyltransferase

3. Enzymes concentrated in the nucleolus
RNA polymerase I
Topoisomerase
RNA methylases
Ribonuclease
Protein kinase
Phosphoprotein phosphatase

4. Enzymes bound to membranes
Glucose 6-phosphatase
Acid phosphatase

For further details see references 8, 9, 13, 15, and 16.

are usually a large number of mitochondria per cell and collectively they occupy a significant proportion of the cell volume; e.g. liver parenchymal cells have 1000–1600 mitochondria per cell and they occupy 10–20 per cent of the volume of the cytoplasm. The mitochondrion has two membranes that have markedly different properties. The outer membrane is smooth in appearance whereas the inner membrane has a number of infoldings called cristae (Fig. 8.3). The inner mitochondrial membrane encloses the matrix, which is 60–70 per cent of the total mitochondrial volume. On the inner surface of the inner mitochondrial membrane are small mushroom-shaped particles containing mitochondrial ATPase, which is driven in the direction of ATP synthesis in oxidative phosphorylation. The number of cristae per mitochondrion varies with the tissue, being highest in those tissues having high potential metabolic rates, e.g. flight muscle (see Chapter 6, Section 6.4.1.1).

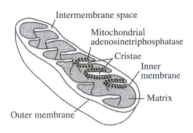

Fig. 8.3 Cross-section of a typical mitochondrion.

The mitochondrial matrix is granular in appearance and has a very high concentration of protein, usually about 500 mg cm^{-3}.[18] About 70 per cent of the mitochondrial protein is in the matrix, 20 per cent is part of the inner mitochondrial membrane, and only 4 per cent is part of the outer membrane. The inner mitochondrial membrane has a ratio of protein:lipid of 75 per cent to 25 per cent, which is high compared to other cell membranes. It is estimated, using differential scanning calorimetry combined with freeze-fracture electron microscopy, that between one-third and one-half of the surface area is occupied by proteins,[19] most of which are involved in the electron-transport chain.

The chemical composition, permeability, and enzyme composition of the inner and outer membranes differ considerably. The outer membrane has a much higher proportion of lipid than the inner membrane, the latter having a very low content of cholesterol. Several of the enzymes present in the outer membrane are also present in the endoplasmic reticulum, e.g. cytochrome b_5 reductase and glycerophosphate acyltransferase, and it is often considered that the outer membrane may be derived from the endoplasmic reticulum. The outer membrane possesses proteins termed porins which act as non-specific pores for solutes of M_r up to 10 000, e.g. ATP, coenzyme A, and carnitine, whereas the inner membrane is generally permeable only to uncharged molecules of M_r up to 150. There are, however, a number of specific proteins in the inner membrane, called translocators, that are responsible for the transport of specific metabolites across the membrane. For example, translocators exist for inorganic phosphate, malate, succinate, citrate, isocitrate, 2-oxoglutarate, glutamate, aspartate, and adenine nucleotides (these are discussed further in Section 8.3.1.4). A list of the principal enzymes located in different regions of the mitochondrion is given in Table 8.3 and some are discussed later in the chapter. A number of the enzymes in the matrix have been shown to associate with one another and with proteins of the inner mitochondrial membranes. These are discussed in Section 8.5.3.2. Although the concentration of protein is high in mitochondria, particularly in the matrix, the overall density of mitochondria is lower than that of nuclei, largely owing to the absence of high concentrations of nucleic acids in the former.

8.2.4 The lysosome

Lysosomes are roughly spherical organelles or vacuoles slightly smaller than mitochondria and having diameters between 0.2 μm and 0.8 μm. They were first discovered by De Duve[21] as particles containing hydrolytic enzymes that sedimented with the mitochondrial fraction in subcellular fractionation. Not only their sizes but their densities (about 1.2 g cm^{-3}) are similar to those of the mitochondria. The

Table 8.3 Principal enzymes or groups of enzymes present in the mitochondrion

1. **Matrix**
   Tricarboxylic-acid cycle enzymes except succinate dehydrogenase
   Enzymes catalysing β-oxidation of fatty acids
   Pyruvate carboxylase
   Phosphoenolpyruvate carboxykinase
   (pigeon liver)
   Carbamoyl-phosphate synthase
   Ornithine carbamoyltransferase
   Glutamate dehydrogenase

2. **Inner membrane**
   Succinate dehydrogenase ⎫
   NADH dehydrogenase ⎬ + associated respiratory chain
   ATP synthase
   3-Hydroxybutyrate dehydrogenase
   Carnitine palmitoyltransferase
   Glycerol 3-phosphate dehydrogenase (FAD enzyme)
   5-Aminolaevulinate synthase
   Hexokinase
   Cytochrome c oxidase

3. **Intermembrane space**
   Adenylate kinase
   Nucleosidediphosphate kinase
   Nucleosidemonophosphate kinase
   L-Xylulose-reductase

4. **Outer membrane**
   NADH dehydrogenase (rotenone insensitive)
   Cytochrome b_5 reductase
   Amine oxidase (flavin-containing)
   Kynureninase
   Acyl CoA synthetase
   Glycerophosphate acyltransferase
   Cholinephosphotransferase
   Adenylate kinase
   Hexokinase
   Phospholipase A_2

For further details see reference 20.

Table 8.4 Enzymes present in lysosomes

1. **Proteolytic enzymes**
   Cathepsins B, D, G, L
   Elastase
   Collagenase

2. **Hydrolysis of glycosides**
   β-D-Glucuronidase
   β-N-Acetyl-D-hexosaminidase
   Hyaluronoglucoasminidase
   Lysozyme
   Neuraminidase

3. **Hydrolysis of esters**
   Deoxyribonuclease II
   Ribonuclease II

4. **Hydrolysis of lipids**
   Phospholipases A_1 and A_2
   Cholesterol esterase

5. **Others**
   Acid phosphatase
   Aryl sulphatase

For a more comprehensive list, see reference 22.

characteristic feature of lysosomes is that they contain about 60 hydrolytic enzymes, most of which have acid pH optima (Table 8.4).

Lysosomes are bounded by a single membrane[23] enclosing a dense granular matrix. The membrane shows limited permeability to solutes, e.g. very little sucrose can penetrate the lysosomal membrane. Transport of many solutes is by carrier-mediated processes. In liver cells the lysosomes only occupy 0.3 per cent of the total cell volume and on average there are about 20 lysosomes per cell. About 35 percent of the total protein is in the membranes and the remainder is in the matrix. The protein concentration in the matrix is about 200 mg cm^{-3} (calculated from reference 24 assuming 35 per cent of lysosomal protein is membrane protein) and the intralysosomal pH is about 1.5 units below that of the cytosol.

Several of the proteins both of the membrane and matrix are glycoproteins and many are acidic. The low intralysosomal pH is thought to be maintained mainly by Donnan equilibrium[25] (an equilibrium that results in the asymmetric distribution of diffusible ions across a membrane, brought about by the presence of a larger ion that is unable to diffuse across the membrane), but there is also some evidence for a membrane-bound adenosinetriphosphatase that may act as a proton pump (a process by which protons are transported across a membrane against the concentration gradient at the expense of metabolic energy). The proteins on the inner surface of the lysosomal membrane contain about 16 μg sialic acid per mg protein and this, together with other glycoproteins in the matrix, is thought to be responsible for the maintenance of the low pH.

Lysosomes are morphologically heterogeneous and this may reflect the way in which they function. Their mode of action is still uncertain, but De Duve and others have proposed the following theory.[22,26] Primary lysosomes that have not participated in intracellular digestion have a full complement of hydrolytic enzymes and are formed by pinching off portions of the smooth endoplasmic reticulum at the Golgi apparatus. The primary lysosomes fuse with other membranous structures containing substrates to be degraded and thereby form secondary lysosomes, i.e. lysosomes in which active digestion is occurring (see also Chapter 9, Section 9.6.3). Some of the low M_r products may diffuse out of the secondary lysosomes.

8.2.5 The peroxisome

Peroxisomes were first discovered in 1954 in kidney tubules as membrane-bound organelles 0.5–1.0 μm in diameter and having a dense granular appearance. In subcellular fractionations De Duve *et al.* isolated them from the light mitochondrial fraction. They were originally called microbodies, but once it became evident that a characteristic of these organelles was that they contained a range of enzymes which generated hydrogen peroxide, they were renamed peroxisomes. Later, these organelles were found to be present in a wide range of eukaryotes, including plants, animals, protozoans, fungi, and tissues. In some plant tissues they were also found to contain enzymes of the glyoxalate cycle, and were called glyoxysomes. All peroxisomes are similar in size (0.1–1.0 μm diameter), are bounded by a single membrane, contain a range of flavin oxidases that generate hydrogen peroxide, and are devoid of DNA.[27,28] The membrane contains a protein[29] thought to render it freely permeable to metabolites of M_r less than 10 000 including glucose, sucrose, urea, uric acid, NAD, CoA, and ATP. In this respect it is like the outer mitochondrial membrane. There is some evidence for a proton-translocating ATPase which may be responsible for maintaining an intra-organelle pH between 5.8 and 6.0. The membrane also contains cytochrome b_5 reductase and long-chain acyl CoA synthase.[29] Peroxisomes are most abundant in tissues that are active in lipid metabolism, e.g. liver, brown body fat. In liver they make up about 2 per cent of the liver volume, and may account for 20 per cent of its oxygen consumption. In yeast grown using methanol as carbon source they make up as much as 80 per cent of the cell volume. The flavin oxidases which they contain include urate oxidase, D-aminoacid oxidase, L-α-hydroxyacid oxidase, acyl CoA oxidase, glutaryl CoA oxidase, polyamine oxidase, pipecolic acid oxidase, oxalate oxidase, trihydroxycholestanoyl CoA oxidase and pristanoyl CoA oxidase. They also contain catalase. Peroxisomes can be detected histochemically using 3,3'-diaminobenzidine which acts as an acceptor for hydrogen peroxide.

Other enzyme systems have been detected in peroxisomes. Catalysis of β-oxidation of fatty acids has been detected in all peroxisomes,[29] the glyoxylate cycle enzymes have been detected in those from some germinating seeds and protozoa, and the enzymes for glycero-ether synthesis required for plasmalogens have been detected in mammalian peroxisomes. Mammalian peroxisomes have also been implicated in cholesterol biosynthesis. The role of the β-oxidation system in peroxisomes in relation to that in mitochondria is particularly interesting. Over 70 per cent of β-oxidation of long-chain fatty acids is believed to occur in rat liver mitochondria. However, the initial stages in the oxidation of very-long-chain fatty acids, including saturated fatty acids greater than C_{24}, and polyunsaturated fatty acids such as arachidonic acid (C_{20}) and larger, appear to occur preferentially in peroxisomes. This is also the case with certain branched-chain fatty acids. An important difference between β-oxidation in peroxisomes and in mitochondria is that in the former the $FADH_2$ formed is reoxidized by molecular oxygen to generate hydrogen peroxide which is acted on by catalase.*

$$^{*}RCH_2CH_2COCoA + FAD \rightleftharpoons$$
$$RCH:CH\text{-}CoA + FADH_2$$

$$FADH_2 + O_2 \rightleftharpoons FAD + H_2O_2$$

$$2H_2O_2 \longrightarrow 2H_2O + O_2$$

*β-oxidation in peroxisomes

All peroxisomal proteins appear to be encoded by nuclear DNA, and are synthesized on free polysomes and have targeting signals to enable them to be imported into the peroxisomes.[30]

8.2.6 The endoplasmic reticulum

Although the endoplasmic reticulum is not generally regarded as an organelle it forms an extensive network throughout the cytoplasm and effectively separates the cytosol into two components; that which is enclosed within the endoplasmic reticulum (the lumen) and that which is outside it (Fig. 8.4). It is not possible to separate these two compartments by present biochemical techniques because when a cell is disrupted prior to carrying out subcellular fractionation, the endoplasmic reticulum is broken down and the membranous fragments form vesicles that constitute the 'microsome fraction' in subcellular fractionations.

The amount of endoplasmic reticulum varies between different cell types but there is a good correlation between the density of reticulum and the extent to which the cell secretes proteins. The principal type of liver cell, the hepatocyte, and the cells of the exocrine pancreas (the cells concerned with its digestive functions rather than its hormonal functions) secrete a number of proteins and both have dense endoplasmic reticula (in the hepatocyte the endoplasmic reticulum occupies about 15 per cent of the total cell volume; 1 g of liver contains about 10 m^2 area of endoplasmic reticulum), whereas a non-secreting cell such as the erythrocyte is devoid of endoplasmic reticulum.

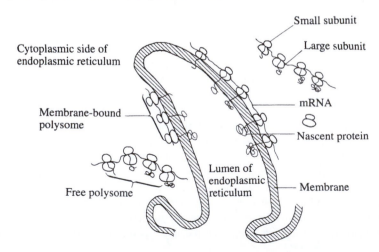

Cytoplasmic side of endoplasmic reticulum

Small subunit

Large subunit

mRNA

Membrane-bound polysome

Nascent protein

Free polysome

Lumen of endoplasmic reticulum

Membrane

Fig. 8.4 The endoplasmic reticulum showing free and membrane-bound polysomes.

The endoplasmic reticulum, which is composed mainly of lipids and proteins, is organized into channels and it is assumed that proteins are collected and secreted into these channels. The proteins of the endoplasmic reticulum have both structural and catalytic roles. It is estimated that there are about 25–50 molecules of phospholipid per polypeptide chain in the endoplasmic reticulum. The membrane has limited permeability: uncharged molecules with M_r values up to about 600 can pass through freely, but charged species with M_r values greater than 90, and macromolecules, cannot pass freely. The amounts and distribution of many of the proteins have been estimated; cytochrome P450 is the most abundant, comprising 4 per cent of the total protein. The majority of the proteins appear to be on the cytosolic surface of the endoplasmic reticulum (Table 8.5).

A large number of different processes occurs on the endoplasmic reticulum including protein synthesis, protein folding, and protein transport, biosynthesis of fatty acids, sterols, phospholipids, and the metabolism of foreign compounds, and this is reflected in the enzyme composition (Table 8.5). Protein synthesis occurs both on polysomes (Fig. 8.4), which are free in the cytosol, and on polysomes bound to the endoplasmic reticulum. The ribosomes that comprise the free and bound polysomes arise from the same ribosome pool. Whether a polysome remains free or bound is determined by the nature of mRNA that is being

Table 8.5 Principal enzymes of the endoplasmic reticulum

Enzyme or group of enzymes	Location
Cholesterol biosynthetic enzymes	Smooth endoplasmic reticulum
Steroid hydroxylation enzymes	Smooth endoplasmic reticulum
Carnitine acyltransferase	
Fatty-acid elongating enzymes (C_{16}–C_{24})	
Glycerolphosphate acyltransferase	
Drug-metabolizing enzymes (aromatic ring hydroxylation, side-chain oxidation, deamination, dealkylation, dehalogenation)	Smooth endoplasmic reticulum
Protein synthesis	Rough endoplasmic reticulum, cytosolic side
Glucose 6-phosphatase	Rough endoplasmic reticulum, luminal side
Adenosinetriphosphatase	Cytosolic side
Cytochrome b_5 reductase	Cytosolic side
NADPH-cytochrome reductase	Cytosolic side
GDPmannose α-D-mannosyltransferase	Cytosolic side
Nucloesidediphosphatase	Luminal side
β-D-Glucuronidase	Luminal side
UDP glucuronosyltransferase	
5' Nucleotidase	Cytosolic side
Cholesterol acyltransferase	Cytosolic side
Proline cis–trans isomerase	Luminal side
Proline cis–trans isomerase	Luminal side
Protein disulphide isomerase	Luminal side
Proline hydroxylase	
Lysyl hydroxylase	

For further details see references 31–34.

translated. Proteins that are destined either to become integral membrane proteins or to be transported across the endoplasmic reticulum on the luminal side contain a signal sequence of between 16 and 30 amino acids, a preponderance of which are hydrophobic. The signal sequence is recognized by a signal-recognition particle which ensures that it becomes membrane bound. The growing polypeptide is extruded through the membrane, eventually either becoming an integral membrane protein or being cleaved to remove the signal peptide and released into the lumen. These proteins may either become secreted out of the cell or be contained within vesicles or dispersed to the lysosomes.

The lumen of the endoplasmic reticulum is the site at which proteins that are destined to be secreted are folded before passing through the cell membrane. Many to these proteins contain disulphide bridges. The lumen provides the environment in which folding can occur efficiently. It is much smaller in volume than the cytosol. The intraluminal volume has been measured as about 1 μl per mg microsomal protein.[35] From this it can be estimated that the lumen occupies about 8 per cent of the cell volume, roughly one-tenth that of the cytosol. Besides having a small volume, it is rich in chaperones and enzymes required for protein folding, e.g. protein disulphide isomerase and proline cis–trans isomerase. It also has a less reducing environment than the cytosol which will favour the formation of disulphide bridges. This is brought about by having a lower ratio of reduced:oxidized glutathione (the tripeptide γ-glutamyl-cysteinyl-glycine); $\approx$ 10:1 compared with $\approx$ 100:1 in the cytosol.[34]

The endoplasmic reticulum is also connected to the Golgi apparatus, which sorts these proteins in a manner dependent on their recognition regions for the lysosomes or for secretion. The enzymes destined for the lysosomes contain a recognition marker in the form of a mannose 6-phosphate residue that is recognized by a receptor (M_r 215 000) that transports the enzyme from the Golgi to the lysosomes (for details, see reference 36).

One of the enzymes that has its active site on the luminal side of the endoplasmic reticulum is glucose-6-phosphatase. It plays a key role in the release of glucose from the liver into the blood, and is therefore important in the maintenance of blood glucose levels. Glucose-6-phosphate is generated in the liver both by gluconeogenesis and by glycogenolysis. Both processes occur in the cytosol. The conversion of cytosolic glucose-6-phosphate to glucose and the release of glucose into the blood requires at least three proteins present in the endoplasmic reticulum and one from the plasma membrane. Glucose-6-phosphate is transported across the endoplasmic reticulum by the transport protein, T1. Once in the lumen it is acted on by glucose-6-phosphatase. The glucose then passes back into the cytosol using another transport protein, T3. From the cytosol it exits into the blood using the plasma membrane glucose transporter GLUT 2.[35]

8.2.7 The cytosol

The cytosol is the aqueous phase in which the cell organelles are suspended.* Operationally, it is the soluble phase that remains as the supernatant after a tissue homogenate has been centrifuged for sufficient '$g \cdot$ min' (Fig. 8.2) to sediment the microsome fraction and free ribosomes and is referred to as the 'high-speed' supernatant. It thus comprises most of the cytosol together with small amounts of solutes that may have leaked from the nucleus, the lumen of the endoplasmic reticulum, and any other organelles disrupted during homogenization. The 'high-speed' supernatant will also differ from the cytosol in that the former will be appreciably diluted with the buffer used for homogenization.

* The term *cytoplasm* refers to the entire contents of the cell outside the nucleus, excluding the cell membrane. The cytoplasm therefore includes subcellular organelles such as mitochondria, lysosomes, peroxisomes, etc.

In recent years, evidence has mounted that suggests that the cytosol is not simply a random mixture of proteins, nucleic acids, and other molecules. High-voltage electron photomicrographs provide evidence for a type of cytomatrix.[37] Cells can be disrupted using dextran sulphate, so that their plasma membrane has holes large enough to allow proteins having M_r up to 400 000 through. These permeabilized cells retain 85–90 per cent of cellular protein after incubation at 37°C for 30 min and all the enzymes required for glycolysis are retained. Similar results have been obtained when cells are partially disrupted using detergents.

Certain groups of enzymes are known to form multienzyme complexes, e.g. pyruvate dehydrogenases (see Chapter 7, Section 7.7), and in the case of glycolytic enzymes there is evidence for a type of association that is less well defined (see Section 8.5.3). It is possible that many other weak associations of macromolecules may occur that have not yet been detected, either because they have not been sought or because the association may be so weak that the dilution of the cytosol during the isolation of a high-speed supernatant may be sufficient to cause dissociation.

In most tissues a substantial proportion of the protein of the cell is in the cytosol. The viscosity of the cytosol is approximately equivalent to that of a 15 per cent solution of sucrose. The cytosol may be regarded as a 'very crowded' protein solution. On average, the distance between globular proteins is of the order of their own size.[14] This crowding causes the rate of diffusion of small molecules to be low (see Section 8.5.2). Measurements of the rates of diffusion of small molecules in the cytosol show that the rates are appreciably lower than in a protein solution of equivalent concentration. This has been interpreted as evidence that the characteristic macromolecules of the cytosol do form defined structures that retard the rates of diffusion of metabolites and other solutes.[38] The importance of diffusion rates in the cytosol is highlighted by evidence that ATP and O_2 gradients exist within the cytosol.[39] The presence of ATP gradients within the cytosol has been demonstrated by showing that the catalytic activities of two ATP-requiring enzymes, one in the cytosol and one on the inside of the plasma membrane, differ in intact cells because their average distances from the sites of ATP generation (mitochondria) differ.

In addition to protein, the cytosol also contains most of the tRNA of the cell. The principal monovalent cations present are Na^+ and K^+ and these are thought to exist predominantly as hydrated ions in free solution, but the principal divalent cations, Mg^{2+} and Ca^{2+}, are thought to be predominantly bound to nucleotides, nucleic acids, and acidic polysaccharides. Much of the intracellular Ca^{2+} is sequestered into the lumen of the endoplasmic reticulum and into the mitochondria, leaving only micromolar concentrations in the cytosol. A large number of enzymes are present in the cytosol, particularly those responsible for glycolysis, gluconeogenesis, fatty-acid synthesis, nucleotide biosynthesis, and aminoacyl-tRNA synthesis (Table 8.6).

Other organelles that are not described here are discussed in the following specific references: chloroplasts, reference 41; Golgi apparatus, references 42 and 43.

8.3 Compartmentation of metabolic pathways

In the intact cell the individual enzymes of a metabolic pathway function together and it is unusual for high concentrations of intermediates to build up. In addition, there are many connections between the main metabolic pathways, since some of the substrates, cofactors, and regulatory molecules, and sometimes the enzymes, are

Table 8.6 Enzymes or groups of enzymes present in the cytosol

1. **Carbohydrate metabolism**
   Glycolytic enzymes including phosphorylase, phosphorylase kinase, and protein kinase
   Glycogen synthase
   Fructose-bisphosphatase
   Phosphoenolpyruvate carboxykinase (rat liver)
   Enzymes of the pentose phosphate pathway
   Malate dehydrogenase
   Isocitrate dehydrogenase
   Lactate dehydrogenase
   Citrate (*pro*-3S)-lyase
   Malate dehydrogenase (oxaloacetate-decarboxylating) (NADP+)
   Glucose-1-phosphate uridylyltransferase

2. **Lipid metabolism**
   Acetyl CoA carboxylase
   Fatty-acid synthase complex
   Glycerol-3-phosphate dehydrogenase (NAD+)

3. **Amino-acid and protein metabolism**
   Aspartate aminotransferase
   Alanine aminotransferase
   Arginase
   Argininosuccinate lyase
   Argininosuccinate synthase
   Aminoacyl-tRNA synthetases

4. **Nucleic-acid synthesis**
   Nucleoside kinase
   Nucleotide kinase

For further details see reference 40.

common to more than one pathway. These complex interactions can only be fully understood when, in addition to studying the isolated systems, some knowledge has been acquired concerning the intracellular locations and concentrations of the enzymes and substrates involved and any permeability barriers separating the components. The functions of some enzymes, e.g. malate dehydrogenase, isocitrate dehydrogenase, and ATP citrate (*pro*-3S)-lyase, cannot be understood fully without a knowledge of their location. We illustrate the importance of compartmentation using two examples: (i) the interrelationships between glycolysis, gluconeogenesis, fatty-acid oxidation and biosynthesis, and the tricarboxylic-acid cycle in the liver cell; and (ii) arginine and ornithine metabolism in the fungus *Neurospora*.

8.3.1 Compartmentation of carbohydrate and fatty-acid metabolism in liver

The processes of glycolysis, gluconeogenesis, β-oxidation of fatty acids, fatty-acid synthesis, and the tricarboxylic-acid cycle are of major importance in many cells. The enzymes concerned constitute a major portion of the protein of the cell. In this section we are primarily concerned with the intracellular location of the processes and their interconnections and we shall not consider the individual reactions in detail.

8.3.1.1 Location of the enzymes catalysing the pathways

The pathways discussed are illustrated in Fig. 8.5. The sequence of glycolytic enzymes from phosphorylase to pyruvate kinase and lactate dehydrogenase is located in the cytosol and in the soluble phase of the nucleus. It is not clear whether the soluble phase of the nucleus is continuous with that of the cytosol, but the concentrations of glycolytic enzymes are slightly higher in the nucleus.[13] In skeletal muscle there is evidence that the glycolytic enzymes are organized in a so-called 'glycogen particle' (see Chapter 7, Section 7.13) present in the cytosol. A similar structure may exist in liver and in other types of cell in which glycogen is stored, although they have not been investigated. Gluconeogenesis, the reversal of glycolysis, uses the same enzymes except for the three steps indicated below.

Step 1 (Mitochondria)

$$\text{ATP} + \text{pyruvate} + CO_2 + H_2O \overset{\text{pyruvate}}{\underset{\text{carboxylase}}{\rightleftharpoons}} \text{ADP} + \text{orthophosphate} + \text{oxaloacetate.}$$

$$\text{GTP} + \text{oxaloacetate} \overset{\text{phosphoenolpyruvate}}{\underset{\text{carboxykinase}}{\rightleftharpoons}} \text{GDP} + \text{phosphoenolpyruvate} + CO_2.$$

Step 2 (Cytosol)

$$\text{D-Fructose 1,6-bisphosphate} + H_2O \overset{\text{fructose}}{\underset{\text{bisphosphatase}}{\rightleftharpoons}} \text{D-fructose 6-phosphate} + \text{orthophosphate.}$$

Step 3 (Cytosol)

$$\text{UTP} + \alpha\text{-D-glucose 1-phosphate} \overset{\text{glucose 1-phosphate}}{\underset{\text{uridylyl transferase}}{\rightleftharpoons}} \text{UDPglucose} + \text{pyrophosphate.}$$

$$\text{UDPglucose} + (1,4\text{-}\alpha\text{-D-glucosyl})_n \overset{\text{glycogen}}{\underset{\text{synthase}}{\rightleftharpoons}} \text{UDP} + (1,4\text{-}\alpha\text{-D-glucosyl})_{n+1}.$$

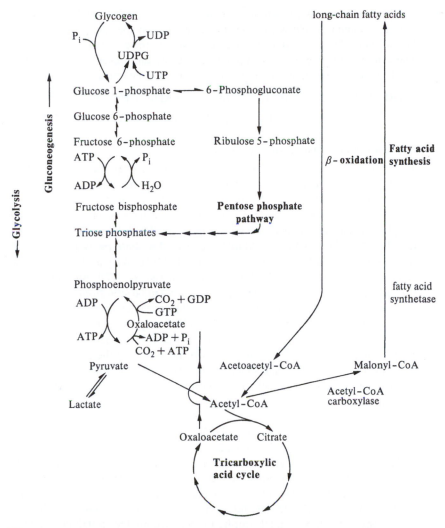

Fig. 8.5 Metabolic pathways for the biosynthesis and degradation of glycogen and of fatty acids. P_i = orthophosphate.

Fructose-bisphosphatase and glycogen synthase are located in the cytosol, whereas pyruvate carboxylase is exclusively in the mitochondrial matrix. The location of phosphoenolpyruvate carboxykinase varies with the species, being totally in the cytosol in rat, > 95 per cent in mitochondria in pigeon and domestic fowl, and distributed in both compartments in other species.[44,45] The alternative pathway for conversion of glucose 6-phosphate (pentose-phosphate pathway) also exists in the cytosol.

The tricarboxylic-acid cycle, β-oxidation of fatty acids, and ketone body formation[46,47] occur in the mitochondria, all enzymes being present in the mitochondrial matrix except succinate dehydrogenase, which is bound to the inner surface of the inner mitochondrial membrane. The electron-transport chain is located on the inner mitochondrial membrane. It now seems probable that other tricarboxylic-acid cycle enzymes form a loose association with the membrane. The fatty acids are synthesized from acetyl CoA using acetyl CoA carboxylase and fatty-acid synthase present in the cytosol. Palmitic acid is the main product of mammalian fatty-acid synthases.[48] Palmitic acid may then act as the substrate for the malonyl CoA-dependent elongase present on the endoplasmic reticulum, where the chain may be extended up to C_{24}. The desaturating enzymes that catalyse the following reaction also occur on the endoplasmic reticulum.[49]

$$CH_3.(CH_2)_x.CH_2.CH_2.(CH_2)_y.CO.SCoA + NADH + H^+ + O_2 \rightleftharpoons$$
$$CH_3.(CH_2)_x.CH=CH.(CH_2)_y.CO.SCoA + 2H_2O + NAD^+.$$

8.3.1.2 Interconnections between pathways

The principal connections between the pathways are through the redox cofactors NAD^+, NADH and $NADP^+$, NADPH; the adenine nucleotides AMP, ADP, ATP; and the metabolite acetyl CoA. Metabolic intermediates of one pathway may also be regulators of other pathways, e.g. citrate, fructose bisphosphate, and acetyl CoA. The degradative pathways generate NADH (glycolysis, β-oxidation, tricarboxylic-acid cycle) and NADPH (pentose-phosphate pathway) and ATP, whereas the biosynthetic pathways require NADPH (fatty-acid biosynthesis, cholesterol biosynthesis) or NADH (gluconeogenesis) and ATP and GTP. Acetyl CoA is the metabolite that links the pathways. It can be seen from the locations of the pathways (Fig. 8.6) that acetyl CoA will be generated in the mitochondria by either pyruvate dehydrogenase (mitochondrial enzyme) or by β-oxidation of fatty acids, and it will be utilized either by the tricarboxylic-acid cycle or ketone body formation[46,47] in the mitochondria or by fatty-acid synthesis in the cytosol. It can also be appreciated that ATP and NADH will be generated principally by oxidative reactions in the mitochondria, whereas ATP, NADH, and NADPH will be required for biosynthesis in the cytosol.

8.3.1.3 Location of metabolites involved in the pathways

Using methods described in Section 8.2.1 it is a relatively straightforward matter to determine the subcellular location of the enzymes catalysing metabolic pathways, especially when compared with the problems of determining the intracellular location and concentrations of metabolites and cofactors. The direct determination of metabolite concentrations is difficult for two reasons: (i) the concentrations often change rapidly during the time required for isolation of subcellular organelles and (ii) substances having low values of M_r readily diffuse out of organelles and redistribute during isolation. For example, tricarboxylic-acid cycle intermediates have a turnover time of milliseconds and so any method of isolating an organelle without a rapid quenching of metabolism would be ineffective.[50]

Most attention has focused on the distribution of adenine nucleotides, the nicotinamide coenzymes, and the intermediates of the tricarboxylic-acid cycle in the mitochondrial compartments and in the cytosol. Little work has been carried out on other organelles. Four methods to determine these distributions have been generally used,[50,51] three direct and one indirect. The direct methods involve isolating mitochondria under conditions that minimize any change in the metabolite concentration or distribution, either by rapidly freeze-drying the tissue and then fractionating using organic solvents, or by disrupting the plasma membrane of isolated cells using digitonin and then rapidly sedimenting mitochondria within 30 s. Using ^{31}P or ^{13}C NMR it has been shown that it is possible to demonstrate compartmentation of metabolites such as orthophosphate and sugar phosphates in intact muscle, and also measure metabolic fluxes *in vivo*.[52,53] This non-destructive technique does not suffer from many of the limitations of the

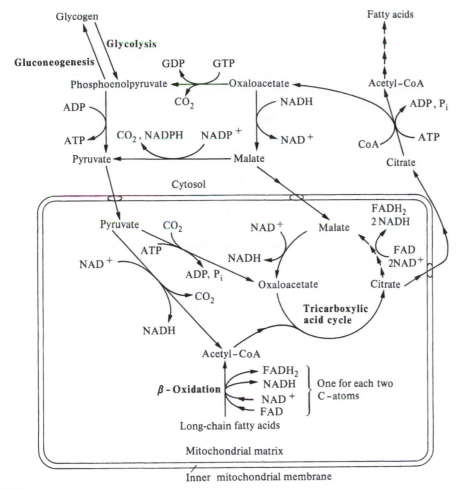

Fig. 8.6 The intracellular location of the principal metabolic pathways leading to the biosynthesis and degradation of carbohydrates and fatty acids. P_i = orthophosphate.

other methods, although it is not sensitive enough for low concentrations of metabolites and a single spectrum can take up to 45 min.[50] The fourth method is an indirect method and makes use of reactions that are known to be near equilibrium in certain cell compartments (see Chapter 6, Section 6.3.3.1). The concentration of certain metabolites may then be determined from a knowledge of the concentrations of other metabolites that are common substrates of reactions at equilibrium (see reference 54).

All the methods have shortcomings. The method using non-polar solvents assumes that the metabolites are deposited during the freeze-drying process in the same location that they have *in vivo* and that they are not extracted by the non-polar solvent. The 'rapid' isolation method is still slow enough for certain changes to have taken place, e.g. it is estimated that a molecule of ATP can diffuse the length of a mitochondrion in < 0.3 s.[55] The indirect method assumes certain reactions are at equilibrium, whereas they may only be close to equilibrium and some may be closer than others. Nevertheless, in spite of these shortcomings there is general qualitative agreement about the distribution of many of the metabolites in the mitochondria and the cytosol, but in some cases there is substantial disagreement about the actual magnitudes of the concentration gradients.

8.3.1.4 Location of nicotinamide cofactors, adenine nucleotides, and coenzyme A

One of the most important groups of substances about which information is necessary is the nicotinamide cofactors, since these cofactors participate in a large number of reactions. It has been known since the 1950s that NAD^+, NADH, $NADP^+$, and NADPH cannot cross the inner mitochondrial membrane. This became apparent when it was found that intact mitochondria oxidized exogenously supplied NADH very slowly compared with the rate of oxidation of NADH by damaged mitochondria.[46] The concentrations of these coenzymes in the mitochondria and the cytosol have been determined by the indirect method (see Section 8.3.1.3). In this method certain enzymes that are present in high concentrations and whose subcellular distribution is known are assumed to catalyse reactions that are near to equilibrium, whereas reactions catalysed by other enzymes or transport processes are assumed to be not at equilibrium. A list of some of the reactions that are considered to be near equilibrium is given in Table 8.7.

The determination of the free $[NAD^+]/[NADH]$ ratio in the cytosol requires the following assumptions and measurements.

1. Lactate dehydrogenase is present exclusively in the cytosol and is present in sufficiently high concentration that the reaction is near equilibrium.

Table 8.7 Enzymes catalysing near equilibrium reactions in vivo which have been used to estimate metabolite compartmentation in rat liver
Activity

Enzyme	Location	Equilibrium definition	K_{eq}	Activity (katal kg^{-1})
Lactate dehydrogenase	Cytosol	$K = \dfrac{[\Sigma\,\text{pyruvate}][NADH][H^+]}{[\Sigma\,\text{lactate}][NAD^+]}$	1.1×10^{-11} mol dm^{-3}	4.1×10^{-3}
Malate dehydrogenase	Cytosol	$K = \dfrac{[\Sigma\,\text{OAA}][NADH][H^+]}{[\Sigma\,\text{malate}][NAD^+]}$	2.9×10^{-2} mol dm^{-3}	6.6×10^{-3}
Glycerol-3-phosphate dehydrogenase	Cytosol	$K = \dfrac{[\Sigma\,\text{DHAP}][NADH][H^+]}{[\Sigma\,\text{G3P}][NAD^+]}$	1.4×10^{-11} mol dm^{-3}	
Isocitrate dehydrogenase	Cytosol	$K = \dfrac{[\Sigma\,\text{oxogl}][CO_2][NADPH]}{[\Sigma\,\text{isocitrate}][NADP^+]}$	1.2 mol dm^{-3}	3.7×10^{-4}
Phosphogluconate dehydrogenase	Cytosol	$K = \dfrac{[\Sigma\,\text{ribulose 5P}][CO_2][NADPH]}{[\Sigma\,\text{phosphogluconate}][NADP^+]}$	0.17 mol dm^{-3}	4.7×10^{-5}
3-Hydroxybutyrate dehydrogenase	Inner mitochondrial Membrane	$K = \dfrac{[\Sigma\,\text{acetoacetate}][NADH][H^+]}{[\Sigma\,\text{3HOB}][NAD^+]}$	4.9×10^{-9} mol dm^{-3}	
Glutamate dehydrogenase	Mitochondrial Matrix	$K = \dfrac{[\text{oxogl}][NH_4^+][NADH][H^+]}{[\text{glutamate}][NAD^+]}$	3.9×10^{-13} mol^2 dm^{-6}	1.97×10^{-3}

Values of K are defined at 311 K (38°C), ionic strength, 0.25, and measured *in vitro* near pH 7.0.
References 55, 56.
OAA = oxaloacetate; DHAP = dihydroxyacetone phosphate; G3P = glycerol 3-phosphate; oxogl = 2-oxoglutarate; and 3HOB = 3-hydroxybutyrate.

$$\text{Lactate} + \text{NAD}^+ \rightleftharpoons \text{pyruvate} + \text{NADH} + \text{H}^+$$

$$K = \frac{[\text{pyruvate}][\text{NADH}][\text{H}^+]}{[\text{lactate}][\text{NAD}^+]}.$$

2. The total concentrations of lactate and pyruvate in the tissue are determined after freeze-clamping (see Chapter 6, Section 6.3.3.1).

3. It is assumed that the total concentration ratio of pyruvate/lactate is approximately the same as that in the cytosol. The mitochondria occupy about 20 per cent of the cell volume in liver and it is assumed that pyruvate and lactate freely diffuse into and out of the mitochondria.

4. It is also assumed that the ratio of the total concentrations of pyruvate and lactate is equal to the ratio of their free concentrations, i.e. no preferential binding occurs.

A further assumption is that the equilibrium constant measured *in vitro* with low enzyme concentrations is applicable *in vivo* (see discussion in Section 8.5.4).

By this method the [NAD$^+$]/[NADH] ratio has been determined in the cytosol. A similar ratio is obtained using the other enzymes catalysing reactions that are near equilibrium, namely glycerol 3-phosphate dehydrogenase and malate dehydrogenase, and these findings give support to the validity of this approach.

The same procedure is used to determine the [NAD$^+$]/[NADH] ratio within the inner membrane of the mitochondria using reactions catalysed by the two enzymes 3-hydroxybutyrate dehydrogenase, which is attached to the inner mitochondrial membrane, and glutamate dehydrogenase, which is in the mitochondrial matrix. It is necessary to measure the tissue contents of 3-hydroxybutyrate, acetoacetate, glutamate, 2-oxoglutarate, and ammonia. The procedure can be extended to determination of the ratio [NADP$^+$]/[NADPH] in the cytosol using glucose 6-phosphate dehydrogenase, isocitrate dehydrogenase, or the 'malic' enzyme (L-malate:NADP$^+$ oxidoreductase (oxaloacetate-decarboxylating) EC 1.1.1.40). The mitochondrial [NADP$^+$]/[NADPH] ratio is more difficult to determine because there is no NADP$^+$-requiring enzyme that is exclusively mitochondrial. Calculations[56] have been made on the assumption that the mitochondrial glutamate dehydrogenase that reacts with either NAD$^+$ or NADP$^+$ *in vitro** must react likewise *in vivo* and establish a near-equilibrium as shown in the following equations, but the results are still regarded as controversial.

> * it is unusual for a dehydrogenase to use NAD$^+$ or NADP$^+$ interchangeably; see discussion in Chapter 5, Section 5.5.4.3

(i) NAD$^+$ + H$_2$O + L-glutamate $\rightleftharpoons$ 2-oxoglutarate + NH$_4^+$ + NADH.

(ii) NADPH + NH$_4^+$ + 2-oxoglutarate $\rightleftharpoons$ L-glutamate + H$_2$O + NADP$^+$.

Sum NAD$^+$ + NADPH $\rightleftharpoons$ NADH + NADP$^+$.

The results of determinations of the ratios [NAD$^+$]/[NADH] and [NADP$^+$]/[NADPH] are given in Table 8.8.

Although the exact values of these ratios vary with the method used and are sensitive to hormonal and nutritional changes, it is clear that the ratio [NAD$^+$]/[NADH] is much higher in the cytosol compared with the mitochondria, and that in the cytosol the ratio [NADP$^+$]/[NADPH] is much lower than the ratio [NAD$^+$]/[NADH]. The principal significance of the different ratios [NAD$^+$]/[NADH] and [NADP$^+$]/[NADPH] is that it permits a further level of compartmentation within the cytosol. Spatial compartmentation occurs between pathways operating in the mitochondria and in the cytosol, but compartmentation based on enzyme specificity also occurs within the cytosol between

Table 8.8 The ratios of the concentrations of oxidized and reduced nicotinamide cofactors in rat liver determined by use of 'near equilibrium' enzymes

	Cytosol	Mitochondria
[NAD+]/[NADH]	700–1000	7–8
[NADP+]/[NADPH]	0.012–0.014	—

For further details see reference 54.

NAD^+- and $NADP^+$-requiring dehydrogenases. The principal reactions generating NADPH are those of the pentose phosphate pathway,

$$\text{Glucose 6-phosphate} + NADP^+ \rightleftharpoons \text{glucono-}\delta\text{-lactone 6-phosphate}$$

$$\text{6-Phospho-D-gluconate} + NADP^+ \rightleftharpoons \text{D-ribulose 5-phosphate} + CO_2 + NADPH$$

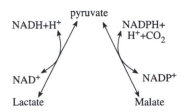

and the principal reactions utilizing NADPH are those of fatty-acid and cholesterol biosynthesis. There is a correlation between tissues synthesizing large amounts of fat, e.g. adipose and mammary tissue, and the activity of the pentosephosphate pathway. Glycolysis and gluconeogenesis, on the other hand, require NAD^+ and NADH, respectively. Although the ratios of the NAD^+–NADH redox couple and of the $NADP^+$–NADPH redox couple are very different, they are nevertheless 'linked' to one another in the cytosol in such a way that changes in one couple will affect the other. The link between them is through enzymes that have common substrates, e.g. lactate dehydrogenase and the 'malic' enzyme, which are both present in the cytosol in large enough amounts to ensure equilibrium.

Since the principal oxidative reactions, e.g. oxidation of NADH, occur in the mitochondria and nicotinamide cofactors do not cross the inner mitochondrial membrane, it is necessary for the reducing power generated by glycolysis to be transferred to the mitochondria. This is brought about by two shuttles, the malate–aspartate shuttle and the glycerol 3-phosphate shuttle (Fig. 8.7). The malate–aspartate shuttle uses two malate dehydrogenases, two aspartate aminotransferases, and two translocators (specific proteins on the inner mitochondrial membrane allowing selective transport of particular ions across the membrane; see Section 8.2.3). NADH in the cytosol is oxidized in the reaction catalysed by malate dehydrogenase and the malate formed enters the mitochondria in exchange for 2-oxoglutarate. There malate is reoxidized to oxaloacetate, which is then transaminated to aspartate that leaves the mitochondria in exchange for glutamate entering. The cycle is completed by transamination of aspartate in the cytosol. The malate–aspartate shuttle seems to be the principal shuttle in liver and heart for the transport of reducing equivalents into the mitochondria. A second shuttle that has been studied principally in insect flight muscle is the glycerol 3-phosphate shuttle. It is also present in other species. As can be seen in Fig. 8.7, it differs from the malate–aspartate shuttle in that *sn*-glycerol 3-phosphate does not enter the mitochondrial matrix, since the flavin-linked glycerol 3-phosphate dehydrogenase present in the inner mitochondrial membrane appears to have its substrate-binding site on the outer face of the membrane but the reduced flavin is generated on the inner face of the membrane.

Equally important in linking the biosynthetic and degradative pathways are the adenine nucleotides. The principal site of generation of ATP is the mitochondrion and of ATP utilization is the cytosol. Adenine nucleotides do not cross

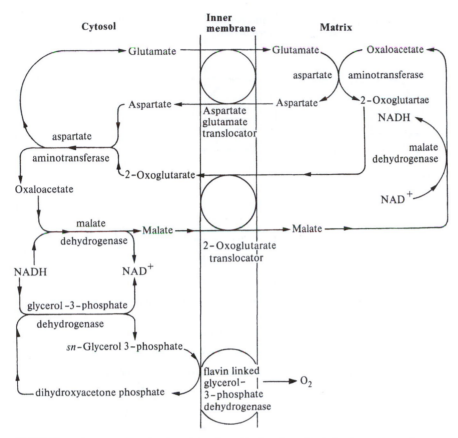

Fig. 8.7 The malate–aspartate shuttle and the glycerol 3-phosphate shuttle for the transport of reducing equivalents into the mitochondria.

the inner mitochondrial membrane freely but are controlled by the adenine nucleotide translocator (Fig. 8.8), which allows the passage of one molecule of ATP in one direction in exchange for one molecule of ADP in the other direction.[57] It can select cytosolic ADP from ATP (whose concentration is much higher) by virtue of its much lower K_m^{ADP} on the cytosolic face of the membrane. On the inside of the mitochondria selectivity is much lower as the K_m^{ADP} and K_m^{ATP} are approximately the same and ATP is exported as a result of its higher concentration. AMP is not transported by the translocator and cannot cross the inner mitochondrial membrane without first being phosphorylated.

Although the adenine nucleotide translocator constitutes as much as 6 per cent of the mitochondrial membrane protein, it appears to be the rate-limiting step in oxidative phosphorylation, limiting the influx of ADP to the site of oxidative phosphorylation. The ratio [ATP]/[ADP] in the cytosol is regulated by two enzymes present in large quantities and operating near equilibrium, namely glyceraldehyde-3-phosphate dehydrogenase and 3-phosphoglycerate kinase:

(i)   Glyceraldehyde 3-phosphate + NAD+ + orthophosphate $\rightleftharpoons$
      3-phosphoglyceroylphosphate + NADH + H+.

(ii) 3-Phosphoglyceroylphosphate + ADP $\rightleftharpoons$ 3-phosphoglycerate + ATP.

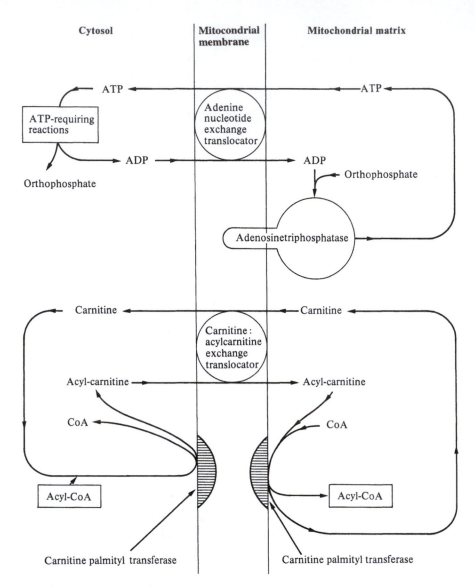

Fig. 8.8 Translocation of adenine nucleotides and of fatty acids across the inner mitochondrial membrane (see references 57–59).

Since these enzymes catalyse reactions near equilibrium, it can be seen that the ratio [NAD$^+$]/[NADH] in the cytosol is linked to the phosphorylation state of the adenine nucleotides in the cytosol. For a review of adenine nucleotide transport, see reference 57.

Besides nicotinamide cofactors and adenine nucleotides, the third important link between metabolism of fat and carbohydrate is acetyl CoA, which can be either the end-product of one pathway (β-oxidation or glycolysis) or the initial substrate of another (tricarboxylic-acid cycle or fatty-acid synthesis). The enzymes involved in acetyl CoA metabolism, together with their intracellular locations, are shown in Fig. 8.9. Coenzyme A and its acylated derivatives are unable to cross the inner mitochondrial membrane. Long-chain fatty acids are

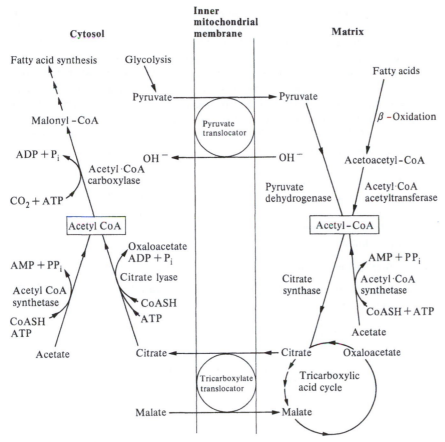

Fig. 8.9 The intracellular location of enzymes concerned with the metabolism of acetyl coenzyme A. P_i = orthophosphate; PP_1 = pyrophosphate.

converted to their carnitine derivatives,[58] which can then cross the membrane (Fig. 8.8). Two separate transferases are involved in the translocation. Carnitine palmitoyltransferase I is located on the outer mitochondrial membrane and catalyses the conversion of acyl CoAs to acylcarnitine. The latter then passes through the intramitochondrial membrane space and the inner membrane where it is converted back to acyl CoA by carnitine palmitoyltransferase II.[58,59] The intracellular concentrations of carnitine in actively metabolizing cells are quite high, e.g. heart 4 mM, liver 2 mM, and highest in rat epididymal fluid, 60 mM.[58]

As can be seen from Fig. 8.9, most of the important steps in generating acetyl CoA, i.e. pyruvate dehydrogenase and acetyl CoA acetyltransferase, are present in the mitochondrial matrix. The principal enzymes using acetyl CoA are citrate (*si*) synthase, which enables acetyl CoA to enter the tricarboxylic-acid cycle, and acetyl CoA carboxylase, which is the first step in fatty-acid biosynthesis. Acetyl CoA generated in the mitochondrial matrix is effectively transferred to the cytosol by conversion to citrate, the latter is transported using the tricarboxylate translocator and then cleaved back to acetyl CoA by ATP citrate (*pro-3S*) lyase. Citrate synthase and ATP citrate (*pro-3S*) lyase are exclusive to the mitochondrial matrix and the cytosol, respectively. ATP citrate (*pro-3S*) lyase is present in high concentrations in the cytosol and catalyses a reaction operating close to equilibrium.

From the three groups of substances we have considered, i.e. nicotinamide cofactors, adenine nucleotides, and coenzyme A derivatives, it is clear that the inner mitochondrial membrane plays an important role in the compartmentation of these compounds, since none can freely cross the membrane. In addition to these coenzymes, several other metabolites are not uniformly distributed across the mitochondrial membrane and there are several translocators that control the movement of intermediates of the tricarboxylic-acid cycle and certain amino acids. It is necessary to know the distribution of these compounds in order to understand the regulation of intermediary metabolism so that calculations of reaction velocities can be made from knowledge of the kinetic properties measured *in vitro*.

It is difficult to measure the distribution of these metabolites directly, for the reasons discussed with regard to nicotinamide cofactors. Indirect measurements have been made using the estimated values for the [NAD⁺]/[NADH] ratio. For example, the concentrations of malate and oxaloacetate in the cytosol are estimated as follows.

(1)   Total cell malate and oxaloacetate is measured.

(2)   The ratio of free $[NAD^+]/[NADH]$ is calculated as described earlier in this section. Now, for the reaction catalysed by malate dehydrogenase (MDH):

$$\text{Malate} + NAD^+ \rightleftharpoons \text{oxaloacetate} + NADH + H^+.$$

$$K_{MDH} = \frac{[\text{oxaloacetate}]\,[NADH]\,[H^+]}{[\text{malate}]\,[NAD^+]}.$$

Therefore

$$\frac{[\text{Malate}]_{\text{cytosol}}}{[\text{Oxaloacetate}]_{\text{cytosol}}} = \frac{[NADH]\,[H^+]}{[NAD^+]\,K_{MDH}}.$$

If $[H^+]$ and the ratio $[NADH]/[NAD^+]$ are known for the cytosol and also K_{MDH}, the ratio [malate]/[oxaloacetate] in the cytosol can be calculated. A similar equation may be written for malate and oxaloacetate in the mitochondria.

Total cell malate = malate$_{\text{cytosol}}$ + malate$_{\text{mitochondria}}$.

Total cell oxaloacetate = oxaloacetate$_{\text{cytosol}}$ + oxaloacetate$_{\text{mitochondria}}$.

Thus, the distribution of both malate and oxaloacetate can be calculated. There is, however, less agreement between workers concerning the concentrations of dicarboxylic and tricarboxylic acids in the two compartments than about the [NAD⁺]/[NADH] ratios. One reason for this is that the determinations are more indirect, relying on more than one reaction being at equilibrium, and thus discrepancies are likely to increase. There is also only qualitative agreement between results from the direct and indirect methods (see Section 8.3.1.3) of determining metabolite distribution. However, the data so far available do suggest that citrate, isocitrate, and possibly 2-oxoglutarate are more concentrated in the mitochondria, whereas glutamate, aspartate, and possibly oxaloacetate are more highly concentrated in the cytosol,[55,60] but the actual concentration gradients obtained by different methods may differ by at least an order of magnitude and the results have to be treated with caution.

8.3.2 Compartmentation of arginine and ornithine metabolism in *Neurospora*

The metabolism of ornithine and arginine in *Neurospora* illustrates the importance of knowing not only the intracellular location of both enzymes and substrates but also the kinetic properties of the enzymes in order to understand their function *in vivo*. *Neurospora*, like many other fungi, is able to grow when supplied with glucose as the sole source of carbon. It is therefore readily able to synthesize both ornithine and arginine. *Neurospora* possesses all the urea-cycle enzymes (see Fig. 8.10) and also an active urease. When it is grown in a medium supplemented with arginine or ornithine it is readily able to assimilate these amino acids and use them in the biosynthesis of other amino acids. Table 8.9 shows some of the activities and K_m values of the principal enzymes present in *Neurospora* that utilize ornithine or arginine. Even when grown on minimal medium, *Neurospora* has substantial intracellular pools of ornithine and arginine, estimated at about 12 and 8 mmol dm^{-3}, respectively.

At first sight it would seem that with a high intracellular pool of arginine and such an active arginase, arginine would be continually degraded at a high rate. It would also appear that provided the supply of carbamoyl phosphate was not limiting, with the presence of all the urea-cycle enzymes (Fig. 8.10) there would be a continuous high rate of synthesis of arginine and hence a substrate cycle (see Chapter 6, Section 6.3.4.1), but this does not occur.

However, a number of additional factors have to be taken into consideration before extrapolations from *in vitro* to *in vivo* conditions can be made. Arginase activity is normally assayed at pH 9.5 after a preincubation with Mn^{2+}. Although

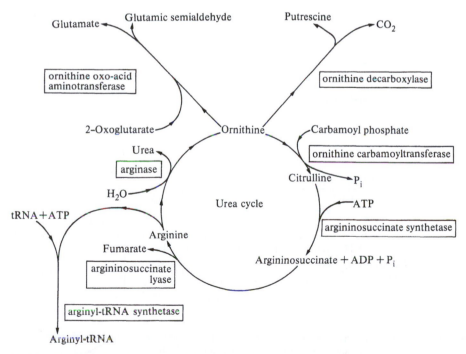

Fig. 8.10 The metabolic pathways for arginine and ornithine metabolism in the fungus *Neurospora crassa*. P_i = orthophosphate.

this treatment results in optimum catalytic activity, it is non-physiological and if the preincubation is omitted and assays are carried out at pH 7.5 the activity is appreciably lower[66] and appears sigmoidal with respect to arginine (for the significance of sigmoidal kinetics see Chapter 6, Section 6.2.2.1). However, the intracellular location of the enzymes and substrates is perhaps the most important factor, as seen in Fig. 8.11. The biosynthesis of ornithine from glutamate, involving acetylglutamate synthase and acetylglutamate kinase, and the first part of the urea cycle occur in the mitochondria, but the later stages of the urea cycle from citrulline to arginine occur in the cytosol, where arginase and also arginyl-tRNA synthetase are located.

It has been shown that approximately 98 per cent of the arginine and ornithine are located in a separate compartment known as a vacuole (Fig. 8.11) and thus the concentrations in the cytosol are much lower. This vacuole, which has been isolated and partially characterized,[61] is capable of storing amino acids, particularly basic amino acids, and similar vacuoles have been found in other fungi and plants.[69] It is capable of storing large phosphate reserves in the form of polyphosphate.[70]

Table 8.9 The activities and Michaelis constants for ornithine- and arginine-metabolizing enzymes in *Neurospora*

Enzyme	K_m ornithine (mmol dm^{-3})	K_m arginine (mmol dm^{-3})	Activity (katal kg^{-1} protein)
Ornithine carbamoyltransferase	1.9		4.3
Ornithine oxyacid aminotransferase	2.0		0.23
Ornithine decarboxylase	0.2		8.3×10^{-4}
Arginase*		20	17.2
Arginyl-tRNA synthetase		0.02	1.3×10^{-5}
Argininosuccinate lyase		0.8	0.18
Argininosuccinate synthase			0.06

*Activity after Mn^{2+} activation.
For further details see references 61–65.

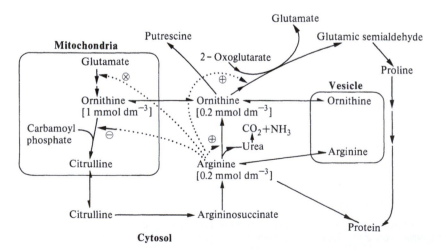

Fig. 8.11 The intracellular location of arginine and ornithine metabolism in *Neurospora crassa:* [ ] = concentrations in compartments indicated. Enzymes that are regulated by the intracellular concentrations of arginine are indicated ⊕ by induction, ⊖ by repression, and ⊗ by feedback inhibition (see references 61, 67, and 68).

The pathways are thus controlled to a large extent by the transport of substrates across both vacuole and mitochondrial membranes. Arginine is taken up into the vacuole by a specific membrane carrier protein[71] having an M_r of 40 000 and a $K_m^{arginine}$ of 0.4 mM,[61] a process coupled to ATP hydrolysis, and it is transported across the inner mitochondrial membrane by facilitated diffusion ($K_m^{arginine} =$ 6.5 mM).[71] In the intact organism the complete urea cycle does not operate to any appreciable extent, instead different halves of the cycle are operative under different physiological conditions. When a mutant deficient in urease is grown on a minimal medium and a small amount of [^{14}C]*guanidino*-arginine of high specific activity is added, most of the label is found in urea. The specific activities of the proteins labelled in short time-periods show that the added [^{14}C]arginine does not equilibrate with the bulk of the arginine pool.[72] The low cytosolic concentration of arginine enables arginyl-tRNA synthetase to compete successfully with arginase. On the basis of their K_m values (Table 8.9) arginase will be operating at about 1 per cent maximum, whereas arginyl-tRNA synthetase will be at more than 90 per cent maximum provided that tRNA^{Arg} and ATP are not limiting. Thus, when grown on minimal medium the intracellular arginine, mainly in the vacuole, is synthesized using the urea-cycle enzymes to convert ornithine to arginine, whilst the arginase catalyses very little arginine degradation. On the other hand, when arginine is added to the growth medium it is the degradative part of the urea cycle that is active. The pool of arginine in the cytosol increases rapidly, e.g. from 0.2 to 15 mmol dm^{-3},[72] and, in addition, higher levels of ornithine-oxyacid aminotransferase are induced and urea production is readily detectable. Also, the higher intracellular arginine inhibits acetylglutamate synthase and acetylglutamate kinase thereby depressing ornithine biosynthesis.[73] The arginine taken into the cells is partly metabolized via arginase and ornithine-oxyacid aminotransferase, leading to the formation of glutamate, proline, and other amino acids, and the remainder is taken up into the vesicles, where it largely displaces ornithine. The latter is transported into the cytosol where it is also metabolized by ornithine-oxyacid aminotransferase. For reviews, see references 61 and 74.

8.4 Vectorial organization of enzymes associated with membranes

The intracellular membranes form the boundaries of subcellular compartments separating enzymes and metabolites in different regions of the cell, as has been discussed in the previous section. Associated with the intracellular membranes are a number of enzymes that are asymmetrically arranged about the membranes. Several of these are concerned with the transport of metabolites and ions between compartments and their functioning can only be understood from studies of the intact membrane or organelles or in reconstituted systems where the enzyme, once isolated, has been reincorporated into membranes or membrane-like structures. In this section we discuss the methods used to study enzymes vectorially arranged in membranes and the progress that has been made in understanding how these enzymes function.

8.4.1 Enzymes in membranes

The 'fluid mosaic' model for membrane structure was first proposed by Singer and Nicolson in 1972.[75] It postulated that integral membrane proteins resemble

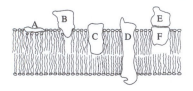

Fig. 8.12 Disposition of enzymes in a membrane.

A. A peripheral protein that can readily be extracted from the membrane by salt solutions, e.g. fructose-bisphosphate aldolase, glyceraldehyde-3-phosphate dehydrogenase in erythrocytes.

B. An integral protein with only a small portion of the polypeptide chain embedded in the bilayer, e.g. α-D-glucosidase and aminopeptidase in the brush border.

C. An integral protein with a large portion of the polypeptide chain embedded in the bilayer, e.g. acyl CoA desaturase.

D. An integral protein spanning the phospholipid bilayer, e.g. ion transporters such as adenosinetriphosphatase (Ca^{2+}-activated).

E. A protein attached to the membrane by a second protein (F) that is embedded in the bilayer, e.g. β-D-glucuronidase from the endoplasmic reticulum attached to a specific integral protein, egasin.

See references 78, 79–81.

'icebergs' floating in a two-dimensional lipid 'sea' and that they could freely diffuse laterally in the lipid matrix unless their movements were restricted by associations with other cell components. The main lipid components, the phospholipids, were arranged in a bilayer. The model has been amply confirmed and there is now much detailed information about the topography of membrane proteins.[76,77] The protein components of the membrane may be either on the surface of the lipid bilayer (peripheral proteins) or embedded in the lipid (integral proteins) as seen in Fig. 8.12. These two groups of proteins can be distinguished by the ease with which they are extracted from the membranes. Peripheral proteins form mainly electrostatic bonds with the charged phospholipid head groups and can be extracted by changes in ionic strength. The integral proteins bind to the hydrocarbon moiety of the phospholipids mainly by hydrophobic interactions and they are extracted from membranes using detergents or organic solvents. The extent to which an integral protein is embedded in the lipid bilayer varies from protein to protein, and this is reflected in the ease with which the protein is extracted. In addition, many of the integral proteins, once extracted, will aggregate once the detergent is removed because they contain a number of non-polar amino-acid side-chains that will take part in hydrophobic interactions. The range of interactions between proteins and the phospholipid bilayer is illustrated in Fig. 8.12.

Some proteins, such as fructose-bisphosphate aldolase and glyceraldehyde-3-phosphate dehydrogenase attached to the inner surface of the plasma membrane of erythrocytes, can readily be extracted by 0.1 mol dm^{-3} EDTA or 1 mmol dm^{-3} ATP.[78] The integral proteins range from those in which only a small portion of the polypeptide chain is embedded in the bilayer, e.g. aminopeptidase from the intestinal brush border where a fragment of polypeptide chain of M_r 8000–10 000 is embedded within the membrane, whereas the remainder (M_r 280 000) is exposed in the lumen of the intestine,[82] to those in which the bulk of the polypeptide chain is presumed to lie in the bilayer, e.g. C_{55}-isoprenoid alcohol phosphokinase.[83] The latter enzyme from *Staphylococcus aureus* catalyses ATP-dependent phosphorylations of C_{55}-isoprenoid alcohols and functions in the biosynthesis of lipopolysaccharide. The enzyme has 58 per cent hydrophobic amino acids and when shaken with 1-butanol–water mixtures it partitions in the butanol phase.

For some integral proteins the hydrophobic regions embedded in the bilayer merely act as an anchoring point on the enzyme. The enzymes may then be completely separated from phospholipid without any appreciable change in their kinetic properties, e.g. α-D-glucosidase and aminopeptidase[84,85] from intestinal brush border, and nucleotide pyrophosphatase[86] from endoplasmic reticulum. Other integral proteins, however, require a minimum amount of phospholipid to retain activity, e.g. amine oxidase (flavin-containing)[87] from the outer mitochondrial membrane, cytochrome *c* oxidase[88] from the inner mitochondrial membrane, and adenosinetriphosphatase (Ca^{2+}-activated) of the sarcoplasmic reticulum.

Transmembrane proteins (see Fig. 8.12D) are usually amphiphilic proteins having a transmembrane hydrophobic segment and two hydrophilic segments, one on either side of the membrane.[76,89] They may traverse the membrane once, e.g. receptor tyrosine kinases, guanylyl cyclases, or more commonly several times, e.g. subunit *c* of F_o ATPase. It is often possible to predict the regions of the polypeptide chain traversing the membrane by assessing the hydrophobicity of segments of the polypeptide chain using a hydropathy plot.[89] Proteins which are anchored to the membrane (see Fig. 8.12B) have hydrophobic tails often in the form of a lipid anchor. The most common form of lipid anchor is glycosylphosphatidylinositol linkage which becomes attached to the C-terminus of the protein. It is found in a number of membrane-anchored enzymes such as carbonic anhydrase,[90] alkaline phosphatase, acetylcholinesterase, 5′-nucleotidase, aminopeptidase P, membrane dipeptidase trehalase, and promastigote surface protease (from *Leishmania major*).[91,92] Other methods of lipid anchorage include myristoylation, e.g. some α-subunits of G-proteins, palmitoylation, e.g. some α-subunits of G proteins, endothelial nitric oxide synthase, adenylyl cyclase, and non-receptor tyrosine kinases, and prenylation, e.g. γ-subunit of G proteins.[93-96] The addition of a single acyl group, e.g. myristoyl group (C_{14}), to the amino terminus of a protein will not, on its own, provide sufficient hydrophobic interactions to hold the protein to a lipid bilayer.[97] It is assumed that additional hydrophobic groups within the polypeptide, or other secondary valance bonds, provide the additional stabilization.

8.4.2 The role of the membrane

The role the membrane plays in relation to the bound enzymes falls into one of four categories.

1. The membrane can act as an anchoring point. For example, the functions of aminopeptidase and α-D-glucosidase in the brush border are to hydrolyse peptides and maltose prior to their uptake across the brush-border membrane. Anchoring the enzymes presumably makes more economical use of the enzymes, since they are retained at the required location (see Section 8.4.3). This is analogous to the use of immobilized enzymes in industrial processes; Chapter 11, Section 11.5.

2. The substrates and products of an enzyme reaction may have only limited solubility in water and may be more soluble in the lipid bilayer, and thus the enzyme will operate more efficiently when lipid bound, e.g. enzymes involved in lipid and glycolipid biosynthesis, in redox reactions involving long-chain ubiquinones, or in glycosyl-transfer reactions involving polyisoprenyl carrier lipids.[98]

3. The phospholipid bilayer may act as the medium in which certain multi-enzyme complexes are organized. An example of this is the acyl CoA desaturase complex,[79] which comprises at least three proteins—cytochrome b_5 reductase, cytochrome b_5, and acyl CoA desaturase—which act together to desaturate fatty acids. All three proteins are localized on the cytosolic face of the endoplasmic reticulum (Fig. 8.13).

4. Membranes may act to separate substrates, products, and effectors in an enzyme reaction, or may allow controlled transport processes to occur. These separations and transport processes cannot readily be studied once the enzyme has been separated from the associated phospholipid membrane components and are usually studied using intact membranes or vesicles obtained from them, or using reconstituted systems. Three examples, each differing in the way in which the enzyme is disposed in relation to the membrane are considered in the next sections.

8.4.3 Hydrolases of the microvillar membrane

Enzymes of this group are good examples with which to illustrate the role of the membrane as an anchoring point for enzymes. Microvillar membranes occur on the luminal surfaces of the epithelium of the small intestine and the proximal tubule of the kidney. The microvillar membranes are highly convoluted to increase their surface area, so as to facilitate efficient absorption—in the case of

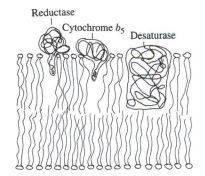

Fig. 8.13 The organization of the acyl CoA desaturase complex in the membrane of the endoplasmic reticulum.

CH$_3$-C(CH$_3$)(CH$_3$)-CH$_2$-C(CH$_3$)(CH$_3$)-⟨C$_6$H$_4$⟩-(OCH$_2$CH$_2$)$_n$OH

Tritron X-100 (isooctyl phenoxy polyethylene oxide)

the small intestine, absorption of nutrients, and in the case of the proximal tubule, reabsorption of solutes filtered by the kidney. The role of hydrolases in the small intestine is obviously to fulfil a digestive function. The proximal tubule contains a similar range of hydrolases but their function is not clear. The hydrolases form the major proportion of the integral membrane proteins of the microvillar membrane. A method that has been important in studying these enzymes is to compare the detergent extracted enzymes, usually using Triton X-100, with the enzymes extracted without detergent but after protease treatment to split off the hydrophobic anchoring region.

With all these hydrolytic enzymes, the protease-extracted enzymes show full activity when compared with the detergent-extracted enzyme. The size of the hydrophobic anchor can be determined from the difference in M_r of the detergent-extracted hydrolase and the protease-extracted hydrolase. In determining the M_r of the detergent-extracted form by gel filtration, allowance has to be made for the micelle of detergent attached to the hydrophobic region necessary for solubilization. Most anchors range in M_r from 3000 to 10 000. The structural arrangement of the hydrolases so far studied falls into one of three types, either homodimers (α_2), heterodimers ($\alpha\beta$), or heterotetramers ($\alpha_2\beta_2$). Aminopeptidase, alkaline phosphatase, and dipeptidyl peptidase are homodimers, whereas sucrase-isomaltase and γ-glutamyl transferase are heterodimers. Endopeptidase-24.18 (EC 3.4.24.18) is a zinc-metallopeptidase present on the brush borders of kidney proximal tubules. It is a heterotetramer in which the β subunits include the anchor region, and the α subunits are attached to the β subunits by disulphide bridges.[99] The probable structural arrangements are shown in Fig. 8.14. All the hydrolases studied are glycoproteins having 13–15 per cent carbohydrate by weight, the carbohydrate moiety being attached to the region of the protein protruding from the membrane. Sucrase-isomaltase is the most abundant protein of the microvilli of the epithelial cells of the small intestine, comprising 10 per cent of the integral membrane proteins. It accounts for all the sucrase activity of the microvilli, 90 per cent of the isomaltase activity, and 70–80 per cent of the maltase activity. It comprises two non-identical subunits, one of which contains the isomaltase catalytic site and is anchored directly to the membrane, the other contains the sucrase activity and also accounts for the maltase activity and is bound to the isomaltase polypeptide, but is not anchored directly to the membrane (Fig. 8.14). The sucrase-isomaltase is synthesized as a single polypeptide precursor that is cleaved during post-translational processing. It is generally assumed that the role of the membrane with all these hydrolases is to ensure efficient hydrolysis coupled with absorption through the associated membrane. For further details, see references 81, 84, and 99–101.

8.4.4 Adenosinetriphosphatase (Ca^{2+}-activated) of the sacroplasmic reticulum

In contrast with the enzymes described in the previous section, for which the membrane acts primarily as an anchoring point, adenosinetriphosphatases (ATPases) are good examples of transmembrane proteins involved in the movement of ions through the membrane. A number of different adenosinetriphosphatase enzymes are known in which the hydrolysis of ATP is coupled to the transport of ions across a membrane, e.g. adenosinetriphosphatase (Na$^+$, K$^+$-activated) of the plasma membrane, mitochondrial adenosinetriphosphatase (also known ATP synthase) associated with oxidative phosphorylation (linked to H$^+$-movement), and adenosinetriphosphatase (Ca^{2+}-activated) of the endoplas-

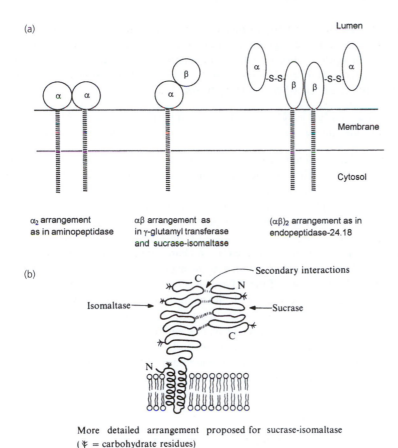

(a)

α_2 arrangement
as in aminopeptidase

$\alpha\beta$ arrangement as
in γ-glutamyl transferase
and sucrase-isomaltase

$(\alpha\beta)_2$ arrangement as in
endopeptidase-24.18

(b)

More detailed arrangement proposed for sucrase-isomaltase
(⊻ = carbohydrate residues)

Fig. 8.14 The different modes of attachment of microvillar enzymes to the cell membrane.[99–101]

mic reticulum and sarcoplasmic reticulum. Of these, the adenosinetriphosphatase (Ca^{2+}-activated) is perhaps the best-understood in molecular terms and therefore is the example chosen here. The sarcoplasmic reticulum is the intracellular membrane surrounding each myofibril in the muscle fibres. The calcium ion concentration in the cytosol is $\approx 0.1\ \mu M$ whereas that within the sarcoplasmic reticulum is > 1 mM. Calcium ions are released from the sarcoplasmic reticulum in response to a nervous impulse. This triggers the process of muscle contraction. Relaxation occurs when Ca^{2+} is removed and this is achieved by adenosinetriphosphatase (Ca^{2+}-activated), which pumps Ca^{2+} from the cytosol back into the sarcoplasmic reticulum at the expense of ATP hydrolysis. Vesicles of the sarcoplasmic reticulum can be prepared and from these adenosinetriphosphatase (Ca^{2+}-activated) can be isolated. There are only four main protein components of this membrane and adenosinetriphosphatase makes up about 90 per cent of this protein. The abundance and relative ease of isolation of this protein has made it a good model system for studying ion transport.

The adenosinetriphosphatase (Ca^{2+}-activated) requires an annulus of 30 molecules of phospholipid for activity. Removal of the phospholipids causes aggregation of the adenosinetriphosphatase; however, the phospholipids can readily be exchanged with added phospholipids if they are mixed in the presence of detergent and the detergent is later removed. Alternatively, the phospholipids may be replaced by certain non-ionic detergents, e.g. dodecyl octaethylene glycol monoether, which provide the amphiphilic environment necessary for maintenance of activity.[102] The adenosinetriphosphatase can be incorporated into vesicles or liposomes (structures composed of multiple phospholipid bilayers completely enclosing an aqueous phase)[103] containing internal phosphate ions. These reconstituted systems catalyse ATP-dependent Ca^{2+} transport. The stoichiometry of a Ca^{2+} transport associated with ATP hydrolysis is as follows:

$$ATP + 2Ca^{2+} + Mg^{2+} + 2K^+ \rightleftharpoons ADP + \text{orthophosphate} + 2Ca^{2+} + Mg^{2+} + 2K^+.$$
    outside inside   inside   outside

Under suitable conditions it is possible to demonstrate the reversibility of this process, i.e. a Ca^{2+} concentration gradient across the membrane can drive the synthesis of ATP. During the process of Ca^{2+} transport, the adenosinetriphosphatase becomes phosphorylated by ATP and then dephosphorylated.[104]

The ATPase comprises a single polypeptide chain (1001 amino-acid residues), the sequence of which has been deduced from the sequence of the cDNA.[105] There are two genes that encode the Ca^{2+} ATPases of fast-twitch and slow-twitch muscles and the sequences show them to be highly conserved.[104] The secondary structure and topography of the ATPase has been predicted on the basis of the likely folding pattern, together with the charged residues involved in the Ca^{2+}-binding and transmembrane regions. In addition, electron micrographs show a globule attached to the membrane by a stalk and helices are observed on freeze-fracture as intramembranous particles. A probable topology of the enzyme is given in Fig. 8.15. As can be seen, there are three distinct domains on the cytoplasmic side of the membrane, eight helices running through the membrane, and five loops on the luminal side of the membrane. The steps in the mechanism are considered as shown in the scheme.

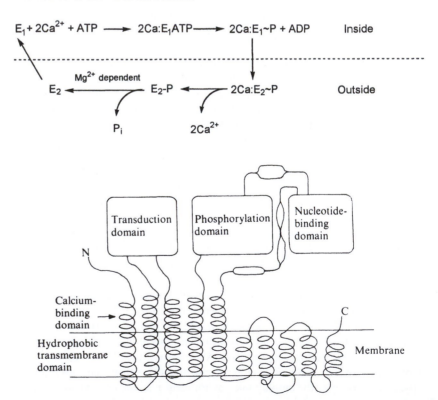

Fig. 8.15 The proposed structure of Ca^{2+}-ATPase.[104]

The initial step involves the binding of two Ca^{2+} and one ATP to the enzyme (E_1). E_1 has a high affinity for Ca^{2+}. On becoming phosphorylased on Asp351 E_1 undergoes a conformational change to E_2. The $2Ca:E_2\sim P$ has a lower affinity for Ca^{2+} than $2Ca:E_1ATP$ and is able to release Ca^{2+} on the luminal side of the membrane, via the channel formed by the helices of the transmembrane region of the protein. The phosphoester becomes hydrolysed and the enzyme reverts to E_1 conformation. The Ca^{2+}-binding sites are believed to be the glutamic acid residues on the stalk of the helices close to the membrane.[104] It is not yet clear how many of the helices are involved in forming the channel, or how they are arranged in relation to one another.

The cycle works as a result of the chemical specificity and vectorial specificity of the intermediates involved. The free enzyme (without Ca^{2+}) is not phosphorylated by ATP, but it binds two Ca^{2+} from the cytosol with very high affinity. This changes the chemical specificity, since the enzyme is now able to become phosphorylated by ATP. The phosphorylated enzyme now has a much lower affinity for Ca^{2+} which it releases into the lumen. The vectorial specificity arises from the different binding affinities for Ca^{2+} of the phosphorylated and dephosphorylated ATPase.

8.4.5 Adenylate cyclase

$3':5'$-Cyclic AMP is perhaps the best-understood of the second messengers that are released inside cells in response to an external signal, generally a hormone (see Chapter 6, Section 6.4.2.1). Other intracellular messengers include Ca^{2+} ions, phosphoinositides, diacylglycerol, and $3':5'$-cyclic GMP. Cyclic AMP is generated by the reaction catalysed by adenylate cyclase:

$$ATP \rightarrow 3':5\text{-cyclic AMP} + \text{pyrophosphate}.$$

The intracellular concentrations of cAMP in unstimulated cells range from 10^{-7} to 10^{-6} mol dm^{-3}. The rises are generally short-lived and the cAMP is degraded by a phosphodiesterase. In this section we focus on the mechanism by which the signal from the hormone is relayed across the membrane to stimulate adenylate cyclase activity. The prototypical hormone-sensitive adenylate cyclase comprises three components: a receptor protein having a 7-helix membrane-spanning structure, a heterotrimeric G protein, and the adenylate cyclase itself. The most thoroughly studied system is that involving the β-adrenergic receptor and adenylate cyclase. There are four types of catecholamine receptor, α_1, α_2, β_1, and β_2, which differ in their relative affinities towards agonists and antagonists*.

* Agonists are structural analogues that mimic the effect of the hormone; antagonists are structural analogues that block the effect of the hormone.

The β receptors belong to a large family of receptors proteins of which over 100 have been cloned and sequenced. The common structural feature is that they contain seven hydrophobic stretches of 20–25 amino acids, which are predicted to form seven transmembrane α-helices connected by alternate extracellular and intracellular loops (Fig. 8.16).[106] The extracellular C-terminal segment is glycosylated and also contains a number of conserved cysteine residues. The receptor binding domain lies within the membrane-spanning helices. There is strong evidence from site-directed mutagenesis that the carboxyl of Asp113 is involved in binding the amino group of catecholamines. A molecular model for the binding of catecholamines to the receptor domain has been proposed.[106]

Hormones show very high affinities for their receptors, since most circulating hormones are present in nanomolar concentrations. In order to measure the strength of binding, physiological concentrations of hormones, or equivalent

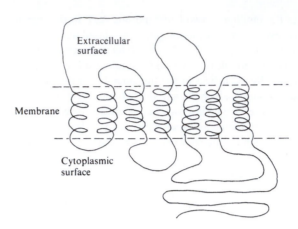

Fig. 8.16 Proposed structure of β-adrenergic receptor.[106]

concentrations of their agonists or antagonists, must be used otherwise non-specific binding to other proteins will confuse the picture. This generally means using radioactively labelled agonists or antagonists having very high specific radioactivity. In the case of the β-receptor, ^{3}H-labelled dihydroalprenolol was used;[107] this has $K_d = 6 \times 10^{-9}$ mol dm^{-3}. The β-receptor has been purified from turkey erythrocytes. The receptors are first solubilized using the detergent digitonin, since the receptor is an integral membrane protein, and it is then purified using a combination of ion-exchange chromatography and affinity chromatography. The latter is performed on alprenolol-agarose and elution is with alprenolol;[108] a 12 000-fold purification can be achieved. The high purification factor is an indication of the small proportion of the total protein in the extract it represents.

The second component of the hormone-sensitive adenylate cyclase system is the G protein (**G**uanine nucleotide binding protein) or transducer. G proteins transduce the signal from the receptor to the adenylate cyclase (Fig. 8.17). All G proteins comprise three subunits, α, β, and γ. There are four subfamilies of α subunits, five of β, and at least six of γ.[109,110] Different α subunits exist that may act positively ($α_s$, stimulating) on the effector protein (adenylate cyclase) or negatively ($α_i$, inhibitory). In the resting state a G protein has GDP bound to the α subunit. On interacting with the hormone receptor complex, the α subunit exchanges GDP for GTP. This brings about a conformational change in the α

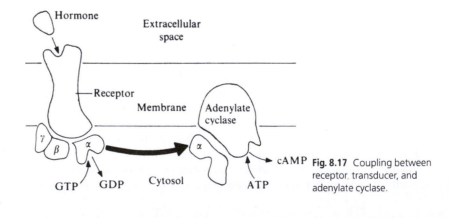

Fig. 8.17 Coupling between receptor, transducer, and adenylate cyclase.

subunit, such that it dissociates from the $\beta\gamma$ dimer. Detached α_s subunits are free to interact with adenylate cyclase causing activation, and detached α_i subunits cause inhibition. The α subunit has GTPase activity, and once the bound GTP is hydrolysed to GDP, the α subunit reverts to its original conformation and re-associates with the $\beta\gamma$ dimer. Of the three polypeptide chains, the α chain differs most among members of the family, whereas the β and γ chains can be exchanged within a family. Two toxins have been useful in studying the mode of action of the G proteins, namely cholera toxin and pertussis toxin. Cholera toxin is able to catalyse the ADP-ribosylation of one of the arginine residues of the α_s subunit, and pertussis toxin catalyses ADP-ribosylation of a cysteine residue four residues from the carboxyl terminus of the α_i subunit.[96]

$$\alpha + NAD^+ \rightarrow ADP\text{-ribosyl-}\alpha + nicotinamide + H^+.$$

In the case of cholera toxin it is the α_s that is modified and this leads to inhibition of GTP hydrolysis and thus to prolonged activation of adenylate cyclase. Pertussis toxin causes the modification of α_i thereby preventing it from exchanging GDP for GTP, and this in turn causes inhibition of hormone-mediated activation of adenylate cyclase. The G proteins are only loosely tethered to the inner face of the plasma membrane mainly by prenylation of the carboxyl terminus of the γ subunit and by myristoylation and palmitoylation of the amino terminus of the α subunit (see Section 8.4.1) and thus are more easily extracted than the receptors and adenylate cyclase. X-ray structures have been obtained of the components of G proteins.[110] The α subunit has two domains, a GTP–GDP binding domain, and an α-helical domain that may be involved in binding the receptor.

The third component, adenylate cyclase, has been purified from rabbit myocardial membranes[111] and from bovine brain (see Chapter 2, Section 2.8.6).[112] In both cases there is an initial solubilization and the main purification step is affinity chromatography on forskolin–agarose. Forskolin is a diterpene isolated from the roots of the aromatic herb *Coleus forskohlii* that causes stimulation of adenylate cyclase. For many years progress on the purification of adenylate cyclase had been slow because of the very small amounts present in the membrane and because of its lability. Since the initial purification and partial sequencing it has been possible to isolate cDNA clones and thereby deduce a complete sequence from which it is possible to deduce a topology within the membrane (Fig. 8.17). It has two hexa-helical transmembrane segments, with two large cytoplasmic domains (C_1 and C_2) and a smaller N-terminal domain.[113] The two cytoplasmic domains contain the catalytic site and also the regulatory sites. A crystal structure has been solved for the complex comprising the catalytic domains of adenylate cyclase together with the α_s and GTPγS.[114] GTPγS is a non-hydrolysable analogue of GTP in which the oxygen linking the β and γ phosphates is replaced by sulphur. The mechanism of activation of adenylate cyclase by α_s binding is thought to involve both C_1 and C_2 domains resulting in an allosteric change.[114]

Forskolin

8.5 The concentrations of enzymes and substrates *in vivo*

When cells or tissues are disrupted and extracts are prepared prior to assay, the suspension containing the enzymes is considerably diluted compared with that existing in the intact cell. This undoubtedly affects many aspects of an enzyme-catalysed reaction. Although many biochemists are aware of this problem when comparing measurements made *in vitro* and *in vivo*, there is little quantitative information that might enable a better extrapolation from *in vitro* to *in vivo* conditions. In this section we highlight some of these differences and discuss what progress has been made in understanding them.

8.5.1 The high concentrations of macromolecules within cells

The protein concentration in all of the intracellular compartments is high. In the mitochondrial matrix the concentration is estimated to be about 500 mg cm^{-3},[18] and in the cytosol concentrations may be in the region of 200 mg cm^{-3}. This has a number of effects on the properties of the medium in which enzymes act. The medium is much more viscous: a solution of albumin at a concentration of 200 mg cm^{-3} has 2–3 times the viscosity of water and one having a concentration of 500 mg cm^{-3} has 5–10 times the viscosity of water.[115] The viscosity of the cytoplasm has been estimated to be about 3.7 times that of water.[116] This will affect the rate of diffusion of solutes, particularly those of macromolecules (see Section 8.5.2). A protein concentration of 500 mg cm^{-3} means that individual protein molecules will be in close proximity to one another and this will tend to promote interactions between protein molecules that would not occur in dilute solutions. A good example of this is given in Section 8.5.3.3. It will also have a profound effect on the state of water molecules in the cell. The water molecules in close proximity to protein molecules differ in some of their properties from those of pure solvent. Apart from a lower freezing point, it can be shown by various physical methods, e.g. NMR spectroscopy, IR spectroscopy,[117] that these water molecules do not rotate as freely as those in pure water. These differences are due to various types of secondary bonds that bind water molecules to protein either directly or indirectly. The water molecules in solutions of protein have been classified into three types[118] based on the differences in their physical properties, as summarized in Table 8.10.

The actual contents of type II and III water will depend on the particular protein, since a higher proportion of polar amino-acid residues will lead to a higher proportion of bound water. The number of molecules of internal water (type III) for various proteins that have been studied by X-ray crystallography[116,120] is as follows: papain, 15; actinidin (a related protease), 15; lysozyme, 4; carboxypeptidase, 24; cytochrome *c*, 3; penicillopepsin, 13; and haemoglobin, 4. A protein the size of haemoglobin (M_r 67 000) has about 1000 molecules of type II water covering its surface.[116] In the case of ribonuclease T1 there are 30 water

Table 8.10 Physical properties of water in protein solutions

Type		Rotational correlation time[a]	Average number of grams of water per gram of protein
I	as in pure water	3×10^{-12} s	
II	loosely bound to protein; freezing point lowered	approx. 10^{-9} s	0.3–0.5
III	tightly bound, motion same as protein molecule, integral part of protein structure	10^{-5}–10^{-7} s	0.003

[a]Rotational correlation time can be considered as the time taken for a molecule to rotate through an angle of 360°. For a more detailed explanation see reference 119.
For reviews on water in biological systems, see references 118 and 120.

molecules tightly bound and their precise location is known. Included in these are a chain of 10 water molecules and two separate water molecules associated with the active site. Their function is probably in the spatial requirement for maintaining catalysis.[121] In the case of lysozyme, where four molecules of internal water are tightly bound, it has been shown that when lysozyme is gradually hydrated, enzyme activity is just detectable when there are between 150 and 250 water molecules per lysozyme. This corresponds approximately to the sum of the type II and type III water molecules. If the mitochondrial proteins behave in the same way as detailed in Table 8.10, then in the mitochondrial matrix with a concentration of 500 mg cm^{-3} protein, 25 per cent of water would be type II and 1.5 per cent type III. Other estimates put the amount of free water (type I) around 50 per cent of the total. Thus, a high proportion of the water in the mitochondrial matrix will have properties different from those of pure water. The same will apply to other subcellular compartments, though in some cases to a lesser degree. Although the nucleus has a slightly lower protein content (approximately 34 g protein per 100 g nuclei)[13] than the mitochondrial matrix, it has a much higher content of nucleic acids, particularly DNA (9 g DNA per 100 g nuclei).[13] Isolated DNA has been shown to have about 20 tightly bound water molecules per nucleotide residue (approximately 11 g water per g nucleotide).[120] Tightly bound water molecules are impermeable to ions and do not form ice-like structures on freezing. It is unlikely that DNA in the nucleus will bind this high proportion of water, since much of its surface will be covered with nucleoprotein. Nevertheless, it would appear that much of the water in the nucleus will be bound to macromolecules.

8.5.2 Diffusion rates within cells

When considering the flux through a metabolic pathway *in vivo*, it is important to know not only the activities of all the enzymes involved, in order to determine the rate-limiting steps, but also the 'transit times' of each step. The 'transit time' is the time required for the product of one enzyme-catalysed reaction to diffuse to the active site of the next enzyme. A number of approximate estimates have been made.[122-124] The two principal parameters required for these calculations are (i) the distance separating two sequential enzymes in a metabolic pathway and (ii) the diffusion coefficient of the product of the first enzyme in the medium of which the cell is composed. The average distance separating two enzymes may be estimated if the number of molecules of each enzyme per cell is known, together with the volume of the relevant subcellular compartment. If an enzyme and substrate are uniformly dispersed at concentrations of 1 μmol dm^{-3}, then the mean distance of travel from one enzyme to the next would be about 100 nm.[116] The diffusion coefficient of the enzyme, substrates, and products can easily be measured in aqueous solution, but it is clear that the diffusion coefficients are significantly lower in the cytosol compared with those in pure water. Various methods have been used to estimate the diffusion coefficients in the cytosol.[31,116] For example, a solution containing 3.7 per cent actin and 12.4 per cent globular protein has been used to simulate cytoplasm.[125] The ratio D_{water}/D_{cells} varies from 2 to 5 for molecules of M_r = 200–400 to 200-fold for protein of M_r = 40 000–150 000.

From the calculations[122,123] made so far it seems that the 'transit time' for the steps in certain metabolic pathways could be rate-limiting, particularly in the larger cell compartments, unless the enzymes are not randomly distributed but are organized. For two enzymes uniformly dispersed at 1 μmol dm^{-3} concentration, the transit time would be of the order of 10^{-5} s. It will depend on the catalytic-centre activities whether diffusion is likely to be rate-limiting. These activities vary over a wide range (1–10^7 s^{-1}); thus for a catalytic-centre activity in the middle of this range, $\approx 10^3$ s^{-1}, only about 1 per cent of the metabolite's lifetime would be spent in diffusion, and the rest of the time bound to a catalytic centre. It seems likely that the diffusion time is seldom rate-limiting in the small prokaryote cells. In eukaryote cells diffusion is more likely to be rate-limiting, because of their larger size, and the barriers within the cytoplasm. Diffusion is also more likely to be rate-limiting either when the metabolite concentration is low, since the rate is directly related to concentration, or when the catalytic-centre activity is high. For example, carbonic anhydrase has very high catalytic activity, and the rate of carbon dioxide diffusion to the catalytic centre will be rate-limiting.[125]

The calculated 'transit time' for fructose 6-phosphate to diffuse to 6-phospho-fructokinase in the cytosol is too long in relation to the known rate at which glycolysis can occur. This suggests that the enzymes may not be randomly arranged in the cytosol but that some kind of ordering exists. There is also kinetic evidence that suggests that an enzyme (aspartate aminotransferase) and its substrate (aspartate) may not be uniformly distributed throughout the mitochondrial matrix, but that the concentration of aspartate may be highest in the centre and the aspartate aminotransferase highest at the periphery of the matrix.[126]

8.5.3 Enzyme associations *in vivo*

It seems probable that in living cells many weak interactions occur that are not easily detected once a cell has been disrupted and the components suspended in buffer solution. In the equilibrium

$$\text{Protein}_A + \text{protein}_B \rightleftharpoons \text{protein}_A \cdot \text{protein}_B$$

dilution will favour the dissociation of the protein$_A \cdot$ protein$_B$ complex, and in many cases when cell disruption is carried out prior to subcellular fractionation, the concentrations of proteins may be lowered by at least one order of magnitude. Thus, the multienzyme complexes discussed in Chapter 7 probably represent only the stronger protein interactions. In recent years attempts have been made to try to demonstrate other weaker interactions. Most attention has focused on the two major metabolic pathways, namely the glycolytic pathway[127] occurring in the cytosol, and the tricarboxylic-acid cycle[128] occurring in the mitochondrial matrix. The enzymes catalysing the steps in both these pathways are usually present in cells in high concentrations, because of their central importance in metabolism. Skeletal muscle is a particularly rich source of glycolytic enzymes since most of the ATP generation in white muscle is through glycolysis. It has been shown that a number of glycolytic enzymes bind to fibrous-actin,[129] one of the major proteins concerned in muscle contraction. The capacity of actin to bind glycolytic enzymes varies with the enzyme; the capacities being greatest with 6-phosphofructokinase, pyruvate kinase, fructose-bisphosphate aldolase, glyceraldehyde-3-phosphate dehydrogenase, glucose-6-phosphate isomerase, and lactate dehydrogenase. The kinetic properties of some of the enzymes change on binding and this, together with the fact that binding to actin is reversible, could be an additional means of regulation, but further work is required before the relevance of these observations to the situation in intact muscles can be assessed.

8.5.3.1 Glycolytic enzymes

Several attempts have also been made to demonstrate associations between individual enzymes of the glycolytic pathway. We have discussed in Chapter 7, Section 7.13, the evidence for the existence of a 'glycogen particle' that contains a number of glycolytic enzymes. Fructose-bisphosphate aldolase and glyceralde-hyde-3-phosphate dehydrogenase catalyse sequential steps in glycolysis:

Fructose-bisphosphate $\rightleftharpoons$ dihydroxyacetone phosphate + glyceraldehyde
3-phosphate.

Glyceraldehyde 3-phosphate + NAD$^+$ + orthophosphate $\rightleftharpoons$
3-phospho-D-glyceroyl phosphate + NADH.

(Triosephosphate isomerase converts dihydroxyacetone phosphate to glyceral-dehyde 3-phosphate.)

Direct methods to demonstrate association between the two enzymes, e.g. by ultracentrifugation or gel filtration, have yielded negative results; however, there is indirect evidence for some form of association.[130] When glyceraldehyde 3-phos-phate is present in aqueous solution, there is normally an equilibrium between the aldehyde and the diol forms (see opposite) (see also Chapter 5, Section 5.5.2).

If the results of kinetic studies on glyceraldehyde-3-phosphate dehydrogenase alone are compared with those from fructose-bisphosphate aldolase coupled with glyceraldehyde-3-phosphate dehydrogenase, the rate of the coupled reaction is such as to suggest that the aldehyde form of glyceraldehyde 3-phosphate is used directly by the dehydrogenase and does not equilibrate with the diol. This in turn suggests a close association of the enzymes and is supported by measurements of polarization of fluorescence.

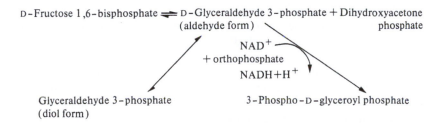

If fluorescein is coupled covalently to either fructose-bisphosphate aldolase or glyceraldehyde-3-phosphate dehydrogenase and the fluorescein-labelled enzyme mixed with varying amounts of the unlabelled form of the second enzyme, a change in fluorescence is observed, suggesting an interaction between the enzymes; the binding constant may be determined from the titration curve. This interaction between glyceraldehyde-3-phosphate dehydrogenase and fructose-bisphosphate aldolase has also been detected by active-enzyme centrifugation[131] (see Chapter 3, Section 3.2.5) and has a dissociation constant in the micromolar range.[132]

There is evidence for the association of a number of glycolytic enzymes with actin and other cytoskeletal proteins, both *in vitro* and *in vivo*.[127] When muscle homogenates containing F-actin are centrifuged, many of the glycolytic enzymes sediment with the actin. The association with actin has also been demonstrated using antigenic probes against aldolase, phosphofructokinase, and glyceralde-hyde-3-phosphate dehydrogenase. This association of glycolytic enzymes with actin is not restricted to muscle cells, but has also been demonstrated in cultured fibroblasts. In red blood cells glycolytic enzymes in the cell membrane are associ-ated with band 3 membrane protein. The effect of association is not only to improve catalytic efficiency but it may also affect the allosteric properties of some

* The concentration of fructose
6-phosphate to give 50% maximum
velocity.

of the enzymes. The active tetrameric form of 6-phosphofructokinase is stabilized by binding to F-actin. This binding virtually abolishes the allosteric regulation by fructose 1,6-bisphosphate and increases the K_i for ATP and decreases the $K_{0.5}$* for fructose 6-phosphate.[127,133]

In a number of mammalian tissues, hexokinase is distributed between the mitochondria and the cytosol. When bound to the mitochondrial membrane its properties are modified, and this may act as a form of control mechanism. In one of the most-studied tissues, the brain, up to 80 per cent of the hexokinase may be bound to the mitochondria and much of this may be released *in vitro* by increasing concentrations of glucose 6-phosphate. Glucose 6-phosphate inhibits the free enzyme more effectively than the bound form. The two processes therefore act synergistically. A higher proportion of hexokinase binds to mitochondria when ATP concentrations are diminished and this may enable hexokinase to compete more favourably with other ATP-requiring processes. There is also evidence that ATP released from the mitochondria is used by the hexokinase in preference to exogenous ATP. This in turn suggests that the initiation of the glycolytic pathway is directly coupled to ATP synthesis in mitochondria.[127,134]

8.5.3.2 The tricarboxylic-acid cycle enzymes and related metabolic pathways

The enzymes catalysing the tricarboxylic-acid cycle, together with aspartate aminotransferase and glutamate dehydrogenase, occur in the matrix of the mitochondria. Apart from succinate dehydrogenase that is covalently linked to inner mitochondrial membrane, all other enzymes are readily released when the inner mitochondrial membrane is disrupted. A number of investigations have been carried out to see whether there is any evidence for associations between the enzymes involved. All of the dehydrogenases which form part of the tricarboxylic-acid cycle have to interact with the electron transport system which is an integral part of the inner mitochondrial membrane. Also when the concentration of each of the enzymes (> 1 mmol dm^{-3})[134] is taken into account together with their size, they are estimated to be able to occupy a surface area approximately equal to that of the inner mitochondrial membrane.[127]

Weak interactions occur between a number of tricarboxylic-acid cycle enzymes, and also with enzymes that feed into or have substrates in common with those of the tricarboxylic-acid cycle.[135] The type of evidence for these interactions is as follows: (i) detection of complexes by M_r determination through light scattering or gel filtration, (ii) cosedimentation of enzyme pairs, (iii) the change in fluorescence when a pair of enzymes interacts, one of which has a fluorescent probe attached, (iv) coprecipitation of enzymes using poly(ethylene glycol), and (v) cross-linking of enzyme pairs.[135] Some examples of these interactions are described next.[127,136,137] During the purification of citrate (*si*) synthase it is difficult to remove the last traces of malate dehydrogenase activity, suggesting a type of interaction or similarity in properties. In the mitochondrial matrix, because of the very high protein concentration, there is a limitation on free water and this could promote interaction between the two enzymes. In order to bring about water limitation *in vitro*, experiments have been carried out in which some of the water has been replaced by poly(ethylene glycol). It has been demonstrated that under these conditions citrate (*si*) synthase will coprecipitate with mitochondrial malate dehydrogenase but not with cytosolic malate dehydrogenase.

In a similar set of experiments, the partitioning of malate dehydrogenase and aspartate aminotransferase in a biphasic system of water and dextran–trimethylaminopropyl poly(ethylene glycol) was studied. If either malate dehydrogenase and aspartate aminotransferase (both mitochondrial) or malate dehydrogenase and aspartate aminotransferase (both cytosolic) were used, the enzymes in each pair appeared in the same fraction after counter current distribution. If, on the other hand, the pair of enzymes were not from the same subcellular fraction, e.g. mitochondrial malate dehydrogenase and cytosolic aspartate aminotransferase, then they appeared in different fractions, suggesting no interaction. Thus, the interaction between the enzymes seems to be specific. Similar specificity is found in the binding of tricarboxylic-acid enzymes to the inner mitochondrial membrane. Mitochondrial malate dehydrogenase and mitochondrial aconitase bind to the inner mitochondrial membrane whereas cytosolic malate dehydrogenase, cytosolic aconitase, and cytosolic (NADP[+]) isocitrate dehydrogenase do not.[127] Further evidence for an association between these enzymes and other enzymes of the tricarboxylic-acid cycle comes from studies on immobilized enzymes.[138] When fumarate hydratase and mitochondrial malate dehydrogenase were immobilized on a gel filtration column, it was possible to demonstrate that the immobilized enzymes were able to bind enzymes catalysing related steps in metabolism, e.g. immobilized fumarate hydratase binds malate dehydrogenase and citrate synthase whereas immobilized malate dehydrogenase binds citrate synthase and immobilized aspartate aminotransferase binds malate dehydrogenase. Another method used to study the effects of enzyme immobilization is to fuse genetically enzymes catalysing sequential steps in the tricarboxylic-acid cycle. This has been done using yeast mitochondrial citrate synthase, malate dehydrogenase, and aconitase.[139] These fused enzymes have been used to show the kinetic advantages of having active sites in close proximity, and molecular models reveal the presence of electrostatically favourable channels that link aconitase with citrate synthase, and citrate synthase with malate dehydrogenase.

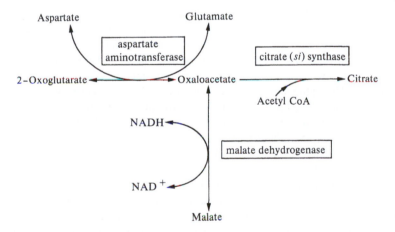

Although associations have been demonstrated only between pairs of tricarboxylic-acid cycle enzymes in mitochondria, there is evidence[140] for a multienzyme aggregate in bacteria containing several of the tricarboxylic-acid cycle enzymes. By preparing spheroplasts and then lysing them in a suitable medium, it was possible to demonstrate by gel filtration that a multienzyme aggregate existed that contained citrate synthase, aconitase, isocitrate dehydrogenase, succinate CoA ligase, and malate dehydrogenase, and was thus able to catalyse the steps from fumarate to oxoglutarate. It is possible that it may also contain oxoglutarate dehydrogenase. Several of these dehydrogenases, e.g. pyruvate dehydrogenase, oxoglutarate dehydrogenase, malate dehydrogenase, and hydroxyacyl CoA dehy-drogenase, are able to bind to complex I, which catalyses the electron-transport chain from NAD^+ to coenzyme Q.[141] Thus there is evidence for a physical link between the tricarboxylic-acid cycle and the electron-transport chain.

Carbamoyl phosphate synthase I is one of the most abundant liver mitochondr-ial proteins,[142] being present in concentrations between 0.4 and 1.0 mmol dm^{-3}. In addition to playing a key role in the urea cycle, it appears to act as a linking enzyme by binding glutamate dehydrogenase and also enhancing the stability of the complex between glutamate dehydrogenase and aspartate aminotransferase. The association between glutamate dehydrogenase and carbamoyl phosphate synthase I has been demonstrated by cross-linking the two in the mitochondrial matrix.[142] Aspartate amonotransferase has been shown to bind carbamyl phosphate syn-thase I, glutamate dehydrogenase, pyruvate carboxylase, and oxoglutarate de-hydrogenase with K_ds of 0.5, 0.1, 0.01, and 0.4 μmol dm^{-3}, respectively.[135] At the concentration at which these enzymes are present in the mitochondrial matrix, these interactions would be stable. Aspartate aminotransferase is also able to bind to the inner mitochondrial membrane.

These experiments illustrate the point that many weak interactions may occur *in vivo* but that special methods may have to be used to detect them.

8.5.3.3 Arginase and ornithine carbamoyltransferase

Another interaction between two enzymes that is not easily demonstrated in dilute solutions is that between arginase and ornithine carbamoyltransferase from yeast.[143,144] The function of ornithine carbamoyltransferase is in the biosyn-thesis of arginine from ornithine (see Fig. 8.10). When arginine is added to the growth medium it acts as a good source of both carbon and nitrogen for the syn-thesis of many compounds, e.g. other amino acids, proteins, and carbohydrates, and also is a preferred source of arginine compared to that synthesized endo-genously. Two mechanisms operate to suppress ornithine carbamoyltransferase activity in this situation. One is the suppression of synthesis of ornithine car-bamoyltransferase and is relatively slow acting. The second mechanism, which is fast acting, involves interaction between ornithine carbamoyltransferase and arginase. This latter mechanism is demonstrated in concentrated suspensions by comparing the activity of ornithine carbamoyltransferase in cell extracts (Fig. 8.18, curve A) with that in permeabilized cells (curve B). In the latter, yeast cells are made permeable to low-M_r substances by use of nystatin (a polyene fungicide produced by *Streptomyces noursei*), but the proteins remain within the cells and thus the protein concentration is unchanged and therefore high.

In both situations A and B in Fig. 8.18 the synthesis of ornithine carbamoyl-transferase is suppressed after arginine is added to the medium, but additionally in B, ornithine carbamoyltransferase becomes bound to arginase in the presence of ornithine and arginine and the ornithine carbamoyltransferase activity is inhibited.

Both arginase and ornithine carbamoyltransferase are trimeric, and the complex formed between them has been shown to have a stoichiometry $\alpha_3\beta_3$. The dissociation constant for the complex, $\alpha_3\beta_3$, in the presence of the substrates ornithine and arginine is 2.3×10^{-8} mol dm^{-3}.[144]

$$\alpha_3\beta_3 = \alpha_3 + \beta_3$$

Ornithine and arginine promote the association of the complex, and it was orig-inally proposed[143] that ornithine bound to the catalytic site on ornithine carba-

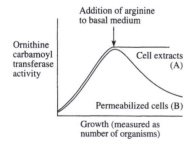

Ornithine carbamoyl transferase activity

Addition of arginine to basal medium

Cell extracts (A)

Permeabilized cells (B)

Growth (measured as number of organisms)

Fig. 8.18 The effects on ornithine carbamoyltransferase activity in yeast of adding arginine to the growth medium.[143]

moyltransferase and also to a regulatory site. However, using a bisubstrate analogue (transition-state analogue), δ-N-(phosphonacetyl)-L-ornithine,[145] it has been shown that a conformational change can be demonstrated on binding the catalytic site but there is no evidence for a separate regulatory site.[146] It seems probable that binding to the active sites of both enzymes induces the necessary conformational changes for their association.

8.5.4 The relative concentrations of enzymes and substrates

The constants K_m and V_{max} (see Chapter 4, Section 4.3), determined by steady-state kinetic studies *in vitro*, are frequently used to estimate the rates at which reactions might occur *in vivo;* e.g. the functions of the various isoenzymes of hexokinase have been deduced from such kinetic data (see Chapter 4, Section 4.3.1.3). There can, however, be a major problem in such extrapolations, namely that in steady-state kinetics it is assumed that $[S] \gg [E]$; this may not always be the case *in vivo.* Table 8.11 shows the intracellular concentrations of some of the glycolytic and tricarboxylic-acid cycle enzymes and their substrates. Although there are often quite wide variations in the calculated concentrations (compare Table 8.11 with, for example, reference 134), it can be seen that for many of these the enzyme concentration is of the same order of magnitude as that of the substrate concentration.

Glycolytic and tricarboxylic-acid cycle enzymes are of major importance in liver and muscle and are therefore present in higher concentrations than the majority of other enzymes in these tissues, but other enzymes are known to be present in high concentrations, e.g. 0.78 mmol dm^{-3} cathepsin D, and 1.54 mmol dm^{-3} cathepsin B in the lysosomal matrix.[153]

Table 8.11 Intracellular concentrations of enzymes and substrates in rat tissues

	Enzyme	Enzyme concentration (μmol dm^{-3})	Substrate concentrations (μmol dm^{-3})	References
1.	Phosphoglucomutase	5	G6P = 450	147, 148
2.	Fructose-bisphosphate aldolase (muscle)	66	FBP = 65, DHAP = 28, G3P = 3	147, 149
3.	Glyceraldehyde-3-phosphate dehydrogenase (muscle)	35	G3P = 3, NAD+ = 600, P_i = 2800	147, 148, 149
4.	Triosephosphate isomerase (muscle)	8	DHAP = 28, G3P = 3	147, 149
5.	Pyruvate carboxylase (liver mitochondria)	10–40	Pyruvate = 50–1000 ATP = 1000–6000	147, 150, 151
6.	Phosphoenolpyruvate carboxykinase (liver cytosol)	15	Oxaloacetate = 5, GTP = 100–600	45, 147
7.	Citrate synthase (liver mitochondria)	35	Oxaloacetate = 0.01–0.1 Acetyl-CoA = 50–1000	147, 150, 151
8.	Malate dehydrogenase (liver mitochondria)	70	Malate = 80–340 NAD+ = 18 000	147, 152
9.	Malate dehydrogenase (liver cytosol)	30	Malate = 300–450 NAD+ = 500	147, 152
10.	Aspartate aminotransferase (liver mitochondria)	50	Glutamate = 380–4800 Oxaloacetate = 0.01–0.1	147, 152
11.	Aspartate aminotransferase (liver cytosol)	1.6	Glutamate = 2500–4400 Oxaloacetate = 5	147, 152

G6P = glucose 6-phosphate; FBP = fructose-bisphosphate; DHAP = dihydroxyacetone phosphate; G3P = glyceraldehyde 3-phosphate; and P_i = orthophosphate.

One of the assumptions in steady-state kinetics is that, since [S] ≫ [E], the total substrate concentration approximates to the free substrate concentration. Clearly this will not apply for enzymes present in high concentrations as an appreciable amount of the substrate will be enzyme-bound. Various equations have been derived for the rate of an enzyme reaction when the substrate is not in great excess.[147,154–156] Perhaps the most important factor is that knowledge of free substrate concentration *in vivo* is required and this cannot easily be determined.

It is not only in the case of enzymes that are present in high concentrations that it is unsafe to assume that the total substrate and free substrate approximate to the same value. An enzyme present in small amounts in a tissue may be in competition with an enzyme present in high concentrations for a common substrate that may be largely bound to the latter enzyme. Thus, the free substrate concentration may, in several cases, be appreciably lower than the total substrate concentration. This situation is well illustrated in the case of oxaloacetate in mitochondria. A number of reactions compete for oxaloacetate and the enzymes concerned are present in high concentrations (compared with substrate); thus a very high proportion of the total oxaloacetate in mitochondria must be enzyme-bound (Fig. 8.19). The kinetic properties (K_m and V_{max}) of citrate (*si*) synthase are such that one would expect any oxaloacetate formed to be converted to citrate, and not to phosphoenolpyruvate for gluconeogenesis; clearly this does not always happen. From the $K_{m, oxaloacetate}$ (≈ 5×10^{-6} mol dm^{-3}) and V_{max} of citrate synthase, together with the free oxaloacetate concentration in the mitochondrial matrix (≈ 10^{-9} mol dm^{-3}), it is possible to calculate the expected reaction rate for citrate synthase. This is at least one order of magnitude lower than the estimated rate of the tricarboxylic-acid cycle in mitochondria.[128] It has been suggested that there must be some type of macromolecular organization (multi-enzyme complex) that prevents the competition between the enzymes for oxaloacetate, and that oxaloacetate may be directly channelled from malate dehydrogenase to citrate synthase.

However, this overlooks another aspect of the difference between experiments carried out *in vitro* and the situation *in vivo*, namely that the latter behaves more like an open system rather than the closed system used *in vitro*. Although the intermediates of most metabolic pathways in cells are present in low concentrations, they are continuously being replaced by new source material and the end-products are often removed from the cell. Thus, although the intracellular concentration of oxaloacetate is very low and undoubtedly much of it is enzyme-bound, changes in the amounts of the enzymes utilizing oxaloacetate do not appreciably affect the intracellular concentration of free oxaloacetate in the 'near-open' system existing in the cell. Mathematical modelling with the aid of

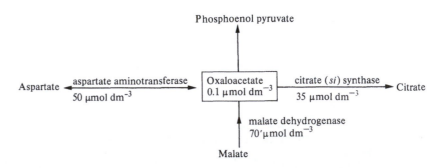

Fig. 8.19 The concentration of enzymes metabolizing oxaloacetate in mitochondria.

computers[157] has been made to determine the effect of a sharp increase in the amount of a protein that binds oxaloacetate (i.e. comparable with an oxaloacetate-metabolizing enzyme). This is found to cause only a transient fall in the intracellular concentration of free oxaloacetate. The reason is that oxaloacetate is rapidly replenished because it is linked metabolically to the other tricarboxylic-acid cycle intermediates that are present in much higher concentrations. In particular, the relatively large malate pool becomes slightly diminished to replenish oxaloacetate. Mathematical modelling techniques for studying the control of metabolic activity (see Chapter 6, Section 6.3.3.2) are still being developed and they require accurate kinetic data and knowledge of pool sizes to give reliable simulations, but they will play an increasing role in our understanding of metabolism in the whole cell, where the complexity of the situation makes computer simulations essential.

Another example in which the proportion of bound ligand is important *in vivo* is that of cAMP-dependent protein kinase.[158] This enzyme is present in muscle and when activated by a sufficiently high concentration of cAMP phosphorylates a number of proteins, amongst them phosphorylase kinase (as shown below).

$$R_2C_2 \text{ (inactive)} + 2cAMP \rightleftharpoons R_2(cAMP)_2 + 2C \text{ (active)}$$

R_2C_2 = inactive cAMP-dependent protein kinase (R = regulatory, C = catalytic)

$$\text{ATP} + \text{protein} \xrightleftharpoons[]{\text{active protein kinase}} \text{ADP} + \text{phosphoprotein}$$

(e.g. inactive   (e.g. active
   phosphorylase kinase)   phosphorylase kinase)

The role of phosphorylase kinase in the control of glycogen metabolism is discussed in Chapter 6 (Section 6.4.2.1). The kinetics of activation of cAMP-dependent protein kinase by cAMP are complex,[159] but the concentration of cAMP that gives 50 per cent activation (apparent K_a) of the enzyme is reported to be in the range 10^{-8} to 10^{-7} mol dm^{-3}. The cAMP content of most tissues is in the range 0.1 to 1.0 μmol kg^{-1} wet weight (i.e. if equally distributed throughout the cell would give 10^{-7} to 10^{-6} mol dm^{-3}) and this would appear to be a sufficient concentration to activate substantially the phosphorylase kinase. However, it is known that in tissues such as resting muscle the rate of glycogenolysis is very low. This led initially to the suggestion that cAMP must be compartmentalized within the muscle so that the concentration to which the protein kinase is exposed is much lower. It is now clear that this explanation is unnecessary. The comparison between the *in vitro* experiment and the *in vivo* situation is false because the concentration of protein kinase in muscle is high (about 0.23 μmol dm^{-3})— approximately the same as that of cAMP. The apparent K_a will be dependent on the concentration of protein kinase, since high concentrations of the latter will favour association of the R and C subunits. Thus, at physiological concentrations of the protein kinase the apparent K_a is 1.5×10^{-6} mol dm^{-3}, i.e. higher concentrations of cAMP are required to effect activation. In addition, an inhibitor has been isolated from skeletal muscle that binds to the C subunit and inactivates it. Both these factors clearly have to be taken into account in explaining the activation *in vivo*.

Another factor to be considered, which arises from the high concentrations of certain enzymes in tissues, is the equilibrium position. When the enzymes in a reaction are present in only catalytic amounts, the equilibrium position for the reaction is unchanged, although in practice it may be difficult to study the equilibrium in the absence of the appropriate enzyme:

$$A + B \rightleftharpoons C + D$$

$$K_{eq} = \frac{[C]\,[D]}{[A]\,[B]}.$$

However, if the enzyme is present in concentrations similar to that of the substrates, so that appreciable amounts of the substrates and products are enzyme-bound, then the equilibrium may be that between the enzyme–substrate complex and the enzyme–product complex. If the enzyme-catalysed reaction proceeds, for example, by ternary complexes (Chapter 4, Section 4.3.5.1), the relevant equilibrium is

$$EAB \rightleftharpoons ECD$$

$$K'_{eq} = \frac{[ECD]}{[EAB]}.$$

A number of examples are given in Table 8.12. It can be seen that most of the values for K_{eq} for the bound enzyme are near unity, and that this makes a very significant difference in a number of the examples cited. Thermodynamic arguments have been proposed (see reference 134) to account for the values of K_{eq} near unity. These suggest that for the evolution of ideal catalysts the free-energy change should be near zero ($K_{eq} \approx 1$) (compare triose phosphate isomerase, Fig. 5.29), since this would give balanced rates of diffusion, chemical transformation, and dissociation steps in a metabolic sequence.

8.6 Conclusions

It can be seen from the preceding section that it has been possible to identify a number of factors that affect enzyme reactions and metabolic pathways *in vivo*, but that are not always apparent in many of the better defined *in vitro* situations. Although many factors have been identified, in many cases it is not possible to be

Table 8.12 Changes in equilibrium by high concentrations of enzymes

Enzyme	$K^{\dagger}_{eq}$	K'_{eq} enzyme bound
1. Lactate dehydrogenase	0.37×10^{-4}	0.25
2. Alcohol dehydrogenase	0.5×10^{-4}	5–10
3. Pyruvate kinase	3×10^{-4}	1.0–2.0
4. Arginine kinase	0.1	1.2
5. Creatine kinase	0.1	≈ 1.0
6. Adenylate kinase	0.4	1.6
7. Phosphoglycerate kinase	8×10^{-4}	0.5–1.5
8. Hexokinase	2000	≈ 1.0
9. Adenosine triphosphatase	1.3×10^{5}	9
10. Triose phosphate isomerase	2.2	0.6
11. Phosphoglucomutase	1.7	0.4

References 141, 160, 161. † Equilibrium constant when catalytic amounts of enzymes are present, in contrast to when enzyme and substrate are present in comparable amounts.

sure of the magnitude of these effects *in vivo*. In some cases this is because research interest in this area has only recently developed, in others it is a question of devising satisfactory techniques, for example there is still much controversy about the intercellular concentrations and distribution of certain metabolites because none of the methods used is ideal. Even for a metabolic pathway, such as glycolysis that has been actively studied since the early 1900s and in which the three-dimensional structures of many of the enzymes are known, there is still much to be learned about its organization in the cytosol and about its control *in vivo*.

References

1. Fulton, A. B., *Cell* **30**, 345 (1982).
2. Srivastava, D. K. and Bernhard, S. A., *Science* **234**, 1081 (1986).
3. De Robertis, E. D. P. and De Robertis, E. M. F., *Cell and molecular biology* (7th edn). Saunders, Philadelphia (1980). See p. 16.
4. Graham, J., in *Centrifugation* (2nd edn) (Rickwood, D., ed.), p. 161. IRL Press, Oxford (1987).
5. Dallner, G., *Methods Enzymol.* **31A**, 191 (1974).
6. Evans, W. H., in *Biological membranes* (Findlay, J. B. C. and Evans, W. H., eds), p. 1. IRL Press, Oxford (1987).
7. Shaw, P. J., *Essays Biochem.* **31**, 77 (1996).
8. Zbarsky, I. B., *Int. Rev. Cytol.* **54**, 295 (1978).
9. Richardson, J. C. W. and Agutter, P. S., *Biochem. Soc. Trans.* **8**, 459 (1980).
10. Gerace, L., *Trends Biochem. Sci.* **11**, 443 (1986).
11. Paine, P. L., *Trends Cell Biol.* **3**, 325 (1993).
12. Davis, L. I., *A. Rev. Biochem.* **64**, 865 (1995)
13. Siebert, G., in *Comprehensive biochemistry*, Vol. 23, (Florkin, M. and Stotz, F. H., eds), p. 1. Elsevier, Amsterdam (1968).
14. Dubochet, J., *Trends Cell Biol.* **3**, 1 (1993).
15. Shaw, P. J. and Jordan, E. G., *A. Rev. Cell Devel. Biol.* **11**, 93 (1995).
16. Fambrough, D. M., in *Handbook of cytology* (Lima de Faria, A., ed.), Chapter 18. North-Holland, Amsterdam (1969).
17. Nelson, W. G., Pienta, K. J., Barrack, E. R., and Coffey, D. S., *A. Rev. Biophys. Bioeng.* **15**, 457 (1986).
18. Srere, P. A., *Trends Biochem. Sci.* **5**, 120 (1980).
19. Hackenbrock, C. R., *Trends Biochem. Sci.* **6**, 151 (1981).
20. Tzagoloff, A., *Mitochondria*, Chapter 2. Plenum Press, New York (1982).
21. De Duve, C. and Wattiaux, R., *A. Rev. Physiol.* **28**, 435 (1966).
22. Holtzman, E., *Lysosomes: cellular organelles*. Plenum Press, London (1989).
23. Lloyd, L. B. and Forster, S., *Trends Biochem. Sci.* **11**, 365 (1986).
24. Goldman, R. and Rottenberg, H., *FEBS Lett.* **33**, 233 (1973).
25. Price, N. C. and Dwek, R. A., *Principles and problems in physical chemistry for biochemists* (2nd edn). Oxford University Press (1979). See p. 66.
26. Schneider, Y. J., Octave, J. N., and Trouet, A., *Curr. Top. Membrane Transport* **24**, 413 (1985).
27. Tolbert, N. E., *A. Rev. Biochem.* **50**, 133 (1981).
28. Van den Bosch, H., Schutgens, R. B. H., Wanders, R. J. A., and Tager, J. M., *A. Rev. Biochem.* **61**, 157 (1992).
29. Osmundsen, H., Bartlett, K., Pourfarzam, M., Eaton, S., and Sleboda, J., in *Channelling in intermediary metabolism* (Agius, L. and Sherratt, H. S. A, eds), Chapter 15. Portland Press, London (1997).
30. Erdmann, R., Veenhuis, M., and Kunau, W.-H., *Trends Cell Biol.* **7**, 400 (1997).
31. De Pierre, J. W. and Ernster, L., *A. Rev. Biochem.* **46**, 201 (1977).
32. De Pierre, J. W. and Dallner, G., *Biochim. biophys. Acta* **415**, 411 (1975).
33. Lichtenstein, A. H. and Breecher, P., *J. biol. Chem.* **255**, 9098 (1980).
34. Helenius, A., Marquardt, T., and Braakman, I., *Trends Cell Biol.* **2**, 227 (1992).
35. Burchell, A., Allan, B. B., and Hume, R., *Mol. Membrane Biol.* **11**, 217 (1994).
36. Dahms, N. M., Lobel, P., and Kornfeld, S., *J. biol. Chem.* **264**, 12115 (1989).
37. Clegg, J. S., *Curr. Top. Cell Regul.* **33**, 3 (1992).
38. Mastro, A. M., Babich, M. A., Taylor, W. D., and Keith, A. D., *Proc. natn. Acad. Sci. USA* **81**, 3414 (1984).
39. Aw, T. and Jones, D. P., *Am. J. Physiol.* **249**, C385 (1984).
40. Anderson, N. and Green, J. G., in *Enzyme cytology* (Roodyn, D. B., ed.), Chapter 8. Academic Press, New York (1967).
41. Halliwell, B., *Chloroplast metabolism*. Clarendon Press, Oxford (1984).
42. Farquhar, M. G. and Palade, G. E., *Trends Cell Biol.* **8**, 2 (1998).
43. Munro, S., *Trends Cell Biol.* **8**, 11 (1998).

44. Hanson, R. W. and Patel, Y. M., *Adv. Enzymol.* **69**, 203 (1994).

45. Denton, R. H. and Halestrap, A. P., *Essays Biochem.* **15**, 37 (1979).

46. Lehninger, A. L., *J. biol. Chem.* **190**, 345 (1951).

47. Quant, P. A., *Essays Biochem.* **28**, 13 (1994).

48. Bloch, K. and Vance, D., *A. Rev. Biochem.* **46**, 263 (1977).

49. Jeffcoat, R., *Essays Biochem.* **15**, 1 (1979).

50. Fell, D., *Understanding the control of metabolism*, pp. 30–7. Portland Press, London (1997).

51. Zuurendonk, P. F., Akerboom, T. P. M., and Tager, J. M., in *Use of isolated liver cells and kidney tubules in metabolic studies* (Tager, J. M. and Williamson, J. R., eds), p. 17. North-Holland, Amsterdam (1976).

52. Brindle, K. M., Davies, S. E. C., and Williams, S.-P., *Biochem. Soc. Trans.* **19**, 997 (1991).

53. Evans, J. N. S., in *Biomolecular NMR spectroscopy*, pp. 259–67. Oxford University Press (1995).

54. Gumaa, K. A., McLean, P., and Greenbaum, A. L., *Essays Biochem.* **7**, 39 (1971).

55. Veech, R. L., in *Microenvironments and metabolic compartmentation* (Srere, P. A. and Estabrook, R. W., eds), p. 17. Academic Press, New York (1978).

56. Krebs, H. A. and Veech, R. L., in *Energy levels and metabolic control in mitochondria* (Papa, S., Tager, J. M., Quagliariello, E., and Slater, E. C., eds), p. 329. Adriatica Editrice, Ban, Italy (1969).

57. Klingenberg, M., *Arch. Biochem. Biophys.* **270**, 1 (1989).

58. Ramsey, R. R., *Essays Biochem.* **28**, 47 (1994).

59. McGarry, J. D. and Brown, N. F., *Eur. J. Biochem.* **244**, 1 (1997).

60. Williamson, J. R., in *Gluconeogenesis: its regulation in mammalian species* (Hanson, R. W. and Mehlman, M. A., eds), Chapter 5. Wiley, New York (1976).

61. Davis, R. H., *Microbiol. Rev.* **50**, 280 (1986).

62. Weiss, R. L. and Davis, R. H., *J. biol. Chem.* **248**, 5403 (1973).

63. Vogel, R. H. and Kopac, M. J., *Biochim. biophys. Acta* **37**, 539 (1960).

64. Nazano, M., *Biochim. biophys. Acta* **145**, 146 (1967).

65. Cohen, B. B. and Bishop, J. O., *Genet. Res.* **8**, 243 (1966).

66. Davis, R. H., Weiss, R. L., and Bowman, B. J., in *Microenvironment and metabolic compartmentation* (Srere, P. A. and Estabrook, R. W., eds), p. 197. Academic Press, New York (1978).

67. Bowman, B. J. and Davis, R. H., *J. Bact.* **130**, 274 (1977).

68. Goodman, I. and Weiss, R. L., *J. Bact.* **141**, 227 (1980).

69. Wiemken, A. and Durr, M., *Arch. Microbiol.* **101**, 45 (1974).

70. Westenberg, B., Boller, T., and Wiemken, A., *FEBS Lett.* **254**, 133 (1989).

71. Paek, Y. L. and Weiss, R. L., *J. biol. Chem.* **264**, 7285 (1989).

72. Weiss, R. L., *J. Bact.* **126**, 1173 (1976).

73. Yu, Y. G. and Weiss, R. L. *J. biol. Chem.* **267**, 15491 (1992).

74. Davis, R. H., *Trends Biochem. Sci.* **13**, 101 (1988).

75. Singer, S. J. and Nicolson, G. L., *Science* **175**, 720 (1972).

76. Jennings, M. L., *A. Rev. Biochem.* **58**, 999 (1989).

77. Levy, D., *Essays Biochem.* **31**, 49 (1996).

78. Shin, B. C. and Carraway, K. L., *J. biol. Chem.* **248**, 1436 (1973).

79. Jeffcoat, R., *Essays Biochem.* **15**, 1 (1979).

80. Tomino, S. and Paigen, K., *J. biol. Chem.* **250**, 1146 (1975).

81. Hooper, N. M., Karran, E. H., and Turner, A. J., *Biochem. J.* **321**, 265 (1997).

82. Benajlba, A. and Maroux, S., *Eur. J. Biochem.* **107**, 381 (1980).

83. Sandermann, H. and Strominger, J. L., *Proc. natn. Acad. Sci. USA* **68**, 2441 (1971).

84. Maroux, S. and Louvard, D., *Biochim. biophys. Acta* **419**, 189 (1976).

85. Kenny, A. J. and Booth, A. G., *Essays Biochem.* **14**, 1 (1978). See Section VII, B.

86. Bischoff, F., Tran-Thi, T., and Decker, K. F. A., *Eur. J. Biochem.* **51**, 353 (1975).

87. Erwin, V. G. and Hellerman, L., *J. biol. Chem.* **242**, 4230 (1967).

88. Tzagoloff, A. and Maclennan, D. H., *Biochim. biophys. Acta* **99**, 476 (1965).

89. Branden, C. and Tooze, J., *Introduction to protein structure*, 2nd edn. Chapter 12. Garland Publishing, New York (1999).

90. Sly, W. S. and Hu, P. Y., *A. Rev. Biochem.* **64**, 375 (1995).

91. Turner, A. J., *Essays Biochem.* **28**, 113 (1994).

92. Udenfriend, S. and Kodukula, K., *A. Rev. Biochem.* **64**, 563 (1995).

93. Milligan, G. and Grassie, M. A., *Essays Biochem.* **32**, 49 (1997).

94. Mumby, S. M., *Curr. Opin. Cell Biol.* **9**, 148 (1997).

95. Zhang, F. L. and Casey, P. J., *A. Rev. Biochem.* **65**, 241 (1996).

96. Fields, T. A. and Casey, P. J., *Biochem. J.* **321**, 561 (1997).

97. McLaughlin, S. and Aderien, A., *Trends Biochem. Sci.* **20**, 272 (1995).

98. Sandermann, H., *Biochim. biophys. Acta* **515**, 209 (1978).

99. Milhiet, P.-E., Corbeil, D., Simon, V., Kenny, A. J., Crine, P., and Boileau, G., *Biochem. J.* **300**, 37 (1994).

100. Brunner, J., Wacker, H., and Semenza, G., *Methods Enzymol.* **96**, 386 (1983).

101. Kenny, A. J. and Maroux, S., *Physiol. Rev.* **62**, 91 (1982).

102. Dean, W. L. and Tanford, C., *Biochemistry* **17**, 1683 (1978).

103. Bangham, A. D., Hill, M. V., and Miller, N. G. M., in *Methods in membrane biology*, Vol. 1. (Korn, F., ed.), Chapter 1. Plenum Press, London (1974).

104. MacLennan, D. H., Rice, W. J., and Green, N. M., *J. biol. Chem.* **272**, 28815 (1997).

105. Maclennan, D. H., Branbl, C. J., Korczak, B., and Green, N. M., *Nature, Lond.* **316**, 696 (1985).

106. Strader, C. D., Fong, T. M., Tota, M. R., and Underwood, D., *A. Rev. Biochem.* **63**, 101 (1994).

107. Strosborg, A. D., Vauguelin, G., Trautman, O., Klutchko, C., Bottari, S., and Andre, C., *Trends Biochem. Sci.* **5**, 11 (1980).

108. Brandt, D. R. and Ross, F. M., *J. biol. Chem.* **261**, 1656 (1986).

109. Neer, E. J., *Cell* **80**, 249 (1995).

110. Coleman, D. E. and Sprang, S. R., *Trends Biochem. Sci.* **21**, 41 (1996).

111. Pfeuffer, E., Drehev, R-M., Metzger, H., and Pfeuffer, T., *Proc. natn. Acad. Sci. USA* **82**, 3086 (1985).

112. Smigel, M. D., *J. biol. Chem.* **261**, 1976 (1986).

113. Tausssig, R. and Gilman, A. G., *J. biol. Chem.* **270**, 1 (1995).

114. Tesmer, J. J. G., Sunahara, R. K., Gilman, A. G., and Sprang, S. R., *Science* **278**, 1907 (1997).

115. Ansari, M. H., MSc Thesis, University of Stirling (1980). See pp. 84–5.

116. West, I. C., in *Channelling in intermediary metabolism* (Agius, L. and Sherratt, H. S. A, eds), Chapter 2. Portland Press, London (1997).

117. Vandermeulen, D. L. and Ressler, N., *Arch. Biochem. Biophys.* **199**, 197 (1980).

118. Cooke, R. and Kuntz, I. D., *A. Rev. Biophys. Bioeng.* **3**, 95 (1974).

119. Knowles, P., *Essays Biochem.* **8**, 79 (1972). See p. 88.

120. Saenger, W., *A. Rev. Biophys. Bioeng.* **16**, 93 (1987).

121. Mailin, R., Zielenkiewicz, P., and Saenger, W., *J. biol. Chem.* **266**, 4848 (1991).

122. Weisz, P. B., *Nature, Lond.* **195**, 772 (1962).

123. Hubscher, G., Mayer, R. L., and Hansen, H. J. M., *Bioenergetics* **2**, 115 (1971).

124. Welch, G. R., *Prog. Biophys. molec. Biol.* **32**, 103 (1977).

125. Luby-Phelps, K., *Curr. Opin. Cell Biol.* **6**, 3 (1994).

126. Duszynski, J., Mueller, G., and La Noue, K., *J. biol. Chem.* **253**, 6149 (1978).

127. Ovadi, J. and Orosz, F., in *Channelling in intermediary metabolism* (Agius, L. and Sherratt, H. S. A, eds), Chapter 13. Portland Press, London (1997).

128. Srere, P. A., Sherry, A. D., Malloy, C. R., and Sumegi, B., in *Channelling in intermediary metabolism* (Agius, L. and Sherratt, H. S. A, eds), Chapter 11. Portland Press, London (1997).

129. Masters, C. J., *Trends Biochem. Sci.* **3**, 206 (1978).

130. Ovadi, J. and Keleti, T., *Eur. J. Biochem.* **85**, 157 (1978).

131. Batke, J., Askoth, G., Lakatos, S., Schmitt, B., and Cohen, R., *Eur. J. Biochem.* **107**, 389 (1980).

132. Tompa, P., Bar, J., and Batke, J., *Eur. J. Biochem.* **159**, 117 (1986).

133. Brooks, S. P. J. and Storey, K. B., *FEBS Lett.* **278**, 135 (1991).

134. Srivastava, D. K. and Bernhardt, S. A., *Curr. Top. Cell. Regul.* **28**, 1 (1986).

135. Fahien, L. A. and Chobanian, M. C., in *Channelling in intermediary metabolism* (Agius, L. and Sherratt, H. S. A, eds), Chapter 12. Portland Press, London (1997).

136. Srere, P. A., Halper, L. A., and Finkelstein, M. B., in *Microenvironment and metabolic compartmentation* (Srere, P. A. and Estabrook, R. W., eds), p. 419. Academic Press, New York (1978).

137. Backman, L. and Johansson, G., *FEBS Lett.* **65**, 39 (1976).

138. Beeckmans, S. and Kanarek, L., *FEBS Lett.* **117**, 527 (1981).

139. Velot, C., Mixon, M. B., Teige, M., and Srere, P. A., *Biochemistry* **36**, 14271 (1997).

140. Barnes, S. J. and Weitzmann, P. D. J., *FEBS Lett.* **201**, 217 (1986).

141. Sumegi, B. and Srere, P. A., *J. biol. Chem.* **259**, 15040 (1984).

142. Fahien, L. A., Kmiotek, F. H., Woldegiorgis, G., Evenson, M., Shrago, F., and Marshall, M., *J. biol. Chem.* **260**, 6069 (1985).

143. Wiame, J. M., *Curr. Top. Cell. Regul.* **4**, 1 (1971).

144. Hensley, P., *Curr. Top. Cell. Regul.* **29**, 35 (1988).

145. Mon, M., Aoyagi, K., Tatibana, M., Ishikawa, T., and Ishii, H., *Biochem. Biophys. Res. Commun.* **76**, 900 (1977).

146. Fisenstein, E. and Hensley, P., *J. biol. Chem.* **261**, 6192 (1986).

147. Sols, A. and Marco, R., *Curr Top. Cell. Regul.* **2**, 227 (1970).

148. Fersht, A. R., in *Enzyme structure and mechanism* (2nd edn). Freeman, New York (1985). See p. 328.

149. Veech, R. L., Rayman, L., Dalziel, K., and Krebs, H. A., *Biochem. J.* **115**, 837 (1969).

150. Barritt, G. J., Zander, G. L., and Utter, M. F., in *Gluconeogenesis: its regulation in mammalian species* (Hanson, R. W. and Mehlman, M. A., eds), p. 3. Wiley, New York (1976).

151. Srere, P. A., in *Gluconeogenesis. its regulation in mammalian species* (Hanson, R. W. and Mehlman, M. A., eds). p. 153. Wiley, New York (1976).

152. Sobboll, S., Scholz, R., Freisl, M., Elbers, R., and Heldt, H. W., in *Use of isolated liver cells and kidney tubules in metabolic studies* (Tager, J. M. and Williamson, J. R., eds), p. 29. North-Holland, Amsterdam (1976).

153. Dean, R. T. and Barrett, A. J., *Essays Biochem.* **12**, 1 (1976).

154. Cha, S., *J. biol. Chem.* **245**, 4814 (1970).

155. Griffiths, J. R., *Biochem. Soc. Trans.* **7**, 15 (1980).

156. Chaplin, M. F., *Trends Biochem. Sci.* **6**, VI (1981).

157. Ottaway, J. H., *Biochem. Soc. Trans.* **7**, 1161 (1979).

158. Nimmo, H. G. and Cohen, P., *Adv. cyclic Nucleotides Res.* **8**, 146 (1977).

159. Swillens, S., Van Cauter, F., and Dumont, J. E., *Biochim. biophys. Acta* **364**, 250 (1974).

160. Nageswara, B. D., Cohn, M., and Scopes, R. K., *J. biol. Chem.* **253**, 8056 (1978).

161. Gutfreund, H., *Prog. Biophys. molec. Biol.* **29**, 161 (1975).

9

Enzyme turnover

9.1 Introduction

Since the early 1940s when Schoenheimer *et al.*[1] first used isotopically labelled precursors to study protein synthesis, it has been realized that most proteins in living cells are continually being replaced whether the organism is fully grown and in nitrogen balance or is growing and increasing its total protein (and hence nitrogen) content. However, it was not until the late 1960s that much information was gained about the rate at which individual enzymes are replaced or turned over in particular tissues and cells and the general importance of enzyme turnover as a control mechanism was appreciated (for reviews see references 2–11).

In any given tissue or cell type there is a wide range of turnover rates for different enzymes, e.g. in rat liver the half-lives of enzymes range from about 15 minutes to over 100 hours.[4,10] The underlying mechanisms that account for this wide range of turnover rates are not fully understood, although considerable progress has now been made in identifying the proteases responsible and their selection mechanisms (see Section 9.6).

At first sight the process of enzyme turnover might seem to be inefficient and wasteful of energy, since every peptide bond formed requires the hydrolysis of ATP and GTP, whereas proteolysis is not coupled to the generation of ATP. However, turnover is important if a cell is to be able to adapt to changes in its environment or if it becomes necessary to remove abnormal enzyme or protein molecules that may arise by mutation or by errors in gene expression. In general, the longer the life of an individual cell the more important is the process of intracellular enzyme turnover. For example, in a bacterium such as *Escherichia coli* growing under optimal conditions, cell division may occur every 20 min. Adaptation occurs largely by induction or repression of enzyme synthesis. If lactose is added to the growth medium, induction of β-D-galactosidase occurs. If lactose is then removed from the medium, existing enzyme molecules will be diluted out rapidly within the cells provided they continue to divide rapidly and β-D-galactosidase is no longer synthesized. Early experiments aimed at detecting turnover of proteins in exponentially growing bacteria suggested that the proteins were completely stable.[4] More recent experiments have shown that turnover does occur but that it is limited largely to abnormal proteins and certain regulatory proteins (see Section 9.6.5).

Increases in the rates of turnover do occur when bacteria approach the stationary phase and the supply of nutrients becomes limiting. In contrast to that of rapidly dividing bacteria, the average life of an adult liver cell is between 160 and 400 days and many enzymes are completely replaced every few days. This turnover is necessary, for example, if a liver cell is to be able to adapt to changes in nutrients that it may encounter throughout its life.

When discussing enzyme turnover in this chapter we are referring to the intracellular processes by which individual enzyme molecules are degraded and replaced by synthesis of new enzyme molecules. The steady-state level of an enzyme in a cell depends on the rates of two opposing processes, namely the rate of enzyme synthesis and the rate of enzyme degradation. The general mechanism by which enzyme synthesis occurs is much better understood than that of enzyme degradation, although, on the whole, enzyme degradation has a more important regulatory influence on enzyme turnover, particularly in eukaryotes. The mechanism by which enzymes are synthesized is no different from that of protein synthesis in general and there are good accounts of this in many textbooks[12-15] and it will not be considered here. Therefore, we shall now consider the kinetics of enzyme turnover and possible mechanisms of enzyme degradation.

9.2 Kinetics of enzyme turnover

In order to understand the importance of enzyme turnover as a regulatory mechanism, it is necessary to make a kinetic analysis of the process. Equations that govern the steady-state levels of enzymes, and the rates of changes to new steady-state levels, have been derived by Schimke *et al.*[2,10,16] From the study of a number of enzymes in which the amounts of enzyme protein present in cells or tissues have been measured, it has been shown that the rate of degradation of an enzyme normally obeys first-order kinetics, i.e. the rate of enzyme degradation $= k[E]$, where [E] is the enzyme concentration. There are, however, some exceptions; e.g. some proteins present in nerve and muscle show biphasic kinetics, that is one fraction of the protein is degraded at a different rate from the remainder of the same protein in that tissue.[17] In most cases, therefore, this means that the degradation of an enzyme is a random process; a newly synthesized enzyme molecule is as likely to be degraded as an old one. The process of enzyme synthesis has been shown generally to obey zero-order kinetics, i.e. the rate of synthesis of an enzyme is independent of the enzyme concentration within the cell. Thus, the rate of change of level of an enzyme in a cell $(d[E]/dt)$ is given by

$$\frac{d[E]}{dt} = k_s - k_d [E],$$

where k_s is the rate constant for enzyme synthesis and k_d the rate constant for enzyme degradation. In the steady-state, $d[E]/dt = 0$ and therefore $k_s = k_d[E]$.

We now consider the expression for the rate of approach to a new steady-state level if either the rate of synthesis or the rate of degradation changes. If the rate constants for the new rates of synthesis and degradation are k'_s and k'_d, respectively, $[E_0]$ is the initial enzyme concentration, and $[E_t]$ is the enzyme concentration at time t after the change in rate of synthesis or degradation, then the equation describing the time course of approach to a new steady-state level is given by

$$\frac{[E_t]}{[E_0]} = \frac{k'_s}{k'_d[E_0]} - \left(\frac{k'_s}{k'_d[E_0]} - 1\right)e^{-k'_d t}.$$

For a derivation of this equation, see Appendix 9.1 and reference 16. This equation is useful for analysing changes in enzyme concentrations resulting from hormonal, nutritional, or other physiological changes. When the rate of degradation of an enzyme is measured, it is more common to express the result in terms of the half-life of the enzyme rather than in terms of the rate constant, k_d. For a first-order reaction the relationship is:

$$t_{\frac{1}{2}} = \frac{\ln 2}{k_d} = \frac{0.69}{k_d}.$$

Figure 9.1 illustrates the effect of differences in the rate of degradation of three enzymes on the rate of attainment of new steady states, when a tenfold increase in the rate of synthesis of the enzyme is induced by a stimulus. The effect of withdrawal of the stimulus 5 hours later is also shown. It can be seen that the enzyme having the shortest half-life, ornithine decarboxylase, responds very rapidly to the changed rate of synthesis and has almost reached the new steady state within 2–3 hours; similarly, when the stimulus is removed the old steady-state level is reached within a further 2–3 hours. In contrast, the amount of the enzyme having the longest half-life, pyruvate kinase, is still increasing almost linearly after 5 hours and would require over 70 hours to achieve 90 per cent of the new steady-state level.

Thus, it is the rate of degradation of an enzyme that primarily determines how rapidly the amount of enzyme changes in response to a given stimulus. The shorter the half-life, the quicker the response, and thus the better the control of the level of enzyme. In deriving these equations it is assumed that the rate of growth of the cells is slow in comparison with the rate of synthesis of the enzymes so that changes in the total volume of the cells can be neglected.

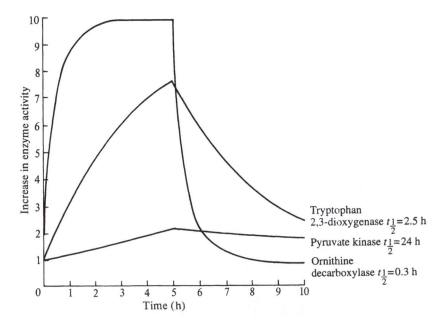

Fig. 9.1 The changes in enzyme activities resulting from a 10-fold increase in the rate of synthesis of each enzyme at zero time, and a 10-fold reduction in the rate of synthesis at 5 h. Note how the differences in half-lives affect the rates of attainment of new steady states.

Tryptophan 2,3-dioxygenase $t_{\frac{1}{2}}$=2.5 h

Pyruvate kinase $t_{\frac{1}{2}}$=24 h

Ornithine decarboxylase $t_{\frac{1}{2}}$=0.3 h

9.3 Methods for measurement of rates of enzyme turnover

There are several methods that can be used to measure the rates of protein synthesis and protein degradation; the most appropriate will depend on the type of information required. In this section we outline the principal methods; there are several reviews and papers in which more detailed treatment is given.[3,18–21] In order to understand how the concentration of a particular enzyme is maintained within a cell or tissue and how the concentration is changed in various physiological conditions, it is most useful to measure k_s and k_d. Knowledge of these rate constants will not give any insight as to the mechanism of synthesis or degradation, e.g. which are the rate-limiting steps in these processes, but they will indicate whether the change in concentration is due to a change in the rate of synthesis or of degradation, or of both.

A change in the enzyme content of a tissue or organism will usually first be observed from the measurement of enzyme activity. It is then necessary to establish whether this change represents a change in the amount of enzyme protein as opposed to some conformational change or covalent modification of the enzyme that alters its activity (see Chapter 6, Section 6.2). To distinguish these it is necessary to purify the enzyme and then measure the amount of enzyme protein. If previously purified enzyme is available it may be used to raise a specific antibody by injecting the enzyme into another species. The serum fractions containing the antibody may then be collected and used to titrate enzyme protein. Alternatively, monoclonal antibodies may be raised against the enzyme (see reference 22). The complex between the enzyme and the antibody forms a precipitate which can be collected by centrifugation and then estimated. Isolation of the enzyme from the tissues at a number of different times, to prove that the amount of enzyme protein is changing, is a very time-consuming process and may require large amounts of tissue. The use of an antibody greatly simplifies the procedure and can be used with smaller amounts of tissue.

Once it has been established that a change in the amount of enzyme protein occurs, the rates of synthesis and degradation are measured. Isotopically labelled precursors are most frequently used for this, as outlined in the following sections.

9.3.1 Measurement of k_s

The rate of enzyme synthesis is usually measured either after giving a single pulse of a suitable radioactively labelled precursor, e.g. an amino acid, or by giving a constant infusion of the labelled precursor. In the first method (involving a single pulse), incorporation of labelled precursor into the enzyme is measured at short time intervals. The specific activity of the amino-acid pool in the tissue or cells concerned is also measured. If short time intervals are used, the initial rate of incorporation is determined (i.e. when the specific activity of label in the enzyme is low and the loss of radioactivity by degradation is negligible). From these data k_s may be determined. One problem arises particularly in experiments with whole animals. The radioactive amino acids enter the plasma and thence pass into the cells, where they equilibrate with endogenous amino acids. The amino acids are then converted to aminoacyl-tRNAs, which are the ultimate precursors of the enzyme (Fig. 9.2).

Therefore, to calculate accurately the rates of synthesis, the specific radioactivity of the isolated aminoacyl-tRNAs should be used.[19] However, in rat liver, the

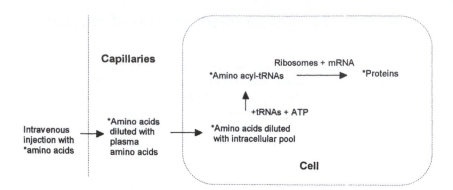

Fig. 9.2 The pathway of incorporation of radioactively labelled (*) amino acids into tissue proteins.

tissue in which many measurements of k_s have been made, there is rapid equilibration between the free amino-acid pool within the cells and the aminoacyl-tRNAs and therefore in this particular tissue the specific radioactivity of the free amino acids may be used.

The second method is that of constant infusion of radioactively labelled amino acids. The amino acids are either infused intravenously or given in the diet until the specific radioactivity of the amino-acid pool within the cells has reached a plateau level. Measurements are then made of the incorporation of the label into the enzyme at the end of the infusion period and of the changes in specific radioactivity of the free amino acid within the tissue. From these measurements k_s may be determined; the mathematical relationships are complex and are described in references 18 and 19.

9.3.2 Measurement of k_d

The rate of degradation can also be measured by use of radioactively labelled precursors. If a pulse of labelled precursor is given and the decay of the specific radioactivity of the labelled enzyme is measured at fixed time intervals, k_d can be calculated. Measurements are made for a longer period than those required for measurement of k_s. Ideally, k_d is measured when the specific radioactivity of the enzyme is high, and the specific radioactivity of the amino-acid pool has declined (Fig. 9.3).

An assumption made in the calculation of k_d is that no significant reutilization of the precursors released by protein degradation takes place. If reutilization takes place, k_d may be considerably underestimated. This has been a major problem in the accurate determination of k_d, but reutilization can often be minimized by a suitable choice of precursor; for example [*guanidino* [14]C]arginine is found to be a better label for this purpose than uniformly labelled arginine or other labelled amino acids when liver protein degradation is measured.[23] This is because the guanidino group is rapidly removed by arginase present in high concentrations in the liver. [^{14}C]carbonate has also been found to be a good precursor for measurement of k_d; much of the [^{14}C]carbonate is incorporated into the carboxyl groups of glutamate and aspartate.[24] The radioactivity of the latter amino acids declines rapidly because of decarboxylations that occur when glutamate and aspartate become transaminated and subsequently oxidized in the tricarboxylic-acid cycle. The $H^{14}CO_3^-$ produced in this way is diluted by the large intracellular pool of HCO_3^- and lost as $^{14}CO_2$.

Another example in which reincorporation does not occur is that of 5-amino-laevulinic acid,[25] which is incorporated specifically into haem-containing pro-

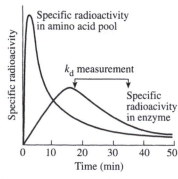

Fig. 9.3 Precursor–product relationship after a single pulse of radioactively labelled amino acid is given to a tissue or cells. Diagram shows the specific radioactivities in the amino-acid pool (precursor) and in the enzyme (product).

teins, e.g. cytochromes, catalase. Haem is degraded by a pathway that differs from that of its synthesis and thus reincorporation does not take place. The validity of this method depends on the assumption that the haem prosthetic group is degraded at the same rate as the protein moiety, which may or may not be the case

The double isotope technique[18] is a convenient method of comparing the rates of turnover of proteins having the same cellular origin. It entails giving two successive pulses of a radioactively labelled amino acid, the first being ^{3}H-labelled and the second ^{14}C-labelled; these are shown as ^{3}H-A and ^{3}H-B, and ^{14}C-A and ^{14}C-B, respectively, in Fig. 9.4, where A and B are two different proteins. In the example in Fig. 9.4, the turnover of A is twice that of B. The proteins that are being studied are isolated at time *Y*. Typically, the ^{3}H pulse would have been given 24 h before isolation of the proteins, whereas the ^{14}C pulse would have been given 30 min before isolation (time X), although the precise times would depend on the proteins and tissue concerned. The times are chosen so that the specific activities of the first label (^{3}H) in the proteins being studied are declining, whereas the specific activities of the second label (^{14}C) are increasing (Fig. 9.3). The proteins are isolated at time *Y* and the ratio ^{14}C/[3]H (*b/a* and *d/c*) in each is measured; these ratios are directly related to their turnover rates. The method has two advantages: (i) that the proteins only have to be isolated at one time point, and (ii) that the differences in turnover rates between proteins are more readily apparent than when a single isotope is used.

A method that has been widely used to determine rates of degradation is to inhibit protein synthesis in the tissue concerned by administration of cycloheximide or puromycin and then to measure the subsequent decay of enzyme activity. This method has the advantage of simplicity in that it does not require the isolation of the enzyme, but it has the disadvantages (a) that it cannot distinguish enzyme inactivation from enzyme degradation and (b) that cycloheximide and puromycin, in addition to inhibiting protein synthesis, may also affect protease activity, thereby affecting the rate of enzyme degradation.[3] Evidence suggests that the latter complication is less serious when studying enzymes that have short half-lives.[26]

9.4 Results from measurements of rates of enzyme turnover

The rates of turnover of a number of enzymes have now been measured in a variety of tissues and organisms and it is possible to correlate these results with the structures and properties of the enzymes concerned. The most widely measured parameter is the enzyme's half-life. Some of the results obtained are given in Table 9.1. It is apparent that a wide range of turnover rates can occur within one tissue, e.g. the half-lives of liver proteins range from about 15 min (ornithine decarboxylase) to 7 days (6-phosphofructokinase). The half-life of a given enzyme may differ from one tissue to another (Table 9.2) or even between different compartments within the same cell, e.g. 5-aminolaevulinate synthase in the cytosol has a half-life of 20 min, whereas in the mitochondria the half-life is 60 min.[34] Some results suggest that in certain organelles or membranes within an organelle, a number of enzymes turn over at similar rates,[35] e.g. cytochrome *c* oxidase and ATP synthase from the inner mitochondrial membrane of yeast turn over at similar rates during sporulation, and also ribosomal proteins turn over at similar rates. However, other

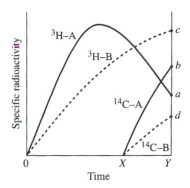

Fig. 9.4 Double isotope technique for measuring protein turnover.

Table 9.1 The half-lives of enzymes in rat liver

Enzyme	Half-life (hours)
Ornithine decarboxylase	0.3
5-Aminolaevulinate synthase (mitochondria)	1.2
RNA polymerase I	1.3
Tyrosine aminotransferase	1.5
Tryptophan 2,3-dioxygenase	2.0
Thymidine kinase	2.6
Hydroxymethylglutaryl-CoA reductase	4.0
Serine dehydratase	5.2
Phosphoenolpyruvate carboxykinase	8.0
Dehydro-orotase	12.0
RNA polymerase II	13.0
Glucose-6-phosphate dehydrogenase	24
Glucokinase	30
Catalase	33
Acetyl-CoA carboxylase	50
Glyceraldehyde-3-phosphate dehydrogenase	74
Pyruvate kinase	84
Arginase	108
Fructose-bisphosphate aldolase	117
Lactate dehydrogenase (LDH-5)	144
6-Phosphofructokinase	168

See references 2, 4, 27–30.

Table 9.2 The half-lives (in days) of enzymes in different tissues of the rat

Enzyme	Liver	Heart	Skeletal muscle	Kidney	Brain
Pyruvate kinase	3.5	4.2	21.6	—	—
Lactate dehydrogenase[a]	1.8–3.8	3.7–9.9	3.0–8.5	3.0–5.0	3.0–7.2
Ornithine oxo-acid aminotransferase	0.95	—	—	4.0	—
Fatty-acid synthase	6.4	—	—	—	2.8

See references 2, 27, 31, and 32.
[a] The half-life of lactate dehydrogenase varies with the particular isoenzyme.[35]

Table 9.3 Variation in half-lives of enzymes within the same organelle of rat liver

	Half-life (hours)
1. Microsomal membrane	
Cytochrome b_5 (outside)	100
Cytochrome b_5 reductase (outside)	140
NADPH-cytochrome reductase (outside)	70
Nucleosidediphosphatase (inside)	30
Carboxylesterase (inside)	95
Hydroxymethylglutaryl-CoA reductase	4
NAD+ nucleosidase	430
2. Mitochondria	
Carbamoyl phosphate synthase	185
Malate dehydrogenase	62
Glutamate dehydrogenase	24
Carnitine acetylase transferase	43
Ornithine oxo-acid aminotransferase	46
Cytochrome c oxidase	>190
Amine oxidase (flavin-containing)	67
Pyruvate dehydrogenase system	
$E_1\alpha$	108
$E_1\beta$	120
E_2	134
E_3	120

See references 37–41.

data indicate that there is a considerable range of turnover rates of enzymes within a single organelle or membrane[36,37] (Table 9.3) and it seems likely that in general organelles do not turn over as units.

From the data available so far, it seems that the polypeptide chains within a multienzyme complex, e.g. pyruvate dehydrogenase, and within an oligomeric enzyme, e.g. cytochrome c oxidase and amine oxidase (flavin-containing), turnover at similar rates (Table 9.3), although with skeletal muscle phosphorylase kinase (subunit composition $\alpha_4\beta_4\gamma_4\delta_4$) the δ-subunit (calmodulin, see Chapter 6, Section 6.4.2.1) is turned over faster than the other subunits: α 38 h, β 41 h, γ 31 h, and δ 10 h.[42]

9.5 Possible correlations between the rates of turnover and the structure and function of enzymes

Towards the end of the 1970s the half-lives of a number of enzymes were known, and as can be seen from Tables 9.1–9.3, it was clear that a wide range of turnover rates (over 1000-fold) exists. Several attempts were made to try to correlate the rate of turnover with particular structural properties of the enzymes with varying degrees of success, e.g. size, pI, hydrophobicity, stability against denaturation, and the binding of substrates, effectors, and coenzymes (Table 9.4).[42,49–53]

One of the difficulties with this approach was that there was considerable variation in the amount of structural information available for different enzymes. Least structural information was generally available for the enzymes with short half-lives. By 1984 Rechsteiner et al.[54] had developed a technique of microinjection of purified enzymes into HeLa cells. Using this method in a study of 35 proteins of known sequence and X-ray structure they found that the half-lives for those proteins did not correlate with size, charge, or proportion of hydrophobic residues. They then decided to examine the amino-acid sequences inferred from DNA sequences of a group of proteins which were known to be rapidly degraded. They found a correlation between the abundance of particular amino-acid sequences in a protein and the turnover rate.[55] Sequences rich in four amino acids, namely proline, glutamate, serine, and threonine (PEST—using the single-letter code) were more abundant in the rapidly degraded enzymes. By 1991, 43 out of 47 short-lived proteins were known to contain PEST regions, whereas in mammalian proteins in general only about 10 per cent have PEST-rich sequences. PEST-rich sequences are hydrophilic and are thus more likely to occur on solvent-exposed loops of proteins, but they do not have a single type of secondary

Table 9.4 Examples of ligands that stabilize enzymes against degradation

Enzyme	Ligand	Effect
Hexokinase	Glucose	Reduces rate of degradation by trypsin
Thymidine kinase	Thymidine	Stabilizes against degradation in cell extracts
Tryptophan 2,3-dioxygenase	Tryptophan or tryptophan analogues	Induces tryptophan 2,3-dioxygenase by stabilizing it against degradation
Serine dehydratase Tyrosine aminotransferase	Pyridoxal phosphate	Protects against group-specific proteases that degrade pyridoxal phosphate-requiring apoenzymes
Aspartate carbamoyltransferase	Aspartate CTP or ATP	Increased susceptibility to trypsin Decreased susceptibility to trypsin
Tetrahydrofolate dehydrogenase	Methotrexate	Decreases rate of intracellular degradation.
Ornithine decarboxylase	α-Methylornithine or diaminobutanone	Decreases rate of intracellular degradation, protects against chymotrypsin
Acetyl CoA carboxylase	Palmityl CoA citrate	Increased susceptibility to lysosomal protease Decreased susceptibility to lysosomal protease
Hydroxymethylglutaryl CoA reductase	Mevinolin	Decreases intracellular degradation

See references 3, 43–48.

structure, some occurring in α-helices and others in β-turns.[56] The importance of peptide motifs in targeting proteins for degradation is discussed in Section 9.6.1.

In addition to making comparisons on purely structural grounds, it is useful to relate turnover with function. Many of the enzymes with short half-lives catalyse rate-limiting steps in metabolic pathways, e.g. ornithine decarboxylase, 5-amino-laevulinate synthase, tyrosine aminotransferase, tryptophan 2,3-dioxygenase, serine dehydratase, hydroxymethylglutaryl CoA reductase, and phosphoenolpyruvate carboxykinase. Other proteins that turnover rapidly include transcriptional factors such as MATα2 and GCN4 from yeast, proteins regulating growth and cell division such as c-*jun* gene product, the tumour supressor protein p53, Mos kinase, cyclins, and the α-subunit of G-proteins.[57] Abnormal proteins, e.g. mutant proteins or proteins containing chemically modified amino acids, are degraded much faster than their normal counterparts. Mutations in β-D-galactosidase[58] from *E. coli* cause the enzyme to be degraded rapidly, and the incorporation of azetidine-2-carboxylate in place of proline into haemoglobin reduces its half-life from 120 days to 20 min.[59] The mechanism by which abnormal proteins are selectively degraded at a faster rate is not yet fully understood.

Azetidine-2-carboxylate

Proline

9.6 The mechanisms of protein degradation

Since the mid-1970s the general features concerning the mechanism of intracellular protein degradation in eukaryotes have emerged, although many details have still to be resolved. The sites at which intracellular protein degradation occur are both lysosomal and non-lysosomal. The process may be energy dependent, requiring ATP, or energy independent, and it may be selective or non-selective. It is convenient to divide proteins into long-lived and short-lived, since this relates to the mode of degradation. Many long-lived enzymes are the so-called 'house-keeping' enzymes. These include enzymes catalysing non-rate-limiting steps in the major metabolic pathways. They are generally degraded by lysosomal proteases. The short-lived enzymes are present in much smaller amounts and are degraded by extralysosomal routes.

Two most important discoveries led to the understanding a mechanism of intracellular protein degradation. In 1980 Hershko *et al.*[60] showed that ATP-dependent proteolysis in rabbit reticulocytes involved conjugation with the protein ubiquitin, and in the same year Wilk and Orlowski[61] isolated a large multicatalytic proteinase complex that corresponds to what is now known as the proteasome. Both the ubiquitin-conjugating system and the proteasomes were found to have wide distribution in eukaryotes and the importance of these in intracellular protein degradation in a wide range of cells and organisms has now become apparent. In 1993 evidence for the existence of novel forms of proteasomes in eubacteria came from the discovery of genes in *E. coli* encoding polypeptides similar to those in eukaryote proteasomes, but these have been studied in much less detail.[11] The evidence for the ubiquitin pathway is strongest in the case of short-lived and abnormal proteins, but there is not yet conclusive evidence for ubiquitination of the majority of enzymes listed in Table 9.1.[42] The proteasome is found in the nucleus and cytosol and is the key enzyme responsible for the non-lysosomal protein degradation. It constitutes up to 1 per cent of the total protein in mammalian cells.[62] The importance of the ubiquitin–proteasome pathway was first revealed by studies using mouse cell-cycle mutant cells known as ts85 cells. These cells have a temperature-sensitive mutation in the first enzyme required for ubiquitin conjugation. When shifted to the non-permissive temperature the turnover of short-lived proteins was inhibited by about 80 per cent.[63] More recently, however, this interpretation has been questioned, since ubiquitin-conjugated proteins have been detected when the cells were incubated at the non-permissive temperature.[42]

The pattern emerging is that most short-lived proteins are degraded by a non-lysosomal pathway involving the proteasome. This pathway in most cases involves tagging the protein by conjugation with ubiquitin prior to its degradation. However, this is not always the case; ornithine decarboxylase, a very short-lived enzyme (Table 9.1) is degraded by proteasomes without prior ubiquitination[64] (see Section 9.6.2.4) as is the protooncogene product c-Jun.[56] The bulk of the long-lived enzymes is degraded by the lysosomes. Lysosomes appear capable of both selective and non-selective protein degradation. The non-selective process involves microautophagy and accounts for the slow degradation of long-lived proteins. The selective process involves selective uptake into the lysosomes by the recognition of certain sequences on the proteins such as KFERQ.[65] Some membrane proteins become conjugated with ubiquitin as a mechanism to target them to the lysosomes.[66]

Enzymes degraded by lysosomal and non-lysosomal routes can often be distinguished by two methods. First, chloroquine and ammonia are able to penetrate the lysosomal membranes and raise the internal pH,[67] thereby inhibiting lysosomal proteases and so affecting the breakdown of enzymes degraded by the lysosomal pathway. Second, the Q_{10} (see Chapter 4, Section 4.3.4.1) for degradation of short-lived proteins degraded by the non-lysosomal route[68] is about 2, whereas degradation via the lysosomes[69] has a Q_{10} of about 4. It is estimated that in the liver about 60 per cent of proteins are degraded in the cytosol and about 40 per cent in the lysosomes.[70] Enzymes degraded by the proteasome pathway can be identified by the use of a specific proteasome inhibitor, lactocystin, which reacts with a threonine residue of the β-subunits of mammalian proteasomes.[71]

A number of different proteases are involved in intracellular protein degradation of which the most important is the ubiquitin–proteasome system. In Section 9.6.1 we discuss the mechanisms by which ubiquitin becomes conjugated to proteins, in Section 9.6.2 the proteasome and its mechanism for degrading proteins,

Chloroquine

in Section 9.6.3 the lysosomal pathway, in Section 9.6.4 proteins degraded in the endoplasmic reticulum, and in Section 9.6.5 protein degradation in prokaryotes.

9.6.1 The conjugation of proteins by ubiquitin as a means of marking them for degradation

Conjugation of proteins with the polypeptide ubiquitin is of crucial importance in understanding how many proteins become marked, prior to degradation. Ubiquitin is a small basic protein (76 residues) found in all eukaryote cells so far examined. Its sequence shows extreme conservation, and this is probably due to the structural requirements for correct folding and stability.[72]

It has a compact globular structure with its C-terminus accessible, and it is with its terminal carboxyl that it forms an isopeptide bond with the amino groups of other proteins.

It is called an isopeptide bond since it is isomeric with the α-peptide formed between the α-amino group and α-carboxyl group that links amino acids in proteins (see Chapter 3, Section 3.3.1.2). Ubiquitin is a very stable protein and this arises from its hydrophobic core comprising a 3.5-turn amphipathic α-helix which intercalates into a five-strand β-sheet, and there are also several β-turns.[72] About 50 per cent or more of the ubiquitin in a cell is coupled to other proteins. Ubiquitin has a number of functions in addition to marking proteins for proteasome degradation. It has been implicated in several cellular processes including cell cycle control, oncoprotein degradation, apoptosis, regulation of transcription, stress responses, maintenance of chromatin structure, DNA repair and antigen presentation, and degradation of abnormal proteins.[66,73] In addition to forming an isopeptide bond with other proteins, ubiquitin is capable of forming large conjugates with itself through isopeptide linkages mainly involving Lys48 and Gly76.[66,74] Polyubiquitination is a prerequisite for efficient targeting and proteasome-mediated degradation. Ubiquitin has a total of seven lysine residues, and thus seven potential sites for the formation of multiple ubiquitin chains, leading to a potentially very large number of possible isomeric ubiquitinations. In addition to linkage through Lys48, other linkages, e.g. to Lys63 and Lys29, have been found.[74] Three enzymes are necessary to catalyse the formation of the complex and are usually referred to as E_1, E_2, and E_3 (Fig. 9.5).

The first is the ubiquitin-activating enzyme, E_1, which initially activates ubiquitin in an ATP-dependent reaction through the formation of a thiol ester bond between the carboxyl terminus of ubiquitin and the thiol group of a specific cysteine residue of E_1. The ubiquitinating enzyme, E_1 binds both ubiquitin and ATP strongly having K_ds < 1 μmol dm^{-3}. The intracellular concentration of ATP is in the millimolar range and that of ubiquitin is above 20 μmol dm^{-3}, so that E_1 would be saturated under normal physiological conditions, and thus ATP and ubiquitin could not control the rates of ubiquitin ligation.[72] Ubiquitin is then transesterified to a specific cysteine residue on one of several ubiquitin-conjugating enzymes (E_2). The E_2 enzymes in turn may transfer the ubiquitin either directly to a substrate or to a third class of enzymes known as ubiquitin protein ligases (E_3). The E_3s also promote the addition of multiple ubiquitins to the substrate proteins, a prerequisite to proteasome degradation.[9] Only one functional E_1 has so far been identified, but over 30 different E_2s and several E_3s are known.[75] The selectivity of ubiquitination arises in part because of the large set of E_3s that exist, each acting on only a few protein substrates, and also the large set of E_2s, each of which interacts with one or a few E_3s. All E_2s have a conserved

Ubiquitin. The isopeptide bond linking ubiquitin to targeted proteins is formed at the C terminus of ubiquitin.

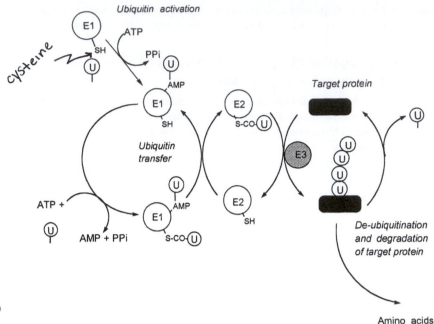

Fig. 9.5 The ubiquitin cycle for protein degradation.

domain of about 130 amino-acid residues containing the active-site cysteine involved in ubiquitin conjugation. Specific complexes between a subset of E_2s and an E_3 ubiquitin protein ligase have been detected.[75] The E_3s are assumed to recognize specific motifs on the protein substrate. The first example of an inherited disease involving ubiquitin is Angelman syndrome, which results from a defect in E_3 and hence ubiquitination.[76]

In addition to the selectivity arising from the multiple ubiquitinating enzymes discussed above, further selectivity and editing is brought about by deubiquitinating enzymes. If ubiquitination is regarded as a post-translational regulatory process rather like protein phosphorylation (see Chapter 6, Section 6.4.2.1), then the deubiquitinating enzymes have a regulatory role comparable to that of protein phosphatases (Chapter 6, Section 6.4.2.4). A large number of deubiquitinating enzymes are now known.[77] They are all thiol proteases with a specificity for cleavage of the carboxyl terminal Gly76 of ubiquitin.

Much work has been carried out with the aim of determining the specific recognition signals on proteins that target them for degradation. Although some have been identified, they appear to account only for a small proportion of proteins degraded. It has been suggested that the total number of different recognition signals could exceed ten.[74] The signals may be quite complex, since they may determine not only which proteins are degraded, but also their rates of degradation. The signals may also be constitutively active or conditionally active.[74] An example of the latter would be signals that are regulated by phosphorylation. In this class are certain cyclins and transcriptional factors which become phosphorylated prior to degradation. Fructose-bisphosphatase from *Saccharomyces cerevisiae* and NAD⁺-requiring glutamate dehydrogenase from *S. cerevisiae* and *Candida utilis* are both phosphorylated prior to their degradation,[78,79] but neither has yet been proven to be ubiquitinated. Two other enzymes that become more susceptible to proteolysis following phosphorylation are hydroxymethylglutaryl

CoA reductase[80] and pyruvate kinase.[81] One of the earliest hypotheses was the 'N-end rule' proposed by Bachmair *et al.*[82] They constructed a chimeric gene consisting of the genes for ubiquitin linked to β-galactosidase. When this gene is expressed it produces a conjugate protein with ubiquitin linked to the N-terminus of β-galactosidase via a peptide bond. The protein became de-ubiquitinated and degraded by the cell's proteolytic system. Site-directed mutagenesis (see Chapter 5, Section 5.4.5) was then used to replace the normal N-terminal amino acid of β-galactosidase (Met) in turn by other amino acids. It was then possible to study the degradation of the fusion products. The half-lives of the fusion products ranged from 2 min to over 20 h depending on the N-terminal amino acid as shown below:

$$H_2N-----------Gly\text{-}Met-----------COOH$$

$$\quad\quad\text{ubiquitin} \quad\quad\quad\quad\quad\quad \beta\text{-galactosidase}$$

Replacement of Met by Ser, Ala, Gly, Thr, or Val has a stabilizing effect. Replacement of Met by Ile, Gln, Glu, or Tyr has a destabilizing effect. Replacement of Met by Leu, Phe, Asp, Lys, or Arg has a strongly destabilizing effect. The N-terminal residue thus appeared an important determinant in protein degradation through the ubiquitin pathway. However, the physiological significance of this mechanism is probably limited to artificial, denatured, and abnormal proteins in certain cell types. There is little evidence that it is important for normal proteins. Approximately 80 per cent of intracellular proteins are N-terminally blocked by acetylation and thus excluded from the rule. Most of the remaining 20 per cent are also likely to be precluded, since they generally have N-termini which stabilize them from degradation by the N-end rule. This is probably due to the specificity of the initiator methionine aminopeptidase which removes the N-terminal methionine if the second amino acid is stabilizing according to the N-end rule.[75] In a survey no relationship could be found between the half-lives of proteins and their N-terminal amino acids.[83]

Other recognition signals so far proposed are the nine-residue 'destruction box' (RTALGDIGN)[84] present in cyclins (see Section 9.7.4), the sequence SSSTDSTP present in the transcriptional activator Gcn4p from yeast, and PEST sequences present in a number of short-lived proteins. Deletion of PEST regions from some proteins caused their partial stabilization.[85] Abnormal, damaged, and denatured proteins may be degraded by the ubiquitin system because their normally buried recognition signals become exposed.

9.6.2 **Proteasome-mediation protein degradation**

In 1980 a large multisubunit protease was isolated from bovine pituitaries,[61] but it was about five years before the wider significance of this particle, which became known as the proteasome, was realized. In its structural complexity, it is like the counterpart to the ribosome in protein synthesis. It is a multicatalytic proteolytic complex that degrades ubiquitin-tagged proteins in an ATP-dependent manner.[66,86] It comprises the 20s proteasome (mass $\approx 700\,000$) together with two particles of a further complex, the 19s regulatory unit (mass $\approx 700\,000$) making up the 26s proteasome.

9.6.2.1 **The structure of the 20s proteasome**

Proteasomes have been found in all eukaryotes studied so far, and they have also been found in archaebacteria. Because the proteasomes isolated from the archaebacterium *Thermoplasma acidophilum* have a simpler structure, they have been

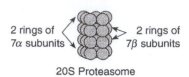

2 rings of 7α subunits → ← 2 rings of 7β subunits

20S Proteasome

Fig. 9.6 26s proteasome structure and assembly.[66]

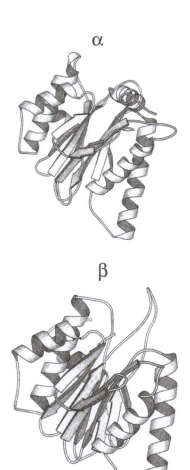

α

β

Fig. 9.7 Structure of α- and β-subunits of the *Thermoplasma* proteasome.[86]

the subject of detailed study using electron microscopy and X-ray crystallography. In 1995 the crystal structure was solved at 0.34 nm resolution.[87] The 20s proteasome is a cylindrical structure comprising 14 α-subunits and 14 β-subunits. These are arranged in four rings with the two rings of α-subunits on the outside and the β-subunits on the inner rings (Fig. 9.6 and also Fig. 3.30).

The proteolytic sites are all in the 20s proteasome. Six out of the 14 β-subunits have functional active sites. Five distinct proteolytic activities can be detected in mammalian 20s proteasomes using artificial chromogenic peptides as substrates. These are peptides linked through an amide bond to a chromophore, e.g. 4-nitroanilides, β-naphthylamides. Proteases that cleave the chromophore from the peptide can be detected since the amine, e.g. 4-nitroaniline or β-naphthylamine, is now able to ionize and has a distinct absorption spectrum. The five proteolytic activities are usually referred to as 'chymotrypsin-like', 'trypsin-like', 'postglutamyl hydrolase', 'branched-chain amino acid' protease, and 'small neutral amino acid' protease. These combined activities have the potential to cleave most types of peptides bonds.

In the 20s proteasomes from *Thermoplasma acidophilum* there are only single types of both α- and β-subunits and only chymotrypsin-like activity is detected. However, even in lower eukaryotes there are seven different types of α- and β-subunits. Seven is the maximum number of different α- and β-subunits that can be present in a single proteasome. In higher eukaryotes there exist both constitutive and inducible subunits. The latter can replace the former in certain physiological situations, e.g. treatment of cells with γ-interferon induces the synthesis of three new distinct β-subunits.[11] The α- and β-subunits from *T. acidophilus* have 26 per cent sequence identity, but their tertiary structure shows much greater similarity in which both have a central five-strand β-sheet sandwich flanked by α-helices (Fig. 9.7). The α-subunits are capable of assembling spontaneously into a seven-membered ring, and they contain the binding sites for the 19s subunits, whereas the β-subunits require the α-subunit as a scaffold on which to assemble.

The barrel-shaped 20s proteasome has three cavities, a central cavity bounded by the two rings of β-subunits, and two outer cavities or 'antechambers', each formed by one α-subunit ring and one β-subunit ring. The entrance to the outer chamber is 1.3 nm diameter and that of the inner chamber 2.2 nm. These entrances are sufficiently small that a polypeptide chain would have to unfold in order to enter the inner cavity. Evidence in support of this comes from using Nanogold-labelled substrate. The Nanogold particle which is electron dense and about 2 nm diameter can easily be identified by electron microscopy. When such a substrate is used the Nanogold can be seen accumulating at the openings of the α-rings.[86]

9.6.2.2 Mechanism of action of the 20s proteasome

20s proteasomes are able to degrade proteins in an ATP-independent manner provided the proteins are unfolded. The β-subunits that have proteolytic activity have their active sites facing inside. This makes them accessible only to polypeptides that can enter the cavities. Individual proteasomes degrade proteins quite slowly, taking 1–2 min per protein.[11] The volume inside the cavity is estimated to be ≈ 84 nm^3 which is enough space to accommodate a single folded polypeptide of $M_r \approx 70\,000$ but a somewhat smaller polypeptide when unfolded.[88] Only one polypeptide chain can enter the inner cavity at a time, and the most abundant degradation products are hepta-, octa-, and nonapeptides. The size of these degradation products corresponds approximately to the distance separating the active sites on the β-subunits.[11] The 20s proteasome does not fit into either the category of a typical serine protease or thiol protease (Chapter 5, Section 5.5.1.7). It is not inhibited by typical thiol protease inhibitors and is not inhibited by diisopropylfluorophosphate (Chapter 5, Section 5.4.4.2).

It is inhibited by a small peptide aldehyde inhibitor, acetyl-leu-leu-norleucenal, which combines with the N-terminal threonine. Lactocystin also inhibits by covalently binding to the N-terminal threonine residue.[89] Replacement of Thr1 by site-directed mutagenesis with alanine caused loss of activity, but activity was retained after replacement with serine. This strongly suggests that threonine is involved in catalysis, and that a hydroxyl group is necessary, but the proteasome lacks the catalytic triad (see Chapter 5, Section 5.5.1.2) typical of serine and thiol proteases.[11] However, Glu17, Lys 33, and Asp166 are required for activity. Lys33 and Glu17 form a salt bridge across the bottom of the active site and may participate in the delocalization of the threonine side-chain proton by forming a charge-relay system.[88]

N-Acetyl-leucyl-leucyl-norleucenal

9.6.2.3 The 26s proteasome

The 26s proteasome is formed by the ATP-dependent association of the 20s proteasome and the 19s complex. It catalyses the ATP-dependent degradation of polyubiquitinated proteins. It occurs in both the cytosol and nucleus of all eukaryotic cells so far studied. The 26s proteasome is labile, and until it was realized that it could be stabilized by glycerol and ATP it was difficult to isolate. It appears dumb-bell shaped on electron micrographs with the 19s complexes fitting on to both ends of the barrel-shaped 20s proteasome (Fig. 9.8). The 19s regulatory complex is also known as PA700 (proteasome activator of M_r 700 000).

The V-shaped invagination on either end of the dumb-bell is thought to be the probable site of entry of polyubiquitinated proteins tagged for degradation. 26s proteasomes have been isolated from a range of different species of eukaryotes, and there is some variation in composition both between species and in different tissues within a single species. This is probably because certain factors may associate reversibly with the fundamental units to act in a regulatory fashion, and so may depend on the physiological status of the cell or tissue. The 19s complex consists of approximately 20 subunits, most of which are present as single copies. Six of these subunits are a particular type of ATPase known as AAA ATPases (*A*TPases *a*ssociated with different cellular *a*ctivities).[90] AAA ATPases are Mg^{2+}-dependent ATPases found in various cell locations and having a variety of functions, including

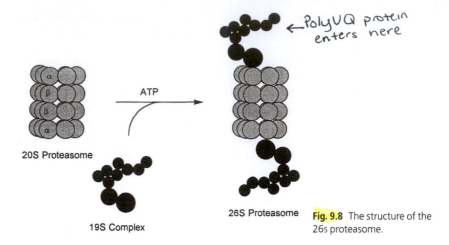

Fig. 9.8 The structure of the 26s proteasome.

vesicle and organelle biogenesis, cell cycle regulation, metalloproteinases, and as a component of 26s proteasomes. The feature they have in common is the AAA motif, a conserved sequence of 230–250 amino-acid residues which includes the ATP binding site. The conserved sequence includes the P-loop which comprises a five-strand parallel β-sheet[91] typical of nucleotide-binding proteins (see Chapter 3, Section 3.3.2.10). In addition, most of them are able to polymerize to form a hexameric ring structure. The six different AAA ATPases which form a heterohexamer in the 19s regulatory complex are designated Sug1p, Sug2p, Yta1p, Yta2p, Yta3p, and Yta5p. All six are essential for proteasomal activity for large substrate molecules. They have similar structures, but distinct functions. For example *sug2* mutants are defective in spindle pole duplication in yeast, and it has been proposed that Sug2 specifically unfolds components of the spindle pole body prior to degradation. The expression of these AAA ATPase genes appears to be regulated by hormones. For example, the expression of *sug1* increases dramatically prior to cell death and after stimulation by certain hormones.[90] Possible functions of these ATPases include the formation of the 26s proteasome from the 20s proteasome and the 19s regulatory complex, unfolding of ubiquitinated proteins, and injection into the 20s proteasome.[11] They resemble bacterial chaperonin 60 in that the central hexameric core could function in the folding and unfolding of proteins. They are considered to be 'reverse chaperones'. This is discussed in Chapter 3, Section 3.6.2.

The functions of the remaining subunits are less clear, but may include capture of substrates, release and disassembly of polyubiquitin, and maintenance of the particle's structure. One of the subunits has been identified as having ubiquitin isopeptidase activity,[92] catalysing the removal of distal residues from polyubiquitinated substrates. It may have an editing function in rescuing slowly degraded substrates.

9.6.2.4 Proteasome-mediated degradation of ornithine decarboxylase and other proteins without prior ubiquitination

Most short-lived proteins present in the cytosol and nucleus of eukaryote cells are degraded after polyubiquitination by 26s proteasomes. However, there are some that are degraded by proteasomes without prior ubiquitination. The most well-studied amongst these is ornithine decarboxylase. Ornithine decarboxylase catalyses the first step in the biosynthesis of polyamines (Fig. 9.9). It is the shortest lived cellular enzyme known (Table 9.1). Polyamines are synthesized most rapidly

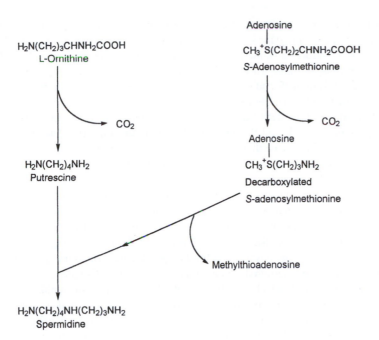

$H_2N(CH_2)_3CHNH_2COOH$
L-Ornithine

$\longrightarrow CO_2$

$H_2N(CH_2)_4NH_2$
Putrescine

Adenosine
|
$CH_3{}^+S(CH_2)_2CHNH_2COOH$

S-Adenosylmethionine

$\longrightarrow CO_2$

Adenosine
|
$CH_3{}^+S(CH_2)_3NH_2$

Decarboxylated

S-adenosylmethionine

Methylthioadenosine

$H_2N(CH_2)_4NH(CH_2)_3NH_2$
Spermidine

Fig. 9.9 The biosynthesis of spermidine.

at the onset of cell proliferation. This association with cell proliferation and the short half-life of ornithine decarboxylase have led to the suggestion that the latter may be an oncogene product.[93] Mammalian ornithine decarboxylase is a dimer of two identical subunits each having 461 amino-acid residues, and active sites that span the two subunits. The enzyme is strongly suppressed by the accumulation of cellular polyamines, which prevent their further build up. This is important since polyamines are cytotoxic.

The main mechanism that effects ornithine decarboxylase degradation is through binding to a specific protein known as an antizyme, and this in turn initiates proteasome-mediated degradation. The antizyme was first discovered in 1976.[94] It is an $M_r \approx 26\,500$ protein that binds reversibly and preferentially to the subunits of ornithine decarboxylase thereby destabilizing the dimer, and initiating its breakdown. Two pieces of evidence show that ornithine decarboxylase is degraded by a proteasome pathway. Its degradation is inhibited by antiproteasome antibodies and by the selective proteasome inhibitor, lactacystin.[95]

When the intracellular concentration of polyamines increases, antizyme production is induced. This induction is under translational control, since it can be inhibited by cycloheximide (a translational inhibitor) but not by actinomycin D (a transcription inhibitor). The antizyme mRNA is present in cells in which ornithine decarboxylase has not been suppressed, and it is an unusually stable mRNA ($t_{1/2} \approx 12\,h$). It is the C-terminus of the antizyme that binds ornithine decarboxylase. Amino-acid residues 106–212 of the antizyme contain the binding region, but residues 55–105 are also required for destabilization of ornithine decarboxylase. Ornithine decarboxylase from trypanosomes is shorter than those from mammalian sources lacking residues 423–461, which contain two PEST regions. It is comparatively stable and unable to bind the antizyme. This led to the identification of the antizyme binding site on mammalian ornithine decarboxylases, and the production of an engineered truncated version lacking residues 423–461, which was unable to bind the antizyme, and was much more

Lactacystin

stable. If *in vitro* translation of the antizyme mRNA is attempted using rabbit reticulocytes, no antizyme is detected unless the polyamine spermidine is added. Spermidine induces a frameshift thereby stimulating antizyme production. This was confirmed by making a variant antizyme mRNA in which a single nucleotide was excised from the 5′ end of the reading frame. This mRNA could be translated in the absence of spermidine.[95]

Purified ornithine decarboxylase is degraded by purified 26s proteasomes in the presence of ATP and the antizyme, but without ubiquitin. The antizyme is like ubiquitin in that neither are degraded and both are recycled. Little is yet known about the details of targeting ornithine decarboxylase to the proteasome.

Ornithine decarboxylase regulates the biosynthesis of the polyamines spermidine and spermine. However, the intracellular concentrations of these amines are also controlled by regulation of their degradation. Spermine/spermidine N^1-acetyltransferase catalyses the rate-limiting first step in the polyamine degradative pathway. Like ornithine decarboxylase it is a short-lived enzyme ($t_{1/2} = 20–40$ min) and its C-terminal region is essential for its rapid degradation, but unlike ornithine decarboxylase the C-terminal region does not contain PEST sequences. It contains a sequence MATEE which is critical for its degradation. Also, unlike ornithine decarboxylase it becomes polyubiquitinated prior to proteasome degradation.[96]

Two other proteins, IκBα and c-Jun, have been shown to be degraded by 26s proteasomes *in vitro* in the absence of ubiquitin.[66] IκBα is the inhibitory subunit of a transcriptional factor NF-κB, and c-Jun is also a transcriptional factor and a proto-oncogene product.

A number of enzymes have been shown to be degraded after they have been oxidized, e.g. glutamine synthetase, glutamine phosphoribosyl pyrophosphate amidotransferase, phosphoenolpyruvate carboxykinase, tyrosine aminotransferase, and fructose bisphosphate aldolase.[97] Oxygen radicals and other activated oxygen species generated as by-products of cellular metabolism or from environmental sources may cause the oxidation of certain amino-acid side-chains. This may result in the inactivation of an enzyme and also cause partial or complete unfolding. Glutamine synthetase is oxidized by mixed function oxidases and this causes the loss of a single histidine residue per subunit. These oxidized enzymes are generally more susceptible to proteasome degradation. In most mammalian cells oxidized proteins are cleaved in an ATP- and ubiquitin-independent pathway by the 20s core proteasome.[98]

9.6.3 Lysosomal protein degradation

Lysosomes play an important role in the turnover of long-lived intracellular proteins. Lysosomes perform at least two types of degradative process: phagocytosis and autophagy. The former is concerned with the uptake of foreign materials and their destruction, whereas the latter is concerned with intracellular turnover, particularly of organelles, and will be considered here. Lysosomes contain at least 50 different hydrolytic enzymes which all have the common feature of an acidic pH optimum. The pH within lysosomes is maintained at about 4.6. These hydrolytic enzymes include at least nine endopeptidases, known as cathepsins, and nine exopeptidases.[99] Together they have the capacity to degrade proteins to small peptides and amino acids. The concentrations of cathepsins D and B within the lysosomes are estimated to be as high as 0.78 and 1.54 mmol dm^{-3}. respectively.[100] Lysosomal proteolysis can be a very significant part of intracellular proteolysis. For example, when isolated hepatocytes are incubated in the absence of amino acids, cellular protein breakdown may be as much as 4–5% h^{-1} and 70–80% of this is due to autophagy by lysosomes; similar rates occur in the livers of whole animals when fasted for 10 h.[101] Lysosomes are thus likely to account for the degradation of organelles such as mitochondria and endoplasmic reticulum and long-lived cytosolic proteins. The process of autophagy entails the sequestration of organelles or cytosol into a particle known as an autophagosome.[102] The autophagosome normally fuses with endosomes and they in turn fuse with primary lysosomes to form active lysosomes. It is at this point that the sequestered proteins encounter the lysosomal proteases. Autophagy is a non-selective process as can be seen from measurements of the rate of sequestration of various cytosolic enzymes. Although the rates of sequestration of the enzymes listed in Table 9.5 are similar, there is a wide range in the rates of enzyme degradation. If the rates of enzyme degradation are measured in the presence of the leupeptin, an inhibitor of most lysosomal proteases, they are reduced only by the rate at which they are sequestered. This suggests that those with the faster degradation rates are degraded largely by the non-lysosomal mechanisms (Section 9.6.2) whereas the more stable enzymes are degraded entirely by the lysosomal route.[102]

Enzymes degraded by lysosomes do not generally require ubiquitination. There is, however, some evidence that certain membrane proteins may be ubiquitinated and degraded by the lysosomal pathway.[66] The growth hormone receptor is a transmembrane protein present on the plasma membrane. It is degraded by ligand-induced endocytosis. Endocytosis occurs after the growth factor receptor has bound growth factor. Using cells that have a thermolabile ubiquitinating system, it is found that when incubated above the permissive temperature degradation is inhibited.[103] There is also a yeast mating factor which is a transmembrane protein in the plasma membrane which has been shown to become ubiquitinated. It normally has a very short half-life (≈ 13 min), but in a vacuolar mutant deficient in a peptidase, the protein is much more stable ($t_{1/2} > 2$ h).[104] In yeast, vacuoles are the equivalent of lysosomes. It is not clear at present whether many membrane proteins are degraded by this route or whether these examples are atypical.

9.6.4 Proteins of the endoplasmic reticulum

Although some proteins present in the endoplasmic reticulum may be degraded by a lysosomal route, at least 25 proteins present in the endoplasmic reticulum are either targeted to the cytosol, where they are degraded by 26s proteasomes, or they are degraded on the luminal side of the endoplasmic reticulum. Many of

Table 9.5 Cytosolic liver enzymes with different half-lives are autophagically sequestered at the same rate

Enzyme	Half-life (h)	Degradation rate (% h⁻¹)	Rate of autophagic sequestration (% h⁻¹)
Aldolase	17.4	3.9±0.2	3.58±0.17
Lactate dehydrogenase	17.0	4.0±0.4	3.64±0.16
Glucokinase	12.7	5.3±0.6	3.38±0.46
Serine dehydratase	10.3	6.5±0.6	3.13±0.15
Tryptophan oxygenase	3.3	19.0±0.5	3.69±0.07
Tyrosine aminotransferase	3.1	20.0±1.1	3.52±0.08
Ornithine decarboxylase	0.9	52.9±1.2	3.28±0.07

Reference 102.

these proteins that have so far been studied are not enzymes, e.g. the cystic fibrosis transmembrane conductance regulator which transports Na^+ and MHC class 1 proteins, but they include misfolded carboxypeptidase Y and hydroxy-methylglutaryl CoA reductase. The major route for degradation of endoplasmic reticular proteins is thought to be through the ubiquitin–proteasome system, preceded by retrograde transport into the cytosol.[105] This applies both to proteins residing in the membrane and proteins on a secretory pathway. The retrograde transport has been studied mainly using mutants of *Saccharomyces cerevisiae*. The apparatus involved is known as a translocon and involves a protein Sec61p and the lumenal chaperone BiP. Those proteins degraded on the luminal side of the endoplasmic reticulum first combine with a luminal chaperone protein, cal-nexin, before degradation.[105,106] At present, it is not clear whether calnexin has a general role in the endoplasmic reticulum degradation process, or whether it is only required for selected groups of proteins.[105] Hydroxymethylglutaryl CoA reductase (HMG CoA reductase) contains two domains, a catalytic domain (548 residues) and an anchoring domain (339 residues). The latter anchors the enzyme to the endoplasmic reticulum. HMG CoA reductase has a short half-life (Table 9.1). It has been shown[107] that if the catalytic domain is cleaved from the complete enzyme, then the rate of degradation of the truncated enzyme (catalytic domain only) is only one-fifth the rate of degradation of the complete enzyme. The truncated enzyme is no longer able to associate with the membrane, which is believed to be involved in initiating the breakdown.

9.6.5 Protein degradation in prokaryotes

As mentioned in Section 9.1 there is little turnover of proteins in prokaryotes during exponential growth. The main proteins that are degraded are either abnormal proteins, which may be non-functional, and certain regulatory proteins. Up to about 80 per cent of such protein degradation is accounted for by the activities of two proteases known as protease Lon and protease Clp.[108–111] These two proteases belong to the broad family of soluble ATP-dependent proteases, which include also the 26s proteasome, although neither require ubiquitinated substrates. Apart from the ATP-dependence of proteolysis, the other feature they have in common is that their structures are barrel-shaped and the protease catalytic sites are within the barrel. Protease Lon (formerly known as protease La) is the product of the *lon* gene. It is a tetramer comprising four identical subunits (M_r 88 000) having separate ATP binding sites and proteolytic sites. The ATP binding site is in the N-terminal half. It is an endoprotease degrading proteins to peptides having between 5 and 20 amino-acid residues. The degradation is dependent on ATP hydrolysis, although small peptides can be degraded slowly in the absence of ATP. It has been proposed that the binding of ATP causes a conformational change activating the protease activity. Protease Lon has a high affinity for ATP ($K_{m(ATP)} \approx 50\ \mu mol\ dm^{-3}$) such that ATP would not be limiting at normal physiological concentrations ($\approx 3\ mmol\ dm^{-3}$). There is evidence that it has a protein binding site in addition to the proteolytic site which involves Ser679. In a mutant in which Ser679 is replaced by alanine, the proteolytic activity is lost but protein substrates are still able to bind to protease Lon and stimulate ATP hydrolysis.[112] Protein binding is thought to cause a conformational change in protease Lon promoting ATPase activation, normally required for proteolysis.[113] A protease homologue of Lon protease, PIM1 protease, is localized in the mitochondrial matrix, but its activity is dependent on the chaperone activity of the mitochondrial Hsp70 system.[114]

Protease Clp (EC 3.4.21.92) catalyses the ATP-dependent degradation of proteins to small peptides. The test substrate used to assay it is usually α-casein, and for this reason it became known as *caseinolytic protease* or *Clp*. It is able to degrade peptides having up to five amino-acid residues in the absence of ATP. It comprises two different subunits, ClpP (protease) and ClpA (ATPase), and its composition is $(ClpP_{14}ClpA_6)_2$.[115] It has many features in common with the 26s proteasome. ClpP exists as a stable tetradecamer consisting of two rings of seven subunits each. It associates with the ClpA in an ATP-dependent manner. The structure of ClpP has been solved at 2.3 Å resolution and the proteolytic sites located facing the central cavity.[108] Ser111 and His135 are essential for catalytic activity. ClpA has nucleotide binding motifs similar to other AAA ATPases and it is a member of the Hsp100 family of chaperones.[115]

A third protease which appears to have a narrower specificity is protease FtsH (*f*ilamentous *t*emperature *s*ensitive cell division mutant).[109] It is a metalloprotease (M_r = 71 000) belonging to the AAA ATPase family, anchored to the cell membrane, and having its ATP binding site on the cytosolic domain. It catalyses the degradation of a membrane translocase subunit, and a transcription factor σ^{30} (see Chapter 7, Section 7.3.1).

9.7 The significance of enzyme turnover

The turnover of specific enzymes has been studied most extensively in eukaryotes in response to nutritional and hormonal changes, and to a lesser extent in the different phases of the cell cycle. These aspects will be considered in turn.

9.7.1 Changes in enzyme-turnover rates in animals in response to changes in diet

Arginase and serine dehydratase present in the liver are concerned indirectly with protein breakdown. If an animal is fed on a high-protein diet, then the surplus protein is not stored as such; it becomes deaminated and the carbon skeletons of the amino acids are converted to precursors of lipid or carbohydrate before being stored in those forms. Non-essential dietary amino acids, of which serine is one, are degraded before the essential amino acids. Serine dehydratase catalyses the first step in the breakdown of serine (see opposite).

The ammonia resulting from deamination of the amino acids eventually enters the urea cycle to be converted to urea. Arginase catalyses the final step in urea formation. The levels of these two enzymes change in response to changing dietary intake. An increase in protein intake causes increases in the activities of both arginase and serine dehydratase in the liver. Both of these increases have been shown to be due to increases in total enzyme protein and not due to activation of pre-existing enzyme. A decrease in dietary protein causes a decrease in the amount of arginase. As can be seen from Table 9.6, changes in the levels of enzymes can be affected by changes in the rate of synthesis, in the rate of degradation, or of both. In the case of arginase a further adaptation has been observed by starving animals that had previously been fed on a low-protein diet. Under these conditions the arginase content of the liver increases in spite of there being no protein intake. Endogenous proteins are now degraded as a source of energy. In contrast to the increase in arginase activity when placed on a high protein diet, under fasting conditions the increase in arginase activity is brought about in a more economical fashion by a decrease in k_d but no change in k_s. An enzyme that resides in the plasma membrane and is believed to be important in amino-acid

$$CH_2OH \cdot CH(NH_2)COOH \rightarrow$$
$$CH_2 = C(NH_2)COOH \rightarrow$$
$$CH_3C(=NH)COOH \rightarrow$$
$$CH_3CO \cdot COOH + NH_3.$$

The reactions catalysed by serine dehydratase.

Acetyl CoA + CO$_2$ + ATP $\longrightarrow$

malonyl CoA + ADP
+ orthophosphate

uptake into cells is γ-glutamyltransferase. The activity of this enzyme is much higher in hepatoma cells than in normal liver cells, and this probably reflects the need for a greater rate of uptake to sustain the more rapid growth of the tumour cells. The increase is brought about by a decrease in the rate of degradation of the enzyme. Its half-life in normal cells is 3 h compared to 24 h in hepatoma cells.[121]

Another important enzyme to be studied in response to dietary change is acetyl CoA carboxylase (see opposite). This enzyme, which catalyses the first (rate-limiting) step in fatty-acid synthesis, is sensitive to changes in dietary fat.

High dietary fat lessens the need for endogenous fatty-acid synthesis and *vice versa*. Again, the changes in enzyme level can be brought about by changing either the rate of synthesis or the rate of degradation (Table 9.6). When animals are fasted, not only acetyl CoA carboxylase but also fatty-acid synthase and L-malate-NADP$^+$ oxidoreductase (oxaloacetate-decarboxylating) are degraded more rapidly than in fed animals. L-Malate-NADP$^+$ oxidoreductase (oxaloacetate-decarboxylating) is one of the enzymes involved in lipogenesis, since it catalyses the formation of NADPH.

9.7.2 Changes in enzyme stability in microorganisms in response to changes in carbon- and nitrogen-containing nutrients

Changes in the carbon and nitrogen sources in the growth medium cause adaptation in microorganisms; this adaptation may involve synthesis of new enzymes and degradation or inactivation of existing enzymes. An enzyme that is no longer necessary for the growth of the microorganisms on a new medium may be lost either passively, i.e. diluted out as the organism continues to grow, or in an active fashion by inactivation or degradation. The active loss of an enzyme that is no longer required may involve reversible or irreversible steps. Reversible inactivation may occur by covalent modification (see Chapter 6, Section 6.2.1). Examples from microorganisms include the following.

(1) Phosphorylation of the pyruvate dehydrogenase complex in *Neurospora crassa*[122] which occurs in the presence of ATP and inactivates the complex.

Table 9.6 The effect of diet on the turnover rates of enzymes from rat liver

Enzyme	Change of diet	Effect on enzyme
Arginase	Low-protein → high-protein	Increased activity mainly due to increased k_s
	High-protein → low-protein	Decreased activity, k_s decreased, k_d ncreased
	Low-protein → fasting	Increased activity, k_d decreased
Serine dehydratase	Protein-free → high-amino-acid	Large increase in activity due to increased k_s
Ornithine oxoglutarate aminotransferase	Basal → high-protein	Increased activity, increase in k_s, k_d unchanged
Acetyl CoA carboxylase	Basal → fat-free diet	Increased activity, increased k_s
	Basal → fasting	Decreased activity, k_s decreased, k_d ncreased
	Basal → high-fat diet	Decreased activity, k_s decreased, k_d ncreased
Fatty acid synthase	Basal → fat-free diet	Increased activity, k_s increased, k_d unchanged
	Basal → fasting	Decreased activity, k_s decreased, k_d increased
Pyruvate kinase	Basal → fasting	Decreased activity, k_s decreased, k_d slight increase
	Fasting → refeeding	Increased activity, k_s increased, k_d unchanged
Phosphoenol pyruvate carboxykinase	Fed → starved	Increased activity, k_s increased, k_d slightly decreased
	Starved → refed	Decreased activity, k_s decreased, k_d slightly increased

See references 2, 31, 40, 116–120.

(2) Inactivation of enzymes required for gluconeogenesis, i.e. phosphoenolpyruvate carboxykinase, malate dehydrogenase, and fructose-bisphosphatase in *S. cerevisiae* when the source of carbon is changed from acetate to glucose. When *S. cerevisiae* is grown with acetate as the source of carbon, it carries out gluconeogenesis to provide it with adequate hexoses and their derivatives. On the other hand, when it is supplied with glucose as the source of carbon, gluconeogenesis is suppressed. These differences are reflected in the stability of fructose-bisphosphatase, which catalyses a rate-limiting step in gluconeogenesis (see Chapter 6, Section 6.4.1). When grown on acetate, the k_d for fructose-bisphosphatase breakdown is 0.008 h^{-1}, whereas when grown on glucose it is 0.42 h^{-1}. The role of phosphorylation is best seen when the acetate-grown cells are switched to a glucose medium. Within 2 min the activity of fructose-bisphosphatase drops very dramatically. This initial fall can be shown to be due to phosphorylation of a serine residue on fructose-bisphosphatase, although no proteolysis occurs. However, by 30 min the fructose-bisphosphatase protein is degraded. Thus, it appears that the initial inactivation is through phosphorylation, but that this then labels the protein for degradation, i.e. a conditionally active mechanism (see Section 9.6.1).

(3) Phosphorylation of NAD$^+$-dependent glutamate dehydrogenase from *S. cerevisiae* and *Candida utilis* to form a much less active enzyme.[79] This enzyme is responsible in yeast for glutamate breakdown and is phosphorylated when glutamate in the medium is exchanged for ammonia and glucose as sources of nitrogen and carbon. The NAD$^+$-requiring glutamate dehydrogenase in these organisms is principally concerned with oxidative deamination of glutamate:

$$\text{Glutamate} + \text{NAD}^+ \rightleftharpoons \text{2-oxoglutarate} + \text{NADH} + \text{NH}_4^+.$$

When *S. cerevisiae* is grown with glutamate as the source of carbon and nitrogen, the NAD$^+$-requiring glutamate dehydrogenase activity is high. If the organism is switched to glucose and ammonia as the sources of carbon and nitrogen, then oxidative deamination of glutamate is not required and the NAD$^+$-requiring glutamate dehydrogenase activity rapidly declines. The mechanism is a conditionally active one similar to that in the previous example, namely the glutamate dehydrogenase initially becomes phosphorylated and is then degraded by proteolysis.

(4) Inactivation of ornithine carbamoyltransferase in yeast by binding arginase. This occurs when arginine is added to the growth medium and is described more fully in Chapter 8, Section 8.5.3.3.

9.7.3 Changes in enzyme turnover in response to hormonal stimuli

The systems most studied under this heading are various liver enzymes concerned with the catabolism of amino acids and their response to hormones, particularly the glucocorticoids. Some studies have been made with intact animals but others have used a strain of hepatoma cells grown in culture which responds very similarly to that of intact liver and which has the advantage that the hormone concentration can be more easily controlled. Glucocorticoids include cortisol, corticosterone, and cortisone, and their principal effects are to raise blood glucose levels, mobilize amino acids, and stimulate gluconeogenesis in the liver. Trytophan 2,3-dioxygenase, the first enzyme in the tryptophan catabolic pathway, can be induced in liver

Table 9.7 Changes in enzyme turnover in response to hormonal treatment

Tissue	Enzyme	Hormonal change	Effect on enzyme
Rat liver	Tryptophan 2,3-dioxygenase	Glucocorticoid administration	Increased activity, k_s increased, k_d unchanged
Rat liver	Tyrosine aminotransferase	Glucocorticoid, glucagon or insulin administration	Increased activity, k_s increased, k_d unchanged
Rat kidney	Ornithine oxo-acid aminotransferase	Oestrogen	Increased activity, k_s increased
Rat liver	Carbamoyl phosphate synthase	Thyroidectomy	Decreased k_d
Rat liver	Malate dehydrogenase	Thyroidectomy	Decreased k_d
Rat liver	Arginase	Glucocorticoid	Increased activity, k_s increased
Rat liver	Serine dehydratase	Glucagon	Increased activity, k_s increased

References 2, 32, 39.

by administration of either cortisol or tryptophan or analogues of tryptophan. In each case the total amount of enzyme protein is increased, but in the case of cortisol administration it is due to an increase in the rate of enzyme synthesis, whereas in response to tryptophan or its analogues the rate of enzyme degradation is decreased, possibly as a result of stabilization of the enzyme in the presence of the substrate. The activity of tyrosine aminotransferase, the first step in tyrosine breakdown, is also induced by cortisol which increases the rate of its biosynthesis. Unlike tryptophan 2,3-dioxygenase, tyrosine aminotransferase is also induced by the pancreatic hormones insulin and glucagon.

Other liver enzymes that have been studied are alanine aminotransferase and ornithine oxoacid aminotransferase. The former increases in response to glucocorticoids, which affect only the rate of synthesis, while the latter has been shown to increase in response to oestrogen, again by an increase in the rate of synthesis; there is no change in the rate of degradation. The effects are summarized in Table 9.7. The effects of hormones on the turnover of specific enzymes have been most studied in liver, but hormone effects on protein turnover in general have also been studied extensively in skeletal muscle. In this tissue insulin, growth hormone, insulin-like growth factor (IGF), testosterone, and adrenaline all favour protein synthesis. This is largely achieved by an increase in the rate of protein synthesis in the case of growth hormone, IGF, and testosterone, but insulin and adrenalin act mainly by decreasing the rate of protein catabolism.[21]

The steroid and thyroid hormones enter the cells of the target tissue, where they bind to their hormone receptors. This induces a conformational change in the receptor proteins, which in turn enables them to bind DNA at the appropriate hormone response element upstream from the gene encoding for the enzymes and stimulating their expression. Other hormones, such as glucagon and adrenalin, bind to receptors on the cell membrane and their effects are thought to be mediated through changes in the intracellular concentrations of cyclic nucleotides (see Chapter 6, Section 6.4.2.1).

9.7.4 The role of proteolysis in the cell cycle

The cycle of events that leads to the formation of two daughter cells from a single parent cell is known as the cell cycle. Understanding the molecular changes that occur and how they are regulated is central to understanding growth, differentiation, and oncogenesis. A prerequisite for mitosis to occur is the replication of DNA. The cell cycle (Fig. 9.10) is divided into four phases: (i) the G_1 phase or

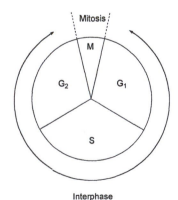

Fig. 9.10 Phases of the cell cycle. Mitosis generally represents about 5 per cent of the cell cycle.

presynthetic gap, the phase before DNA is replicated, (ii) the S phase in which DNA is replicated, (iii) the G_2 phase or post-synthetic gap, and (iv) the M phase, or mitotic phase, in which the nuclear membrane disappears, the chromosomes segregate to opposite poles and the nuclear membrane reforms around the daughter nuclei, and the cells divide into two daughter cells.

Since the late 1970s much progress has been made in understanding key enzyme changes that occur during the cycle. Much of the understanding has come from the study of mutants of budding yeast (*Saccharomyces cerevisiae*) and fission yeast (*Schizosaccharomyces pombe*) and many of the important protein factors involved have names which relate to these mutations, giving rise to a very complex nomenclature. In this section the aim is restricted to explaining the role of proteolysis in the context of the cell cycle. For further information on the cell cycle and the nomenclature see reference 123.

Two types of polypeptide that have key roles as regulators of the cell cycle are the p34 protein and a family of proteins known as cyclins. The p34 is a polypeptide, M_r 34 000, that associates with one of the cyclins to form a heterodimer known as cyclin-dependent protein kinase. The p34 protein is the product of the *cdc28* gene in *Saccharomyces cerevisiae* but the homologous gene in *Schizosaccharomyces pombe* is known as *cdc2*. There are number of different, but related, p34s and also a number of different cyclins, giving rise to a large number of potential heterodimers, each with slightly different properties. The catalytic activities of cyclin-dependent protein kinases are regulated by phosphorylation at three sites on p34: Thr-14, Tyr-15, and Thr-161. A large number of substances have been suggested as potential physiological substrates of cyclin-dependent protein kinases including lamins, vimentin, caldesmon, nucleolin, histone H1, various tyrosine kinases, and RNA polymerase II.[124]

Cyclins are synthesized periodically during the cell cycle, appearing at highest concentration during the M phase. No enzyme activity has been attributed to them; it has been suggested that their role is to determine the subcellular location and/or specificity of p34. The different families of cyclins are designated A, B, C, D, and E, but they can also be divided into those synthesized in late G_1 phase and those synthesized in S phase. The appearance and disappearance of particular cyclin-dependent protein kinases during the cell cycle is regulated by the synthesis and proteolytic degradation of specific cyclins (Fig. 9.11).[125] The half-life of cyclin B may change over a period of 1–2 min from greater than 1 h to less than 1 min.[125] Two *Saccharomyces* cyclins Cln1 and Cln2, required for progression from G_1, are synthesized in late G_1 phase and disappear in S phase as a result of protein degradation.[85] Cln2 has been shown to become polyubiquitinated by the E_2 enzyme, and genetic evidence indicates that proteasomes are responsible. Conditional mutants in a non-ATPase regulatory subunit of the 26s proteasome are blocked in both the G_1–S transition and the G_2–M transition at the restrictive temperature. Cyclins required during the S phase such as cyclins Clb5 and Clb2 are also degraded by a ubiquitin–proteasome pathway. PEST sequences are thought to provide the degradation signal for Cln2, and the 'destruction box' (RTALGDIGN)[84] for Clb5 and Clb2. The destruction box was originally identified when a truncated cyclin lacking the first 90 amino-acid residues was found not to be degraded at mitosis and the cells remained in a persistent metaphase state.[126] When the first 90 amino-acid residues of the sequence of a number of cyclins from different species were compared, a common feature was the RTALGDIGN motif. Studies with further mutants showed that residues 54–66 appeared to be the determinants of proteolysis and these contained the destruction box.

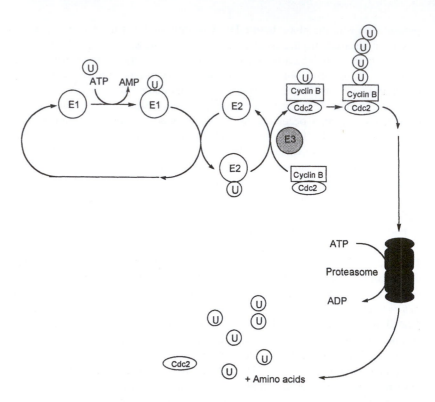

Fig. 9.11 Steps in the degradation of cyclin B.

In addition to cyclins that are required for cyclin-dependent kinase activity, there are proteins that inhibit cyclin-dependent kinase activity. One such protein present in *Saccharomyces cerevisiae* is Sic1. It is also degraded by a ubiquitin–proteasome pathway, so that proteolysis can have the effects of both turning on and turning off activity.[85] Although proteasomes may be the principal proteolytic agents regulating progress through the cell cycle, other proteases such as calpains may be important. A calpain-selective inhibitor, ZLLY-DMK, proved to be more effective in inhibiting the progress of fibroblasts in late G_1 into S phase than the proteasome inhibitor lactocystin.[127]

9.8 Other processes in which intracellular proteolysis is important

There are a number of processes in which intracellular proteolysis is important, involving degradation of proteins but not necessarily enzymes.

9.8.1 Apoptosis

Apoptosis is the term used to describe programmed cell death. It is important in the regulation of cell turnover during development, tissue homeostasis, senescence, and in cancer. It involves an irreversible step involving a cascade of proteases which eventually cause cell disintegration.[128] The main proteases that have so far been implicated in apoptosis are caspases, but the proteasome pathway may also be important. Caspases (from **C**ysteine **asp**artate) are a family of cysteine proteases structurally related to interleukin-1β converting enzymes sometimes abbreviated to ICE and having a near-absolute requirement for an aspartate residue in the P_1 site (Chapter 5, Fig. 5.9). The family has been divided into three groups based on sequence relationships.[128] Caspases are synthesized as zymogens, in the case of caspase 1 the zymogen (M_r 45 000) which becomes activated to its active form comprising two subunits, one of M_r 20 000 and the other 10 000.[129] From a study of developmental mutants it has been shown that 14 genes are involved in apoptosis in the nematode *Caenorhabditis elegans*. The gene product of one of these CED-3 is related to mammalian caspases.[129]

Proteasomes have been shown to be involved in thyroxine-induced apoptosis of the tadpole tail. By the use of proteasome inhibitors there is evidence that proteasomes are required for apoptosis in proliferating cells but not in quiescent non-dividing cells.[130]

9.8.2 Oncogenes and transcriptional factors

Proto-oncogenes and transcriptional factors are amongst the short-lived proteins that are degraded by a proteasome pathway. The proto-oncogene c-*jun* encodes transcriptional activator c-Jun which is degraded by 26s proteasomes. A ubiquitinated form of c-Jun has been detected although ubiquitination may not be an absolute requirement for degradation.[132,132] The viral oncogene product v-Jun lacks the δ domain present in c-Jun and is a relatively long-lived protein. The δ domain appears to be necessary for ubiquitination.[131] Other proto-oncogene products including N-Myc, c-Myc, and c-Fos are rapidly degraded after ubiquitination.[133] The p53 tumour suppressor protein is one of a group of proteins that show sequence-specific DNA binding. It becomes destabilized and degraded after interaction with a gene product from the oncogenic papilloma virus. This gene product interacts with the ubiquitin ligase E_3 to promote the degradation of p53.[134] p53 is the most commonly mutated gene in human cancer with most mutations occurring in the DNA contact residues between 102 and 292 resulting in loss of DNA binding.[135] This renders it more susceptible to proteolysis.

The NFκB transcription factor family regulates the expression of a large number of different genes including those involved in the immune response. It is formed from a precursor (M_r 105 000) by release of a fragment (M_r 50 000). The p50 combines with another subunit, p65, to form a heterodimer. This NFκB heterodimer is capable of binding a decameric sequence on the immunoglobulin light chain enhancer DNA. It can also bind other enhancers. A third polypeptide IκBα regulates the actions of the heterodimer. When it is bound to the heterodimer, the complex is located in the cytoplasm. A variety of extracellular signals stimulate the breakdown of the regulatory polypeptide IκBα by a ubiquitin-dependent process. This enables the heterodimer to translocate to the nucleus and interact with the enhancer region and activate transcription. The formation of the p50 from its precursor is also a ubiquitin-dependent process.[134,136]

9.8.3 Phytochrome turnover

Phytochrome is a major photoreceptor in plants involved in red light-mediated morphogenesis. It is a polypeptide (M_r 124 000) having a tetrapyrrole chromophore. A pulse of light converts the red light-absorbing form to a far-red light-absorbing form, which then aggregates. The latter has a half-life of ≈ 2 h compared with > 100 h for the former. It appears that the conformational change, which involves isomerization of the chromophore, causes increased ubiquitination and degradation of the far-red light-absorbing form.[137] This allows plants to respond to darkness by quickly removing active phytochrome.

9.9 Conclusions

Enzyme turnover is an important method of regulation of enzyme activity and is controlled by regulation of either enzyme synthesis or enzyme inactivation/degradation, or in some cases by regulation of both steps. The general mechanism of enzyme synthesis is well understood. Considerable progress has been made since the early 1980s in the general understanding of protein degradation, particularly by the ubiquitin–proteasome pathway. The resolution of the proteasome structure has been a major advance in this respect. There are still many details concerning the recognition signals to be resolved. These not only have to explain how different proteins are recognized by the ubiquitinating enzymes, but also the factors that determine the wide range in half-lives of different enzymes. Also although a general mechanism for protein degradation via ubiquitination has been formulated, for the vast majority of enzymes whose turnover rates have been measured (see Table 9.1) their mechanism of degradation has not been proven. For a critical account of enigmatic aspects of intracellular protein degradation the reader is referred to reference 42. The hypothesis that there are two basic mechanisms of intracellular protein degradation; the ubiquitin–proteasome pathway for short-lived proteins and the non-ubiquitin lysosomal pathway for long-lived proteins at present appear to be an oversimplification; a number of proteins do not appear to fit neatly into either category.

References

1. Schoenheimer, R., *The dynamic state of body constituents.* Harvard University Press, Cambridge, Massachusetts (1942).

2. Schimke, R. T., *Adv. Enzymol.* **37**, 135 (1973).

3. Goldberg, A. L. and Dice, J. F., *A. Rev. Biochem.* **43**, 835 (1974).

4. Goldberg, A. L. and St John, A. C., *A. Rev. Biochem.* **45**, 747 (1976).

5. Beynon, R. J. and Bond, J. S., *Am. J. Physiol.* **251**, C141 (1986).

6. Doherty, F. and Mayer, R. J., *Intracellular protein degradation.* IRL Press, Oxford (1992).

7. Peters, J.-M., Harris, J. R. and Finley, D., *Ubiquitin and the biology of the cell.* Plenum, New York (1998).

8. Rivett, A. J., *Curr. Top. cell. Regul.* **28**, 291 (1986).

9. Hershko, A. and Ciechanover, A., *A. Rev. Biochem.* **61**, 761 (1992).

10. Walker, R., *The molecular biology of enzyme synthesis*, Chapter 1. Wiley, New York (1983).

11. Coux, O., Tanaka, K., and Goldberg, A. L., *A. Rev. Biochem.* **65**, 801 (1996).

12. Stryer, L., *Biochemistry* (4th edn), Chapter 34. Freeman, New York (1995).

13. Voet, D. and Voet, J. G., *Biochemistry* (2nd edn), Chapter 30. Wiley, New York (1995).

14. Mathews, C. K. and Van Holde, K. E., *Biochemistry* (2nd edn), Chapter 27. Benjamin/Cummings, Menlo Park, California (1995).

15. Adams, R. L. P., Knowler, J. T., and Leader, D. P., *The biochemistry of nucleic acids* (11th edn), Chapter 12. Chapman and Hall, London (1992).

16. Berlin, C. M. and Schimke, R. T., *Molec. Pharmacol.* **1**, 149 (1965).

17. Chiu, F. C. and Goldman, J. E., *J. Neurochem.* **42**, 166 (1984).

18. Zak, R., Martin, A. F., and Blough, R., *Physiol. Rev.* **59**, 407 (1979).

19. Garlick, P. J., in *Comprehensive biochemistry* (Florkin, M. and Stotz, E. H., eds), Vol. **19B**, Chapter 2. Elsevier, Amsterdam (1980).

20. Wilde, C. J., Paskin, N., Saxton, J., and Mayer, R. J., *Biochem. J.* **192**, 311 (1980).

21. Rooyackers, O. E. and Nair, K. S., *A. Rev. Nutr.* **17**, 457 (1997).

22. Voet, D. and Voet, J. G., *Biochemistry* (2nd edn), pp. 1214–15. Wiley, New York (1995).

23. Schimke, R. T., *J. biol. Chem.* **239**, 3808 (1964).

24. Millward, D. J., *Clin. Sci.* **39**, 577 (1970).

25. Druyan, R., De Bernard, B., and Rabinowitz, M., *J. biol. Chem.* **244**, 5874 (1969).

26. Gunn, J. M., Ballard, J. F., and Hanson, R. W., *J. biol. Chem.* **251**, 3586 (1976).

27. Schimke, R. T. and Katanuma, N., *Intracellular protein turnover*, p. 143. Academic Press, New York (1975).

28. Schimke, R. T. and Katanuma, N., *Intracellular protein turnover*, p. 181. Academic Press, New York (1975).

29. Schimke, R. T. and Katanuma, N., *Intracellular protein turnover*, p. 283. Academic Press, New York (1975).

30. Hopgood, M. F. and Ballard, F. J., *Biochem. J.* **144**, 371 (1974).

31. Volpe, J. J., Lyles, T. O., Runcari, D. A. K., and Vagelos, P. R., *J. biol. Chem.* **248**, 2502 (1978).

32. Kobayasbi, K., Kito, K., and Katunuma, N., *J. Biochem.* **79**, 787 (1976).

33. Masters, C., *Curr. Top. cell. Regul.* **21**, 205 (1982).

34. Goldberg, A. L. and St John, A. C., *A. Rev. Biochem.* **45**, 751 (1976).

35. Gear, A. R. L., *Biochem. J.* **120**, 577 (1970).

36. Walker, J. H., Burgess, R. J., and Mayer, R. J., *Biochem. J.* **176**, 927 (1978).

37. Russell, S. M., Burgess, R. J., and Mayer, R. J., *Biochem. J.* **192**, 321 (1980).

38. Schimke, R. T. and Katanuma, N., *Intracellular protein turnover*, p. 325. Academic Press, New York (1975).

39. Nicoletti, M., Guerri, C., and Grisolia, S., *Eur. J. Biochem.* **75**, 583 (1977).

40. Ip, M. M., Chee, P. Y., and Swick, R. W., *Biochem. biophys. Acta* **354**, 29 (1974).

41. Miyazawa, S., Ozasa, H., Furuta, S., Osumi, T., Hashimoto, T., Miura, S. *et al.*, *J. Biochem.* **93**, 453 (1983).

42. Jennissen, H. P., *Eur. J. Biochem.* **231**, 1 (1995).

43. McLintock, D. K. and Markus, G., *J. biol. Chem.* **243**, 2855 (1968).

44. Hillcoat, B. L., Swett, V., and Bertino, J. R., *Proc. natn. Acad. Sci. USA* **58**, 1632 (1967).

45. McCann, P. P., Tardiff, C., Duchesne, M., and Mamont, P. S., *Biochem. Biophys. Res. Commun.* **76**, 893 (1977).

46. Stevens, L. and McKinnon, I. M., *Biochem. J.* **166**, 635 (1977).

47. Taneke, T., Wade, K., Ogiwara, H., and Numa, S., *FEBS Lett.* **82**, 95 (1977).

48. Sinensky, M. and Logel, J., *J. biol. Chem.* **258**, 8547 (1983).

49. Dice, J. F. and Goldberg, A. L., *Arch. Biochem. Biophys.* **170**, 213 (1975).

50. Bohley, P., Wollert, H-G., Riemann, D., and Riemann, S., *Acta Biol. Med. Ger.* **40**, 1655 (1981).

51. Dice, J. F. and Goldberg, A. L., *Proc. natn. Acad. Sci. USA* **72**, 3893 (1975).

52. Katunuma, N., Kominami, E., Banno, Y., Kito, K., Aoki, Y., and Urata, G., *Adv. Enz. Regul.* **14**, 325 (1976).

53. Woodbury, R. G., Everett, M., Sanada, Y., Katanuma, N., Lagunoff, D., and Neurath, H., *Proc. natn. Acad. Sci. USA* **75**, 5311 (1978).

54. Rechsteiner, M., Chin, D., Hough, R., McGarry, T., Rogers, S., Rote, K. V. *et al.*, *Cell Fusion* (Ciba Symposium 103) (Evered, D. and Whelan, J., eds), pp. 181–201 (1984).

55. Rogers, S., Wells, R., and Rechsteiner, M., *Science* **234**, 364 (1986).

56. Rechsteiner, M. and Rogers, S. W., *Trends Biochem. Sci.* **21**, 267 (1996).

57. Jentsch, S. and Schlenker, S., *Cell* **82**, 881 (1995).

58. Lin, S. and Zabin, I., *J. biol. Chem.* **247**, 2205 (1972).

59. Ballard, F. J., *Essays Biochem.* **13**, 22 (1977).

60. Hershko, A., Ciechanover, A., Heller, H., Haas, A. L., and Rose, I. A., *Proc. natn. Acad. Sci. USA* **77**, 1783 (1980).

61. Wilk, S. and Orlowski, M., *J. Neurochem.* **35**, 1172 (1980)

62. Baumeister, W. and Lupas, A., *Curr. Opin. Struct. Biol.* **7**, 273 (1997).

63. Ciechanover, A., Finley, D., and Varshavsky, A., *Cell* **37**, 57 (1984).

64. Murakami, Y., Tanahashi, N., Tanaka, K., Omura, S., and Hayashi, S., *Biochem. J.* **317**, 77 (1996).

65. Dice, J. F., *Trends Biochem. Sci.* **15**, 305 (1990).

66. Pickart, C. M., *FASEB J.* **11**, 1055 (1997).

67. Poole, B., and Wibo, M., *J. biol. Chem.* **248**, 6221 (1973).

68. Neff, N. T., De Martino, G. N., and Goldberg, A. L., *J. Cell Physiol.* **101**, 439 (1979).

69. Hough, R., and Rechsteiner, M., *Proc. natn. Acad. Sci. USA* **81**, 90 (1984).

70. McElligott, M. A., Miao, P., and Dice J. F., *J. biol. Chem.* **260**, 11986 (1985).

71. Fenteany, G., Standaert, R. F., Lane, W. S., Choi, S., Corey, E. J., and Schreiber, S. L., *Science* **268**, 726 (1995).

72. Haas, A. L., and Siepmann, T. J., *FASEB J.* **11**, 1257 (1997).

73. Johnston, S. C., Larsen, C. N., Cook, W. J., Wilkinson, K. D., and Hill, C. P., *EMBO J.* **16**, 3787 (1997).

74. Varshavsky, A., *Trends Biochem. Sci.* **22**, 383 (1997).

75. Kumar, S., Kao, W. H., and Howley, P. M., *J. biol. Chem.* **272**, 13548 (1997).

76. Haas, A. L., *FASEB J.* **11**, 1053 (1997).

77. Wilkinson, K. D., *FASEB J.* **11**, 1245 (1997).

78. Purwin, C., Leidig, F., and Holzer, H., *Biochem. Biophys. Res. Commun.* **107**, 1482 (1982).

79. Hemmings, B. A., in *Molecular aspects of cellular regulation* (Cohen, P., ed.), Vol. 3, p. 155. Elsevier, Amsterdam (1984).

80. Parker, R. A., Miller, S. J., and Gibson, D. M., *Biochem. Biophys. Res. Commun.* **125**, 629 (1984).

81. Bergström, G., Ekman, P., Humble, E., and Engström, L., *Biochem. Biophys. Acta* **532**, 259 (1978).

82. Bachmair, A., Finley, D., and Varshavsky, A., *Science* **234**, 179 (1986).

83. Rogers, S. W. and Rechsteiner, M., *J. Biol. Chem.* **263**, 19850 (1988).

84. Seufert, W., Futcher, B., and Jentsch, S., *Nature* **373**, 78 (1995).

85. Hilt, W. and Wolf, D. H., *Trends Biochem. Sci.* **21**, 96 (1996).

86. Baumeister, W., Cejka, Z., Kania, M., and Seemuller, E., *Biol. Chem.* **378**, 121 (1997).

87. Löwe, J., Stock, D., Jap, B., Zwicki, P., Baumeister, W., and Huber, R., *Science* **268**, 533 (1995).

88. Baumeister, W., Walz, J., Zühl, F., and Seemüller, E., *Cell* **92**, 367 (1998).

89. Fenteany, G. and Schreiber, S. L., *J. biol. Chem.* **273**, 8545 (1998).

90. Patel, S. and Latterich, M., *Trends Cell Biol.* **8**, 65 (1998).

91. Lupas, A., Flanagan, J. M., Tamura, T., and Baumeister, W., *Trends Biochem. Sci.* **22**, 399 (1997).

92. Lam, Y. A., Xu, W., DeMartino, G., and Cohen, R. E., *Nature* **385**, 737 (1997).

93. Hayashi, S., Murakami, Y., and Matsufuji, S., *Trends Biochem. Sci.* **21**, 27 (1996).

94. Fong, W. F., Heller, J. S., and Canellakis, E. S., *Biochim. Biophys. Acta* **428**, 456 (1976).

95. Murakami, Y., Tanahashi, N., Tanaka, K., Omura, S., and Hayashi, S., *Biochem. J.* **317**, 77 (1996).

96. Coleman, C. S. and Pegg, A. E., *J. biol. Chem.* **272**, 12164 (1997).

97. Rivett, A. J. and Levine, R. L., *Biochem. Soc. Trans.* **15**, 816 (1987).

98. Grune, T., Reinheckel, T., and Davies, K. J. A., *FASEB J.* **11**, 526 (1997).

99. Kirschke, H., and Barrett, A. J., in *Lysosomes: their role in protein breakdown* (Glaumann, H. and Ballard, J. F., eds), Chapter 6. Academic Press, London (1987).

100. Dean, R. T. and Barrett, A. J., *Essays Biochem.* **12**, 1 (1976).

101. Seglen, P. O., in *Lysosomes: their role in protein breakdown* (Glaumann, H. and Ballard, J. F., eds), Chapter 10. Academic Press, London (1987).

102. Seglen, P.O., Gordon, P. B., and Holen, I., *Seminars Cell Biol.* **1**, 441 (1990).

103. Strous, G. J., van Kerkhof, P., Govers, R., Ciechanover, A., and Schwartz, A. L., *EMBO J.* **15**, 3806 (1996).

104. Kölling, R. and Hollenberg, C. P., *EMBO J.* **13**, 3261 (1994).

105. Sommer, T. and Wolf, D. H., *FASEB J.* **11**, 1227 (1997).

106. Brodsky, J. L. and McCracken, A. A., *Trends Cell Biol.* **7**, 151 (1997).

107. Gill, G., Faust, J. R., Chiu, D. J., Goldstein, J. L., and Brown, M. S., *Cell* **41**, 249 (1985).

108. Wang, J., Hartling, J. A., and Flanagan, J. M., *Cell* **91**, 447 (1997).

109. Suziki, C. K., Rep, M., van Dijl, J. M., Suda, K., Grivell, L. A., and Schatz, G., *Trends Biochem. Sci.* **22**, 118 (1997).

110. Goldberg, A. L., *Eur. J. Biochem.* **203**, 9 (1992).

111. Gottesman, S. and Maurizi, M. R., *Microbiol. Rev.* **56**, 592 (1992).

112. Fischer, H. and Glockshuber, R., *J. biol. Chem.* **268**, 22502 (1993).

113. Goldberg, A. L., Moerschell, R. P., Chung, C. H., and Maurizi, R., *Methods Enzymol.* **244**, 350 (1994).

114. Langer, T. and Neupert, W., *Experientia* **52**, 1069 (1996).

115. Barrett, A. J., *Eur. J. Biochem.* **237**, 1 (1996).

116. Schimke, R. T., and Katunuma, N., *Intracellular protein turnover*, p. 127. Academic Press, New York (1975).

117. Cladaras, C. and Cottam, G. L., *Arch. Biochem. Biophys.* **200**, 426 (1980).

118. Gunn, J. M., Hanson, R. W., Meynhas, O., Reshef, L., and Ballard, F. J., *Biochem. J.* **150**, 195 (1975).

119. Hopgood, M. F., Ballard, F. J., Reshef, L., and Hanson, R. W., *Biochem. J.* **134**, 445 (1973).

120. Numa, S. and Yamashita, S., *Curr. Top. cell. Regul.* **8**, 197 (1974). See pp. 206–20.

121. Ding, J. L., Smith, G. D., and Peters, T. J., *FEBS Lett.* **142**, 207 (1982).

122. Wieland, O. H., Hartmann, U., and Siess, E. A., *FEBS Lett.* **27**, 240 (1972).

123. Murray, A. and Hunt, T., *The cell cycle.* Freeman, New York (1993).

124. Nigg, E. E., *Trends Cell Biol.* **3**, 296 (1993).

125. Murray, A., *Cell* **81**, 149 (1995).

126. Glotzer, M., Murray, A. W., and Kirschner, M. W., *Nature* **349**, 132 (1991).

127. Mellgren, R. L., *Biochem. Biophys. Res. Commun.* **236**, 555 (1997).

128. Nagata, S., *Curr. Biol.* **6**, 1241 (1996).

129. Nicholson, D. W. and Thornberry, N. A.. *Trends Biochem. Sci.* **22**, 299 (1997).

130. Drexler, H. C. A., *Proc. natn. Acad. Sci. USA.* **94**, 855 (1997).

131. Treier, M., Staszewski, L. M., and Bohmann, D., *Cell* **78**, 787 (1994).

132. Jariel-Encontre, I., Pariat, M., Martin. F., Carillo, S., Salvat,C., and Piechaczyk, M., *J. biol. Chem.* **270**, 16623 (1995).

133. Ciechanover, A., DiGuissseppe, J. A., Bercovich, B., Orian, A., Richter, J. D., Schwartz, A. L. *et al.*, *Proc. natn. Acad. Sci. USA.* **88**, 139 (1991).

134. Wilkinson, K. D., *A. Rev. Nutr.* **15**, 161 (1995).

135. Cho, Y., Gorina, S., Jeffrey, P. D., and Pavletich, N. P., *Science* **265**, 346 (1994).

136. Chiao, P. J., Miyamoto, S., and Verma, I. M., *Proc. natn. Acad. Sci. USA* **91**, 28 (1994).

137. Jabben, M., Shanklin, J., and Vierstra, R. D., *J. biol. Chem.* **264**, 4998 (1989).

Appendix 9.1

Derivation of equation describing the time course of approach to a new steady-state level of enzyme after a change in its rate of synthesis and/or degradation has occurred.

The steady-state level of an enzyme in a cell or tissue is given by the equation:

$$\frac{d[E]}{dt} = k_s - k_d[E] \qquad \text{(see Section 9.3)}$$

where k_s is the rate constant for its synthesis (zero order with respect to [E]) and k_d is the rate constant for its degradation (first order with respect to [E]).

If there is a change in the rate of synthesis and/or degradation of the enzyme as a result of some hormonal, dietary, etc. influence then the enzyme concentration will change to a new steady-state level.

Let

k_s' be the new rate constant for enzyme synthesis,
k_d' be the new rate constant for enzyme degradation,
$[E_0]$ be the initial steady-state concentration,
$[E_t]$ be the enzyme concentration at time t after change in rate constants from k_s to k_s' and k_d to k_d'.

$$\frac{d[E]}{dt} = k_s' - k_d'[E] \qquad (A.9.1)$$

$$\frac{d[E]}{(k_s' - k_d'[E])} = dt. \qquad (A.9.2)$$

By integration of eqn (A.9.2),

$$\ln(k_s' - k_d'[E]) = -k_d't + c. \qquad (A.9.3)$$

At time $t = 0$, $[E] = [E_0]$, therefore $c = \ln(k_s' - k_d'[E_0])$. Thus at time t,

$$\ln(k_d' - k_d'[E_t]) - \ln(k_s' - k_d'[E_0]) = -k_d't. \qquad (A.9.4)$$

Taking antilogarithms of eqn (A.9.4):

$$\frac{k_s' - k_d'[E_t]}{k_s' - k_d'[E_0]} = e^{-k_d't} \qquad (A.9.5)$$

This can be rearranged to give

$$\frac{[E_t]}{[E_0]} = \frac{k_s'}{k_d'[E_0]} - \left(\frac{k_s'}{k_d'[E_0]} - 1\right)e^{-k_d't}.$$

See also reference 16.

10

Clinical aspects of enzymology

10.1 Introduction

Since the mid-1950s there has been a considerable increase in both the measurement of enzyme activities and the use of purified enzymes, in clinical practice. Before 1940 the only enzymes whose activities were measured for clinical diagnoses were the hydrolytic enzymes lipase, amylase, phosphatases, trypsin, and pepsin, and their measurement constituted less than 5 per cent of the analyses carried out in the average clinical chemistry laboratory. At the present time up to 25 per cent of the work of an average clinical chemistry laboratory may consist of enzyme assays for diagnoses and up to about 20 different enzymes are assayed routinely. The reason for the change in emphasis is largely due to increased understanding and awareness of the molecular details of metabolism and to the development of rapid and reliable methods of enzyme assay. The discovery in 1954 that serum aspartate aminotransferase activity increased shortly after myocardial infarction greatly stimulated the search for other enzymes in serum which could be used as indicators of tissue damage. Understanding the function of particular enzymes has enabled the rational design of drugs to inhibit specific enzymes. In addition, during the last forty years many enzymes have been isolated and purified (see Chapter 2) and this has made it possible to use enzymes to determine the concentrations of substrates of clinical importance. The availability of highly specific antibodies, often from single clones (monoclonal antibodies), in conjunction with purified enzymes to amplify detection has enabled a wide range of biological compounds to be measured by enzyme immunoassay (see Section 10.8.7). A further development arising from the increased availability of purified enzymes is the potential for enzyme therapy (see Section 10.9).

The measurement of enzyme activities in serum is of major importance as an aid to diagnosis. In recent years, however, it has also become important as a means of monitoring progress after therapy, recovery after surgery, and detection of transplant rejection. We shall therefore discuss measurement of enzyme activities for clinical diagnoses first (Sections 10.2–10.6). In Section 10.7 we discuss some examples of the use of exogenous enzyme inhibitors to reduce the activities of certain endogenous enzymes. In Section 10.8 we shall describe the use of enzymes as a means of estimating concentrations of substrates, and finally in Section l0.9 we consider enzyme-replacement therapy. There are a number of books and reviews on clinical applications of enzymology in which more detailed information can be obtained.[1–7]

10.2 Determination of enzyme activities for clinical diagnosis

For the measurement of an enzyme activity to be useful as a routine diagnostic clinical method the following conditions should be fulfilled.

1. The enzyme should be present in blood, urine, or some readily available tissue fluid. Tissue biopsies should not be used routinely, although they may be used if the diagnostic value is sufficiently important.

2. The enzyme should be easy to assay and it is advantageous if the method can readily be automated.

3. The differences in the ranges of enzyme activities obtained from normal and diseased subjects should be diagnostically significant and there should be a good correlation between the level of enzyme activities and the pathological state.

4. It is also useful if the enzyme is sufficiently stable so that the sample may be stored at least for a limited time.

Serum is the fluid in which most measurements of enzyme activities are made. Urine can be used for a few enzymes that are cleared by the kidney. Enzymuria also occurs as a result of kidney damage or failure, when a range of kidney enzymes are detectable in urine.[8] Other body fluids such as those from the pleura, peritoneum, pericardium, cerebrospinal canal, synovia, stomach, duodenum, semen, vagina, and within the amnion are occasionally used.[8] Red blood cells and white blood cells, although they are relatively accessible, have not so far been used extensively in diagnosis. The enzymes present in serum can be considered in one of two categories: (i) plasma-specific enzymes and (ii) non-plasma-specific enzymes. The former are enzymes whose normal function is in the plasma, e.g. enzymes concerned with blood coagulation,[9,10] complement activation,[11,12] and lipoprotein metabolism.[13] The latter are enzymes that have no physiological function in the plasma, enzymes for which the cofactors or even substrates may be lacking. This category includes enzymes that are secreted by tissues, e.g. amylase, lipase, and phosphatases, and also enzymes associated with cellular metabolism, whose presence in normal serum at low levels may be due to turnover of cells within the tissue causing release of the enzyme content. The rationale of enzyme measurement for diagnosis is that if a tissue is broken down or is producing an abnormally high amount of intracellular enzyme that is then released, there is an elevation of enzyme activity in the serum. For several enzymes there is a very high concentration gradient (a factor of the order of 10^3 or greater) across the cell membrane to the extracellular fluid, e.g. for skeletal muscle creatine kinase 50 000-fold,[6] and for many enzymes more than one thousand-fold, and thus a small amount of tissue damage may affect considerably the serum concentration of certain enzymes.[14] However, not all increases in the activities of enzyme in serum are due to tissue damage. They may be due to other factors such as increased cell turnover, cellular proliferation as in neoplasia, or decreased clearance by the kidney.

* Necrosis is the death of a portion of tissue or organ.

In the diseased state a tissue may become inflamed or it may become necrotic.* If the latter occurs there is likely to be fairly complete release of the enzyme content from the dead cells. However, the pattern of enzymes detectable in the serum may not completely resemble that of the tissue from which it arises, since enzymes may be inactivated at different rates. Inflammation of a tissue may result in a change in the permeability of the cells, so that release of enzymes may occur from the cytosol but not from organelles,[15] e.g. glutamate dehydrogenase which is present in high concentrations in the mitochondrial matrix is released only when cell destruction is fairly complete.[16]

Many of the enzymes released into the serum seem to be removed at a fairly rapid rate, e.g. glutathione S-transferase α-isoenzyme has a half-life of 90 min, but others are lost more slowly (Table 10.1). Mitochondrial enzymes are released more slowly than cytosolic enzymes. The mechanism of their removal has not been studied appreciably; it may be due to inactivation and degradation occurring in the serum or it may be due to clearance by the kidney. Some experiments, involving intravenous injections of ^{125}I-labelled lactate dehydrogenase (LDH-5) into rabbits, have been carried out to study the fate of serum enzymes. The results suggest that LDH-5 undergoes denaturation in the plasma and that the products are excreted into the small intestine where they are further degraded and then reabsorbed, as amino acids and small peptides, back into the circulation.[19] The quite rapid removal of most released enzymes means that monitoring particular enzymes present in the serum of a diseased subject gives an up-to-date picture of the release of enzymes from diseased tissue. Monitoring enzymes in the serum is thus useful not only in the initial diagnosis but also in studying the course of the disease and the response to treatment.

Ideally, it would be desirable to study tissue-specific enzymes for diagnostic purposes, since these would identify the tissue from which they arose. However, there are relatively few enzymes that are entirely tissue-specific, although there are several that are much more abundant in one tissue than another, e.g. acid

Table 10.1 Examples of the half-lives of enzymes found in serum

Enzyme and tissue of origin	Half-life in serum (h)
Lactate dehydrogenase (muscle and liver)	
Isoenzyme 1 (LDH-1)	50–110
Isoenzyme 5 (LDH-5)	8–14
Aspartate aminotransferase (liver)	
Mitochondria	6–7
Cytosol	12–17
Alanine aminotransferase	40–50
Creatine kinase	
Isoenzyme MM (CK-1)	10–20
Isoenzyme MB (CK-2)	7–17
Isoenzyme BB (CK-3)	3
Alkaline phosphatase	
Liver	140–230
Bone	30–50
Intestine	< 1
Placenta	170
Glutathione S-transferase (liver)	1.5

See references 2, 3, 17, 18.

phosphatase in the prostate and acetylcholinesterase in erythrocytes. Sometimes the measurement of more than one enzyme in serum helps to establish a pattern characteristic for the tissue of origin. Although a particular enzyme activity may not be specific to one tissue, there may exist isoenzymes (see Chapter 1, Section 1.5.2) that show a different pattern in different tissues. The best-studied example is that of lactate dehydrogenase (further information on lactate dehydrogenase is given in Chapter 5, Section 5.5.4). Lactate dehydrogenase is composed of four subunits. There are two types of subunit, which combine in lactate dehydrogenase to give five possible combinations: α_4, $\alpha_3\beta$, $\alpha_2\beta_2$, $\alpha\beta_3$, and β_4. These five types may be separated by electrophoresis and it has been found that their distribution varies with the tissue (Fig. 10.1). Thus, although by measurements of lactate dehydrogenase activity in serum the tissue of origin could not be identified, identification might be possible if the isoenzyme distribution were determined by electrophoresis. Several enzymes that may be present in serum are known to exist in multiple forms, e.g. alkaline phosphatase, amylase, creatine kinase, ceruloplasmin, glucose 6-phosphate dehydrogenase, and aspartate aminotransferase. Apart from lactate dehydrogenase, creatine kinase and alkaline phosphatase are the isoenzymes most studied in diagnostic enzymology. Microscale electrophoretic separations of some isoenzymes can be performed within 30 min. Some isoenzymes may be differentiated by criteria other than electrophoretic mobility, e.g. by substrate specificity, heat stability, or sensitivity to inhibitors (Table 10.2). Antibodies may also be used to differentiate the different subunits of isoenzymes. This method has been used to distinguish the different isoenzymes of human phosphofructokinase, and to identify which form is absent in inherited deficiencies of phosphofructokinase.[20] The tissues of origin are often distinguished by the proportion of different forms of isozymes, rather than a complete absence of one or other.

In many cases where measurements of enzyme activity are used as an aid to diagnosis, more than one activity is measured in order to permit a greater degree of discrimination. Perhaps the two examples in which enzyme monitoring has been most useful and widely applied are those of liver and heart diseases. It must be emphasized that the measurement of enzyme activity itself is not sufficient evidence on which to make a diagnosis but must be taken in conjunction with the

electrophoretic mobility:
mobility of molecule due to size, charge, and shape

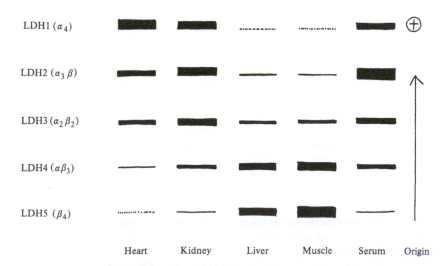

Fig. 10.1 Electrophoretic separation of isoenzymes of lactate dehydrogenase from different tissues.

LDH1 (α_4)

LDH2 ($\alpha_3\beta$)

LDH3 ($\alpha_2\beta_2$)

LDH4 ($\alpha\beta_3$)

LDH5 (β_4)

Heart   Kidney   Liver   Muscle   Serum   Origin

Table 10.2 Methods used to determine the tissue of origin of serum enzymes

Enzyme	Differentiation by electrophoresis	Differentiation by other methods
Lactate dehydrogenase	LDH-1 predominant in heart, erythrocytes, brain, and kidney	Differential heat stability, LDH-4 and LDH-5 show greater heat stability
	LDH-3 predominant in leucocytes	LDH-1 is more sensitive to inhibition by excess pyruvate than LDH-5
	LDH-5 predominant in skeletal muscle and liver	LDH-1 and LDH-2 only can use 2-hydroxybutyrate as an alternative substrate
Acid phosphatase		Prostate enzyme inhibited by tartrate. Erythrocyte enzyme inhibited by formaldehyde. Differential activities towards substrates, phenyl phosphate, glycerol 2-phosphate, and thymolphthalein phosphate.
Alkaline phosphatase	Complex electrophoretic patterns. Liver enzyme most cationic	Placental enzyme shows greater heat stability. Phenylalanine preferentially inhibits the placental and intestinal enzyme. Urea preferentia ly inhibits bone enzyme. Levamisole preferentially inhibits bone and liver enzymes.
Creatine kinase	Brain, lung, and thyroid mostly CK-1. Skeletal muscle almost exclusively CK-3. Cardiac tissue CK-3 + about 15% CK-1	Anti-M antibody inhibits M- but not B-subunits.

References 1, 2.

clinical symptoms and other types of evidence, e.g. electrocardiographic and ultrasonography evidence in cases of heart and liver disease.

Increased activities of certain enzymes in serum are commonly found in malignant diseases, but no enzyme is specific for cancer and so far none has been very useful in the early detection of cancer. Besides the use of isoenzyme patterns to identify the tissue of origin, isoenzyme patterns are also used in forensic science. Since a number of enzymes in serum and red blood cells occur as different isoenzymes, the particular pattern in a blood sample may provide supporting evidence as to its origin. Isoenzyme patterns that are currently used by the Metropolitan Police Science Laboratory include adenosine deaminase, adenylate kinase, carbonate dehydratase, acid phosphatase, esterase, phosphoglucomutase, aminopeptidase, and lactoylglutathione lyase.[21]

10.3 Clinical enzymology of liver disease

The enzymes most used in diagnosis of liver disease are serum aspartate aminotransferase (now abbreviated to AST but previously referred to as serum glutamate oxaloacetate transaminase, SGOT), alanine aminotransferase (now abbreviated to ALT but previously referred to as serum glutamate pyruvate transaminase, SPGT), alkaline phosphatase, and γ-glutamyltransferase. Others that are also sometimes measured are lactate dehydrogenase, isocitrate dehydrogenase 5′ nucleotidase, and glutathione transferase. Ornithine carbamoyltransferase is an enzyme that is almost exclusive to the liver and although its liver:blood ratio is 10^5:1 it has not been widely used as a diagnostic aid because the assay is complex compared to those of other liver enzymes and is not easily automated.[22] The aminotransferases that catalyse the following reactions are usually assayed by coupling to the reactions catalysed by malate and lactate dehydrogenases so that the change in A_{340} due to oxidation of NADH can be measured continuously (see Chapter 4, Section 4.2).

↓

detection at 340nm

1. Aspartate + 2-oxoglutarate $\rightleftharpoons$ oxaloacetate + glutamate
   Oxaloacetate + NADH + H$^+$ $\rightleftharpoons$ malate + NAD$^+$

2. Alanine + 2-oxoglutarate $\rightleftharpoons$ pyruvate + glutamate
   Pyruvate + NADH + H$^+$ $\rightleftharpoons$ lactate + NAD$^+$

Alkaline phosphatase is assayed by measuring the hydrolysis of 4-nitrophenyl phosphate. γ-Glutamyltransferase is assayed by measuring the release of 4-nitroaniline when γ-glutamyl-4-nitroanilide is used as one of the substrates. 4-Nitroaniline absorbs at 400 nm.

HOOC.CH (NH$_2$)CH$_2$CH$_2$CO +glycylglycine $\longrightarrow$ γ-glutamylglycylglycine

$+$ H$_2$N$-$\<benzene ring>$-$NO$_2$

HN$-$\<benzene ring>$-$NO$_2$

Although this is a most sensitive substrate, it has a poor solubility in water, limiting the concentrations that can be used. Recently, the more soluble carboxylated derivative, L-γ-glutamyl-3-carboxy-4-nitroanilide, has been used[23] and it is at least as sensitive.

The most commonly occurring liver diseases are viral or toxic hepatitis, cirrhosis, primary and secondary liver tumours, and obstruction of the bile-duct due to stones, tumours, or stricture. The conditions may be acute, intermittent, or chronic. In addition, the liver plays a central role in the metabolism of many drugs. Many of the enzymes involved are located on the smooth endoplasmic reticulum. Liver damage may not only be caused by drug overdose, but in some cases by therapeutic doses. Liver surgery also causes damage. The measurement of serum enzyme activities is thus not only used in diagnosis, but also in monitoring drug treatment, recovery from surgery, and as an indicator of possible transplant rejection. The two aminotransferases are in the category of non-plasma-specific enzymes, i.e. they are enzymes concerned with intracellular metabolism and have no known function in the plasma. They are released from the liver when parenchymal cells (the principal cells of the liver) become necrotic as in, for example, viral or toxic hepatitis or cirrhosis; the increase in their activity in the plasma is related to the extent of cell breakdown. Aminotransferases in serum may also arise from cardiac tissue after myocardial infarction, but the proportions of aspartate aminotransferase and alanine aminotransferase differ. This ratio was used to differentiate between various conditions, although it is only rarely used now. In cases of chronic hepatitis or cirrhosis, aminotransferases are often monitored over a period of several months to follow the progress of the disease and to provide an indication of the prognosis.

Measurements of serum alkaline phosphatase are also important in diagnosis of liver disease. Serum alkaline phosphatase arises mainly from the sinusoidal surface (i.e. that adjacent to the blood space) of liver cells and bile duct canaliculi and from the osteoblasts (bone-forming cells). The enzymes from these sources differ in their isoenzyme patterns. Liver alkaline phosphatase is a non-plasma-specific enzyme that is secreted into the serum at low levels in the absence of liver cell damage. Damage to parenchymal liver cells does not generally increase serum alkaline phosphatase levels, but when the bile duct epithelium is affected there is a pronounced increase. In growing children where there is higher osteoblastic activity, there may be uncertainty as to the origin of elevated serum alkaline

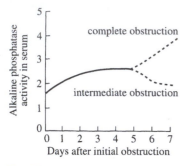

Fig. 10.2 Typical changes in the activity of serum alkaline phosphatase after complete and intermittent obstructive jaundice (see reference 18).

phosphatase. The measurement of serum 5′ nucleotidase activity can then be used since it arises mainly from the liver endoplasmic reticulum. It is possible to differentiate obstructive jaundice from hepatitis by measurement of aminotransferases and alkaline phosphatase, since there is a proportionately greater increase in alkaline phosphatase activity in obstructive jaundice than in hepatitis, so that the ratios of activities of serum alkaline phosphatase/serum alanine aminotransferase one week after the onset of viral hepatitis might be 0.2–0.6 compared with 10–20 one week after the onset of obstructive jaundice.[18] Whether the bile-duct obstruction is intermittent or complete can be assessed by monitoring the alkaline phosphatase activity over a period of a few days (Fig. 10.2). Glutamate dehydrogenase can also be used in place of alkaline phosphatase to distinguish hepatic and biliary disease. The ratio of glutamate dehydrogenase/alanine aminotransferase is higher in cases of obstructive jaundice than, for example, in viral hepatitis.[16]

γ-Glutamyltransferase is now also being used extensively in the diagnosis of liver disease. Although it is not present exclusively in the liver, the proportion present in the liver is much higher than is the case for either aminotransferases. It is a microsomal enzyme also found in bile canaliculi, kidney, and pancreas.[24] The activity is greatly elevated in the serum in cases of chronic alcoholism, when it causes cirrhosis of the liver, and in cases of metastatic invasion of the liver. The increased serum level in cases of chronic alcoholism is due to both increased synthesis in the liver and increased release into the serum; γ-glutamyltransferase activity follows a similar pattern to that of alkaline phosphatase in obstructive jaundice.

Glutathione S-transferase is emerging as another alternative to the aminotransferases for assessing liver damage. It catalyses detoxification reactions between glutathione and electrophiles. These electrophiles include, for example, carcinogens, chemotherapeutic drugs, paracetamol, and a range of xenobiotics. The enzyme is present in very high concentrations in liver, comprising about 5 per cent of the cytosol protein, and this makes it a highly sensitive indicator of liver damage.[17,25] Since it has a short half-life in serum (≈ 90 min) it is also a rapidly responding marker. It is present in other tissues, and unique isoenzymes exist in certain tumours, which may reduce their susceptibility to chemotherapy. The enzyme can be measured by a spectrophotometric assay which can be automated, although certain substances present in serum may interfere with the assay. The immunoassay is more costly but is able to distinguish the different isoenzymes.[25] Because of its high concentration in liver, the enzyme is a sensitive and rapidly responding marker of liver injury.[17]

The enzymes discussed so far increase in activity in serum when cells are damaged and their contents leak out into the blood. None of these assess how well the liver is functioning. For this the prothrombin time is one of the most useful measures. The factors required for blood clotting are synthesized in the liver and secreted into the blood. The one-stage prothrombin time measures the time required for prothrombin activation and the conversion of fibrinogen to fibrin which is catalysed by thrombin. If any of the factors required for thrombin activation is present below a critical concentration, then the prothrombin time is increased and this could be an indication of reduced synthetic capacity of the liver.[5,17]

10.4 Clinical enzymology of heart disease

Enzymological methods have been particularly useful in providing supporting evidence for myocardial infarction (the process of formation of an area of dead heart muscle) and also monitoring the course of the infarct. In myocardial infarction a coronary artery becomes obstructed and this leads to irreversible damage and necrosis of the heart tissue. If a definitive diagnosis cannot be made by an electrocardiogram, then enzyme tests may help. In addition, serum enzyme activities may be used in monitoring progress and recovery after heart surgery. Enzymes are released from the necrotic tissue into the plasma; the three enzymes most commonly assayed are creatine kinase, aspartate aminotransferase, and lactate dehydrogenase. Each enzyme shows a different time-course for release into the plasma and subsequent disappearance. These differences depend on the concentration of each present in cardiac tissue, their rate of release, and their subsequent rate of clearance or degradation.[1-5,18] Their half-lives in the plasma range from 12 to 120 h (Table 10.1). Creatine kinase is the earliest to be detectable, rising 4–6 h after the onset of pain, reaching a peak at 24–36 h, and then rapidly declining. Aspartate aminotransferase reaches a peak between 48 and 60 h and lactate dehydrogenase between 45 and 72 h; the latter declines more slowly than the former (Fig. 10.3). Because blood specimens may be taken from times ranging from 4 h to 10 days after the onset of the infarct, the enzyme test most appropriate in a particular situation may vary. Creatine kinase is the test of choice unless the blood sample had been taken a long time after the suspected occurrence of the infarct, in which case lactate dehydrogenase activity is measured. Creatine kinase has the advantage of greater specificity than either aspartate aminotransferase or lactate dehydrogenase, both of which are present in high concentration in liver. Although aspartate aminotransferase was the first enzyme test to be used in diagnosis of heart disease in the 1950s, it is now the least used, because of its low specificity. Both creatine kinase and lactate dehydrogenase have the advantage that further discrimination is possible if the isoenzyme patterns are examined. To measure the isoenzymes generally entails electrophoretic separation, but the individual isozymes can be detected *in situ* on the gels and quantitated by scanning fluorimetry or densitometry.[5] The prognosis and progress can be assessed by the increase in the activities of these enzymes.

Creatine kinase does not occur in the liver, unlike aspartate aminotransferase and lactate dehydrogenase, but it is present in skeletal muscle and brain. In skeletal muscle it has about eight times the concentration on a gram wet weight basis

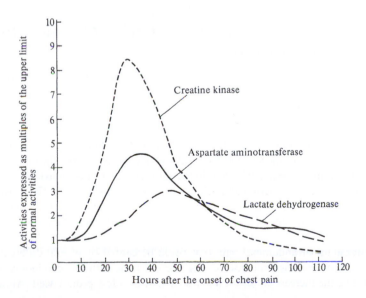

Fig. 10.3 Typical changes in the activities of serum creatine kinase, aspartate aminotransferase (ALT), and lactate dehydrogenase following myocardial infarction. Activities are expressed as multiples of the upper limit of the normal level (see reference 1).

compared to that of cardiac muscle.[26] An increase in creatine kinase activity in serum is most frequently associated with damage to cardiac or skeletal muscle, and less frequently to brain damage. There are two different types of creatine kinase subunit, referred to as M (muscle) and B (brain). The enzyme is a dimer and may exist as MM (CK-MM), MB (CK-MB), or BB (CK-BB).* Skeletal muscle has $\approx 99\%$ MM and $\approx 1\%$ MB, whereas cardiac muscle has 80–85% MM and 15–20% MB, while brain has predominantly BB. In normal serum there are low levels of creatine kinase (94–98% MM, 2–4% MB, BB undetectable). If it is not possible to differentiate on clinical grounds whether a rise in serum creatine kinase is due to damage to skeletal or cardiac muscle, then resolution from studying the isoenzyme pattern is usually possible. The measurement of CK-MB is a more sensitive indicator of myocardial infarction than that of total creatine kinase activity. With a minor myocardial infarction, total serum creatine kinase activities may still lie within the normal range, although the CK-MB shows a clinically significant increase.[5] Antibodies against the B and M subunits can also be used to distinguish the isoenzymes. Anti-M subunit antibody inhibits CK-MM and the M subunit of CK-MB. This provides a means for a less labour-intensive method of assessing the isoenzyme forms present in serum.[27]

Lactate dehydrogenases from liver and heart may also be distinguished by their isoenzyme patterns (Fig. 10.1). This generally requires the separation of the five isoenzymes and assay *in situ* in the gel, although it is possible to measure LDH-1 using α-hydroxybutyrate as an alternative substrate. The LDH-1 isoenzyme shows much greater catalytic activity with α-hydroxybutyrate than the other isoenzymes. In normal serum the ratio of LDH-1:LDH-2 activities is ≈ 0.85 but after a myocardial infarction it increases to greater than 1.0.[5] Measurement of both serum aspartate aminotransferase and alanine aminotransferase can be used to differentiate between liver and heart disease. The ratio of activities of aspartate aminotransferase/alanine aminotransferase in heart tissue (20–25) is higher than the ratio in liver (3–5) and this is reflected in the ratios of enzymes released into the serum, but this method has now been generally superseded by measurement of either creatine kinase or lactate dehydrogenase activities together with their isoenzyme distribution.

10.5 Other enzyme activities that become elevated in serum in disease

A number of other enzymes are assayed in serum and urine for diagnostic purposes; the more frequently measured ones are discussed below.

10.5.1 α-Amylase

α-Amylase is an endoamylase that hydrolyses the 1,4-α link in amylose and amylopectin. In humans the enzyme occurs in a variety of tissues, but the highest concentrations are in the pancreas and in the salivary glands. Low amylase activities can be detected in the serum and urine of normal healthy subjects. The most pronounced increases in serum and urinary amylase occur in acute pancreatitis, when the activities may rise to 20–30 times the normal levels. Acute pancreatitis can be difficult to diagnose without enzyme tests as the patient usually complains of intense upper abdominal pain that could be caused by several different disorders. In other pancreatic diseases, e.g. chronic pancreatitis or carcinoma of the pancreas, the increase in serum amylase activities is less pronounced. Although the highest concentration of amylase is in the pancreas, the increased excretion of amylase may occur in a number of intra-abdominal disorders in which the tissue affected is in close proximity to the pancreas, e.g. gastric or duodenal ulcers, peritonitis, and mesenteric thrombosis. Amylase activity in urine may also be increased in mumps when the salivary glands become inflamed. Salivary and pancreatic amylase can be distinguished by their isoenzyme pattern. An unusual feature of amylase secretion compared with other enzymes is that it is efficiently cleared by the kidneys and readily detected in urine. Only enzymes having M_r values less than about 55 000 are cleared by the kidney and appear in urine. Larger serum enzymes are generally inactivated and removed by a receptor-mediated endocytotic mechanism present on the reticuloendothelial system in bone marrow, spleen, or liver.[3] Most enzymes of importance in clinical diagnosis have M_r values between 70 000 and 160 000; exceptions are uropepsinogen, adenylate kinase (21 000), and α-amylase (49 000).[28] During an attack of acute pancreatitis the level of α-amylase in the serum may be raised for 2–3 days but persist at a higher level in urine for 7–10 days.[5]

The exocrine pancreas also secretes most of the lipase used in the digestive system and serum lipase activity may be raised 10–40 times the normal level in acute pancreatitis. The older method of assaying lipase involved titrating the fatty acid released after hydrolysis of an olive oil emulsion. This method was time-consuming and not very reproducible and this was one of the reasons amylase activity was usually measured. Now that a simpler and more reliable turbimetric method is available, lipase activities may be used increasingly for diagnosis of pancreatic diseases.[5]

10.5.2 Creatine kinase and fructose-bisphosphate aldolase

These two enzymes are useful as indicators of skeletal muscle disorders. Creatine kinase occurs predominantly in skeletal muscle, cardiac muscle, and brain, but in very low activities in all other tissues (see Section 10.4). The serum activity is raised in all forms of muscular dystrophy, the level depending on the mass of muscle that is becoming dystrophic. The greatest increases, which may be as much as 50-fold, are seen in Duchenne-type muscular dystrophy. Creatine kinase is also raised in muscle injury and may even be raised after strenuous exercise or after an intramuscular injection.[5] Although creatine kinase activity is also raised in myocardial infarction, the rise is short-lived and is usually easy to distinguish from skeletal muscle defects. Serum levels of creatine kinase may also be raised in cases of brain damage, e.g. stroke. The tissue of origin of creatine kinase can generally be determined from the isoenzyme pattern (see Section 10.4).

Fructose-bisphosphate aldolase is also present in high concentrations in skeletal muscle and thus could be used to support the conclusions reached from the assay of creatine kinase. However, fructose-bisphosphate aldolase is also present in many other tissues, although not at such high concentrations, and also the increases in serum concentrations are not generally as great as those found with creatine kinase. Particularly important in this respect is the high fructose-bisphosphate aldolase content of red blood cells; it is therefore essential that a sample of serum to be assayed does not contain lysed red blood cells.

10.5.3 Alkaline phosphatase

Alkaline phosphatase is a phosphatase having low specificity, hydrolysing a variety of organic phosphate esters, and having a pH optimum between 9 and 10.

NO₂

OPO₃²⁻

4-nitophenyl phosphate

Its activity is detectable in a number of different tissues, notably intestinal epithelium, kidney tubules, bone (osteoblasts), leucocytes, liver, and placenta. The true substrates of the enzyme are not known and it is usually assayed with 4-nitrophenyl phosphate.

Apart from its application in the diagnosis of obstructive jaundice, it is also very useful in the detection of bone disease. Its activity in serum is raised in those bone diseases in which there is increased activity of the bone-forming cells or osteoblasts as in, for example, osteomalacia, rickets, Paget's disease, and hyperparathyroidism. Its activity is lower than normal in a rare form of rickets known as hypophosphatasia. This disease was first described in 1948 and to date about 300 cases have been reported. It is a severe form of rickets in which alkaline phosphatase levels in serum, bone, and other tissues are subnormal, in contrast to other forms of rickets. Evidence from subjects suffering from hypophosphatasia, and from the use of inhibitors of alkaline phosphatase, has led to the hypothesis that a physiological substrate of alkaline phosphatase is inorganic pyrophosphate, and that the enzyme shows pyrophosphatase activity. Pyrophosphate has been shown to inhibit the growth of hydroxyapatite crystals, a process occurring in normal bone formation.[29] Also hypophosphatasia patients have increased pyrophosphate concentrations in their urine. Other possible roles have been suggested for alkaline phosphatase in skeletal mineralization, e.g. increasing local levels of inorganic phosphate, transport of inorganic phosphate, and acting as a calcium-binding protein.[30]

The alkaline phosphatase present in normal serum is believed to be derived mainly from the liver and bone cells with a small amount from the intestine.[31] It is useful to be able to differentiate between increased activities in the serum due to secretion from liver, from bone, or from other tissues. The isoenzyme patterns of alkaline phosphatase from various tissues are quite complex, multiple bands being obtained from most tissues. The alkaline phosphatases from liver, kidney, and bone are coded for by the same gene, and are therefore not strictly isozymes according to the definition given in Chapter 1, Section 1.5.2, but are the result of post-translational modification.[29] Differentiation of the tissue of origin by electrophoresis is not an easy method to use in a routine investigation in the case of alkaline phosphatases. Other properties of the alkaline phosphatases from different tissues that have been useful are the effects of heat and various inhibitors (see Table 10.2). The results suggest that alkaline phosphatases are of three distinctive types, namely, placental, intestinal, and those from other tissues. There is an increase in alkaline phosphatase activity during the third trimester of pregnancy and this is due to release of placental enzyme. The latter can readily be differentiated from the bone and liver enzymes because it is quite stable to heating for 30 min at 329 K (56°C) while the others are labile.[5] The liver enzyme also has a longer half-life in serum compared with that from bone (Table 10.1).

10.5.4 Acid phosphatase

This enzyme, like alkaline phosphatase, is a phosphatase of low specificity and the natural substrates are unknown. The pH optimum is between 4.0 and 5.5. The enzyme occurs in a variety of tissues, e.g. liver, spleen, erythrocyte, prostate, and the osteoclasts of bone.[29] The highest concentration is present in the prostate and detection of prostatic carcinoma is the main purpose of clinical assays of this enzyme.

Five isozymes contribute to the serum levels of acid phosphatase from each of the tissues listed above.[29] Most of the acid phosphatase in normal serum is derived from blood platelets, but approximately one-third in adult males is derived from the prostate. The enzyme is very labile and it must be assayed as soon as possible after collecting the serum. It is clearly important to have an early and unequivocal diagnosis of prostatic carcinoma; however, small deviations in the level of acid phosphatase in the serum could be due to release from other tissues or lysis of erythrocytes during serum collection. Two methods are available for differentiating erythrocyte and prostatic enzymes. The first relies on the differing sensitivities of the enzymes towards formaldehyde and tartrate; 0.5 per cent formaldehyde almost completely inhibits the erythrocyte enzyme but hardly affects that of the prostatic enzyme, whereas 0.01 mol dm^{-3} tartrate has almost no effect on the erythrocyte enzyme whilst inhibiting the prostatic enzyme. The second is to use either thymolphthalein monophosphate or 1-naphthyl phosphate which is much more readily degraded by the prostate enzyme.

Elevated serum levels of acid phosphatase can be used to help diagnose carcinoma of the prostate after metastasis has occurred, but whilst the tumour is confined within the prostate significant increases may not be evident. Attempts to find markers that are more specific for the prostate gland and that are more sensitive for early detection are being investigated. A glycoprotein prostate-specific antigen and steroid 5α reductase appear more promising alternatives.[4,32]

1-Naphthyl phosphate

Thymolphthalein phosphate

10.6 The detection and significance of enzyme deficiencies

The previous section has been concerned with the detection of elevated enzyme activities in serum and urine and the diagnosis of the associated diseases. These represent the large majority of enzyme assays performed for clinical diagnosis. Much less frequently encountered are those subjects who suffer from enzyme deficiencies due to inborn errors in metabolism. More than 400 different inborn errors in metabolism are now known although none of them occur very frequently (Table 10.3), e.g. phenylketonuria, 1 in 12 000 births (about 50 births per year in the UK); cystinuria, 1 in 14 000; galactosaemia, 1 in 57 000; maple syrup disease, 1 in 180 000; fructosuria, 1 in 130 000; alcaptonuria, 1 in 200 000.[33-35] In many cases the enzyme deficiency has been established (Table 10.3).

The initial identification of the disease is rarely the result of an enzyme assay and in many cases the presence of a metabolite is used in detection of the disease; thus, phenylketonuria is usually detected by measurement of phenylpyruvate in urine or phenylalanine in blood. The determination of the enzyme defect may require a tissue biopsy. For some genetic diseases, e.g. phenylketonuria, DNA probes are now available which can be used on small amounts of blood or cells or amniotic fluid.[5] Some inborn errors in metabolism are relatively harmless, e.g. albinism, alkaptonuria, but others must be detected early if the defect is to be circumvented. Two important ones that fall in the latter category are phenylketonuria and galactosaemia. In phenylketonuria there is a lack of phenylalanine 4-monooxygenase, an enzyme required for the breakdown of phenylalanine:

+ Tetrahydropteridine   + Dihydropteridine + H_2O

+ O_2

Table 10.3 Some examples of inborn errors in metabolism due to enzyme deficiencies[33]

Inborn error	Enzyme deficiency	Frequency (1 in X)[a]
Alkaptonuria	Homogentisate 1,2-dioxygenase	10^5–10^6
Phenylketonuria	Phenylalanine 4-monooxygenase	10 000–20 000
Maple syrup disease	Branched-chain oxo-acid oxidative decarboxylases	180 000
Galactosaemia	Galactose 1-phosphate uridylyltransferase	35 000–60 000
Glycogen storage disease type Ia	Glucose 6-phosphatase	100 000
Glycogen storage disease type V (McArdle disease)	Muscle glycogen phosphorylase	$\approx 10^6$
Pentosuria	L-Xylulose reductase	2500 Ashkenazi Jews
Fructosuria	Fructokinase	130 000
Gaucher disease	Glucocerebrosidase	60 000–360 000
Tay-Sachs	β-N-Acetyl-D-hexosaminidase	300 000
Butyrylcholinesterase deficiency	Butyrylcholinesterase	3000 Caucasians
Wilson disease	P-type ATPase (Cu^{2+} transporter)	50 000
Acatalasaemia	Catalase	25 000–250 000
Cystic fibrosis	Cystic fibrosis transmembrane conductance regulator	2000–3000 Caucasians, rare in other ethnic groups
Hypophosphatasia	Alkaline phosphatase	$\approx 10^5$
Xeroderma pigmentosa	DNA binding proteins and helicases	250 000

[a]Frequencies are expressed as 1 in X (the number given). In some cases a range is given which reflects variation between populations.

Phenylalanine is being produced continuously in a normal subject as a result of protein turnover and dietary intake and is largely broken down via tyrosine to homogentisic acid. In the phenylketonuric subject, phenylalanine has to be degraded by a minor pathway via phenylpyruvic acid (Fig. 10.4).[33–35]

Phenylketonuria is associated with mental retardation: it is not clear whether this is due to the build up of high concentrations of phenylalanine or its metabolites. The mental disturbance can be prevented if the subjects from birth to 7 years are fed a diet with a considerably reduced phenylalanine content, although there is evidence that the diet must be maintained continuously throughout adulthood to prevent the accumulation of deleterious effects.[36] Phenylalanine is an essential amino acid and so its dietary content must be reduced to the minimum required to maintain growth and protein turnover. If the protein in a normal diet is reduced so that there is no surplus phenylalanine, then it will become deficient in other essential amino acids. Special semisynthetic diets have been compiled. These usually contain protein hydrolysate from which the phenylalanine content has been reduced but the other essential amino acids are in adequate amounts.[33] One problem which has arisen recently is the use in foodstuffs of the artificial sweetener, aspartame (see Chapter 11, Section 11.5.3), which has replaced about 12 per cent of the total sweeteners consumed by the average American. In the intestine this releases L-phenylalanine, L-aspartic acid, and methanol.[33] Early detection of phenylketonuria is thus very important and infants are now screened at birth in many countries.

Galactosaemia may result from one of three enzyme deficiencies: galactose 1-phosphate uridylyltransferase, galactokinase, or uridine diphosphate galactose 4-epimerase. The first of these occurs most frequently. As can be seen from Fig. 10.5, this results in a block in the conversion of galactose 1-phosphate to

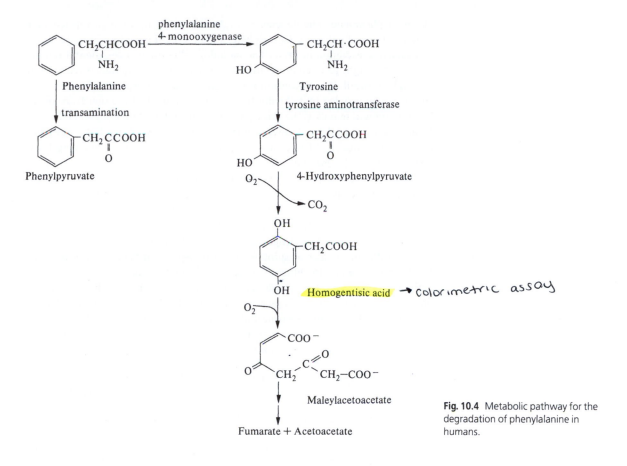

Fig. 10.4 Metabolic pathway for the degradation of phenylalanine in humans.

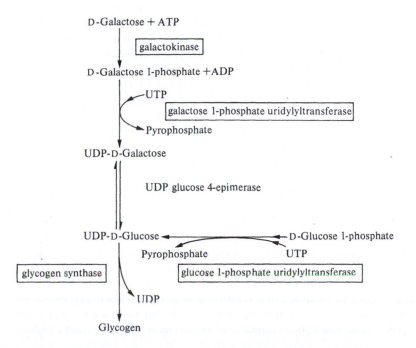

Fig. 10.5 The pathways of galactose metabolism.

glucose 1-phosphate. The deficiency results in a build up of high intracellular concentrations of galactose 1-phosphate, a high level of galactose in body fluids, and galactose excretion in urine. The pathological consequences, mental retardation, slow weight gain, enlarged liver, and cataracts, are generally attributed to the high intracellular concentrations of galactose 1-phosphate. The harmful effects can be avoided by reducing the galactose content of the diet from birth. (Lactose present in milk is the principal source of galactose in infants.) Galactose 1-phosphate uridylyltransferase is normally present in many human tissues including erythrocytes. Thus, galactosaemia may be detected by assaying lysed erythrocytes for the enzyme.[33]

The most frequently occurring hereditary disease known is glucose-6-phosphate dehydrogenase deficiency, in which the erythrocytes are affected.[37] It is more common in African Americans than in most other populations. The disease is manifested as haemolytic anaemia and occurs after the subjects have been treated with antimalarial drugs such as pamaquine or primaquine, or have received other drugs such as sulphonamides or aspirin, or have eaten fava beans. Erythrocytes metabolize glucose using the glycolytic and pentose phosphate pathways. The agents that cause haemolysis appear to inhibit glucose-6-phosphate dehydrogenase which is present only in low levels in these subjects. This in turn causes reduction in the intracellular NADPH concentration and the latter is required for glutathione reductase. It is thought that haemolysis is induced by lowering of the levels of glutathione:

$$\text{D-Glucose 6-phosphate} + \text{NADP}^+ \rightleftharpoons \text{D-glucono-}\delta\text{-lactone 6-phosphate} + \text{NADPH} + \text{H}^+.$$

$$\text{H}^+ + \text{NADPH} + \text{oxidized glutathione} \rightleftharpoons \text{NADP}^+ + 2 \text{ glutathione}.$$

Another example of drug sensitivity being associated with a genetically inherited deficiency occurs in the case of serum cholinesterase (EC 3.1.1.8) deficiency. Cholinesterase is normally present in serum at quite high levels. The enzyme is considered to be synthesized by the liver and lower levels of activity are detected in serum in certain liver diseases, e.g. acute hepatitis and advanced carcinoma of the liver, when the rate of synthesis of the enzyme is reduced. The enzyme was originally called pseudocholinesterase since it hydrolyses a variety of esters of choline, in contrast to the acetylcholinesterase (EC 3.1.1.7) of nervous tissue and erythrocytes, which has a more restricted specificity. However, it has now been renamed butyrylcholinesterase since it hydrolyses butyrylcholine more rapidly than acetylcholine. The relative deficiency of butyrylcholinesterase in some individuals became apparent in the 1950s as a result of the use of suxamethonium (succinyl choline) as a muscle relaxant in surgery.

$$(CH_3)_3N^+(CH_2)_2O_2C(CH_2)_2CO_2(CH_2)_2N^+(CH_3)_3$$
Succinyl choline

In normal subjects the effects of suxamethonium are short-lived and this is due to its breakdown by serum butyrylcholinesterase. However, in certain individuals the effects of the drug were much more long-lasting; this was associated with low butyrylcholinesterase levels. Most of these individuals have a mutant form of cholinesterase in which Asp70 is substituted by glycine. When the cholinesterase from suxamethonium-sensitive individuals was further studied it became apparent that the enzyme differed from that of normal individuals in having a high K_m

with all substrates tested. This enabled a simple screening test to be carried out on serum to determine whether a subject is likely to be sensitive to suxamethonium. The test involves determining the percentage inhibition of cholinesterase by a particular concentration of dibucaine measured under standard conditions. The cholinesterase from suxamethonium-sensitive individuals is inhibited by only about 20 per cent under conditions that reduce the normal enzyme activity by about 80 per cent.[33,38] For reviews on genetically inherited enzyme deficiencies see references 33 and 37.

Dibucaine

10.7 **Enzyme inhibitors and drug design**

The diseases described in the previous section are all genetically inherited diseases in which there is an enzyme deficiency. Other clinical situations arise as the result of enzyme overactivity. In this section we consider some examples of exogenous inhibitors or drugs that have been used to reduce or inhibit the activities of endogenous enzymes. For more detailed accounts see references 39 and 40.

10.7.1 **Inhibition of hydroxymethylglutaryl CoA reductase**

An important genetically inherited disease that involves a protein defect, although not an enzyme defect, is familial hypercholesterolaemia. Familial hypercholesterolaemia is characterized by high circulating cholesterol and cholesterol ester levels in blood. Subjects homozygous for the disease may have 4–6 times the normal circulating levels of cholesterol, and generally have symptoms of coronary disease in childhood or adolescence, with the deposition of arterial plaques. The manifestation of the disease is less severe in the more common heterozygous state, with cholesterol and cholesterol ester levels of about twice the normal range and increased frequency of coronary disease at middle age. The condition has been shown to arise from a deficiency in the low-density lipoprotein receptors that are responsible for the uptake of cholesterol and cholesterol esters by the tissues: hence the higher circulating levels of cholesterol and cholesterol esters.

It has been found that the circulating levels of cholesterol and cholesterol esters can be reduced, and therefore so can the tendency to form arterial plaques, if the subjects are given a competitive inhibitor of hydroxymethylglutaryl CoA reductase (HMG CoA reductase). HMG CoA reductase catalyses the rate-limiting step in the formation of cholesterol. It is an enzyme having a short half-life (see Chapter 9, Section 9.6.4).

Mevinolin

Compactin   Monacol K
R = H   R = CH$_3$

$$\text{HMG CoA} + 2\text{NADPH} + 2\text{H}^+ \rightarrow \text{mevalonate} + \text{CoASH} + 2\text{NADP}^+.$$

Three competitive inhibitors[41] of HMG CoA reductase have been found, namely, mevinolin, compactin (K_i = 1 nmol dm^{-3}),[39] and monacol K, which can be used to inhibit the enzyme. All three are fungal products and have a structural resemblance to HMG CoA. Mevalonate is a precursor not only of steroids but also of a number of other important cell constituents such as ubiquinone, dolichol, and isopentenyl adenine (present as one of the modified bases in tRNA). However, sterol biosynthesis appears to be selectively inhibited at low concentrations.

HMG CoA

Competitive inhibitors of HMG CoA reductase; the regions resembling HMG CoA are outlined.

10.7.2 **Inhibition of xanthine oxidase** (proteins)

Gout is characterized by higher than normal levels of urate in the blood. Urate acid is the end-product of purine breakdown in humans; it is mainly derived from nucleic acids, and is excreted by the kidney. Sodium urate has a low solubility and uric acid is lower still. If the rate of urate formation is higher than normal, uric acid is liable to deposit in tissues, particularly where the pH is slightly acidic. Deposition of uric acid crystals in the joints is particularly painful, but more serious is the possibility of renal failure due to deposition in the kidney. The tendency to gout is familial and this has led to the suggestion that it may be a metabolic defect. Dietary factors and alcohol ingestion may exacerbate the situation. One form of treatment is to reduce uric acid production by inhibition of the last step in the metabolic pathway in which xanthine is oxidized to uric acid. Xanthine is more soluble than uric acid and is also excreted by the kidney. The inhibitor used is allopurinol. Allopurinol is also a substrate for xanthine oxidase and becomes hydroxylated on C-2 to form alloxanthine. The latter binds strongly to the active site ($K_{di} \approx 0.5$ nmol dm^{-3}) and so inhibits further catalytic activity. It effectively acts as a suicide substrate for the enzyme[42] (see Chapter 5, Section 5.4.4.3).

10.7.3 Inhibition of HIV-1 protease

The replication of human immunodeficiency virus-1 (HIV-1), the cause of AIDS, requires the proteolytic cleavage of a virus-encoded polyprotein which also contains several important viral proteins. The HIV-1 protease that catalyses this proteolytic cleavage is itself part of the polyprotein sequence. It has to be excised from the rest of the polyprotein before it can hydrolyse the peptide bonds necessary for release of other key proteins. HIV-1 protease is an aspartyl protease having a sequence Asp-Thr-Gly at its active site. It resembles other aspartic acid proteases, e.g. pepsin ($M_r \approx 33\,000$), except that it is much smaller ($M_r \approx 11\,000$). It has been suggested that pepsin and related proteases have arisen by gene duplication, each having two aspartic acid residues at the active site. A number of retroviral proteases, e.g. Rous sarcoma virus protease, are homodimers, with an aspartic acid residue from each monomer forming part of the active site. They also show inhibition by pepstatin, characteristic of aspartic acid proteases. X-ray crystallographic studies have shown that HIV-1 protease is a symmetrical dimer. The protease is essential for virus multiplication, so potential inhibitors have been designed as transition-state analogues of the protease (Fig. 10.6).

Many inhibitors have been designed that are peptide substrate analogues in which the scissile P1–P1′ amide bond (see Chapter 5, Fig. 5.9) has been replaced by a non-hydrolysable isostere with tetrahedral geometry (Fig. 10.6). An important consideration in designing an HIV-1 protease inhibitor to be used as a drug is that it must not appreciably inhibit other human aspartic acid proteases such as, for example, renin.[43] Renin, secreted by the kidney, in combination with angiotensin I-converting enzyme, secreted by the kidney and the lung, catalyses the two steps in the conversion of angiotensinogen via angiotensin I to angiotensin II. Angiotensin II is the most potent endogenous substance controlling blood pressure. Many of the strategies used in designing HIV-1 protease inhibitors had been previously developed for renin inhibitors designed for controlling blood pressure.[39] For a review of HIV-1 protease inhibitors see references 29 and 44.

10.8 The use of enzymes to determine the concentrations of metabolites of clinical importance

With the wider availability of purified enzymes it has become possible in recent years to measure a variety of metabolites by enzymatic methods. The principle is

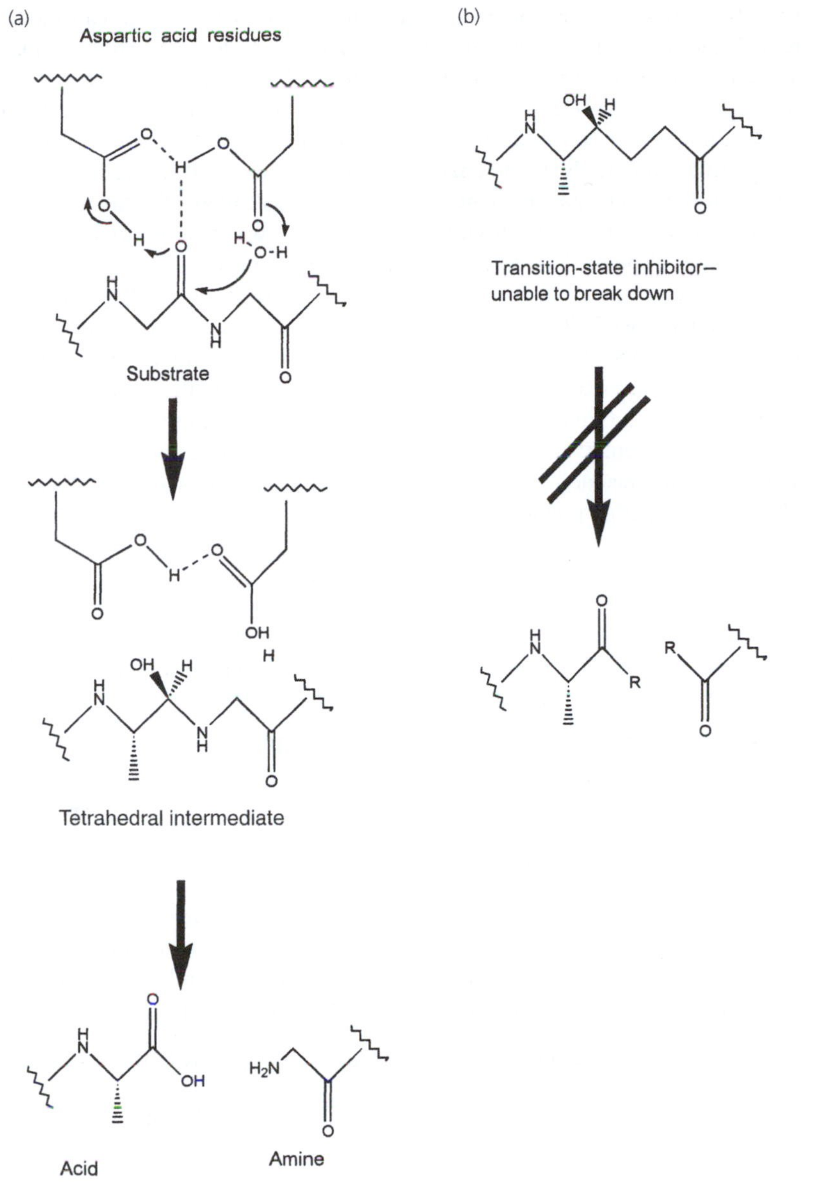

(a)

Aspartic acid residues

Substrate

Tetrahedral intermediate

Acid

Amine

(b)

Transition-state inhibitor—
unable to break down

Fig. 10.6 (a) The proposed mechanism of action of aspartyl proteases such as HIV protease, and (b) the structure of a transition-state inhibitor.[43]

to use an enzyme to transform a metabolite into its product and then to estimate the amount transformed by utilizing the difference in the properties of the substrate and product. There are a number of advantages in using enzymatic methods to estimate metabolites in body fluids such as serum or urine. These methods can make use of the high specificity of the enzyme (see Chapter 1, Section 1.3.2) to estimate the metabolite in the presence of many other substances, thus avoiding purification steps that may be necessary prior to a chemical analysis. Enzymatic methods can sometimes be used to measure concentrations of labile substances that would be degraded by harsher chemical methods. There may also be certain disadvantages. The cost of purified enzymes may be too high for routine use. This will depend on the amount required to carry out each assay in a reasonably short time. In some cases it is possible to use immobilized enzymes that can then be reused, e.g. urease, hexokinase, trypsin, α-amylase, and leucine aminopeptidase. These can be immobilized on the inner walls of plastic tubing and used in continuous flow analysis[3] (see Chapter 11, Section 11.5.4). Enzymes are susceptible to inactivation by a variety of agents. An inhibitor present in serum or urine might, for example, completely inhibit enzyme activity or it may only cause partial inhibition. In the latter case we might obtain a false result unless the time-course of the reaction were studied to check that complete reaction had occurred.

A number of points have to be checked before an enzyme can be reliably used in the measurement of the concentration of a metabolite. (For a detailed discussion see reference 45.) The equilibrium position of the reaction is important and it is best, if possible, to use a reaction in which the equilibrium is far to the right in order to obtain a quantitative conversion of the metabolite to the product of the reaction. If the equilibrium is not so favourable, this can sometimes be overcome either by removing the products by chemical trapping or by coupling to a second enzyme reaction. For example, the equilibrium constant for glucose-6-phosphate isomerase is close to unity,

$$K = \frac{[\text{fructose 6-phosphate}]}{[\text{glucose 6-phosphate}]} = 0.3,$$

but by coupling this to glucose-6-phosphate dehydrogenase the product of the isomerase is effectively removed.

1. D-Fructose 6-phosphate $\rightleftharpoons$ D-glucose 6-phosphate

2. D-glucose 6-phosphate + $NADP^+$ $\rightleftharpoons$ D-glucono-δ-lactone 6-phosphate + NADPH + H^+

3. D-Glucono-δ-lactone 6-phosphate + H_2O $\rightleftharpoons$ 6-phospho-D-gluconate.

The third step, which is the spontaneous non-enzymic hydrolysis of D-glucono-δ-lactone 6-phosphate, ensures that reactions 1 and 2 proceed to virtual completion.

The equilibrium for the glutamate dehydrogenase reaction lies to the left, but glutamate can be estimated in terms of the NADH produced,

L-Glutamate + NAD^+ + H_2O $\rightleftharpoons$ 2-oxoglutarate + NADH + NH_4^+,

provided the 2-oxoglutarate is trapped by reaction with hydrazine:

$$\begin{array}{ccc}
\text{COO}^- & & \text{COO}^- \\
| & & | \\
\text{C}=\text{O} & +\,H_2NNH_2 \;\rightarrow\; & \text{C}=\text{NNH}_2 \;+\; H_2O \\
| & & | \\
\text{CH}_2 & & \text{CH}_2 \\
| & & | \\
\text{CH}_2 & & \text{CH}_2 \\
| & & | \\
\text{COO}^- & & \text{COO}^-
\end{array}$$

A sufficiently large amount of enzyme must be used so that the substrate is almost completely (≥ 99 per cent) converted to the product in a short time (usually a few minutes). This will depend on the V_{max} for the enzyme and also on the K_m in relation to the amount of metabolite to be estimated. If the latter is, for example, approximately equal to the K_m, then when the substrate has become 99 per cent converted to the products the rate will only be 1 per cent of V_{max}.

The specificity of the enzyme should be known, since some enzymes that do not show absolute specificity will convert related substrates or homologues to their products, often at a reduced rate. If the enzyme preparation is not pure, then side-reactions will occur. Some examples of the use of enzymes to assay metabolites that are important in clinical chemistry are given below. For the estimation of certain metabolites, e.g. blood glucose and plasma triglycerides, enzyme methods are now used almost exclusively, but for others, e.g. serum creatinine, non-enzymic methods are still being used in the majority of clinical chemistry laboratories.[5]

10.8.1 Blood glucose

The oral glucose tolerance test is still the most widely used diagnostic procedure in the investigation of glucose metabolism in humans, e.g. in the investigation of diabetes. A subject is usually given an oral test dose of glucose and the blood glucose concentration is measured for the subsequent 3–4 h. In a normal subject the blood glucose concentration rises during the initial 30–60 min and then falls rapidly back to the basal level due to the increased utilization by the tissues. In a diabetic subject the increase in the concentration of blood glucose is more pronounced and the decline to normal levels is much slower. The extent of the rise and the rate of the subsequent fall depends on the severity of the diabetes. The test requires several measurements of blood glucose. Chemical methods depend on the reducing properties of glucose and therefore are not specific for glucose. There are three enzymatic methods that are generally available: hexokinase coupled with glucose-6-phosphate dehydrogenase, glucose oxidase coupled with peroxidase or measuring the oxygen consumed directly using an oxygen electrode, and glucose dehydrogenase.[46] Since glucose is taken up and metabolized by erythrocytes and leucocytes, it is important that the blood sample is either cooled rapidly or has sodium fluoride added (2 mg ml^{-1}) to prevent glycolysis.

1. D-Glucose + ATP $\rightleftharpoons$ D-glucose 6-phosphate + ADP.
   D-Glucose 6-phosphate + NADP$^+$ $\rightleftharpoons$ D-glucono-δ-lactone 6-phosphate + NADPH + H$^+$.

2. D-Glucose + H$_2$O + O$_2$ $\rightleftharpoons$ D-gluconic acid + H$_2$O$_2$.
   H$_2$O$_2$ + dye$_{reduced}$ $\rightleftharpoons$ H$_2$O + dye$_{oxidized}$.

3. D-Glucose + NAD$^+$ $\rightleftharpoons$ D-glucono-δ-lactone + NADH + H$^+$.

In the first method the NADPH produced would be measured by the change in A_{340}. Although hexokinase is not specific for glucose, glucose-6-phosphate dehydrogenase is highly specific for glucose 6-phosphate and thus the specificity is built into the second step. The second method using glucose oxidase is more frequently used because the enzymes and substrates are less expensive and the final absorption is in the visible range, although it is more susceptible to interference. Glucose oxidase specifically oxidizes β-D-glucose, and α-D-glucose is oxidized 150 times slower. Blood glucose contains an equilibrium mixture of both isomers. Most glucose oxidase preparations contain some aldose 1-epimerase as an impurity that enables rapid epimerization to occur and the total glucose to be estimated. The hydrogen peroxide produced in the first step oxidizes an acceptor, which then has a changed absorption spectrum. The most commonly used acceptor was o-dianisidine; however, various substances present in blood, e.g. uric acid, bilirubin, haemoglobin, and ascorbic acid, interfere by competing with *o*-dianisidine for hydrogen peroxide. Alternative acceptors such as 3-methyl-2-benzothiazoline hydrazone or 4-aminophenazine that have been used are not affected by these agents.[47]

The third method utilizes glucose dehydrogenase (EC 1.1.1.47) from *Bacillus cereus* and has the advantages that it does not require any additional coupling enzymes and also that the glucose dehydrogenase can be immobilized on nylon tubing and used in a biosensor (see reference 48; for further discussion on biosensors, see Chapter 11, Section 11.5.5).

10.8.2 Uric acid

Chemical

Uric acid + $H_3PW_{12}O_{40} \rightarrow$ allantoin
$\qquad$ + CO_2 + tungsten blue

Enzymatic

Uric acid + $2H_2O + O_2 \rightleftharpoons$
$\qquad$ allantoin + CO_2 + H_2O_2

Uric acid is the end-product of purine metabolism in humans. Measurement of uric acid in blood and urine is important in the diagnosis of gout, when the levels of uric acid are raised. They are also raised in other conditions, e.g. in leukaemia, and after ingestion of food rich in nucleoproteins. Uric acid may be estimated by either chemical or enzymatic methods. The chemical method involves the oxidation of uric acid to allantoin using phosphotungstic acid, whereas in the enzymatic method urate oxidase is used to catalyse a similar oxidation.

The chemical method has the disadvantage of lack of specificity, whereas the enzymatic method has the advantage that urate oxidase shows absolute specificity. It has now almost completely replaced the chemical method. The enzyme measurement uses the difference in absorbance at 293 nm between uric acid and allantoin. Alternatively, the hydrogen peroxide formed could be coupled to peroxidase as in the glucose oxidase assay to enable measurements to be made in the visible part of the spectrum, or the oxygen consumption can be measured using an oxygen electrode.

10.8.3 Urea

Determinations of the levels of urea in serum are frequently carried out as a test of renal function, especially in conjunction with urinary urea estimation, which enables the glomerular filtration rate to be assessed. Urea is hydrolysed by urease to ammonium carbonate. This reaction is most frequently coupled with glutamate dehydrogenase, so that the ammonia, together with oxoglutarate, causes the oxidation of NADH which can be monitored at 340 nm. An alternative method is to have the urease immobilized and allow the ammonia formed to diffuse through a semipermeable layer where it reacts with an indicator dye.[49]

10.8.4 Cholesterol, cholesterol esters, and triglycerides

1. Cholesterol ester + $H_2O \rightleftharpoons$
$\qquad$ cholesterol + $RCOO^-$

2. Cholesterol + $O_2 \rightleftharpoons$
$\qquad$ 4-cholesten-3-one + H_2O_2

3. $2H_2O_2$ + 4-aminophenazine $\rightleftharpoons$
$\qquad$ quinoneimine dye + $4H_2O$

The measurement of the concentrations of cholesterol and cholesterol esters in the serum is important in assessing the risk factors in arteriosclerosis and in myocardial infarction. High circulating levels of cholesterol and cholesterol esters are indicative of high risk. Cholesterol can be measured by a colorimetric method that is not highly specific and thus not very satisfactory. It also involves the use of concentrated sulphuric acid, which can be a nuisance in routine analysis. The introduction of straightforward enzymatic methods of high specificity has greatly simplified procedures. Cholesterol can be measured on between 5 and 100 μl serum using a single enzyme, cholesterol oxidase, and cholesterol esters can be measured by using this enzyme together with cholesterol esterase. The product of cholesterol oxidase action, 4-cholesten-3-one, can be estimated by its absorption at 240 nm, but it is more satisfactory to use horseradish peroxidase to catalyse the oxidation of 4-aminophenazine by hydrogen peroxide to a quinoneimine dye that absorbs at 500 nm. At the higher wavelength there are fewer interfering substances.

The measurement of serum triglycerides in the plasma or serum is useful in following the course of diabetes mellitus, nephrosis, and biliary obstruction. It can also be measured enzymatically, by hydrolysing the triglyceride and then phosphorylating the glycerol formed, and eventually coupling to a dehydrogenase so that oxidation of NADH can be monitored either by the decrease in absorbance at 340 nm, or by the decrease in fluorescence at 460 nm after excitation at 355 nm:

Lipase
1. Triglyceride + $3H_2O \rightarrow$ glycerol + 3 fatty acids

Glycerol kinase
2. Glycerol + ATP $\rightleftharpoons$ glycerol-3-phosphate + ADP

Pyruvate kinase
3. ADP + PEP $\rightleftharpoons$ ATP + Pyruvate

Lactate dehydrogenase
4. Pyruvate + NADH + H^+ $\rightleftharpoons$ Lactate + NAD^+

10.8.5 **Other metabolites**

A number of other metabolites that are measured less frequently can also be measured by enzymatic methods. Creatine may be measured by the decrease in A_{340} using creatine kinase coupled to pyruvate kinase and lactate dehydrogenase:

$$\text{Creatine + ATP} \xrightleftharpoons{\text{Mg}^{2+}} \text{phosphocreatine + ADP.}$$

$$\text{ADP + phosphoenolpyruvate} \xrightleftharpoons{\text{Mg}^{2+}} \text{ATP + pyruvate.}$$

$$\text{Pyruvate + NADH + H}^+ \rightleftharpoons \text{L-lactate + NAD}^+.$$

The measurement of creatinine is important in the creatinine-clearance test, which is used to test glomerular filtration occurring in the kidney. The colorimetric method using picric acid has been used for many years but is now being replaced by an enzymatic method using creatininase. It is used in conjunction with the coupling enzymes mentioned above for the assay of creatine:

$$\text{Creatinine + H}_2\text{O} \rightleftharpoons \text{creatine.}$$

Ammonia may be estimated using glutamate dehydrogenase:

$$\text{NH}_4^+ + \text{NADH + 2-oxoglutarate} \rightleftharpoons \text{L-glutamate + NAD}^+.$$

Ethanol may be estimated using alcohol dehydrogenase,

$$\text{CH}_3\text{CH}_2\text{OH + NAD}^+ \rightleftharpoons \text{CH}_3\text{CHO + NADH + H}^+,$$

and lactate in a similar way by using lactate dehydrogenase.

10.8.6 **Thiamin diphosphate in erythrocytes**

This example is rather different from the previous ones, since it is a case where an enzyme assay is used to detect a vitamin deficiency. The enzyme transketolase, which catalyses the following reaction, requires thiamin diphosphate as a cofactor for activity:

$$\text{D-Ribose 5-phosphate} + \text{D-xylulose 5-phosphate} \rightleftharpoons \text{sedoheptulose}$$
$$\text{7-phosphate} + \text{D-glyceraldehyde 3-phosphate}.$$

Normally, thiamin deficiency rarely occurs in developed countries, where there is very little malnutrition. However, the deficiency is common in areas where there is severe malnutrition and it also occurs in the case of Wernicke's encephalopathy. This syndrome occurs in alcoholics, who become thiamin deficient as a result of consuming little other than alcohol. The early stages of this syndrome may be detected as thiamin deficiency. Thiamin is normally converted into thiamin diphosphate, the cofactor for transketolase and also for pyruvate and oxoglutarate dehydrogenases (see Chapter 7, Section 7.7):

$$\text{Thiamin} + \text{ATP} \rightleftharpoons \text{thiamin monophosphate} + \text{ADP}.$$

$$\text{Thiamin monophosphate} + \text{ATP} \rightleftharpoons \text{thiamin diphosphate} + \text{ADP}.$$

The erythrocytes are a good and readily available source of enzymes of the pentose-phosphate pathway such as transketolase. The transketolase activity is thus measured in the presence and absence of added thiamin diphosphate. In normal subjects the activity is not increased by addition of the cofactor, in contrast to the situation in thiamin-deficient subjects.

10.8.7 Enzyme immunoassay

In addition to the direct use of enzymes to measure the concentrations of their substrates in body fluids, there is also an indirect method in which enzymes are used as a type of amplification system in what is known as enzyme immunoassay. Since the first description of the method of radioimmunoassay by Yalow and Berson in 1960, a large number of methods have been developed for measurement of hormones, drugs, and a variety of small biological molecules that are based on antigen–antibody interactions. These methods rely on the specificity of antigen–antibody interactions for binding the species to be assayed, together with some method of increasing the sensitivity of detection by use of either a radioactive label or an enzyme label in enzyme immunoassays. The first enzyme immunoassay was developed in 1971.[50]

Small molecules such as hormones and drugs do not induce antibody formation when injected into a foreign species. However, if they are chemically combined to a protein such as albumin, and then are injected into an animal, they will induce antibody formation. The antibodies so formed will combine specifically with the antigen, forming an insoluble antigen–antibody complex. The antibodies will also combine with the original small molecule (known as a hapten) although they will not give rise to a precipitate. Immunoassays can thus be devised for any molecule (acting as a hapten) that, when coupled to a protein, is capable of inducing antibody formation. The method thus has potentially very wide application.

The second stage of the method is to detect the antigen–antibody or antibody–hapten complexes, preferably when they are present in low concentrations. Two of the methods used are either to incorporate a radioactive label onto the antigen or antibody (radioimmunoassay), or to incorporate an enzyme molecule onto the antigen or antibody (enzyme immunoassay). Radioimmunoassays do not in general entail the use of enzymes and thus are outside the scope of this book (a good review of the method is given in reference 51). Radioimmunoassays are, however, being used in the detection of a number of the isoenzymes listed in Table 10.2. A further development in enzyme immunoassays is to have the antibody adsorbed onto a solid support such as a plastic tube. This facilitates the separation of the free antigen from the antibody–antigen complex, and the technique is then referred to as enzyme-linked immunosorbant assay (ELISA).

A variety of enzyme immunoassays techniques have been developed; two of the most widely used are described here, but for more details see reference 52. The two different methods are competitive binding assay and sandwich assay outlined in Fig. 10.7. In the competitive binding assay the substance to be measured (the antigen) is first used to raise antibodies. If this is a small molecule (L) then it would first be coupled to a protein to make it antigenic before being injected into a rabbit to raise the necessary antibodies. The antibodies are adsorbed onto a suitable surface. A conjugate has to be prepared containing the substance L and the enzyme being used in the detection procedure. A number of enzymes have been used in this type of coupling, e.g. alkaline phosphatase, β-galactosidase, horseradish peroxidase, lysozyme, malate dehydrogenase, and glucose-6-phosphate dehydrogenase. The coupling is usually carried out using a bifunctional reagent, e.g. glutaraldehyde. An ideal coupling is one in which the linkage is stable and complete, and in which the properties of the enzyme and the antigen are minimally affected. To measure an unknown concentration of L, a known amount of the antibody is adsorbed to the support. A mixture containing the unknown concentration of L and a known amount of the enzyme conjugate (L-Enz) is mixed with the immobilized antibody. The latter is then washed and the amount of enzyme activity attached to the support is measured. The competition between L and L-Enz for the immobilized antibody means that the more of L present the smaller the amount of L-Enz attached to the support. If a standard set of known concentrations of L is assayed in the same way then the amount present in the unknown can be determined. The method is very sensitive since very small amounts of enzyme activity can be detected. The method can be used for any antigenic substance, or any hapten that can be made antigenic by conjugation with a protein.

The sandwich method (Fig. 10.7) requires a second antibody that recognizes a different antigenic determinant on L. The second antibody is then coupled to the detecting enzyme (Ab-Enz). A fixed amount of the immobilized antibody is mixed with the unknown concentration of L and an excess of Ab-Enz. After washing the support only Ab-Enz bound to L will remain fixed, and this can be measured by assaying for enzyme activity .

These methods have been applied to measurement of a wide range of clinically important substances. For example, the ELISA method has been used to detect the AIDS virus. In this case serum suspected of containing the AIDS virus is incubated on a polystyrene dish, to which any AIDS virus would adhere. The dish is washed and an inert protein is added which binds to any remaining surface sites on the polystyrene. Then antibody against AIDS viral protein coat (Ab1) is added together with an Ab-Enz which recognizes Ab1. The amount of Ab-Enz sticking to the plate is directly related to the concentration of AIDS viral proteins. This method has been developed to measure a variety of drugs, including opiates, barbiturates, amphetamine, cocaine, anticancer drugs, and hormones such as thyroxine, and to screen antigens from pathogenic organisms, and is being used increasingly in clinical laboratories. It has three advantages over radioimmunoassays in that the reagents are stable, it avoids the need to use radioactive materials, and it is also capable of greater sensitivity.

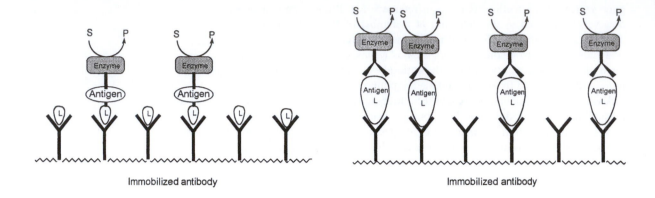

Competitive Enzyme Immunoassay   Sandwich Enzyme Immunoassay

Fig. 10.7 Enzyme-linked immunoassay.

10.9 **Enzyme therapy**

Although the attempts at enzyme therapy in clinical trials have so far met with limited success, the technique of administering enzymes for therapeutic purposes seems sufficiently promising that may become a feasible approach for treatment of certain diseases in the near future. The principal areas in which work on enzyme therapy has so far concentrated are the removal of toxic substances from the blood, genetic deficiency diseases, and cancer. In addition, enzyme therapy may be used in the following areas:

(1) degradation of necrotic tissue by use of proteolytic enzymes such as trypsin, chymotrypsin, or bromelain;

(2) removal of blood clots using streptokinase or urokinase (see Chapter 11, Section 1 1.4.3.8); and

(3) treatment of pancreatic insufficiency, which occurs in cystic fibrosis,[53] by administration of suitably entrapped enzymes, e.g. proteolytic enzymes.

How the enzyme is administered depends on which of the above objectives is being considered. If the enzyme is used to remove or metabolize a substance present in the blood, e.g. a toxic metabolite or a blood clot, then it is not necessary for the enzyme to enter the intracellular compartments; it is only necessary for the enzyme to be present in the blood. This type of application may be either intra- or extracorporeal. The latter, which involves the use of a bypass as in kidney dialysis, has the advantage that immobilized enzymes may be used and these can be retained outside the body and are not cleared by the kidney. Immobilized enzymes also have the advantage of being less immunogenic. There are several examples of this type of application,[54] such as removal of circulating urea by urease in cases of kidney failure, removal of bilirubin by bilirubin oxidase, removal of fibrin from blood clots by fibrolysin, and removal of asparagine by asparaginase (see Section l0.9.2). For intracellular enzyme therapy, it is necessary for the enzyme to be taken up by the appropriate target cells. For more detailed accounts of enzyme therapy, see references 53, 55, and 56.

10.9.1 **Treatment of genetic deficiency disease**

There are now a large number of genetic deficiency diseases in which the enzyme deficiency has been identified (Table 10.3), although these only afflict small proportions of the population. The principal treatment that is at present available for a number of these diseases is careful regulation of the diet to reduce the substrate loading and thereby minimize the harmful effects of the defect, e.g. for phenylketonuria, a diet low in phenylalanine; for galactosaemia, a diet low in galactose; for maple syrup disease, a diet low in leucine, isoleucine, and valine.

The ultimate cure for these diseases would be to insert the gene that codes for the deficient enzyme into chromosomes. Although there are still many problems to be solved before this may be feasible, significant progress has been made. The introduction of a gene into somatic cells (somatic replacement) as opposed to germ cells would limit any benefit to the individual being treated and would not extend to future generations. In a sense it is like a complex drug delivery system. The genetic diseases that are the best potential candidates for gene therapy are those in which there is a single gene defect. To introduce genes into somatic cells, techniques must be available to manipulate the cells *in vitro*. At present this includes fibroblasts, lymphocytes, endothelial cells, muscle cells, hepatocytes, and keratinocytes.[57] For example, cDNA encoding phenylalanine hydroxylase, the enzyme deficient in phenylketonuria, has been cloned and efficiently transduced into phenylalanine-deficient hepatocytes *in vitro*, but so far transduction efficiency *in vivo* is very low.[58] The first use of gene therapy to cure a genetic disorder was in adenosine deaminase deficiency. Individuals lacking this enzyme are born with severe combined immunodeficiency syndrome, and they lack a functional immune system. Gene therapy trials were begun in 1990. T lymphocytes were removed from deficient individuals and were infected with a modified virus carrying the adenosine deaminase gene and then infused into the individuals' circulatory system. Because the lymphocytes have a limited life-span, reinfusion is repeated every 3 to 5 months. The patients have so far responded well to the treatment.[57,59]

An alternative to gene therapy is enzyme therapy. Although purified enzymes are available for many of these deficiency diseases, there are many problems in delivering the enzyme to the required site under such conditions that it will remain stable and functional for a reasonable time. When considering enzyme therapy to alleviate genetic deficiency disease, the prospects of a successful treatment are likely to be best in cases in which the enzyme can be most easily targeted to the tissue or organelle required, and also in cases in which an enzyme treatment for a limited period is likely to be of value. The latter is a consideration because any enzyme applied is likely to be turned over and eventually cleared from the system. These restrictions have made lysosomal storage diseases the prime candidates for enzyme therapy.[60] These are a group of diseases in which one of the normal lysosomal enzymes is deficient. Various substances are normally taken up into the lysosomes by the process of endocytosis (in which a portion of the membrane surrounds the substance and is then involuted and pinched off within the lysosome) and are subsequently degraded. However, in lysosomal storage diseases the substances are endocytosed but then accumulate. Because the lysosomes have the capability to endocytose foreign material, it would seem reasonable that they should take up added enzymes. There is also quite a good understanding of the signals that cause lysosomal proteins made in the Golgi apparatus to be taken up by the lysosomes. The presence of a mannose-6-phosphate residue on the protein targets it to the lysosomes. Work has been in progress since 1964 on attempts at replacement therapy in lysosomal storage diseases and a number of technical problems have been identified. Quite large amounts of the appropriate enzyme are required in a high state of purity and in a non-immunogenic form. Some of the early problems arose from immunological reactions, especially when the enzyme was obtained from a different organism. Many enzymes when administered are either inactivated or degraded fairly rapidly, although some experiments carried out using fibroblasts grown in culture have shown that enzymes can be taken up from the culture medium and will function within the cells. In the earlier work little was known about the signals needed for receptor-mediated endocytosis.[60]

The lysosomal storage disease most studied is Gauchers disease. In this disease a glucocerebroside accumulates largely in macrophages because of the defective activity of glucosylceramidase.[55] These macrophages distribute to various organs of the body including liver, spleen, lungs, and lymph modes, causing their enlargement. The results of enzyme therapy in over 1200 patients with Gaucher disease show that after one year there is a significant regression of hepatic, splenic, bony, and haematological abnormalities. The β-glucosidase used is generally either of human placental origin, or human recombinant enzyme from cells grown in culture, and the high cost at present limits treatment.

Enzyme therapy has also been tested for pancreatic insufficiency, cystic fibrosis, and xeroderma pigmentosa. Pancreatic insufficiency arises in pancreatitis, where premature activation of the pancreatic enzymes leads to self-digestion and loss of part of the pancreas. It can also arise after excessive drug consumption, including alcohol. Cystic fibrosis is a genetically inherited disease occurring with a frequency of about 1 in 2000 births. The genetic defect is in a transmembrane protein known as cystic fibrosis transmembrane conductance regulator. It is a disabling and fatal disorder in which the main focus is the exocrine glands such as the sweat glands and digestive glands. Thick mucus often leads to blockage of the pancreatic duct and also the bronchial tubes leading to lung infections. Pancreatic insufficiency can be alleviated by giving pancreatic extract containing lipase, amylase, and proteases. The most efficient form of delivery to date is in enteric-coated microspheres. A special polymer coating protects the enzymes at low pH, such as in the stomach, but releases them at pHs above about 6.[54] The lung infection associated with cystic fibrosis has been alleviated by application of recombinant deoxyribonuclease.[61] Undegraded DNA in combination with mucopolysaccharide forms a viscous gel in the lungs. The deoxyribonuclease acts by degrading DNA released from the dead human and bacterial cells that accumulate.

Xeroderma pigmentosa, a genetic defect in which individuals are extremely sensitive to solar UV radiation, is due to a defective DNA repair mechanism. In trials in which a prokaryotic DNA repair enzyme was delivered to the skin in topically applied liposomes, there was evidence of fewer thymine dimers present in DNA and less erythema in the skin.[62]

10.9.2 Cancer therapy

The enzyme therapy that has so far proved most useful in anticancer treatment has been the administration of asparaginase in cases of acute lymphoblastic leukaemia and to a lesser extent in acute granulocytic leukaemia.[63] The initial observation that led to its discovery was made in the 1950s and the first treatments were given in the 1970s. It was observed that injection of normal guinea-pig serum caused complete regression of a lymphosarcoma in mice and delayed the appearance of other tumours in experimental animals. These effects occurred only when guinea-pig serum was used. Extensive analysis showed that the active factor present in guinea pig was asparaginase. This enzyme is widely distributed in plants, animals, and bacteria, but it is absent from the serum of most common mammals. Other sources of asparaginase were tested and also found to cause regression of certain tumours. Asparaginase catalyses the breakdown of asparagine which is present in human serum at a concentration of about 40 μmol dm^{-3}.

$$\text{L-Asparagine} + H_2O \rightleftharpoons \text{L-aspartate} + NH_3$$

The tumours have a nutrient requirement for asparagine, and asparaginase, by lowering the serum asparagine concentration, retards the growth of the tumour. Asparaginase has been most effective when administered in combination with other chemotherapeutic agents. Microorganisms are the richest sources of asparaginase, but can cause liver damage and immunological responses, and repeated treatment leads to resistance. Human asparaginase might be the ideal source, but large quantities are required, which may become available through recombinant technology in the future.[64] Other enzymes, e.g. glutaminase, serine dehydratase, arginase, phenylalanine ammonia-lyase, and leucine dehydrogenase, are also being tested as possible anticancer agents; the rationale is that certain types of cancer cells may have lost the ability to synthesize amino acids and are thus dependent on uptake from the serum for supply, but so far none are in general use.

A rather different way in which enzymes may be used in cancer therapy is in the activation of prodrugs (Fig. 10.8). Prodrugs are drugs that have been chemically modified so that they remain inactive until specifically activated in much the same way as zymogens become activated (see Chapter 6, Section 6.2.1.1). Many drugs that are used in cancer chemotherapy inhibit particular steps in metabolic pathways. For example, methotrexate is a structural analogue of folic acid and a competitive inhibitor of dihydrofolate reductase ($K_i \approx 0.1$ nmol dm^{-3}). Tetrahydrofolate is required for the synthesis of thymidylate, a precursor of DNA. Methotrexate is therefore most toxic to cells that divide most rapidly, e.g. cancer cells. The effectiveness of such an inhibitor could be improved if it could be delivered specifically to the tumours cells. It could then be given in higher doses without killing non-cancer cells. Two strategies have been developed for the activation of prodrugs at tumour targets; the first is known as antibody-directed enzyme prodrug therapy (ADEPT), and the second gene-directed enzyme prodrug therapy (GDEPT). The first of these will be described briefly here, but for reviews of both see references 65–67. The prodrug to be used has to be synthesized in inert form; this generally involves some form of conjugation. An enzyme is required to activate the drug at the target site. The enzyme used is generally an exogenous enzyme and it is targeted using an antibody that recognizes specific antigenic sites on the tumour. Some examples of prodrugs and their activating enzymes are given in Table 10.4. The enzymes used have to be fairly stable and be able to release the drug at a rate (determined by K_m and k_{cat}) at which it

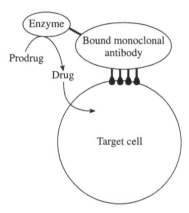

Fig. 10.8 The use of a monoclonal antibody–enzyme conjugate together with a prodrug to target a drug to tumour cells.

Table 10.4 Examples of prodrugs and their activating enzymes

Prodrug	Drug	Activating enzyme
Methotrexate α-phenylalanine	Methotrexate	Carboxypeptidase-A
Etoposide phosphate	Etoposide	Alkaline phosphatase
p-N-bis (2-Chloroethyl) aminobenzoylglutamic acid	Benzoic acid mustard	Carboxypeptidase-G$_2$
Mitomycin phosphate	Mitomycin	Alkaline phosphatase
Aniline mustard glucuronide	Aniline mustard	β-Glucuronidase
5-Fluorocytosine	5-Fluorouracil	Cytosine deaminase
2-(L-α-Aminoacyl) methotrexate	Methotrexate	Aminopeptidase

can be taken up at the tumour site. The enzyme has to be coupled to the antibody using suitable cross-linking agents. The technology developed for enzyme-linked immunosorbant assay can be used (see Section 10.8.7). An alternative will be to use fusion proteins produced by constructs linking genes for the enzyme and antibody which could be made in quantity using bacterial expression systems.

Gene-directed enzyme prodrug therapy differs from antibody-directed enzyme prodrug therapy in that the enzyme used to activate the prodrug is directed to the tumour target in the form of its gene linked to a virus vector. The enzyme then becomes expressed in the tumour target cells.[66]

10.10 Conclusions

Measurements of enzyme activity are being used increasingly in clinical diagnoses and enzyme preparations are being used increasingly to estimate metabolites and for enzyme therapy. These trends are likely to continue to increase for several reasons. Sophisticated equipment is becoming more widely available in clinical laboratories and many assays can now be performed on automated equipment. A wider variety of pure enzymes is becoming commercially available and many of these have been successfully immobilized to support materials. Many enzymes previously only available in very small quantities can now be made in larger quantities by recombinant DNA technology using bacterial and other expression systems (see Chapter 2, Section 2.4.1). These factors should enable the enzymes to be used more economically both in the assay of metabolites and in enzyme-replacement therapy.

In the field of clinical diagnosis, the assay of more than one enzyme or the combination of enzyme assay and isoenzyme distribution can give a reasonable degree of discrimination between different diseases. Many enzymes which are difficult to assay in an automated procedure can now be measured as enzyme protein by radioimmunoassays or enzyme immunoassays, and this will greatly increase the range of possibilities. This, in turn, could widen the scope of enzyme diagnoses.

References

1. Horder, M. and Wilkinson, J. H., in *Chemical diagnosis of disease* (Brown, S. S., Mitchell, F. L., and Young, D. S., eds), Chapter 7. Elsevier/North-Holland, Amsterdam (1979).

2. Moss, D. W. and Rosalki, S. B., *Enzyme tests in diagnosis* (2nd edn). Arnold, London (1995).

3. Burtis, C. A. and Ashwood, E. R., *Tietz textbook of clinical chemistry* (2nd edn). W B Saunders, Philadelphia (1993).

4. Marshall, W. J., *Clinical chemistry* (3rd edn). Mosby, London (1995).

5. Kaplan, A., Jack, R., Opheim, K. E., Toivola, B., and Lyon, A. W., *Clinical chemistry: interpretation and techniques* (4th edn). Williams and Williams, Baltimore (1995).

6. Kaplan, L. A., Pesce, A. J., and Kazmierczak, S. C., *Clinical chemistry: theory, analysis and correlation* (3rd edn), Chapters 26–31, 54, and 55. Mosby, St Louis, USA (1996).

7. Bergmeyer, H., *Methods in enzymatic analysis* (3rd English edn). Verland Chemie Weinheim, Academic Press, New York (1983). See Vol. 1.

8. Moss, D. W. and Rosalki, S. B., *Enzyme tests in clinical diagnosis* (2nd edn). Chapter 10. Arnold, London (1995).

9. Furie, B. and Furie, B. C., *Cell* **53**, 505 (1988).

10. Scully, M. F., *Essays Biochem.* **27**, 17 (1992).

11. Law, S. K. A. and Reid, K. B. M., *Complement*. IRL Press, Oxford (1988).

12. Golub, E. S. and Green, D. R., *Immunology: a synthesis*, Chapter 8. Sinauer, Sunderland, USA (1991).

13. Devlin, T. M., *Textbook of biochemistry with clinical correlations* (3rd edn), pp. 406–7. Wiley-Liss, New York (1992).

14. Schmidt, E. and Schmidt, F. W., *FEBS Lett.* **62**, Suppl. E62 (1976).

15. Horder, M. and Wilkinson, J. H., in *Chemical diagnosis of disease* (Brown, S. S., Mitchell, F. L., and Young, D. S., eds), Chapter 7, pp. 378–9. Elsevier/North-Holland, Amsterdam (1979).

16. Schmidt, E. S. and Schmidt, F. W., *Clin. Chim. Acta* **43**, 43 (1988).

17. Tredger, J. M. and Sherwood, R. A., *Ann. Clin. Biochem.* **34**, 121 (1997).

18. Schmidt, E. and Schmidt, F. W., *Brief guide to practical enzyme diagnosis* (2nd edn). Boehringer Mannheim, Mannheim (1976).

19. Wilkinson, J. H., *The principles and practice of diagnostic enzymology*, pp. 225–9. Edward Arnold, London (1976).

20. Vora, S., *Analyt. Biochem.* **144**, 307 (1985).

21. Divall, G. B., *Electrophoresis* **6**, 249 (1985).

22. Burtis, C. A. and Ashwood, E. R., *Tietz textbook of clinical chemistry* (2nd edn), pp. 786–7. W B Saunders, Philadelphia (1993).

23. Burtis, C. A. and Ashwood, E. R., *Tietz textbook of clinical chemistry* (2nd edn), p. 849. W B Saunders, Philadelphia (1993).

24. Moss, D. W. and Rosalki, S. B., *Enzyme tests in diagnosis* (2nd edn), Chapter 3. Arnold, London (1995).

25. Beckett, G. J. and Hayes, J. D., *Adv. Clin. Chem.* **30**, 281 (1993).

26. Kaplan, L. A., Pesce, A. J., and Kazmierczak, S. C., *Clinical chemistry: theory, analysis and correlation* (3rd edn), p. 596. Mosby, St Louis, USA (1996).

27. Moss, D. W. and Rosalki, S. B., *Enzyme tests in diagnosis* (2nd edn), p. 98. Arnold, London (1995).

28. Hall, C. L. and Hardwicke, J., *A. Rev. Med.* **30**, 199 (1979).

29. Christenson, R. H., *Clin. Biochem.* **30**, 573 (1997).

30. Whyte, M. P., in *Principles of bone biology* (Bilezikian, J. P., Raisz, L. G., and Rodan, G. A., eds), Chapter 68. Academic Press, New York (1996).

31. Moss, D. W. and Rosalki, S. B., *Enzyme tests in diagnosis* (2nd edn), p. 43. Arnold, London (1995).

32. Lombardo, M. E. and Hudson, P. B., *Steroids* **61**, 651 (1996).

33. Scriver, C. R., Beaudet, A. L., Sly, W. S., Valle, D., Stanbury, J. B., Wyngaarden, J. B. *et al.*, *The metabolic and molecular bases of inherited disease* (7th edn), Vol. 1. McGraw-Hill, New York (1995).

34. Cummings, M. R., *Human heredity: principles and issues* (3rd edn), Chapter 10. West Publishing, St Paul, USA (1994).

35. Woo, S. L. C., *Biochemistry* **28**, 1 (1989).

36. Ledley, F. D., Dilella, A. G., and Woo, S. L. C., *Trends in Genetics* **1**, 309 (1985).

37. Scriver, C. R., Beaudet, A. L., Sly, W. S., Valle, D., Stanbury, J. B., Wyngaarden, J. B. *et al.*, *The metabolic and molecular bases of inherited disease* (7th edn),Vol. 2. McGraw-Hill, New York (1995).

38. Jokanovic, M. and Maksimovic, M., *Eur. J. Clin. Chem. Clin. Biochem.* **35**, 11 (1997).

39. Edwards, P. D., Hesp, B., Trainor, D. A., and Willard, A. K., in *Enzyme chemistry: impact and applications* (2nd edn), (Suckling, C. J., ed.), Chapter 5. Chapman and Hall, London (1990).

40. Palfreyman, M. G., McCann, P. P., Lovenberg, W., Temple, J. G., and Sjoerdsma, A., *Enzymes as targets for drug design.* Academic Press, New York (1989).

41. Endo, A., *Trends Biochem. Sci.* **6**, 10 (1981).

42. Massey, V., Komai, H., Palmer, G., and Elion, G., *J. biol. Chem.* **245**, 2837 (1970).

43. Vacca, J. P. and Condra, J. H., *Drug Disc. Today.* **2**, 261 (1997).

44. Wlodawer, A. and Erickson, J. W., *A. Rev. Biochem.* **62**, 543 (1993).

45. Bergmeyer, H. U., *Methods in enzymatic analysis* (2nd English edn). Verland Chemie Weinheim, Academic Press, New York (1974). See Vol. 1, Section A. 11.3.

46. Burrin, J. M. and Price, C. P., *Ann. Clin. Biochem.* **22**, 327 (1985).

47. Burtis, C. A. and Ashwood, E. R., *Tietz textbook of clinical chemistry* (2nd edn), p. 961. W B Saunders, Philadelphia (1993).

48. Chaplin, M. F., in *Molecular biology and biotechnology* (3rd edn), (Walker, J. M. and Gingold, E. B., eds), Chapter 19. Royal Society of Chemistry, London (1993).

49. Kaplan, L. A., Pesce, A .J., and Kazmierczak, S. C., *Clinical chemistry: theory, analysis and correlation* (3rd edn), p. 499. Mosby, St Louis, USA (1996).

50. Engvall, E. and Perlman, P., *Immunochemistry* **8**, 871 (1971).

51. Orth, D. N., *Methods Enzymol.* **37**, 22 (1975).

52. Kenney, D. M., *A practical guide to ELISA.* Pergamon, Oxford (1991).

53. Klein, M. D. and Langer, R., *Trends Biotech.* **4**, 179 (1986).

54. Lebenthal, E., Rolston, D. D. K., and Holsclaw, D. S., *Pancreas* **9**, 1 (1994).

55. Grabowski, G. A., *Drugs* **52**, 159 (1996).

56. Kaye, E. M., *Curr Opinion Pediat.* **7**, 650 (1995).

57. Morgan, R. A. and Anderson, W. F., *A. Rev. Biochem.* **62**, 191 (1993).

58. Eisensmith, R. C. and Woo, S. L. C., *Eur. J. Pediat.* **155**, S15 (1996).

59. Cummings, M. R., *Human heredity: principles and issues* (3rd edn), p. 222. West Publishing, St Paul, USA (1994).

60. Neufeld, E. F., *A. Rev. Biochem.* **60**, 257 (1991).

61. Edgington, S. M., *Biotechnology* **11**, 580 (1993).

62. Yarosh, D., Klein, J., Kibitel, J., Alas, L., O'Connor, A., Cummings, B. *et al.*, *Photoderm. Photoimmun. Photomed.* **12**, 122 (1996).

63. Gallagher, M. P., Marshall, R. D., and Wilson, R., *Essays Biochem.* **24**, 1 (1989).

64. Wang, M. L., Rollence, M. L., Filpula, D., and Shorr, R. G. L., *FASEB J.* **9**, A94 (1995).

65. Huennekens, F. M., *Trends Biotech.* **12**, 234 (1994).

66. Connors, T. and Knox, R. J., *Stem Cells* **13**, 501 (1995).

67. Sherwood, R. F., *Adv. Drug Delivery Rev.* **22**, 269 (1996).

11

Enzyme technology

11.1 Introduction

Enzymes have remarkable catalytic properties (see Chapter 1, Section 1.3.1), especially when compared with other types of catalysts. In recent years they have been exploited on an increasing scale in food, pharmaceutical, and chemical industries. The advantages of high catalytic activity, lack of undesirable side reactions, and operation under mild conditions are often highly desirable. Also, with the advent of genetic engineering, a wider range of enzymes has become available on a larger scale and this has increased the scope of enzyme technology (Fig. 11.1). Genetic engineering itself creates a demand for certain highly purified enzymes (see Section 11.5.7) and the whole area of enzyme technology is expanding rapidly. Between 1983 and 1995 the sales of enzymes for industrial use have doubled, and a further doubling is predicted by 2005[1].

The enzymes present in a number of bacteria and fungi have been used for many years by man, but it is only since the 1970s that isolated enzymes, as opposed to enzymes present in whole organisms, have been used in industrial processes. There are now over 60 enzymes in commercial use, of which three-quarters are hydrolytic enzymes[1] (Fig. 11.2) although costwise they represent less than 50%.

We first consider the main applications of reactions that are catalysed by enzymes in intact organisms, then consider the use of isolated enzymes, and follow this by discussion of the advantages of either immobilized enzymes or immobilized whole cells. This short chapter serves to illustrate some of the new and traditional technological applications of enzymology. There are many articles and books on enzyme technology (see, for example, references 1–6).

11.2 Use of microorganisms in brewing and cheesemaking

The oldest industries in which enzymes in living organisms have been used are those of brewing and cheesemaking. These processes were carried out centuries before there was any understanding of their biochemical basis. In fact, much of the emphasis in biochemistry at the end of the nineteenth century and in the early

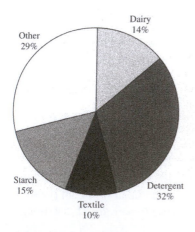

Fig. 11.1 Distribution of enzyme sales (1994).[1]

part of the twentieth century was aimed at understanding these fermentation processes.

As can be seen from Table 11.1, enzymes may be involved in either hydrolysis of polysaccharides to monosaccharides or in glycolysis leading to ethanol production. In the case of brewing, the starting material is the polysaccharide starch, whereas in winemaking the starting material is mainly mono- or disaccharides. The largest carbohydrates that most yeasts are able to ferment are trisaccharides. Therefore, when polysaccharides are used as the principal source of carbohydrates they must be degraded before fermentation can occur. The principal raw material used in brewing is the starch obtained from barley. Barley seeds are allowed to germinate and this leads to release of amylases that cause the breakdown of starch. The starch grains are contained within the cells of the endosperm, the walls of which are surrounded by a number of polymers including 1,3-β-D-glucans. It is necessary for these glucans to be degraded so that the amylases can be in contact with the starch granules (Fig. 11.3). The release of a variety of peptidases is also important, since they catalyse the degradation of endogenous protein, releasing peptides and amino acids. The amino acids and peptides are essential for the growth of yeast which occurs at the fermentation stage. The starch granules, dispersed into water, have to be kept above 60°C to prevent gelatinization. The amylase enzymes have to be stabilized to remain active at this temperature. This is achieved by maintaining the pH between 5 and 6, adding Ca^{2+}, and having a high substrate concentration. The amylases hydrolyse the α-1,4 links present in amylose and amylopectin* to produce a mixture of glucose, maltose, maltotriose, and higher unfermentable sugars called dextrins. α-Amylase is an endoamylase that randomly hydrolyses starch to dextrins, while β-amylase, an exoamylase, attacks both starch and dextrins from the non-reducing end removing maltose units. Neither attack the α-1,6 links present in amylopectin.

Anaerobic glycolysis is the second stage in the process in which ethanol is the end-product. (For further details of the brewing process see references 7–10.) New strains of yeasts that possess amylolytic activity have been constructed by recombinant DNA technology.[10–12] Glucoamylase genes from *Aspergillus* spp., β-1,3/1,4-glucanase genes from *Bacillus subtilis*, and cellulase (endo-β-1,4-glucanase) and β-galactosidase genes from *Trichoderma* have all been expressed in brewer's yeast.

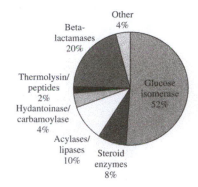

Fig. 11.2 The percentage of total sales of enzymes used in biotechnology.[1]

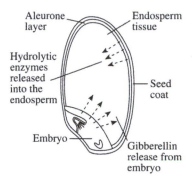

Fig. 11.3 The release of hydrolytic enzymes from a germinating barley seed.

* Amylopectin is polymer of D-glucose units linked α-1,6 as well as α-1,4 and makes up between 17 and 30 per cent of starch, the remainder being amylose which only has α-1,4 links.

Table 11.1 The use of whole organisms as a source of enzymes in the food industry

Process	Organism used	Enzyme step or steps involved
Brewing and winemaking	Barley (*Hordeum*)	α-Amylase and β-amylase, endo-1,3-β-D-glucanase
	Saccharomyces spp	Limited proteolysis; oligosaccharidases (oligosaccharides → monosaccharides) Glycolytic enzymes (monosaccharides → ethanol)
Cheesemaking	*Streptococcus lactis* *Streptococcus cremoris*	Lactose → lactate + limited lipolysis and proteolysis
	Propionobacteria spp	Conversion of lactate → propionate + acetate + CO_2 + H_2O
	Penicillium camemberti and *Penicillium roqueforti*	Limited lipolysis and proteolysis plus other undefined activities
	Calf stomach	Chymosin (rennin); caseinogen → casein
Vinegar production	*Acetobacter*	Ethanol to acetate

The resulting recombinant strains have the potential advantage of enabling the genetically engineered yeast to degrade starch, dextrins, cellulose, and lactose, respectively. However, they have a number of disadvantages over the normal strains, which still outweigh their potential advantages, but they may replace traditional yeasts in the future.[10,11]

Ethanol is not only produced as a beverage but is an important solvent and fuel. There are considerable differences between the production of alcoholic beverages and of industrial ethanol. For the former it is important that all components contributing to the flavour and appearance must be maintained. For industrial ethanol, however, the other components are unimportant; the main consideration is low-cost production of ethanol. Much industrial alcohol is produced by catalytic hydration of ethene derived from crude oil. However, with the increased cost of petrochemicals there is a greater interest in the fermentation process, particularly if a cheap source of raw material is available. Thus, much effort is being expended on devising efficient methods of degrading the most abundant polysaccharide, cellulose. It is also important to use a strain of yeast that will allow maximum fermentation to occur. For example, if sugar beet is used as the raw material, since it contains the sugar raffinose, it is important to use a raffinose-metabolizing strain of yeast. The possibility of production of ethanol using a fermentative bacterium such as *Zymomonas* is being explored. It has the advantage of catalysing a faster fermentation.[13]

For cheesemaking the main process is the fermentation of lactose to lactic acid and this is most frequently carried out by *Streptococcus lactis* or *Streptococcus cremoris*. These organisms are used primarily because they have limited metabolic diversity and do not produce large quantities of other products. However, it is the small quantities of other metabolites that are partially responsible for the characteristic flavours of cheeses. Other species of bacteria are used additionally in the case of certain cheeses that have gaseous cavities. These are the result of decarboxylations occurring late in the fermentation process, usually brought about by *Streptococcus diacetyllactis*, *Leuconostoc* spp., and *Propionobacteria shermanii*.

In Swiss cheese the flavour is partly due to propionic acid formation and the release of free proline by *Propionobacteria* spp. *Propionobacteria* only begins to thrive once appreciable lactic acid formation has occurred. This second fermentation, in which lactic acid is converted to propionic acid, does not result in a high-energy yield compared to the energy yield in lactate formation and thus only develops in the later stages of fermentation. The ATP is generated in the decarboxylation of pyruvate.

$$3CH_3.CHOH.COO^- \xrightarrow{-6[H]} 3CH_3.CO.COO^- \xrightarrow{+6[H]}$$

Lactate   Pyruvate

$$\rightarrow 2CH_3.CH_2.COO^- + CH_3COO^- + CO_2 + H_2O$$

Propionate

The second stage in the production of Camembert, Brie, or blue cheeses is the ripening brought about by the fungi *Penicillium camemberti* and *Penicillium roqueforti*. The *P. camemberti* grows on the surface of the ripening Camembert or Brie, whereas *P. roqueforti* has the ability to grow under fairly anaerobic conditions within the cheese. Although chemical analyses of the blue cheeses have shown the presence of a variety of carboxylic acids, alcohols, and esters , very little is known about the enzymes involved. Lipases are undoubtedly important.

The characteristic flavour and aroma of many cheeses are due to the release of free fatty acids, particularly saturated fatty acids having 4 to 12 carbon atoms from triglycerides by the action of lipases., and also compounds derived from fatty acids such as methyl ketones, e.g. heptanone, and unsaturated alcohols, e.g. oct-1-en-3-ol. A more detailed account of the organisms used in cheesemaking is given in references 14–16.

11.3 Use of microorganisms in the production of organic chemicals

Microorganisms are also used to produce a variety of complex organic compounds. In many cases the enzymes involved have not been characterized. The most widely used organisms for these processes are fungi, and several organic acids are synthesized by species of *Aspergillus*. Citric acid production is the most important in economic terms. The growth conditions for *Aspergillus niger* are arranged to stimulate the overproduction of citric acid. The growth medium contains high concentrations of sugars and low concentrations of Mn^{2+} and Fe^{3+}. The underlying mechanism for this is not fully understood, but it appears that Mn^{2+} deficiency represses all the TCA cycle enzymes except citrate synthase, allowing citric acid accumulation. Citrate is an allosteric inhibitor of phosphofructokinase (see Chapter 6, Section 6.4.1.1). Mn^{2+} deficiency also causes the accumulation of NH_4^+ ions, and the latter relieves citrate inhibition.[17]

In general, it is not economical to produce simple organic compounds using microorganisms, since they can be produced more cheaply by chemical syntheses. However, this is not the case with more complex compounds such as antibiotics, for which microorganisms are used. Table 11.2 summarizes the main processes in which the enzyme complements of certain microorganisms are used to synthesize organic compounds; further details are given in references 17–19.

11.4 Use of isolated enzymes in industrial processes

Although several industrial processes are carried out using whole organisms as a source of enzymes, the efficiency of these processes can often be improved by using isolated enzymes. In addition, enzymes are now being used in a wider variety of processes. The advantages of using isolated enzymes compared to intact organisms are: (i) it may be possible to obtain higher catalytic activity; (ii) undesirable side-reactions may be avoided; and (iii) the processes may be able to be carried out more reproducibly. The principal disadvantage is likely to be the higher cost of using isolated enzymes and also that they may be more sensitive to inactivation than the enzymes within an intact organism. It is advantageous to use enzymes from thermophilic bacteria or fungi (when a suitable enzyme is available) so as to reduce the problem of heat inactivation.[20]

11.4.1 Sources of isolated enzymes

Approximately 90 per cent of isolated enzymes used in industrial processes are from bacteria and fungi, although enzymes from plant and animal tissues are also used.[1] Fungi and bacteria have the advantage that they can be easily grown and it is usually not difficult to scale up a production process (for reviews, see references 21–23). In addition, the sources are not subject to seasonal or other factors, e.g. chymosin (rennin) production is seasonal because it is dependent on

Table 11.2 The production of some organic compounds by use of microorganisms

Compound	Organism used for synthesis	Enzyme steps if known	Uses of compound
Citric acid	*Aspergillus niger*	Monosaccharides → citrate Glycolysis and tricarboxylic-acid cycle involved	Soft drinks, jams, jellies, flavouring, blood transfusion, sequestering agent in electroplating and leather tanning
Itaconic acid	*Aspergillus* spp.	As with citrate formation plus citrate ⇌ aconitate → itaconate + CO_2	Copolymer in acrylic resins
Gluconic acid	*Aspergillus niger*	Glucose → gluconolactone → gluconate (glucose oxidase)	Calcium gluconate is used in Ca therapy, as a sequestering agent, food additive, and in the pharmaceutical industry
Fumaric acid	*Rhizopus*	Tricarboxylic-acid cycle	Paper sizing and polyester resins
Amphotericin B	*Streptomyces* spp.		Antibiotics
Chloramphenicol	*Streptomyces* spp.		Antibiotics
Cycloheximide	*Streptomyces* spp.		Antibiotics
Erythromycin	*Streptomyces* spp.		Antibiotics
Novobiocin	*Streptomyces* spp.		Antibiotics
Streptomycin	*Streptomyces* spp.		Antibiotics
Griseofulvin	*Penicillium* spp.		Antibiotics
Penicillin	*Penicillium* spp.		Antibiotics
Cephalosporin	*Acremonium* spp.		Antibiotic
Cyclosporins	*Trichoderma polysporum*		Antibiotic, antifungal agent, and immunosuppressant
Lovastatin	*Aspergillus terreus*		Treatment of hypercholesterolaemia
Bacitracin	*Bacillus subtilis*		Antibiotic
Gramicidin	*Bacillus brevis*		Antibiotic
Carotenoid pigments	*Blakeslea trispora*	Synthesis from acetate	Colouring agent
Riboflavin	*Ashbya gossypi*	Synthesis from glycine, formate, and CO_2	Nutrient

the supply of calf stomachs. The majority of enzymes that have so far been used are hydrolytic enzymes and many of these are produced extracellularly by fungi. The possibility of producing larger quantities of enzymes from animal and plant sources by use of tissue culture methods is now being explored. Some proteins such as vaccines are already being produced in tissue culture.

With microbial enzymes it is often possible to increase the yields by changes in the growth conditions, addition of inducers, or strain selection, including increasing the number of gene copies by genetic engineering.[24,25] With enzymes from animal and plant sources, the yields may be increased by the introduction of the appropriate genes and their promoter regions into the more rapidly growing microorganisms. However, there are often problems, such as the formation of inclusion bodies through incorrect folding, the lack of glycosylation, or degradation of the recombinant protein, which have to be overcome before a satisfactory product is obtained. There are also strict controls on the use of recombinant proteins in the food industry. The enzyme chymosin produced in three separate transgenic microorganisms (see Chapter 2, Section 2.8.7) was approved for food use in the UK in 1991, but it is not permitted in Germany and The Netherlands although the protein was shown to be identical to that produced from calves.[22]

11.4.2 Isolation of enzymes

The methods available for isolating enzymes on a laboratory scale (Chapter 2) or an industrial scale are largely the same. However, the criteria for selection of a particular method vary according to use. Enzymes required for food-processing and related industries and for detergents are generally required in large quantities at relatively low purity. Those enzymes required for clinical diagnosis and related areas are generally required in smaller quantities at a higher level of purity. Although a high state of purity is not generally required in food-processing, it may be necessary to exclude certain contaminating enzymes; e.g. in biscuit manufacture, where proteases are used, it is necessary that the extract containing protease is low in amylase activity. Many of the basic methods of enzyme purification require some adaptation for scaling up.[26]

As mentioned previously, many of the fungal enzymes used are extracellular enzymes and they can readily be separated from the mycelium by filtration. It is usually necessary to concentrate the extract and this is done by spray-drying, which is less costly than precipitation by ammonium sulphate or organic solvents. If the enzyme is intracellular, the mycelium has to be disrupted. Enzymatic methods of breaking the cell wall would be too costly and difficult to scale up, so high-pressure homogenization methods are usually used.

11.4.3 Enzymes isolated on an industrial scale and their applications

The principal isolated enzyme preparations that are used industrially are those concerned with carbohydrate metabolism (Table 11.3) and protein hydrolysis (Table 11.4) and these account for over three-quarters of current sales. Other

Table 11.3 Carbohydrate-metabolizing enzymes used in industry

Enzyme	Reaction or substrate	Sources of enzyme	Utilization
Amylase	Endohydrolysis of 1,4-α-D-glucosidic linkages in polysaccharides	*Bacillus subtilis* *Aspergillus niger* *Aspergillus oryzae*	Production of sugars and oligosaccharides from starch Enables starch to be fermented
Exo-1,4-α-D-glucosidase (amyloglucosidase)	Hydrolysis of terminal 1,4-linked α-D-glucose residues successively from non-reducing end of polysaccharides	*Aspergillus niger* *Aspergillus oryzae* *Rhizopus* sp	Production of glucose from starch
Cellulase	Endohydrolysis of 1,4-β-D-glycosidic linkages in cellulose and β-D-glucans	*Trichoderma viride* *Aspergillus niger*	Cellulose $\rightarrow$ cellobiose for fermentation (cellulase preparations having very high activity have yet not been obtained)
Polygalacturonase	Random hydrolysis of 1,4-α-D-galactosiduronic linkages in pectin	*Mucor, Botrytis, Pencillium,* and *Aspergillus*	Extraction of fruit juice from pulp, clarification of wines and fruit juices
β-D-Galactosidase	Lactose $\rightarrow$ glucose + galactose	*Aspergillus niger* *Aspergillus oryzae*	Hydrolysis gives sweeter, more-soluble sugars
β-D-Fructofuranosidase	Sucrose $\rightarrow$ glucose + fructose	*Aspergillus oryzae* *Saccharomyces*	Hydrolysis gives sweeter, more-soluble sugars
Glucose oxidase	Glucose + O_2 $\rightarrow$ gluconolactone + H_2O_2	*Aspergillus niger* *Penicillum* spp	Analytical reagent for glucose; desugaring egg products, removing O_2 from mayonnaise and fruit juices susceptible to oxidation
Xylose isomerase (usually referred to as glucose isomerase)	Glucose $\rightleftharpoons$ fructose	*Streptomyces* spp *Lactobacillus brevis*	Production of high-fructose syrup

Table 11.4 Protein-metabolizing enzymes used in industry

Enzyme	Reaction or substrate	Source of enzyme	Utilization
Papain	Peptide-bond cleavage	*Papaya* latex	Removal of turbity (due to protein) in beer, meat tenderizing
Chymosin (rennin)	Clotting of milk	Calf stomach	Production of cheese
Trypsin Chymotrypsin	Peptide-bond cleavage	Animal pancreas	Meat tenderizers, medical uses
Fungal proteases	Peptide-bond hydrolysis	*Aspergillus oryzae* *Aspergillus niger*	Meat tenderizers, breadmaking to improve viscoelastic properties of doughs, chill-proofing of beer (to prevent haze due to protein–tannin interaction)
		Mucor pusillus, Rhizomucor miehei, Cryptonectria parasitica	Used as a substitute for chymosin in cheese making since they cause only limited proteolysis
Bacterial proteases		*Bacillus subtilis*	Detergents, removal of gelatin from films

enzymes that have been used less extensively include AMP deaminase, steroid 11β-isomerase, aminoacylase, streptokinase, penicillinase, nitrile hydratase, hyaluronoglucosaminidase, hyaluronoglucuronidase, triacylglycerol lipase, catalase, keratinase, pullulanase, pectinase, anthocyanase, and phosphodiesterase. In some instances isolated enzymes are replacing whole organisms in the processing of foods, but in many cases enzymes are being used in completely new processes. In the next section we consider briefly the principal uses of enzymes in industry.

11.4.3.1 Alcoholic beverages

In the traditional brewing process, where the starch from barley is hydrolysed prior to fermentation, the barley is allowed to germinate to an extent that ensures sufficient α- and β-amylases are produced for the hydrolytic steps not to be rate limiting. The extent to which the glucose moieties present in starch become completely hydrolysed and fermented varies with the particular beverage. In a typical beer 2–3 per cent carbohydrate may remain unfermented. For low-alcohol beers the mash may be held at 80°C for a short time, denaturing the heat labile β-amylase, and leaving a higher than normal proportion of unfermentable carbohydrates. However, the addition of enzymes that cleave the α-1,6 bonds is sometimes used to produce 'lite' low-carbohydrate beers suitable for diabetics, and if the product is then diluted to have the normal alcohol level a low-calorie beer for slimming drinkers results.[27] For the production of spirits as complete a conversion of the glucose moieties to alcohol is desirable. The definition of Scotch whisky prohibits the use of exogenous enzymes, but for the production of lesser spirits and fuel alcohol where sorghum, potatoes, wheat, or rye may be used as the source of starch, the addition of exogenous enzymes is permitted.[10] The enzymes most used are amylases from *Bacillus subtilis*, *Aspergillus oryzae*, or *A. niger*. The α-amylase from *B. subtilis* is more heat stable and is used when harsh conditions are required to liquefy particular cereals. However, the extract lacks α-1,6 glucosidase activity and therefore only degrades the amylopectin component of starch to oligosaccharides. The enzymes from *Aspergillus* are also able to cleave α-1,6 glucoside links and therefore degrade the starch more completely into fermentable sugars.[10,27]

11.4.3.2 Breadmaking

In breadmaking the carbohydrate for fermentation is also derived from the hydrolysis of starch. The aeration (CO_2 production) of bread during its manufacture will depend on adequate α- and β-amylase activity. In some countries, where harvesting conditions do not promote germination, amylase activity becomes a limiting factor. In this case, addition of the fungal amylases (see Section 11.4.3.1) is preferable on account of their greater heat lability. It is important that only limited starch hydrolysis occurs. The fungal enzymes are rapidly inactivated during baking. Fungal proteases may also be used in breadmaking to improve the viscoelastic properties of the dough.

11.4.3.3 Cheesemaking

The increased demand for cheese coupled with the reduced availability of chymosin (rennin), which is obtained principally from calf stomachs, led to a search for substitutes for chymosin. Chymosin, an aspartic acid protease, causes the clotting of milk, a process which involves cleavage of a single peptide bond in κ-casein between Phe105 and Met106, releasing the acidic C-terminal peptide. The high specificity of chymosin does not depend solely on the recognition of residues 105 and 106; all the residues from 98 to 111 appear to be involved in the recognition process.[16] The release of the C-terminal peptide is followed by Ca^{2+}-induced aggregation of the modified micelles to form a gel[16,28] (Fig. 11.4).

Proteolytic enzymes such as trypsin also cause clotting, but then further degrade casein. If this occurs in cheesemaking it leads to undesirable flavours. Proteins do not contribute significantly to the flavour of foods; it is the peptides and amino acids that often account for the sweet, sour, bitter, and salty flavours. Peptides having chain lengths of 3–15 amino-acid residues and rich in hydrophobic amino acids are the main agents causing bitterness.[29] In the search for a substitute for chymosin, enzymes that cause clotting but only limited proteolysis are desirable.

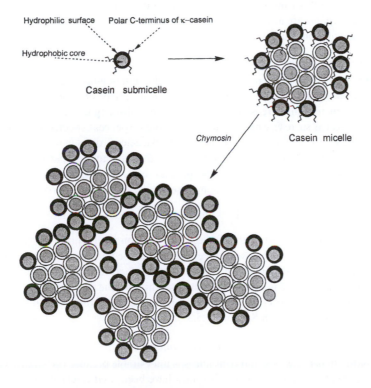

Hydrophilic surface   Polar C-terminus of κ–casein

Hydrophobic core

Casein submicelle

Chymosin

Casein micelle

Aggregation of casein micelles

Fig. 11.4 Ca^{2+}-induced aggregation of micelles. The filled annuli represent κ casein and the open annuli α and β caseins.

Chymosin is able to cleave the Phe105–Met106 peptide bond in κ-casein over 100 times faster than any other peptide bonds in κ-casein.[16] The fungal aspartate proteases from *Mucor pusillus*, *Rhizomucor miehei*, and *Cryphonectria parasitica* can substitute for chymosin, although they do not have quite such a good clotting/ proteolysis ratio. However, with the increase in the number of vegetarians, particularly in the UK, there is a demand for cheeses in which a microbial coagulant has been used.[30] Recombinant DNA chymosin, in which the chymosin gene has been cloned into expression systems such as *Aspergillus oryzae* and *Kluyveromyces lactis* (see Chapter 2, Section 2.8.7), has also been used in the manufacture of certain cheeses since 1994.[30] The other exogenous enzyme to be used in cheese manufacture is lipase. Lipolysis is important in ripening, and the extent to which it occurs depends on the particular cheese. In traditional cheese manufacture the endogenous lipase present in unpasteurized milk was the source. With pasteurized milk in which the endogenous lipase is inactivated, a source of lipase is present in the rennet paste. Exogenous lipase is added for cheese varieties such as Mozzarella, Parmesan, Samso, and Romi, when an enhancement or acceleration of the lipolytic flavour is required.[15,16,30]

11.4.3.4 Meat tenderizing and marination

Toughness in meat is due primarily to collagen and elastin and also actomyosin. Lower-quality cuts of meat are just as nutritious as prime cuts. Thus, to make maximum use of the carcass, efforts have been made to tenderize the lower-quality cuts. Also the flavour of meat is to some extent related to the concentrations of peptides and amino acids present (see Section 11.4.3.3). Proteases are used in both tenderizing and marination. Papain is the traditional enzyme used, but others such as bromelain, trypsin, chymotrypsin, and microbial proteases from *Aspergillus* are also used.[31] Over 5 per cent of United States beef is subject to tenderizing procedures by packers.

11.4.3.5 Sweeteners

A very important use of enzymes is in the production of glucose syrups which are used as sweetening agents on a large scale in a variety of processed foods. A glucose–fructose mixture is sweeter than glucose alone (glucose is about 70 per cent as sweet as sucrose, whereas fructose is about 60 per cent sweeter than sucrose) and cheaper to produce than sucrose that is extracted from sugar cane or sugar beet. The glucose–fructose mixture is produced from starches, particularly corn starch. Two enzymes are used in the process: exo-1,4-α-D-glucosidase obtained from *Aspergillus niger* and xylose isomerase mainly from *Streptomyces* spp.*

$$\text{Starch} \rightarrow \text{glucose} \rightleftharpoons \text{fructose}$$

This process is discussed further in Section 11.5.1 on immobilized enzymes.

Another example of the use of an enzyme to cause an increase in sweetness is the use of β-D-galactosidase in ice-cream manufacture. Since the mixture of glucose and galactose is sweeter than lactose, this enzyme can be used to catalyse an increase in sweetness of a product. The manufacture of the artificial sweetener aspartame is discussed in Section 11.5.3.

11.4.3.6 Clarification of beers, wines, and fruit juices

Turbidity in beers, wines, and fruit juices is undesirable because turbid drinks are less acceptable to the consumer. Enzymes have been used to clarify a variety of beverages. In the case of fruit juices they are more easily prepared from extracted

* The enzyme used for isomerization is xylose isomerase, EC 5.3.1.5, although it is usually referred to as glucose isomerase, since it will also isomerize glucose;[32]
$K_{m(\text{D-glucose})} = 0.17$ mmol dm^{-3} and $K_{m(\text{D-xylose})} = 0.01$ mmol dm^{-3}.

pulp if the viscosity is not too high. In the case of beer a haze or turbidity may develop on cooling and this appears to be due to a complex between tannin and protein. It can be successfully removed by addition of small amounts of protease, usually papain. A careful balance has to be struck, because not only does the protein cause the haze, its foaming properties account for the head; therefore, too much proteolysis removes not only the chill haze but also the good head (the froth on top of the beer).[10] In the case of wines and fruit juices the turbidity may be due to starch or pectin and is removed by addition of amylases or polygalacturonases, which also cause a decrease in viscosity. A detailed discussion of the enzymes used is given in reference 10.

11.4.3.7 **Detergents**

In 1913 the first patent was described for the use of enzymes in detergents. It was a crude extract of pancreas (pancreatin) in sodium carbonate solution. However, the large-scale development of detergents containing enzymes took place during the 1960s, so that by 1969 about half of the detergents marketed contained enzymes. Following the indication of possible health hazards, primarily skin allergies, the use of such detergents declined sharply in the early 1970s; since then methods have been developed for encapsulation of enzyme particles using nonionic surfactants and this has reduced dust inhalation problems with powdered enzymes. The greater importance attached to 'environmental friendliness' since the 1980s has led to the reduction of the detergent component of 'detergents' with an increase in the enzyme content, and also a lowering of the washing temperatures used. The principal enzymes used in 'biological' detergents are mixtures of amylase and neutral and alkaline proteases that are active in the pH range 6.5 to 10 and at temperatures from 303 K (30°C) to 333 K (60°C). Detergents often contain oxidizing and chelating agents and so the enzymes must withstand these also. All proteases used in detergents today are from strains of *Bacillus*, usually either neutral or alkaline protease. They are serine proteases of low specificity having temperature optima about 60°C and alkaline pH optima. The carbohydrate-hydrolysing enzymes are α-amylases from either *B. licheniformis* or *B. amyloliquifaciens* having high temperature optima. A mixture of carbohydrate-hydrolysing and protein-hydrolysing enzymes of low specificity is most useful, since polysaccharides and proteins require partial degradation before they are removed from garments. By contrast, lipids can be removed without any prior degradation by ionic detergents at alkaline pHs; however, since 1988 lipases have begun to be introduced into detergents. During the same period cellulases have also been added to detergents. Their role is somewhat different, acting mainly on damaged fibres in cotton garments. The macroscopic effects of this microscopic 'shaving' are that it enhances the softness, brightens the colours, and improves the removal of particulate material soiling the garments.[33]

11.4.3.8 **Medical applications**

Isolated enzymes are used in medicine as reagents both in analysis and in therapy. About 50 different enzymes are used in different aspects of clinical diagnoses, and for most of these much higher levels of purity are required than for most industrial enzymes. Two of the major enzymes used are peroxidase from horseradish and alkaline phosphatase from beef intestinal mucosa, both being required for immunoassays. The enzymes may be used in test strips, ELISA, biosensors, and autoanalysers. The use of enzymes, e.g. glucose oxidase, in clinical analysis was discussed in Chapter 10, Section 10.8.1.

The use of enzymes as therapeutic agents can be subdivided into two categories: (i) the topical application of an enzyme as an extracellular agent, and (ii) the intracellular application of enzymes to treat metabolic deficiency and related diseases. The latter were discussed in Chapter 10, Section 10.9.1.

A variety of enzymes have found extracellular application in medicine, including proteolytic enzymes in clot removal, collagenase in skin ulcers, hyaluronidase in drug administration, trypsin as an anti-inflammatory agent and a wound cleanser, rhodanese (thiosulphate sulphurtransferase) in cyanide poisoning, and uricase in gout.

Blood clots are liquefied under physiological conditions by the action of plasmin. Plasmin is produced slowly in the plasma by the action of fibrinokinase on plasminogen (plasmin precursor). Streptokinase (an enzyme from *Streptococcus haemolyticus*) and urokinase (an enzyme isolated from human urine) are effective activators of plasminogen and appear to be useful in relieving peripheral thrombosis. Urokinase, although more expensive to produce, does not cause the immunological complications found with streptokinase. Tissue plasminogen activators are potentially more useful, since they bring about clot-specific plasminogen activation by binding to fibrin. Human tissue plasminogen activator has been cloned in *E. coli* and is now being produced in transformed mammalian cells; it has been approved for treatment of myocardial infarction.[34]

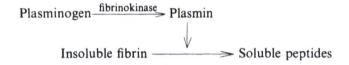

Hyaluronidase (EC 3.2.1.35 and 3.2.1.36) obtained from beef testes is a useful agent to aid diffusion when coinjected with other drugs such as antibiotics, adrenalin, heparin, and local anaesthetics. Hyaluronidase hydrolyses polyhyaluronic acid, the main polysaccharide component of connective tissue, and thus aids the diffusion of other substances by reducing the viscosity in the locality of the injection.

For further details of the medical applications of isolated enzymes, see reference 34.

11.5 Immobilized enzymes

Attaching an enzyme to an insoluble support will permit its reuse and continuous use without a difficult recovery process. The attachment to the support may also stabilize the enzyme. If two or more enzymes catalysing a sequence of reactions are immobilized in close proximity to each other, an efficient immobilized multienzyme protein analogue can be produced. There are therefore potential advantages in immobilizing enzymes that are required on a large scale. Much research effort since about 1970 has been devoted to producing satisfactory immobilized enzymes for large-scale application. For details of the methods used for immobilization, see references 35–40. A number of immobilized enzymes are being used currently in industry (Table 11.5) and others are at the research and development stage (see reference 41).

There are a number of ways in which an enzyme may be immobilized, by adsorption, covalent linkage, crosslinking, matrix entrapment, or encapsulation, as summarized in Fig. 11.5. Adsorption is one of the simplest methods, but has the disadvantage that the enzyme may become partially desorbed under the

Table 11.5 Examples of industrial applications of immobilized enzymes and whole cells

Enzyme or whole cells	Carrier	Methods of Immobilization	Application
Aminoacylase	DEAE-Sephadex	Adsorption	Preparation of L-amino acids from a racemic DL-amino acid mixture
Xylose isomerase (glucose isomerase)	Duolite A7 (ceramic beads) Amberlite IRA904 exchange resin	Adsorption	Preparation of fructose–glucose mixture
Thermolysin		Two-phase system	Preparation of aspartame
Lactase (β-galactosidase)	Silica particles	Adsorption	Preparation of lactose-free milk
Glucose oxidase		Membrane entrapped	Measurement of blood glucose
E. coli containing aspartate ammonia-lyase	Polyacrylamide	Entrapment	Preparation of aspartate
Pseudomonas putida containing arginine deaminase	Polyacrylamide	Entrapment	Preparation of citrulline
Achromobacter liquidum containing histidine ammonia-lyase	Polyacrylamide	Entrapment	Preparation of uraconic acid
E. coli containing penicillin amidase	Polyacrylamide, cellulose	Entrapment	Preparation of 6-aminopenicillanic acid
Brevibacterium ammoniagenes containing fumarate hydratase	Polyacrylamide	Entrapment	Preparation of malate and fumarate
Rhodococcus rhodochrous containing nitrile hydratase	Polyacrylamide	Entrapment	Preparation of acrylamide for polymers

conditions used. Covalent linkage is generally more costly than crosslinking, since it usually requires the activation of the matrix. Crosslinking is most frequently carried out using glutaraldehyde which is relatively cheap and available in industrial quantities. The uses of immobilized enzymes can be broadly divided into preparative and analytical. Examples of the former are large-scale isomerization of glucose and production of L-amino acids from racemic mixtures. Use of immobilized glucose oxidase is an example of the latter. The state of purification of enzymes used for immobilization ranges from complete purification down to the use of whole cells. If a sequence of enzymes is required, then it is often more convenient and thus more economical to use whole cells.[36] However, where side-reactions would be detrimental to the process, then purified enzymes have to be used. We discuss in the following sections some current applications of immobilized enzymes.

11.5.1 Production of syrups from corn starch

The production of syrups from starches was mentioned in Section 11.4.3.5. Syrups (sugar solutions) are used extensively as sweetening agents in a variety of manufactured foods. Several million tonnes of high-fructose corn syrup are now produced annually and this is displacing sucrose in traditional applications. It represents one of the biggest developments in food science since 1974. Fructose is more satisfactory than glucose because it is sweeter and does not crystallize so readily from concentrated solutions. The overall process requires two main enzyme-catalysed steps:

$$\text{Starch} \xrightarrow{\text{exo-1,4-}\alpha\text{-D-glucosidase}} \text{glucose} \underset{\text{xylose isomerase}}{\rightleftharpoons} \text{fructose.}$$

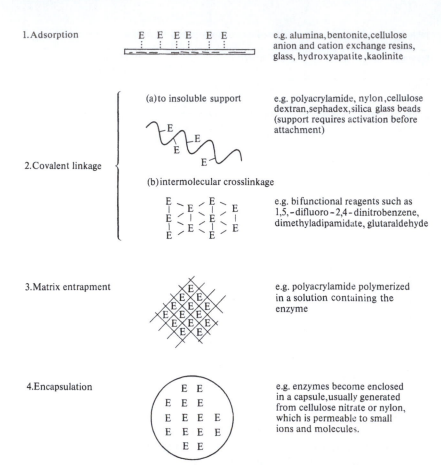

1. Adsorption

e.g. alumina, bentonite, cellulose
anion and cation exchange resins,
glass, hydroxyapatite, kaolinite

2. Covalent linkage

(a) to insoluble support

e.g. polyacrylamide, nylon, cellulose
dextran, sephadex, silica glass beads
(support requires activation before
attachment)

(b) intermolecular crosslinkage

e.g. bifunctional reagents such as
1,5,-difluoro-2,4-dinitrobenzene,
dimethyladipamidate, glutaraldehyde

3. Matrix entrapment

e.g. polyacrylamide polymerized
in a solution containing the
enzyme

4. Encapsulation

e.g. enzymes become enclosed
in a capsule, usually generated
from cellulose nitrate or nylon,
which is permeable to small
ions and molecules.

Fig. 11.5 Methods used for enzyme immobilization (for details see references 35–38).

The first of these steps is catalysed by exo-1,4-α-D-glucosidase obtained from *Aspergillus niger*. This enzyme has been immobilized on porous silica. The second step is catalysed by xylose isomerase obtained from *Streptomyces* spp. This enzyme has been immobilized by crosslinking with glutaraldehyde. The isomerization generally takes place on a packed bed reactor in which the substrate solution flows through the reactor containing a plug of the immobilized enzyme. Under these conditions the enzyme is exposed to high-substrate and low-product concentration. Other methods of immobilization, including the use of ion exchange resins, have been used. The equilibrium constant for the enzyme at 323 K (50°C), the operating temperature of the process, is approximately unity, so that approximately equal amounts of fructose and glucose are present at equilibrium. However, to avoid using excessively long reaction times a conversion of about 45 per cent is used. Immobilization of the enzyme increases its half-life up to as much as 100 days, and the enzyme is usually replaced after three half-lives.[42]

11.5.2 Production of L-amino acids from racemic mixtures

Over 5×10^7 kg of DL-amino acids are produced each year. Most of this amount has been produced by synthetic methods, and since most biological systems utilize only L-amino acids, a method for resolving the isomers was required for many purposes. The production of chiral chemicals is particularly important in the pharmaceutical industry. Although a number of pharmaceuticals are sold as racemates, it is important to know beforehand whether there is a difference in the pharmacological action of the different isomers. Resolution of optically active acids by combining with optically active bases is time-consuming and expensive. Resolution by an enzymatic method that uses an immobilized enzyme is now carried out commercially in Japan. It was the first reported commercial use of an immobilized enzyme. The procedure is shown in Fig. 11.6. The method is based on the specificity of aminoacylase for L-N-acetylated amino acids and the differences in solubility of free amino acids compared with their N-acetylated derivatives. This allows the L-amino acid to crystallize out and the D-N-acetyl amino acid to be racemized and recycled. The enzyme is immobilized by binding to DEAE-Sephadex columns. The columns can easily be recharged when necessary, simply by passing more enzyme down the column.

11.5.3 Immobilized whole cells for use in amino-acid interconversions

Besides immobilization of single enzymes it is also possible to immobilize whole cells.[36] The usual method is to entrap the cells by polymerizing acrylamide and N,N'-methylenebisacrylamide around them. Whole cells from strains of *E. coli* and *Pseudomonas putida* containing the appropriate enzymes have been successfully used for the production of aspartate, citrulline, phenylalanine, tryptophan, serine, urocanic acid, and 6-aminopenicillanic acid by entrapment.[43]

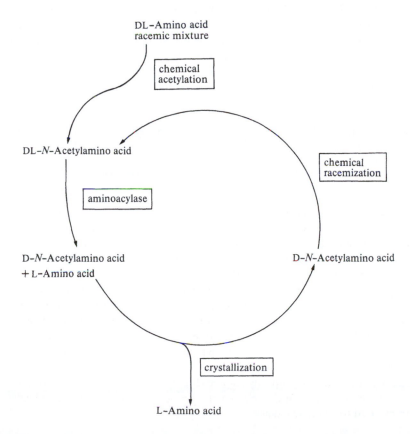

Fig. 11.6 Production of L-amino acids from racemic mixtures of amino acids by use of aminoacylase.

Immobilized *E. coli* cells have been found to have an operational half-life of up to two years.[44] Aspartate and citrulline are used in the pharmaceutical industry, and urocanic acid is used as an ultraviolet filter in suntan preparations.

$$^-OOC-CH=CH-COO^- + NH_3 \xrightarrow{\text{aspartate ammonia–lyase}} {}^-OOCCH_2CHCOO^-$$

Fumarate   Aspartate

$$H_2N-C(=NH_2^+)-NH(CH_2)_3CHCOO^- \xrightarrow{\text{L–arginine deiminase}} H_2N-C(=O)-NH(CH_2)_3CHCOO^- + NH_4^+$$

Arginine   Citrulline

Histidine   Urocanic acid

6-Aminopenicillanic acid is used as a precursor of the semisynthetic penicillins, e.g. methicillin, oxacillin, carbenicillin, and ampicillin. Benzylpenicillins and phenoxypenicillins (penicillins G and V) are the naturally occurring penicillins produced by *Penicillium* spp. The semisynthetic penicillins have advantages over the natural penicillins as antibiotics and account for more than 85 per cent of penicillins sold for therapeutic use. Chemical hydrolysis of penicillins G or V not only cleaves the amide bond but also opens the β-lactam ring. The greater selectivity of penicillin amidase catalyses the formation of 6-aminopenicillanic acid. Semisynthetic cephalosporins are made in a similar way.

Penicillin   6–Aminopenicillanic acid

The artificial sweetening agent aspartame (*N*-L-α-aspartyl-L-phenylalanine methyl ester), discovered in 1969, is about 200 times sweeter than sucrose. It is formed from the amino acids phenylalanine and aspartate, but it is important to have the correct stereoisomers, since α-aspartame is sweet, but β-aspartame is bitter. An enzyme-catalysed synthesis has been devised using a proteolytic enzyme, thermolysin. In water, the equilibrium of the thermolysin-catalysed reaction favours hydrolysis, but when the amino group of L-aspartate is protected as *N*-(benzyloxycarbonyl)-L-aspartate, and the carboxyl group of phenylalanine is methylated, it catalyses the reverse reaction, namely the formation of the peptide bond between L-phenylalanine methyl ester and *N*-(benzyloxycarbonyl)-L-aspartate, and the *N*-protected aspartame formed crystallizes out as an insoluble salt with the remaining D-phenylalanine (Fig. 11.7).[45] Alternatively, a two-phase system can be used in which the product is formed in water and then immediately transferred to an organic phase.

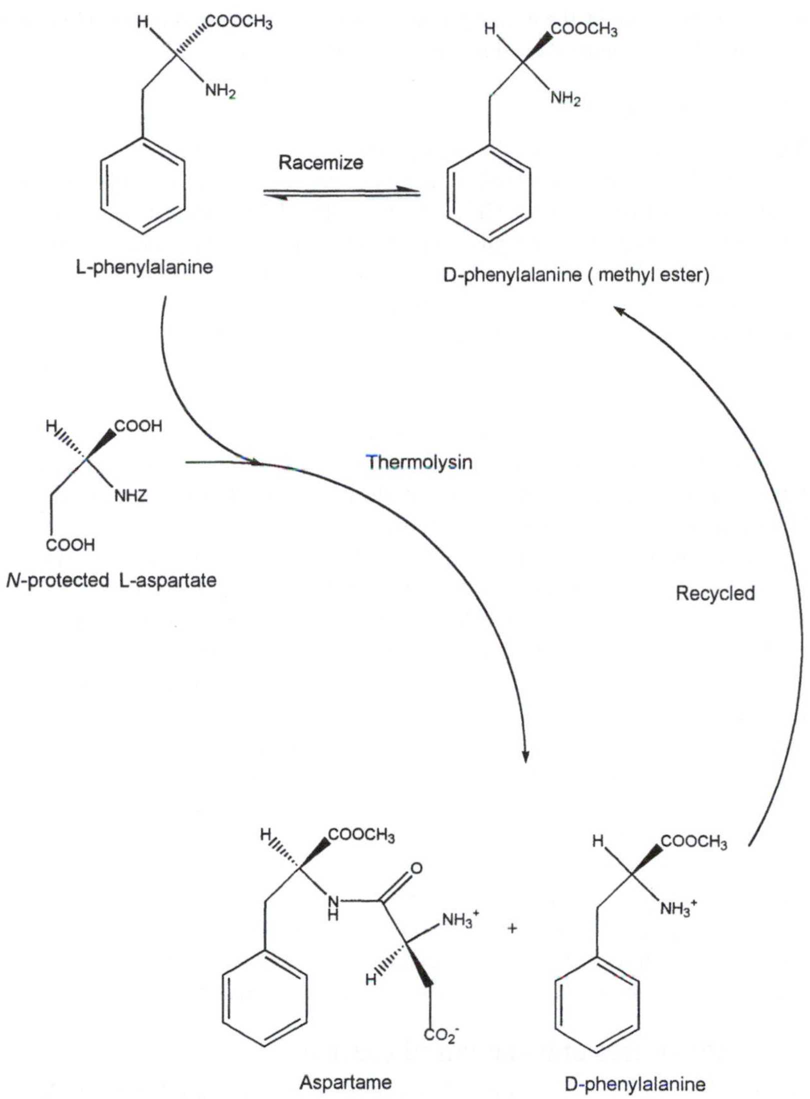

Fig. 11.7 The production of aspartame using thermolysin. For clarity the protecting group (Z = benzyloxy carbonyl) is not shown in the aspartame structure.

11.5.4 Enzymic synthesis of acrylamide

Acrylamide is widely used as a starting material for the production of a number of polymers, as an ingredient in coagulators and soil conditioners, and for adhesives and paints. It can be produced by reaction of acrylonitrile and water in the presence of suitable inorganic catalysts such as copper, but this results in unwanted side-reactions. Nitriles occur naturally in some plants, bacteria, and fungi, e.g. in antibiotics such as toyocamycin and treponemycin, and certain microorganisms are capable of using nitriles both as a carbon and nitrogen source. Aliphatic nitriles are degraded in two stages, the first catalysed by nitrile hydratase, and the second by an amidase. When *Rhodococcus rhodochrous* is grown on the optimum culture medium for inducing nitrile hydratase, the latter accounts for more than 50 per cent of the total soluble protein. It is not necessary to purify the enzyme, but simply to entrap the organism in a polyacrylamide

$$RCN + H_2O \rightleftharpoons RCONH_2$$
$$RCONH_2 + H_2O \rightleftharpoons RCOOH + NH_3$$

gel.[46] It is now one of the major products obtained using immobilized enzymes with a world-wide annual production of over 15 000 tons.[41]

11.5.5 Biosensors

The term 'biosensor' is generally used to describe an analytical device which has a biological recognition mechanism (most commonly enzyme and substrate) and transduces it into a signal, which is usually electrical. The electrical signal can be detected as a change in current (amperometric biosensors), change in voltage (potentiometric biosensors), or change in conductivity of a solution (conductimetric biosensors). A typical arrangement would be to have an enzyme immobilized using one of the methods described in Fig. 11.5, in close proximity to a suitable electrode as illustrated for the measurement of glutamate (Fig. 11.8). For this, NAD^+ is immobilized to an insoluble dextran, and, together with glutamate dehydrogenase and lactate dehydrogenase, entrapped in a dialysis membrane surrounding the ammonium-ion-sensitive electrode. The NH_4^+ generated in close proximity to the electrode results in an electrode potential proportional to the logarithm of the glutamate concentration.

Different types of electrodes would be used depending on the enzyme-catalysed reaction, e.g. for production or utilization of protons, a pH electrode would be used, for O_2 an oxygen electrode, for other ions an appropriate ion-selective electrode. For some enzyme-catalysed reactions where there are changes in the concentrations of ionic species, e.g. urease, the change in conductivity could be measured.

Biosensors are potentially capable of measuring a wide variety of substances in solution. They have the advantage over spectrophotometric methods that they can be used with fluids that are optically opaque, such as blood or culture media. Interference by other substances is generally not a problem because of the enzyme specificity. Enzyme electrodes have been devised for measurement of a wide range of substances including amino acids, lactic acid, acetic acid, formic acid, lipids, penicillin, and alcohols. Over half of the present applications of biosensors are in the field of clinical diagnostics, the most widely used being that used to measure blood glucose. For further details see reference 48.

11.5.6 Other uses of immobilized enzymes

A number of other applications of immobilized enzymes are being developed. Lactose is a major by-product of cheese manufacture, since much of the lactose occurs in the whey. Immobilized β-galactosidase generally from *Aspergillus oryzae* is used to hydrolyse lactose to yield a mixture of glucose and galactose, the mixture being more useful as a sweetening agent. Lactose malabsorption is quite common throughout the world population, although not generally in Europe. This has led to a demand for lactose-free milk, which can be prepared by hydrolysing the lactose to glucose and galactose. Cortisol is a useful drug for arthritis treatment and it can be made from the cheap precursor 11-deoxycortisol. Immobilized steroid 11β-monooxygenase and a Δ'-dehydrogenase are used to produce prednisolone, which is a superior therapeutic agent to cortisone.

Immobilized enzymes can also be used in analytical procedures. The most developed application to date is in the analysis of blood glucose using immobilized glucose dehydrogenase. The glucose dehydrogenase from *Bacillus cereus* oxidizes only D-glucose and can be used specifically in the determination of this sugar:

$$\text{D-Glucose} + NAD^+ \rightleftharpoons \text{D-glucono-}\delta\text{-lactone} + NADH + H^+$$

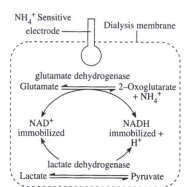

Fig. 11.8 Diagrammatic arrangement of an electrode used to monitor the concentration of glutamate (see reference 47).

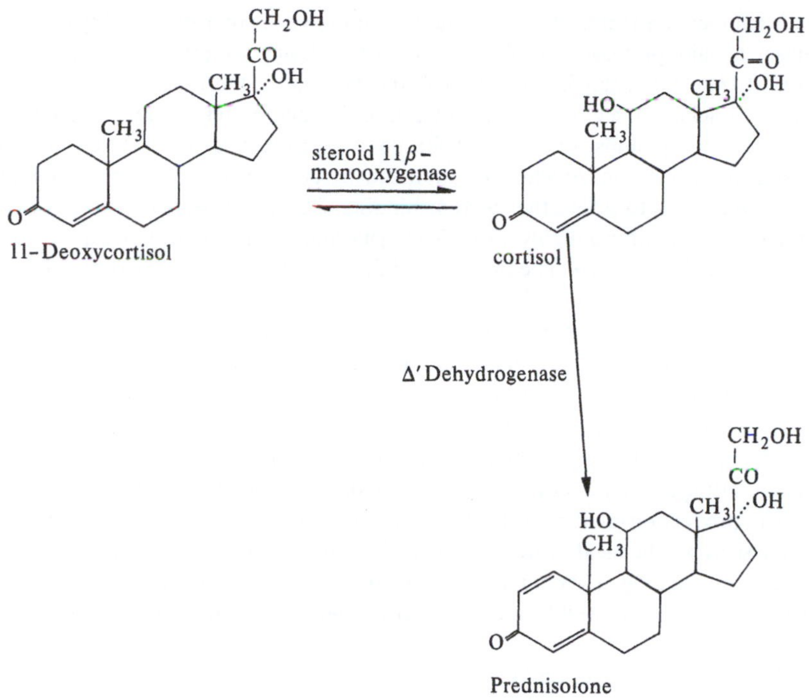
CH2OH
CO
CH3
OH
CH3
O
11-Deoxycortisol
steroid 11β-monooxygenase
CH2OH
C=O
CH3
OH
HO
CH3
O
cortisol
Δ' Dehydrogenase
CH2OH
CO
CH3
OH
HO
CH3
O
Prednisolone

The NADH formed is measured spectrophotometrically or spectrofluorimetrically. The enzyme has been immobilized onto nylon tubing and thus can be reused, so that as little as 0.18 mg can be used for 1500 glucose determinations.

11.5.7 **Future developments**

The range of enzymes used in industrial processes is still only a small proportion of the total enzymes known. Most of those used are extracellular hydrolytic enzymes. Extracellular enzymes are generally more stable than intracellular enzymes and can more easily be purified and therefore more readily used for industrial application. There is evidence that in some cases immobilization increases the stability of enzymes, although the reasons are often not clear. Within the last decade the structures of many enzymes have been solved, and site-directed mutagenesis has been used increasingly to modify the properties of enzymes, the classic cases being tyrosyl-tRNA ligase, β-lactamase, dihydrofolate reductase, and subtilisin. With the aid of new computational approaches to energy calculation in combination with powerful computer-driven molecular graphics it will become possible to predict the likely effects of amino-acid substitution of three-dimensional structures and lead to a rational design of biocatalysts. Amongst the enzymes used in industrial processes, xylose isomerase (glucose isomerase) is one in which improvement has been attempted by site-directed mutagenesis. The optimum temperature at which xylose isomerase can be used is limited by its rate of denaturation. One of the processes that leads to the instability of the enzyme is that at the high concentrations of glucose used, the side-chain of Lys 253 becomes glycosylated. When this is replaced by an arginine residue by site-directed mutagenesis, the rate of amino-group glycosylation is reduced. The modified enzyme is more stable; at 60°C its half-life is increased from 607 to 1550 h.[49]

A number of attempts have been made to improve the thermal stability of sub-tilisin, a major protease in 'biological' detergents. Using site-directed mutagenesis (see Chapter 5, Section 5.4.5) only small increases in the thermal stability have so far been achieved, e.g. an increase of 3°C in its melting temperature. Calcium-chelating compounds are required for good performance detergents, but can result in autolytic inactivation of subtilisin by complexing the enzyme-bound Ca^{2+}. Attempts to delete the calcium binding site of subtilisin by site-directed mutagenesis result in a highly unstable enzyme and further changes are necessary to stabilize the enzyme.[50] The thermal stability of a thermolysin-like protease has been dramatically improved by replacement of eight of its amino-acid residues by site-directed mutagenesis. The eight replacements include the insertion of a disul-phide bridge and a glycine replacement with alanine, and an alanine with proline. These have the effect of increasing the half-life of the enzyme at 100°C in the pres-ence of denaturing agents from < 30 s to 170 min.[51]

Attempts are also being made by site-directed mutagenesis to improve the milk-clotting activity of chymosin and to modify its general proteolytic activity.[16]

The production of the herbicide glyphosate uses a combination of enzyme-catalysed and chemical synthesis. Glyoxal is formed from oxidation of glycolic acid, catalysed by a recombinant glycolate oxidase overexpressed in *Pichia pastoris*. This is followed by a Schiff's base condensation and reduction to form glyphosate.[50]

Approximately 10 per cent of industrially produced enzymes are not hydrolytic enzymes, and the range of these is gradually increasing. In future, an increase in the production of chiral drugs using enzyme-catalysed reactions can be expected. The discovery that some oxidoreductases, transferases, hydrolases, and iso-merases can function in apolar solvents such as toluene, benzene, hexane, and cyclohexane has widened the potential scope of enzymes used in industry. Enzyme catalysis in an organic medium often shifts the position of equilibrium, so that, for example, hydrolytic enzymes can be used for synthesis. Also, recovery of certain products from non-aqueous media may be easier. Potential applica-tions are in steroid oxidation and hydroxylation, phenolic polymerization, and ester and peptide synthesis.[41]

A factor which greatly restricts the use of oxidoreductases and synthetases is the recovery of cofactors such as $NAD(P)^+$, FAD, and adenine nucleotides. These are expensive ingredients, and in order to reduce costs must be able to be recy-cled. This can be achieved by immobilizing the cofactors to polymers such as poly(ethylene glycol).[41]

For the use of enzymes in industrial processes, much larger quantities are often needed than in research laboratories. It can seen from Chapter 2, Section 2.4.1, that when an enzyme is not readily available in sufficient quantity, then using recombinant DNA technology is the answer. This has already occurred in the case of chymosin (see Section 11.4.3.3) and many other enzymes used in indus-try are now recombinant products.[50] Molecular biological techniques themselves place a demand for a wide range of highly purified enzymes. There are currently over 1500 restriction endonucleases known, and many of these, together with DNA polymerases, RNA polymerases, reverse transcriptase, a range of nucleases, DNA ligase, and polynucleotide kinase, are 'tools of the trade' in molecular biology and are sold by the major fine-chemical companies.[50] There is every reason to expect that not only the amounts but also the range of enzymes used in industrial processes will increase in the future.

References

1. Godfrey, T. and West, S. (eds), *Industrial enzymology*. Macmillan, London (1996).

2. Fogarty, W. M. and Kelly, C. T., *Microbial enzymes and biotechnology*. Chapman and Hall, London (1990).

3. Brown, C. M., Campbell, I., and Priest, F. G., *Introduction to biotechnology*. Blackwell Scientific, Oxford (1987).

4. Chaplin, M. F. and Bucke, C., *Enzyme technology*. Cambridge University Press (1990).

5. Walker, J. M. and Gingold, E. B., *Molecular biology and biotechnology* (3rd edn). Royal Society of Chemistry, London (1993).

6. Gerhartz, W., *Enzymes in industry*. VCH, Weinheim, Germany (1990).

7. Pollock, J., *Brewing science*. Academic Press, New York (1987).

8. Brown, C. M., Campbell, I., and Priest, F. G., as reference 3, Chapter 9.

9. Godfrey, T. and West, S., as reference 1, pp. 103-32.

10. Slaughter, J. C. and Priest, F. G., in *Food enzymology* (Fox, P. F., ed.), Vol. 2, Chapter 18. Elsevier Applied Science, London (1991).

11. Tubb, R. S., *Trends Biotech.* **4**, 98 (1986).

12. Finkelstein, D. B. and Ball, C., *Biotechnology of filamentous fungi*, pp. 395–7. Butterworth-Heinemann, Boston (1992).

13. Buchholz, S. E., Dooley, M. M., and Eveleigh, D. E., *Trends Biotech.* **5**, 199 (1987).

14. Gripon, J.-C., Monnet, V., Lamberet, G., and Desmazeaud, M. J., in *Food enzymology* (Fox, P. F., ed.), Vol. 1, Chapter 3. Elsevier Applied Science, London (1991).

15. Law, B., *The Biochemist* **19**(4), 6 (1997).

16. Fox, P., *The Biochemist* **19**(4), 12 (1997).

17. Zidwick, M. J., in *Biotechnology of filamentous fungi* (Finkelstein, D. B. and Ball, C., eds), Chapter 11. Butterworth-Heinemann, Boston (1992).

18. Masurekar, P. S., in *Biotechnology of filamentous fungi* (Finkelstein, D. B. and Ball, C., eds), Chapter 10. Butterworth-Heinemann, Boston (1992).

19. Rambosek, J. A., in *Biotechnology of filamentous fungi* (Finkelstein, D. B. and Ball, C., eds), Chapter 9. Butterworth-Heinemann, Boston (1992).

20. Herbert, R. A., *Trends Biotech.* **10**, 395 (1992).

21. Chaplin, M. F. and Bucke, C., as reference 4, Chapter 2.

22. Scawen, M. D., Atkinson, T., Hammond, P. M., and Sherwood, R. F., in *Molecular biology and biotechnology* (Walker, J. M. and Gingold, E. B., eds), (3rd edn), Chapter 17. Royal Society of Chemistry, London (1993).

23. Gerhartz, W., as reference 6, Chapter 3.

24. Old, R. W. and Primrose, S. B., *Principles of gene manipulation: an introduction to genetic engineering* (5th edn). Blackwell Science (UK), Oxford (1994).

25. Brown, T. A., *Gene cloning: an introduction* (3rd edn). Chapman and Hall, London (1995).

26. Thatcher, D. R., in *Proteins Labfax* (Price, N. C., ed.), Chapter 13. Bios Scientific, Oxford (1996).

27. O'Rourke, T., in *Industrial enzymology* (Godfrey, T. and West, S., eds), Chapter 2.6. Macmillan, London (1996).

28. Fox, P. and Grufferty, M. B., in *Food Enzymology* (Fox, P. F. ed.), Vol. 1, Chapter 5. Elsevier Applied Science, London (1991).

29. Pawlett, D. and Bruce, G., in *Industrial enzymology* (Godfrey, T. and West, S., eds), Chapter 2.9. Macmillan, London (1996).

30. Wigley, R. C., in *Industrial enzymology* (Godfrey, T. and West, S., eds), Chapter 2.7. Macmillan, London (1996).

31. Bailey, M. E. and Murdock, J. A., in *Food enzymology* (Fox, P. F., ed.), Vol. 2, Chapter 26. Elsevier Applied Science, London (1991).

32. Gerhatz, W., as reference 6, Section 4.3.

33. Eriksen, N., in *Industrial enzymology* (Godfrey, T. and West, S., eds), Chapter 2.10. Macmillan, London (1996).

34. Gerhatz, W., as reference 6, Section 5.4.

35. Mosbach, K., *Methods Enzymol.* **44** (1976).

36. Bickerstaff, G., *Immobilization of enzymes and cells*. Humana Press, London (1997).

37. Mosbach, K., *Methods Enzymol.* **135** (1987).

38. Mosbach, K., *Methods Enzymol.* **136** (1987).

39. Gerhatz, W, as reference 6, Section 3.3.

40. Powell, L. W., in *Industrial enzymology* (Godfrey, T. and West, S., eds), Chapter 2.14. Macmillan, London (1996).

41. Katchalski-Katzir, E., *Trends Biotech.* **11**, 471 (1993).

42. Gerhatz, W., as reference 6, Section 4.3.

43. Hamilton, B. K., Hsiao, H-Y., Swann, W. E., Anderson, D. M., and Delente, J. J., *Trends Biotech.* **3**, 64 (1985).

44. Wood, L. L., *Biotechnology* **2**, 1081 (1984)

45. Turner, M. K., *Trends Biotech.* **13**, 253 (1995).

46. Kobayashi, M., Nagasawa, T., and Yamada, H., *Trends Biotech.* **10**, 402 (1992).

47. Bohak, Z. and Sharon, N., *Biotechnological applications of proteins and enzymes*, p. 141. Academic Press, New York (1977).

48. Chaplin, M. F., in *Molecular biology and biotechnology* (Walker, J. M. and Gingold, E. B., eds), (3rd edn), Chapter 19. Royal Society of Chemistry, London (1993).

49. Law, B. A., in *Industrial enzymology* (Godfrey, T. and West, S., eds), Chapter 3.2. Macmillan, London (1996).

50. Cowan, D., *Trends Biotech* **14**, 177 (1996).

51. Vandenburg, B., Vriend, G., Veltman, O. R., Venema, G., and Eijsink, V. G. H., *Proc. Natn. Acad. Sci. USA* **95**, 2056 (1998).

Appendix

Enzymes referred to in Chapters 1–11

The enzymes referred to in the 11 chapters of this book are listed below in alphabetical order generally under their recommended names as in *Enzyme nomenclature* (Recommendations (1992) of the Nomenclature Committee of the International Union of Biochemistry and Molecular Biology), except in a few cases where trivial names are more frequently used. The complete Enzyme Nomenclature Database can be found on the web site http://expasy.hcuge.ch/sprot/enzyme.html. A few of the enzymes listed have not yet been classified, usually because insufficient information is at present available for this to be done. In these cases we have indicated what is known about the probable nature of the reaction concerned. The specificities of a number of proteases are given in an abbreviated form, e.g. Arg- means the amide bond cleaved is that on the carboxyl side of arginine, whereas -Leu means that cleavage occurs on the amino side of leucine.

	EC number
Acetoin dehydrogenase	1.1.1.5

Acetoin + NAD^+ $\rightleftharpoons$ diacetyl + NADH

	EC number
Acetylcholinesterase	3.1.1.7

Acetylcholine + H_2O $\rightleftharpoons$ choline + acetate

Acetyl-CoA acetyltransferase — 2.3.1.9

Acetyl-CoA + acetyl-CoA $\rightleftharpoons$ CoA + acetoacetyl-CoA

Acetyl-CoA carboxylase — 6.4.1.2

ATP + acetyl-CoA + CO_2 + H_2O $\rightleftharpoons$ ADP + orthophosphate + malonyl-CoA

Acetylglutamate kinase — 2.7.2.8

ATP + *N*-acetyl-L-glutamate $\rightleftharpoons$ ADP + *N*-acetyl-L-glutamate 5-phosphate

β-N-acetyl-D-hexosaminidase — 3.2.1.52

Hydrolysis of terminal non-reducing *N*-acetyl-D-hexosamine residues in *N*-acetyl-β-D-hexosaminides

Acetylornithine aminotransferase — 2.6.1.11

N^2-Acetyl-L-ornithine + 2-oxoglutarate $\rightleftharpoons$ *N*-acetyl-L-glutamate 5-semialdehyde + L-glutamate

Acid phosphatase — 3.1.3.2

An orthophosphoric monoester + H_2O $\rightleftharpoons$ an alcohol + orthophosphate

Aconitase 4.2.1.3
Citrate $\rightleftharpoons$ *cis*-aconitate + H_2O
The preferred name is aconitate hydratase

Actinidin (*Actinidain*) 3.4.22.14
Hydrolysis of proteins with broad specificity for peptide bonds similar
to that of papain

[Acyl-carrier protein] S-acetyltransferase 2.3.1.38
Acetyl-CoA + [acyl-carrier protein] $\rightleftharpoons$ CoA + acetyl-[acyl-carrier protein]

[Acyl-carrier protein] S-malonyltransferase 2.3.1.39
Malonyl-CoA + [acyl-carrier protein] $\rightleftharpoons$ CoA + malonyl-[acyl carrier
protein]

Acyl-CoA desaturase 1.14.99.5
Stearyl-CoA + AH_2 + O_2 $\rightleftharpoons$ oleyl-CoA + A + $2H_2O$

Acyl-CoA oxidase 1.3.3.6
Acyl-CoA + O_2 $\rightleftharpoons$ *trans*-2,3-dehydroacyl-CoA + H_2O_2

Acyl-CoA synthetase
See Long-chain-fatty-acid-CoA ligase

Adenosinetriphosphatase 3.6.1.3
ATP + H_2O $\rightleftharpoons$ ADP + orthophosphate

Adenylate cyclase 4.6.1.1
ATP $\rightleftharpoons$ 3′:5′-cyclic AMP + pyrophosphate

Adenylate kinase 2.7.4.3
ATP + AMP $\rightleftharpoons$ ADP + ADP

Alanine aminotransferase 2.6.1.2
L-Alanine + 2-oxoglutarate $\rightleftharpoons$ pyruvate + L-glutamate

Alcohol dehydrogenase 1.1.1.1
Alcohol + NAD^+ $\rightleftharpoons$ aldehyde or ketone + NADH

Aldose 1-epimerase 5.1.3.3
α-D-Glucose $\rightleftharpoons$ β-D-glucose

Alkaline phosphatase 3.1.3.1
An orthophosphate monoester + H_2O $\rightleftharpoons$ an alcohol + orthophosphate

Amidophosphoribosyltransferase 2.4.2.14
5-Phospho-β-D-ribosylamine + pyrophosphate + L-glutamate $\rightleftharpoons$
L-glutamine + 5-phospho-α-D-ribose 1-diphosphate + H_2O

Amine oxidase (flavin-containing) 1.4.3.4
RCH_2NH_2 + H_2O + O_2 $\rightleftharpoons$ RCHO + NH_3 + H_2O_2

Amino-acid N-acetyltransferase 2.3.1.1
Acetyl-CoA + L-glutamate $\rightleftharpoons$ CoA + *N*-acetyl-L-glutamate

D-Amino-acid oxidase 1.4.3.3
D-amino acid + H_2O + O_2 $\rightleftharpoons$ 2-oxo-acid + NH_3 + H_2O_2

Aminoacylase 3.5.1.14
An *N*-acyl-L-aminoacid + H_2O $\rightleftharpoons$ a carboxylate + an L-aminoacid

Aminoacyl-tRNA ligases 6.1.1.1–22
ATP + L-amino acid + tRNA^{AA} $\rightleftharpoons$ AMP + pyrophosphate
+ L-aminoacyl-tRNA^{AA}
One enzyme for each of the amino acids

5-Aminolaevulinate synthase   2.3.1.37
  Succinyl-CoA + glycine $\rightleftharpoons$ aminolaevulinate + CoA + CO_2

Aminopeptidase (cytosol)   3.4.11.1
  Aminoacyl-peptide + H_2O $\rightleftharpoons$ amino acid + peptide

AMP deaminase   3.5.4.6
  AMP + H_2O $\rightleftharpoons$ IMP + NH_3

α-Amylase   3.2.1.1
  Endohydrolysis of 1,4-α-D-glucosidic linkages in polysaccharides
  containing three or more 1,4-α-linked D-glucose units

β-Amylase   3.2.1.2
  Hydrolysis of 1,4-α-D-glucosidic linkages in polysaccharides so as to
  remove successive maltose units from the non-reducing ends of the chains

Amylo-1,6-glucosidase   3.2.1.33
  Endohydrolysis of 1,6-α-D-glucosidic linkages at points of branching in
  chains of 1,4-linked α-D-glucose residues. Together with EC 2.4.1.25,
  these two activities constitute the glycogen *debranching* system

Angiotensin I-converting enzyme (Peptidyl-dipeptidase)   3.4.15.1
  Release of a C-terminal dipeptide, Xaa-Xbb-Xcc, when neither Xaa nor
  Xbb is Pro. Thus conversion of angiotensin I to angiotensin II

Anthranilate phosphoribosyltransferase   2.4.2.18
  N-(5′-Phospho-D-ribosyl)-anthranilate + pyrophosphate $\rightleftharpoons$
                            anthranilate + 5-phospho-α-D-ribose 1-diphosphate
  In many organisms this enzyme exists as a complex with anthranilate
  synthase (EC 4.1.3.27)

Anthranilate synthase   4.1.3.27
  Chorismate + L-glutamine $\rightleftharpoons$ anthranilate + pyruvate + L-glutamate
  In many organisms this enzyme exists as a complex with anthranilate
  phosphoribosyltransferase (EC 2.4.2.18)

Arginase   3.5.3.1
  L-Arginine + H_2O $\rightleftharpoons$ L-ornithine + urea

Arginine deiminase   3.5.3.6
  L-Arginine + H_2O $\rightleftharpoons$ L-citrulline + NH_3

Argininosuccinate lyase   4.3.2.1

N-(L-Arginino)succinate $\rightleftharpoons$ fumarate + L-arginine

Argininosuccinate synthase   6.3.4.5
  ATP + L-citrulline + L-aspartate $\rightleftharpoons$ AMP + pyrophosphate
                                                     + (L-arginino)succinate

Arginyl-tRNA ligase   6.1.1.19
  ATP + L-arginine + tRNA^{Arg} $\rightleftharpoons$ AMP + pyrophosphate
                                                     + L-arginyl-tRNA^{Arg}

Aromatic-L-amino acid decarboxylase   4.1.1.28
  L-Tryptophan $\rightleftharpoons$ tryptamine + CO_2

Arylsulphatase   3.1.6.1
  A phenol sulphate + H_2O $\rightleftharpoons$ a phenol + sulphate

Asparaginase   3.5.1.1
  L-Asparagine + H_2O $\rightleftharpoons$ L-aspartate + NH_3

Aspartate aminotransferase   2.6.1.1
L-Aspartate + 2-oxoglutarate $\rightleftharpoons$ oxaloacetate + L-glutamate

Aspartate ammonia-lyase   4.3.1.1
L-Aspartate $\rightleftharpoons$ fumarate + NH_3

Aspartate carbamoyltransferase   2.1.3.2
Carbamoyl phosphate + L-aspartate $\rightleftharpoons$ orthophosphate
$\qquad\qquad\qquad\qquad\qquad$ + *N*-carbamoyl-L-aspartate

Aspartate kinase   2.7.2.4
ATP + L-aspartate $\rightleftharpoons$ ADP + 4-phospho-L-aspartate

Aspergillopepsin I   3.4.23.18
Hydrolyses proteins with broad specificity. Generally favours hydrophobic
residues in P1 and P1′, but accepts Lys in P1, which leads to activation of
trypsinogen. Does not clot milk

ATP citrate (pro-3S)-lyase   4.1.3.8
ATP + citrate + CoA $\rightleftharpoons$ ADP + orthophosphate + acetyl-CoA
$\qquad\qquad\qquad\qquad\qquad\qquad$ + oxaloacetate

Often referred to as citrate cleavage enzyme

Branched-chain 2-oxoacid dehydrogenase   1.2.4.4
(3-Methyl-2-oxobutanoate dehydrogenase)
3-Methyl-2-oxobutanoate + lipoamide $\rightleftharpoons$ *S*-(2-methylpropanoyl)-
$\qquad\qquad\qquad\qquad\qquad\qquad$ dihydrolipoamide + CO_2

Part of multienzyme complex

Branching enzyme
See 1,4-α-Glucan branching enzyme

Bromelain   3.4.22.4
A thiol protease, cleaving preferentially at Lys-, Ala-, Tyr-, and Gly-

Carbamoyl-phosphate synthase (ammonia)   6.3.4.16
2ATP + NH_3 + CO_2 + H_2O $\rightleftharpoons$ 2ADP + orthophosphate + carbamoyl
$\qquad\qquad\qquad\qquad\qquad\qquad\qquad$ phosphate

This enzyme is present in the livers of ureotelic vertebrates

Carbamoyl-phosphate synthase (glutamine-hydrolysing)   6.3.5.5
2ATP + L-glutamine + CO_2 + H_2O $\rightleftharpoons$ 2ADP + orthophosphate
$\qquad\qquad\qquad\qquad\qquad$ + L-glutamate + carbamoyl phosphate

This enzyme is present in bacteria, fungi, and mammals. Both carbamoyl
phosphate synthases from fungi are EC 6.3.5.5, as is the mammalian
enzyme that participates in pyrimidine biosynthesis

Carbonate dehydratase   4.2.1.1
H_2CO_3 $\rightleftharpoons$ CO_2 + H_2O
Trivial name often used is carbonic anhydrase

Carbon monoxide dehydrogenase   1.2.99.2
CO + H_2O + acceptor $\rightleftharpoons$ CO_2 + reduced acceptor

Carboxylesterase   3.1.1.1
A carboxylic ester + H_2O $\rightleftharpoons$ an alcohol + a carboxylic acid anion

Carboxypeptidase A   3.4.17.1
Peptidyl-L-amino acid + H_2O $\rightleftharpoons$ peptide + L-amino acid

Carboxypeptidase B   3.4.17.2
Peptidyl-L-lysine (-L-arginine) + H_2O $\rightleftharpoons$ peptide + L-lysine (or L-arginine)

Carboxypeptidase Y   3.4.16.5
  Release of a C-terminal amino acid with broad specificity

Carnitine palmitoyltransferase   2.3.1.21
  Palmitoyl-CoA + L-carnitine $\rightleftharpoons$ CoA + L-palmitoylcarnitine

Caspase-1   3.4.22.92
  Release of interleukin 1β by specific cleavage of 116Asp–Ala117 and
  27Asp–Gly28

Catalase   1.11.1.6
  $2H_2O_2 \rightleftharpoons O_2 + 2H_2O$

Cathepsin B   3.4.22.1
  Hydrolyses proteins with a specificity resembling that of papain

Cathepsin D   3.4.23.5
  Protease with specificity similar to, but narrower than, that of pepsin A

Cathepsin G   3.4.21.20
  Protease with specificity similar to that of chymotrypsin

Cathepsin L   3.4.22.15
  Hydrolysis of proteins: no action on acylamino acid esters

Cellulase   3.2.1.4
  Endohydrolysis of $1,4$-β-D-glycosidic linkages in cellulose, lichenin, and
  cereal β-D-glucans

Chitinase   3.2.1.14
  Random hydrolysis of N-acety-β-D-glucosaminide $1,4$-β-linkages in chitin
  and chitodextrin

Chitin synthase   2.4.1.16
  UDP-N-acetyl-D-glucosamine + $[1,4$-$(N$-acetyl-β-D-glucosaminyl)]$_n$ $\rightleftharpoons$
                                 UDP + $[1,4$-$(N$-acetyl-β-D-glucosaminyl)]$_{n+1}$

Chloramphenicol O-acetyltransferase   2.3.1.28
  Acetyl-CoA + chloramphenicol $\rightleftharpoons$ CoA + chloramphenicol 3-acetate

Cholesterol acyltransferase   2.3.1.26
  Acyl-CoA + cholesterol $\rightleftharpoons$ CoA + cholesterol ester

Cholesterol esterase   3.1.1.13
  A cholesterol ester + H_2O $\rightleftharpoons$ cholesterol + a fatty-acid anion

Cholesterol oxidase   1.1.3.6
  Cholesterol + O_2 $\rightleftharpoons$ 4-cholesten-3-one + H_2O_2

Choline phosphotransferase   2.7.8.2
  CDPcholine + 1,2-diacylglycerol $\rightleftharpoons$ CMP + a phosphatidylcholine

Cholinesterase   3.1.1.8
  An acylcholine + H_2O $\rightleftharpoons$ choline + a carboxylic-acid anion

Chorismate synthase   4.6.1.4
  5-O-(1-Carboxyvinyl)-3-phosphoshikimate $\rightleftharpoons$ chorismate
                                                            + orthophosphate

Chymosin   3.4.23.4
  Protease, cleaves a single bond in casein κ
  Trivial name often used is rennin

Chymotrypsin   3.4.21.1
  Protease, preferential cleavage at Tyr-, Trp-, Phe-, and Leu

Citrate (si) synthase   4.1.3.7
Citrate + CoA $\rightleftharpoons$ acetyl-CoA + H_2O + oxaloacetate

Clostripain   3.4.22.8
Protease, with preferential cleavage of peptides: Arg-, especially Arg–Pro
bond but not Lys-

Coagulation Factor IXa   3.4.21.22
Selective cleavage of Arg–Ile in Factor X to form Factor Xa

Collagenase (vertebrate)   3.4.24.7
Cleaves preferentially one bond in native collagen leaving an N-terminal
(75%) and a C-terminal (25%) fragment

Creatine kinase   2.7.3.2
ATP + creatine $\rightleftharpoons$ ADP + phosphocreatine

Creatininase   3.5.2.10
Creatinine + H_2O $\rightleftharpoons$ creatine

Crotonyl-[acyl-carrier protein] hydratase   4.2.1.58
(3R)-3-Hydroxybutanoyl-[acyl-carrier protein] $\rightleftharpoons$ but-2-enoyl-[acyl-carrier
protein] + H_2O

3′:5′-Cyclic-nucleotide phosphodiesterase   3.1.4.17
Nucleoside 3′:5′-cyclic phosphate + H_2O $\rightleftharpoons$ nucleoside 5′-phosphate

Cystathionine-γ-lyase   4.4.1.1
L-Cystathionine + H_2O $\rightleftharpoons$ L-cysteine + NH_3 + 2-oxobutyrate

Cytidylate kinase   2.7.4.14
ATP + (d)CMP $\rightleftharpoons$ ADP + (d)CDP

Cytochrome b_5 reductase   1.6.2.2
NADH + 2 ferricytochrome b_5 $\rightleftharpoons$ NAD^+ + 2-ferrocytochrome b_5

Cytochrome c oxidase   1.9.3.1
4 Ferrocytochrome c + O_2 $\rightleftharpoons$ 4 ferricytochrome c + $2H_2O$

Cytosine deaminase   3.5.4.1
Cytosine + H_2O $\rightleftharpoons$ uracil + NH_3

Debranching enzyme
Consists of two activities:
α-dextrin endo-1,6-α-glucosidase
Hydrolysis of 1,6-α-D-glucosidic linkages in pullulan, amylopectin, and
glycogen, and in α- and β-amylase limit dextrins of amylopectin and
glycogen and isoamylase   3.2.1.41
Hydrolysis of 1,6-α-D-glucosidic linkages in glycogen, amylopectin, and
their β-limit dextrins   3.2.1.68

2-Dehydro-3-deoxyphosphogluconate aldolase   4.1.2.14
2-Dehydro-3-deoxy-D-gluconate 6-phosphate $\rightleftharpoons$ pyruvate
+ D-glyceraldehyde 3-phosphate

2-Dehydro-3-deoxyphosphoheptanoate aldolase   4.1.2.15
2-Dehydro-3-deoxy-D-*arabino*-heptanoate 7-phosphate + orthophosphate
$\rightleftharpoons$ phospho*enol*pyruvate + D-erythrose 4-phosphate + H_2O

3-Dehydroquinate dehydratase—often known as dehydroquinase   4.2.1.10
3-Dehydroquinate $\rightleftharpoons$ 3-dehydroshikimate + H_2O

3-Dehydroquinate synthase   4.6.1.3
  3-Deoxy-*arabino*-heptulosonate 7-phosphate $\rightleftharpoons$ 3-dehydroquinate
                                                              + orthophosphate

Deoxycytidine triphosphate pyrophosphatase   3.6.1.12
  dCTP + H_2O $\rightleftharpoons$ dCMP + pyrophosphate

Deoxycytidylic acid deaminase   3.5.4.12
  dCMP + H_2O $\rightleftharpoons$ dUMP + NH_3

Deoxycytidylate 5-hydroxymethyltransferase   2.1.2.8
  5,10-Methylenetetrahydrofolate + H_2O + deoxcytidylate $\rightleftharpoons$
                                tetrahydrofolate + 5-hydroxymethyldeoxycytidylate

Deoxyribonuclease I   3.1.21.1
  Endonucleolytic cleavage of DNA to 5′-phospho-dinucleotide and
  5′-phospho-oligonucleotide end-products

Deoxyribonuclease II   3.1.22.1
  Endonucleolytic cleavage of DNA to 3′-phospho-mononucleotides and
  3′-phospho-oligonucleotides

Dihydrofolate reductase   1.5.1.3
  5,6,7,8-Tetrahydrofolate + NADP $\rightleftharpoons$ 7,8-dihydrofolate + NADPH

Dihydrolipoamide S-acetyltransferase   2.3.1.12
  Acetyl-CoA + dihydrolipoamide $\rightleftharpoons$ CoA + *S*-acetylhydrolipoamide
  A component of the pyruvate dehydrogenase multienzyme system

Dihydrolipoamide dehydrogenase   1.8.1.4
  Dihydrolipoamide + NAD^+ $\rightleftharpoons$ lipoamide + NADH
  A component of the pyruvate, oxoglutarate, and branched chain 2-oxoacid
                                dehydrogenase multienzyme complexes

Dihydrolipoamide succinyltransferase   2.3.1.61
  Succinyl-CoA + dihydrolipoamide $\rightleftharpoons$ CoA + *S*-succinyldihydrolipoamide
  A component of the oxoglutarate dehydrogenase multienzyme system

Dihydro-orotase   3.5.2.3
  (*S*)-Dihydro-orotate + H_2O $\rightleftharpoons$ *N*-carbamoyl-L-aspartate

Dipeptidyl peptidase   3.4.14.1–5
  A number of enzymes of differing specificities catalysing the reaction
  Dipeptidyl-polypeptide + H_2O $\rightleftharpoons$ dipeptide + polypeptide

Diphosphate-fructose-6-phosphate 1-phosphotransferase   2.7.1.90
  Diphosphate + D-fructose 6-phosphate = phosphate
                                              + D-fructose 1,6-bisphosphate

DNA ligase (ATP)   6.5.1.1
  ATP + (deoyribonucleotide)$_n$ + (deoyribonucleotide)$_m$ $\rightleftharpoons$ AMP
                              + pyrophosphate + (deoyribonucleotide)$_{m+n}$

DNA ligase (NAD$^+$)   6.5.1.2
  NAD^+ + (deoyribonucleotide)$_n$ + (deoyribonucleotide)$_m$ $\rightleftharpoons$ AMP
              + nicotinamide mononucleotide + (deoyribonucleotide)$_{m+n}$

DNA-methyltransferase (adenine-specific)   2.1.1.72
  S-Adenosyl-L-methionine + DNA adenine $\rightleftharpoons$ *S*-adenosyl-L-homocysteine
                                              + DNA 6-methylaminopurine

DNA-methyltransferase (cytosine-specific)   2.1.1.73
  S-Adenosyl-L-methionine + DNA cytosine $\rightleftharpoons$ *S*-adenosyl-L-homocysteine
                                              + DNA 5-methylcytosine

DNA nucleotidylexotransferase   2.7.7.31

Deoxynucleoside triphosphate + $DNA_n \rightleftharpoons$ pyrophosphate + DNA_{n+1}

Also known as terminal transferase or terminal addition enzyme

DNA-directed DNA polymerase   2.7.7.7

Deoxynucleoside triphosphate + $DNA_n \rightleftharpoons$ pyrophosphate + DNA_{n+1}

Often referred to as DNA polymerase

Dodecenoyl-CoA Δ-isomerase (Δ³-cis-Δ²-trans-enoyl-CoA isomerase)   5.3.3.8

3-*cis*-Dodecenyl-CoA + 2-*trans*-dodecenyl-CoA

dUTP pyrophosphatase   3.6.1.23

dUTP + $H_2O \rightleftharpoons$ dUMP + pyrophosphate

Elastase (pancreatic)   3.4.21.36

Protease preferentially cleaving bonds involving the carbonyl groups of amino acids bearing uncharged non-aromatic side-chains

Hydrolysis of elastin

Endodeoxyribonuclease

See Type II site-specific deoxyribonuclease

Endo-1,3-β-D-glucanase   3.2.1.39

Hydrolysis of 1,3-β-D-glucosidic linkages in 1,3-β-D-glucans

Endoproteases A and B (yeast)

These enzymes are classified as EC 3.4.23.18 (*Aspergillopepsin I*) and EC 3.4.21.48 (*Cerevisin*), respectively

Endoprotease Glu-C

See Glutamyl endopeptidase

Enolase   4.2.1.11

2-Phospho-D-glycerate $\rightleftharpoons$ phosphoenolpyruvate + H_2O

Enoyl-[acyl-carrier protein] reductase (NADPH)   1.3.1.10

Acyl-[acyl-carrier protein] + $NADP^+ \rightleftharpoons$ *trans*- 2,3-dehydroacyl-[acyl-carrier protein] + NADPH

Enoyl-CoA hydratase   4.1.2.17

(3*S*)-3-Hydroxyacyl-CoA $\rightleftharpoons$ *trans*-2(or 3)-enoyl-CoA + H_2O

Enteropeptidase   3.4.21.9

Selective cleavage of Lys[6]–Ile bond in trypsinogen

Often referred to as enterokinase

Erythronolide synthase   2.3.1.94

6 Malonyl-CoA + propionyl-CoA $\rightleftharpoons$ 7 CoA + 6-deoxyerythronolide *b*

Exodeoxyribonuclease I   3.1.11.1

Exonucleolytic cleavage of DNA in the 3′ to 5′ direction to yield 5′-phosphomononucleotides

Exo-1,4-α-D-glucosidase   3.2.1.3

Hydrolysis of terminal 1,4-linked α-D-glucose residues successively from the non-reducing ends of the chains with release of β-D-glucose

Fatty-acyl-CoA synthase   2.3.1.86

Acetyl-CoA + $n$malonyl-CoA + $2n$NADH + $2n$NADPH $\rightleftharpoons$
long-chain-acyl-CoA + nCoA + nCO$_2$ + $2n$NAD$^+$ + $2n$NADP$^+$

The enzyme is a multifunctional polypeptide catalysing reactions of EC 2.3.1.38, 2.3.1.39, 2.3.1.41, 1.1.1.100, 4.2.1.61, and 1.3.1.9.

See Fig. 7.21 for details

Ficin   3.4.22.3
A thiol protease preferentially cleaving Lys-, Ala-, Tyr-, Gly-, Asn-, Leu-,
and Val-

β-D-Fructofuranosidase   3.2.1.26
Hydrolysis of terminal non-reducing β-D-fructofuranoside residues in
β-D-fructofuranosides

Fructokinase   2.7.1.4
ATP + D-fructose $\rightleftharpoons$ ADP + D-fructose 6-phosphate

Fructose-bisphosphatase   3.1.3.11
D-Fructose 1,6-bisphosphate + H_2O $\rightleftharpoons$ D-fructose 6-phosphate
                                          + orthophosphate

Fructose-2,6-bisphosphate 2-phosphatase   3.1.3.46
D-Fructose 2,6-bisphosphate + H_2O $\rightleftharpoons$ D-fructose 6-phosphate
                                          + orthophosphate

Fructose-bisphosphate aldolase   4.1.2.13
D-Fructose 1,6-bisphosphate $\rightleftharpoons$ glycerone phosphate + D-glyceraldehyde
                                                       3-phosphate
Glycerone phosphate was formerly known as dihydroxyacetone phosphate

Fumarate hydratase   4.2.1.2
L-Malate $\rightleftharpoons$ fumarate + H_2O
Often referred to as fumarase

Galactokinase   2.7.1.6
ATP + D-galactose $\rightleftharpoons$ ADP + α-D-galactose 1-phosphate

Galactose 1-phosphate uridylyltransferase   2.7.7.10
UTP + α-D-galactose 1-phosphate $\rightleftharpoons$ pyrophosphate + UDPgalactose

α-D-Galactosidase   3.2.1.22
Hydrolysis of terminal, non-reducing α-D-galactose residues in
α-D-galactosides, including galactose oligosaccharides, galactomannans, and
galactolipids

β-D-Galactosidase   3.2.1.23
Hydrolysis of terminal non-reducing β-D-galactose residues in
β-D-galactosides

GDPmannose α-D-mannosyltransferase   2.4.1.48
GDPmannose + heteroglycan $\rightleftharpoons$ GDP + 1,2(1,3)-α-D-mannosyl-
                                                       heteroglycan

1,4-α-Glucan branching enzyme   2.4.1.18
Transfers a segment of 1,4-α-D-glucan chain to a primary hydroxyl group
in a similar glucan chain

4-α-D-Glucanotransferase   2.4.1.25
Transfers a segment of 1,4-α-D-glucan to a new position in an acceptor,
which may be glucose or a 1,4-α-D-glucan
This activity forms part of the glycogen debranching system

3-Glucanase   3.2.1.39
Now known as Glucan endo-1,3-β-D-glucosidase
See Endo-1,3-β-D-glucanase

Glucokinase   2.7.1.2
ATP + D-glucose $\rightleftharpoons$ ADP + D-glucose 6-phosphate

Glucose 1-dehydrogenase   1.1.1.47
β-D-Glucose + NAD(P)$^+$ $\rightleftharpoons$ D-glucono-1,5-lactone + NAD(P)H

Glucose oxidase   1.1.3.4
   β-D-Glucose + O_2 $\rightleftharpoons$ D-glucono-1,5-lactone + H_2O_2

Glucose-6-phosphatase   3.1.3.9
   D-Glucose 6-phosphate + H_2O $\rightleftharpoons$ D-glucose + orthophosphate

Glucose 6-phosphate 1-dehydrogenase   1.1.1.49
   D-Glucose 6-phosphate + $NADP^+$ $\rightleftharpoons$ D-glucono-1,5-lactone 6-phosphate
                                                                + NADPH

Glucose 6-phosphate isomerase   5.3.1.9
   D-Glucose 6-phosphate $\rightleftharpoons$ D-fructose 6-phosphate

Glucose 1-phosphate uridylyltransferase   2.7.7.9
   (UTP-glucose 1-phosphate uridylyltransferase)
   UTP + α-D-glucose 1-phosphate $\rightleftharpoons$ pyrophosphate + UDPglucose

α-D-Glucosidase   3.2.1.20
   Hydrolysis of terminal, non-reducing 1,4-linked α-D-glucose residues with
   release of α-D-glucose
   Often referred to as maltase

Glucosylcerebrosidase   3.2.1.45
   D-glucosyl-N-acylsphingosine + H_2O $\rightleftharpoons$ D-glucose + N-acylsphingosine

β-D-Glucuronidase   3.2.1.31
   A β-D-glucuronide + H_2O $\rightleftharpoons$ an alcohol + D-glucuronate

Glucuronosyltransferase   2.4.1.17
   UDPglucuronate + acceptor $\rightleftharpoons$ UDP + acceptor β-D-glucuronide

Glutamate dehydrogenase   1.4.1.2
   L-Glutamate + H_2O + NAD^+ $\rightleftharpoons$ 2-oxoglutarate + NH_3 + NADH

Glutamate dehydrogenase (NAD(P)$^+$)   1.4.1.3
   L-Glutamate + H_2O + $NAD(P)^+$ $\rightleftharpoons$ 2-oxoglutarate + NH_3 + NAD(P)H

Glutaminase   3.5.1.2
   L-Glutamine + H_2O $\rightleftharpoons$ L-glutamate + NH_3

Glutamine phosphoribosyl pyrophosphate amidotransferase   2.4.2.14
   5-Phospho-β-D-ribosylamine + pyrophosphate + L-glutamate $\rightleftharpoons$
              L-α-glutamine + 5-phospho-α-D-ribose 1-diphosphate + H_2O

Glutamine synthetase (glutamate–ammonia ligase)   6.3.1.2
   ATP + L-glutamate + NH_3 $\rightleftharpoons$ ADP + orthophosphate + L-glutamine

Glutamate 5-kinase (γ-glutamyl kinase)   2.7.2.11
   ATP + L-glutamate $\rightleftharpoons$ ADP + L-glutamate 5-phosphate

Glutamate-5-semialdehyde dehydrogenase   1.2.1.41
   L-Glutamate-5-semialdehyde + orthophosphate + $NADP^+$ $\rightleftharpoons$ L-glutamyl
                                                    5-phosphate + NADPH

Glutamyl endopeptidase   3.4.21.19
   Preferential cleavage: Glu-, Asp-
   Also known as V8 protease

γ-Glutamyltransferase   2.3.2.2
   (5-L-Glutamyl)-peptide + an amino acid $\rightleftharpoons$ peptide + 5-L-glutamyl-amino
                                                                acid

Glutaryl-CoA oxidase   1.3.99.7
   Glutaryl-CoA + acceptor $\rightleftharpoons$ crotonyl-CoA + CO_2 + reduced acceptor

Glutathione reductase (NADPH)   1.6.4.2
   NADPH + oxidized glutathione $\rightleftharpoons$ NADP$^+$ + 2 glutathione

Glutathione-S-transferase   2.5.1.45
   RX + GSH $\rightleftharpoons$ HX + R-S-G
   (GSH = reduced glutathione)

Glyceraldehyde 3-phosphate dehydrogenase (phosphorylating)   1.2.1.12
   D-Glyceraldehyde 3-phosphate + orthophosphate + NAD$^+$ $\rightleftharpoons$
                              3-phospho-D-glyceroylphosphate + NADH

Glycerol kinase   2.7.1.30
   ATP + glycerol $\rightleftharpoons$ ADP + *sn*-glycerol 3-phosphate

Glycerol 3-phosphate dehydrogenase   1.1.99.5
   sn-Glycerol 3-phosphate + acceptor $\rightleftharpoons$ glycerone phosphate + reduced
                                                                        acceptor

Glycerol 3-phosphate dehydrogenase (NAD$^+$)   1.1.1.8
   sn-Glycerol 3-phosphate + NAD$^+$ $\rightleftharpoons$ glycerone phosphate + NADH

Glycerol 3-phosphate O-acyltransferase   2.3.1.15
   Acyl-CoA + *sn*-glycerol 3-phosphate $\rightleftharpoons$ CoA + 1-acyl-*sn*-glycerol
                                                                   3-phosphate

Glycine decarboxylase (glycine dehydrogenase-decarboxylating)   1.4.4.2
   Glycine + lipoylprotein $\rightleftharpoons$ *S*-aminomethyldihydrolipoylprotein + CO$_2$

Glycogen (starch) synthase   2.4.1.11
   UDPglucose + (1,4-α-D-glucosyl)$_n$ $\rightleftharpoons$ UDP + (1,4-α-D-glucosyl)$_{n+1}$

Glycolate oxidase (hydroxy-acid oxidase)   1.1.3.15
   (*S*)-2-hydroxy-acid + O$_2$ $\rightleftharpoons$ 2-oxo-acid + H$_2$O$_2$

Guanylate cyclase   4.6.1.2
   GTP $\rightleftharpoons$ 3′,5′-cyclic GMP + pyrophosphate

Hexokinase   2.7.1.1
   ATP + D-hexose $\rightleftharpoons$ ADP + D-hexose 6-phosphate

Histidine ammonia-lyase   4.3.1.3
   L-Histidine $\rightleftharpoons$ urocanate + NH$_3$

HIV protease
   See Retropepsin

Homogentisate 1,2-dioxygenase   1.13.11.5
   Homogentisate + O$_2$ $\rightleftharpoons$ 4-maleylacetoacetate

Homoserine dehydrogenase   1.1.1.3
   L-Homoserine + NAD(P)$^+$ $\rightleftharpoons$ L-aspartate 4-semialdehyde + NAD(P)H

Hyaluronoglucosaminidase   3.2.1.35
   Random hydrolysis of 1,4-linkages between *N*-acetyl-β-D-glucosamine and
   D-glucuronate residues in hyaluronate

Hyaluronoglucuronidase   3.2.1.36
   Random hydrolysis of 1,3-linkages between β-D-glucuronate and
   N-acetyl-D-glucosamine residues in hyaluronate

Hydrogenase   1.18.99.1
   2 Reduced ferredoxin + 2H$^+$ $\rightleftharpoons$ 2 oxidized ferredoxin + H$_2$

L-α-Hydroxy-acid oxidase (S-2-hydroxy-acid oxidase)   1.1.3.15
   (*S*)-2-hydroxy-acid + O$_2$ $\rightleftharpoons$ 2-oxo-acid + H$_2$O$_2$

3-Hydroxyacyl-CoA dehydrogenase   1.1.1.35
(S)-3-hydroxyacyl-CoA + NAD$^+$ $\rightleftharpoons$ 3-oxoacyl-CoA + NADH

3-Hydroxybutyrate dehydrogenase   1.1.1.30
(R)-3-Hydroxybutyrate + NAD$^+$ $\rightleftharpoons$ acetoacetate + NADH

Hydroxymethylglutaryl-CoA reductase   1.1.1.88
(S)-Mevalonate + CoA + 2NAD$^+$ $\rightleftharpoons$ 3-hydroxy-3-methylglutaryl-CoA
+ 2NADH

Indole-3-glycerol-phosphate synthase   4.1.1.48
1-(2-Carboxyphenylamino)-1-deoxy-D-ribulose 5-phosphate $\rightleftharpoons$
1-(ind-3-yl)-glycerol 3-phosphate + CO_2 + H_2O

Iodide peroxidase   1.11.1.8
Iodide + H_2O_2 $\rightleftharpoons$ iodine + 2H_2O
Used to iodinate tyrosine residues

Isocitrate dehydrogenase (NAD$^+$)   1.1.1.41
Isocitrate + NAD$^+$ $\rightleftharpoons$ 2-oxoglutarate + CO_2 + NADH

Isocitrate lyase   4.1.3.1
Isocitrate $\rightleftharpoons$ succinate + glyoxylate

Isoleucyl-tRNA ligase   6.1.1.5
ATP + L-isoleucine + tRNA^{Ile} $\rightleftharpoons$ AMP + pyrophosphate
+ L-isoleucyl-tRNA^{Ile}

Isomaltase (Oligo-1,6-glucosidase)   3.2.1.10
Hydrolysis of 1,6-α-D-glucosidic linkages in isomaltose and dextrins
produced from starch and glycogen by α-amylase
The enzyme from intestinal mucosa also catalyses the hydrolysis of sucrose
and maltose (see Chapter 8, Section 8.4.3)

C_{55} isoprenoid-alcohol kinase   2.7.1.66
Also known as undecaprenol kinase
ATP + undecaprenol $\rightleftharpoons$ ADP + undecaprenyl phosphate

Kynureninase   3.7.1.3
L-Kynurenine + H_2O $\rightleftharpoons$ anthranilate + L-alanine

β-Lactamase (penicillinase)   3.5.2.6
A β-lactam + H_2O $\rightleftharpoons$ a substituted β-amino acid

L-Lactate dehydrogenase   1.1.1.27
(S)-Lactate + NAD$^+$ $\rightleftharpoons$ pyruvate + NADH

Lactose synthase   2.4.1.22
UDPgalactose + D-glucose $\rightleftharpoons$ UDP + lactose

Lactoylglutathione lyase   4.4.1.5
(R)-S-Lactoylglutathione $\rightleftharpoons$ glutathione + methylglyoxal

Leucine dehydrogenase   1.4.1.9
L-Leucine + H_2O + NAD$^+$ $\rightleftharpoons$ 4-methyl-2-oxopentanoate + NH_3 + NADH

Long-chain acyl-CoA synthase   6.2.1.3
ATP + long-chain carboxylic acid + CoA $\rightleftharpoons$ AMP + pyrophosphate
+ acyl-CoA

Lysine hydroxylase   1.14.11.4
Procollagen-L-lysine + 2-oxoglutarate + O_2 $\rightleftharpoons$ procollagen-5-hydroxy-
L-lysine + succinate + CO_2

Lysozyme   3.2.1.17
Hydrolysis of 1,4-β-linkages between N-acetylmuramic acid and
N-acetyl-D-glucosamine residues in a peptidoglycan and between
N-acetyl-D-glucosamine residues in chitodextrins

Malate dehydrogenase   1.1.1.37
(S)-Malate + NAD^+ $\rightleftharpoons$ oxaloacetate + NADH

Malate dehydrogenase (oxaloacetate-decarboxylating) (NADP$^+$)   1.1.1.40
(S)-Malate + $NADP^+$ $\rightleftharpoons$ pyruvate + CO_2 + NADPH
Often referred to as the 'malic' enzyme

Malonyl-CoA dependent elongase system   2.3.1.119
Referred to as Icosanoyl-CoA synthase
A microsomal membrane enzyme system that catalyses the elongation of
acyl-CoAs (C_{16}–C_{22})
Stearyl-CoA + malonyl-CoA + 2 NAD(P)H $\rightleftharpoons$ Icosanoyl-CoA + CO_2
                                                            + $NAD(P)^+$

Methionine adenosyltransferase   2.5.1.6
ATP + L-methionine + H_2O $\rightleftharpoons$ orthophosphate + pyrophosphate
                                        + S-adenosyl-L-methionine

Multicatalytic endopeptidase complex   3.4.99.46
Cleavage at Xaa– bonds in which Xaa carries a hydrophobic, basic, or
acidic side-chain

NADH dehydrogenase   1.6.99.3
NADH + acceptor $\rightleftharpoons$ NAD^+ + reduced acceptor

NAD$^+$ nucleosidase   3.2.2.5
NAD^+ + H_2O $\rightleftharpoons$ nicotinamide + ADPribose

NADP$^+$-cytochrome reductase   1.6.2.4
$NADP^+$ +2 ferrocytochrome $\rightleftharpoons$ NADPH + 2 ferricytochrome

Neuraminidase   3.2.1.18
Also known as exo-α-sialidase
Hydrolysis of α-2,3-, α-2,6-, and α-2,8-glycosidic linkages (at a decreasing
rate respectively) of terminal sialic residues in oligosaccharides,
glycoproteins, glycolipids, colominic acid and synthetic substrates

Nicotinamide-nucleotide adenylyltransferase   2.7.7.1
ATP + nicotinamide ribonucleotide $\rightleftharpoons$ pyrophosphate + NAD

Nitric oxide synthase   1.14.13.39
L-Arginine + nNADH + mO$_2$ $\rightleftharpoons$ citrulline + NO + nNADP$^+$

Nitrile hydratase   4.2.1.84
An aliphatic amide $\rightleftharpoons$ a nitrile + H_2O

Nitrogenase   1.18.6.1
3 Reduced ferredoxin + $6H^+$ + N_2 + nATP $\rightleftharpoons$ 3 oxidized ferredoxin
                                        + $2NH_3$ + nADP + n orthophosphate

Nitrogenase (flavodoxin)   1.19.6.1
6 Reduced flavodoxin + $6H^+$ + N_2 + nATP $\rightleftharpoons$ 6 oxidized flavodoxin
                                        + $2NH_3$ + nADP + n orthophosphate

Nucleoside-diphosphatase   3.6.1.6
A nucleoside diphosphate + H_2O $\rightleftharpoons$ a nucleotide + orthophosphate

Nucleoside-diphosphate kinase   2.7.4.6
ATP + nucleoside diphosphate $\rightleftharpoons$ ADP + nucleoside triphosphate

Nucleoside-phosphate kinase   2.7.4.4
ATP + nucleoside phosphate $\rightleftharpoons$ ADP + nucleoside diphosphate

Nucleoside-triphosphatase   3.6.1.15
NTP + H_2O $\rightleftharpoons$ NDP + orthophosphate

5'-Nucleotidase   3.1.3.5
A 5'-ribonucleotide + H_2O $\rightleftharpoons$ a ribonucleoside + orthophosphate

Nucleotide pyrophosphatase   3.6.1.9
A dinucleotide + H_2O $\rightleftharpoons$ 2 mononucleotides
Substrates include NAD^+, $NADP^+$, FAD, CoA, ATP, and ADP

D-Octopine dehydrogenase   1.5.1.11
N^2-(D-1-carboxyethyl)-L-arginine + NAD^+ + H_2O $\rightleftharpoons$ L-arginine
                                               + pyruvate + NADH

Ornithine carbamoyltransferase   2.1.3.3
Carbamoylphosphate + L-ornithine $\rightleftharpoons$ orthophosphate + L-citrulline

Ornithine decarboxylase   4.1.1.17
L-Ornithine $\rightleftharpoons$ putrescine + CO_2

Ornithine-oxo-acid transaminase   2.6.1.13
L-Ornithine + a 2-oxoacid $\rightleftharpoons$ L-glutamate 5-semialdehyde + an L-amino acid

Orotate phosphoribosyltransferase   2.4.2.10
Orotidine 5'-phosphate + pyrophosphate $\rightleftharpoons$ orotate
                                    + 5-phospho-α-D-ribose 1-diphosphate

Orotate reductase (NADP$^+$)   1.3.1.14
(S)-Dihydroorotate + NAD^+ $\rightleftharpoons$ orotate + NADH

Orotidine-5'-phosphate decarboxylase   4.1.1.23
Orotidine 5-phosphate $\rightleftharpoons$ UMP + CO_2

3-Oxoacyl-[acyl-carrier protein] reductase   1.1.1.100
(R)-3-Hydroxyacyl-[acyl-carrier protein] + $NADP^+$ $\rightleftharpoons$
                         3-oxoacyl-[acyl-carrier protein] + NADPH

3-Oxoacyl-[acyl-carrier protein] synthase   2.3.1.41
Acyl-[acyl-carrier protein] + malonyl-[acyl-carrier protein] $\rightleftharpoons$
                    3-oxoacyl-[acyl-carrier protein] + CO_2 + acyl-carrier protein

Oxoglutarate dehydrogenase (also known as oxoglutarate decarboxylase)   1.2.4.2
2-Oxoglutarate + lipoamide $\rightleftharpoons$ S-succinyldihydrolipoamide + CO_2
A component of the oxoglutarate dehydrogenase multienzyme system

Oxalate oxidase   1.2.3.4
Oxalate + O_2 $\rightleftharpoons$ $2CO_2$ + H_2O_2

4-Oxalocrotonate tautomerase   No EC number yet assigned

Papain   3.4.22.2
Hydrolysis of proteins with broad specificity for peptide bonds, with
preference for a residue bearing a large hydrophobic side-chain at the P2
position. Dose not accept Val at P1'

Pectinase (polygalacturonase)   3.5.1.15
Hydrolyses random 1,4 α-D-galactosiduronic linkages in pectate and other
galacturonans

Penicillin amidase 3.5.1.11

Penicillin + H_2O $\rightleftharpoons$ a carboxylate + 6-aminopenicillinate

Penicillinase

See under recommended name β-Lactamase

Penicillopepsin

See Aspergillopepsin I

Pepsin A 3.4.23.1

Preferential cleavage: hydrophobic, preferably aromatic residues in P1 and P1′ positions. Cleaves Phe[1]–Val, Gln[4]–His, Glu[13]–Ala, Ala[14]–Leu, Leu[15]–Tyr, Tyr[16]–Leu, Gly[23]–Phe, Phe[24]–Phe, and Phe[25]–Tyr bonds in the B chain of insulin

Peptidylprolyl isomerase 5.2.1.8

Peptidylproline ($\omega = 180$) $\rightleftharpoons$ peptidylproline ($\omega = 0$)

Peroxidase 1.11.1.7

Donor + H_2O_2 $\rightleftharpoons$ oxidized donor + $2H_2O$

Phenylalanine ammonia-lyase 4.3.1.5

L-Phenylalanine $\rightleftharpoons$ *trans*-cinnamate + NH_3

Phenylalanine hydroxylase

See Phenylalanine 4-monooxygenase

Phenylalanine 4-monooxygenase 1.14.16.1

L-Phenylalanine + tetrahydrobiopterin + O_2 $\rightleftharpoons$ L-tyrosine
+ dihydrobiopterin + H_2O

Phosphodiesterase I 3.1.4.1

Hydrolytically removes 5′-nucleotides successively from the 3′-hydroxy termini of 3′-hydroxy-terminated oligonucleotides

Phosphoenolpyruvate carboxykinase (GTP) 4.1.1.32

GTP + oxaloacetate $\rightleftharpoons$ GDP + phospho*enol*pyruvate + CO_2

6-Phosphofructokinase 2.7.1.11

ATP + D-fructose 6-phosphate $\rightleftharpoons$ ADP + D-fructose 1,6-bisphosphate

6-Phosphofructo-2-kinase 2.7.1.105

ATP + D-fructose 6-phosphate $\rightleftharpoons$ ADP + D-fructose 2,6-bisphosphate

Phosphoglucomutase 5.4.2.2

α-D-Glucose 1-phosphate $\rightleftharpoons$ α-D-glucose 6-phosphate

Phosphogluconate dehydrogenase (decarboxylating) 1.1.1.44

6-Phospho-D-gluconate + NAD(P)$^+$ $\rightleftharpoons$ D-ribulose 5-phosphate + CO_2
+ NAD(P)H

Phosphoglycerate kinase 2.7.2.3

ATP + 3-phospho-D-glycerate $\rightleftharpoons$ ADP + 3-phospho-D-glyceroyl phosphate

Phosphoglycerate mutase 5.4.2.1

2-Phospho-D-glycerate $\rightleftharpoons$ 3-phospho-D-glycerate

Phospho-2-keto-3-deoxy-gluconate aldolase

See 2-Dehydro-3-deoxyphosphogluconate aldolase

Phospho-2-keto-3-deoxy-heptonate aldolase

See 2-Dehydro-3-deoxyphosphoheptanoate aldolase

Phospholipase A_1   3.1.1.32
A phosphatidylcholine + H_2O ⇌ 2-acylglycerophosphocholine
+ a carboxylate

Phospholipase A_2   3.1.1.4
A phosphatidylcholine + H_2O ⇌ 1-acylglycerophosphocholine
+ a carboxylate

Phospholipase C   3.1.4.3
A phosphatidylcholine + H_2O ⇌ 1, 2-diacylglycerol + choline phosphate

Phosphoribosylaminoimidazole carboxylase   4.1.1.21
1-(5-Phospho-D-ribosyl)-5-amino-4-imidazolecarboxylate ⇌
1-(5-Phospho-D-ribosyl)-5-amino-4-imidazole + CO_2

Phosphoribosylaminoimidazolesuccinocarboxamide synthase   6.3.2.6
ATP + 1-(5-phospho-D-ribosyl)-5-amino-4-carboxyimidazole + L-aspartate
⇌ ADP + orthophosphate + 1-(5-phospho-D-ribosyl)-5-amino-
4-(*N*-succinocarboxyamide)-imidazole

Phosphorylase   2.4.1.1
$(1,4-\alpha$-D-Glucosyl$)_n$ + orthophosphate ⇌ $(1,4-\alpha$-D-glucosyl$)_{n+1}$
+ α-D-glucose 1-phosphate

Phosphorylase kinase   2.7.1.38
4ATP + 2 phosphorylase *b* ⇌ 4ADP + phosphorylase *a*

[Phosphorylase] phosphatase   3.1.3.17
[Phosphorylase *a*] + $4H_2O$ ⇌ 2[phosphorylase *b*] + 4 orthophosphate

3-Phosphoshikimate 1-carboxyvinyltransferase   2.5.1.19
Phospho*enol*pyruvate + 3-phosphoshikimate ⇌ orthophosphate
+ 5-*O*-(1-carboxyvinyl)-3-phosphoshikimate

Pipecolate oxidase   1.5.3.7
L-pipecolate + O_2 ⇌ 2,3,4,5-tetrahydropyridine-2-carboxylate + H_2O_2

Plasmin   3.4.21.7
Preferential cleavage: Lys- > Arg-; higher selectivity than trypsin
Converts fibrin into soluble products

Polyamine oxidase   1.5.3.11
N^1-Acetylspermine + O_2 + H_2O ⇌ N^1-acetylspermidine
+ 2-aminopropanal + H_2O_2

Polynucleotide 5′-hydroxyl-kinase   2.7.1.78
ATP + 5′-dephospho-DNA ⇌ ADP + 5′-phospho-DNA

Polygalacturonase   3.2.1.15
Random hydrolysis of $1,4-\alpha$-D-galactosiduronic linkages in pectate and other
galacturonans

Proline hydroxylase   1.14.11.2
Procollagen L-proline + 2-oxoglutarate + O_2 ⇌ procollagen
trans-4-hydroxy-L-proline + succinate + CO_2

Proline racemase   5.1.1.4
L-Proline ⇌ D-proline

Protease Clp   3.4.22.36
Hydrolyses proteins to small peptides in the presence of ATP and Mg^{2+}

Proteasome

See Multicatalytic endopeptidase complex

Protein disulphide-isomerase   5.3.4.1

Catalyses the rearrangement of both intrachain and interchain disulphide
bonds in proteins

Protein disulphide reductase (glutathione)   1.8.4.2

2 Glutathione + protein-disulphide $\rightleftharpoons$ oxidized glutathione + protein-dithiol

Protein kinase

ATP + a protein $\rightleftharpoons$ ADP + a phosphoprotein
The enzyme from rat tissues is stimulated by cyclic AMP and will activate
phosphorylase kinase. Other enzymes will phosphorylate other proteins.
Some enzymes are activated by cyclic AMP, some by cyclic GMP, and
some by neither (see Chapter 6, Section 6.4.2)

Protein phosphatase

See [Phosphorylase phosphatase]

Protein-tyrosine kinase   2.7.1.112

ATP + protein-tyrosine $\rightleftharpoons$ ADP + protein-tyrosine phosphate

Pullulanase   3.2.1.41

Starch debranching enzyme hydrolyses 1,6-α glucosidic links in pullulanin
and starch

Pyruvate carboxylase   6.4.1.1

ATP + pyruvate + HCO_3^- $\rightleftharpoons$ ADP + orthophosphate + oxaloacetate

Pyruvate dehydrogenase (lipoamide) (also known as pyruvate decarboxylase)   1.2.4.1

Pyruvate + lipoamide $\rightleftharpoons$ S-acetyldihydrolipoamide + CO_2
A component of the pyruvate dehydrogenase multienzyme system

[Pyruvate dehydrogenase (lipoamide)] kinase   2.7.1.99

ATP + [pyruvate dehydrogenase (lipoamide)] $\rightleftharpoons$ ADP + [pyruvate
                                        dehydrogenase (lipoamide)] phosphate

[Pyruvate dehydrogenase (lipoamide)]-phosphatase   3.1.3.43

[Pyruvate dehydrogenase (lipoamide)] phosphate + H_2O $\rightleftharpoons$
                                [pyruvate dehydrogenase (lipoamide)] + orthophosphate

Pyruvate kinase   2.7.1.40

ATP + pyruvate $\rightleftharpoons$ ADP + phospho*enol*pyruvate

Renin   3.4.23.15

Cleaves at Leu- bond in angiotensinogen to generate angiotensin I
Also known as angiotensinogenase
Restriction endonuclease
See Type II site-specific deoxyribonuclease

Retropepsin   3.4.23.16

Protease specific for a P1 residue that is hydrophobic, and P1′ variable.
but often Pro

Reverse transcriptase   2.7.7.49

Deoxynucleoside triphosphate + DNA_n $\rightleftharpoons$ pyrophosphate + DNA_{n+1}
Also known as DNA nucleotidyltransferase (RNA-directed)

Rhodanese

See Thiosulphate sulphurtransferase

Riboflavin synthase (lumazine synthase)　　　　2.5.1.9
2,6,7-Dimethyl-8-(1-D-ribityl)lumazine $\rightleftharpoons$ riboflavin
　　　　+ 4-(1-D-ribitylamino)-5-amino-2,6-dihydroxypyrimidine

Ribonuclease H　　　　3.1.26.4
Endonucleolytic cleavage to 5′-phosphomonoester

Ribonuclease II　　　　3.1.13.1
Exonucleotic cleavage in the 3′ to 5′ direction to yield
3′-phosphomononucleotides
Also known as exoribonuclease II

Ribonuclease (pancreatic)　　　　3.1.27.5
Endonucleolytic cleavage of RNA to 3′-phosphomononucleotides and
3′-phosphooligonucleotides ending in Cp or Up with 2′,3′-cyclic
phosphate intermediates

Ribonucleotide reductase　　　　1.17.4.2
2′-deoxyribonucleotide + oxidized thioredoxin + $H_2O \rightleftharpoons$
　　　　　　　　ribonucleotide + reduced thioredoxin

Ribulose-bisphosphate carboxylase　　　　4.1.1.39
D-Ribulose 1,5-bisphosphate + $CO_2 \rightleftharpoons$ 2 3-phospho-D-glycerate

RNA-methyltransferases　　　　2.1.29–2.1.1.37
S-Adenosyl-L-methionine + tRNA $\rightleftharpoons$ S-adenosyl-L-homocysteine
　　　　　　　　+ tRNA containing methyl groups
A group of enzymes which methylate tRNA on different positions
on the purine or pyrimidine bases

RNA polymerase　　　　2.7.7.6
Nucleoside triphosphate + $RNA_n \rightleftharpoons$ pyrophosphate + RNA_{n+1}
Also known as DNA-directed RNA polymerase

L-Serine dehydratase　　　　4.2.1.13
L-Serine + $H_2O \rightleftharpoons$ pyruvate + NH_3 + H_2O

Shikimate 5-dehydrogenase　　　　1.1.1.25
Shikimate + $NADP^+ \rightleftharpoons$ 5-dehydroshikimate + NADPH

Shikimate kinase　　　　2.7.1.71
ATP + shikimate $\rightleftharpoons$ ADP + shikimate 3-phosphate

Steroid 11β-monooxygenase　　　　1.14.15.4
A steroid + reduced adrenal ferredoxin + $O_2 \rightleftharpoons$
　　　　an 11β-hydroxysteroid + oxidized adrenal ferredoxin + H_2O

Streptokinase　　　　No EC number assigned
Although this protein has been purified, it is not clear whether it
possesses a catalytic site. Streptokinase interacts with plasminogen
and the complex formed shows plasmin activity (see under plasmin).
The nature of activation process has to be clarified

Subtilisin　　　　3.4.21.61
Hydrolysis of proteins with broad specificity for peptide bonds, and a
preference for a large uncharged residue in P1
Hydrolyses peptide amides

Succinate-CoA ligase (GDP-forming)　　　　6.2.1.4
GTP + succinate + CoA $\rightleftharpoons$ GDP + orthophosphate + succinyl-CoA
Sometimes misleadingly referred to as succinate thiokinase

Succinate dehydrogenase   1.3.99.1
Succinate + acceptor $\rightleftharpoons$ fumarate + reduced acceptor

Sucrose α-glucosidase   3.2.1.48
Hydrolysis of sucrose and maltose by an α-D-glucosidase-type action,
still sometimes referred to as sucrase or invertase

Superoxide dismutase   1.15.1.1
2 Peroxide radical + 2H$^+$ $\rightleftharpoons$ O$_2$ + H$_2$O$_2$

Tetrahydrofolate dehydrogenase
See Dihydrofolate reductase

Thermolysin   3.4.24.27
A metalloprotease cleaving preferentially -Leu > -Phe
Also known as *Bacillus thermoproteolyticus* neutral

Thiosulphate sulphurtransferase   2.8.1.1
Sometimes known as rhodanase
Thiosulphate + cyanide $\rightleftharpoons$ sulphite + thiocyanate

Threonine dehydratase   4.2.1.16
L-Threonine + H$_2$O $\rightleftharpoons$ 2-oxobutyrate + NH$_3$ + H$_2$O

Thrombin   3.4.21.5
Protease cleaving preferentially Arg–Gly bonds in fibrinogen to form fibrin
and release fibrinopeptides A and B

Thymidine kinase   2.7.1.21
ATP + thymidine $\rightleftharpoons$ ADP + thymidine 5′-phosphate

Thymidylate synthase   2.1.1.45
5,10-Methylenetetrahydrofolate + dUMP $\rightleftharpoons$ dihydrofolate + dTMP

Topoisomerase I   5.99.1.2
ATP-independent breakage of single-stranded DNA followed by passage
and rejoining

Transketolase   2.2.1.1
Sedoheptulose 7-phosphate + D-glyceraldehyde 3-phosphate $\rightleftharpoons$
                                    D-ribose 5-phosphate + D-xylulose 5-phosphate

Trehalase   3.2.1.28
αα-Trehalose + H$_2$O $\rightleftharpoons$ 2 D-glucose

Triacylglycerol lipase   3.1.1.3
Triacylglycerol + H$_2$O $\rightleftharpoons$ diacylglycerol + a carboxylate

Triose phosphate isomerase   5.3.1.1
D-Glyceraldehyde 3-phosphate $\rightleftharpoons$ glycerone phosphate

Trypsin   3.4.21.4
A protease cleaving preferentially Arg-, Lys-

Tryptophan 2,3-dioxygenase   1.13.11.11
L-Tryptophan + O$_2$ $\rightleftharpoons$ L-formylkynurenine

Tryptophan synthase   4.2.1.20
L-Serine + 1-(indol-3-yl)glycerol 3-phosphate $\rightleftharpoons$ L-tryptophan
                                                  + glyceraldehyde 3-phosphate

Type II site-specific deoxyribonuclease   3.1.21.4

A group of endodeoxyribonucleases often referred to as restriction endonucleases. The Type II enzymes catalyse sequence-specific cleavage. There are over 400 enzymes of differing specificity known

Tyrosine 3-monooxygenase   1.14.16.2

L-Tyrosine + tetrahydrobiopterin + O_2 $\rightleftharpoons$ 3,4-dihydroxy-L-phenylalanine + dihydrobiopterin + H_2O

Also known as tyrosine 3-hydroxylase

Tyrosine kinase   2.7.1.112

A protein tyrosine + ATP $\rightleftharpoons$ ADP + a protein tyrosine phosphate

Tyrosine transaminase   2.6.1.5

L-Tyrosine + 2-oxoglutarate $\rightleftharpoons$ 4-hydroxyphenylpyruvate + L-glutamate

Tyrosyl-tRNA ligase   6.1.1.1

ATP + L-tyrosine + tRNA^{Tyr} $\rightleftharpoons$ AMP + pyrophosphate + L-tyrosyl-tRNA^{Tyr}

Ubiquitin-protein ligase   6.3.2.19

ATP + ubiquitin + protein-lysine $\rightleftharpoons$ AMP + diphosphate + protein-N-ubiquityl-lysine

Ubiquitin thiolesterase   3.1.2.15

Ubiquitin-C-terminal thioester + H_2O $\rightleftharpoons$ ubiquitin + a thiol

UDPglucose 4-epimerase   5.1.3.2

UDPglucose $\rightleftharpoons$ UDPgalactose

UDPglucose-hexose-1-phosphate uridylyltransferase   2.7.7.12

UDPglucose + α-D-galactose 1-phosphate $\rightleftharpoons$ α-D-glucose 1-phosphate + UDPgalactose

UDPglucuronosyltransferase

See Glucuronosyltransferase

UDP-N-acetylglucosamine 2-epimerase   5.1.3.14

UDP-N-acetyl-D-glucosamine $\rightleftharpoons$ UDP-N-acetyl-D-mannosamine

Urate oxidase   1.7.3.3

Urate + O_2 $\rightleftharpoons$ unidentified products

Urease   3.5.1.5

Urea + H_2O $\rightleftharpoons$ CO_2 + $2NH_3$

Urokinase (recommended name: U-plasminogen activator)   3.4.21.73

Cleavage of Arg–Val in plasminogen to form plasmin

Uropepsinogen   No EC number assigned

The enzyme, pepsin, which is secreted in the stomach in mammals, is subsequently absorbed by the intestine into the blood, whence it is excreted by the kidney as an inactive protease, uropepsinogen. Uropepsinogen is activated to uropepsin at pH 1–2

Xanthine oxidase   1.1.3.22

Xanthine + H_2O + O_2 $\rightleftharpoons$ urate + H_2O_2

Xylose isomerase   5.3.1.5

D-Xylose $\rightleftharpoons$ D-xylulose

Some enzymes also convert D-glucose to D fructose

L-*Xylulose reductase*   1.1.1.10

Xylitol + NADP$^+$ $\rightleftharpoons$ L-xylulose + NADPH

Index

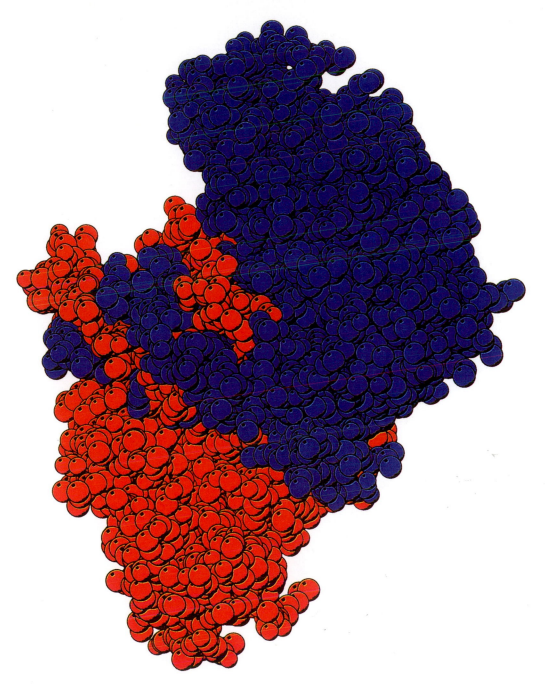

Plate 1 Structure of the open form of citrate synthase (5csc). The two subunits of the dimeric protein are shown in space filling format as red and blue. (See Fig. 3.31 (a).)

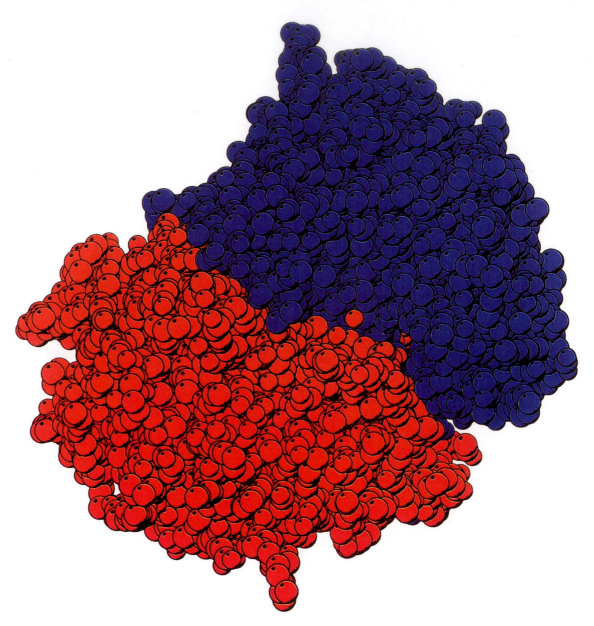

Plate 2 Structure of closed form of citrate synthase (4cts). The two subunits of the dimeric protein are shown in space filling format as red and blue. (See Fig. 3.31 (b).)

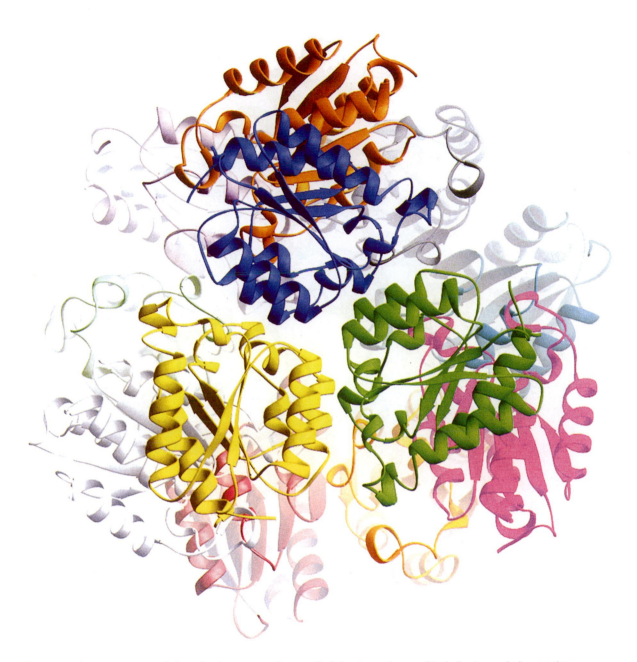

Plate 3 The structure of the dodecamer of type II dehydroquinase (1 doi) viewed down the three-fold axis of a trimeric unit; each monomer is individually coloured. The overall molecule is roughly spherical in shape with a diameter of 10 nm, and possesses tetrahedral symmetry with four trimers packing together. The structure is drawn using the programs Molscript and Ribbons. (See Fig. 5.33 (a).)

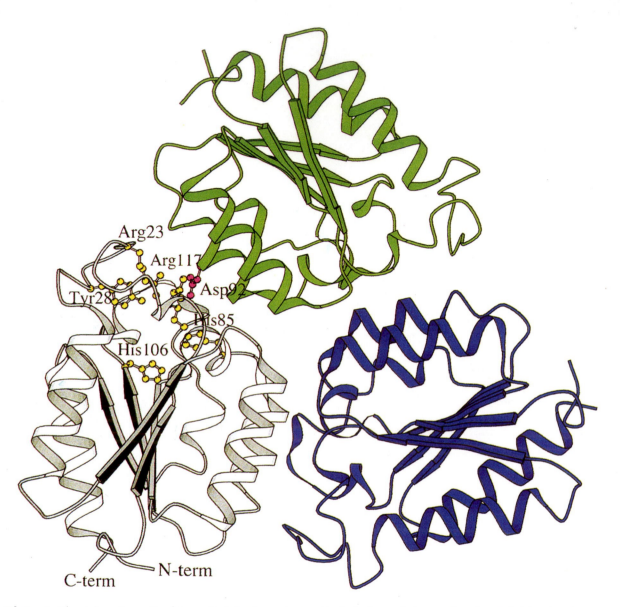

Plate 4 The trimeric unit of type II dehydroquinase (1 doi) viewed down the three-fold axis. The residues implicated in the catalytic mechanism are highlighted in one of the subunits. The structure is drawn using the programs Molscript and Ribbons. (See Fig. 5.34.)

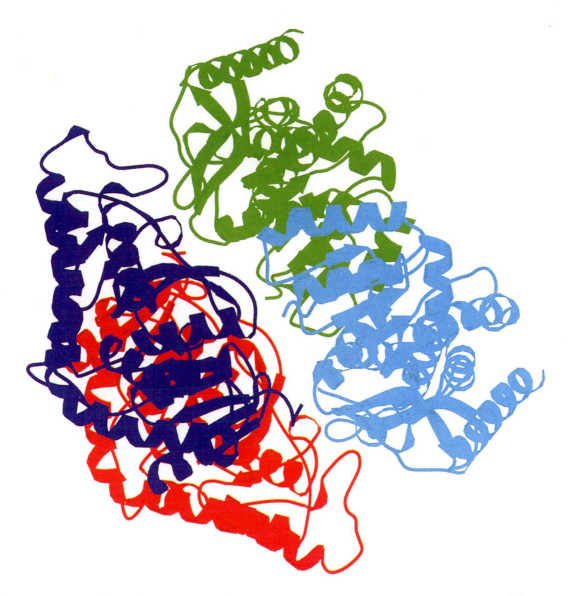

Plate 5 The structure of lactate dehydrogenase shown as a cartoon representation of the tetramer (1 ldm) drawn in Molscript. The arrangement of the tetramer as a dimer of dimers is clearly brought out. (See Fig. 5.39 (a).)

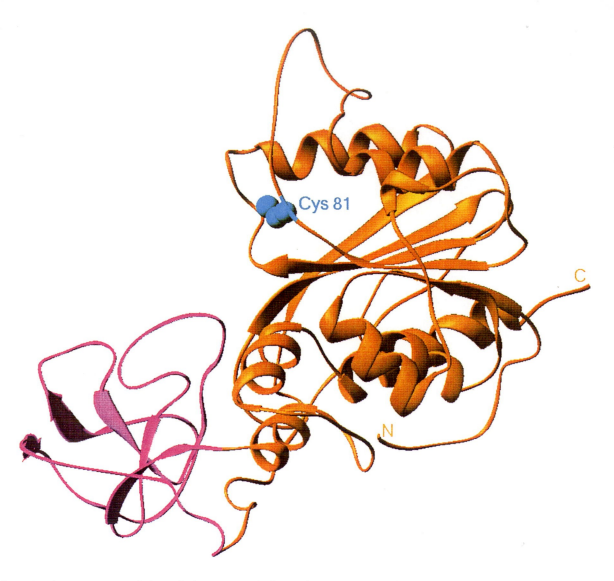

Plate 6 The structure of the HhaI DNA methyltransferase enzyme (1 hmy), drawn using the graphics program Ribbons. The large domain (brown) contains the binding site for S-adenosylmethionine and the catalytically essential Cys 81 (blue). The small domain is shown in magenta. The binding site for DNA lies between the domains. (See Fig. 5.43.)

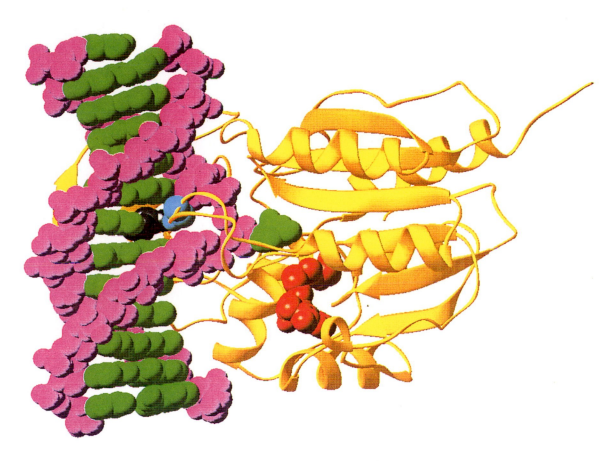

Plate 7 The structure of the DNA methyltransferase-DNA complex (1 mht), showing the movement of the active site loop and distortion of DNA so as to flip out the target cytosine base. The DNA is shown in space filling form (sugar-phosphate backbone in magenta and bases in green) and the enzyme (yellow) in a cartoon format using the program Ribbons. The location of the S-adenosylmethionine (red, space filling) binding site in the enzyme is indicated. The side chains of Ser 87 (blue) and Gln 287 (black) are seen to occupy the position in the double helix vacated by the target cytosine. (See Fig. 5.44 (a).)